Table 2. Exponential Functions

x	e^x	e^{-x}	x	e^x	e^{-x}
0.00	1.0000	1.0000	2.5	12.182	0.0821
0.05	1.0513	0.9512	2.6	13.464	0.0743
0.10	1.1052	0.9048	2.7	14.880	0.0672
0.15	1.1618	0.8607	2.8	16.445	0.0608
0.20	1.2214	0.8187	2.9	18.174	0.0550
0.25	1.2840	0.7788	3.0	20.086	0.0498
0.30	1.3499	0.7408	3.1	22.198	0.0450
0.35	1.4191	0.7047	3.2	24.533	0.0408
0.40	1.4918	0.6703	3.3	27.113	0.0369
0.45	1.5683	0.6376	3.4	29.964	0.0334
0.50	1.6487	0.6065	3.5	33.115	0.0302
0.55	1.7333	0.5769	3.6	36.598	0.0273
0.60	1.8221	0.5488	3.7	40.447	0.0247
0.65	1.9155	0.5220	3.8	44.701	0.0224
0.70	2.0138	0.4966	3.9	49.402	0.0202
0.75	2.1170	0.4724	4.0	54.598	0.0183
0.80	2.2255	0.4493	4.1	60.340	0.0166
0.85	2.3396	0.4274	4.2	66.686	0.0150
0.90	2.4596	0.4066	4.3	73.700	0.0136
0.95	2.5857	0.3867	4.4	81.451	0.0123
1.0	2.7183	0.3679	4.5	90.017	0.0111
1.1	3.0042	0.3329	4.6	99.484	0.0101
1.2	3.3201	0.3012	4.7	109.95	0.0091
1.3	3.6693	0.2725	4.8	121.51	0.0082
1.4	4.0552	0.2466	4.9	134.29	0.0074
1.5	4.4817	0.2231	5	148.41	0.0067
1.6	4.9530	0.2019	6	403.43	0.0025
1.7	5.4739	0.1827	7	1096.6	0.0009
1.8	6.0496	0.1653	8	2981.0	0.0003
1.9	6.6859	0.1496	9	8103.1	0.0001
2.0	7.3891	0.1353	10	22026	0.00005
2.1	8.1662	0.1225			
2.2	9.0250	0.1108			
2.3	9.9742	0.1003			
2.4	11.023	0.0907			

Undergraduate Texts in Mathematics

Undergraduate Texts in Mathematics

Apostol: Introduction to Analytic Number Theory.

Armstrong: Basic Topology.

Bak/Newman: Complex Analysis.

Banchoff/Wermer: Linear Algebra Through Geometry.

Childs: A Concrete Introduction to Higher Algebra.

Chung: Elementary Probability Theory with Stochastic Processes.

Croom: Basic Concepts of Algebraic Topology.

Curtis: Linear Algebra: An Introductory Approach.

Dixmier: General Topology.

Driver: Why Math?

Ebbinghaus/Flum/Thomas Mathematical Logic.

Fischer: Intermediate Real Analysis.

Fleming: Functions of Several Variables. Second edition.

Foulds: Optimization Techniques: An Introduction.

Foulds: Combination Optimization for Undergraduates.

Franklin: Methods of Mathematical Economics.

Halmos: Finite-Dimensional Vector Spaces. Second edition.

Halmos: Naive Set Theory.

Iooss/Joseph: Elementary Stability and Bifurcation Theory.

Jänich: Topology.

Kemeny/Snell: Finite Markov Chains.

Klambauer: Aspects of Calculus.

Lang: Undergraduate Analysis.

Lang: A First Course in Calculus. Fifth Edition.

Lang: Calculus of One Variable. Fifth Edition.

Lang: Introduction to Linear Algebra. Second Edition.

Lax/Burstein/Lax: Calculus with Applications and Computing, Volume 1. Corrected Second Printing.

LeCuyer: College Mathematics with APL.

Lidl/Pilz: Applied Abstract Algebra.

Macki/Strauss: Introduction to Optimal Control Theory.

Malitz: Introduction to Mathematical Logic.

Marsden/Weinstein: Calculus I, II, III. Second edition.

continued after Index

Murray H. Protter
Charles B. Morrey, Jr.

Intermediate Calculus

Second Edition

With 266 Illustrations

Springer-Verlag
New York Berlin Heidelberg Tokyo

Murray H. Protter
Department of Mathematics
University of California
Berkeley, CA 94720
U.S.A.

Charles B. Morrey, Jr.
Formerly of:
Department of Mathematics
University of California
Berkeley, CA 94720
U.S.A.

AMS Classifications: 06-01, 08-01

Library of Congress Cataloging in Publication Data
Protter, Murray H.
Intermediate calculus.
(Undergraduate texts in mathematics)
Rev. ed. of: Calculus with analytic geometry. 1971.
Includes index.
1. Calculus. 2. Geometry, Analytic. I. Morrey, Charles Bradfield. II. Protter, Murray H. Calculus with analytic geometry. III. Title. IV. Series.
QA303.P974 1985 515′.15 84-14118

This is the second edition of *Calculus with Analytic Geometry: A Second Course*, the first edition of which was published by Addison-Wesley Publishing Company, Inc., © 1971.

Typeset by Asco Trade Typesetting, Ltd., Hong Kong.
Printed and bound by R. R. Donnelley & Sons, Harrisonburg, Virginia.
Printed in the United States of America.

9 8 7 6 5 4 3 2 (Corrected Second Printing, 1986)

ISBN 0-387-96058-9 Springer-Verlag New York Berlin Heidelberg Tokyo
ISBN 3-540-96058-9 Springer-Verlag Berlin Heidelberg New York Tokyo

Preface

Analytic geometry and calculus at a college or university almost always consists of a three-semester course. Typically, the first two semesters cover plane analytic geometry and the calculus of functions of one variable. The third semester usually deals with three-dimensional analytic geometry, partial differentiation, multiple integration, and a selection of other topics which depend on the book used. Some courses may even include a small amount of linear algebra. Most texts for such a three-semester sequence run to an unwieldy 1,000 pages or more.

We believe that an instructor can add a great deal of flexibility to the calculus program by separating the text materials used in the third semester from those used in the first year. Such a division makes for a greater choice in the selection of topics taken up in the third semester. Moreover, at many universities there is a fourth semester of analysis in the lower division program. In such a case it is desirable to have one book which carries through the entire year, as this text does.

In recent years the percentage of students who enter college after completing a year of calculus in high school has been increasing; by now, the number is substantial. These students, many of whom have taken the Advanced Placement program, have mastered the calculus of functions of one variable from a variety of texts and are ready to begin the third semester of calculus with analytic geometry with a text suited to their needs.

In the first five chapters in this book we present the material which is most frequently taught in the third semester of calculus. We suppose that the student has completed the usual two semesters of plane analytic geometry and one-variable calculus from any standard text. Chapters 6 through 10 provide additional material which can be used either to replace some of the traditional third-semester course or to fill out a fourth semester of analysis. The latter option would give students a thorough preparation for a junior-level course in real analysis.

One of the main features of our text is the flexibility which results from the relative independence of the chapters. For example, if an instructor wishes to teach Chapter 6 on Fourier series and if the students have already had the standard topics on infinite series which we present in Sections 1 through 10 of Chapter 3, then the instructor need only present the advanced material on uniform convergence of series in Sections 11, 12, and 13 as preparation for Fourier series. On the other hand, if the instructor chooses to skip Chapter 6, there is no inconvenience in presenting the remainder of the book.

We also wish to emphasize the flexibility of our treatment of both vector field theory and Green's and Stokes' theorems in Chapters 9 and 10. A minimum of preparation from Chapters 2, 4, and 5 is needed for this purpose. We first establish Green's theorem for simple domains, a result which is adequate for most applications. Here the presentation is quite elementary. Then we continue with a section on orientable surfaces, as well as proofs of Green's and Stokes' theorems, which use a partition of unity. The serious student will benefit greatly from these sections, since the methods we use are straightforward, detailed, and sufficiently general so that, for example, it can be shown that Cauchy's theorem for complex analytic functions in general domains is a corollary of Green's theorem.

Chapter 7, on the implicit function theorem and the inverse function theorem, provides an excellent preparation for those students who intend to go on in mathematics. However, it may be skipped with little or no inconvenience by those instructors who prefer to concentrate on the last two chapters of the text. Chapter 8, on differentiation under the integral sign and improper integrals, treats a useful topic, especially for those planning to work in applied mathematics or related fields of technology. It is worth noting that the material in Chapter 8 is seldom presented in texts at the lower division level. As with Chapter 7, the omission of this chapter will not affect the continuity of the remainder of the book.

Many students are not familiar with the simple properties of matrices and determinants. Also, they are usually not aware of Cramer's rule for solving m linear equation in n unknowns when m and n are different integers. In an appendix we provide an introduction to matrices and determinants sufficient to establish Cramer's rule. The instructor may wish to use this material as optional independent reading for those interested students who are unfamiliar with linear algebra. We include illustrative examples and exercises in this appendix so that a good student can easily learn the material without help.

Berkeley, California
October 1984

MURRAY H. PROTTER

Contents

CHAPTER 1

Analytic Geometry in Three Dimensions

1. The Number Space R^3. Coordinates. The Distance Formula

Analytic geometry in three dimensions makes essential use of coordinate systems. To introduce a coordinate system, we consider triples (a, b, c) of real numbers, and we call the set of all such triples of real numbers the **three-dimensional number space**. We denote this space by R^3. Each individual triple is a **point** in R^3. The three elements in each number triple are called its **coordinates**. We now show how three-dimensional number space may be represented on a geometric or Euclidean three-dimensional space.

In three-dimensional space, consider three mutually perpendicular lines which intersect in a point O. We designate these lines the **coordinate axes** and, starting from O, set up identical number scales on each of them. If the positive directions of the x, y, and z axes are labeled x, y, and z, as shown in Fig. 1-1, we say the axes form a **right-handed system**. Figure 1-2 illustrates the axes in a **left-handed system**. We shall use a right-handed coordinate system throughout.

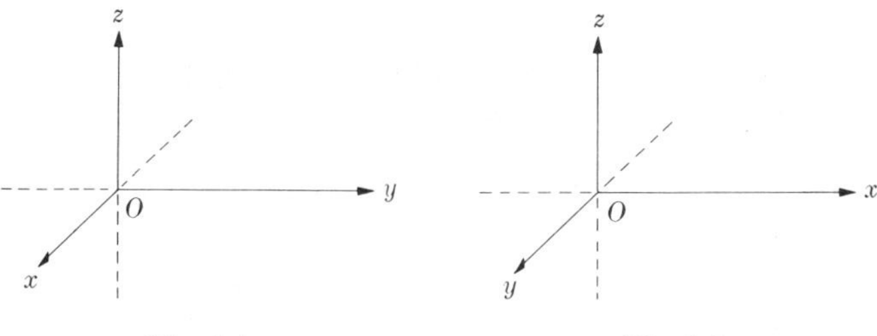

Fig. 1-1

Fig. 1-2

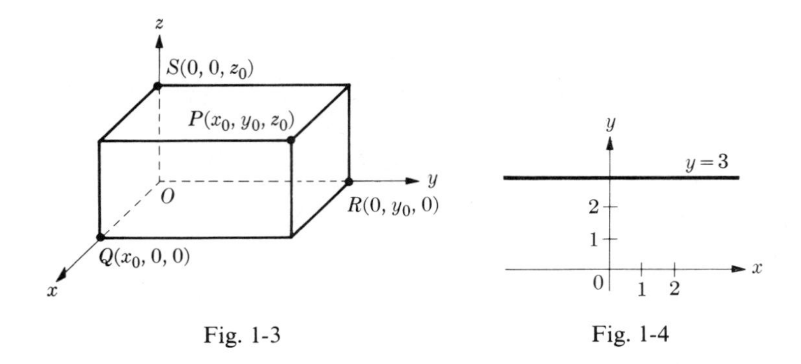

Fig. 1-3

Fig. 1-4

Any two intersecting lines in space determine a plane. A plane containing two of the coordinate axes is called a **coordinate plane**. Clearly, there are three such planes.

To each point P in three-dimensional space we can assign a point in R^3 in the following way. Through P construct three planes, each parallel to one of the coordinate planes as shown in Fig. 1-3. We label the intersections of the planes through P with the coordinate axes Q, R, and S, as shown. Then, if Q is x_0 units from the origin O, R is y_0 units from O, and S is z_0 units from O, we assign to P the number triple (x_0, y_0, z_0) and say that the point P has **Rectangular coordinates** (x_0, y_0, z_0). To each point in space there corresponds exactly one ordered number triple and, conversely, to each ordered number triple there is associated exactly one point in three-dimensional space. We have just described a **rectangular** coordinate system. In Section 7 we shall discuss other coordinate systems.

In studying analytic geometry in the plane, an equation such as

$$y = 3$$

repesents all points lying on a line parallel to the x axis and three units above it (Fig. 1-4).

We shall use set notation to describe sets of points. A set of points is determined by its properties. If P is a generic element of a set described by properties, say A and B, we write this fact in *set notation*:

$$\{P : P \text{ has properties } A \text{ and } B\}.$$

The symbol before the colon is a generic element in the domain under discussion, while the properties are described after the colon. For sets in the plane a typical point is denoted (x, y), and we write the line $y = 3$ in the form

$$\{(x, y) : y = 3\}.$$

The equation $y = 3$ is the property that the generic point (x, y) must possess, i.e., the set consists of all points in the plane with coordinates $(x, 3)$.

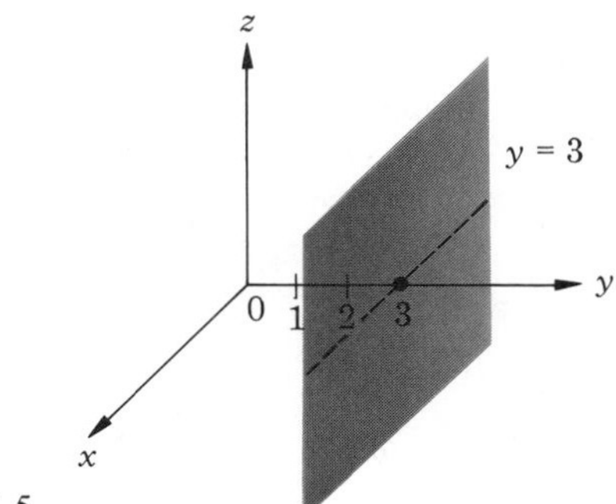

Fig. 1-5

The equation

$$y = 3$$

in the context of three-dimensional geometry represents something entirely different. The graph of the points satisfying this equation is a plane parallel to the xz plane (the xz plane is the coordinate plane determined by the x axis and the z axis) and three units from it (Fig. 1-5). In set notation, we represent this plane by writing

$$\{(x, y, z) : y = 3\}.$$

We see that set notation, by use of the symbols (x, y) or (x, y, z), indicates clearly when we are dealing with two- or three-dimensional geometry. The equation $y = 3$ by itself is ambiguous unless we know in advance the dimension of the geometry.

In three dimensions the plane represented by $y = 3$ is perpendicular to the y axis and passes through the point $(0, 3, 0)$. Since there is exactly one plane which is perpendicular to a given line and which passes through a given point, we see that the graph of the equation $y = 3$ consists of one and only one such plane. Conversely, from the very definition of a rectangular coordinate system every point with y coordinate 3 must lie in this plane. Equations such as $x = a$ or $y = b$ or $z = c$ always represent planes parallel to the coordinate planes.

We recall from Euclidean geometry that *any two nonparallel planes intersect in a straight line*. Therefore, the graph of all points which simultaneously satisfy the equations

$$x = a \qquad \text{and} \qquad y = b$$

is a line parallel to (or coincident with) the z axis. Conversely, any such line is the graph of a pair of equations of the above form. Since the plane $x = a$ is parallel to the z axis, and the plane $y = b$ is parallel to the z axis, the line of intersection must be parallel to the z axis also. (Corresponding statements hold with the axes interchanged.)

A plane separates three-dimensional space into two parts, each of which

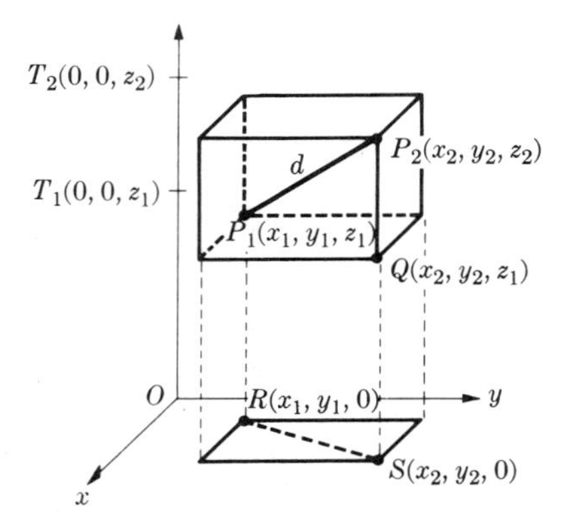

Fig. 1-6

is called a **half-space**. The inequality

$$x > 5$$

represents all points with x coordinate greater than 5. In set notation, we write

$$\{(x, y, z) : x > 5\}.$$

The set of such points comprises a half-space. Two intersecting planes divide three-space into 4 regions which we call **infinite wedges**. Three intersecting planes divide space into 8 regions (or possibly fewer), four planes into 15 regions (or possibly fewer), and so on. The inequality

$$|y| \leq 4 \Leftrightarrow -4 \leq y \leq 4$$

represents all points between (and on) the planes $y = -4$ and $y = 4$. Regions in space defined by inequalities are more difficult to visualize than those in the plane. However, **polyhedral domains**—that is, those bounded by a number of planes—are frequently simple enough to be sketched. A polyhedron with six faces in which opposite faces are congruent parallelograms is called a **parallelepiped**. Cubes and rectangular bins are particular cases of parallelepipeds.

Theorem 1. *The distance d between the points $P_1(x_1, y_1, z_1)$ and $P_2(x_2, y_2, z_2)$ is*

$$d = \sqrt{(x_2 - x_1)^2 + (y_2 - y_1)^2 + (z_2 - z_1)^2}.$$

PROOF. We make the construction shown in Fig. 1-6. By the Pythagorean theorem we have

$$d^2 = |P_1 Q|^2 + |QP_2|^2.$$

Where the symbol $|P_1 Q|$ denotes the length of the line segment with end points P_1 and Q, and similarly for $|QP_2|$.

Noting that $|P_1Q| = |RS|$, we use the formula for distance in the xy plane to get

$$|P_1Q|^2 = |RS|^2 = (x_2 - x_1)^2 + (y_2 - y_1)^2.$$

Furthermore, since P_2 and Q are on a line parallel to the z axis, we see that

$$|QP_2|^2 = |T_1T_2|^2 = (z_2 - z_1)^2.$$

Therefore

$$d^2 = (x_2 - x_1)^2 + (y_2 - y_1)^2 + (z_2 - z_1)^2.$$

The midpoint P of the line segment connecting the point $P_1(x_1, y_1, z_1)$ and $P_2(x_2, y_2, z_2)$ has coordinates $P(\bar{x}, \bar{y}, \bar{z})$ given by the formulas

$$\bar{x} = \frac{x_1 + x_2}{2}, \qquad \bar{y} = \frac{y_1 + y_2}{2}, \qquad \bar{z} = \frac{z_1 + z_2}{2}.$$

If P_1 and P_2 lie in the xy plane—that is, if $z_1 = 0$ and $z_2 = 0$—then so does the midpoint P, and we recognize the formula as the one we learned in plane analytic geometry. The above formula for $\bar{x}$ is proved as in the case of plane geometry, by passing planes through P_1, P, and P_2 perpendicular to the x axis. The formulas for $\bar{y}$ and $\bar{z}$ are established by analogy.

Example 1. Find the coordinates of the point Q which divides the line segment from $P_1(1, 4, -2)$ to $P_2(-3, 6, 7)$ in the proportion 3 to 1.

Solution. The midpoint P of the segment P_1P_2 has coordinates $P(-1, 5, \frac{5}{2})$. When we find the midpoint of PP_2 we get $Q(-2, \frac{11}{2}, \frac{19}{4})$.

Example 2. One endpoint of a segment P_1P_2 has coordinates $P_1(-1, 2, 5)$. The midpoint P is known to lie in the xz plane, while the other endpoint is known to lie on the intersection of the planes $x = 5$ and $z = 8$. Find the coordinates of P and P_2.

Solution. For $P(\bar{x}, \bar{y}, \bar{z})$ we note that $\bar{y} = 0$, since P is in the xz plane. Similarly, for $P_2(x_2, y_2, z_2)$ we have $x_2 = 5$ and $z_2 = 8$. From the midpoint formula we get

$$\bar{x} = \frac{-1 + 5}{2}, \qquad 0 = \bar{y} = \frac{2 + y_2}{2}, \qquad \bar{z} = \frac{5 + 8}{2}.$$

Therefore the points have coordinates $P(2, 0, \frac{13}{2})$, $P_2(5, -2, 8)$.

Problems

In Problems 1 through 5, find the lengths of the sides of triangle ABC and state whether the triangle is a right triangle, an isosceles triangle, or both.

1. $A(2, 1, 3)$, $B(3, -1, -2)$, $C(0, 2, -1)$

2. $A(4, 3, 1)$, $B(2, 1, 2)$, $C(0, 2, 4)$
3. $A(3, -1, -1)$, $B(1, 2, 1)$, $C(6, -1, 2)$
4. $A(1, 2, -3)$, $B(4, 3, -1)$, $C(3, 1, 2)$
5. $A(0, 0, 0)$, $B(4, 1, 2)$, $C(-5, -5, -1)$

In Problems 6 and 7, find the midpoint of the segment joining the given points A, B.

6. $A(4, -2, 6)$, $B(-2, 8, 1)$
7. $A(-2, 3, 5)$, $B(-6, 0, 4)$

In Problems 8 and 9, in each case find the coordinates of the three points which divide the given segment AB into four equal parts.

8. $A(3, 4, -1)$, $B(7, -2, 5)$
9. $A(1, -6, 0)$, $B(6, 12, 7)$

In Problems 10 through 13, find the lengths of the medians of the given triangles ABC.

10. $A(2, 1, 3)$, $B(3, -1, -2)$, $C(0, 2, -1)$
11. $A(4, 3, 1)$, $B(2, 1, 2)$, $C(0, 2, 4)$
12. $A(3, -1, -1)$, $B(1, 2, 1)$, $C(6, -1, 2)$
13. $A(1, 2, -3)$, $B(4, 3, -1)$, $C(3, 1, 2)$
14. One endpoint of a line segment is at $P(4, 6, -3)$ and the midpont is at $Q(2, 1, 6)$. Find the other endpoint.
15. One endpoint of a line segment is at $P_1(-2, 1, 6)$ and the midpoint Q lies in the plane $y = 3$. The other endpoint, P_2, lies on the intersection of the planes $x = 4$ and $z = -6$. Find the coordinates of P_2 and Q.

In Problems 16 through 19, determine whether or not the three given points lie on a line.

16. $A(1, -1, 2)$, $B(-1, -4, 3)$, $C(3, 2, 1)$
17. $A(2, 3, 1)$, $B(4, 6, 5)$, $C(-2, -2, -7)$
18. $A(1, -1, 2)$, $B(3, 3, 4)$, $C(-2, -6, -1)$
19. $A(-4, 5, -6)$, $B(-1, 2, -1)$, $C(3, -3, 6)$
20. Describe the set of points in space given by $\{(x, y, z) : x = 2, y = -4\}$.
21. Describe the set of points in space given by $\{(x, y, z) : -2 \leq y \leq 5\}$.
22. Describe the set of points in space given by $\{(x, y, z) : x \geq 0, y \geq 0, z \geq 0\}$.
23. Describe the set of points in space given by $\{(x, y, z) : x^2 + y^2 + z^2 < 1\}$.
24. Derive the formula for determining the midpoint of a line segment.
25. The formula for the coordinates of a point $Q(x_0, y_0, z_0)$ which divides the line segment from $P_1(x_1, y_1, z_1)$ to $P_2(x_2, y_2, z_2)$ in the ratio p to q is

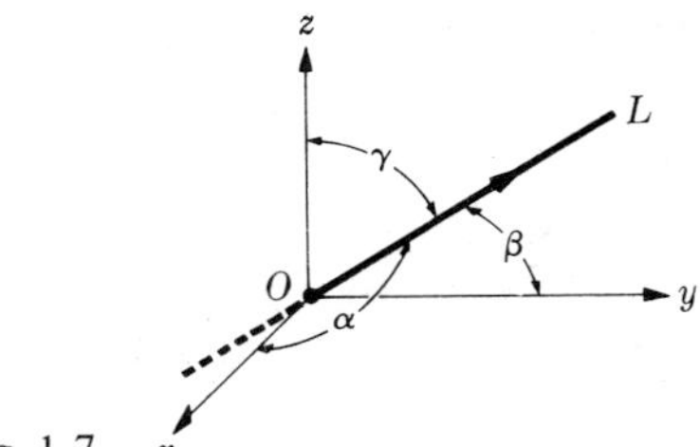

Fig. 1-7

$$x_0 = \frac{px_2 + qx_1}{p+q}, \qquad y_0 = \frac{py_2 + qy_1}{p+q}, \qquad z_0 = \frac{pz_2 + qz_1}{p+q}.$$

Derive this formula.

26. Find the equation of the graph of all points equidistant from the points $(2, -1, 3)$ and $(3, 1, -1)$. Can you describe the graph?

27. Find the equation of the graph of all points equidistant from the points $(5, 1, 0)$ and $(2, -1, 4)$. Can you describe the graph?

28. Find the equation of the graph of all points such that the sum of the distances from $(1, 0, 0)$ and $(-1, 0, 0)$ is always equal to 4. Describe the graph.

29. The points $A(0, 0, 0)$, $B(1, 0, 0)$, $C(\frac{1}{2}, \frac{1}{2}, 1/\sqrt{2})$, $D(0, 1, 0)$ are the vertices of a four-sided figure. Show that $|AB| = |BC| = |CD| = |DA| = 1$. Prove that the figure is not a rhombus.

30. Prove that the diagonals joining opposite vertices of a rectangular parallelepiped (there are four of them which are interior to the parallelepiped) bisect each other.

2. Direction Cosines and Numbers

In three-dimensional space we consider a line passing through the origin O and place an arrow on it so that one of the two possible directions on the line is distinguished (Fig. 1-7). We call such a line a **directed line**. If no arrow is placed, then L is called an **undirected line**. We use the symbol $\vec{L}$ to indicate a directed line while the letter L without the arrow over it indicates an undirected line. We denote by α, β, and γ the angles made by the directed line $\vec{L}$ and the positive directions of the x, y, and z axes, respectively. We define these angles to be the **direction angles** of the directed line $\vec{L}$. The *undirected* line L will have two possible sets of direction angles according to the ordering chosen. The two sets are

$$\alpha, \beta, \gamma \qquad \text{and} \qquad 180° - \alpha, 180° - \beta, 180° - \gamma.$$

The term "line" without further specification shall mean undirected line.

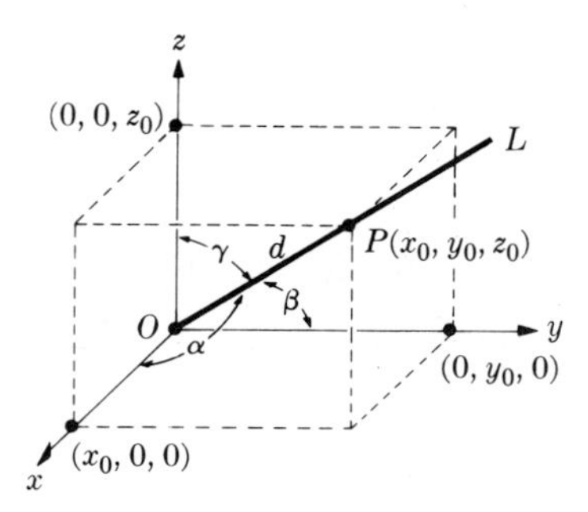

Fig. 1-8

Definition. If α, β, γ are direction angles of a directed line $\vec{L}$, then $\cos\alpha$, $\cos\beta$, $\cos\gamma$ are called the **direction cosines** of $\vec{L}$.

Since $\cos(180° - \theta) = -\cos\theta$, we see that if λ, μ, ν are direction cosines of a directed line $\vec{L}$, then λ, μ, ν and $-\lambda$, $-\mu$, $-\nu$ are the two sets of direction cosines of the undirected line L.

We shall show that the direction cosines of any line L satisfy the relation

$$\cos^2\alpha + \cos^2\beta + \cos^2\gamma = 1.$$

Let $P(x_0, y_0, z_0)$ be a point on a line L which goes through the origin. Then the distance d of P from the origin is

$$d = \sqrt{x_0^2 + y_0^2 + z_0^2},$$

and (see Fig. 1-8) we have

$$\cos\alpha = \frac{x_0}{d}, \qquad \cos\beta = \frac{y_0}{d}, \qquad \cos\gamma = \frac{z_0}{d}.$$

Squaring and adding, we get the desired result.

To define the direction cosines of any line L in space, we simply consider the line L' parallel to L which passes through the origin, and assert that *by definition* L **has the same direction cosines as** L'. Thus *all parallel lines in space have the same direction cosines.*

Definition. Two sets of number triples, a, b, c and a', b', c', neither all zero, are said to be **proportional** if there is *a* number k such that

$$a' = ka, \qquad b' = kb, \qquad c' = kc.$$

REMARK. The number k may be positive or negative but not zero, since by hypothesis neither of the number triples is 0, 0, 0. If none of the numbers a, b, and c is zero, we may write the proportionality relations as

$$\frac{a'}{a} = k, \qquad \frac{b'}{b} = k, \qquad \frac{c'}{c} = k$$

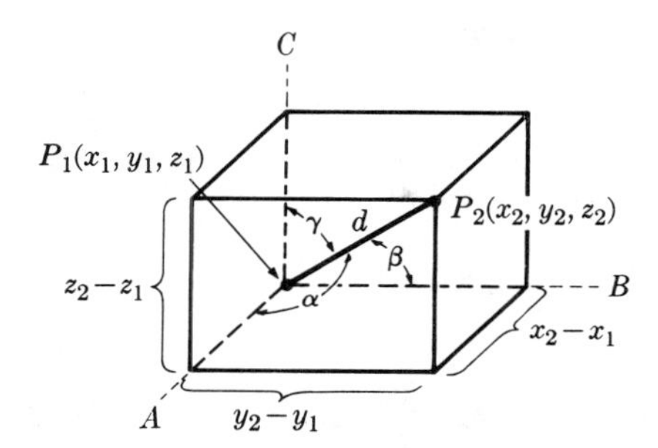

Fig. 1-9

or, more simply,

$$\frac{a'}{a} = \frac{b'}{b} = \frac{c'}{c}.$$

Definition. Suppose that *a* line L has direction cosines λ, μ, ν. Then *a* set of numbers a, b, c is called *a* **set of direction numbers** for L if a, b, c and λ, μ, ν are proportional.

A line L has unlimited sets of direction numbers.

Theorem 2. *If $P_1(x_1, y_1, z_1)$ and $P_2(x_2, y_2, z_2)$ are two points on a line L, then*

$$\lambda = \frac{x_2 - x_1}{d}, \qquad \mu = \frac{y_2 - y_1}{d}, \qquad \nu = \frac{z_2 - z_1}{d}$$

is a set of direction cosines of L where d is the distance from P_1 to P_2.

Proof. In Fig. 1-9 we note that the angles α, β, and γ are equal to the direction angles, since the lines P_1A, P_1B, P_1C are parallel to the coordinate axes. We read off from the figure that

$$\cos \alpha = \frac{x_2 - x_1}{d}, \qquad \cos \beta = \frac{y_2 - y_1}{d}, \qquad \cos \gamma = \frac{z_2 - z_1}{d},$$

which is the desired result.

Corollary 1. *If $P_1(x_1, y_1, z_1)$ and $P_2(x_2, y_2, z_2)$ are two points on a line L, then*

$$x_2 - x_1, \qquad y_2 - y_1, \qquad z_2 - z_1$$

constitute a set of direction numbers for L.

Multiplying λ, μ, ν of Theorem 2 by the constant d, we obtain the result of the Corollary.

Example 1. Find direction numbers and direction cosines for the line L passing through the points $P_1(1, 5, 2)$ and $P_2(3, 7, -4)$.

Solution. From the Corollary, 2, 2, -6 form a set of direction numbers. We compute

$$d = |P_1 P_2| = \sqrt{4 + 4 + 36} = \sqrt{44} = 2\sqrt{11},$$

and so

$$\frac{1}{\sqrt{11}}, \quad \frac{1}{\sqrt{11}}, \quad -\frac{3}{\sqrt{11}}$$

form a set of direction cosines. Since L is undirected, it has two such sets, the other being $-\frac{1}{\sqrt{11}}, -\frac{1}{\sqrt{11}}, \frac{3}{\sqrt{11}}$.

Example 2. Do the three points $P_1(3, -1, 4)$, $P_2(1, 6, 8)$, and $P_3(9, -22, -8)$ lie on the same straight line?

Solution. A set of direction numbers for the line L_1 through P_1 and P_2 is -2, 7, 4. A set of direction numbers for the line L_2 through P_2 and P_3 is 8, -28, -16. Since the second set is proportional to the first (with $k = -4$), we conclude that L_1 and L_2 have the same direction cosines. Therefore the two lines are parallel. However, they have the point P_2 in common and so must coincide.

From Theorem 2 and the statements in Example 2, we easily obtain the next result.

Corollary 2. *A line L_1 is parallel to a line L_2 if and only if a set of direction numbers of L_1 is proportional to a set of direction numbers of L_2.*

The angle between two intersecting lines in space is defined in the same way as the angle between two lines in the plane. It may happen that two lines L_1 and L_2 in space are neither parallel nor intersecting. Such lines are said to be **skew** to each other. Nevertheless, the angle between L_1 and L_2 can still be defined. Denote by L_1' and L_2' the lines passing through the origin and parallel to L_1 and L_2, respectively. The **angle between L_1 and L_2 is defined to be the angle between the intersecting lines L_1' and L_2'.**

Theorem 3. *If L_1 and L_2 have direction cosines λ_1, μ_1, ν_1 and λ_2, μ_2, ν_2, respectively, and if θ is the angle between L_1 and L_2, then*

$$\cos\theta = \lambda_1\lambda_2 + \mu_1\mu_2 + \nu_1\nu_2.$$

Proof. From the way we defined the angle between two lines we may consider L_1 and L_2 as lines passing through the origin. Let $P_1(x_1, y_1, z_1)$ be a point on L_1 and $P_2(x_2, y_2, z_2)$ a point on L_2, neither O (see Fig. 1-10). Denote by d_1 the distance of P_1 from O, by d_2 the distance of P_2 from O; let $d =$

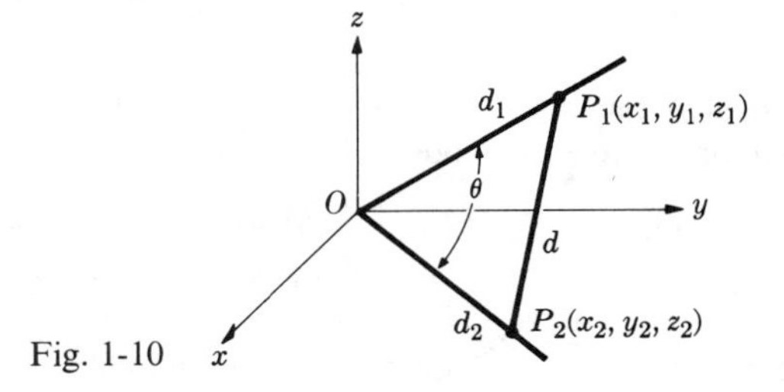

Fig. 1-10

$|P_1P_2|$. We apply the Law of Cosines to triangle OP_1P_2, getting

$$d^2 = d_1^2 + d_2^2 - 2d_1d_2 \cos\theta \qquad \text{or} \qquad \cos\theta = \frac{d_1^2 + d_2^2 - d^2}{2d_1d_2}$$

and $\cos\theta$

$$= \frac{x_1^2 + y_1^2 + z_1^2 + x_2^2 + y_2^2 + z_2^2 - (x_2 - x_1)^2 - (y_2 - y_1)^2 - (z_2 - z_1)^2}{2d_1d_2}.$$

After simplification we obtain

$$\cos\theta = \frac{x_1x_2 + y_1y_2 + z_1z_2}{d_1d_2} = \frac{x_1}{d_1}\cdot\frac{x_2}{d_2} + \frac{y_1}{d_1}\cdot\frac{y_2}{d_2} + \frac{z_1}{d_1}\cdot\frac{z_2}{d_2}$$

$$= \lambda_1\lambda_2 + \mu_1\mu_2 + \nu_1\nu_2.$$

Corollary. *Two lines L_1 and L_2 with direction numbers a_1, b_1, c_1 and a_2, b_2, c_2, respectively, are perpendicular if and only if*

$$a_1a_2 + b_1b_2 + c_1c_2 = 0.$$

EXAMPLE 3. Find the cosine of the angle between the line L_1, passing through the points $P_1(1, 4, 2)$ and $P_2(3, -1, 3)$, and the line L_2, passing through the points $Q_1(3, 1, 2)$ and $Q_2(2, 1, 3)$.

SOLUTION. A set of direction numbers for L_1 is 2, -5, 1. A set for L_2 is -1, 0, 1. Therefore direction cosines for the two lines are

$$L_1: \quad \frac{2}{\sqrt{30}}, \quad \frac{-5}{\sqrt{30}}, \quad \frac{1}{\sqrt{30}}; \qquad L_2: \quad \frac{-1}{\sqrt{2}}, \quad 0, \quad \frac{1}{\sqrt{2}}.$$

We obtain

$$\cos\theta = -\frac{1}{\sqrt{15}} + 0 + \frac{1}{2\sqrt{15}} = -\frac{1}{2\sqrt{15}}.$$

We observe that two lines always have two possible supplementary angles of intersection. If $\cos\theta$ is negative, we have obtained the obtuse angle and, if it is positive, the acute angle of intersection.

Problems

In Problems 1 through 4, find a set of direction numbers and a set of direction cosines for the line passing through the given points.

1. $A(2, 1, 4)$, $B(3, 5, -2)$
2. $A(4, 2, -3)$, $B(1, 0, 5)$
3. $A(-2, 1, -4)$, $B(0, -5, -7)$
4. $A(6, 7, -2)$, $B(8, -5, 1)$

In each of Problems 5 through 8, a point P_1 and a set of direction numbers are given. Find the coordinates of another point on the line L determined by P_1 and the given direction numbers.

5. $P_1(1, 4, -2)$, direction numbers 2, 1, 4
6. $P_1(3, 5, -1)$, direction numbers 2, 0, 4
7. $P_1(0, 4, -3)$, direction numbers 0, 0, 5
8. $P_1(1, 2, 0)$, direction numbers 4, 0, 0

In each of Problems 9 through 12, determine whether or not the three given points lie on a line.

9. $A(3, 1, 0)$, $B(2, 2, 2)$, $C(0, 4, 6)$
10. $A(2, -1, 1)$, $B(4, 1, -3)$, $C(7, 4, -9)$
11. $A(4, 2, -1)$, $B(2, 1, 1)$, $C(0, 0, 2)$
12. $A(5, 8, 6)$, $B(-2, -3, 1)$, $C(4, 2, 8)$

In each of Problems 13 through 15, determine whether or not the line through the points P_1, P_2 is parallel to the line through the points Q_1, Q_2.

13. $P_1(4, 8, 0)$, $P_2(1, 2, 3)$; $Q_1(0, 5, 0)$, $Q_2(-3, -1, 3)$
14. $P_1(2, 1, 1)$, $P_2(3, 2, -1)$; $Q_1(0, 1, 4)$, $Q_2(2, 3, 0)$
15. $P_1(3, 1, 4)$, $P_2(-3, 2, 5)$; $Q_1(4, 6, 1)$, $Q_2(0, 5, 8)$

In each of Problems 16 through 18, determine whether or not the line through the points P_1, P_2 is perpendicular to the line through the points Q_1, Q_2.

16. $P_1(2, 1, 3)$, $P_2(4, 0, 5)$; $Q_1(3, 1, 2)$, $Q_2(2, 1, 6)$
17. $P_1(2, -1, 0)$, $P_2(3, 1, 2)$; $Q_1(2, 1, 4)$, $Q_2(4, 0, 4)$
18. $P_1(0, -4, 2)$, $P_2(5, -1, 0)$; $Q_1(3, 0, 2)$, $Q_2(2, 1, 1)$

In each of Problems 19 through 21, find $\cos\theta$ where θ is the angle between the line L_1 passing through P_1, P_2 and L_2 passing through Q_1, Q_2.

19. $P_1(2, 1, 4)$, $P_2(-1, 4, 1)$; $Q_1(0, 5, 1)$, $Q_2(3, -1, -2)$
20. $P_1(4, 0, 5)$, $P_2(-1, -3, -2)$; $Q_1(2, 1, 4)$, $Q_2(2, -5, 1)$
21. $P_1(0, 0, 5)$, $P_2(4, -2, 0)$; $Q_1(0, 0, 6)$, $Q_2(3, -2, 1)$

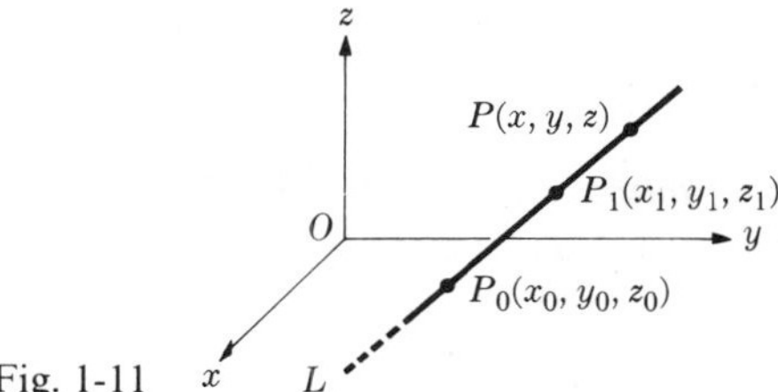

Fig. 1-11

22. A **regular tetrahedron** is a 4-sided figure each side of which is an equilateral triangle. Find 4 points in space which are the vertices of a regular tetrahedron with each edge of length 2 units.

23. A **regular pyramid** is a 5-sided figure with a square base and sides consisting of 4 congruent isosceles triangles. If the base has a side of length 4 and if the height of the pyramid is 6 units, find the area of each of the triangular faces.

24. The points $P_1(1, 2, 3)$, $P_2(2, 1, 2)$, $P_3(3, 0, 1)$, $P_4(5, 2, 7)$ are the vertices of a plane quadrilateral. Find the coordinates of the midpoints of the sides. What kind of quadrilateral do these four midpoints form?

25. Prove that the four interior diagonals of a parallelepiped bisect each other.

26. Given the set $S = \{(x, y, z) : x^2 + y^2 + z^2 = 1\}$. Find the points of $S \cap L$ where L is the line passing through the origin and having direction numbers 2, 1, 3.

27. Let A, B, C, D be the vertices of any quadrilateral in three-dimensional space. Prove that the lines joining the midpoints of the sides form a parallelogram.

3. Equations of a Line

A line in three-dimensional space is determined by two points. If $P_0(x_0, y_0, z_0)$ and $P_1(x_1, y_1, z_1)$ are given points, we seek an analytic method of representing the line L determined by these points. The result is obtained by solving a geometric problem. A point $P(x, y, z)$, different from P_0, is on L if and only if the direction numbers determined by P and P_0 are proportional to those determined by P_1 and P_0 (Fig. 1-11). Calling the proportionality constant t, the proportionality conditions are

$$x - x_0 = t(x_1 - x_0), \qquad y - y_0 = t(y_1 - y_0), \qquad z - z_0 = t(z_1 - z_0).$$

Thus we obtain the **two-point form of the parametric equations of a line**:

$$\begin{aligned} x &= x_0 + (x_1 - x_0)t, \\ y &= y_0 + (y_1 - y_0)t, \\ z &= z_0 + (z_1 - z_0)t. \end{aligned} \tag{1}$$

Definition. Given two points $P_0(x_0, y_0, z_0)$ and $P_1(x_1, y_1, z_1)$, the point which is t **of the way from** P_0 **to** P_1 is the point $P(x, y, z)$ whose coordinates are given by equations (1).

EXAMPLE 1. Find the parametric equations of the line through the points $A(3, 2, -1)$ and $B(4, 4, 6)$. Locate three additional points on the line.

SOLUTION. Substituting in (1) we obtain

$$x = 3 + t, \qquad y = 2 + 2t, \qquad z = -1 + 7t.$$

To get an additional point on the line we let $t = 2$ and obtain $P_1(5, 6, 13)$; $t = -1$ yields $P_2(2, 0, -8)$ and $t = 3$ gives $P_3(6, 8, 20)$.

Theorem 4. *The parametric equations of a line L through the point $P_0(x_0, y_0, z_0)$ with direction numbers a, b, c are given by*

$$x = x_0 + at, \qquad y = y_0 + bt, \qquad z = z_0 + ct. \tag{2}$$

PROOF. The point $P_1(x_0 + a, y_0 + b, z_0 + c)$ must be on L, since the direction numbers formed by P_0 and P_1 are just a, b, c. Using the two-point form (1) for the equations of a line through P_0 and P_1, we get (2) precisely.

If $\cos\alpha$, $\cos\beta$, $\cos\gamma$ are the direction cosines of a line L passing through the point (x_0, y_0, z_0), the equations of the line are

$$x = x_0 + t\cos\alpha, \qquad y = y_0 + t\cos\beta, \qquad z = z_0 + t\cos\gamma.$$

EXAMPLE 2. Find the parametric equations of the line L through the point $A(3, -2, 5)$ with direction numbers 4, 0, -2. What is the relation of L to the coordinate planes?

SOLUTION. Substituting in (2), we obtain

$$x = 3 + 4t, \qquad y = -2, \qquad z = 5 - 2t.$$

Since all points on the line must satisfy all three of the above equations, L must lie in the plane $y = -2$. This plane is parallel to the xz plane. Therefore L is parallel to the xz plane. Setting $x = 0$ in the first equation, we get $t = -3/4$. From the third equation, $z = 5 - 2(-3/4) = 13/2$. Hence the line L intersects the yz plane in the point $(0, -2, 13/2)$. Similarly, setting $z = 0$, we find that $t = 5/2$, and L intersects the xy plane in the point $(13, -2, 0)$.

If none of the direction numbers is zero, the parameter t may be eliminated from the system of equations (2). We may write

$$\frac{x - x_0}{a} = \frac{y - y_0}{b} = \frac{z - z_0}{c} \tag{3}$$

for the equations of a line. For any value of t in (2) the ratios in (3) are equal.

Conversely, if the ratios in (3) are all equal we may set the common value equal to t and (2) is satisfied.

If one of the direction numbers is zero, the form (3) may still be used if the zero in the denominator is interpreted properly. The equations

$$\frac{x - x_0}{a} = \frac{y - y_0}{b} = \frac{z - z_0}{0}$$

are understood to stand for the equations

$$\frac{x - x_0}{a} = \frac{y - y_0}{b} \qquad \text{and} \qquad z = z_0.$$

The system

$$\frac{x - x_0}{0} = \frac{y - y_0}{b} = \frac{z - z_0}{0}$$

stands for

$$x = x_0 \qquad \text{and} \qquad z = z_0.$$

We recognize these last two equations as those of planes parallel to coordinate planes. In other words, *a line is represented as the intersection of two planes.* This fact will be discussed further in the next section.

The two-point form for the equations of a line may be written *symmetrically*. The equations

$$\frac{x - x_0}{x_1 - x_0} = \frac{y - y_0}{y_1 - y_0} = \frac{z - z_0}{z_1 - z_0}.$$

are called the **symmetric form for the equations of a line.**

EXAMPLE 3. Find the point of intersection of the line

$$L = \{(x, y, z) : x = 3 - t, \quad y = 2 + 3t, \quad z = -1 + 2t\}$$

with the plane $S = \{(x, y, z) : z = 5\}$.

SOLUTION. Denoting by $P(x_0, y_0, z_0)$ the point of intersection of L and S, we see that z_0 must have the value 5. From the equation $z = -1 + 2t$, we conclude that at the point of intersection, $5 = -1 + 2t$ or $t = 3$. Then $x_0 = 3 - 3 = 0$ and $y_0 = 2 + 3(3) = 11$. Hence* $P = L \cap S$ has coordinates (0, 11, 5).

EXAMPLE 4. Find the equations of the line through the point $P(2, -1, 3)$ and parallel to the line through the points $Q(1, 4, -6)$ and $R(-2, -1, 5)$.

* The symbol $\cap$ denotes *intersection*; thus $L \cap S$ is the intersection of the line L and the plane S.

SOLUTION. The line through Q and R has direction numbers: $-2-1=-3$, $-1-4=-5$, $5-(-6)=11$. Therefore the parallel line through P is given by the equations

$$x = 2 - 3t, \quad y = -1 - 5t, \quad z = 3 + 11t.$$

PROBLEMS

In each of Problems 1 through 4, find the equations of the line going through the given points.

1. $A(1,3,2)$, $B(2,-1,4)$
2. $A(1,0,5)$, $B(-2,0,1)$
3. $A(4,-2,0)$, $B(3,2,-1)$
4. $A(5,5,-2)$, $B(6,4,-4)$

In each of Problems 5 through 9, find the equations of the line passing through the given point with the given direction numbers.

5. $P_1(1,0,-1)$, direction numbers 2, 1, -3
6. $P_1(-2,1,3)$, direction numbers 3, -1, -2
7. $P_1(4,0,0)$, direction numbers 2, -1, -3
8. $P_1(1,2,0)$, direction numbers 0, 1, 3
9. $P_1(3,-1,-2)$, direction numbers 2, 0, 0

In each of Problems 10 through 14, decide whether or not L_1 and L_2 are perpendicular.

10. $L_1: \dfrac{x-2}{2} = \dfrac{y+1}{-3} = \dfrac{z-1}{4}$; $\quad L_2: \dfrac{x-2}{-3} = \dfrac{y+1}{2} = \dfrac{z-1}{3}$

11. $L_1: \dfrac{x}{1} = \dfrac{y+1}{2} = \dfrac{z+1}{3}$; $\quad L_2: \dfrac{x-3}{0} = \dfrac{y+1}{-3} = \dfrac{z+4}{2}$

12. $L_1: \dfrac{x+2}{-1} = \dfrac{y-2}{2} = \dfrac{z+3}{3}$; $\quad L_2: \dfrac{x+2}{1} = \dfrac{y-2}{2} = \dfrac{z+3}{-1}$

13. $L_1: \dfrac{x+5}{4} = \dfrac{y-1}{3} = \dfrac{z+8}{5}$; $\quad L_2: \dfrac{x-4}{3} = \dfrac{y+7}{2} = \dfrac{z+4}{1}$

14. $L_1: \dfrac{x+1}{0} = \dfrac{y-2}{1} = \dfrac{z+8}{0}$; $\quad L_2: \dfrac{x-3}{1} = \dfrac{y+2}{0} = \dfrac{z-1}{0}$

15. Find the equations of the medians of the triangle with vertices at $A(4,0,2)$, $B(3,1,4)$, $C(2,5,0)$.
16. Find the equations of any line through $P_1(2,1,4)$ and perpendicular to any line having direction numbers 4, 1, 3.
17. Find the equations of any line through $P_1(2,-1,5)$ and perpendicular to any line having direction numbers 2, -3, 1.
18. Find the points of intersection of the line

$$L = \{(x, y, z) : x = 3 + 2t, \quad y = 7 + 8t, \quad z = -2 + t\}$$

with each of the coordinate planes.

19. Find the points of intersection of the line

$$L = \left\{(x, y, z) : \frac{x+1}{-2} = \frac{y+1}{3} = \frac{z-1}{7}\right\}$$

with each of the coordinate planes.

20. Show that the following lines are coincident:

$$L_1 : \frac{x-1}{2} = \frac{y+1}{-3} = \frac{z}{4}; \qquad L_2 : \frac{x-5}{2} = \frac{y+7}{-3} = \frac{z-8}{4}.$$

21. Find the equations of the line through (3, 1, 5) which is parallel to the line

$$L = \{(x, y, z) : x = 4 - t, \quad y = 2 + 3t, \quad z = -4 + t.\}$$

22. Find the equations of the line through (3, 1, −2) which is perpendicular to and intersects the line

$$L = \left\{(x, y, z) : \frac{x+1}{1} = \frac{y+2}{1} = \frac{z+1}{1}\right\}.$$

[*Hint:* Let (x_0, y_0, z_0) be the point of intersection and determine its coordinates.]

23. A triangle has vertices at $A(2, 1, 6)$, $B(-3, 2, 4)$ and $C(5, 8, 7)$. Perpendiculars are drawn from these vertices to the xz plane. Locate the points A', B', and C' which are the intersections of the perpendiculars through A, B, C and the xz plane. Find the equations of the sides of the triangle $A'B'C'$.

Let $P_1(x_1, y_1, z_1)$ and $P_2(x_2, y_2, z_2)$ be two points and suppose $P(\bar{x}, \bar{y}, \bar{z})$ is on the line segment joining P_1 to P_2. If P is h of the way from P_1 to P_2, then the coordinates $(\bar{x}, \bar{y}, \bar{z})$ are given by

$$\bar{x} = x_1 + h(x_2 - x_1), \qquad \bar{y} = y_1 + h(y_2 - y_1), \qquad \bar{z} = z_1 + h(z_2 - z_1). \quad (4)$$

The above equations are called the **point of division formula**.

24. Find the point 1/5 of the way from

$$P_1(2, -1, 3) \text{ to } P_2(6, 2, -5).$$

25. What is the relation of the points P_1, P, and P_2 when $h > 1$? when h is negative?

26. Show that the medians of any triangle in three-dimensional space intersect at a point which divides each median in the proportion 2 : 1.

4. The Plane

Any three points not on a straight line determine a plane. While this characterization of a plane is quite simple, it is not convenient for beginning the study of planes. Instead we use the fact that *there is exactly one plane which passes through a given point and is perpendicular to a given line.*

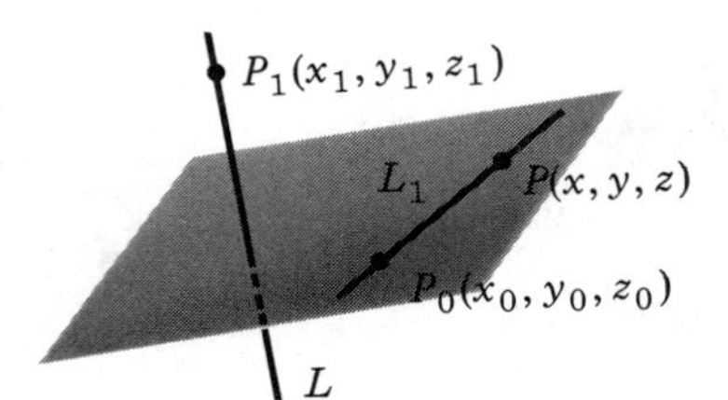

Fig. 1-12

Let $P_0(x_0, y_0, z_0)$ be a given point, and suppose that a given line L goes through the point $P_1(x_1, y_1, z_1)$ and has direction numbers A, B, C.

Theorem 5. *The equation of the plane passing through P_0 and perpendicular to L is*

$$A(x - x_0) + B(y - y_0) + C(z - z_0) = 0.$$

PROOF. We establish the result by solving a geometric problem. Let $P(x, y, z)$ be a point on the plane (Fig. 1-12). From Euclidean geometry we recall that if a line L_1 through P_0 and P is perpendicular to L, then P must be in the desired plane. A set of direction numbers for the line L_1 is

$$x - x_0, \qquad y - y_0, \qquad z - z_0.$$

Since L has direction numbers A, B, C, we conclude that the two lines L and L_1 are perpendicular if and only if their direction numbers satisfy the relation

$$A(x - x_0) + B(y - y_0) + C(z - z_0) = 0,$$

which is the equation we seek.

REMARK. Note that only the direction of L—and not the coordinates of P_1—enters the above equation. We obtain the same plane and the same equation if any line parallel to L is used in its stead.

EXAMPLE 1. Find the equation of the plane through the point $P_0(5, 2, -3)$ which is perpendicular to the line through the points $P_1(5, 4, 3)$ and $P_2(-6, 1, 7)$.

SOLUTION. The line through the points P_1 and P_2 has direction numbers -11, -3, 4. The equation of the plane is

$$-11(x - 5) - 3(y - 2) + 4(z + 3) = 0$$

$$\Leftrightarrow \qquad 11x + 3y - 4z - 73 = 0.$$

All lines perpendicular to the same plane are parallel and therefore have proportional direction numbers.

Definition. A set of **attitude numbers** of a plane is any set of direction numbers of a line perpendicular to the plane.

In Example 1 above, 11, 3, -4 form a set of attitude numbers of the plane.

EXAMPLE 2. What are sets of attitude numbers for planes parallel to the coordinate planes?

SOLUTION. A plane parallel to the yz plane has an equation of the form $x - c = 0$, where c is a constant. A set of attitude numbers for this plane is 1, 0, 0. A plane parallel to the xz plane has attitude numbers 0, 1, 0, and any plane parallel to the xy plane has attitude numbers 0, 0, 1.

Since lines perpendicular to the same or parallel planes are themselves parallel, we get at once the next theorem.

Theorem 6. *Two planes are parallel if and only if their attitude numbers are proportional.*

Theorem 7. *If A, B, and C are not all zero, the graph of an equation of the form*

$$Ax + By + Cz + D = 0 \tag{1}$$

is a plane.

PROOF. Suppose that $C \neq 0$, for example. Then the point $P_0(0, 0, -D/C)$ is on the graph as its coordinates satisfy the above equation. Therefore we may write

$$A(x - 0) + B(y - 0) + C\left(z + \frac{D}{C}\right) = 0,$$

and the graph is the plane passing through P_0 perpendicular to any line with direction numbers A, B, C.

An equation of the plane through three points not on a line can be found by assuming that the plane has an equation of the form (1), substituting in turn the coordinates of the three points, and solving simultaneously the three resulting equations. The fact that there are four constants, A, B, C, D, and only three equations is illusory, since we may divide through by one of them (say D) and obtain three equations in the unknowns A/D, B/D, C/D. This is equivalent to setting D (or one of the other constants) equal to some convenient value. An example illustrates the procedure.

EXAMPLE 3. Find an equation of the plane passing through the points (2, 1, 3), (1, 3, 2), $(-1, 2, 4)$.

SOLUTION. Since the three points lie in the plane, each of them satisfies equation (1). We have

$$\begin{aligned}(2, 1, 3)&: \quad 2A + B + 3C + D = 0,\\ (1, 3, 2)&: \quad A + 3B + 2C + D = 0,\\ (-1, 2, 4)&: \quad -A + 2B + 4C + D = 0.\end{aligned}$$

Solving for A, B, C in terms of D, we obtain

$$A = -\tfrac{3}{25}D, \qquad B = -\tfrac{4}{25}D, \qquad C = -\tfrac{5}{25}D.$$

Setting $D = -25$, we get the equation

$$3x + 4y + 5z - 25 = 0.$$

PROBLEMS

In each of Problems 1 through 4, find the equation of the plane which passes through the given point P_0 and has the given attitude numbers.

1. $P_0(1, 4, 2)$; 3, 1, -4
2. $P_0(2, 1, -5)$; 3, 0, 2
3. $P_0(4, -2, -5)$; 0, 3, -2
4. $P_0(-1, -2, -3)$; 4, 0, 0

In each of Problems 5 through 8, find the equation of the plane which passes through the three points.

5. $(1, -2, 1), (2, 0, 3), (0, 1, -1)$
6. $(2, 2, 1), (-1, 2, 3), (3, -5, -2)$
7. $(3, -1, 2), (1, 2, -1), (2, 3, 1)$
8. $(-1, 3, 1), (2, 1, 2), (4, 2, -1)$

In each of Problems 9 through 12, find the equation of the plane passing through P_1 and perpendicular to the line L_1.

9. $P_1(2, -1, 3)$; $\quad L_1: x = -1 + 2t, y = 1 + 3t, z = -4t$
10. $P_1(1, 2, -3)$; $\quad L_1: x = t, y = -2 - 2t, z = 1 + 3t$
11. $P_1(2, -1, -2)$; $\quad L_1: x = 2 + 3t, y = 0, z = -1 - 2t$
12. $P_1(-1, 2, -3)$; $\quad L_1: x = -1 + 5t, y = 1 + 2t, z = -1 + 3t$

In each of Problems 13 through 16, find the equations of the line through P_1 and perpendicular to the given plane M_1.

13. $P_1(-2, 3, 1)$; $\quad M_1: 2x + 3y + z - 3 = 0$
14. $P_1(1, -2, -3)$; $\quad M_1: 3x - y - 2z + 4 = 0$
15. $P_1(-1, 0, -2)$; $\quad M_1: x + 2z + 3 = 0$
16. $P_1(2, -1, -3)$; $\quad M_1: x = 4$

In each of Problems 17 through 20, find an equation of the plane through P_1 and parallel to the plane Φ.

17. $P_1(1, -2, -1)$; $\Phi: 3x + 2y - z + 4 = 0$

18. $P_1(-1, 3, 2)$; $\Phi: 2x + y - 3z + 5 = 0$

19. $P_1(2, -1, 3)$; $\Phi: x - 2y - 3z + 6 = 0$

20. $P_1(3, 0, 2)$; $\Phi: x + 2y + 1 = 0$

In each of Problems 21 through 23, find the equations of the line through P_1 parallel to the given line L.

21. $P_1(2, -1, 3)$; $L: \dfrac{x-1}{3} = \dfrac{y+2}{-2} = \dfrac{z-2}{4}$

22. $P_1(0, 0, 1)$; $L: \dfrac{x+2}{1} = \dfrac{y-1}{3} = \dfrac{z+1}{-2}$

23. $P_1(1, -2, 0)$; $L: \dfrac{x-2}{2} = \dfrac{y+2}{-1} = \dfrac{z-3}{4}$

In each of Problems 24 through 28, find the equation of the plane containing L_1 and L_2.

24. $L_1: \dfrac{x+1}{2} = \dfrac{y-2}{3} = \dfrac{z-1}{1}$; $L_2: \dfrac{x+1}{1} = \dfrac{y-2}{-1} = \dfrac{z-1}{2}$

25. $L_1: \dfrac{x-1}{3} = \dfrac{y+2}{2} = \dfrac{z-2}{2}$; $L_2: \dfrac{x-1}{1} = \dfrac{y+2}{1} = \dfrac{z-2}{0}$

26. $L_1: \dfrac{x+2}{1} = \dfrac{y}{0} = \dfrac{z+1}{2}$; $L_2: \dfrac{x+2}{2} = \dfrac{y}{3} = \dfrac{z+1}{1}$

27. $L_1: \dfrac{x}{2} = \dfrac{y-1}{3} = \dfrac{z+2}{-1}$; $L_2: \dfrac{x-2}{2} = \dfrac{y+1}{3} = \dfrac{z}{-1} (L_1 \| L_2)$

28. $L_1: \dfrac{x-2}{2} = \dfrac{y+1}{-1} = \dfrac{z}{3}$; $L_2: \dfrac{x+1}{2} = \dfrac{y}{-1} = \dfrac{z+2}{3} (L_1 \| L_2)$

In Problems 29 and 30, find the equation of the plane through P_1 and the given line L.

29. $P_1(3, -1, 2)$; $L = \left\{(x, y, z): \dfrac{x-2}{2} = \dfrac{y+1}{3} = \dfrac{z}{-2}\right\}$

30. $P_1(1, -2, 3)$; $L = \{(x, y, z): x = -1 + t, y = 2 + 2t, z = 2 - 2t\}$

31. Show that the plane $2x - 3y + z - 2 = 0$ is parallel to the line

$$\frac{x-2}{1} = \frac{y+2}{1} = \frac{z+1}{1}.$$

32. Show that the plane $5x - 3y - z - 6 = 0$ contains the line

$$x = 1 + 2t, \qquad y = -1 + 3t, \qquad z = 2 + t.$$

33. A plane has attitude numbers A, B, C, and a line has direction numbers a, b, c. What condition must be satisfied in order that the plane and line be parallel?

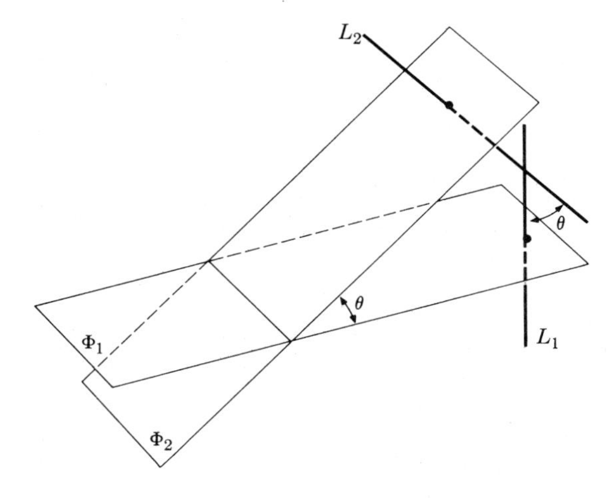

Fig. 1-13

34. Show that the three planes $7x - 2y - 2z - 5 = 0$, $3x + 2y - 3z - 10 = 0$, $7x + 2y - 5z - 16 = 0$ all contain a common line. Find the coordinates of two points on this line.

*35. Find a condition that three planes $A_1x + B_1y + C_1z + D_1 = 0$, $A_2x + B_2y + C_2z + D_2 = 0$, $A_3x + B_3y + C_3z + D_3 = 0$ either have a line in common or have no point in common.

5. Angles. Distance from a Point to a Plane

The angle between two lines was defined in Section 2. We recall that if line L_1 has direction cosines λ_1, μ_1, ν_1 and line L_2 has direction cosines λ_2, μ_2, ν_2, then

$$\cos\theta = \lambda_1\lambda_2 + \mu_1\mu_2 + \nu_1\nu_2,$$

where θ is the angle between L_1 and L_2.

Definition. Let Φ_1 and Φ_2 be two planes, and let L_1 and L_2 be two lines which are perpendicular to Φ_1 and Φ_2, respectively. Then the **angle between Φ_1 and Φ_2** is, by definition, the angle between L_1 and L_2. (See Fig. 1-13.) Furthermore, we make the convention that we always select the acute angle between these lines as the angle between Φ_1 and Φ_2.

Theorem 8. *The angle θ between the planes $A_1x + B_1y + C_1z + D_1 = 0$ and $A_2x + B_2y + C_2z + D_2 = 0$ is given by*

$$\cos\theta = \frac{|A_1A_2 + B_1B_2 + C_1C_2|}{\sqrt{A_1^2 + B_1^2 + C_1^2}\sqrt{A_2^2 + B_2^2 + C_2^2}}.$$

PROOF. From the definition of attitude numbers for a plane, we know that they are direction numbers of any line perpendicular to the plane. Converting to direction cosines, we get the above formula.

Corollary. *Two planes with attitude numbers A_1, B_1, C_1 and A_2, B_2, C_2 are perpendicular if and only if*

$$A_1 A_2 + B_1 B_2 + C_1 C_2 = 0.$$

EXAMPLE 1. Find $\cos\theta$ where θ is the angle between the planes $3x - 2y + z = 4$ and $x + 4y - 3z - 2 = 0$.

SOLUTION. Substituting in the formula of Theorem 8, we have

$$\cos\theta = \frac{|3 - 8 - 3|}{\sqrt{9 + 4 + 1}\sqrt{1 + 16 + 9}} = \frac{4}{\sqrt{91}}.$$

Two nonparallel planes intersect in a line. Every point on the line satisfies the equations of both planes and, conversely, every point which satisfies the equations of both planes must be on the line. *Therefore we may characterize any line in space by finding two planes which contain it.* Since every line has an unlimited number of planes which pass through it and since *any* two of them are sufficient to determine the line uniquely, we see that there is an unlimited number of ways of writing the equations of a line. The next example shows how to transform one representation into another.

EXAMPLE 2. The two planes

$$2x + 3y - 4z - 6 = 0 \qquad \text{and} \qquad 3x - y + 2z + 4 = 0$$

intersect in a line. (That is, the points which satisfy *both* equations constitute the line.) Find a set of parametric equations of the line of intersection.

SOLUTION. We solve the above equations for x and y in terms of z, getting

$$x = -\tfrac{2}{11}z - \tfrac{6}{11}, \qquad y = \tfrac{16}{11}z + \tfrac{26}{11} \qquad \text{and} \qquad \frac{x + \frac{6}{11}}{-\frac{2}{11}} = \frac{y - \frac{26}{11}}{\frac{16}{11}} = \frac{z}{1}.$$

We can therefore write

$$x = -\tfrac{6}{11} - \tfrac{2}{11}t, \qquad y = \tfrac{26}{11} + \tfrac{16}{11}t, \qquad z = t,$$

which are the desired parametric equations.

Three planes may be parallel, may pass through a common line, may have no common points, or may have a unique point of intersection. If they have a unique point of intersection, the intersection point may be found by solving simultaneously the three equations of the planes. If they have no common point, an attempt to solve simultaneously will fail. A further examination will show whether or not two or more of the planes are parallel.

EXAMPLE 3. Determine whether or not the planes $\Phi_1 : 3x - y + z - 2 = 0$; $\Phi_2 : x + 2y - z + 1 = 0$; $\Phi_3 : 2x + 2y + z - 4 = 0$ intersect. If so, find the point of intersection.

SOLUTION. Eliminating z between Φ_1 and Φ_2, we have

$$4x + y - 1 = 0. \tag{1}$$

Eliminating z between Φ_2 and Φ_3, we find

$$3x + 4y - 3 = 0. \tag{2}$$

We solve equations (1) and (2) simultaneously to get

$$x = \tfrac{1}{13}, \qquad y = \tfrac{9}{13}.$$

Substituting in the equation for Φ_1, we obtain $z = \frac{32}{13}$. Therefore the single point of intersection of the three planes is $(\frac{1}{13}, \frac{9}{13}, \frac{32}{13})$.

EXAMPLE 4. Find the point of intersection of the plane

$$3x - y + 2z - 3 = 0$$

and the line

$$\frac{x+1}{3} = \frac{y+1}{2} = \frac{z-1}{-2}.$$

SOLUTION. We write the equations of the line in parametric form:

$$x = -1 + 3t, \qquad y = -1 + 2t, \qquad z = 1 - 2t.$$

The point of intersection is given by a value of t; call it t_0. This point must satisfy the equation of the plane. We have

$$3(-1 + 3t_0) - (-1 + 2t_0) + 2(1 - 2t_0) - 3 = 0 \quad \Leftrightarrow \quad t_0 = 1.$$

The desired point is $(2, 1, -1)$.

We now derive an important formula which tells us how to find the perpendicular distance from a point in space to a plane.

Theorem 9. *The distance d from the point $P_1(x_1, y_1, z_1)$ to the plane*

$$Ax + By + Cz + D = 0$$

is given by

$$d = \frac{|Ax_1 + By_1 + Cz_1 + D|}{\sqrt{A^2 + B^2 + C^2}}.$$

PROOF. We write the equations of the line L through P_1 which is perpendicular to the plane. They are

$$L: \quad x = x_1 + At, \qquad y = y_1 + Bt, \qquad z = z_1 + Ct.$$

Denote by (x_0, y_0, z_0) the point of intersection of L and the plane. Then

$$d^2 = (x_1 - x_0)^2 + (y_1 - y_0)^2 + (z_1 - z_0)^2. \tag{3}$$

Also (x_0, y_0, z_0) is on both the line and the plane. Therefore, we have for some value t_0

$$\begin{aligned} x_0 &= x_1 + At_0, \\ y_0 &= y_1 + Bt_0, \\ z_0 &= z_1 + Ct_0, \end{aligned} \tag{4}$$

and

$$\begin{aligned} &Ax_0 + By_0 + Cz_0 + D \\ &\qquad = 0 = A(x_1 + At_0) + B(y_1 + Bt_0) + C(z_1 + Ct_0) + D. \end{aligned}$$

Thus, from (3) and (4), we write

$$d = \sqrt{A^2 + B^2 + C^2}\,|t_0|,$$

and now, inserting the relation

$$t_0 = \frac{-(Ax_1 + By_1 + Cz_1 + D)}{A^2 + B^2 + C^2}$$

in the preceding expression for d, we obtain the desired formula.

EXAMPLE 5. Find the distance from the point $(2, -1, 5)$ to the plane

$$3x + 2y - 2z - 7 = 0.$$

SOLUTION.

$$d = \frac{|6 - 2 - 10 - 7|}{\sqrt{9 + 4 + 4}} = \frac{13}{\sqrt{17}}.$$

PROBLEMS

In each of Problems 1 through 4, find $\cos\theta$ where θ is the angle between the given planes.

1. $2x - y + 2z - 3 = 0, \quad 3x + 2y - 6z - 11 = 0$
2. $x + 2y - 3z + 6 = 0, \quad x + y + z - 4 = 0$
3. $2x - y + 3z - 5 = 0, \quad 3x - 2y + 2z - 7 = 0$
4. $x + 4z - 2 = 0, \quad y + 2z - 6 = 0$

In each of Problems 5 through 8, find the equations in parametric form of the line of intersection of the given planes.

5. $3x + 2y - z + 5 = 0, \quad 2x + y + 2z - 3 = 0$

6. $x + 2y + 2z - 4 = 0, \quad 2x + y - 3z + 5 = 0$

7. $x + 2y - z + 4 = 0, \quad 2x + 4y + 3z - 7 = 0$

8. $2x + 3y - 4z + 7 = 0, \quad 3x - 2y + 3z - 6 = 0$

In each of Problems 9 through 12, find the point of intersection of the given line and the given plane.

9. $3x - y + 2z - 5 = 0, \qquad \dfrac{x-1}{2} = \dfrac{y+1}{3} = \dfrac{z-1}{-2}$

10. $2x + 3y - 4z + 15 = 0, \qquad \dfrac{x+3}{2} = \dfrac{y-1}{-2} = \dfrac{z+4}{3}$

11. $x + 2z + 3 = 0, \qquad \dfrac{x+1}{1} = \dfrac{y}{0} = \dfrac{z+2}{2}$

12. $2x + 3y + z - 3 = 0, \qquad \dfrac{x+2}{2} = \dfrac{y-3}{3} = \dfrac{z-1}{1}$

In each of Problems 13 through 16, find the distance from the given point to the given plane.

13. $(2, 1, -1), \quad x - 2y + 2z + 5 = 0$

14. $(3, -1, 2), \quad 3x + 2y - 6z - 9 = 0$

15. $(-1, 3, 2), \quad 2x - 3y + 4z - 5 = 0$

16. $(0, 4, -3), \quad 3y + 2z - 7 = 0$

17. Find the equation of the plane through the line

$$\frac{x+1}{3} = \frac{y-1}{2} = \frac{z-2}{4}$$

which is perpendicular to the plane

$$2x + y - 3z + 4 = 0.$$

18. Find the equation of the plane through the line

$$\frac{x-2}{2} = \frac{y-2}{3} = \frac{z-1}{-2}$$

which is parallel to the line

$$\frac{x+1}{3} = \frac{y-1}{2} = \frac{z+1}{1}.$$

19. Find the equation of the plane through the line

$$\frac{x+2}{3} = \frac{y}{-2} = \frac{z+1}{2}$$

which is parallel to the line

$$\frac{x-1}{2} = \frac{y+1}{3} = \frac{z-1}{4}.$$

20. Find the equations of any line through the point $(1, 4, 2)$ which is parallel to the plane

$$2x + y + z - 4 = 0.$$

21. Find the equation of the plane through $(3, 2, -1)$ and $(1, -1, 2)$ which is parallel to the line

$$\frac{x-1}{3} = \frac{y+1}{2} = \frac{z}{-2}.$$

In each of Problems 22 through 26, find all the points of intersection of the three given planes. If the three planes pass through a line, find the equations of the line in parametric form.

22. $2x + y - 2z - 1 = 0, \quad 3x + 2y + z - 10 = 0, \quad x + 2y - 3z + 2 = 0$

23. $x + 2y + 3z - 4 = 0, \quad 2x - 3y + z - 2 = 0, \quad 3x + 2y - 2z - 5 = 0$

24. $3x - y + 2z - 4 = 0, \quad x + 2y - z - 3 = 0, \quad 3x - 8y + 7z + 1 = 0$

25. $2x + y - 2z - 3 = 0, \quad x - y + z + 1 = 0, \quad x + 5y - 7z - 3 = 0$

26. $x + 2y + 3z - 5 = 0, \quad 2x - y - 2z - 2 = 0, \quad x - 8y - 13z + 11 = 0$

In each of Problems 27 through 29, find the equations in parametric form of the line through the given point P_1 which intersects and is perpendicular to the given line L.

27. $P_1(3, -1, 2); \quad L: \dfrac{x-1}{2} = \dfrac{y+1}{-1} = \dfrac{z}{3}$

28. $P_1(-1, 2, 3); \quad L: \dfrac{x}{2} = \dfrac{y-2}{0} = \dfrac{z+3}{-3}$

29. $P_1(0, 2, 4); \quad L: \dfrac{x-1}{3} = \dfrac{y-2}{1} = \dfrac{z-3}{4}$

30. If $A_1x + B_1y + C_1z + D_1 = 0$ and $A_2x + B_2y + C_2z + D_2 = 0$ are two intersecting planes, what is the graph of all points which satisfy

$$A_1x + B_1y + C_1z + D_1 + k(A_2x + B_2y + C_2z + D_2) = 0,$$

where k is a constant?

31. Find the equation of the plane passing through the point $(2, 1, -3)$ and the intersection of the planes $3x + y - z - 2 = 0$, $2x + y + 4z - 1 = 0$. [*Hint:* See Problem 30.]

32. Given a regular tetrahedron with each edge 2 units in length. (See Problem 22 at the end of Section 2.) Find the distance from a vertex to the opposite face.

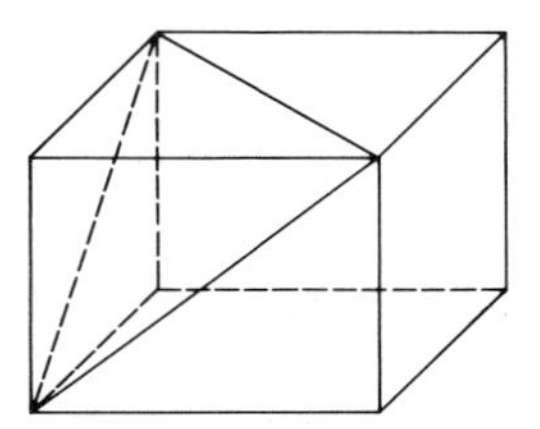

Fig. 1-14

33. Given a regular pyramid with each edge of the base 2 units in length and each lateral edge 4 units in length. (See Problem 23 at the end of Section 2.) Find the distance from the top of the pyramid to the base. Also find the distance from one of the vertices of the base to a face opposite to that vertex.

34. A slice is made in a cube by cutting through a diagonal of one face and proceeding through one of the vertices of the opposite face as shown in Fig. 1-14. Find the angle between the planar slice and the first face which is cut.

6. The Sphere. Cylinders

A **sphere** is the graph of all points at a given distance from a fixed point. The fixed point is called the **center** and the fixed distance is called the **radius**.

If the center is at the point (h, k, l), the radius is r, and (x, y, z) is any point on the sphere, then, from the formula for the distance between two points, we obtain the relation

$$(x - h)^2 + (y - k)^2 + (z - l)^2 = r^2. \tag{1}$$

Equation (1) is the **equation of a sphere**. If it is multiplied out and the terms collected we have the equivalent form

$$x^2 + y^2 + z^2 + Dx + Ey + Fz + G = 0 \tag{2}$$

with $D = -2h$, $E = -2k$, $F = -2l$, $G = h^2 + k^2 + l^2 - r^2$.

Example 1. Find the center and radius of the sphere with equation

$$x^2 + y^2 + z^2 + 4x - 6y + 9z - 6 = 0.$$

Solution. We complete the square by first writing

$$x^2 + 4x \qquad + y^2 - 6y \qquad + z^2 + 9z \qquad = 6;$$

then, adding the appropriate quantities to both sides, we have

$$(x^2 + 4x + 4) + (y^2 - 6y + 9) + (z^2 + 9z + \tfrac{81}{4}) = 6 + 4 + 9 + \tfrac{81}{4}$$

$$\Leftrightarrow \qquad (x + 2)^2 + (y - 3)^2 + (z + \tfrac{9}{2})^2 = \tfrac{157}{4}.$$

The center is at $(-2, 3, -\frac{9}{2})$ and the radius is $\frac{1}{2}\sqrt{157}$.

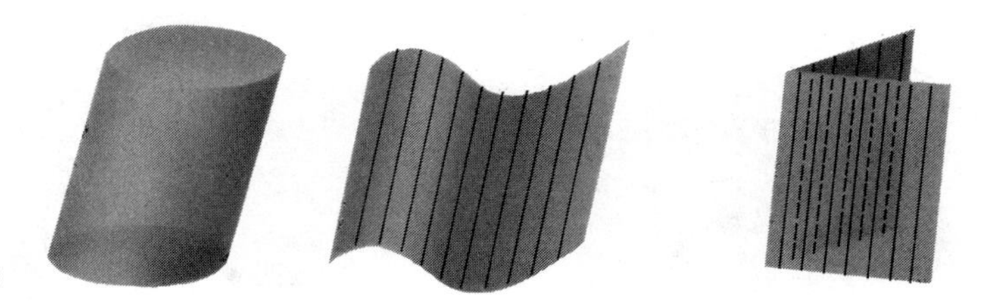
Fig. 1-15

EXAMPLE 2. Find the equation of the sphere which passes through (2, 1, 3), (3, 2, 1), (1, −2, −3), (−1, 1, 2).

SOLUTION. Substituting these points in the form (2) above for the equation of a sphere, we obtain

$$\begin{aligned}
(2,1,3)&: \quad 2D + E + 3F + G = -14,\\
(3,2,1)&: \quad 3D + 2E + F + G = -14,\\
(1,-2,-3)&: \quad D - 2E - 3F + G = -14,\\
(-1,1,2)&: \quad -D + E + 2F + G = -6.
\end{aligned}$$

Solving these by elimination (first G, then D, then F) we obtain, successively,

$$D + E - 2F = 0, \qquad D + 3E + 6F = 0, \qquad 3D + F = -8,$$

and

$$2E + 8F = 0, \qquad 3E - 7F = 8; \qquad \text{and so} \qquad -38E = -64.$$

Therefore

$$E = \tfrac{32}{19}, \qquad F = -\tfrac{8}{19}, \qquad D = -\tfrac{48}{19}, \qquad G = -\tfrac{178}{19}.$$

The desired equation is

$$x^2 + y^2 + z^2 - \tfrac{48}{19}x + \tfrac{32}{19}y - \tfrac{8}{19}z - \tfrac{178}{19} = 0.$$

A **cylindrical surface** is a surface which consists of a collection of parallel lines. Each of the parallel lines is called a **generator** of the **cylinder** or cylindrical surface.

The customary right circular cylinder of elementary geometry is clearly a special case of the type of cylinder we are considering. Figure 1-15 shows some examples of cylindrical surfaces. Note that a plane is a cylindrical surface.

Theorem 10. *An equation of the form*

$$f(x, y) = 0$$

is a cylindrical surface with generators all parallel to the z axis. The surface intersects the xy plane in the curve

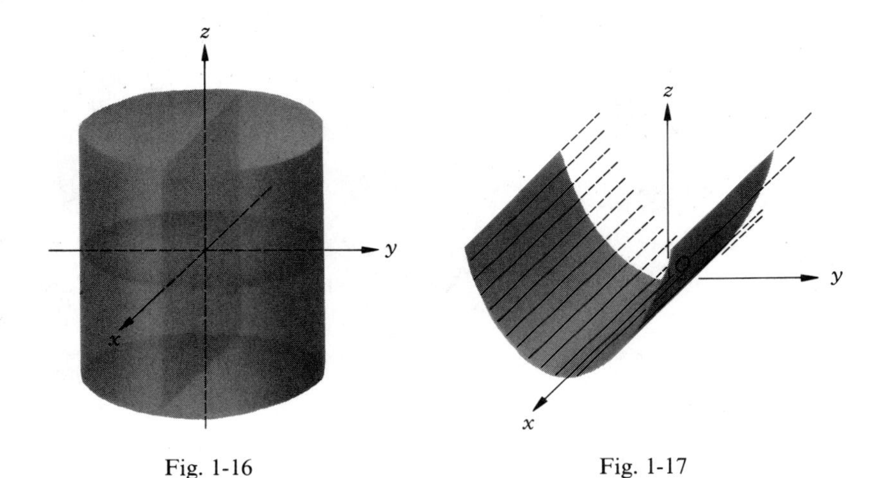

Fig. 1-16 Fig. 1-17

$$f(x, y) = 0, \qquad z = 0.$$

A similar result holds with axes interchanged.

PROOF. Suppose that x_0, y_0 satisfies $f(x_0, y_0) = 0$. Then any point (x_0, y_0, z) for $-\infty < z < \infty$ satisfies the same equation, since z is absent. Therefore the line parallel to the z axis through $(x_0, y_0, 0)$ is a generator.

EXAMPLE 3. Describe and sketch the graph of the equation $x^2 + y^2 = 9$.

SOLUTION. The graph is a right circular cylinder with generators parallel to the z axis (Theorem 10). It is sketched in Fig. 1-16.

EXAMPLE 4. Describe and sketch the graph of the equation $y^2 = 4z$.

SOLUTION. According to Theorem 10 the graph is a cylindrical surface with generators parallel to the x axis. The intersection with the yz plane is a parabola. The graph, called a **parabolic cylinder**, is sketched in Fig. 1-17.

PROBLEMS

In each of Problems 1 through 4, find the equation of the sphere with center at C and given radius r.

1. $C(1, 4, -2)$, $r = 3$
2. $C(2, 0, -3)$, $r = 5$
3. $C(0, 1, 4)$, $r = 6$
4. $C(-2, 1, -3)$, $r = 1$

In each of Problems 5 through 9, determine the graph of the equation. If it is a sphere, find its center and radius.

5. $x^2 + y^2 + z^2 + 2x - 4z + 1 = 0$

6. $x^2 + y^2 + z^2 - 4x + 2y + 6z - 2 = 0$

7. $x^2 + y^2 + z^2 - 2x + 4y - 2z + 7 = 0$

8. $x^2 + y^2 + z^2 + 4x - 2y + 4z + 9 = 0$

9. $x^2 + y^2 + z^2 - 6x + 4y + 2z + 10 = 0$

10. Find the equation of the graph of all points which are twice as far from $A(3, -1, 2)$ as from $B(0, 2, -1)$.

11. Find the equation of the graph of all points which are three times as far from $A(2, 1, -3)$ as from $B(-2, -3, 5)$.

12. Find the equation of the graph of all points whose distances from the point $(0, 0, 4)$ are equal to their perpendicular distances from the xy plane.

In each of Problems 13 through 24, describe and sketch the graph of the given equation.

13. $x = 3$
14. $x^2 + y^2 = 16$
15. $2x + y = 3$
16. $x + 2z = 4$
17. $x^2 = 4z$
18. $z = 2 - y^2$
19. $4x^2 + y^2 = 16$
20. $4x^2 - y^2 = 16$
21. $y^2 + x^2 = 9$
22. $z^2 = 2 - 2x$
23. $x^2 + y^2 - 2x = 0$
24. $z^2 = y^2 + 4$

In each of Problems 25 through 28, describe the curve of intersection, if any, of the given surface S and the given plane Φ. That is, describe the set $S \cap \Phi$.

25. $S: x^2 + y^2 + z^2 = 25$; $\quad \Phi: z = 3$

26. $S: 4x = y^2 + z^2$; $\quad \Phi: x = 4$

27. $S: x^2 + 2y^2 + 3z^2 = 12$; $\quad \Phi: y = 4$

28. $S: x^2 + y^2 = z^2$; $\quad \Phi: 2x + z = 4$

29. Verify that the graph of the equation $(x - 2)^2 + (y - 1)^2 = 0$ is a straight line. Show that every straight line parallel to one of the coordinate axes can be represented by a *single* equation of the *second* degree.

30. If

$$S_1 = \{(x, y, z) : x^2 + y^2 + z^2 + A_1x + B_1y + C_1z + D_1 = 0\}$$

and

$$S_2 = \{(x, y, z) : x^2 + y^2 + z^2 + A_2x + B_2y + C_2z + D_2 = 0\}$$

are two spheres, then the **radical plane** is obtained by subtraction of the equations for S_1 and S_2. It is

$$\Phi = \{(x, y, z) : (A_1 - A_2)x + (B_1 - B_2)y + (C_1 - C_2)z + (D_1 - D_2) = 0\}.$$

Show that the radical plane is perpendicular to the line joining the centers of the spheres S_1 and S_2.

31. Show that the three radical planes (see Problem 30) of three spheres interest in a common line or are parallel.

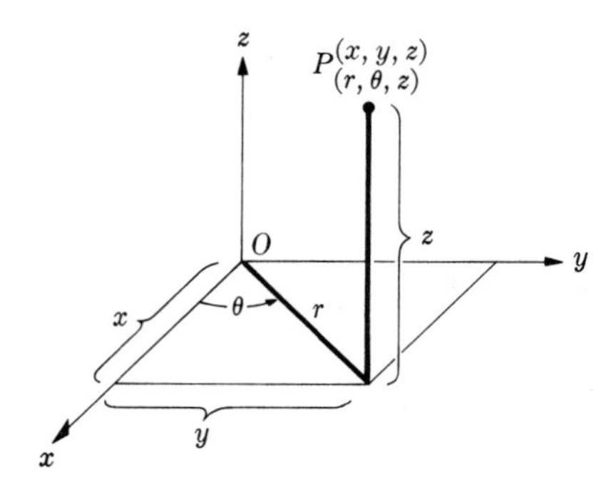

Fig. 1-18

7. Other Coordinate Systems

In plane analytic geometry we employ a rectangular coordinate system for certain types of problems and a polar coordinate system for others. We know that there are circumstances in which one system is more convenient than the other. A similar situation prevails in three-dimensional geometry, and we now take up systems of coordinates other than the rectangular one which we have studied exclusively so far. One such system, known as **cylindrical coordinates**, is described in the following way. A point P in space with rectangular coordinates (x, y, z) may also be located by replacing the x and y values with the corresponding polar coordinates r, θ and by allowing the z value to remain unchanged. In other words, to each ordered number triple of the form (r, θ, z), there is associated a point in space. The transformation from cylindrical to rectangular coordinates is given by the equations

$$x = r\cos\theta, \qquad y = r\sin\theta, \qquad z = z.$$

The transformation from rectangular to cylindrical coordinates is given by

$$r^2 = x^2 + y^2, \qquad \tan\theta = y/x, \qquad z = z.$$

If the coordinates of a point are given in one system, the above equations show how to get the coordinates in the other. Figure 1-18 exhibits the relation between the two systems. It is always assumed that the origins of the systems coincide and that $\theta = 0$ corresponds to the xz plane. We see that the graph of $\theta =$ const consists of all points in a plane containing the z axis. The graph of $r =$ const consists of all points on a right circular cylinder with the z axis as its central axis. (The term "cylindrical coordinates" comes from this fact.) The graph of $z =$ const consists of all points in a plane parallel to the xy plane.

EXAMPLE 1. Find cylindrical coordinates of the points whose rectangular coordinates are $P(3, 3, 5)$, $Q(2, 0, -1)$, $R(0, 4, 4)$, $S(0, 0, 5)$, $T(2, 2\sqrt{3}, 1)$.

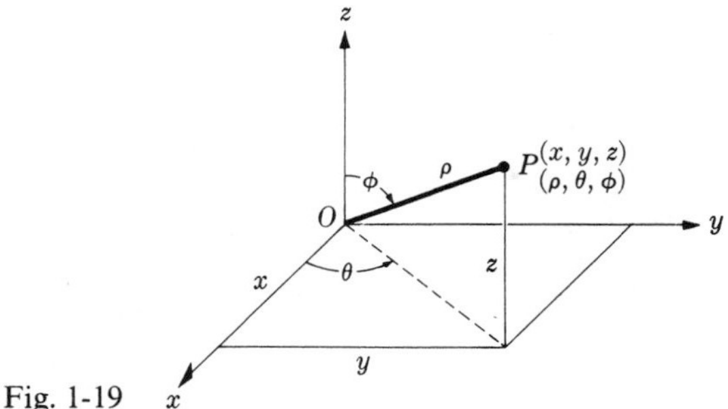

Fig. 1-19

SOLUTION. For the point P we have $r = \sqrt{9+9} = 3\sqrt{2}$, $\tan\theta = 1$, $\theta = \pi/4$, $z = 5$. Therefore one set of coordinates is $(3\sqrt{2}, \pi/4, 5)$. For Q we have $r = 2$, $\theta = 0$, $z = -1$. The coordinates are $(2, 0, -1)$. For R we get $r = 4$, $\theta = \pi/2$, $z = 4$. The result is $(4, \pi/2, 4)$. For S we see at once that the coordinates are $(0, \theta, 5)$ for any θ. For T we get $r = \sqrt{4+12} = 4$, $\tan\theta = \sqrt{3}$, $\theta = \pi/3$. The answer is $(4, \pi/3, 1)$.

REMARK. Just as polar coordinates do not give a one-to-one correspondence between ordered number pairs and points in the plane, so cylindrical coordinates do not give a one-to-one correspondence between ordered number triples and points in space.

A **spherical coordinate system** is defined in the following way. A point P with rectangular coordinates (x, y, z) has spherical coordinates (ρ, θ, ϕ) where ρ is the distance of the point P from the origin, θ is the same quantity as in cylindrical coordinates, and ϕ is the angle that the directed line $\overrightarrow{OP}$ makes with the positive z direction. Figure 1-19 exhibits the relation between rectangular and spherical coordinates. The transformation from spherical to rectangular coordinates is given by the equations

$$x = \rho\sin\phi\cos\theta, \qquad y = \rho\sin\phi\sin\theta, \qquad z = \rho\cos\phi.$$

The transformation from rectangular to spherical coordinates is given by

$$\rho^2 = x^2 + y^2 + z^2,$$

$$\tan\theta = \frac{y}{x},$$

and

$$\cos\phi = \frac{z}{\sqrt{x^2+y^2+z^2}}.$$

We note that the graph of $\rho = \text{const}$ is a sphere with center at the origin (from which is derived the term "spherical coordinates"). The graph of $\theta =$

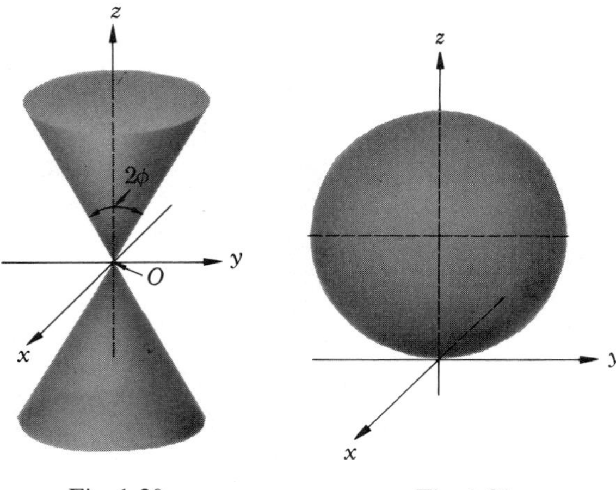

Fig. 1-20 Fig. 1-21

const is a plane through the z axis, as in cylindrical coordinates. The graph of $\phi = \text{const}$ is a **cone** with vertex at the origin and angle opening 2ϕ if $0 < \phi < \pi/2$. (See Fig. 1-20.) The lower nappe of the cone in Fig. 1-20 is or is not included according as negative values of ρ are or are not allowed.

EXAMPLE 2. Find an equation in spherical coordinates of the sphere

$$x^2 + y^2 + z^2 - 2z = 0.$$

Sketch the graph.

SOLUTION. Since $\rho^2 = x^2 + y^2 + z^2$ and $z = \rho \cos \phi$, we have

$$\rho^2 - 2\rho \cos \phi = 0 \quad \Leftrightarrow \quad \rho(\rho - 2 \cos \phi) = 0.$$

The graph of this equation is the graph of $\rho = 0$ and $\rho - 2 \cos \phi = 0$. The graph of $\rho = 0$ is on the graph of $\rho - 2 \cos \phi = 0$ (with $\phi = \pi/2$). Plotting the surface

$$\rho = 2 \cos \phi,$$

we get the surface shown in Fig. 1-21.

If ρ is constant, then the quantities (θ, ϕ) form a coordinate system on the surface of a sphere. Latitude and longitude on the surface of the earth also form a coordinate system. If we restrict θ so that $-\pi < \theta \leq \pi$, then θ is called the **longitude** of the point in spherical coordinates. If ϕ is restricted so that $0 \leq \phi \leq \pi$, then ϕ is called the **colatitude** of the point. That is, ϕ is $(\pi/2) -$ latitude, where latitude is taken in the ordinary sense—i.e., positive north of the equator and negative south of it.

PROBLEMS

1. Find a set of cylindrical coordinates for each of the points whose rectangular coordinates are

   a) $(3, 3, 7)$, b) $(4, 8, 2)$, c) $(-2, 3, 1)$.

2. Find the rectangular coordinates of the points whose cylindrical coordinates are

   a) $(2, \pi/3, 1)$, b) $(3, -\pi/4, 2)$, c) $(7, 2\pi/3, -4)$.

3. Find a set of spherical coordinates for each of the points whose rectangular coordinates are

   a) $(2, 2, 2)$, b) $(2, -2, -2)$, c) $(-1, \sqrt{3}, 2)$.

4. Find the rectangular coordinates of the points whose spherical coordinates are

   a) $(4, \pi/6, \pi/4)$, b) $(6, 2\pi/3, \pi/3)$, c) $(8, \pi/3, 2\pi/3)$.

5. Find a set of cylindrical coordinates for each of the points whose spherical coordinates are

   a) $(4, \pi/3, \pi/2)$, b) $(2, 2\pi/3, 5\pi/6)$, c) $(7, \pi/2, \pi/6)$.

6. Find a set of spherical coordinates for each of the points whose cylindrical coordinates are

   a) $(2, \pi/4, 1)$, b) $(3, \pi/2, 2)$, c) $(1, 5\pi/6, -2)$.

In each of Problems 7 through 16, find an equation in cylindrical coordinates of the graph whose (x, y, z) equation is given. Sketch.

7. $x^2 + y^2 + z^2 = 9$
8. $x^2 + y^2 + 2z^2 = 8$
9. $x^2 + y^2 = 4z$
10. $x^2 + y^2 - 2x = 0$
11. $x^2 + y^2 = z^2$
12. $x^2 + y^2 + 2z^2 + 2z = 0$
13. $x^2 - y^2 = 4$
14. $xy + z^2 = 5$
15. $x^2 + y^2 - 4y = 0$
16. $x^2 + y^2 + z^2 - 2x + 3y - 4z = 0$

In each of Problems 17 through 22, find an equation in spherical coordinates of the graph whose (x, y, z) equation is given. Sketch.

17. $x^2 + y^2 + z^2 - 4z = 0$
18. $x^2 + y^2 + z^2 + 2z = 0$
19. $x^2 + y^2 = z^2$
20. $x^2 + y^2 = 4$
21. $x^2 + y^2 = 4z + 4$ (Solve for ρ in terms of ϕ.)
22. $x^2 + y^2 - z^2 + z - y = 0$
23. Theorem 10 on page 29 describes a surface when the equation of the surface has one of the rectangular coordinates absent. Describe, in the form of a theorem, the nature of a surface with equation $f(r, z) = 0$ where r, θ, z are cylindrical coordinates. Do the same when the equation is of the form $f(\theta, z) = 0$.
24. Same as Problem 23 for spherical coordinates when the equation is of the form $f(\rho, \phi) = 0$.

CHAPTER 2

Vectors

1. Directed Line Segments and Vectors in the Plane

Let A and B be two points in a plane. The length of the line segment joining A and B is denoted $|AB|$.

Definition. The directed segment from A **to** B is defined as the line segment AB which is ordered so that A precedes B. We use the symbol $\overrightarrow{AB}$ to denote such a directed segment. We call A its **base** and B its **head**.

If we select the opposite ordering of the line segment AB, then we get the directed segment $\overrightarrow{BA}$. In this case B precedes A and we call B the base and A the head. To distinguish geometrically the base and head of a directed line segment, we usually draw an arrow at the head, as shown in Fig. 2-1.

Definitions. The **magnitude** of a directed line segment $\overrightarrow{AB}$ is its length $|AB|$. Two directed line segments $\overrightarrow{AB}$ and $\overrightarrow{CD}$ are said to **have the same magnitude and direction** if and only if either one of the following two conditions holds:

i) $\overrightarrow{AB}$ and $\overrightarrow{CD}$ are both on the same line L, their magnitudes are equal,

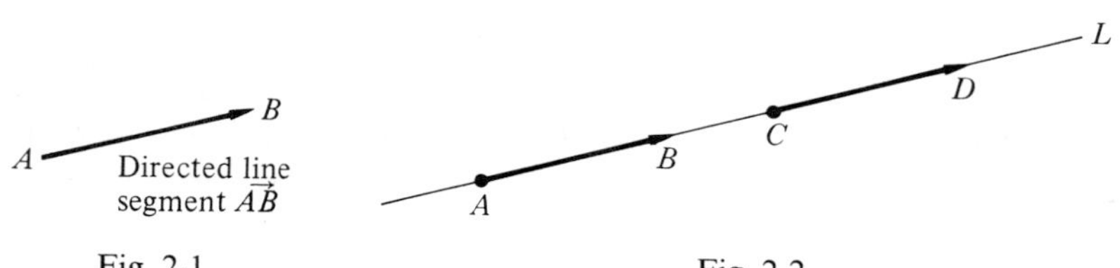

Fig. 2-1

Fig. 2-2

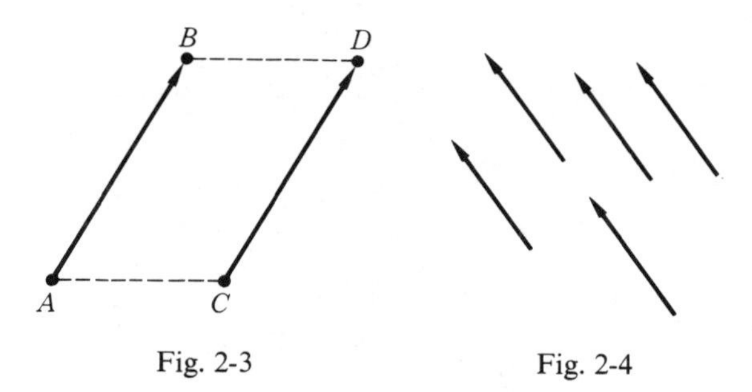

Fig. 2-3 Fig. 2-4

and the heads B and D are pointing in the same direction, as shown in Fig. 2-2.

ii) The points A, C, D, and B are the vertices of a parallelogram, as shown in Fig. 2-3.

Figure 2-4 shows several line segments having the same magnitude and direction. Whenever two directed line segments $\overrightarrow{AB}$ and $\overrightarrow{CD}$ have the same magnitude and direction, we say that they are **equivalent** and write

$$\overrightarrow{AB} \approx \overrightarrow{AD}.$$

We now prove a theorem which expresses this equivalence relationship in terms of coordinates.

Theorem 1. *Suppose that A, B, C, and D are points in a plane. Denote the coordinates of A, B, C, and D by (x_A, y_A), (x_B, y_B), (x_C, y_C), (x_D, y_D), respectively. (See Fig. 2-5.)*

i) *If the coordinates above satisfy the equations*

$$x_B - x_A = x_D - x_C \qquad \text{and} \qquad y_B - y_A = y_D - y_C, \tag{1}$$

then

$$\overrightarrow{AB} \approx \overrightarrow{CD}.$$

ii) *Conversely, if $\overrightarrow{AB}$ is equivalent to $\overrightarrow{CD}$, then the coordinates of the four points satisfy Eqs. (1).*

Proof. First, we suppose that Eqs. (1) hold, and we wish to show that $\overrightarrow{AB} \approx \overrightarrow{CD}$. If $x_B - x_A$ and $y_B - y_A$ are positive, as shown in Fig. 2-5, then both $\overrightarrow{AB}$ and $\overrightarrow{CD}$ are directed segments which are pointing upward and to the right. It is a simple result from plane geometry that ΔABE is congruent to ΔCDF; hence $|AB| = |CD|$ and the line through AB is parallel to (or coincides with) the line through CD. Therefore A, B, D, C are the vertices of a parallelogram or the line through AB coincides with the line through CD. In either case, $\overrightarrow{AB}$ is equivalent to $\overrightarrow{CD}$. The reader can easily draw the

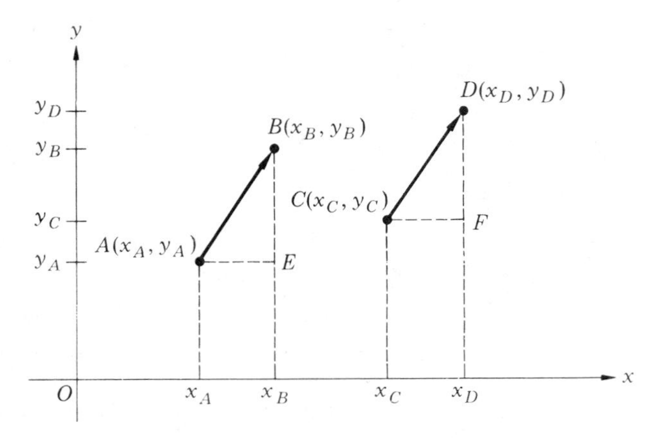

Fig. 2-5

same conclusion if either $x_B - x_A$ or $y_B - y_A$ is negative or if both quantities are negative. See Problem 27 at the end of Section 2.

To prove the converse, we assume that $\overrightarrow{AB} \approx \overrightarrow{CD}$. That is, we suppose that $|AB| = |CD|$ and that the line through AB is parallel to (or coincides with) the line through CD. If the directed line segments appear as in Fig. 2-5, we conclude from plane geometry that ΔABE is congruent to ΔCDF. Consequently $x_B - x_A = x_D - x_C$ and $y_B - y_A = y_D - y_C$. The reader may verify that the result holds generally if $\overrightarrow{AB}$ and $\overrightarrow{CD}$ are pointing in directions other than upward and to the right.

If we are given a directed line segment $\overrightarrow{AB}$, we see at once that there is an unlimited number of equivalent ones. In fact, if C is any given point in the plane, we can use Eqs. (1) of Theorem 1 to find the coordinates of the unique point D such that $\overrightarrow{CD} \approx \overrightarrow{AB}$. Theorem 1 also yields various simple properties of the relation $\approx$. For example, if $\overrightarrow{AB} \approx \overrightarrow{CD}$, then $\overrightarrow{CD} \approx \overrightarrow{AB}$; also if $\overrightarrow{AB} \approx \overrightarrow{CD}$ and $\overrightarrow{CD} \approx \overrightarrow{EF}$, then $\overrightarrow{AB} \approx \overrightarrow{EF}$.

The definition of a vector involves an abstract concept—that of a collection of directed line segments.

Definition. A **vector** is a collection of all directed line segments having a given magnitude and a given direction. We shall use boldface letters to denote vectors; thus when we write $\mathbf{v}$ for a vector, it stands for an entire collection of directed line segments. A particular directed line segment in the collection is called a **representative** of the vector* $\mathbf{v}$. Any member of the collection may be used as a representative.

Figure 2-4 shows five representatives of the same vector. Since any two representative directed line segments of the same vector are equivalent, the collection used to define a vector is called an **equivalence class**.

* We also say that the directed line segment *determines* the vector $\mathbf{v}$.

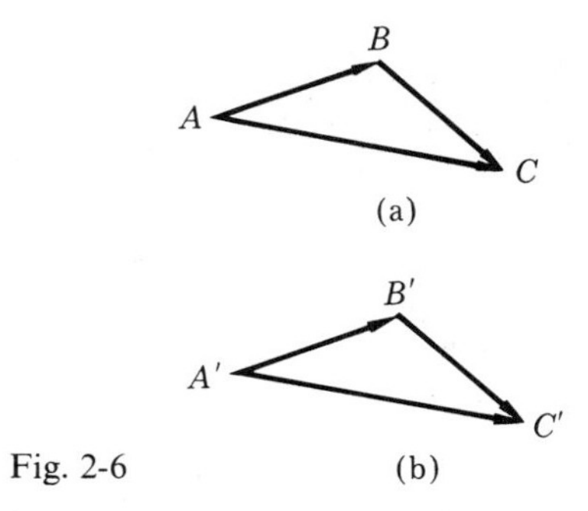

Fig. 2-6

The vector as we have defined it is sometimes called a **free vector**. There are other ways of introducing vectors; one is to call a directed line segment a vector. We then would make the convention that directed line segments with the same magnitude and direction (i.e., equivalent) are equal vectors. When there is no danger of confusion we shall identify directed line segments and vectors.

Vectors occur with great frequency in various branches of physics and engineering. Problems in mechanics, especially those involving forces, are concerned with "lines of action," i.e., the lines along which forces act. In such problems it is convenient (but not necessary) to define a vector as the equivalence class of all directed line segments which lie along a given straight line and which have a given magnitude.

Definition. The **length** of a vector is the common length of all its representative segments. A **unit vector** is a vector of length one. Two vectors are said to be **orthogonal** (or **perpendicular**) if any representative of one vector is perpendicular to any representative of the other (i.e., the representatives lie along perpendicular lines).

For convenience, we consider directed line segments of zero length; these are simply points. The **zero vector**, denoted by $\mathbf{0}$, is the class of directed line segments of zero length. We make the convention that the zero vector is orthogonal to every vector.

2. Operations with Vectors

Vectors may be added to yield other vectors. Suppose $\mathbf{u}$ and $\mathbf{v}$ are vectors, i.e., each is a collection of directed line segments. To add $\mathbf{u}$ and $\mathbf{v}$, first select a representative of $\mathbf{u}$, say $\overrightarrow{AB}$, as shown in Fig. 2-6(a). Next take the particular representative of $\mathbf{v}$ which has its base at the point B, and label it $\overrightarrow{BC}$. Then draw the directed line segment $\overrightarrow{AC}$. The sum $\mathbf{w}$ of $\mathbf{u}$ and $\mathbf{v}$ is the class of directed line segments of which $\overrightarrow{AC}$ is a representative. We write

$$\mathbf{u} + \mathbf{v} = \mathbf{w}.$$

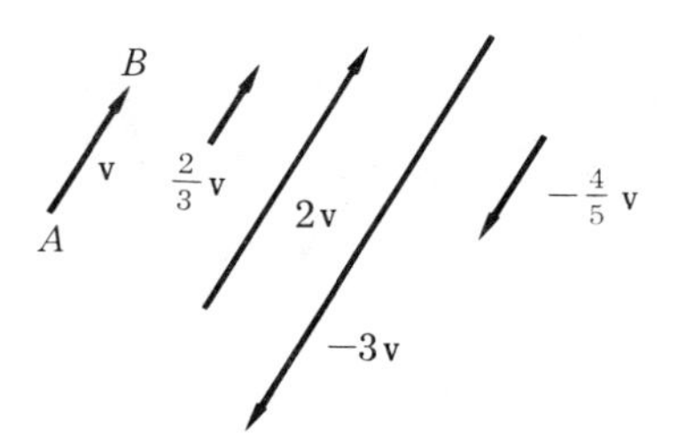

Fig. 2-7

It is important to note that we could have started with any representative of **u**, say $\overrightarrow{A'B'}$ in Fig. 2-6(b). Then we could have selected the representative of **v** with base at B'. The directed line segment $\overrightarrow{A'C'}$ and $\overrightarrow{AC}$ are representatives of the same vector, as is easily seen from Theorem 1.

Vectors may be multiplied by numbers to yield new vectors. If **v** is a vector and c is a *positive* real number, then $c\mathbf{v}$ is a vector with its representatives having the same direction as those of **v** but with magnitudes c times as long as the representatives of **v**. If c is a negative number, then the representatives of $c\mathbf{v}$ have the opposite direction to those of **v** and their magnitudes are $|c|$ times as long as those of **v**. If c is zero, we get the vector **0**. Figure 2-7 shows various multiples of a representative line segment $\overrightarrow{AB}$ of a vector **v**. We write $-\mathbf{v}$ for the vector $(-1)\mathbf{v}$.

Definitions. Suppose we are given a rectangular coordinate system in the plane. We call I the point with coordinates $(1,0)$ and J the point with coordinates $(0,1)$ as shown in Fig. 2-8. The **unit vector i** is defined as the vector which has $\overrightarrow{OI}$ as a representative. The **unit vector j** is defined as the vector which has $\overrightarrow{OJ}$ as a representative.

Theorem 2. *Suppose a vector* **w** *has* $\overrightarrow{AB}$ *as a representative. Denote the coordinates of A and B by* (x_A, y_A) *and* (x_B, y_B), *respectively. Then* **w** *may be expressed in the form*

$$\boxed{\mathbf{w} = (x_B - x_A)\mathbf{i} + (y_B - y_A)\mathbf{j}.}$$

PROOF. From Eqs. (1) in Theorem 1, we know that **w** is given by $\overrightarrow{OP}$ where P has coordinates $(x_B - x_A, y_B - y_A)$. (See Fig. 2-8.) Let $Q(x_B - x_A, 0)$ and $R(0, y_B - y_A)$ be the points on the coordinates axes as shown in Fig. 2-8. It is clear geometrically that $\overrightarrow{OQ}$ is $(x_B - x_A)$ times as long as $\overrightarrow{OI}$ and that $\overrightarrow{OR}$ is $(y_B - y_A)$ times as long as $\overrightarrow{OJ}$. We denote by **u** the vector given by $\overrightarrow{OQ}$ and by **v** the vector given by $\overrightarrow{OR}$. Using the rule for addition of vectors, we obtain

$$\mathbf{w} = \mathbf{u} + \mathbf{v},$$

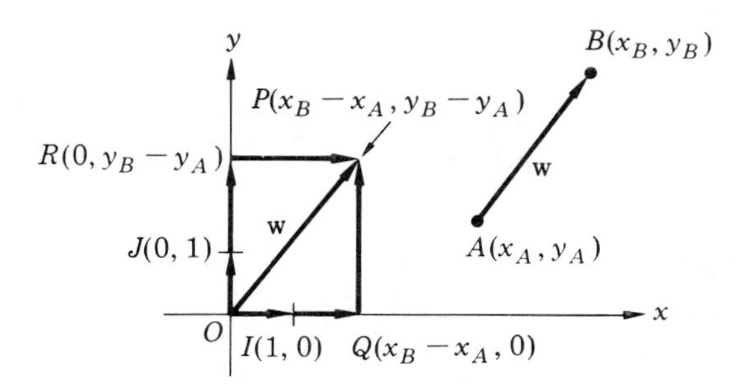

Fig. 2-8

since $\overrightarrow{OP}$ as well as $\overrightarrow{OR}$ determines **v**. Since $\mathbf{u} = (x_B - x_A)\mathbf{i}$ and $\mathbf{v} = (y_B - y_A)\mathbf{j}$, the result of the theorem is established.

EXAMPLE 1. A vector **v** is determined by $\overrightarrow{AB}$. Given that A has coordinates $(3, -2)$ and B has coordinates $(1, 1)$, express **v** in terms of **i** and **j**. Draw a figure.

SOLUTION. Using the formula in Theorem 2, we have (see Fig. 2-9)

$$\mathbf{v} = (1 - 3)\mathbf{i} + (1 + 2)\mathbf{j} = -2\mathbf{i} + 3\mathbf{j}.$$

NOTATION. The length of a vector **v** will be denoted by $|\mathbf{v}|$.

Theorem 3. *If* $\mathbf{v} = a\mathbf{i} + b\mathbf{j}$, *then*

$$\boxed{|\mathbf{v}| = \sqrt{a^2 + b^2}.}$$

Therefore $\mathbf{v} = \mathbf{0}$ *if and only if* $a = b = 0$.

PROOF. By means of Theorem 2, we know that **v** is determined by the directed

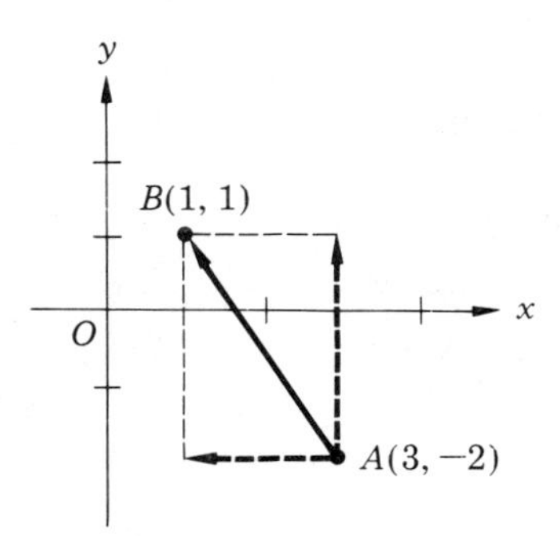

Fig. 2-9

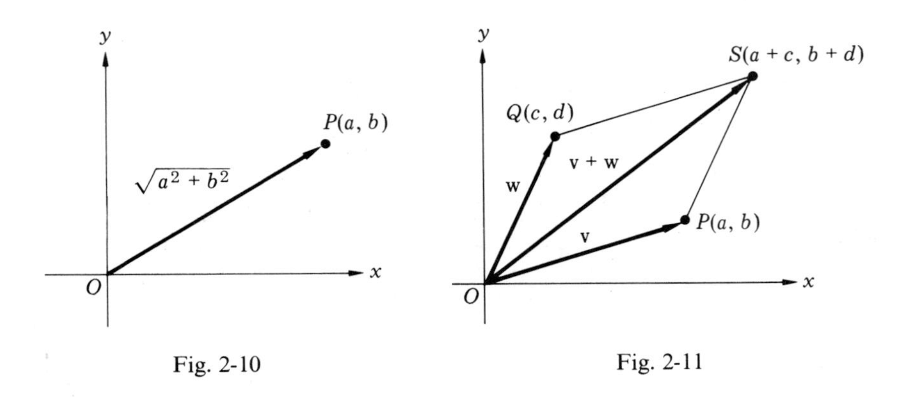

Fig. 2-10 Fig. 2-11

segment $\overrightarrow{OP}$ where P has coordinates (a, b). (See Fig. 2-10.) Then $|\overrightarrow{OP}| = \sqrt{a^2 + b^2}$; since, by definition, the length of a vector is the length of any of its representatives, the result follows.

The next theorem is useful for problems concerned with the addition of vectors and the multiplication of vectors by numbers.

Theorem 4. *If* $\mathbf{v} = a\mathbf{i} + b\mathbf{j}$ *and* $\mathbf{w} = c\mathbf{i} + d\mathbf{j}$, *then*

$$\boxed{\mathbf{v} + \mathbf{w} = (a + c)\mathbf{i} + (b + d)\mathbf{j}.}$$

Further, if h is any number, then

$$\boxed{h\mathbf{v} = (ha)\mathbf{i} + (hb)\mathbf{j}.}$$

PROOF. Let P, Q, and S have coordinates as shown in Fig. 2-11. Then $\overrightarrow{OP}$ and $\overrightarrow{OQ}$ determine $\mathbf{v}$ and $\mathbf{w}$, respectively. Since $\overrightarrow{PS} \approx \overrightarrow{OQ}$, we use the rule for addition of vectors to find that $\overrightarrow{OS}$ determines $\mathbf{v} + \mathbf{w}$. The point R (see Fig. 2-12) is h of the way from O to P. Hence $\overrightarrow{OR}$ is a representative of $h\mathbf{v}$.

We conclude from the theorems above that the addition of vectors and their multiplication by numbers satisfy the following laws:

$$\left.\begin{aligned} \mathbf{u} + (\mathbf{v} + \mathbf{w}) &= (\mathbf{u} + \mathbf{v}) + \mathbf{w} \\ c(d\mathbf{v}) &= (cd)\mathbf{v} \end{aligned}\right\} \quad \text{Associative laws}$$

$$\mathbf{u} + \mathbf{v} = \mathbf{v} + \mathbf{u} \qquad \text{Commutative law}$$

$$\left.\begin{aligned} (c + d)\mathbf{v} &= c\mathbf{v} + d\mathbf{v} \\ c(\mathbf{u} + \mathbf{v}) &= c\mathbf{u} + c\mathbf{v} \end{aligned}\right\} \quad \text{Distributive laws}$$

$$1 \cdot \mathbf{u} = \mathbf{u}, \qquad 0 \cdot \mathbf{u} = \mathbf{0}, \qquad (-1)\mathbf{u} = -\mathbf{u}$$

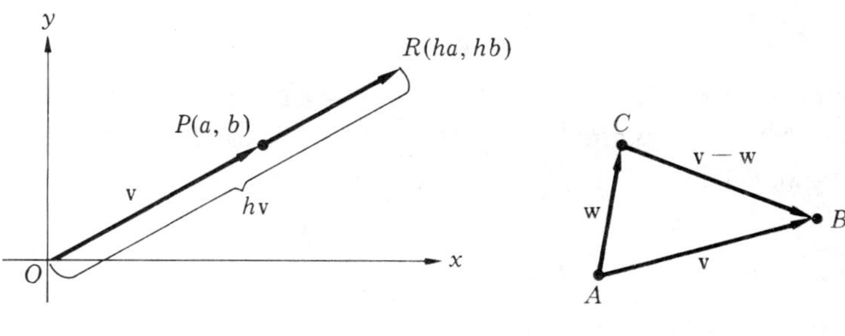

Fig. 2-12 Fig. 2-13

where $-\mathbf{u}$ denotes that vector such that $\mathbf{u} + (-\mathbf{u}) = \mathbf{0}$. These laws hold for all $\mathbf{u}$, $\mathbf{v}$, $\mathbf{w}$ and all numbers c and d. It is important to note that multiplication and division of vectors in the ordinary sense is not (and will not be) defined. However, subtraction is defined. If $\overrightarrow{AB}$ is a representative of $\mathbf{v}$ and $\overrightarrow{AC}$ is one of $\mathbf{w}$, then $\overrightarrow{CB}$ is a representative of $\mathbf{v} - \mathbf{w}$ (see Fig. 2-13).

EXAMPLE 2. Given the vectors $\mathbf{u} = 2\mathbf{i} - 3\mathbf{j}$, $\mathbf{v} = -4\mathbf{i} + \mathbf{j}$. Express the vector $2\mathbf{u} - 3\mathbf{v}$ in terms of $\mathbf{i}$ and $\mathbf{j}$.

SOLUTION. $2\mathbf{u} = 4\mathbf{i} - 6\mathbf{j}$ and $-3\mathbf{v} = 12\mathbf{i} - 3\mathbf{j}$. Adding these vectors, we get $2\mathbf{u} - 3\mathbf{v} = 16\mathbf{i} - 9\mathbf{j}$.

Definition. Let $\mathbf{v}$ be any vector except $\mathbf{0}$. The unit vector $\mathbf{u}$ in the direction of $\mathbf{v}$ is defined by

$$\mathbf{u} = \left(\frac{1}{|\mathbf{v}|}\right)\mathbf{v}.$$

EXAMPLE 3. Given the vector $\mathbf{v} = -2\mathbf{i} + 3\mathbf{j}$, find a unit vector in the direction of $\mathbf{v}$.

SOLUTION. We have $|\mathbf{v}| = \sqrt{4+9} = \sqrt{13}$. The desired vector $\mathbf{u}$ is

$$\mathbf{u} = \frac{1}{\sqrt{13}}\mathbf{v} = -\frac{2}{\sqrt{13}}\mathbf{i} + \frac{3}{\sqrt{13}}\mathbf{j}.$$

EXAMPLE 4. Given the vector $\mathbf{v} = 2\mathbf{i} - 4\mathbf{j}$. Find the directed line segment $\overrightarrow{AB}$ of $\mathbf{v}$, given that A has coordinates $(3, -5)$.

SOLUTION. Denote the coordinates of B by x_B, y_B. Then we have (by Theorem 2)

$$x_B - 3 = 2 \quad \text{and} \quad y_B + 5 = -4.$$

Therefore $x_B = 5$, $y_B = -9$.

Problems

In Problems 1 through 5, express $\mathbf{v}$ in terms of $\mathbf{i}$ and $\mathbf{j}$, given that the endpoints A and B of the directed segment $\overrightarrow{AB}$ of $\mathbf{v}$ have the given coordinates. Draw a figure.

1. $A(3, -2)$, $B(1, 5)$
2. $A(-4, 1)$, $B(2, -1)$
3. $A(5, -6)$, $B(0, -2)$
4. $A(4, 4)$, $B(8, -3)$
5. $A(7, -2)$, $B(-5, -6)$

In Problems 6 through 9, in each case find a unit vector $\mathbf{u}$ in the direction of $\mathbf{v}$. Express $\mathbf{u}$ in terms of $\mathbf{i}$ and $\mathbf{j}$.

6. $\mathbf{v} = 4\mathbf{i} + 3\mathbf{j}$
7. $\mathbf{v} = -5\mathbf{i} - 12\mathbf{j}$
8. $\mathbf{v} = 2\mathbf{i} - 2\sqrt{3}\mathbf{j}$
9. $\mathbf{v} = -2\mathbf{i} + 5\mathbf{j}$

In Problems 10 through 15, find the representative $\overrightarrow{AB}$ of the vector $\mathbf{v}$ from the information given. Draw a figure.

10. $\mathbf{v} = 7\mathbf{i} - 3\mathbf{j}$, $A(2, -1)$
11. $\mathbf{v} = -2\mathbf{i} + 4\mathbf{j}$, $A(6, 2)$
12. $\mathbf{v} = 3\mathbf{i} + 2\mathbf{j}$, $B(-2, 1)$
13. $\mathbf{v} = -4\mathbf{i} - 2\mathbf{j}$, $B(0, 5)$
14. $\mathbf{v} = 3\mathbf{i} + 2\mathbf{j}$, midpoint of segment AB has coordinates $(3, 1)$
15. $\mathbf{v} = -2\mathbf{i} + 3\mathbf{j}$, midpoint of segment AB has coordinates $(-4, 2)$
16. Find a representative of the vector $\mathbf{v}$ of unit length making an angle of 30° with the positive x direction. Express $\mathbf{v}$ in terms of $\mathbf{i}$ and $\mathbf{j}$.
17. Find the vector $\mathbf{v}$ (in terms of $\mathbf{i}$ and $\mathbf{j}$) which has length $2\sqrt{2}$ and makes an angle of 45° with the positive y axis (two solutions).
18. Given that $\mathbf{u} = 3\mathbf{i} - 2\mathbf{j}$, $\mathbf{v} = 4\mathbf{i} + 3\mathbf{j}$. Find $\mathbf{u} + \mathbf{v}$ in terms of $\mathbf{i}$ and $\mathbf{j}$. Draw a figure.
19. Given that $\mathbf{u} = -2\mathbf{i} + 3\mathbf{j}$, $\mathbf{v} = \mathbf{i} - 2\mathbf{j}$. Find $\mathbf{u} + \mathbf{v}$ in terms of $\mathbf{i}$ and $\mathbf{j}$. Draw a figure.
20. Given that $\mathbf{u} = -3\mathbf{i} - 2\mathbf{j}$, $\mathbf{v} = 2\mathbf{i} + \mathbf{j}$. Find $3\mathbf{u} - 2\mathbf{v}$ in terms of $\mathbf{i}$ and $\mathbf{j}$. Draw a figure.
21. Show that if $\overrightarrow{AB} \approx \overrightarrow{CD}$ and $\overrightarrow{CD} \approx \overrightarrow{EF}$, then $\overrightarrow{AB} \approx \overrightarrow{EF}$.
22. Show that if $\overrightarrow{AB} \approx \overrightarrow{DE}$ and $\overrightarrow{BC} \approx \overrightarrow{EF}$, then $\overrightarrow{AC} \approx \overrightarrow{DF}$. Draw a figure.
23. Show that if $\overrightarrow{AB} \approx \overrightarrow{DE}$, c is any real number, C is the point c of the way from A to B, and F is the point c of the way from D to E, then $\overrightarrow{AC} \approx \overrightarrow{DF}$. Draw a figure.
24. Show that the vectors $\mathbf{v} = 2\mathbf{i} + 4\mathbf{j}$ and $\mathbf{w} = 10\mathbf{i} - 5\mathbf{j}$ are orthogonal.
25. Show that the vectors $\mathbf{v} = -3\mathbf{i} + \sqrt{2}\mathbf{j}$ and $\mathbf{w} = 4\sqrt{2}\mathbf{i} + 12\mathbf{j}$ are orthogonal.
26. Let $\mathbf{u}$ and $\mathbf{v}$ be two nonzero vectors. Show that they are orthogonal if and only if the equation

$$|\mathbf{u} + \mathbf{v}|^2 = |\mathbf{u}|^2 + |\mathbf{v}|^2$$

holds.

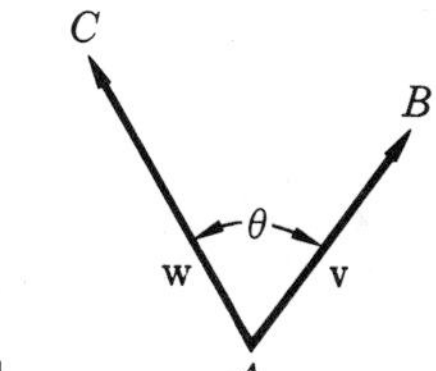

Fig. 2-14

27. Vectors in the plane may be divided into categories as follows: pointing upward to the right, upward to the left, downward to the right, downward to the left; also those pointing vertically and those horizontally. Theorem 1, part (i) was established for vectors pointing upward to the right (Fig. 2-5). Write a proof of Theorem 1(i) for three of the remaining five possibilities. Draw a figure and devise, if you can, a general proof.

28. The same as Problem 27 for part (ii) of Theorem 1.

29. Write out a proof establishing the associative, commutative, and distributive laws for vectors. (See page 42.)

3. Operations with Plane Vectors, Continued. The Scalar Product

Two vectors $\mathbf{v}$ and $\mathbf{w}$ are said to be **parallel** or **proportional** when each is a scalar multiple of the other (and neither is zero). Parallel vectors have parallel directed line segments.

By the **angle between two vectors** $\mathbf{v}$ and $\mathbf{w}$ (neither $= \mathbf{0}$), we mean the measure of the angle between two representatives of $\mathbf{v}$ and $\mathbf{w}$ having the same base (see Fig. 2-14). Two parallel vectors make an angle either of 0 or of π, depending on whether they are pointing in the same or opposite directions.

Suppose that $\mathbf{i}$ and $\mathbf{j}$ are the usual unit vectors pointing in the direction of the x and y axes, respectively. Then we have the following theorem.

Theorem 5. *If θ is the angle between the vectors*

$$\mathbf{v} = a\mathbf{i} + b\mathbf{j} \qquad \textit{and} \qquad \mathbf{w} = c\mathbf{i} + d\mathbf{j},$$

then

$$\boxed{\cos\theta = \frac{ac + bd}{|\mathbf{v}|\,|\mathbf{w}|}.}$$

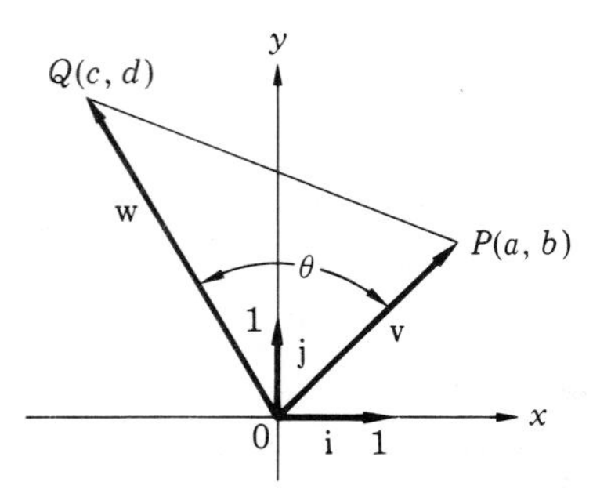

Fig. 2-15

PROOF. We draw the directed line segments of **v** and **w** with base at the origin of the coordinate system, as shown in Fig. 2-15. Using Theorem 2 of Section 2, we see that the coordinates of P are (a, b) and those of Q are (c, d). The length of **v** is $|OP|$ and the length of **w** is $|OQ|$. We apply the law of cosines to $\triangle OPQ$, obtaining

$$\cos\theta = \frac{|OP|^2 + |OQ|^2 - |QP|^2}{2|OP||OQ|}.$$

Therefore

$$\cos\theta = \frac{a^2 + b^2 + c^2 + d^2 - (a-c)^2 - (b-d)^2}{2|OP||OQ|} = \frac{ac + bd}{|\mathbf{v}||\mathbf{w}|}.$$

EXAMPLE 1. Given the vectors $\mathbf{v} = 2\mathbf{i} - 3\mathbf{j}$ and $\mathbf{w} = \mathbf{i} - 4\mathbf{j}$, compute the cosine of the angle between **v** and **w**.

SOLUTION. We have

$$|\mathbf{v}| = \sqrt{4+9} = \sqrt{13}, \qquad |\mathbf{w}| = \sqrt{1+16} = \sqrt{17}.$$

Therefore

$$\cos\theta = \frac{2\cdot 1 + (-3)(-4)}{\sqrt{17}\sqrt{13}} = \frac{14}{\sqrt{221}}.$$

Suppose that we have two directed line segments $\overrightarrow{AB}$ and $\overrightarrow{CD}$, as shown in Fig. 2-16. The **projection of $\overrightarrow{AB}$ in the direction $\overrightarrow{CD}$** is defined as the directed length, denoted $\overrightarrow{EF}$, of the line segment EF obtained by dropping perpendiculars from A and B to the line containing $\overrightarrow{CD}$. We can find the projection of a vector **v** along a vector **w** by first taking a representative of **v** and finding its projection in the direction of a representative of **w**.

If θ is the angle between two vectors **v** and **w**, the quantity

$$|\mathbf{v}|\cos\theta$$

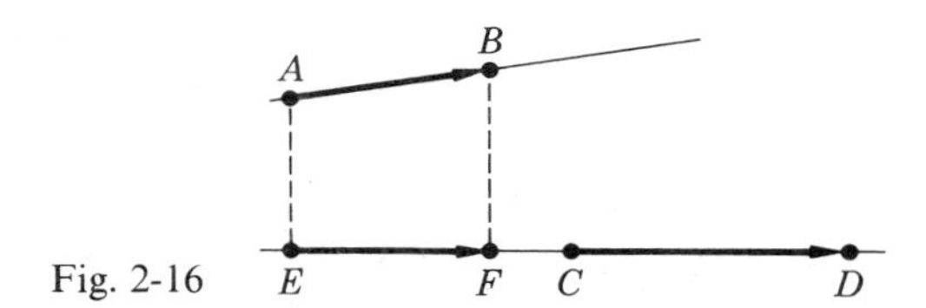

Fig. 2-16

is positive if θ is an acute angle and the projection of $\mathbf{v}$ on $\mathbf{w}$ is positive. If θ is an obtuse angle, then $|\mathbf{v}|\cos\theta$ is negative, and the projection of $\mathbf{v}$ on $\mathbf{w}$ is negative.

Definition. The quantity $|\mathbf{v}|\cos\theta$ is called **the projection of v on w**. We write $\text{Proj}_{\mathbf{w}}\,\mathbf{v}$ for this quantity. If $\mathbf{v} = a\mathbf{i} + b\mathbf{j}$ and $\mathbf{w} = c\mathbf{i} + d\mathbf{j}$, then

$$\boxed{\text{Proj}_{\mathbf{w}}\,\mathbf{v} = |\mathbf{v}|\cos\theta = \frac{ac + bd}{|\mathbf{w}|}.}$$

EXAMPLE 2. Find the projection of $\mathbf{v} = 3\mathbf{i} - 2\mathbf{j}$ on $\mathbf{w} = -2\mathbf{i} - 4\mathbf{j}$.

SOLUTION. We obtain the desired projection by using the formula

$$\text{Proj}_{\mathbf{w}}\,\mathbf{v} = \frac{-6 + 8}{\sqrt{20}} = \frac{1}{\sqrt{5}}.$$

The **scalar product** of two nonzero vectors $\mathbf{v}$ and $\mathbf{w}$, written $\mathbf{v} \cdot \mathbf{w}$, is defined by the formula

$$\boxed{\mathbf{v} \cdot \mathbf{w} = |\mathbf{v}|\,|\mathbf{w}|\cos\theta,}$$

where θ is the angle between $\mathbf{v}$ and $\mathbf{w}$. If one of the vectors is $\mathbf{0}$, the scalar product is defined to be 0. The terms **dot product** and **inner product** are also used to designate scalar product. It is evident from the definition that scalar product satisfies the relations

$$\mathbf{v} \cdot \mathbf{w} = \mathbf{w} \cdot \mathbf{v}, \qquad \mathbf{v} \cdot \mathbf{v} = |\mathbf{v}|^2.$$

Furthermore, if $\mathbf{v}$ and $\mathbf{w}$ are orthogonal, then

$$\mathbf{v} \cdot \mathbf{w} = 0,$$

and conversely. If $\mathbf{v}$ and $\mathbf{w}$ are parallel, we have $\mathbf{v} \cdot \mathbf{w} = \pm|\mathbf{v}|\,|\mathbf{w}|$, and conversely. In terms of the orthogonal unit vectors $\mathbf{i}$ and $\mathbf{j}$, vectors $\mathbf{v} = a\mathbf{i} + b\mathbf{j}$ and $\mathbf{w} = c\mathbf{i} + d\mathbf{j}$ have as their scalar product (see Theorem 5)

$$\boxed{\mathbf{v} \cdot \mathbf{w} = ac + bd.}$$

In addition, it can be verified that the **distributive law**

$$\mathbf{u} \cdot (\mathbf{v} + \mathbf{w}) = \mathbf{u} \cdot \mathbf{v} + \mathbf{u} \cdot \mathbf{w}$$

holds for any three vectors.

EXAMPLE 3. Given the vectors $\mathbf{u} = 3\mathbf{i} + 2\mathbf{j}$ and $\mathbf{v} = 2\mathbf{i} + a\mathbf{j}$. Determine the number a so that $\mathbf{u}$ and $\mathbf{v}$ are orthogonal. Determine a so that $\mathbf{u}$ and $\mathbf{v}$ are parallel. For what value of a will $\mathbf{u}$ and $\mathbf{v}$ make an angle of $\pi/4$?

SOLUTION. If $\mathbf{u}$ and $\mathbf{v}$ are orthogonal, we have

$$3 \cdot 2 + 2 \cdot a = 0 \qquad \text{and} \qquad a = -3.$$

For $\mathbf{u}$ and $\mathbf{v}$ to be parallel, we must have

$$\mathbf{u} \cdot \mathbf{v} = \pm |\mathbf{u}|\,|\mathbf{v}|$$

or

$$6 + 2a = \pm\sqrt{13} \cdot \sqrt{4 + a^2}.$$

Solving, we obtain $a = \frac{4}{3}$.

Employing the formula in Theorem 5, we find

$$\tfrac{1}{2}\sqrt{2} = \cos\frac{\pi}{4} = \frac{6 + 2a}{\sqrt{13} \cdot \sqrt{4 + a^2}},$$

so that $\mathbf{u}$ and $\mathbf{v}$ make an angle of $\pi/4$ when

$$a = 10, \ -\tfrac{2}{5}.$$

Because they are geometric quantities which are independent of the coordinate system, vectors are well suited for establishing certain types

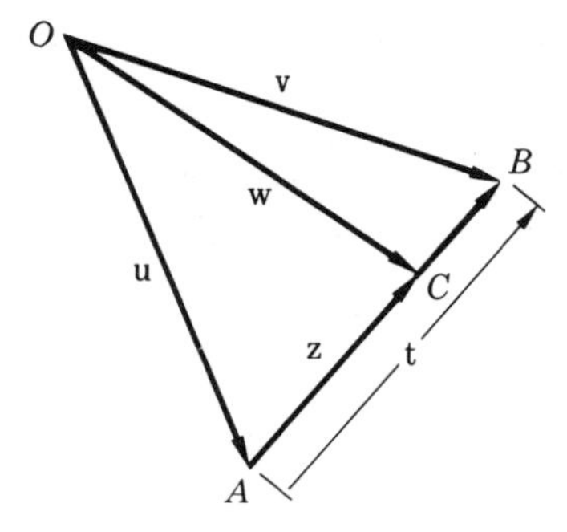

Fig. 2-17

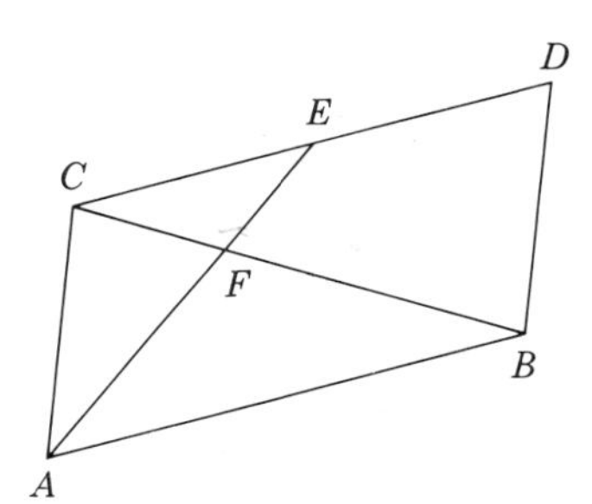

Fig. 2-18

of theorems in plane geometry. We give two examples to exhibit the technique.

EXAMPLE 4. Let $\overrightarrow{OA}$ be a representative of $\mathbf{u}$ and $\overrightarrow{OB}$ a representative of $\mathbf{v}$. Let C be the point on line AB which is $\frac{2}{3}$ of the way from A to B (Fig. 2-17). Express in terms of $\mathbf{u}$ and $\mathbf{v}$ the vector $\mathbf{w}$ which has $\overrightarrow{OC}$ as representative.

SOLUTION. Let $\mathbf{z}$ be the vector with $\overrightarrow{AC}$ as representative, and $\mathbf{t}$ the vector with $\overrightarrow{AB}$ as representative. We have

$$\mathbf{w} = \mathbf{u} + \mathbf{z} = \mathbf{u} + \tfrac{2}{3}\mathbf{t}.$$

Also we know that $\mathbf{t} = \mathbf{v} - \mathbf{u}$, and so

$$\mathbf{w} = \mathbf{u} + \tfrac{2}{3}(\mathbf{v} - \mathbf{u}) = \tfrac{1}{3}\mathbf{u} + \tfrac{2}{3}\mathbf{v}.$$

Let $\overrightarrow{AB}$ be a directed line segment. We introduce a convenient symbol for the vector which has $\overrightarrow{AB}$ as a representative.

NOTATION. The symbol $\mathbf{v}[\overrightarrow{AB}]$ denotes the vector which has $\overrightarrow{AB}$ as a representative.

EXAMPLE 5. Let $ABDC$ be a parallelogram, as shown in Fig. 2-18. Suppose that E is the midpoint of CD and F is $\frac{2}{3}$ of the way from A to E on AE. Show that F is $\frac{2}{3}$ of the way from B to C.

SOLUTION. Let $\overrightarrow{AB}$, $\overrightarrow{AC}$, $\overrightarrow{AF}$, $\overrightarrow{AE}$, $\overrightarrow{BF}$, $\overrightarrow{BC}$, and $\overrightarrow{CE}$ determine vectors. Then $\mathbf{v}[\overrightarrow{AB}]$, $\mathbf{v}[\overrightarrow{AC}]$, $\mathbf{v}[\overrightarrow{AF}]$, etc., are the vectors which determine the directed line segments shown in brackets. By hypothesis, we have

$$\mathbf{v}[\overrightarrow{AB}] = \mathbf{v}[\overrightarrow{CD}] \qquad \text{and} \qquad \mathbf{v}[\overrightarrow{CE}] = \tfrac{1}{2}\mathbf{v}[\overrightarrow{CD}].$$

The rule for addition of vectors gives us

$$\mathbf{v}[\overrightarrow{AE}] = \mathbf{v}[\overrightarrow{AC}] + \mathbf{v}[\overrightarrow{CE}] = \mathbf{v}[\overrightarrow{AC}] + \tfrac{1}{2}\mathbf{v}[\overrightarrow{AB}].$$

Also, since $\mathbf{v}[\overrightarrow{AF}] = \frac{2}{3}\mathbf{v}[\overrightarrow{AE}]$, we obtain

$$\mathbf{v}[\overrightarrow{AF}] = \tfrac{2}{3}\mathbf{v}[\overrightarrow{AC}] + \tfrac{1}{3}\mathbf{v}[\overrightarrow{AB}].$$

The rule for subtraction of vectors yields

$$\mathbf{v}[\overrightarrow{BF}] = \mathbf{v}[\overrightarrow{AF}] - \mathbf{v}[\overrightarrow{AB}] = \tfrac{2}{3}\mathbf{v}[\overrightarrow{AC}] - \tfrac{2}{3}\mathbf{v}[\overrightarrow{AB}] = \tfrac{2}{3}(\mathbf{v}[\overrightarrow{AC}] - \mathbf{v}[\overrightarrow{AB}]).$$

Since $\mathbf{v}[\overrightarrow{BC}] - \mathbf{v}[\overrightarrow{AC}] - \mathbf{v}[\overrightarrow{AB}]$, we conclude that

$$\mathbf{v}[\overrightarrow{BF}] = \tfrac{2}{3}\mathbf{v}[\overrightarrow{BC}],$$

which is the desired result.

Problems

In Problems 1 through 6, given that θ is the angle between $\mathbf{v}$ and $\mathbf{w}$, find $|\mathbf{v}|$, $|\mathbf{w}|$, $\cos\theta$, and the projection of $\mathbf{v}$ on $\mathbf{w}$.

1. $\mathbf{v} = 4\mathbf{i} - 3\mathbf{j}$, $\mathbf{w} = -4\mathbf{i} - 3\mathbf{j}$
2. $\mathbf{v} = 2\mathbf{i} + \mathbf{j}$, $\mathbf{w} = \mathbf{i} - \mathbf{j}$
3. $\mathbf{v} = 3\mathbf{i} - 4\mathbf{j}$, $\mathbf{w} = 5\mathbf{i} + 12\mathbf{j}$
4. $\mathbf{v} = 4\mathbf{i} + \mathbf{j}$, $\mathbf{w} = 6\mathbf{i} - 8\mathbf{j}$
5. $\mathbf{v} = 3\mathbf{i} + 2\mathbf{j}$, $\mathbf{w} = 2\mathbf{i} - 3\mathbf{j}$
6. $\mathbf{v} = 6\mathbf{i} + 5\mathbf{j}$, $\mathbf{w} = 2\mathbf{i} + 5\mathbf{j}$

In Problems 7 through 12, find the projection of the vector $\overrightarrow{AB}$ on the vector $\overrightarrow{CD}$. Draw a figure in each case.

7. $A(1,0)$, $B(2,3)$, $C(1,1)$, $D(-1,1)$
8. $A(0,0)$, $B(1,4)$, $C(0,0)$, $D(2,7)$
9. $A(3,1)$, $B(5,2)$, $C(-2,-1)$, $D(-1,3)$
10. $A(2,4)$, $B(4,7)$, $C(6,-1)$, $D(2,2)$
11. $A(2,-1)$, $B(1,3)$, $C(5,2)$, $D(9,3)$
12. $A(1,6)$, $B(2,5)$, $C(5,2)$, $D(9,3)$

In Problems 13 through 17, find $\cos\theta$ and $\cos\alpha$, given that $\theta = \angle ABC$ and $\alpha = \angle BAC$. Use vector methods and draw figures.

13. $A(-1,1)$, $B(3,-1)$, $C(3,4)$
14. $A(2,1)$, $B(-1,2)$, $C(1,3)$
15. $A(3,4)$, $B(5,1)$, $C(4,1)$
16. $A(4,1)$, $B(1,-1)$, $C(3,3)$
17. $A(0,0)$, $B(3,-5)$, $C(6,-10)$

In Problems 18 through 24, determine the number a (if possible) such that the given condition for $\mathbf{v}$ and $\mathbf{w}$ is satisfied.

18. $\mathbf{v} = 2\mathbf{i} + a\mathbf{j}$, $\mathbf{w} = \mathbf{i} + 3\mathbf{j}$, $\mathbf{v}$ and $\mathbf{w}$ orthogonal
19. $\mathbf{v} = \mathbf{i} - 3\mathbf{j}$, $\mathbf{w} = 2a\mathbf{i} + \mathbf{j}$, $\mathbf{v}$ and $\mathbf{w}$ orthogonal
20. $\mathbf{v} = 3\mathbf{i} - 4\mathbf{j}$, $\mathbf{w} = 2\mathbf{i} + a\mathbf{j}$, $\mathbf{v}$ and $\mathbf{w}$ parallel
21. $\mathbf{v} = a\mathbf{i} + 2\mathbf{j}$, $\mathbf{w} = 2\mathbf{i} - a\mathbf{j}$, $\mathbf{v}$ and $\mathbf{w}$ parallel
22. $\mathbf{v} = a\mathbf{i}$, $\mathbf{w} = 2\mathbf{i} - 3\mathbf{j}$, $\mathbf{v}$ and $\mathbf{w}$ parallel

23. $\mathbf{v} = 5\mathbf{i} + 12\mathbf{j}$, $\mathbf{w} = \mathbf{i} + a\mathbf{j}$, $\mathbf{v}$ and $\mathbf{w}$ make an angle of $\pi/3$

24. $\mathbf{v} = 4\mathbf{i} - 3\mathbf{j}$, $\mathbf{w} = 2\mathbf{i} + a\mathbf{j}$, $\mathbf{v}$ and $\mathbf{w}$ make an angle of $\pi/6$

25. Prove the distributive law for the scalar product, as stated on page 48.

26. Let $\mathbf{i}$ and $\mathbf{j}$ be the usual unit vectors of one coordinate system, and let $\mathbf{i}_1$ and $\mathbf{j}_1$ be the unit orthogonal vectors corresponding to another rectangular system of coordinates. Given that

$$\mathbf{v} = a\mathbf{i} + b\mathbf{j}, \qquad \mathbf{w} = c\mathbf{i} + d\mathbf{j}, \qquad \mathbf{v} = a_1\mathbf{i}_1 + b_1\mathbf{j}_1, \qquad \mathbf{w} = c_1\mathbf{i}_1 + d_1\mathbf{j}_1,$$

show that

$$ac + bd = a_1c_1 + b_1d_1.$$

In Problems 27 through 30, the quantity $|AB|$ denotes (as is customary) the length of the line segment AB, the quantity $|AC|$, the length of AC, etc.

27. Given $\triangle ABC$, in which $\angle A = 120°$, $|AB| = 4$, and $|AC| = 7$. Find $|BC|$ and the projections of $\overrightarrow{AB}$ and $\overrightarrow{AC}$ on $\overrightarrow{BC}$. Draw a figure.

28. Given $\triangle ABC$, with $\angle A = 45°$, $|AB| = 8$, $|AC| = 6\sqrt{2}$. Find $|BC|$ and the projections of $\overrightarrow{AB}$ and $\overrightarrow{AC}$ on $\overrightarrow{BC}$. Draw a figure.

29. Given $\triangle ABC$, with $|AB| = 10$, $|AC| = 9$, $|BC| = 7$. Find the projections of $\overrightarrow{AC}$ and $\overrightarrow{BC}$ on $\overrightarrow{AB}$. Draw a figure.

30. Given $\triangle ABC$, with $|AB| = 5$, $|AC| = 7$, $|BC| = 9$. Find the projections of $\overrightarrow{AB}$ and $\overrightarrow{AC}$ on $\overrightarrow{CB}$. Draw a figure.

31. Given the line segments AB and AC, with D on AB $\frac{2}{3}$ of the way from A to B. Let E be the midpoint of AC. Express $\mathbf{v}[\overrightarrow{DE}]$ in terms of $\mathbf{v}[\overrightarrow{AB}]$ and $\mathbf{v}[\overrightarrow{AC}]$. Draw a figure.

32. Suppose that $\mathbf{v}[\overrightarrow{AD}] = \frac{1}{4}\mathbf{v}[\overrightarrow{AB}]$ and $\mathbf{v}[\overrightarrow{BE}] = \frac{1}{2}\mathbf{v}[\overrightarrow{BC}]$. Find $\mathbf{v}[\overrightarrow{DE}]$ in terms of $\mathbf{v}[\overrightarrow{AB}]$ and $\mathbf{v}[\overrightarrow{BC}]$. Draw a figure.

33. Given $\square ABDC$, a parallelogram, with E $\frac{2}{3}$ of the way from B to D, and F as the midpoint of segment CD. Find $\mathbf{v}[\overrightarrow{EF}]$ in terms of $\mathbf{v}[\overrightarrow{AB}]$ and $\mathbf{v}[\overrightarrow{AC}]$.

34. Given parallelogram $ABDC$, with E $\frac{1}{4}$ of the way from B to C, and F $\frac{1}{4}$ of the way from A to D. Find $\mathbf{v}[\overrightarrow{EF}]$ in terms of $\mathbf{v}[\overrightarrow{AB}]$ and $\mathbf{v}[\overrightarrow{AC}]$.

35. Given parallelogram $ABDC$, with E $\frac{1}{3}$ of the way from B to D, and F $\frac{1}{4}$ of the way from B to C. Show that F is $\frac{3}{4}$ of the way from A to E.

36. Suppose that on the sides of $\triangle ABC$, $\mathbf{v}[\overrightarrow{BD}] = \frac{2}{3}\mathbf{v}[\overrightarrow{BC}]$, $\mathbf{v}[\overrightarrow{CE}] = \frac{2}{3}\mathbf{v}[\overrightarrow{CA}]$, and $\mathbf{v}[\overrightarrow{AF}] = \frac{2}{3}\mathbf{v}[\overrightarrow{AB}]$. Draw figure and show that

$$\mathbf{v}[\overrightarrow{AD}] + \mathbf{v}[\overrightarrow{BE}] + \mathbf{v}[\overrightarrow{CF}] = \mathbf{0}.$$

37. Show that the conclusion of Problem 36 holds when the fraction $\frac{2}{3}$ is replaced by any real number h.

38. Let $\mathbf{a} = \mathbf{v}[\overrightarrow{OA}]$, $\mathbf{b} = \mathbf{v}[\overrightarrow{OB}]$, and $\mathbf{c} = \mathbf{v}[\overrightarrow{OC}]$. Show that the medians of $\triangle ABC$ meet at a point P, and express $\mathbf{v}[\overrightarrow{OP}]$ in terms of $\mathbf{a}$, $\mathbf{b}$, and $\mathbf{c}$. Draw a figure.

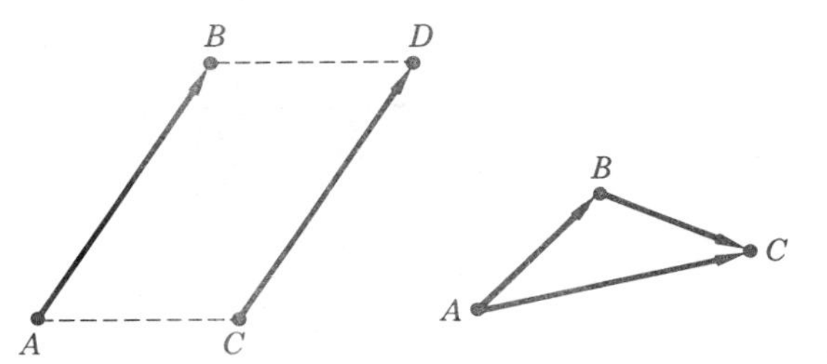

Fig. 2-19

4. Vectors in Three Dimensions

The development of vectors in three-dimensional space is a direct extension of the theory of vectors in the plane as given in Sections 1–3. This section, which is completely analogous to Section 1, may be read quickly. The distinction between two- and three-dimensional vectors appears in Sections 5, 6, and 7.

A **directed line segment** $\overrightarrow{AB}$ is defined as before, except that now the **base** A and the **head** B may be situated anywhere in three-space. The **magnitude** of a directed line segment is its length. Two directed line segments $\overrightarrow{AB}$ and $\overrightarrow{CD}$ are said to **have the same magnitude and direction** if and only if either one of the following two conditions holds:

i) $\overrightarrow{AB}$ and $\overrightarrow{CD}$ are both on the same directed line $\vec{L}$ and their directed lengths are equal; or
ii) the points A, C, D, and B are the vertices of a parallelogram as shown in Fig. 2-19.

We note that the above definition is the same as that given in Section 1.

Whenever two directed line segments $\overrightarrow{AB}$ and $\overrightarrow{CD}$ have the same magnitude and direction, we say they are **equivalent** and write

$$\overrightarrow{AB} \approx \overrightarrow{CD}.$$

We shall next state a theorem which is a direct extension of Theorem 1 in Section 1.

Theorem 6. *Suppose that A, B, C, and D are points in space. Denote the coordinates of A, B, C, and D by (x_A, y_A, z_A), (x_B, y_B, z_B), and so forth.* (i) *If the coordinates satisfy the equations*

$$x_B - x_A = x_D - x_C, \qquad y_B - y_A = y_D - y_C, \qquad z_B - z_A = z_D - z_C, \quad (1)$$

then $\overrightarrow{AB} \approx \overrightarrow{CD}$. (ii) *Conversely, if $\overrightarrow{AB} \approx \overrightarrow{CD}$, the coordinates satisfy the equations in* (1).

Sketch of Proof. (i) We assume that the equations in (1) hold. Then also,

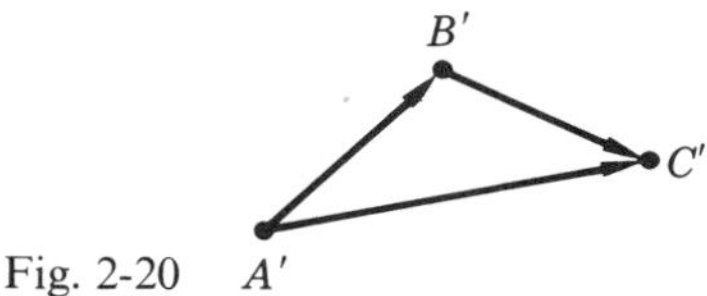

Fig. 2-20

$$x_C - x_A = x_D - x_B, \qquad y_C - y_A = y_D - y_B, \qquad z_C - z_A = z_D - z_B. \quad (2)$$

As in the proof of Theorem 1 in Section 1, we can conclude that the lines AB and CD are parallel because their direction numbers are equal. Similarly, AC and BD are parallel. Therefore $\overrightarrow{AB} \approx \overrightarrow{CD}$ or all the points are on a directed line $\vec{L}$. The proof that $\overrightarrow{AB} \approx \overrightarrow{CD}$ in this latter case is given in Appendix 2. (ii) To prove the converse, assume that $\overrightarrow{AB} \approx \overrightarrow{CD}$. Then either $ACDB$ is a parallelogram or all the points are on a directed line $\vec{L}$. When $ACDB$ is a parallelogram there is a unique point E such that

$$x_E - x_C = x_B - x_A, \qquad y_E - y_C = y_B - y_A, \qquad z_E - z_C = z_B - z_A.$$

As in the proof of Theorem 1 in Section 1, we conclude from part (i) that $ACEB$ is a parallelogram and $D = E$; hence equations (1) hold. The proof that equations (1) hold when the points are on a directed line $\vec{L}$ is given in Appendix 2.

If we are given a directed line segment $\overrightarrow{AB}$, it is clear that there is an unlimited number of equivalent ones. In fact, if C is any given point in three-space, we can use equations (1) of Theorem 6 to find the coordinates of the unique point D such that $\overrightarrow{CD} \approx \overrightarrow{AB}$.

Definitions. A **vector** is the collection of all directed line segments having a given magnitude and direction. We shall use boldface letters to denote vectors. A particular directed line segment in a collection $\mathbf{v}$ is called a **representative** of the vector* $\mathbf{v}$. The **length** of a vector is the common length of all its representatives. A **unit vector** is a vector of length one. Two vectors are said to be **orthogonal** (or **perpendicular**) if any representative of one vector is perpendicular to any representative of the other. The **zero vector**, denoted by $\mathbf{0}$, is the class of directed line "segments" of zero length (i.e., simply points).

As in Section 2, we can define the sum of two vectors. Given $\mathbf{u}$ and $\mathbf{v}$, let $\overrightarrow{AB}$ be a representative of $\mathbf{u}$ and let $\overrightarrow{BC}$ be that representative of $\mathbf{v}$ which has its base at B. Then $\mathbf{u} + \mathbf{v}$ is the vector which has representative $\overrightarrow{AC}$ as shown in Fig. 2-20. If $\overrightarrow{A'B'}$ and $\overrightarrow{B'C'}$ are other representatives of $\mathbf{u}$ and $\mathbf{v}$, respectively, it follows from Theorem 6 that $\overrightarrow{A'C'} \approx \overrightarrow{AC}$. Therefore $\overrightarrow{A'C'}$ is

* We also say that the directed line segment *determines* the vector v.

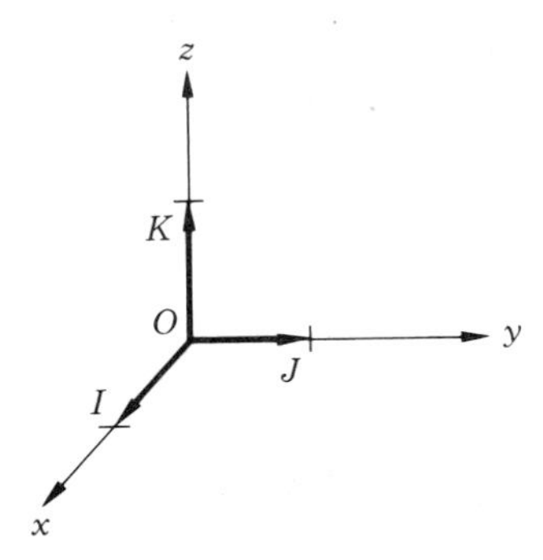

Fig. 2-21

also a representative of $\mathbf{u} + \mathbf{v}$. In other words, the rule for forming the sum of two vectors does not depend on the particular representatives we select in making the calculation.

Vectors may be multiplied by numbers (scalars). Given a vector $\mathbf{u}$ and a number c, let $\overrightarrow{AB}$ determine $\mathbf{u}$ and let C be the point c of the way from A to B. Then $\overrightarrow{AC}$ also determines $c\mathbf{u}$. It follows easily from Theorem 6 that if $\overrightarrow{A'B'}$ determines $\mathbf{u}$ and C' is c of the way from A' to B', then $\overrightarrow{A'C'} \approx \overrightarrow{AC}$ and so $\overrightarrow{A'C'}$ determines $c\mathbf{u}$.

Definitions. Suppose that a rectangular coordinate system is given. Figure 2-21 shows such a system with the points $I(1, 0, 0)$, $J(0, 1, 0)$, and $K(0, 0, 1)$ identified. The **unit vector i** is defined as the vector which has $\overrightarrow{OI}$ as one of its representatives. The **unit vector j** is defined as the vector which has $\overrightarrow{OJ}$ as one of its representatives. The **unit vector k** is defined as the vector which has $\overrightarrow{OK}$ as one of its representatives.

We now establish a direct extension of Theorem 2 in Section 2.

Theorem 7. *Suppose a vector* $\mathbf{w}$ *has* $\overrightarrow{AB}$ *as a representative. Denote the coordinates of A and B by* (x_A, y_A, z_A) and (x_B, y_B, z_B), *respectively. Then* $\mathbf{w}$ *may be expressed in the form*

$$\boxed{\mathbf{w} = (x_B - x_A)\mathbf{i} + (y_B - y_A)\mathbf{j} + (z_B - z_A)\mathbf{k}.}$$

PROOF. From equations (1) of Theorem 6, we know that $\mathbf{w}$ is determined by $\overrightarrow{OP}$ where P has coordinates $(x_B - x_A, y_B - y_A, z_B - z_A)$. Let

$$Q(x_B - x_A, 0, 0), \qquad R(0, y_B - y_A, 0),$$

$$S(0, 0, z_B - z_A), \qquad T(x_B - x_A, y_B - y_A, 0)$$

be as shown in Fig. 2-22. Then Q is $x_B - x_A$ of the way from O to $I(1, 0, 0)$,

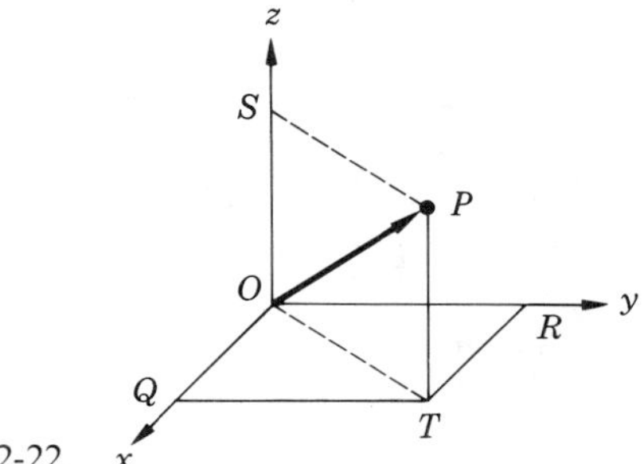

Fig. 2-22

and similarly for R and S with regard to J and K. Therefore,

$$\mathbf{v}(\overrightarrow{OQ}) = (x_B - x_A)\mathbf{i},$$
$$\mathbf{v}(\overrightarrow{QT}) = \mathbf{v}(\overrightarrow{OR}) = (y_B - y_A)\mathbf{j},$$
$$\mathbf{v}(\overrightarrow{TP}) = \mathbf{v}(\overrightarrow{OS}) = (z_B - z_A)\mathbf{k}.$$

Using the rule for addition of vectors, we find

$$\mathbf{v}(\overrightarrow{OP}) = \mathbf{v}(\overrightarrow{OQ}) + \mathbf{v}(\overrightarrow{QT}) + \mathbf{v}(\overrightarrow{TP}),$$

and the proof is complete.

EXAMPLE 1. A vector $\mathbf{v}$ has $\overrightarrow{AB}$ as a representative. If A and B have coordinates $(3, -2, 4)$ and $(2, 1, 5)$, respectively, express $\mathbf{v}$ in terms of $\mathbf{i}$, $\mathbf{j}$, and $\mathbf{k}$.

SOLUTION. From Theorem 7, we obtain

$$\mathbf{v}(\overrightarrow{AB}) = (2 - 3)\mathbf{i} + (1 + 2)\mathbf{j} + (5 - 4)\mathbf{k} = -\mathbf{i} + 3\mathbf{j} + \mathbf{k}.$$

The next two theorems are direct extensions of Theorems 3 and 4 in Section 2. The proofs are left to the reader.

The **length** of a vector $\mathbf{v}$ is denoted by $|\mathbf{v}|$.

Theorem 8. *If* $\mathbf{v} = a\mathbf{i} + b\mathbf{j} + c\mathbf{k}$, *then*

$$\boxed{|\mathbf{v}| = \sqrt{a^2 + b^2 + c^2}.}$$

Theorem 9. *If* $\mathbf{v} = a_1\mathbf{i} + b_1\mathbf{j} + c_1\mathbf{k}$, $\mathbf{w} = a_2\mathbf{i} + b_2\mathbf{j} + c_2\mathbf{k}$, *then*

$$\boxed{\mathbf{v} + \mathbf{w} = (a_1 + a_2)\mathbf{i} + (b_1 + b_2)\mathbf{j} + (c_1 + c_2)\mathbf{k}.}$$

If h is any number, then

$$h\mathbf{v} = ha_1\mathbf{i} + hb_1\mathbf{j} + hc_1\mathbf{k}.$$

In complete analogy with vectors in the plane, we conclude from the theorems above that the addition of vectors and their multiplication by numbers satisfy the following laws:

$$\left.\begin{aligned} \mathbf{u} + (\mathbf{v} + \mathbf{w}) &= (\mathbf{u} + \mathbf{v}) + \mathbf{w} \\ c(d\mathbf{v}) &= (cd)\mathbf{v} \end{aligned}\right\} \quad \text{Associative laws}$$

$$\mathbf{u} + \mathbf{v} = \mathbf{v} + \mathbf{u} \qquad \text{Commutative law}$$

$$\left.\begin{aligned} (c + d)\mathbf{v} &= c\mathbf{v} + d\mathbf{v} \\ c(\mathbf{u} + \mathbf{v}) &= c\mathbf{u} + c\mathbf{v} \end{aligned}\right\} \quad \text{Distributive laws}$$

$$1 \cdot \mathbf{u} = \mathbf{u}, \qquad 0 \cdot \mathbf{u} = \mathbf{0}, \qquad (-1)\mathbf{u} = -\mathbf{u}.$$

Definition. Let $\mathbf{v}$ be any vector except $\mathbf{0}$. The **unit vector u in the direction of v** is defined by

$$\mathbf{u} = \frac{1}{|\mathbf{v}|}\mathbf{v}.$$

EXAMPLE 2. Given the vectors $\mathbf{u} = 3\mathbf{i} - 2\mathbf{j} + 4\mathbf{k}$ and $\mathbf{v} = 6\mathbf{i} - 4\mathbf{j} - 2\mathbf{k}$, express the vector $3\mathbf{u} - 2\mathbf{v}$ in terms of $\mathbf{i}$, $\mathbf{j}$, and $\mathbf{k}$.

SOLUTION. $3\mathbf{u} = 9\mathbf{i} - 6\mathbf{j} + 12\mathbf{k}$ and $-2\mathbf{v} = -12\mathbf{i} + 8\mathbf{j} + 4\mathbf{k}$. Adding these vectors, we get $3\mathbf{u} - 2\mathbf{v} = -3\mathbf{i} + 2\mathbf{j} + 16\mathbf{k}$.

EXAMPLE 3. Given the vector $\mathbf{v} = 2\mathbf{i} - 3\mathbf{j} + \mathbf{k}$, find a unit vector in the direction of $\mathbf{v}$.

SOLUTION. We have $|\mathbf{v}| = \sqrt{4 + 9 + 1} = \sqrt{14}$. The desired vector $\mathbf{u}$ is

$$\mathbf{u} = \frac{1}{\sqrt{14}}\mathbf{v} = \frac{2}{\sqrt{14}}\mathbf{i} - \frac{3}{\sqrt{14}}\mathbf{j} + \frac{1}{\sqrt{14}}\mathbf{k}.$$

EXAMPLE 4. Given the vector $\mathbf{v} = 2\mathbf{i} + 4\mathbf{j} - 3\mathbf{k}$, find the representative $\overrightarrow{AB}$ of $\mathbf{v}$ if the point A has coordinates $(2, 1, -5)$.

SOLUTION. Denote the coordinates of B by x_B, y_B, z_B. Then we have

$$x_B - 2 = 2, \qquad y_B - 1 = 4, \qquad z_B + 5 = -3.$$

Therefore, $x_B = 4$, $y_B = 5$, $z_B = -8$.

Problems

In Problems 1 through 6, express $\mathbf{v}$ in terms of $\mathbf{i}$, $\mathbf{j}$, and $\mathbf{k}$, given that the endpoints P and Q of the representative $\overrightarrow{PQ}$ of $\mathbf{v}$ have the given coordinates. Also find another directed line segment for the same vector $\mathbf{v}$.

1. $P(2,0,3)$, $Q(1,4,-3)$
2. $P(1,1,0)$, $Q(-1,2,0)$
3. $P(-4,-2,1)$, $Q(1,-3,4)$
4. $P(3,2,1)$, $Q(3,3,3)$
5. $P(2,0,0)$, $Q(0,0,-3)$
6. $P(4,-5,-1)$, $Q(-2,1,-3)$

In Problems 7 through 10, in each case find a unit vector $\mathbf{u}$ in the direction of $\mathbf{v}$. Express $\mathbf{u}$ in terms of $\mathbf{i}$, $\mathbf{j}$, and $\mathbf{k}$.

7. $\mathbf{v} = 3\mathbf{i} + 2\mathbf{j} - 4\mathbf{k}$
8. $\mathbf{v} = \mathbf{i} - \mathbf{j} + \mathbf{k}$
9. $\mathbf{v} = 2\mathbf{i} - 4\mathbf{j} - \mathbf{k}$
10. $\mathbf{v} = -2\mathbf{i} + 3\mathbf{j} + 5\mathbf{k}$

In Problems 11 through 17, find the directed line segment $\overrightarrow{AB}$ of the vector $\mathbf{v}$ from the information given.

11. $\mathbf{v} = 2\mathbf{i} + \mathbf{j} - 3\mathbf{k}$, $A(1,2,-1)$
12. $\mathbf{v} = -\mathbf{i} + 3\mathbf{j} - 2\mathbf{k}$, $A(2,0,4)$
13. $\mathbf{v} = 3\mathbf{i} + 2\mathbf{j} - 4\mathbf{k}$, $B(2,0,-4)$
14. $\mathbf{v} = -2\mathbf{i} + 4\mathbf{j} + \mathbf{k}$, $B(0,0,-5)$
15. $\mathbf{v} = \mathbf{i} - 2\mathbf{j} + 2\mathbf{k}$; the midpoint of the segment AB has coordinates $(2,-1,4)$.
16. $\mathbf{v} = 3\mathbf{i} + 4\mathbf{k}$; the midpoint of the segment AB has coordinates $(1,2,-5)$.
17. $\mathbf{v} = -\mathbf{i} + \mathbf{j} - 2\mathbf{k}$; the point three-fourths of the distance from A to B has coordinates $(1,0,2)$.
18. Find a vector $\mathbf{u}$ in the direction of $\mathbf{v} = -\mathbf{i} + \mathbf{j} - \mathbf{k}$ and having half the length of $\mathbf{v}$.
19. Given $\mathbf{u} = \mathbf{i} + 2\mathbf{j} - 4\mathbf{k}$, $\mathbf{v} = 3\mathbf{i} - 7\mathbf{j} + 5\mathbf{k}$, find $\mathbf{u} + \mathbf{v}$ in terms of $\mathbf{i}$, $\mathbf{j}$, and $\mathbf{k}$. Sketch a figure.
20. Given that $\mathbf{u} = -3\mathbf{i} + 7\mathbf{j} - 4\mathbf{k}$, $\mathbf{v} = 2\mathbf{i} + \mathbf{j} - 6\mathbf{k}$, find $3\mathbf{u} - 7\mathbf{v}$ in terms of $\mathbf{i}$, $\mathbf{j}$, and $\mathbf{k}$.
21. Let a and b be any real numbers. Show that the vector $\mathbf{k}$ is orthogonal to $a\mathbf{i} + b\mathbf{j}$.
22. Show that the vector $\mathbf{u} = \mathbf{i} + 2\mathbf{j} - 3\mathbf{k}$ is orthogonal to $a\mathbf{v} + b\mathbf{w}$ where $\mathbf{v} = 2\mathbf{i} + 2\mathbf{j} + 2\mathbf{k}$, $\mathbf{w} = -\mathbf{i} + 2\mathbf{j} + \mathbf{k}$ and a, b are any real numbers. Interpret this statement geometrically.

23. Discuss the relationship between the direction numbers of a line and the representation of a vector $\mathbf{v}$ in terms of the vectors $\mathbf{i}$, $\mathbf{j}$, and $\mathbf{k}$.

24. Suppose two representatives of $\mathbf{u}$ and $\mathbf{v}$ determine a plane. Discuss the relationship between the attitude numbers of this plane and the representations of $\mathbf{u}$ and $\mathbf{v}$ in terms of $\mathbf{i}$, $\mathbf{j}$, and $\mathbf{k}$ as stated in Theorem 7.

5. Linear Dependence and Independence*

Two vectors $\mathbf{u}$ and $\mathbf{v}$, neither zero, are said to be **proportional** if and only if there is a number c such that $\mathbf{u} = c\mathbf{v}$; that is, each vector is a scalar multiple of the other. If $\mathbf{v}_1, \mathbf{v}_2, \ldots, \mathbf{v}_k$ are any vectors and $c_1, c_2, \ldots, c_k$ are numbers, we call an expression of the form

$$c_1\mathbf{v}_1 + c_2\mathbf{v}_2 + \cdots + c_k\mathbf{v}_k$$

a **linear combination** of the vectors $\mathbf{v}_1, \mathbf{v}_2, \ldots, \mathbf{v}_k$. If two vectors $\mathbf{u}$ and $\mathbf{v}$ are proportional, the definition shows that a linear combination of them is the zero vector. In fact, $\mathbf{u} - c\mathbf{v} = \mathbf{0}$. A set of vectors $\{\mathbf{v}_1, \mathbf{v}_2, \ldots, \mathbf{v}_k\}$ is **linearly dependent** if and only if there is a set of constants $\{c_1, c_2, \ldots, c_k\}$, *not all zero*, such that

$$c_1\mathbf{v}_1 + c_2\mathbf{v}_2 + \cdots + c_k\mathbf{v}_k = \mathbf{0}. \tag{1}$$

If no such set of constants exists, then the set $\{\mathbf{v}_1, \mathbf{v}_2, \ldots, \mathbf{v}_k\}$ is said to be **linearly independent**.

It is clear that any two proportional vectors are linearly dependent. As another example, the vectors

$$\mathbf{v}_1 = 2\mathbf{i} + 3\mathbf{j} - \mathbf{k}, \qquad \mathbf{v}_2 = -2\mathbf{i} - \mathbf{j} + \mathbf{k}, \qquad \mathbf{v}_3 = 2\mathbf{i} + 7\mathbf{j} - \mathbf{k}$$

form a linearly dependent set since the selection $c_1 = 3$, $c_2 = 2$, $c_3 = -1$ shows that

$$c_1\mathbf{v}_1 + c_2\mathbf{v}_2 + c_3\mathbf{v}_3 = 3(2\mathbf{i} + 3\mathbf{j} - \mathbf{k}) + 2(-2\mathbf{i} - \mathbf{j} + \mathbf{k}) - (2\mathbf{i} + 7\mathbf{j} - \mathbf{k}) = \mathbf{0}.$$

A set $\{\mathbf{v}_1, \mathbf{v}_2, \ldots, \mathbf{v}_k\}$ is linearly dependent if and only if one member of the set can be expressed as a linear combination of the remaining members. To see this, we observe that in Eq. (1) one of the terms on the left-hand side, say $\mathbf{v}_i$, must have a nonzero coefficient and so may be transferred to the right-hand side. Dividing by the coefficient $-c_i$, we express this particular $\mathbf{v}_i$ as a linear combination of the remaining $\mathbf{v}$'s. Conversely, if some $\mathbf{v}_i$ is expressible in terms of the others, it follows by transposing $\mathbf{v}_i$ that $\mathbf{v}_1, \mathbf{v}_2, \ldots, \mathbf{v}_k$ are linearly dependent.

* For an understanding of Sections 5, 6, and 7, we assume that the reader is acquainted with determinants of the second and third order. For those unfamiliar with the subject a discussion is provided in Appendix 1.

The following statement, a direct consequence of the definition of linear dependence, is often useful in proofs of theorems. If $\{\mathbf{v}_1, \mathbf{v}_2, \ldots, \mathbf{v}_k\}$ is a linearly independent set and if

$$c_1\mathbf{v}_1 + c_2\mathbf{v}_2 + \cdots + c_k\mathbf{v}_k = \mathbf{0},$$

then it follows that $c_1 = c_2 = \cdots = c_k = 0$.

The set $\{\mathbf{i}, \mathbf{j}, \mathbf{k}\}$ *is linearly independent.* To show this observe that the equation

$$c_1\mathbf{i} + c_2\mathbf{j} + c_3\mathbf{k} = \mathbf{0} \tag{2}$$

holds if and only if $|c_1\mathbf{i} + c_2\mathbf{j} + c_3\mathbf{k}| = 0$. But

$$|c_1\mathbf{i} + c_2\mathbf{j} + c_3\mathbf{k}| = \sqrt{c_1^2 + c_2^2 + c_3^2},$$

and this last expression is zero if and only if $c_1 = c_2 = c_3 = 0$. Thus no nonzero constants satisfying (2) exist and $\{\mathbf{i}, \mathbf{j}, \mathbf{k}\}$ is a linearly independent set.

The proof of the next theorem employs the tools on determinants given in Appendix 1. The details of the proof of Theorem 10 are carried out in Appendix 2.

Theorem 10. *Let*

$$\mathbf{u} = a_{11}\mathbf{i} + a_{12}\mathbf{j} + a_{13}\mathbf{k},$$
$$\mathbf{v} = a_{21}\mathbf{i} + a_{22}\mathbf{j} + a_{23}\mathbf{k},$$
$$\mathbf{w} = a_{31}\mathbf{i} + a_{32}\mathbf{j} + a_{33}\mathbf{k},$$

and denote by D the determinant

$$D = \begin{vmatrix} a_{11} & a_{12} & a_{13} \\ a_{21} & a_{22} & a_{23} \\ a_{31} & a_{32} & a_{33} \end{vmatrix}.$$

Then the set $\{\mathbf{u}, \mathbf{v}, \mathbf{w}\}$ *is linearly independent if and only if* $D \neq 0$.

EXAMPLE 1. Determine whether or not the vectors

$$\mathbf{u} = 2\mathbf{i} - \mathbf{j} + \mathbf{k}, \qquad \mathbf{v} = \mathbf{i} + 2\mathbf{j} + \mathbf{k}, \qquad \mathbf{w} = -\mathbf{i} + \mathbf{j} + 3\mathbf{k}$$

form a linearly independent set.

SOLUTION. Expanding D by its first row, we have

$$D = \begin{vmatrix} 2 & -1 & 1 \\ 1 & 2 & 1 \\ -1 & 1 & 3 \end{vmatrix} = 2\begin{vmatrix} 2 & 1 \\ 1 & 3 \end{vmatrix} + \begin{vmatrix} 1 & 1 \\ -1 & 3 \end{vmatrix} + \begin{vmatrix} 1 & 2 \\ -1 & 1 \end{vmatrix}.$$

Therefore $D = 2(5) + 4 + 3 = 17 \neq 0$. The set is linearly independent.

Theorem 11. *If* $\{\mathbf{u}, \mathbf{v}, \mathbf{w}\}$ *is a linearly independent set and* $\mathbf{r}$ *is any vector, then there are constants* A_1, A_2, *and* A_3 *such that*

$$\mathbf{r} = A_1\mathbf{u} + A_2\mathbf{v} + A_3\mathbf{w}. \tag{3}$$

PROOF. According to Theòrem 7, *every* vector can be expressed as a linear combination of $\mathbf{i}$, $\mathbf{j}$, and $\mathbf{k}$. Therefore

$$\begin{aligned}
\mathbf{u} &= a_{11}\mathbf{i} + a_{12}\mathbf{j} + a_{13}\mathbf{k},\\
\mathbf{v} &= a_{21}\mathbf{i} + a_{22}\mathbf{j} + a_{23}\mathbf{k},\\
\mathbf{w} &= a_{31}\mathbf{i} + a_{32}\mathbf{j} + a_{33}\mathbf{k},\\
\mathbf{r} &= b_1\mathbf{i} + b_2\mathbf{j} + b_3\mathbf{k}.
\end{aligned}$$

When we insert all these expressions in (3) and collect all terms on one side, we get a linear combination of $\mathbf{i}$, $\mathbf{j}$, and $\mathbf{k}$ equal to zero. Since $\{\mathbf{i}, \mathbf{j}, \mathbf{k}\}$ is a linearly independent set, the coefficients of $\mathbf{i}$, $\mathbf{j}$, and $\mathbf{k}$ are equal to zero separately. Computing these coefficients, we get the equations

$$\begin{aligned}
a_{11}A_1 + a_{21}A_2 + a_{31}A_3 &= b_1,\\
a_{12}A_1 + a_{22}A_2 + a_{32}A_3 &= b_2,\\
a_{13}A_1 + a_{23}A_2 + a_{33}A_3 &= b_3.
\end{aligned} \tag{4}$$

We have here three equations in the three unknowns A_1, A_2, A_3. The determinant D' of the coefficients in (4) differs from the determinant D of Theorem 10 in that the rows and columns are interchanged. Since $\{\mathbf{u}, \mathbf{v}, \mathbf{w}\}$ is an independent set, we know that $D \neq 0$; also Theorem 6 of Appendix 1 proves that $D = D'$, and so $D' \neq 0$. We now use Cramer's rule (Theorem 11, Appendix 1) to solve for A_1, A_2, A_3.

Note that the proof of Theorem 11 gives the method for finding A_1, A_2, A_3. We work an example.

EXAMPLE 2. Given the vectors

$$\mathbf{u} = 2\mathbf{i} + 3\mathbf{j} + \mathbf{k}, \qquad \mathbf{v} = -\mathbf{i} + \mathbf{j} + 2\mathbf{k},$$
$$\mathbf{w} = 3\mathbf{i} - \mathbf{j} + 3\mathbf{k}, \qquad \mathbf{r} = \mathbf{i} + 2\mathbf{j} - 6\mathbf{k},$$

show that $\mathbf{u}$, $\mathbf{v}$, and $\mathbf{w}$ are linearly independent and express $\mathbf{r}$ as a linear combination of $\mathbf{u}$, $\mathbf{v}$, and $\mathbf{w}$.

SOLUTION. Expanding D by its first row, we obtain

$$D = \begin{vmatrix} 2 & 3 & 1 \\ -1 & 1 & 2 \\ 3 & -1 & 3 \end{vmatrix} = 2\begin{vmatrix} 1 & 2 \\ -1 & 3 \end{vmatrix} - 3\begin{vmatrix} -1 & 2 \\ 3 & 3 \end{vmatrix} + \begin{vmatrix} -1 & 1 \\ 3 & -1 \end{vmatrix}$$

$$= 2(5) - 3(-9) + (-2) = 35.$$

Hence $D \neq 0$ and so $\{\mathbf{u}, \mathbf{v}, \mathbf{w}\}$ is a linearly independent set. Using equations (4), we now obtain the equations

$$2A_1 - A_2 + 3A_3 = 1,$$
$$3A_1 + A_2 - A_3 = 2,$$
$$A_1 + 2A_2 + 3A_3 = -6.$$

Solving these, we find that $A_1 = 1$, $A_2 = -2$, $A_3 = -1$. Finally,

$$\mathbf{r} = \mathbf{u} - 2\mathbf{v} - \mathbf{w}.$$

PROBLEMS

In Problems 1 through 5 state whether or not the given vectors are linearly independent.

1. $\mathbf{u} = 2\mathbf{i} + \mathbf{j} - \mathbf{k}$, $\mathbf{v} = \mathbf{i} - 2\mathbf{j} + 5\mathbf{k}$, $\mathbf{w} = 2\mathbf{i} - 7\mathbf{j} + \mathbf{k}$
2. $\mathbf{u} = \mathbf{i} + 2\mathbf{j} + 3\mathbf{k}$, $\mathbf{v} = 2\mathbf{i} + \mathbf{j} + 4\mathbf{k}$, $\mathbf{w} = 3\mathbf{j} + 2\mathbf{k}$
3. $\mathbf{u} = 2\mathbf{i} + 3\mathbf{j}$, $\mathbf{v} = \mathbf{i} - 4\mathbf{j}$, $\mathbf{w} = \mathbf{i} + 2\mathbf{j}$
4. $\mathbf{u} = -\mathbf{i} + 2\mathbf{j}$, $\mathbf{v} = \mathbf{i} + \mathbf{j} + \mathbf{k}$, $\mathbf{w} = -2\mathbf{j} + 6\mathbf{k}$
5. $\mathbf{u} = \mathbf{i} + \mathbf{j}$, $\mathbf{v} = 2\mathbf{i} - 6\mathbf{j} + 3\mathbf{k}$, $\mathbf{w} = -\mathbf{i} + \mathbf{j}$, $\mathbf{r} = 4\mathbf{k}$

In Problems 6 through 11, show that **u**, **v**, and **w** are linearly independent and express **r** in terms of **u**, **v**, and **w**.

6. $\mathbf{u} = 2\mathbf{i} - \mathbf{j} + \mathbf{k}$, $\mathbf{v} = -\mathbf{i} + \mathbf{j} - 2\mathbf{k}$, $\mathbf{w} = 2\mathbf{i} - \mathbf{j} + 2\mathbf{k}$, $\mathbf{r} = 3\mathbf{i} - \mathbf{j} + 2\mathbf{k}$
7. $\mathbf{u} = \mathbf{i} - \mathbf{j} + \mathbf{k}$, $\mathbf{v} = -\mathbf{i} + 2\mathbf{j} - \mathbf{k}$, $\mathbf{w} = 2\mathbf{i} - \mathbf{j} + \mathbf{k}$, $\mathbf{r} = 2\mathbf{i} + 3\mathbf{j} + 4\mathbf{k}$
8. $\mathbf{u} = 3\mathbf{i} + \mathbf{j} - 2\mathbf{k}$, $\mathbf{v} = 2\mathbf{i} - \mathbf{k}$, $\mathbf{w} = -\mathbf{i} + 2\mathbf{j} + \mathbf{k}$, $\mathbf{r} = \mathbf{i} + 2\mathbf{j} - 3\mathbf{k}$
9. $\mathbf{u} = 2\mathbf{i} - \mathbf{j} + \mathbf{k}$, $\mathbf{v} = \mathbf{i} + \mathbf{j}$, $\mathbf{w} = -\mathbf{i} + \mathbf{j} + 2\mathbf{k}$, $\mathbf{r} = 2\mathbf{i} - \mathbf{j} - 2\mathbf{k}$
10. $\mathbf{u} = \mathbf{i} - 2\mathbf{j} - 3\mathbf{k}$, $\mathbf{v} = 2\mathbf{i} - \mathbf{j} - 2\mathbf{k}$, $\mathbf{w} = -\mathbf{i} + \mathbf{j} + \mathbf{k}$, $\mathbf{r} = 2\mathbf{i} + 3\mathbf{j} + 4\mathbf{k}$
11. $\mathbf{u} = 2\mathbf{i} - 3\mathbf{k}$, $\mathbf{v} = \mathbf{i} + 4\mathbf{j} - \mathbf{k}$, $\mathbf{w} = -2\mathbf{i} + 5\mathbf{j} + 3\mathbf{k}$, $\mathbf{r} = -\mathbf{i} + 20\mathbf{j} + 3\mathbf{k}$
12. Prove Theorem 8.
13. Prove Theorem 9.
14. Show that in three dimensions any set of four vectors must be linearly dependent.
15. Show that if $\overrightarrow{OA}$, $\overrightarrow{OB}$, and $\overrightarrow{OC}$ are directed line segments of **u**, **v**, and **w**, respectively, and if $\{\mathbf{u}, \mathbf{v}, \mathbf{w}\}$ is a linearly dependent set, then the three line segments lie in one plane.
16. Find the equations of the line passing through the point $P(2, 1, -3)$ and parallel to $\mathbf{v} = 3\mathbf{i} - 2\mathbf{j} + 7\mathbf{k}$.
17. Find the equations of the line through the point $A(1, -4, 0)$ and perpendicular to any plane determined by $\mathbf{v} = \mathbf{i} + 2\mathbf{j} - \mathbf{k}$ and $\mathbf{w} = 3\mathbf{i} - \mathbf{j} + \mathbf{k}$.

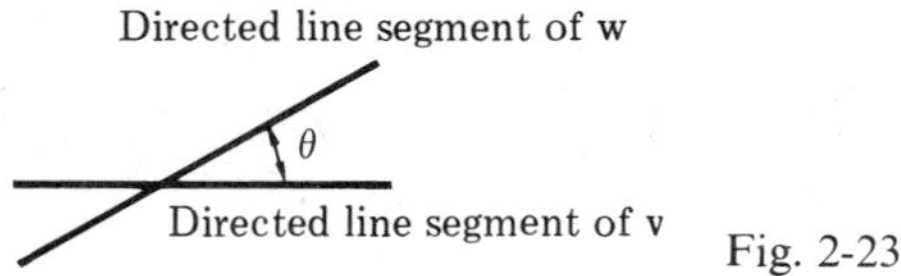

Fig. 2-23

6. The Scalar (Inner or Dot) Product

Two vectors are said to be **parallel** or **proportional** when each is a scalar multiple of the other (and neither is zero). The representatives of parallel vectors are all parallel directed line segments.

By the **angle between two vectors v and w** (neither $= \mathbf{0}$), we mean the measure of the angle between any directed line segment of $\mathbf{v}$ and an intersecting directed line segment of $\mathbf{w}$ (Fig. 2-23). Two parallel vectors make an angle of 0 or π, depending on whether they are pointing in the same or in the opposite direction.

Theorem 12. *If θ is the angle between the vectors*

$$\mathbf{v} = a_1\mathbf{i} + a_2\mathbf{j} + a_3\mathbf{k} \qquad \textit{and} \qquad \mathbf{w} = b_1\mathbf{i} + b_2\mathbf{j} + b_3\mathbf{k},$$

then

$$\boxed{\cos\theta = \frac{a_1b_1 + a_2b_2 + a_3b_3}{|\mathbf{v}|\cdot|\mathbf{w}|}.}$$

The proof is a straightforward extension of the proof of the analogous theorem in the plane (Theorem 5 of Section 3) and will therefore be omitted.

EXAMPLE 1. Given the vectors $\mathbf{v} = 2\mathbf{i} + \mathbf{j} - 3\mathbf{k}$ and $\mathbf{w} = -\mathbf{i} + 4\mathbf{j} - 2\mathbf{k}$, find the cosine of the angle between $\mathbf{v}$ and $\mathbf{w}$.

SOLUTION. We have

$$|\mathbf{v}| = \sqrt{4+1+9} = \sqrt{14}, \qquad |\mathbf{w}| = \sqrt{1+16+4} = \sqrt{21}.$$

Therefore

$$\cos\theta = \frac{-2+4+6}{\sqrt{14}\cdot\sqrt{21}} = \frac{8}{7\sqrt{6}}.$$

Definitions. Given the vectors $\mathbf{u}$ and $\mathbf{v}$, the **scalar (inner** or **dot) product** $\mathbf{u}\cdot\mathbf{v}$ is defined by the formula

$$\boxed{\mathbf{u} \cdot \mathbf{v} = |\mathbf{u}|\,|\mathbf{v}| \cos\theta,}$$

where θ is the angle between the vectors. If either $\mathbf{u}$ or $\mathbf{v}$ is $\mathbf{0}$, we define $\mathbf{u} \cdot \mathbf{v} = 0$. Two vectors $\mathbf{u}$ and $\mathbf{v}$ are **orthogonal** if and only if $\mathbf{u} \cdot \mathbf{v} = 0$.

Thus we see that $\mathbf{0}$ is orthogonal to every vector. These definitions are identical with those for plane vectors.

Theorem 13. *The scalar product satisfies the laws*
a) $\mathbf{u} \cdot \mathbf{v} = \mathbf{v} \cdot \mathbf{u}$; b) $\mathbf{u} \cdot \mathbf{u} = |\mathbf{u}|^2$.
c) *If* $\mathbf{u} = a_1\mathbf{i} + b_1\mathbf{j} + c_1\mathbf{k}$ *and* $\mathbf{v} = a_2\mathbf{i} + b_2\mathbf{j} + c_2\mathbf{k}$, *then*

$$\boxed{\mathbf{u} \cdot \mathbf{v} = a_1a_2 + b_1b_2 + c_1c_2.}$$

Proof. Parts (a) and (b) are direct consequences of the definition; part (c) follows from Theorem 12 since

$$\mathbf{u} \cdot \mathbf{v} = |\mathbf{u}|\,|\mathbf{v}| \cos\theta = |\mathbf{u}|\,|\mathbf{v}| \frac{a_1a_2 + b_1b_2 + c_1c_2}{|\mathbf{u}|\,|\mathbf{v}|}.$$

Corollary. (a) *If c and d are any numbers and if $\mathbf{u}$, $\mathbf{v}$, $\mathbf{w}$ are any vectors, then*

$$\mathbf{u} \cdot (c\mathbf{v} + d\mathbf{w}) = c(\mathbf{u} \cdot \mathbf{v}) + d(\mathbf{u} \cdot \mathbf{w}).$$

b) *We have*

$$\mathbf{i} \cdot \mathbf{i} = \mathbf{j} \cdot \mathbf{j} = \mathbf{k} \cdot \mathbf{k} = 1, \qquad \mathbf{i} \cdot \mathbf{j} = \mathbf{i} \cdot \mathbf{k} = \mathbf{j} \cdot \mathbf{k} = 0.$$

Example 2. Find the scalar product of the vectors

$$\mathbf{u} = 3\mathbf{i} + 2\mathbf{j} - 4\mathbf{k} \quad \text{and} \quad \mathbf{v} = -2\mathbf{i} + \mathbf{j} + 5\mathbf{k}.$$

Solution. $\mathbf{u} \cdot \mathbf{v} = 3(-2) + 2 \cdot 1 + (-4)(5) = -24$.

Example 3. Express $|3\mathbf{u} + 5\mathbf{v}|^2$ in terms of $|\mathbf{u}|^2$, $|\mathbf{v}|^2$, and $\mathbf{u} \cdot \mathbf{v}$.

Solution.
$$\begin{aligned} |3\mathbf{u} + 5\mathbf{v}|^2 &= (3\mathbf{u} + 5\mathbf{v}) \cdot (3\mathbf{u} + 5\mathbf{v}) \\ &= 9(\mathbf{u} \cdot \mathbf{u}) + 15(\mathbf{u} \cdot \mathbf{v}) + 15(\mathbf{v} \cdot \mathbf{u}) + 25(\mathbf{v} \cdot \mathbf{v}) \\ &= 9|\mathbf{u}|^2 + 30(\mathbf{u} \cdot \mathbf{v}) + 25|\mathbf{v}|^2. \end{aligned}$$

Definition. Let $\mathbf{v}$ and $\mathbf{w}$ be two vectors which make an angle θ. We denote by $|\mathbf{v}| \cos\theta$ the **projection of v on w**. We also call this quantity the **component of v along w**. As before we denote this quantity by $\text{Proj}_{\mathbf{w}}\mathbf{v}$.

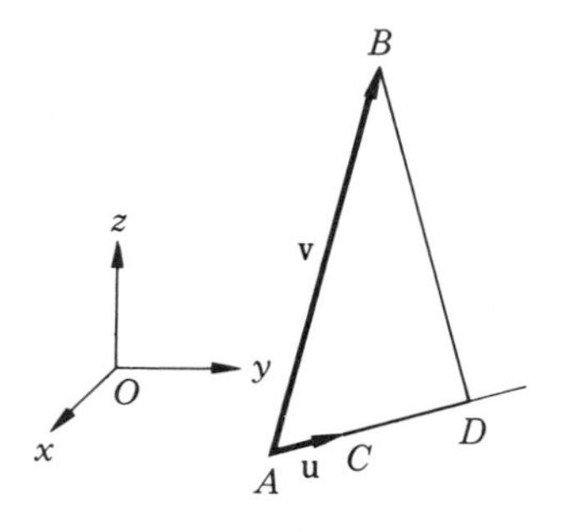

Fig. 2-24

From the formula for $\cos\theta$, we may write

$$\operatorname{Proj}_{\mathbf{w}}\mathbf{v} = |\mathbf{v}|\cos\theta = |\mathbf{v}|\frac{\mathbf{v}\cdot\mathbf{w}}{|\mathbf{v}|\,|\mathbf{w}|} = \frac{\mathbf{v}\cdot\mathbf{w}}{|\mathbf{w}|}.$$

EXAMPLE 4. Find the projection of $\mathbf{v} = -\mathbf{i} + 2\mathbf{j} + 3\mathbf{k}$ on $\mathbf{w} = 2\mathbf{i} - \mathbf{j} - 4\mathbf{k}$.

SOLUTION. We have $\mathbf{v}\cdot\mathbf{w} = (-1)(2) + (2)(-1) + (3)(-4) = -16$; $|\mathbf{w}| = \sqrt{21}$. Therefore, he projection of $\mathbf{v}$ on $\mathbf{w}$ is $|\mathbf{v}|\cos\theta = -16/\sqrt{21}$.

Theorem 14. *If* $\mathbf{u}$ *and* $\mathbf{v}$ *are not* $\mathbf{0}$, *there is a unique number* k *such that* $\mathbf{v} - k\mathbf{u}$ *is orthogonal to* $\mathbf{u}$. *In fact,* k *can be found from the formula*

$$k = \frac{\mathbf{u}\cdot\mathbf{v}}{|\mathbf{u}|^2}.$$

PROOF. $(\mathbf{v} - k\mathbf{u})$ is orthogonal to $\mathbf{u}$ if and only if $\mathbf{u}\cdot(\mathbf{v} - k\mathbf{u}) = 0$. But

$$\mathbf{u}\cdot(\mathbf{v} - k\mathbf{u}) = \mathbf{u}\cdot\mathbf{v} - k|\mathbf{u}|^2 = 0.$$

Therefore, selection of $k = \mathbf{u}\cdot\mathbf{v}/|\mathbf{u}|^2$ yields the result.

Figure 2-24 shows geometrically how k is to be selected. We drop a perpendicular from the head of $\mathbf{v}$ (point B) to the line containing $\mathbf{u}$ (point D). The directed segment $\overrightarrow{AD}$ gives the proper multiple of $\mathbf{u}[\overrightarrow{AC}]$, and the directed segment $\overrightarrow{DB}$ represents the orthogonal vector.

EXAMPLE 5. Find a linear combination of

$$\mathbf{u} = 2\mathbf{i} + 3\mathbf{j} - \mathbf{k} \qquad \text{and} \qquad \mathbf{v} = \mathbf{i} + 2\mathbf{j} + \mathbf{k}$$

which is orthogonal to $\mathbf{u}$.

SOLUTION. We select $k = (2 + 6 - 1)/14 = \frac{1}{2}$, and the desired vector is $\frac{1}{2}\mathbf{j} + \frac{3}{2}\mathbf{k}$.

Problems

In each of Problems 1 through 5, find $\cos\theta$ where θ is the angle between $\mathbf{v}$ and $\mathbf{w}$.

1. $\mathbf{v} = \mathbf{i} + 3\mathbf{j} - 2\mathbf{k}, \quad \mathbf{w} = 2\mathbf{i} + 4\mathbf{j} - \mathbf{k}$
2. $\mathbf{v} = -\mathbf{i} + 2\mathbf{j} + 3\mathbf{k}, \quad \mathbf{w} = 3\mathbf{i} - 2\mathbf{j} - 2\mathbf{k}$
3. $\mathbf{v} = 4\mathbf{i} - 3\mathbf{j} + 5\mathbf{k}, \quad \mathbf{w} = 2\mathbf{i} + \mathbf{j} + \mathbf{k}$
4. $\mathbf{v} = 2\mathbf{i} + 3\mathbf{k}, \quad \mathbf{w} = \mathbf{j} + 4\mathbf{k}$
5. $\mathbf{v} = -2\mathbf{i} - 3\mathbf{j} - 4\mathbf{k}, \quad \mathbf{w} = 2\mathbf{i} - 3\mathbf{j} + 4\mathbf{k}$

In each of Problems 6 through 10, find the projection of the vector $\mathbf{v}$ on $\mathbf{u}$.

6. $\mathbf{u} = 2\mathbf{i} - 6\mathbf{j} + 3\mathbf{k}, \quad \mathbf{v} = \mathbf{i} + 2\mathbf{j} - 2\mathbf{k}$
7. $\mathbf{u} = 6\mathbf{i} + 2\mathbf{j} - 3\mathbf{k}, \quad \mathbf{v} = -\mathbf{i} + 8\mathbf{j} + 4\mathbf{k}$
8. $\mathbf{u} = 12\mathbf{i} + 3\mathbf{j} + 4\mathbf{k}, \quad \mathbf{v} = 4\mathbf{i} + 8\mathbf{j} + \mathbf{k}$
9. $\mathbf{u} = 3\mathbf{i} + 5\mathbf{j} - 4\mathbf{k}, \quad \mathbf{v} = 4\mathbf{i} - 3\mathbf{j} + 5\mathbf{k}$
10. $\mathbf{u} = 2\mathbf{i} - 5\mathbf{j} + 3\mathbf{k}, \quad \mathbf{v} = -\mathbf{i} + 2\mathbf{j} + 7\mathbf{k}$

In each of Problems 11 through 13, find a unit vector in the direction of $\mathbf{u}$.

11. $\mathbf{u} = 2\mathbf{i} - 6\mathbf{j} + 3\mathbf{k}$
12. $\mathbf{u} = -\mathbf{i} + 2\mathbf{k}$
13. $\mathbf{u} = 3\mathbf{i} - 2\mathbf{j} + 7\mathbf{k}$

In each of Problems 14 through 17, find the value of k so that $\mathbf{v} - k\mathbf{u}$ is orthogonal to $\mathbf{u}$. Also, find the value h so that $\mathbf{u} - h\mathbf{v}$ is orthogonal to $\mathbf{v}$.

14. $\mathbf{u} = 2\mathbf{i} - \mathbf{j} + 2\mathbf{k}, \quad \mathbf{v} = 3\mathbf{i} + \mathbf{j} + 2\mathbf{k}$
15. $\mathbf{u} = 2\mathbf{i} - 3\mathbf{j} + 6\mathbf{k}, \quad \mathbf{v} = 7\mathbf{i} + 14\mathbf{k}$
16. $\mathbf{u} = 3\mathbf{i} + 4\mathbf{j} - 5\mathbf{k}, \quad \mathbf{v} = 9\mathbf{i} + 12\mathbf{j} - 5\mathbf{k}$
17. $\mathbf{u} = \mathbf{i} + 3\mathbf{j} - 2\mathbf{k}, \quad \mathbf{v} = 6\mathbf{i} + 10\mathbf{j} - 3\mathbf{k}$
18. Write a detailed proof of Theorem 12.
19. Show that if $\mathbf{u}$ and $\mathbf{v}$ are any vectors ($\neq \mathbf{0}$), then $\mathbf{u}$ and $\mathbf{v}$ make equal angles with $\mathbf{w}$ if
$$\mathbf{w} = \left(\frac{|\mathbf{v}|}{|\mathbf{u}| + |\mathbf{v}|}\right)\mathbf{u} + \left(\frac{|\mathbf{u}|}{|\mathbf{u}| + |\mathbf{v}|}\right)\mathbf{v}.$$
20. Show that if $\mathbf{u}$ and $\mathbf{v}$ are any vectors, the vectors $|\mathbf{v}|\mathbf{u} + |\mathbf{u}|\mathbf{v}$ and $|\mathbf{v}|\mathbf{u} - |\mathbf{u}|\mathbf{v}$ are orthogonal.

In each of Problems 21 through 23, determine the relation between g and h so that $g\mathbf{u} + h\mathbf{v}$ is orthogonal to $\mathbf{w}$.

21. $\mathbf{u} = 3\mathbf{i} - 2\mathbf{j} + \mathbf{k}, \quad \mathbf{v} = \mathbf{i} + 2\mathbf{j} - 3\mathbf{k}, \quad \mathbf{w} = -\mathbf{i} + \mathbf{j} + 2\mathbf{k}$
22. $\mathbf{u} = 2\mathbf{i} + \mathbf{j} - 2\mathbf{k}, \quad \mathbf{b} = \mathbf{i} - \mathbf{j} + \mathbf{k}, \quad \mathbf{w} = -\mathbf{i} + 2\mathbf{j} + 3\mathbf{k}$
23. $\mathbf{u} = \mathbf{i} + 2\mathbf{j} - 3\mathbf{k}, \quad \mathbf{v} = 3\mathbf{i} + \mathbf{j} - \mathbf{k}, \quad \mathbf{w} = 4\mathbf{i} - \mathbf{j} + 2\mathbf{k}$

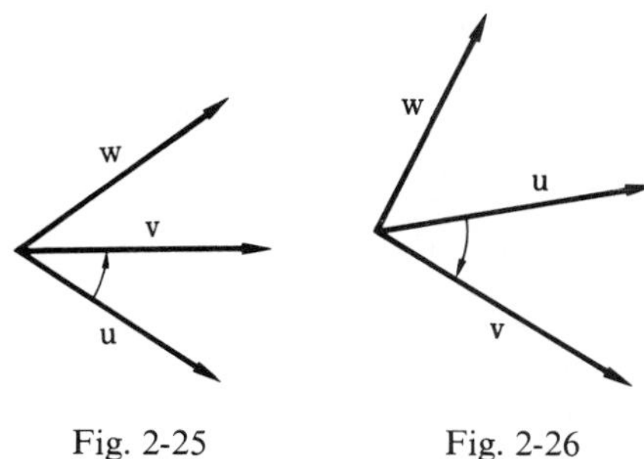

Fig. 2-25 Fig. 2-26

In each of Problems 24 through 26, determine g and h so that $\mathbf{w} - g\mathbf{u} - h\mathbf{v}$ is orthogonal to both $\mathbf{u}$ and $\mathbf{v}$.

24. $\mathbf{u} = 2\mathbf{i} - \mathbf{j} + \mathbf{k}$, $\mathbf{v} = \mathbf{i} + \mathbf{j} + 2\mathbf{k}$, $\mathbf{w} = 2\mathbf{i} - \mathbf{j} + 4\mathbf{k}$
25. $\mathbf{u} = \mathbf{i} + \mathbf{j} - 2\mathbf{k}$, $\mathbf{v} = -\mathbf{i} + 2\mathbf{j} + 3\mathbf{k}$, $\mathbf{w} = 5\mathbf{i} + 8\mathbf{k}$
26. $\mathbf{u} = 3\mathbf{i} - 2\mathbf{j}$, $\mathbf{v} = 2\mathbf{i} - \mathbf{k}$, $\mathbf{w} = 4\mathbf{i} - 2\mathbf{k}$
27. If $\mathbf{u}$ and $\mathbf{v}$ are nonzero vectors, under what conditions is it true that
$$|\mathbf{u} + \mathbf{v}| = |\mathbf{u}| + |\mathbf{v}|?$$
28. Suppose that $\overrightarrow{AB}$, $\overrightarrow{AC}$, and $\overrightarrow{AD}$ are directed line segments of $\mathbf{u}$, $\mathbf{v}$, and $\mathbf{u} + \mathbf{v}$, respectively, with $|\mathbf{u}| = |\mathbf{v}|$. Show that $\overrightarrow{AD}$ bisects the angle between $\overrightarrow{AB}$ and $\overrightarrow{AC}$.
29. Let P be a vertex of a cube. Draw a diagonal of the cube from P and a diagonal of one of the faces from P. Use vectors to find the cosine of the angle between these two diagonals.
30. Use vectors to find the cosine of the angle between two faces of a regular tetrahedron. (See Problem 22, Chapter 1, Section 2.)

7. The Vector or Cross Product

We saw in Section 6 that the scalar product of two vectors $\mathbf{u}$ and $\mathbf{v}$ associates an ordinary number, i.e., a scalar, with each pair of vectors. The vector or cross product, on the other hand, associates a *vector* with each ordered pair of vectors. However, before defining the cross product, we shall discuss the notion of "right-handed" and "left-handed" triples of vectors.

An **ordered triple** $\{\mathbf{u}, \mathbf{v}, \mathbf{w}\}$ of linearly independent vectors is said to be **right-handed** if the vectors are situated as in Fig. 2-25. If the ordered triple is situated as in Fig. 2-26, the vectors are said to form a **left-handed triple**. The notion of left-handed and right-handed triple is not defined if the vectors form a linearly dependent set.

Definition. Two sets of ordered triples of vectors are said to be **similarly**

oriented if and only if both sets are right-handed or both are left-handed. Otherwise they are **oppositely oriented**.

Suppose $\{\mathbf{u}_1, \mathbf{v}_1, \mathbf{w}_1\}$ and $\{\mathbf{u}_2, \mathbf{v}_2, \mathbf{w}_2\}$ are ordered linearly independent sets of triples. From Theorem 11, it follows that we may express $\mathbf{u}_2$, $\mathbf{v}_2$, and $\mathbf{w}_2$ in terms of $\mathbf{u}_1$, $\mathbf{v}_1$, and $\mathbf{w}_1$ by equations of the form

$$\mathbf{u}_2 = a_{11}\mathbf{u}_1 + a_{12}\mathbf{v}_1 + a_{13}\mathbf{w}_1,$$

$$\mathbf{v}_2 = a_{21}\mathbf{u}_1 + a_{22}\mathbf{v}_1 + a_{23}\mathbf{w}_1,$$

$$\mathbf{w}_2 = a_{31}\mathbf{u}_1 + a_{32}\mathbf{v}_1 + a_{33}\mathbf{w}_1.$$

We denoted by D the determinant

$$D = \begin{vmatrix} a_{11} & a_{12} & a_{13} \\ a_{21} & a_{22} & a_{23} \\ a_{31} & a_{32} & a_{33} \end{vmatrix}.$$

Although the proof will not be given, it is a fact that the two triples above are similarly oriented if and only if $D > 0$; they are oppositely oriented if and only if $D < 0$. Note that the determinant cannot be zero, for then $\mathbf{u}_2$, $\mathbf{v}_2$, and $\mathbf{w}_2$ would not be linearly independent. (See the proof of Theorem 10′ in Appendix 2.

It is also true that if $\{\mathbf{u}_1, \mathbf{v}_1, \mathbf{w}_1\}$ and $\{\mathbf{u}_2, \mathbf{v}_2, \mathbf{w}_2\}$ are similarly oriented and if $\{\mathbf{u}_2, \mathbf{v}_2, \mathbf{w}_2\}$ and $\{\mathbf{u}_3, \mathbf{v}_3, \mathbf{w}_3\}$ are similarly oriented, then $\{\mathbf{u}_1, \mathbf{v}_1, \mathbf{w}_1\}$ and $\{\mathbf{u}_3, \mathbf{v}_3, \mathbf{w}_3\}$ are similarly oriented.

The facts above lead to the following result.

Theorem 15. *If* $\{\mathbf{u}, \mathbf{v}, \mathbf{w}\}$ *is a right-handed triple, then* (i) $\{\mathbf{v}, \mathbf{u}, -\mathbf{w}\}$ *is a right-handed triple, and* (ii) $\{c_1\mathbf{u}, c_2\mathbf{v}, c_3\mathbf{w}\}$ *is a right-handed triple provided that* $c_1c_2c_3 > 0$.

To prove (i) we apply the above determinant condition on similar orientation by regarding $\{\mathbf{u}, \mathbf{v}, \mathbf{w}\}$ as $\{\mathbf{u}_1, \mathbf{v}_1, \mathbf{w}_1\}$ and $\{\mathbf{v}, \mathbf{u}, -\mathbf{w}\}$ as $\{\mathbf{u}_2, \mathbf{v}_2, \mathbf{w}_2\}$. To establish (ii) we regard $\{c_1\mathbf{u}, c_2\mathbf{v}, c_3\mathbf{w}\}$ as $\{\mathbf{u}_2, \mathbf{v}_2, \mathbf{w}_2\}$. The details are left to the reader.

Definition. Given the vectors **u** and **v**, the **vector** or **cross product** $\mathbf{u} \times \mathbf{v}$ is defined as follows:

i) if either **u** or **v** is **0**, then

$$\mathbf{u} \times \mathbf{v} = \mathbf{0};$$

ii) if **u** is proportional to **v**, then

$$\mathbf{u} \times \mathbf{v} = \mathbf{0};$$

iii) otherwise,

$$\mathbf{u} \times \mathbf{v} = \mathbf{w}$$

where $\mathbf{w}$ has the three properties: (a) it is orthogonal to both $\mathbf{u}$ and $\mathbf{v}$; (b) it has magnitude $|\mathbf{w}| = |\mathbf{u}|\,|\mathbf{v}| \sin\theta$, where θ is the angle between $\mathbf{u}$ and $\mathbf{v}$, and (c) it is directed so that $\{\mathbf{u}, \mathbf{v}, \mathbf{w}\}$ is a right-handed triple.

REMARK. We shall always assume that any coordinate triple $\{\mathbf{i}, \mathbf{j}, \mathbf{k}\}$ is right-handed. (We have assumed this up to now without pointing out this fact specifically.)

The proofs of the next two theorems are given in Appendix 2.

Theorem 16. *Suppose that* $\mathbf{u}$ *and* $\mathbf{v}$ *are any vectors, that* $\{\mathbf{i}, \mathbf{j}, \mathbf{k}\}$ *is a right-handed triple, and that* t *is any number. Then*

i) $\mathbf{v} \times \mathbf{u} = -(\mathbf{u} \times \mathbf{v})$,
ii) $(t\mathbf{u}) \times \mathbf{v} = t(\mathbf{u} \times \mathbf{v}) = \mathbf{u} \times (t\mathbf{v})$,
iii) $\mathbf{i} \times \mathbf{j} = -\mathbf{j} \times \mathbf{i} = \mathbf{k}$,
$\mathbf{j} \times \mathbf{k} = -\mathbf{k} \times \mathbf{j} = \mathbf{i}$,
$\mathbf{k} \times \mathbf{i} = -\mathbf{i} \times \mathbf{k} = \mathbf{j}$,
iv) $\mathbf{i} \times \mathbf{i} = \mathbf{j} \times \mathbf{j} = \mathbf{k} \times \mathbf{k} = \mathbf{0}$.

Theorem 17. *If* $\mathbf{u}$, $\mathbf{v}$, $\mathbf{w}$ *are any vectors, then*

i) $\mathbf{u} \times (\mathbf{v} + \mathbf{w}) = (\mathbf{u} \times \mathbf{v}) + (\mathbf{u} \times \mathbf{w})$ *and*
ii) $(\mathbf{v} + \mathbf{w}) \times \mathbf{u} = (\mathbf{v} \times \mathbf{u}) + (\mathbf{w} \times \mathbf{u})$.

With the aid of Theorems 16 and 17, the next theorem, an extremely useful one, is easily established.

Theorem 18. *If*

$$\mathbf{u} = a_1\mathbf{i} + a_2\mathbf{j} + a_3\mathbf{k} \qquad \text{and} \qquad \mathbf{v} = b_1\mathbf{i} + b_2\mathbf{j} + b_3\mathbf{k},$$

then

$$\boxed{\mathbf{u} \times \mathbf{v} = (a_2b_3 - a_3b_2)\mathbf{i} + (a_3b_1 - a_1b_3)\mathbf{j} + (a_1b_2 - a_2b_1)\mathbf{k}.} \tag{1}$$

PROOF. By using the laws in Theorems 16 and 17 we obtain (being careful to keep the order of the factors)

$$\begin{aligned}\mathbf{u} \times \mathbf{v} = {} & a_1b_1(\mathbf{i} \times \mathbf{i}) + a_1b_2(\mathbf{i} \times \mathbf{j}) + a_1b_3(\mathbf{i} \times \mathbf{k}) \\ & + a_2b_1(\mathbf{j} \times \mathbf{i}) + a_2b_2(\mathbf{j} \times \mathbf{j}) + a_2b_3(\mathbf{j} \times \mathbf{k}) \\ & + a_3b_1(\mathbf{k} \times \mathbf{i}) + a_3b_2(\mathbf{k} \times \mathbf{j}) + a_3b_3(\mathbf{k} \times \mathbf{k}).\end{aligned}$$

The result follows from Theorem 16, parts (iii) and (iv) by collecting terms.

The formula (1) above is useful in calculating the cross product. The

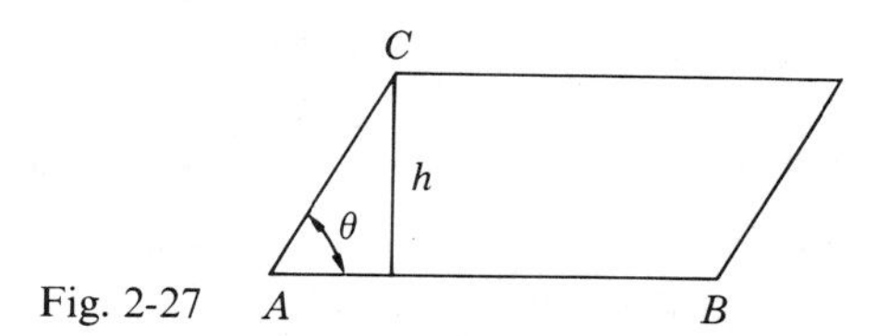

Fig. 2-27

following *symbolic form* is a great aid in remembering the formula. We write

$$\mathbf{u} \times \mathbf{v} = \begin{vmatrix} \mathbf{i} & \mathbf{j} & \mathbf{k} \\ a_1 & a_2 & a_3 \\ b_1 & b_2 & b_3 \end{vmatrix},$$

where it is understood that this "determinant" is to be expanded formally according to its first row. The reader may easily verify that when the above expression is expanded, it is equal to (1).

EXAMPLE 1. Find $\mathbf{u} \times \mathbf{v}$ if $\mathbf{u} = 2\mathbf{i} - 3\mathbf{j} + \mathbf{k}$, $\mathbf{v} = \mathbf{i} + \mathbf{j} - 2\mathbf{k}$.

SOLUTION. Carrying out the formal expansion, we obtain

$$\begin{vmatrix} \mathbf{i} & \mathbf{j} & \mathbf{k} \\ 2 & -3 & 1 \\ 1 & 1 & -2 \end{vmatrix} = \begin{vmatrix} -3 & 1 \\ 1 & -2 \end{vmatrix}\mathbf{i} - \begin{vmatrix} 2 & 1 \\ 1 & -2 \end{vmatrix}\mathbf{j} + \begin{vmatrix} 2 & -3 \\ 1 & 1 \end{vmatrix}\mathbf{k} = 5\mathbf{i} + 5\mathbf{j} + 5\mathbf{k}.$$

REMARKS. In mechanics the cross product is used for the computation of the vector moment of a force **F** applied at a point B, about a point A. There are also applications of cross product to problems in electricity and magnetism. However, we shall confine our attention to applications in geometry.

Theorem 19. *The area of a parallelogram with adjacent sides AB and AC is given by**

$$|\mathbf{v}(\overrightarrow{AB}) \times \mathbf{v}(\overrightarrow{AC})|.$$

The area of $\triangle ABC$ is then $\frac{1}{2}|\mathbf{v}(\overrightarrow{AB}) \times \mathbf{v}(\overrightarrow{AC})|$.

PROOF. From Fig. 2-27, we see that the area of the parallelogram is

$$|AB|h = |AB|\,|AC|\sin\theta.$$

The result then follows from the definition of cross product.

* The notation $\mathbf{v}(\overrightarrow{AB})$ to indicate a vector $\mathbf{v}$ with representative $\overrightarrow{AB}$ was introduced in Section 3.

EXAMPLE 2. Find the area of $\triangle ABC$ with $A(-2,1,3)$, $B(1,-1,1)$, $C(3,-2,4)$.

SOLUTION. We have $\mathbf{v}(\overrightarrow{AB}) = 3\mathbf{i} - 2\mathbf{j} - 2\mathbf{k}$, $\mathbf{v}(\overrightarrow{AC}) = 5\mathbf{i} - 3\mathbf{j} + \mathbf{k}$. From Theorem 19 we obtain

$$\mathbf{v}(\overrightarrow{AB}) \times \mathbf{v}(\overrightarrow{AC}) = -8\mathbf{i} - 13\mathbf{j} + \mathbf{k}$$

and

$$\tfrac{1}{2}|-8\mathbf{i} - 13\mathbf{j} + \mathbf{k}| = \tfrac{1}{2}\sqrt{64 + 169 + 1} = \tfrac{3}{2}\sqrt{26}.$$

The vector product may be used to find the equation of a plane through three points. The next example illustrates the technique.

EXAMPLE 3. Find the equation of the plane through the points $A(-1,1,2)$, $B(1,-2,1)$, $C(2,2,4)$.

SOLUTION. A vector normal to the plane will be perpendicular to both the vectors

$$\mathbf{v}(\overrightarrow{AB}) = 2\mathbf{i} - 3\mathbf{j} - \mathbf{k} \qquad \text{and} \qquad \mathbf{v}(\overrightarrow{AC}) = 3\mathbf{i} + \mathbf{j} + 2\mathbf{k}.$$

One such vector is the cross product

$$\mathbf{v}(\overrightarrow{AB}) \times \mathbf{v}(\overrightarrow{AC}) = -5\mathbf{i} - 7\mathbf{j} + 11\mathbf{k}.$$

Therefore the numbers -5, -7, 11 form a set of *attitude numbers* (see Chapter 1, Section 4) of the desired plane. Using $A(-1,1,2)$ as a point on the plane, we get for the equation

$$-5(x+1) - 7(y-1) + 11(z-2) = 0$$

or

$$5x + 7y - 11z + 20 = 0.$$

EXAMPLE 4. Find the perpendicular distance between the skew lines

$$L_1 : \frac{x+2}{2} = \frac{y-1}{3} = \frac{z+1}{-1}, \qquad L_2 : \frac{x-1}{-1} = \frac{y+1}{2} = \frac{z-2}{4}.$$

SOLUTION. The vector

$$\mathbf{v}_1 = 2\mathbf{i} + 3\mathbf{j} - \mathbf{k}$$

is a vector along L_1. The vector

$$\mathbf{v}_2 = -\mathbf{i} + 2\mathbf{j} + 4\mathbf{k}$$

is a vector along L_2. A vector perpendicular to both $\mathbf{v}_1$ and $\mathbf{v}_2$ (i.e., to both L_1 and L_2) is

$$\mathbf{v}_1 \times \mathbf{v}_2 = 14\mathbf{i} - 7\mathbf{j} + 7\mathbf{k}.$$

Call this common perpendicular **w**. The desired length may be obtained as a *projection*. Select any point on L_1 (call it P_1) and any point on L_2 (call it P_2). Then the desired length is the projection of the vector $\mathbf{v}(\overrightarrow{P_1P_2})$ on **w**. To get this, we select $P_1(-2, 1, -1)$ on L_1 and $P_2(1, -1, 2)$ on L_2; and so

$$\mathbf{v}(\overrightarrow{P_1P_2}) = 3\mathbf{i} - 2\mathbf{j} + 3\mathbf{k}.$$

Therefore,

$$\text{Projection of } \mathbf{v}(\overrightarrow{P_1P_2}) \text{ on } \mathbf{w} = \frac{\mathbf{v}(\overrightarrow{P_1P_2}) \cdot \mathbf{w}}{|\mathbf{w}|}$$

$$= \frac{3 \cdot 14 + (-2)(-7) + 3(7)}{7\sqrt{6}} = \frac{11}{\sqrt{6}}.$$

Problems

In each of Problems 1 through 6, find the cross product $\mathbf{u} \times \mathbf{v}$.

1. $\mathbf{u} = \mathbf{i} + 3\mathbf{j} - \mathbf{k}$, $\quad \mathbf{v} = 2\mathbf{i} - \mathbf{j} + \mathbf{k}$
2. $\mathbf{u} = 4\mathbf{i} - 2\mathbf{j} + 3\mathbf{k}$, $\quad \mathbf{v} = -\mathbf{i} - 2\mathbf{j} - \mathbf{k}$
3. $\mathbf{u} = -\mathbf{i} + 2\mathbf{j}$, $\quad \mathbf{v} = \mathbf{i} + 3\mathbf{j} - 2\mathbf{k}$
4. $\mathbf{u} = 3\mathbf{j} + 2\mathbf{k}$, $\quad \mathbf{v} = 2\mathbf{i} - 3\mathbf{j}$
5. $\mathbf{u} = -2\mathbf{i} + 4\mathbf{j} + 5\mathbf{k}$, $\quad \mathbf{v} = 4\mathbf{i} + 5\mathbf{k}$
6. $\mathbf{u} = 2\mathbf{i} - 3\mathbf{j} + \mathbf{k}$, $\quad \mathbf{v} = 4\mathbf{k}$

In Problems 7 through 11, find in each case the area of $\triangle \mathbf{ABC}$ and the equation of the plane through A, B, and C. Use vector methods.

7. $A(1, -2, 3)$, $\quad B(3, 1, 2)$, $\quad C(2, 3, -1)$
8. $A(3, 2, -2)$, $\quad B(4, 1, 2)$, $\quad C(1, 2, 3)$
9. $A(2, -1, 1)$, $\quad B(3, 2, -1)$, $\quad C(-1, 3, 2)$
10. $A(1, -2, 3)$, $\quad B(2, -1, 1)$, $\quad C(4, 2, -1)$
11. $A(-2, 3, 1)$, $\quad B(4, 2, -2)$, $\quad C(2, 0, 1)$

In Problems 12 through 14, find in each case the perpendicular distance between the given lines.

12. $\dfrac{x+1}{2} = \dfrac{y-3}{-3} = \dfrac{z+2}{4}$; $\quad \dfrac{x-2}{3} = \dfrac{y+1}{2} = \dfrac{z-1}{5}$
13. $\dfrac{x-1}{3} = \dfrac{y+1}{2} = \dfrac{z-1}{5}$; $\quad \dfrac{x+2}{4} = \dfrac{y-1}{3} = \dfrac{z+1}{-2}$
14. $\dfrac{x+1}{2} = \dfrac{y-1}{-4} = \dfrac{z+2}{3}$; $\quad \dfrac{x}{3} = \dfrac{y}{5} = \dfrac{z-2}{-2}$

In Problems 15 through 19, use vector methods to find, in each case, the equations in symmetric form of the line through the given point P and parallel to the two given planes.

15. $P(-1,3,2)$, $3x - 2y + 4z + 2 = 0$, $2x + y - z = 0$

16. $P(2,3,-1)$, $x + 2y + 2z - 4 = 0$, $2x + y - 3z + 5 = 0$

17. $P(1,-2,3)$, $3x + y - 2z + 3 = 0$, $2x + 3y + z - 6 = 0$

18. $P(-1,0,-2)$, $2x + 3y - z + 4 = 0$, $3x - 2y + 2z - 5 = 0$

19. $P(3,0,1)$, $x + 2y = 0$, $3y - z = 0$

In Problems 20 and 21, find in each case equations in symmetric form of the line of intersection of the given planes. Use the method of vector products.

20. $2(x - 1) + 3(y + 1) - 4(z - 2) = 0$

$3(x - 1) - 4(y + 1) + 2(z - 2) = 0$

21. $3(x + 2) - 2(y - 1) + 2(z + 1) = 0$

$(x + 2) + 2(y - 1) - 3(z + 1) = 0$

In each of Problems 22 through 26, find an equation of the plane through the given point or points and parallel to the given line or lines.

22. $(1,3,2)$; $\dfrac{x+1}{2} = \dfrac{y-2}{-1} = \dfrac{z+3}{3}$; $\dfrac{x-2}{1} = \dfrac{y+1}{-2} = \dfrac{z+2}{2}$

23. $(2,-1,-3)$; $\dfrac{x-1}{3} = \dfrac{y+2}{2} = \dfrac{z}{-4}$; $\dfrac{x}{2} = \dfrac{y-1}{-3} = \dfrac{z-2}{2}$

24. $(2,1,-2)$; $(1,-1,3)$; $\dfrac{x+1}{3} = \dfrac{y-1}{2} = \dfrac{z-2}{2}$

25. $(1,-2,3)$; $(-1,2,-1)$; $\dfrac{x-2}{2} = \dfrac{y+1}{3} = \dfrac{z-1}{4}$

26. $(0,1,2)$; $(2,0,1)$; $\dfrac{x-1}{3} = \dfrac{y+1}{0} = \dfrac{z+1}{1}$

In Problems 27 through 29, find in each case the equation of the plane through the line L_1 which also satisfies the additional condition.

27. $L_1: \dfrac{x-1}{2} = \dfrac{y+1}{3} = \dfrac{z-2}{1}$; through $(2,1,1)$

28. $L_1: \dfrac{x-2}{2} = \dfrac{y-2}{3} = \dfrac{z-1}{-2}$; parallel to $\dfrac{x+1}{3} = \dfrac{y-1}{2} = \dfrac{z+1}{1}$

29. $L_1: \dfrac{x+1}{1} = \dfrac{y-1}{2} = \dfrac{z-2}{-2}$; perpendicular to $2x + 3y - z + 4 = 0$

In Problems 30 and 31, find the equation of the plane through the given points and perpendicular to the given planes.

30. $(1,2,-1)$; $2x - 3y + 5z - 1 = 0$; $3x + 2y + 4z + 6 = 0$

31. $(-1,3,2)$; $(1,6,1)$; $3x - y + 4z - 7 = 0$

In Problems 32 and 33, find equations in symmetric form of the line through the given point P, which is perpendicular to and intersects the given line. Use the cross product.

32. $P(3,3,-1)$; $\frac{x}{-1} = \frac{y-3}{1} = \frac{z+1}{1}$

33. $P(3,-2,0)$; $\frac{x-4}{3} = \frac{y-4}{-4} = \frac{z-5}{-1}$

34. Suppose that $\mathbf{a}$ is a nonzero vector. If $\mathbf{a} \cdot \mathbf{x} = \mathbf{a} \cdot \mathbf{y}$ and $\mathbf{a} \times \mathbf{x} = \mathbf{a} \times \mathbf{y}$, is it true that $\mathbf{x} = \mathbf{y}$? Justify your answer.

35. Given that $\mathbf{u} + \mathbf{v} + \mathbf{w} - \mathbf{r} = \mathbf{0}$ and $\mathbf{u} - \mathbf{v} + \mathbf{w} + 2\mathbf{r} = \mathbf{0}$. Prove that $\mathbf{r} \times \mathbf{v} = \mathbf{0}$, that $\mathbf{v} \times \mathbf{w} = \frac{3}{2}\mathbf{r} \times \mathbf{w}$, and that

$$2\mathbf{u} \times \mathbf{w} = \mathbf{w} \times \mathbf{r} = \mathbf{r} \times \mathbf{u}.$$

8. Products of Three Vectors

Since two types of multiplication, the scalar product and the cross product, may be performed on vectors, we can combine three vectors in several ways. For example, we can form the product

$$(\mathbf{u} \times \mathbf{v}) \cdot \mathbf{w}$$

and the product

$$(\mathbf{u} \times \mathbf{v}) \times \mathbf{w}.$$

Also, we can consider the combinations

$$\mathbf{u} \cdot (\mathbf{v} \times \mathbf{w}) \qquad \text{and} \qquad \mathbf{u} \times (\mathbf{v} \times \mathbf{w}).$$

The next theorem gives a simple rule for computing $(\mathbf{u} \times \mathbf{v}) \cdot \mathbf{w}$ and also an elegant geometric interpretation of the quantity $|(\mathbf{u} \times \mathbf{v}) \cdot \mathbf{w}|$.

Theorem 20. *Suppose that $\mathbf{u}_1, \mathbf{u}_2, \mathbf{u}_3$, are vectors and that the points A, B, C, D are chosen so that*

$$\mathbf{v}(\overrightarrow{AB}) = \mathbf{u}_1, \qquad \mathbf{v}(\overrightarrow{AC}) = \mathbf{u}_2, \qquad \mathbf{v}(\overrightarrow{AD}) = \mathbf{u}_3.$$

Then

i) *the quantity $|(\mathbf{u}_1 \times \mathbf{u}_2) \cdot \mathbf{u}_3|$ is the volume of the parallelepiped with one vertex at A and adjacent vertices at B, C, and D.* (See Fig. 2-28.) *This*

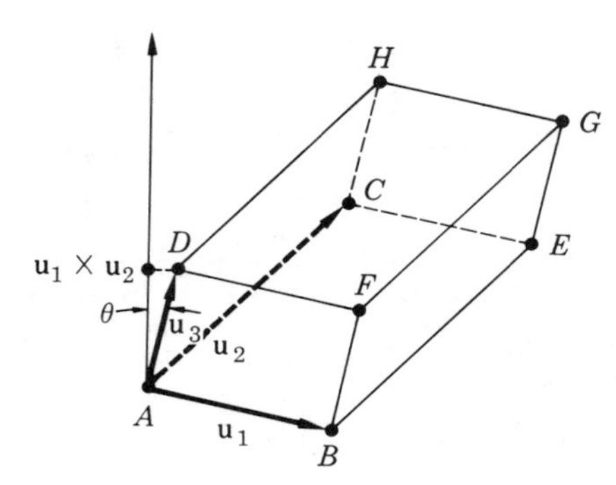

Fig. 2-28

volume is zero if and only if the four points A, B, C, D lie in a plane;

ii) *if* $\{\mathbf{i}, \mathbf{j}, \mathbf{k}\}$ *is the usual right-handed coordinate triple and if*

$$\mathbf{u}_1 = a_1\mathbf{i} + b_1\mathbf{j} + c_1\mathbf{k}, \qquad \mathbf{u}_2 = a_2\mathbf{i} + b_2\mathbf{j} + c_2\mathbf{k},$$

$$\mathbf{u}_3 = a_3\mathbf{i} + b_3\mathbf{j} + c_3\mathbf{k},$$

then

$$\boxed{(\mathbf{u}_1 \times \mathbf{u}_2) \cdot \mathbf{u}_3 = \begin{vmatrix} a_1 & b_1 & c_1 \\ a_2 & b_2 & c_2 \\ a_3 & b_3 & c_3 \end{vmatrix};}$$

iii) $(\mathbf{u}_1 \times \mathbf{u}_2) \cdot \mathbf{u}_3 = \mathbf{u}_1 \cdot (\mathbf{u}_2 \times \mathbf{u}_3)$.

Proof. To prove (i), note that $|\mathbf{u}_1 \times \mathbf{u}_2|$ is the area of the parallelogram $ABEC$ and that

$$|(\mathbf{u}_1 \times \mathbf{u}_2) \cdot \mathbf{u}_3| = |\mathbf{u}_1 \times \mathbf{u}_2| |\mathbf{u}_3| |\cos \theta|,$$

where θ is the angle between the two vectors $\mathbf{u}_3$ and $\mathbf{u}_1 \times \mathbf{u}_2$. The quantity $|\mathbf{u}_3| |\cos \theta|$ is the length of the projection of $\mathbf{u}_3$ on the normal to the plane of $ABEC$. Clearly, $(\mathbf{u}_1 \times \mathbf{u}_2) \cdot \mathbf{u}_3 = 0 \Leftrightarrow \mathbf{u}_1 \times \mathbf{u}_2 = \mathbf{0}$ or $\mathbf{u}_3 = \mathbf{0}$ or $\cos \theta = 0$. If $\cos \theta = 0$, then $\mathbf{u}_3$ is parallel to the plane of $\mathbf{u}_1$ and $\mathbf{u}_2$ and all four points lie in a plane. The proof of parts (ii) and (iii) follow from Theorems 13 and 18 and are left to the reader.

Theorem 21. *If* $\mathbf{u}$, $\mathbf{v}$, *and* $\mathbf{w}$ *are any vectors, then*

i) $(\mathbf{u} \times \mathbf{v}) \times \mathbf{w} = (\mathbf{u} \cdot \mathbf{w})\mathbf{v} - (\mathbf{v} \cdot \mathbf{w})\mathbf{u}$,
ii) $\mathbf{u} \times (\mathbf{v} \times \mathbf{w}) = (\mathbf{u} \cdot \mathbf{w})\mathbf{v} - (\mathbf{u} \cdot \mathbf{v})\mathbf{w}$.

Proof. If $\mathbf{u}$ and $\mathbf{v}$ are proportional or if $\mathbf{w}$ is orthogonal to both $\mathbf{u}$ and $\mathbf{v}$, then both sides of (i) are zero. Otherwise, we see that $(\mathbf{u} \times \mathbf{v}) \times \mathbf{w}$ is orthogonal to the perpendicular to the plane determined by **u** and **v**. Hence $(\mathbf{u} \times \mathbf{v}) \times \mathbf{w}$ is in the plane of **u** and **v**. We choose a right-handed coordinate triple $\{\mathbf{i}, \mathbf{j}, \mathbf{k}\}$ so that **i** is in the direction of **u** and **j** is in the plane of **u** and **v**. Then there are numbers a_1, a_2, b_2, etc., so that

$$\mathbf{u} = a_1\mathbf{i}, \qquad \mathbf{v} = a_2\mathbf{i} + b_2\mathbf{j}, \qquad \mathbf{w} = a_3\mathbf{i} + b_3\mathbf{j} + c_3\mathbf{k}.$$

The reader may now compute both sides of (i) to see that they are equal. The proof of (ii) is left to the reader.

EXAMPLE 1. Given $A(3, -1, 2)$, $B(1, 2, -2)$, $C(2, 1, -2)$, and $D(-1, 3, 2)$, find the volume of the parallelepiped having AB, AC, and AD as edges.

SOLUTION. We have

$$\begin{aligned} \mathbf{u}_1 &= \mathbf{v}(\overrightarrow{AB}) = -2\mathbf{i} + 3\mathbf{j} - 4\mathbf{k}, \\ \mathbf{u}_2 &= \mathbf{v}(\overrightarrow{AC}) = -\mathbf{i} + 2\mathbf{j} - 4\mathbf{k}, \\ \mathbf{u}_3 &= \mathbf{v}(\overrightarrow{AD}) = -4\mathbf{i} + 4\mathbf{j}. \end{aligned}$$

We compute $\mathbf{u}_2 \times \mathbf{u}_3 = 16\mathbf{i} + 16\mathbf{j} + 4\mathbf{k}$. Therefore

$$|\mathbf{u}_1 \cdot (\mathbf{u}_2 \times \mathbf{u}_3)| = |-32 + 48 - 16| = 0.$$

Hence the four points are in a plane. The volume is zero.

EXAMPLE 2. Find the equations of the line through the point $(3, -2, 1)$ perpendicular to the line L (and intersecting it) given by

$$L = \left\{(x, y, z) : \frac{x-2}{2} = \frac{y+1}{-2} = \frac{z}{1}\right\}.$$

SOLUTION. Let $P_0(3, -2, 1)$ and $P_1(2, -1, 0)$,

$$\begin{aligned} \mathbf{u} &= 2\mathbf{i} - 2\mathbf{j} + \mathbf{k}, \\ \mathbf{v} &= \mathbf{v}(\overrightarrow{P_0P_1}) = -\mathbf{i} + \mathbf{j} - \mathbf{k}. \end{aligned}$$

The plane containing L and P_0 has a normal perpendicular to **u** and **v**. Hence this normal is proportional to $\mathbf{u} \times \mathbf{v}$. The desired line is in this plane and perpendicular to L. Therefore it has a direction **w** perpendicular to **u** and $\mathbf{u} \times \mathbf{v}$. Thus for some number c, we have

$$\begin{aligned} c\mathbf{w} &= \mathbf{u} \times (\mathbf{u} \times \mathbf{v}) = (\mathbf{u} \cdot \mathbf{v})\mathbf{u} - (\mathbf{u} \cdot \mathbf{u})\mathbf{v} \\ &= -5(2\mathbf{i} - 2\mathbf{j} + \mathbf{k}) - 9(-\mathbf{i} + \mathbf{j} - \mathbf{k}) = -\mathbf{i} + \mathbf{j} + 4\mathbf{k}. \end{aligned}$$

Consequently, the desired line has equations

$$\frac{x-3}{-1} = \frac{y+2}{1} = \frac{z-1}{4}.$$

Problems

In Problems 1 through 4, find the volume of the parallelepiped having edges AB, AC, and AD, or else show that A, B, C, and D lie on a plane or on a line. If they lie on a plane, find its equation; if they lie on a line, find its equations.

1. $A = (2, -1, 3)$, $B = (-1, 2, 2)$, $C = (1, 0, 1)$, $D = (4, 1, -1)$
2. $A = (3, 1, -2)$, $B = (1, 2, 1)$, $C = (2, -1, 3)$, $D = (4, 3, -7)$
3. $A = (1, 2, -3)$, $B = (3, 1, -2)$, $C = (-1, 3, 1)$, $D = (-3, 4, 3)$
4. $A = (-1, -2, 2)$, $B = (2, -1, 1)$, $C = (0, 1, 3)$, $D = (3, 2, -1)$
5. Prove Theorem 20, parts (ii) and (iii).
6. Complete the proof of Theorem 21.

In Problems 7 through 10, compute $(\mathbf{u} \times \mathbf{v}) \times \mathbf{w}$ directly and by using Theorem 21.

7. $\mathbf{u} = 2\mathbf{i} + 3\mathbf{j} - \mathbf{k}$, $\mathbf{v} = \mathbf{i} - 2\mathbf{j} + \mathbf{k}$, $\mathbf{w} = -\mathbf{i} + \mathbf{j} + 2\mathbf{k}$
8. $\mathbf{u} = 3\mathbf{i} - 2\mathbf{j} + \mathbf{k}$, $\mathbf{v} = \mathbf{i} + \mathbf{j} + 2\mathbf{k}$, $\mathbf{w} = 2\mathbf{i} - \mathbf{j} + 3\mathbf{k}$
9. $\mathbf{u} = \mathbf{i} + 2\mathbf{j} - 3\mathbf{k}$, $\mathbf{v} = -\mathbf{i} + \mathbf{j} - 2\mathbf{k}$, $\mathbf{w} = 3\mathbf{i} - \mathbf{j} + \mathbf{k}$
10. $\mathbf{u} = 2\mathbf{i} - \mathbf{j} + 3\mathbf{k}$, $\mathbf{v} = \mathbf{i} + 2\mathbf{j} + \mathbf{k}$, $\mathbf{w} = 3\mathbf{i} - 2\mathbf{j} - \mathbf{k}$
11. Show that every vector $\mathbf{v}$ satisfies the identity
$$\mathbf{i} \times (\mathbf{v} \times \mathbf{i}) + \mathbf{j} \times (\mathbf{v} \times \mathbf{j}) + \mathbf{k} \times (\mathbf{v} \times \mathbf{k}) = 2\mathbf{v}.$$

In Problems 12 through 14, find, in each case, the equations of the line through the given point and perpendicular to the given line and intersecting it.

12. $(1, 3, -2)$, $\dfrac{x-2}{3} = \dfrac{y+1}{-2} = \dfrac{z-1}{4}$
13. $(2, -1, 3)$, $\dfrac{x+1}{2} = \dfrac{y-2}{3} = \dfrac{z+1}{-5}$
14. $(-1, 2, 4)$, $\dfrac{x-1}{4} = \dfrac{y+2}{-3} = \dfrac{z-1}{2}$

In Problems 15 and 16, express $(\mathbf{t} \times \mathbf{u}) \times (\mathbf{v} \times \mathbf{w})$ in terms of $\mathbf{v}$ and $\mathbf{w}$.

15. $\mathbf{t} = \mathbf{i} + \mathbf{j} - 2\mathbf{k}$, $\mathbf{u} = 3\mathbf{i} - \mathbf{j} + 2\mathbf{k}$, $\mathbf{v} = 2\mathbf{i} + 2\mathbf{j} - \mathbf{k}$, $\mathbf{w} = -\mathbf{i} + \mathbf{j} + 2\mathbf{k}$
16. $\mathbf{t} = 2\mathbf{i} - \mathbf{j} + \mathbf{k}$, $\mathbf{u} = \mathbf{i} + 2\mathbf{j} - 3\mathbf{k}$, $\mathbf{v} = 3\mathbf{i} + \mathbf{j} + 2\mathbf{k}$, $\mathbf{w} = -\mathbf{i} + 2\mathbf{j} - 2\mathbf{k}$
17. Derive a formula expressing $(\mathbf{t} \times \mathbf{u}) \times (\mathbf{v} \times \mathbf{w})$ in terms of $\mathbf{v}$ and $\mathbf{w}$.
18. Given $\mathbf{a} = \mathbf{i} - \mathbf{j} + \mathbf{k}$, $\mathbf{b} = 2\mathbf{i} + 3\mathbf{j} + \mathbf{k}$, $p = 1$. Solve the equations $\mathbf{a} \cdot \mathbf{v} = p$, $\mathbf{a} \times \mathbf{v} = \mathbf{b}$ for $\mathbf{v}$.
19. Given that $\mathbf{a} \cdot \mathbf{b} = 0$, $\mathbf{a} \neq \mathbf{0}$, $\mathbf{b} \neq \mathbf{0}$, find a formula for the solution $\mathbf{v}$ of the equations
$$\mathbf{a} \cdot \mathbf{v} = p, \qquad \mathbf{a} \times \mathbf{v} = \mathbf{b}.$$
[*Hint:* Note that $\mathbf{a}$, $\mathbf{b}$, and $\mathbf{a} \times \mathbf{b}$ are mutually orthogonal.]

20. Show that for any vectors $\mathbf{u}$, $\mathbf{v}$, $\mathbf{w}$, we have

    i) $(\mathbf{u} \pm \mathbf{v}) \cdot [(\mathbf{u} + \mathbf{v}) \times (\mathbf{u} - \mathbf{v})] = 0$

    ii) $(\mathbf{u} + \mathbf{v}) \cdot [(\mathbf{u} \times \mathbf{w}) \times (\mathbf{u} + \mathbf{v})] = 0.$

21. If $\mathbf{u}$, $\mathbf{v}$, $\mathbf{w}$ are any vectors show that

$$[\mathbf{u} \times (\mathbf{v} \times \mathbf{w})] + [\mathbf{v} \times (\mathbf{w} \times \mathbf{u})] + [\mathbf{w} \times (\mathbf{u} \times \mathbf{v})] = \mathbf{0}.$$

*22. Let $\mathscr{F}$ be a collection of objects with $\mathbf{A}$, $\mathbf{B}$, $\mathbf{C}$ members of $\mathscr{F}$. Let $\oplus$ be an operation between members of $\mathscr{F}$ satisfying the relation

$$(\mathbf{A} \oplus \mathbf{B}) \oplus \mathbf{C} = \alpha\mathbf{B} - \beta\mathbf{A}.$$

State general conditions on the numbers α and β such that the formula

$$[(\mathbf{A} \oplus \mathbf{B}) \oplus \mathbf{C}] + [(\mathbf{B} \oplus \mathbf{C}) \oplus \mathbf{A}] + [(\mathbf{C} \oplus \mathbf{A}) \oplus \mathbf{B}] = \mathbf{0}$$

should hold.

9. Vector Functions and Their Derivatives

A **function** of one variable is a set of ordered pairs (x, y) of real numbers in which no two pairs have the same first element. That is, to each value of x (the first element of the pair) there corresponds exactly one value of y (the second element). The set of all values of x which occur is the **domain** of the function, and the set of all y which occur is the **range** of the function.

We denote the collection of all real numbers by R^1, and the set of all ordered pairs of real numbers (a, b) by R^2. We call R^2 the **number plane** and observe that it forms the basis for a rectangular coordinate system in the geometric plane.

For a real-valued function of one real variable, say f, we use the notation $f\colon R^1 \to R^1$ to indicate its domain and range.

We now extend the definition of function to the case where the range is a vector. In this section we restrict ourselves to vectors in the plane. The more general situation in which the range consists of vectors in three-space is discussed in Section 10.

We consider the collection of all vectors in the plane and we denote this set by V_2. That is, any vector in the plane is a member of V_2.

Definition. A **vector function** is the collection of ordered pairs $(t, \mathbf{v})$ in which t is a real number and $\mathbf{v}$ is a vector in V_2; this collection of ordered pairs must have the property that no two pairs have the same first element. The **domain** consists of all possible values of t in the collection, and the **range** consists of all vectors which occur.

In terms of mappings, a vector function is a mapping from a set in R^1 into a set in V_2.

Vector functions are more complicated than ordinary functions, since the elements of the range, namely vectors, are themselves equivalence classes of directed line segments. However, if we concentrate on the representation—the directed line segments—the concept becomes more concrete. Also, many of the properties of ordinary functions extend easily to vector functions when suitably interpreted. We shall use boldface letters such as **f**, **g**, **v**, **F**, **G** to represent vector functions. If the dependence on the independent variable is to be indicated, we shall write $\mathbf{f}(t)$, $\mathbf{g}(s)$, $\mathbf{F}(x)$, and so forth for the function values.

Definition. A vector function **f** is continuous at $t = a$ if $\mathbf{f}(a)$ is defined and if for each $\varepsilon > 0$ there is a $\delta > 0$ such that

$$|\mathbf{f}(t) - \mathbf{f}(a)| < \varepsilon \qquad \text{for all } t \text{ such that} \qquad 0 < |t - a| < \delta.$$

We note that the form of the definition of continuity for vector functions is identical with that for ordinary functions. We must realize, however, that $\mathbf{f}(t) - \mathbf{f}(a)$ is a *vector*, and that the symbol $|\mathbf{f}(t) - \mathbf{f}(a)|$ stands for the length of a vector, whereas in the case of ordinary functions, $|f(t) - f(a)|$ is the absolute value of a number. In words, the continuity of a vector function asserts that as $t \to a$ the vector $\mathbf{f}(t)$ approaches $\mathbf{f}(a)$, in *both* length and direction. When **f** is continuous at a, we also write

$$\lim_{t \to a} \mathbf{f}(t) = \mathbf{f}(a).$$

If **i** and **j** are the customary unit vectors associated with a rectangular coordinate system in the plane, a vector function **f** can be written in the form

$$\mathbf{f}(t) = f_1(t)\mathbf{i} + f_2(t)\mathbf{j},$$

where f_1 and f_2 are functions in the ordinary sense. Statements about vector functions **f** may always be interpreted as statements about a pair of functions (f_1, f_2).

Theorem 22. *A function* **f** *is continuous at* $t = a$ *if and only if* f_1 *and* f_2 *are continuous at* $t = a$.

PROOF. We have $\mathbf{f}(a) = f_1(a)\mathbf{i} + f_2(a)\mathbf{j}$ and

$$|\mathbf{f}(t) - \mathbf{f}(a)| = |(f_1(t) - f_1(a))\mathbf{i} + (f_2(t) - f_2(a))\mathbf{j}|,$$

Also, (see Fig. 2-29)

$$|\mathbf{f}(t) - \mathbf{f}(a)| = \sqrt{|f_1(t) - f_1(a)|^2 + |f_2(t) - f_2(a)|^2}.$$

If

$$\lim_{t \to a} f_1(t) = f_1(a) \qquad \text{and} \qquad \lim_{t \to a} f_2(t) = f_2(a),$$

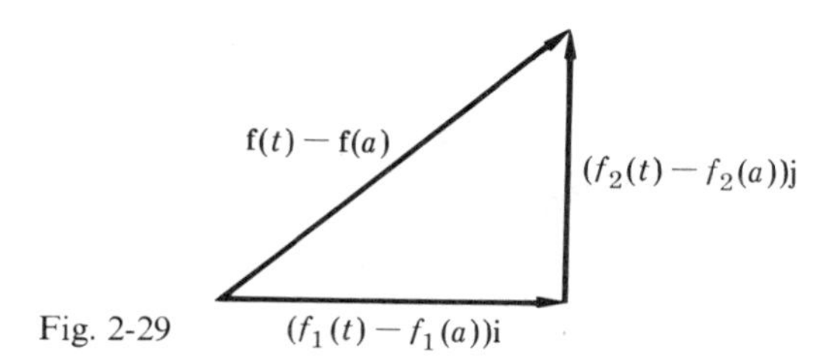

Fig. 2-29

it follows that

$$\lim_{t\to a} \mathbf{f}(t) = \mathbf{f}(a).$$

On the other hand, the inequalities

$$|f_1(t) - f_1(a)| \le |\mathbf{f}(t) - \mathbf{f}(a)|, \qquad |f_2(t) - f_2(a)| \le |\mathbf{f}(t) - \mathbf{f}(a)|$$

show that if

$$\lim_{t\to a} \mathbf{f}(t) = \mathbf{f}(a),$$

then *both*

$$\lim_{t\to a} f_1(t) = f_1(a) \qquad \text{and} \qquad \lim_{t\to a} f_2(t) = f_2(a).$$

Suppose that f is a function from R^1 to R^1. We recall from elementary calculus the definition of the derivative of f at a point x_0, denoted $f'(x_0)$:

$$f'(x_0) = \lim_{h\to 0} \frac{f(x_0 + h) - f(x_0)}{h}.$$

The definition of the derivative of a vector function is completely analogous.

Definition. If $\mathbf{f}$ is a vector function, we define the **derivative** $\mathbf{f}'$ as

$$\mathbf{f}'(t) = \lim_{h\to 0} \frac{\mathbf{f}(t + h) - \mathbf{f}(t)}{h},$$

whenever the limit exists. If $\mathbf{f}$ is given in terms of functions f_1 and f_2 by

$$\mathbf{f}(t) = f_1(t)\mathbf{i} + f_2(t)\mathbf{j},$$

then the derivative may be computed by the simple formula

$$\boxed{\mathbf{f}'(t) = f_1'(t)\mathbf{i} + f_2'(t)\mathbf{j}.} \tag{1}$$

The quantities $f_1'(t)$ and $f_2'(t)$ are derivatives in the ordinary sense. Formula

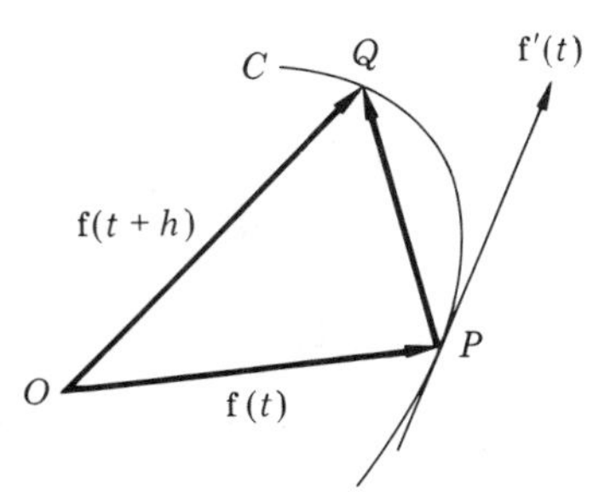

Fig. 2-30

(1) follows directly from the definition of derivative and the theorem on the limit of a sum.

EXAMPLE 1. Find the derivative $\mathbf{f}'(t)$ if $\mathbf{f}(t) = (t^2 + 2t - 1)\mathbf{i} + (3t^3 - 2)\mathbf{j}$.

SOLUTION. $\mathbf{f}'(t) = (2t + 2)\mathbf{i} + 9t^2\mathbf{j}$.

EXAMPLE 2. Given that $\mathbf{f}(t) = (\sin t)\mathbf{i} + (3 - 2\cos t)\mathbf{j}$, find $\mathbf{f}''(t)$.

SOLUTION. $\mathbf{f}'(t) = \cos t\mathbf{i} + 2\sin t\mathbf{j}$. Hence $\mathbf{f}''(t) = -\sin t\mathbf{i} + 2\cos t\mathbf{j}$.

EXAMPLE 3. Given $\mathbf{f}(t) = (3t - 2)\mathbf{i} + (2t^2 + 1)\mathbf{j}$. Find the value of

$$\mathbf{f}'(t) \cdot \mathbf{f}''(t).$$

SOLUTION. $\mathbf{f}'(t) = 3\mathbf{i} + 4t\mathbf{j}$, $\mathbf{f}''(t) = 4\mathbf{j}$, and

$$\mathbf{f}'(t) \cdot \mathbf{f}''(t) = (3\mathbf{i} + 4t\mathbf{j}) \cdot (4\mathbf{j}) = 16t.$$

The derivative of a vector function has a simple geometric interpretation in terms of directed line segments. Draw the particular directed line segment of $\mathbf{f}(t)$ which has its base at the origin of the coordinate system. Then the head of this directed line segment will trace out a curve C as t takes on all possible values in its domain (Fig. 2-30). The directed line segment $\overrightarrow{OP}$ represents $\mathbf{f}(t)$. Let $\overrightarrow{OQ}$ represent $\mathbf{f}(t + h)$. Then $\mathbf{f}(t + h) - \mathbf{f}(t)$ has $\overrightarrow{PQ}$ as one of its directed line segments. Multiplying $\mathbf{f}(t + h) - \mathbf{f}(t)$ by $1/h$ gives a vector in the direction of $\overrightarrow{PQ}$, but $1/h$ times as long. As h tends to zero, the quantity

$$\frac{\mathbf{f}(t + h) - \mathbf{f}(t)}{h}$$

tends to a vector, with its directed line segment tangent to the curve C at the point P. Using vector terminology instead of directed line segments, we conclude that $\mathbf{f}'(t)$ is the vector tangent to the curve $\mathbf{f}(t)$.

PROBLEMS

In Problems 1 through 6, calculate $\mathbf{f}'(t)$ and $\mathbf{f}''(t)$.

1. $\mathbf{f}(t) = (t^2 + 1)\mathbf{i} + (t^3 - 3t)\mathbf{j}$
2. $\mathbf{f}(t) = \left(\dfrac{t-3}{t+1}\right)\mathbf{i} - \dfrac{t^2+1}{t^2+t+1}\mathbf{j}$
3. $\mathbf{f}(t) = (\tan 3t)\mathbf{i} + (\cos \pi t)\mathbf{j}$
4. $\mathbf{f}(t) = \dfrac{1}{t}\mathbf{i} - \dfrac{1}{t^2+1}\mathbf{j}$
5. $\mathbf{f}(t) = e^{2t}\mathbf{i} + e^{-2t}\mathbf{j}$
6. $\mathbf{f}(t) = (e^t + e^{-t})\mathbf{i} + (e^t - e^{-t})\mathbf{j}$
7. Find $\mathbf{f}(t) \cdot \mathbf{f}'(t)$ if $\mathbf{f}(t) = \dfrac{t^2+1}{t^2+2}\mathbf{i} + 2t\mathbf{j}$.
8. Find $\dfrac{d}{dt}(\mathbf{f}(t) \cdot \mathbf{f}'(t))$ if $\mathbf{f}(t) = 2t\mathbf{i} + \dfrac{1}{t+1}\mathbf{j}$.
9. Find $\dfrac{d}{dt}|\mathbf{f}(t)|$ if $\mathbf{f}(t) = \sin 2t\mathbf{i} + \cos 3t\mathbf{j}$.
10. Find $\dfrac{d}{dt}|\mathbf{f}(t)|$ if $\mathbf{f}(t) = (t^2 + 1)\mathbf{i} + (3 + 2t^2)\mathbf{j}$.
11. Find $\dfrac{d}{dt}|\mathbf{f}(t)|$ if $\mathbf{f}(t) = \cos\dfrac{1}{t}\mathbf{i} + \sin\dfrac{1}{t}\mathbf{j}$.
12. Find $\dfrac{d}{dt}(\mathbf{f}'(t) \cdot \mathbf{f}''(t))$ if $\mathbf{f}(t) = (3t + 1)\mathbf{i} + (2t^2 - t^3)\mathbf{j}$.
13. Find $\dfrac{d}{dt}(\mathbf{f}'(t) \cdot \mathbf{f}''(t))$ if $\mathbf{f}(t) = (\log t)\mathbf{i} + \dfrac{2}{t}\mathbf{j}$.
14. Find $\dfrac{d}{dt}(\mathbf{f}''(t) \cdot \mathbf{f}'''(t))$ if $\mathbf{f}(t) = e^{3t}\mathbf{i} + e^{-3t}\mathbf{j}$.
15. Given that $\mathbf{f}(t) = (2t + 1)\mathbf{i} + 3t\mathbf{j}$, $\mathbf{g}(t) = 4t\mathbf{j}$. Find $d\theta/dt$, where $\theta = \theta(t)$ is the angle between $\mathbf{f}$ and $\mathbf{g}$.
16. Write out a proof establishing Formula (1) on page 79.
17. Prove the formula
$$\frac{d}{dt}(F(t)\mathbf{f}(t)) = F(t)\mathbf{f}'(t) + F'(t)\mathbf{f}(t).$$
18. Show that the following Chain Rule holds:
$$\frac{d}{dt}\{\mathbf{f}[g(t)]\} = \mathbf{f}'[g(t)]g'(t).$$
19. Given the vector $\mathbf{f}(t) = (2t + 1)\mathbf{i} + 2t\mathbf{j}$. Describe the curve traced out by the tip of the directed line segment which has its base at the origin.
20. Given the vector $\mathbf{g}(t) = \cos t\mathbf{i} + \sin t\mathbf{j}$. Describe the curve traced out by the tip of the directed line segment which has its base at the origin.
21. Prove that if $\mathbf{f}(t) = \sin 2t\mathbf{i} + \cos 2t\mathbf{j}$, then $\mathbf{f}(t) \cdot \mathbf{f}'(t) = 0$. What is the geometric interpretation of this result?

22. Prove that if

$$\mathbf{f}(t) = \frac{\mathbf{g}(t)}{h(t)}, \qquad \text{then} \qquad \mathbf{f}'(t) = \frac{h(t)\mathbf{g}'(t) - \mathbf{g}(t)h'(t)}{h^2(t)}$$

23. Show that if

$$\mathbf{f}(t) = f_1(t)\mathbf{i} + f_2(t)\mathbf{j}, \qquad \text{then} \qquad \frac{d}{dt}(\mathbf{f}(t)\cdot\mathbf{f}(t)) = 2\mathbf{f}(t)\cdot\mathbf{f}'(t).$$

24. Show that

$$\frac{d}{dt}(\mathbf{f}(t)\cdot\mathbf{g}(t)) = \mathbf{f}(t)\cdot\mathbf{g}'(t) + \mathbf{f}'(t)\cdot\mathbf{g}(t).$$

[*Hint:* Write $\mathbf{f}(t) = f_1(t)\mathbf{i} + f_2(t)\mathbf{j}$ and $\mathbf{g}(t) = g_1(t)\mathbf{i} + g_2(t)\mathbf{j}$.]

10. Vector Velocity and Acceleration in the Plane

The vector function

$$\mathbf{f}(t) = x(t)\mathbf{i} + y(t)\mathbf{j}$$

is equivalent to the pair of parametric equations

$$x = x(t), \qquad y = y(t),$$

since the head of the directed line segment with base at the origin, which represents **f**, traces out a curve which is identical with the curve given in parametric form by these equations.

We recall from elementary calculus that the arc length of a curve in the plane, denoted s, satisfies the relation

$$\left(\frac{ds}{dt}\right)^2 = \left(\frac{dx}{dt}\right)^2 + \left(\frac{dy}{dt}\right)^2. \tag{1}$$

For $\mathbf{f}(t) = x(t)\mathbf{i} + y(t)\mathbf{j}$, we have $\mathbf{f}'(t) = x'(t)\mathbf{i} + y'(t)\mathbf{j}$ and the length $|\mathbf{f}'(t)|$ is given by

$$|\mathbf{f}'(t)| = \sqrt{(x'(t))^2 + (y'(t))^2}.$$

Combining this formula with (1), we get

$$\frac{ds}{dt} = |\mathbf{f}'(t)|.$$

We consider the motion of a particle along a curve C in the plane. Suppose that C is given by the parametric equations

$$C: x = x(t), \qquad y = y(t).$$

Letting t denote the time, we define the **velocity vector v** at the time t as

$$\mathbf{v}(t) \equiv \frac{d\mathbf{f}}{dt} \equiv \mathbf{f}'(t) = x'(t)\mathbf{i} + y'(t)\mathbf{j}.$$

According to the geometrical interpretation of the derivative of a vector function given in the preceding section, the velocity vector is always tangent to the path describing the motion. We define the **speed** of the particle to be *the magnitude of the velocity vector*. The speed is

$$|\mathbf{v}(t)| = |\mathbf{f}'(t)| = \sqrt{(dx/dt)^2 + (dy/dt)^2},$$

which tells us that the speed is identical with the quantity ds/dt; in other words, the speed measures the rate of change in arc length s with respect to time t.

The **acceleration vector** $\mathbf{a}(t)$ is defined as the derivative of the velocity vector, or

$$\mathbf{a}(t) = \mathbf{v}'(t) = \mathbf{f}''(t).$$

EXAMPLE 1. Suppose that a particle P moves according to the law

$$\mathbf{f}(t) = (3\cos 2t)\mathbf{i} + (3\sin 2t)\mathbf{j}.$$

Find $\mathbf{v}(t)$, $\mathbf{a}(t)$, $s'(t)$, $s''(t)$, $|\mathbf{a}(t)|$, and $\mathbf{v}(t)\cdot\mathbf{a}(t)$.

SOLUTION. We have

$$\begin{aligned}
\mathbf{v}(t) &= \mathbf{f}'(t) = (-6\sin 2t)\mathbf{i} + (6\cos 2t)\mathbf{j},\\
\mathbf{a}(t) &= (-12\cos 2t)\mathbf{i} - (12\sin 2t)\mathbf{j},\\
s'(t) &= [(-6\sin 2t)^2 + (6\cos 2t)^2]^{1/2} = 6,\\
s''(t) &= 0,\\
|\mathbf{a}(t)| &= [(-12\cos 2t)^2 + (12\sin 2t)^2]^{1/2} = 12,\\
\mathbf{v}(t)\cdot\mathbf{a}(t) &= 0.
\end{aligned}$$

REMARK. We note that P is moving around a circle with center at O and radius 3, with a constant speed but a changing velocity vector! Since

$$\mathbf{v}(t)\cdot\mathbf{a}(t) = 0,$$

and since $\mathbf{v}(t)$ is always tangent to the circle, we conclude that the acceleration vector is always pointing toward the center of the circle.

EXAMPLE 2. Suppose that a particle P moves according to the law

$$x(t) = t\cos t, \qquad y(t) = t\sin t,$$

or, equivalently,

$$\mathbf{f}(t) = (t\cos t)\mathbf{i} + (t\sin t)\mathbf{j}.$$

Find $\mathbf{v}(t)$, $\mathbf{a}(t)$, $s'(t)$, $s''(t)$, and $|\mathbf{a}(t)|$.

SOLUTION
$$\mathbf{v}(t) = (-t\sin t + \cos t)\mathbf{i} + (t\cos t + \sin t)\mathbf{j},$$
$$\mathbf{a}(t) = (-t\cos t - 2\sin t)\mathbf{i} + (-t\sin t + 2\cos t)\mathbf{j}.$$

Therefore

$$s'(t) = |\mathbf{v}(t)| = \sqrt{1+t^2},$$

$$s''(t) = \tfrac{1}{2}(1+t^2)^{-1/2}(2t) = \frac{t}{\sqrt{1+t^2}},$$

$$|\mathbf{a}(t)| = \sqrt{4+t^2}.$$

PROBLEMS

In Problems 1 through 8, assume that a particle P moves according to the given law, t denoting the time. Compute $\mathbf{v}(t)$, $|\mathbf{v}(t)|$, $\mathbf{a}(t)$, $|\mathbf{a}(t)|$, $s'(t)$, and $s''(t)$.

1. $\mathbf{f}(t) = t^2\mathbf{i} - 3t\mathbf{j}$
2. $\mathbf{f}(t) = \frac{1}{2}t^2\mathbf{i} + \frac{1}{3}t^3\mathbf{j}$
3. $\mathbf{f}(t) = 3t\mathbf{i} + (1 - t^{-1})\mathbf{j}$
4. $\mathbf{f}(t) = (t - \frac{1}{2}t^2)\mathbf{i} + 2t^{1/2}\mathbf{j}$
5. $\mathbf{f}(t) = (\log\sec t)\mathbf{i} + t\mathbf{j}$
6. $\mathbf{f}(t) = (3\cos t)\mathbf{i} + (2\sin t)\mathbf{j}$
7. $x = 2e^t$, $y = 3e^{-t}$
8. $x = e^{-t}\cos t$, $y = e^{-t}\sin t$

In Problems 9 through 13, assume that a particle P moves according to the given law. Find $\mathbf{v}(t)$, $\mathbf{a}(t)$, $s'(t)$, and $s''(t)$ at the given time t.

9. $x = t^2 - t - 1$, $y = t^2 - 2t$, $t = 1$
10. $\mathbf{f}(t) = (5\cos t)\mathbf{i} + (4\sin t)\mathbf{j}$, $t = 2\pi/3$
11. $x = 3\sec t$, $y = 2\tan t$, $t = \pi/6$
12. $\mathbf{f}(t) = 4(\pi t - \sin\pi t)\mathbf{i} + 4(1 - \cos\pi t)\mathbf{j}$, $t = \frac{3}{4}$
13. $x = \log(1+t)$, $y = 3/t$, $t = 2$
14. Suppose that P moves according to the law

$$\mathbf{f}(t) = (R\cos wt)\mathbf{i} + (R\sin wt)\mathbf{j},$$

where $R > 0$ and w are constants. Find $\mathbf{v}(t)$, $\mathbf{a}(t)$, $s'(t)$, and $s''(t)$. Show that

$$|\mathbf{a}(t)| = \frac{1}{R}|\mathbf{v}(t)|^2.$$

15. Let $\mathbf{T}(t)$ be a vector one unit long and parallel to the *velocity vector*. Show that

$$\mathbf{T}(t) = \frac{\mathbf{v}(t)}{ds/dt} = \frac{x'(t)}{s'(t)}\mathbf{i} + \frac{y'(t)}{s'(t)}\mathbf{j}.$$

16. Using the formula of Problem 15, compute $\mathbf{T}(t)$ if the law of motion is

$$\mathbf{f}(t) = (3t - 1)\mathbf{i} + (t^2 + 2)\mathbf{j}.$$

17. Using the formula of Problem 15, compute $\mathbf{T}(t)$ if the law of motion is

$$\mathbf{f}(t) = (4e^{2t})\mathbf{i} + (3e^{-2t})\mathbf{j}.$$

18. A formula for $\mathbf{T}(t)$ is given in Problem 15. Compute $\mathbf{T}'(t)$ and show that $\mathbf{T}(t) \cdot \mathbf{T}'(t) = 0$ [*Hint:* Use the fact that $x'^2 + y'^2 = s'^2$.]

11. Vector Functions in Space. Space Curves. Tangents and Arc Length

The collection of all vectors in three-space is denoted by V_3. We now extend the definition of vector function given in Section 9 to the case where the range is an element of V_3.

Definition. A vector function in three-space, or simply a vector function, is the collection of ordered pairs $(t, \mathbf{v})$ in which t is a real number (an element of R^1) and $\mathbf{v}$ is a vector in V_3. As usual, no two pairs have the same first element.

The definition of continuity for a vector function in three-space is identical with that given in Section 9 for vector functions in the plane. Part (b) of the next theorem is proved exactly as is Theorem 22 in Section 9. The proofs of the remaining parts are left to the reader.

Theorem 23. *Suppose that*

$$\mathbf{f}(t) = f_1(t)\mathbf{i} + f_2(t)\mathbf{j} + f_3(t)\mathbf{k}$$

is a vector function and that the vector $\mathbf{c} = c_1\mathbf{i} + c_2\mathbf{j} + c_3\mathbf{k}$ *is a constant. Then*

a) $\mathbf{f}(t) \to \mathbf{c}$ *as* $t \to a$ *if and only if*

$$f_1(t) \to c_1 \quad \text{and} \quad f_2(t) \to c_2 \quad \text{and} \quad f_3(t) \to c_3.$$

b) $\mathbf{f}$ *is continuous as* a *if and only if* f_1, f_2, *and* f_3 *are.*
c) $\mathbf{f}'(t)$ *exists if and only if* $f_1'(t), f_2'(t)$, *and* $f_3'(t)$ *do.*
d) *We have the formula*

$$\boxed{\mathbf{f}'(t) = f_1'(t)\mathbf{i} + f_2'(t)\mathbf{j} + f_3'(t)\mathbf{k}.}$$

e) *If* $\mathbf{v}(t) = a\mathbf{w}(t)$, *then* $\mathbf{v}'(t) = a\mathbf{w}'(t)$ *where* a *is a constant.*
f) *If* $\mathbf{v}(t) = c(t)\mathbf{w}(t)$, *then* $\mathbf{v}'(t) = c(t)\mathbf{w}'(t) + c'(t)\mathbf{w}(t)$.
g) *If* $\mathbf{v}(t) = \mathbf{w}(t)/c(t)$, *then*

$$\mathbf{v}'(t) = \frac{c(t)\mathbf{w}'(t) - c'(t)\mathbf{w}(t)}{[c(t)]^2}.$$

EXAMPLE 1. Given $\mathbf{f}(t) = 3t^2\mathbf{i} - 2t^3\mathbf{j} + (t^2 + 3)\mathbf{k}$, find $\mathbf{f}'(t)$, $\mathbf{f}''(t)$, $\mathbf{f}'''(t)$.

SOLUTION

$$\mathbf{f}'(t) = 6t\mathbf{i} - 6t^2\mathbf{j} + 2t\mathbf{k},$$
$$\mathbf{f}''(t) = 6\mathbf{i} - 12t\mathbf{j} + 2\mathbf{k},$$
$$\mathbf{f}'''(t) = -12\mathbf{j}.$$

The next theorem shows how to differentiate functions involving scalar and vector products. The proofs are left to the reader.

Theorem 24. *If $\mathbf{u}(t)$ and $\mathbf{v}(t)$ are differentiable, then the derivative of $f(t) = \mathbf{u}(t) \cdot \mathbf{v}(t)$ is given by the formula*

$$\boxed{f'(t) = \mathbf{u}(t) \cdot \mathbf{v}'(t) + \mathbf{u}'(t) \cdot \mathbf{v}(t).} \tag{1}$$

The derivative of $\mathbf{w}(t) = \mathbf{u}(t) \times \mathbf{v}(t)$ is given by the formula

$$\boxed{\mathbf{w}'(t) = \mathbf{u}(t) \times \mathbf{v}'(t) + \mathbf{u}'(t) \times \mathbf{v}(t).} \tag{2}$$

The proofs of formulas (1) and (2) follow from the corresponding differentiation formulas for ordinary functions. Note that in formula (2) it is essential to retain the order of the factors in each vector product.

EXAMPLE 2. Find the derivative $f'(t)$ and $\mathbf{w}'(t)$ of

$$f(t) = \mathbf{u}(t) \cdot \mathbf{v}(t) \qquad \text{and} \qquad \mathbf{w}(t) = \mathbf{u}(t) \times \mathbf{v}(t)$$

when

$$\mathbf{u}(t) = (t + 3)\mathbf{i} + t^2\mathbf{j} + (t^3 - 1)\mathbf{k}$$

and

$$\mathbf{v}(t) = 2t\mathbf{i} + (t^4 - 1)\mathbf{j} + (2t + 3)\mathbf{k}.$$

SOLUTION. By formula (1) we have

$$\begin{aligned} f(t) &= [(t + 3)\mathbf{i} + t^2\mathbf{j} + (t^3 - 1)\mathbf{k}] \cdot (2\mathbf{i} + 4t^3\mathbf{j} + 2\mathbf{k}) \\ &\quad + (\mathbf{i} + 2t\mathbf{j} + 3t^2\mathbf{k}) \cdot [2t\mathbf{i} + (t^4 - 1)\mathbf{j} + (2t + 3)\mathbf{k}] \\ &= (2t + 6) + 4t^5 + 2t^3 - 2 + 2t + 2t^5 - 2t + 6t^3 + 9t^2 \\ &= 6t^5 + 8t^3 + 9t^2 + 2t + 4. \end{aligned}$$

According to (2), we have

$$\begin{aligned}\mathbf{w}'(t) = {} & [(t+3)\mathbf{i} + t^2\mathbf{j} + (t^3-1)\mathbf{k}] \times (2\mathbf{i} + 4t^3\mathbf{j} + 2\mathbf{k}) \\ & + (\mathbf{i} + 2t\mathbf{j} + 3t^2\mathbf{k}) \times [2t\mathbf{i} + (t^4-1)\mathbf{j} + (2t+3)\mathbf{k}].\end{aligned}$$

Computing the two cross products on the right, we get

$$\begin{aligned}\mathbf{w}'(t) = {} & (-7t^6 + 4t^3 + 9t^2 + 6t)\mathbf{i} + (8t^3 - 4t - 11)\mathbf{j} \\ & + (5t^4 + 12t^3 - 6t^2 - 1)\mathbf{k}.\end{aligned}$$

Consider a rectangular coordinate system and a directed line segment from the origin O to a point P in space. As in two dimensions, we denote the vector $\mathbf{v}(\overrightarrow{OP})$ by $\mathbf{r}$. We define an **arc** C in space in a way completely analogous to that in which an arc in the plane was defined. The vector equation

$$\mathbf{r}(t) = x(t)\mathbf{i} + y(t)\mathbf{j} + z(t)\mathbf{k} \tag{3}$$

is considered to be equivalent to the parametric equations

$$x = x(t), \qquad y = y(t), \qquad z = z(t). \tag{4}$$

The arc length of a curve in three-space, denoted s, satisfies a relation similar to the one for arc length in the plane given in Section 10. When a curve is given by (4), we have

$$\left(\frac{ds}{dt}\right)^2 = \left(\frac{dx}{dt}\right)^2 + \left(\frac{dy}{dt}\right)^2 + \left(\frac{dz}{dt}\right)^2.$$

The actual length of an arc is determined by integrating ds/dt between two values of the parameter t. We find for the length l of such an arc

$$l = \int_{t_0}^{t_1} \sqrt{\left(\frac{dx}{dt}\right)^2 + \left(\frac{dy}{dt}\right)^2 + \left(\frac{dz}{dt}\right)^2}\, dt \tag{5}$$

If the curve is given in the vector form (3), we obtain

$$s'(t) = |\mathbf{r}'(t)|, \qquad l = \int_{t_0}^{t_1} |\mathbf{r}'(t)|\, dt. \tag{6}$$

Definitions. If $\mathbf{r}'(t) \neq \mathbf{0}$, we define the vector $\mathbf{T}(t) = \mathbf{r}'(t)/|\mathbf{r}'(t)|$ as the **unit tangent vector** to the path corresponding to the value t. The line through the point P_0 corresponding to $\mathbf{r}(t_0)$ and parallel to $\mathbf{T}(t_0)$ is called the **tangent line to the arc** at t_0; the line directed in the same way as $\mathbf{T}(t_0)$ is called the **directed tangent line at** t_0 (Fig. 2-31).

EXAMPLE 3. The graph of the equations

$$x = a\cos t, \qquad y = a\sin t, \qquad z = bt \tag{7}$$

is called a **helix**.

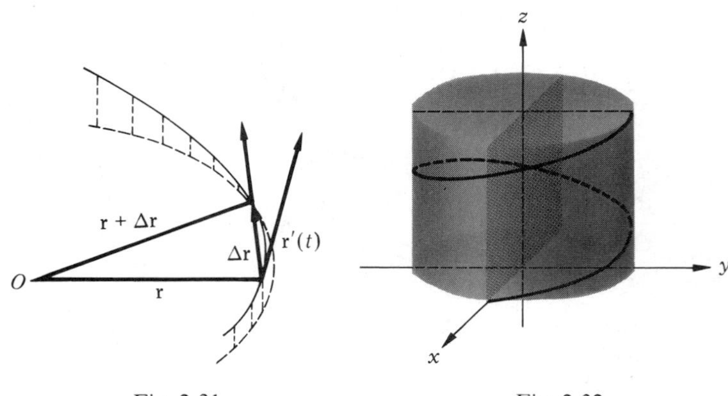

Fig. 2-31 Fig. 2-32

i) Find $s'(t)$.
ii) Find the length of that part of the helix for which $0 \leq t \leq 2\pi$.
iii) Show that the unit tangent vector makes a constant angle with the z axis.

SOLUTION. (See Fig. 2-32 for the graph.)

i) $x'(t) = -a \sin t$, $y'(t) = a \cos t$, $z'(t) = b$. Therefore,

$$s'(t) = \sqrt{a^2 + b^2}.$$

ii) $l(C) = 2\pi\sqrt{a^2 + b^2}$.

iii) $\mathbf{T}(t) = \dfrac{-y\mathbf{i} + x\mathbf{j} + b\mathbf{k}}{\sqrt{a^2 + b^2}}$.

Letting ϕ be the angle between $\mathbf{T}$ and $\mathbf{k}$, we get

$$\cos \phi = b/\sqrt{a^2 + b^2}.$$

REMARK. We note that the helix (7) winds around the cylinder $x^2 + y^2 = a^2$.

Definitions. If t denotes time in the parametric equations of an arc $\mathbf{r}(t)$, then $\mathbf{r}'(t)$ is the **velocity vector** $\mathbf{v}(t)$, and $\mathbf{r}''(t) = \mathbf{v}'(t)$ is called the **acceleration vector**. The quantity $s'(t)$ (a scalar) is called the **speed** of a particle moving according to the law

$$\mathbf{r}(t) = x(t)\mathbf{i} + y(t)\mathbf{j} + z(t)\mathbf{k}.$$

PROBLEMS

In each of Problems 1 through 8, find the derivatives $\mathbf{f}'(t)$ and $\mathbf{f}''(t)$.

1. $\mathbf{f}(t) = t^2\mathbf{i} + (t^2 + 1)\mathbf{j} + (2 - 3t)\mathbf{k}$
2. $\mathbf{f}(t) = (2t^3 + t - 1)\mathbf{i} + (t^{-1} + 1)\mathbf{j} + (t^2 + t^{-2})\mathbf{k}$

3. $\mathbf{f}(t) = (\cos 2t)\mathbf{i} + (\sin 2t)\mathbf{j} + 2t\mathbf{k}$

4. $\mathbf{f}(t) = e^{2t}\mathbf{i} + t^2\mathbf{j} + e^{-2t}\mathbf{k}$

5. $\mathbf{f}(t) = \dfrac{t^2}{t^2+1}\mathbf{i} + \dfrac{1}{t^2+1}\mathbf{j} + (t^2+1)\mathbf{k}$

6. $\mathbf{f}(t) = (\log 2t)\mathbf{i} + e^{3t}\mathbf{j} + (t \log t)\mathbf{k}$

7. $\mathbf{f}: t \to (\cos t)\mathbf{i} + (\tan t)\mathbf{j} + (\sin t)\mathbf{k}$

8. $\mathbf{f}: t \to [(t^2+1)(t^2-2)]\mathbf{i} + (t^3 + 2t^{-3})\mathbf{j} + [\log(t^2+1)]\mathbf{k}$

In Problems 9 through 11, find in each case either $f'(t)$ or $\mathbf{f}'(t)$, whichever is appropriate.

9. $f(t) = \mathbf{u}(t) \cdot \mathbf{v}(t)$, where

$$\mathbf{u}(t) = 3t\mathbf{i} + 2t^2\mathbf{j} + \frac{1}{t}\mathbf{k}, \qquad \mathbf{v}(t) = t^2\mathbf{i} + \frac{1}{t}\mathbf{j} + t^3\mathbf{k}.$$

10. $\mathbf{f}(t) = \mathbf{u}(t) \times \mathbf{v}(t)$, where

$$\mathbf{u}(t) = (\cos t)\mathbf{i} + (\sin t)\mathbf{j} + t\mathbf{k}, \qquad \mathbf{v}(t) = (\sin t)\mathbf{i} + (\cos t)\mathbf{j} + t^2\mathbf{k}.$$

11. $f(t) = \mathbf{u}(t) \cdot [\mathbf{v}(t) \times \mathbf{w}(t)]$, where

$$\mathbf{u}(t) = t\mathbf{i} + (t+1)\mathbf{k}, \qquad \mathbf{v}(t) = t^2\mathbf{j}, \qquad \mathbf{w}(t) = \frac{1}{t^2}\mathbf{k}.$$

In Problems 12 through 15, find the length of the arc C.

12. $C: x = t, \quad y = t^2/\sqrt{2}, \quad z = t^3/3; \qquad 0 \le t \le 2$

13. $C: x = t, \quad y = 3t^2/2, \quad z = 3t^3/2; \qquad 0 \le t \le 2$

14. $C: x = t, \quad y = \log(\sec t + \tan t), \quad z = \log \sec t; \qquad 0 \le t \le \pi/4$

15. $C: x = t\cos t, \quad y = t \sin t, \quad z = t; \qquad 0 \le t \le \pi/2$

In Problems 16 through 18, the parameter t is the time in seconds. Taking s to be the length in meters, in each case find the velocity, speed, and acceleration of a particle moving according to the given law.

16. $\mathbf{r}(t) = t^2\mathbf{i} + 2t\mathbf{j} + (t^3 - 1)\mathbf{k}$

17. $\mathbf{r}(t) = (t \sin t)\mathbf{i} + (t \cos t)\mathbf{j} + t\mathbf{k}$

18. $\mathbf{r}(t) = e^{3t}\mathbf{i} + e^{-3t}\mathbf{j} + te^{3t}\mathbf{k}$

19. a) Given $\mathbf{f}(t) = (3\cos t)\mathbf{i} + (4\cos t)\mathbf{j} + (5 \sin t)\mathbf{k}$, show that

$$\mathbf{f}(t) \cdot \mathbf{f}'(t) = 0$$

for all t.

b) Let $\mathbf{f}(t)$ be a vector function such that $|\mathbf{f}(t)| = 1$ for all t. Show that $\mathbf{f}(t)$ and $\mathbf{f}'(t)$ are orthogonal for all t.

20. Suppose that $\mathbf{f}$: $t \to \alpha(t)\mathbf{i} + \beta(t)\mathbf{j} + \gamma(t)\mathbf{k}$ has the properties: $\mathbf{f}(t) = \mathbf{f}'(t)$ for all t and $\alpha(0) = \beta(0) = \gamma(0) = 2$. Find $\mathbf{f}(t)$.

21. Show that every differentiable vector function $\mathbf{v}(t)$ satisfies the identity

$$2\mathbf{v}'(t) = \mathbf{i} \times (\mathbf{v}' \times \mathbf{i}) + \mathbf{j} \times (\mathbf{v}' \times \mathbf{j}) + \mathbf{k} \times (\mathbf{v}' \times \mathbf{k}).$$

22. Write out a complete proof of Theorem 23.

23. Write out a complete proof of Theorem 24.

CHAPTER 3

Infinite Series

1. Indeterminate Forms

Suppose f and F are real-valued functions defined on some interval of R^1 containing the number a. If $f(x)$ and $F(x)$ both approach 0 as x tends to the value a, the quotient

$$\frac{f(x)}{F(x)}$$

may approach a limit, may become infinite, or may fail to have any limit. In the definition of derivative it is the evaluation of just such expressions that leads to the usual differentiation formulas. We are aware that the expression

$$\frac{f(a)}{F(a)} = \frac{0}{0}$$

is in itself a meaningless one, and we use the term **indeterminate form** for the ratio 0/0.

If $f(x)$ and $F(x)$ both tend to infinity as x tends to a, the ratio $f(x)/F(x)$ may or may not tend to a limit. We use the same term, **indeterminate form**, for the expression ∞/∞, obtained by direct substitution of $x = a$ into the quotient $f(x)/F(x)$.

We recall the Theorem of the Mean, which is established in elementary calculus.

Theorem 1 (Theorem of the Mean). *Suppose that f is continuous for*

$$a \leq x \leq b$$

and that $f'(x)$ exists for each x between a and b. Then there is an x_0 between a and b (that is, $a < x_0 < b$) such that

$$\frac{f(b) - f(a)}{b - a} = f'(x_0).$$

REMARK. **Rolle's Theorem** is the special case $f(a) = f(b) = 0$.

The evaluation of indeterminate forms requires an extension of the Theorem of the Mean which we now prove.

Theorem 2 (Generalized Theorem of the Mean). *Suppose that f and F are continuous for $a \leq x \leq b$, and $f'(x)$ and $F'(x)$ exist for $a < x < b$ with $F'(x) \neq 0$ there. Then $F(b) - F(a) \neq 0$ and there is a number ξ with*

$$a < \xi < b$$

such that

$$\boxed{\frac{f(b) - f(a)}{F(b) - F(a)} = \frac{f'(\xi)}{F'(\xi)}} \tag{1}$$

PROOF. The fact that $F(b) - F(a) \neq 0$ is obtained by applying the Theorem of the Mean (Theorem 1) to F. For then, $F(b) - F(a) = F'(x_0)(b - a)$ for some x_0 such that $a < x_0 < b$. By hypothesis, the right side is different from zero.

For the proof of the main part of the theorem, we define the function $\phi(x)$ by the formula

$$\phi(x) = f(x) - f(a) - \frac{f(b) - f(a)}{F(b) - F(a)}[F(x) - F(a)].$$

We compute $\phi(a)$, $\phi(b)$, and $\phi'(x)$, getting

$$\phi(a) = f(a) - f(a) - \frac{f(b) - f(a)}{F(b) - F(a)}[F(a) - F(a)] = 0,$$

$$\phi(b) = f(b) - f(a) - \frac{f(b) - f(a)}{F(b) - F(a)}[F(b) - F(a)] = 0,$$

$$\phi'(x) = f'(x) - \frac{f(b) - f(a)}{F(b) - F(a)}F'(x).$$

Applying the Theorem of the Mean (i.e., in the special form of Rolle's Theorem) to $\phi(x)$ in the interval $[a, b]$, we find

$$\frac{\phi(b) - \phi(a)}{b - a} = 0 = \phi'(\xi) = f'(\xi) - \frac{f(b) - f(a)}{F(b) - F(a)}F'(\xi)$$

for some ξ between a and b. Dividing by $F'(\xi)$, we obtain formula (1).

The next theorem, known as **l'Hôpital's Rule**, is useful in the evaluation of indeterminate forms.

Theorem 3 (l'Hôpital's Rule). *Suppose that*

$$\lim_{x\to a} f(x) = 0, \qquad \lim_{x\to a} F(x) = 0, \qquad \textit{and} \qquad \lim_{x\to a} \frac{f'(x)}{F'(x)} = L,$$

and that the hypotheses of Theorem 2 hold in some deleted interval about a. Then*

$$\boxed{\lim_{x\to a} \frac{f(x)}{F(x)} = \lim_{x\to a} \frac{f'(x)}{F'(x)} = L.}$$

PROOF. For some h we apply Theorem 2 in the interval $a < x < a + h$. Then

$$\frac{f(a+h) - f(a)}{F(a+h) - F(a)} = \frac{f(a+h)}{F(a+h)} = \frac{f'(\xi)}{F'(\xi)}, \qquad a < \xi < a + h,$$

where we have taken $f(a) = F(a) = 0$. As h tends to 0, ξ tends to a, and so

$$\lim_{h\to 0} \frac{f(a+h)}{F(a+h)} = \lim_{\xi\to a} \frac{f'(\xi)}{F'(\xi)} = L.$$

A similar proof is valid for x in the interval $a - h < x < a$.

EXAMPLE 1. Evaluate

$$\lim_{x\to 3} \frac{x^3 - 2x^2 - 2x - 3}{x^2 - 9}.$$

SOLUTION. We set $f(x) = x^3 - 2x^2 - 2x - 3$ and $F(x) = x^2 - 9$. We see at once that $f(3) = 0$ and $F(3) = 0$, and we have an indeterminate form. We calculate

$$f'(x) = 3x^2 - 4x - 2, \qquad F'(x) = 2x.$$

By Theorem 3 (l'Hôpital's Rule):

$$\lim_{x\to 3} \frac{f(x)}{F(x)} = \lim_{x\to 3} \frac{f'(x)}{F'(x)} = \frac{3(9) - 4(3) - 2}{2(3)} = \frac{13}{6}.$$

REMARKS. It is essential that $f(x)$ and $F(x)$ *both* tend to zero as x tends to a before applying l'Hôpital's Rule. If either or both functions tend to finite limits $\neq 0$, or if one tends to zero and the other does not, then the limit of the quotient is found by the method of direct substitution.

* Let I be an interval which has a as an interior point. The interval I with a removed from it is called a **deleted interval** about a.

It may happen that $f'(x)/F'(x)$ is an indeterminate form as $x \to a$. Then l'Hôpital's Rule may be applied again, and the limit $f''(x)/F''(x)$ may exist as x tends to a. In fact, for some problems l'Hôpital's Rule may be required a number of times before the limit is actually determined. Example 3 below exhibits this point.

EXAMPLE 2. Evaluate

$$\lim_{x \to a} \frac{x^p - a^p}{x^q - a^q}, \qquad a > 0.$$

SOLUTION. We set $f(x) = x^p - a^p$, $F(x) = x^q - a^q$. Then $f(a) = 0$, $F(a) = 0$. We compute $f'(x) = px^{p-1}$, $F'(x) = qx^{q-1}$. Therefore

$$\lim_{x \to a} \frac{f(x)}{F(x)} = \lim_{x \to a} \frac{f'(x)}{F'(x)} = \lim_{x \to a} \frac{px^{p-1}}{qx^{q-1}} = \frac{p}{q} a^{p-q}.$$

EXAMPLE 3. Evaluate

$$\lim_{x \to 0} \frac{x - \sin x}{x^3}.$$

SOLUTION. We set $f(x) = x - \sin x$, $F(x) = x^3$. Since $f(0) = 0$, $F(0) = 0$, we apply l'Hôpital's Rule and get

$$\lim_{x \to 0} \frac{f(x)}{F(x)} = \lim_{x \to 0} \frac{1 - \cos x}{3x^2}.$$

But we note that $f'(0) = 0$, $F'(0) = 0$, and so we apply l'Hôpital's Rule again:

$$f''(x) = \sin x, \qquad F''(x) = 6x.$$

Therefore

$$\lim_{x \to 0} \frac{f(x)}{F(x)} = \lim_{x \to 0} \frac{f''(x)}{F''(x)}.$$

Again we have an indeterminate form: $f''(0) = 0$, $F''(0) = 0$. We continue, to obtain $f'''(x) = \cos x$, $F'''(x) = 6$. Now we find that

$$\lim_{x \to 0} \frac{f(x)}{F(x)} = \lim_{x \to 0} \frac{f'''(x)}{F'''(x)} = \frac{\cos 0}{6} = \frac{1}{6}.$$

L'Hôpital's Rule can be extended to the case where *both* $f(x) \to \infty$ and $F(x) \to \infty$ as $x \to a$. The proof of the next theorem, which we omit, is analogous to the proof of Theorem 3.

Theorem 4 (l'Hôpital's Rule). *Suppose that*

$$\lim_{x \to a} f(x) = \infty, \qquad \lim_{x \to a} F(x) = \infty, \qquad \textit{and} \qquad \lim_{x \to a} \frac{f'(x)}{F'(x)} = L.$$

Then

$$\lim_{x\to a}\frac{f(x)}{F(x)}=\lim_{x\to a}\frac{f'(x)}{F'(x)}=L.$$

REMARK. Theorems 3 and 4 hold for one-sided limits as well as for ordinary limits. In many problems a one-sided limit is required even though this statement is not made explicity. The next example illustrates such a situation.

EXAMPLE 4. Evaluate

$$\lim_{x\to 0}\frac{\log x}{\log(2e^x-2)}$$

SOLUTION. We first note that x must tend to zero through positive values since otherwise the logarithm function is not defined. We set

$$f(x)=\log x,\qquad F(x)=\log(2e^x-2).$$

Then $f(x)\to-\infty$ and $F(x)\to-\infty$ as $x\to 0^+$. Therefore

$$\lim_{x\to 0^+}\frac{f(x)}{F(x)}=\lim_{x\to 0^+}\frac{1/x}{e^x/(e^x-1)}=\lim_{x\to 0^+}\frac{e^x-1}{xe^x}$$

$$=\lim_{x\to 0^+}\frac{1-e^{-x}}{x}.$$

We still have an indeterminate form, and we take derivatives again. We obtain

$$\lim_{x\to 0^+}\frac{\log x}{\log(2e^x-2)}=\lim_{x\to 0^+}\frac{e^{-x}}{1}=1.$$

REMARK. Theorems 3 and 4 are valid when $a=+\infty$ or $-\infty$. That is, if we have an indeterminate expression for $f(\infty)/F(\infty)$, then

$$\lim_{x\to+\infty}\frac{f(x)}{F(x)}=\lim_{x\to+\infty}\frac{f'(x)}{F'(x)},$$

and a similar statement holds when $x\to-\infty$. The next example exhibits this type of indeterminate form.

EXAMPLE 5. Evaluate

$$\lim_{x\to+\infty}\frac{8x}{e^x}.$$

SOLUTION.

$$\lim_{x\to+\infty}\frac{8x}{e^x}=\lim_{x\to+\infty}\frac{8}{e^x}=0.$$

REMARKS. Indeterminate forms of the type $0 \cdot \infty$ or $\infty - \infty$ can often be evaluated by transforming the expression into a quotient of the form 0/0 or ∞/∞. Limits involving exponential expressions may often be evaluated by taking logarithms. Of course, algebraic or trigonometric reductions may be made at any step. The next examples illustrate the procedure.

EXAMPLE 6. Evaluate

$$\lim_{x\to\pi/2} (\sec x - \tan x).$$

SOLUTION. We employ trigonometric reduction to change $\infty - \infty$ into a standard form. We have

$$\lim_{x\to\pi/2} (\sec x - \tan x) = \lim_{x\to\pi/2} \frac{1 - \sin x}{\cos x} = \lim_{x\to\pi/2} \frac{-\cos x}{-\sin x} = 0.$$

EXAMPLE 7. Evaluate

$$\lim_{x\to 0} [(1 + x)^{1/x}].$$

SOLUTION. We have 1^∞, which is indeterminate. Set $y = (1 + x)^{1/x}$ and take logarithms. Then

$$\log y = \log [(1 + x)^{1/x}] = \frac{\log(1 + x)}{x}.$$

By l'Hôpital's Rule,

$$\lim_{x\to 0} \frac{\log(1 + x)}{x} = \lim_{x\to 0} \frac{1}{1 + x} = 1.$$

Therefore, $\lim_{x\to 0} \log y = 1$, and we conclude that

$$\lim_{x\to 0} y = \lim_{x\to 0} (1 + x)^{1/x} = e.$$

PROBLEMS

In each of Problems 1 through 42, find the limit.

1. $\lim_{x\to -2} \frac{2x^2 + 5x + 2}{x^2 - 4}$

2. $\lim_{x\to 2} \frac{x^3 - x^2 - x - 2}{x^3 - 8}$

3. $\lim_{x\to 1} \frac{x^3 - 3x + 2}{x^3 - x^2 - x + 1}$

4. $\lim_{x\to 2} \frac{x^4 - 3x^2 - 4}{x^3 + 2x^2 - 4x - 8}$

5. $\lim_{x\to +\infty} \frac{2x^3 - x^2 + 3x + 1}{3x^3 + 2x^2 - x - 1}$

6. $\lim_{x\to 4} \frac{x^3 - 8x^2 + 2x + 1}{x^4 - x^2 + 2x - 3}$

7. $\lim_{x\to +\infty} \frac{x^3 - 3x + 1}{2x^4 - x^2 + 2}$

8. $\lim_{x\to +\infty} \frac{x^4 - 2x^2 - 1}{2x^3 - 3x^2 + 3}$

9. $\lim_{x\to 0} \frac{\tan 3x}{\sin x}$

10. $\lim_{x\to 0} \frac{\sin 7x}{x}$

11. $\lim_{x\to 0} \frac{e^{2x} - 2x - 1}{1 - \cos x}$

12. $\lim_{x\to 0} \frac{e^{3x} - 1}{1 - \cos x}$

13. $\lim_{x\to 0} \frac{\log x}{e^x}$

14. $\lim_{x\to +\infty} \frac{\log x}{x^h}, \quad h > 0$

15. $\lim_{x\to 0} \frac{\log(1 + 2x)}{3x}$

16. $\lim_{x\to 0} \frac{3^x - 2^x}{x}$

17. $\lim_{x\to 0} \frac{3^x - 2^x}{x^2}$

18. $\lim_{x\to 0} \frac{3^x - 2^x}{\sqrt{x}}$

19. $\lim_{x\to \pi/2} \frac{1 - \sin x}{\cos x}$

20. $\lim_{x\to 2} \frac{\sqrt{2x} - 2}{\log(x - 1)}$

21. $\lim_{x\to \pi/2} \frac{\log \sin x}{1 - \sin x}$

22. $\lim_{x\to \pi/2} \frac{\cos x}{\sin^2 x}$

23. $\lim_{x\to +\infty} \frac{x^3}{e^x}$

24. $\lim_{x\to \pi/2} \frac{\tan x}{\log \cos x}$

25. $\lim_{x\to +\infty} \frac{\sin x}{x}$

26. $\lim_{x\to \pi/2} \frac{\sin x}{x}$

27. $\lim_{x\to 0} \frac{x - \arctan x}{x - \sin x}$

28. $\lim_{x\to 0} \sqrt{x} \log x$

29. $\lim_{x\to 0} x \cot x$

30. $\lim_{x\to \pi/2} (x - \pi/2) \sec x$

31. $\lim_{x\to +\infty} \frac{\arctan x}{x}$

32. $\lim_{\theta\to 0} \left(\csc \theta - \frac{1}{\theta}\right)$

33. $\lim_{x\to 0} \left(\cot^2 x - \frac{1}{x^2}\right)$

34. $\lim_{x\to 0} x^x$

35. $\lim_{x\to 0} x^{4x}$

36. $\lim_{x\to +\infty} \left(1 + \frac{k}{x}\right)^x$

37. $\lim_{x\to 0} x^{(x^2)}$

38. $\lim_{x\to 0} (\cot x)^x$

39. $\lim_{x\to 0} x^{(1/\log x)}$

40. $\lim_{x\to +\infty} \frac{x^p}{e^x}, \quad p > 0$

41. $\lim_{x\to 0^+} (\sin x)^{\tan x}$

42. $\lim_{x\to 0} \frac{\log x}{x^h}$, h is a real number

43. a) Prove that

$$\lim_{x\to 0^+} x^3 e^{1/x} = +\infty.$$

b) Prove that for every positive integer n,

$$\lim_{x \to 0^+} x^n e^{1/x} = +\infty.$$

c) Find the result if in part (b) we have $x \to 0^-$ instead of $x \to 0^+$.

44. By direct methods, find the value of $\lim_{x \to +\infty} (x \sin x)/(x^2 + 1)$. What happens if l'Hôpital's Rule is used? Explain.

45. Prove the following form of l'Hôpital's Rule:
If

$$\lim_{x \to +\infty} f(x) = 0, \quad \lim_{x \to +\infty} g(x) = 0 \quad \text{and} \quad \lim_{x \to +\infty} \frac{f'(x)}{g'(x)} = L,$$

then

$$\lim_{x \to +\infty} \frac{f(x)}{g(x)} = L.$$

[*Hint*: Consider $\lim_{x \to 0^+} f(1/x)/g(1/x)$.]

46. Suppose that

$$\lim_{x \to 0} f(x) = \lim_{x \to 0} f'(x) = \lim_{x \to 0} f''(x) = \lim_{x \to 0} f'''(x) = 0,$$

and that

$$\lim_{x \to 0} \frac{x^2 f'''(x)}{f''(x)} = 2.$$

Find $\lim_{x \to 0} x^2 f'(x)/f(x)$.

47. If the second derivative f'' of a function f exists at a value x_0, show that

$$\lim_{h \to 0} \frac{f(x_0 + h) - 2f(x_0) + f(x_0 - h)}{h^2} = f''(x_0).$$

48. Let $P(x)$ and $Q(x)$ be polynomials of degree m and n, respectively. Analyze

$$\lim_{x \to +\infty} \frac{P(x)}{Q(x)}$$

according as $m > n$ or $m = n$ or $m < n$.

2. Convergent and Divergent Series

The numbers

$$b_1, b_2, b_3, \ldots, b_{12}, b_{13}, b_{14}$$

form a sequence of fourteen numbers. Since this set contains both a first and last element, the sequence is termed **finite**. In all other circumstances it is called **infinite**. The subscripts not only identify the location of each element but also serve to associate a positive integer with each member of the sequence. In other words, *a* **sequence is a function** with *domain* a portion (or all) of the positive integers and with *range* in the collection of real numbers. If we use J to denote the collection of positive integers and R the set of real numbers, then a sequence is a function $f: J \to R$.

If the domain is an infinite collection of positive integers, e.g., all positive integers, we write

$$a_1, a_2, \ldots, a_n, \ldots,$$

the final dots indicating the never-ending character of the sequence. Simple examples of infinite sequences are

$$1, \frac{1}{2}, \frac{1}{3}, \frac{1}{4}, \ldots, \frac{1}{n}, \ldots \tag{1}$$

$$\frac{1}{2}, \frac{2}{3}, \frac{3}{4}, \ldots, \frac{n}{n+1}, \ldots \tag{2}$$

$$2, 4, 6, \ldots, 2n, \ldots \tag{3}$$

Definition. Given the infinite sequence

$$a_1, a_2, \ldots, a_n, \ldots,$$

we say that **this sequence has the limit** c if, for each $\varepsilon > 0$, there is a positive integer N (the size of N depending on ε) such that

$$|a_n - c| < \varepsilon \qquad \text{for all } n > N.$$

We also write $a_n \to c$ as $n \to \infty$ and, equivalently,

$$\lim_{n\to\infty} a_n = c.$$

In the sequence (1) above, we have

$$a_1 = 1, \quad a_2 = \frac{1}{2}, \quad \ldots, \quad a_n = \frac{1}{n}, \quad \ldots$$

and $\lim_{n\to\infty} a_n = 0$. The sequence (2) has the form

$$a_1 = \frac{1}{2}, \quad a_2 = \frac{2}{3}, \quad \ldots, \quad a_n = \frac{n}{n+1}, \quad \ldots$$

and $\lim_{n\to\infty} a_n = 1$. The sequence (3) does not tend to a limit.

An expression such as

$$u_1 + u_2 + u_3 + \cdots + u_{24}$$

is called a *finite series*. The **sum** of such a series is obtained by adding the 24 terms. We now extend the notion of a finite series by considering an expression of the form

$$u_1 + u_2 + u_3 + \cdots + u_n + \cdots$$

which is nonterminating and which we call an **infinite series.*** Our first task is to give a meaning, if possible, to such an infinite succession of additions.

Definition. Given the infinite series $u_1 + u_2 + u_3 + \cdots + u_n + \cdots$, the quantity $s_k = u_1 + u_2 + \cdots + u_k$ is called the k**th partial sum** of the series. That is,

$$s_1 = u_1, \qquad s_2 = u_1 + u_2, \qquad s_3 = u_1 + u_2 + u_3,$$

etc. Each partial sum is obtained by a **finite** number of additions.

Definition. Given the series

$$u_1 + u_2 + u_3 + \cdots + u_n + \cdots \tag{4}$$

with the sequence of partial sums

$$s_1, s_2, s_3, \ldots, s_n, \ldots,$$

we define the **sum of the series** (4) to be

$$\lim_{n\to\infty} s_n \tag{5}$$

whenever the limit exists.

Using the Σ notation for sum, we can also write

$$\sum_{n=1}^{\infty} u_n = \lim_{n\to\infty} s_n.$$

If the limit (5) does not exist, then the sum (4) *is not defined.*

Definitions. If the limit (5) exists, the series $\Sigma_{n=1}^{\infty} u_n$ is said to **converge** to that limit; otherwise the series is said to **diverge**.

REMARK. The expression $\Sigma_{n=1}^{\infty} u_n$ is a shorthand notation for the formal series expression (4). However, the symbol $\Sigma_{n=1}^{\infty} u_n$ is also used as a synonym for the numerical value of the series when it converges. There will be no difficulty in recognizing which meaning we are employing in any particular case. We could obtain more precision by using the (more cumbersome) notation described in the footnote below.

* The definition given here is informal. A more formal definition is as follows: An **infinite series** is an ordered pair $(\{u_n\}, \{s_n\})$ of infinite sequences in which $s_k = u_1 + \cdots + u_k$ for each k. The infinite series $(\{u_n\}, \{s_n\})$ is denoted by $u_1 + u_2 + \cdots + u_n + \cdots$ or $\Sigma_{n=1}^{\infty} u_n$. When no confusion can arise we also denote by $\Sigma_{n=1}^{\infty} u_n$ the limit of the sequence $\{s_n\}$ when it exists.

The sequence of terms

$$a, ar, ar^2, ar^3, \ldots, ar^{n-1}, ar^n, \ldots$$

forms a *geometric progression*. Each term (except the first) is obtained by multiplication of the preceding term by r, the *common ratio*. The partial sums of the **geometric series**

$$a + ar + ar^2 + ar^3 + \cdots + ar^n + \cdots$$

are

$$s_1 = a,$$

$$s_2 = a + ar,$$

$$s_3 = a + ar + ar^2,$$

$$s_4 = a + ar + ar^2 + ar^3,$$

and, in general,

$$s_n = a(1 + r + r^2 + \cdots + r^{n-1}).$$

For example, with $a = 2$ and $r = \frac{1}{2}$,

$$s_n = 2\left(1 + \frac{1}{2} + \frac{1}{4} + \cdots + \frac{1}{2^{n-1}}\right).$$

The identity

$$(1 + r + r^2 + \cdots + r^{n-1})(1 - r) = 1 - r^n,$$

which may be verified by straightforward multiplication, leads to the formula

$$s_n = a\frac{1 - r^n}{1 - r}$$

for the nth partial sum. The example $a = 2$, $r = \frac{1}{2}$ gives

$$s_n = 2\frac{1 - 2^{-n}}{\frac{1}{2}} = 4 - \frac{1}{2^{n-2}}.$$

In general, we may write

$$s_n = a\frac{1 - r^n}{1 - r} = \frac{a}{1 - r} - \frac{a}{1 - r}r^n, \qquad r \neq 1. \tag{6}$$

The next theorem is a direct consequence of formula (6).

Theorem 5. *A geometric series*

$$a + ar + ar^2 + \cdots + ar^n + \cdots$$

converges if $-1 < r < 1$ *and diverges if* $|r| \geq 1$. *In the convergent case we have*

$$\sum_{n=1}^{\infty} ar^{n-1} = \frac{a}{1-r}. \tag{7}$$

PROOF. From (6) we see that $r^n \to 0$ if $|r| < 1$, yielding (7); also, $r^n \to \infty$ if $|r| > 1$. For $r = 1$, the partial sum s_n is na, and s_n does not tend to a limit as $n \to \infty$. If $r = -1$, the partial sum s_n is a if n is odd and 0 if n is even.

The next theorem is useful in that it exhibits a limitation on the behavior of the terms of a convergent series.

Theorem 6. *If the series*

$$\sum_{k=1}^{\infty} u_k = u_1 + u_2 + u_3 + \cdots + u_n + \cdots \tag{8}$$

converges, then

$$\lim_{n\to\infty} u_n = 0.$$

PROOF. Writing

$$s_n = u_1 + u_2 + \cdots + u_n,$$

$$s_{n-1} = u_1 + u_2 + \cdots + u_{n-1},$$

we have, by subtraction, $u_n = s_n - s_{n-1}$. Letting c denote the sum of the series, we see that $s_n \to c$ as $n \to \infty$; also, $s_{n-1} \to c$ as $n \to \infty$. Therefore

$$\lim_{n\to\infty} u_n = \lim_{n\to\infty} (s_n - s_{n-1}) = \lim_{n\to\infty} s_n - \lim_{n\to\infty} s_{n-1} = c - c = 0.$$

REMARK. The converse of Theorem 6 is not necessarily true. Later we shall show (by example) that it is possible both for u_n to tend to 0 and for the series to diverge.

The following corollary, a restatement of Theorem 6, is useful in establishing the divergence of infinite series.

Corollary. *If u_n does not tend to zero as $n \to \infty$, then the series $\Sigma_{n=1}^{\infty} u_n$ is divergent.*

Convergent series may be added, subtracted, and multiplied by constants, as the next theorem shows.

Theorem 7. *If $\Sigma_{n=1}^{\infty} u_n$ and $\Sigma_{n=1}^{\infty} v_n$ both converge and c is any number, then the series*

$$\sum_{n=1}^{\infty} (cu_n), \qquad \sum_{n=1}^{\infty} (u_n + v_n), \qquad \sum_{n=1}^{\infty} (u_n - v_n)$$

all converge, and

$$\sum_{n=1}^{\infty} (cu_n) = c \sum_{n=1}^{\infty} u_n,$$

$$\sum_{n=1}^{\infty} (u_n \pm v_n) = \sum_{n=1}^{\infty} u_n \pm \sum_{n=1}^{\infty} v_n.$$

PROOF. For each n, we have the following equalities for the partial sums:

$$\sum_{j=1}^{n} (cu_j) = c \sum_{j=1}^{n} u_j;$$

$$\sum_{j=1}^{n} (u_j \pm v_j) = \sum_{j=1}^{n} u_j \pm \sum_{j=1}^{n} v_j.$$

The results now follow from the elementary theorems on limits of sequences which the reader will recall from a first course in calculus.

EXAMPLE. Express the repeating decimal $A = 0.151515 \ldots$ as the ratio of two integers.

SOLUTION. We write A in the form of a geometric series:

$$A = 0.15(1 + 0.01 + (0.01)^2 + (0.01)^3 + \cdots),$$

in which $a = 0.15$ and $r = 0.01$. This series is convergent and has sum

$$s = \frac{0.15}{1 - 0.01} = \frac{0.15}{0.99} = \frac{5}{33} = A.$$

PROBLEMS

In Problems 1 through 8, express each repeating decimal as the ratio of two integers.

1. $0.717171 \cdots$
2. $0.464646 \cdots$
3. $0.013013013 \cdots$
4. $2.718718718 \cdots$
5. $0.000141414 \cdots$
6. $32.46513513513 \cdots$
7. $3.614361436143 \cdots$
8. $42.000100010001 \cdots$
9. Find the sum of the geometric series if $a = 3$, $r = -\frac{1}{3}$.
10. Find the sum of the geometric series if $a = -2$, $r = \frac{1}{4}$.
11. The first term of a geometric series is 3 and the fifth term is $\frac{16}{27}$. Find the sum of the infinite series.

12. The fourth term of a geometric series is -1 and the seventh term is $\frac{1}{8}$. Find the sum of the infinite series.

13. A ball is dropped from a height of 6 meters. Upon each bounce the ball rises to $\frac{1}{3}$ of its previous height. Find the total distance travelled by the ball.

In Problems 14 through 18, write the first five terms of the series given. Use the corollary to Theorem 6 to show that the series is divergent.

14. $\sum_{n=1}^{\infty} \frac{n^2}{n+1}$

15. $\sum_{n=1}^{\infty} \frac{2n}{3n+5}$

16. $\sum_{n=1}^{\infty} \frac{n^2-2n+3}{2n^2+n+1}$

17. $\sum_{n=1}^{\infty} (-1)^{n+1}\frac{e^n}{n^3}$

18. $\sum_{n=1}^{\infty} \frac{n^2+n+2}{\log(n+1)}$

In Problems 19 through 22, *assume* that the series

$$\sum_{n=1}^{\infty} \frac{1}{n^2}, \quad \sum_{n=1}^{\infty} \frac{1}{n^3}, \quad \sum_{n=1}^{\infty} \frac{1}{n^4}$$

all converge. In each case use Theorem 7 to show that the given series is convergent.

19. $\sum_{n=1}^{\infty} \frac{3n+2}{n^3}$

20. $\sum_{n=1}^{\infty} \frac{n-2}{n^3}$

21. $\sum_{n=1}^{\infty} \frac{3n^2+4}{n^4}$

22. $\sum_{n=1}^{\infty} \frac{3n^2-2n+4}{n^4}$

23. Suppose that the series $\Sigma_{k=1}^{\infty} u_k$ converges. Show that any series obtained from this one by deleting a finite number of terms also converges. [*Hint:* Since $s_n = \Sigma_{k=1}^{n} u_k$ converges to a limit, find the value to which S_n, the partial sums of the deleted series, must tend.]

24. Show that any number of the form

$$0.a_1 a_2 a_3 a_1 a_2 a_3 a_1 a_2 a_3 \cdots,$$

where a_1, a_2, and a_3 are digits between 0 and 9, is expressible as the ratio of two integers and therefore is a rational number.

25. Show that any repeating decimal is a rational number.

26. Let $a_1, a_2, \ldots, a_n, \ldots$ be a sequence of positive terms such that $a_{n+1} < a_n$ for every n. Give an example to show that $\Sigma_{n=1}^{\infty} a_n$ may not converge.

3. Series of Positive Terms

Except in very special cases, it is not possible to tell if a series converges by finding whether or not s_n, the nth partial sum, tends to a limit. (The geometric series, however, is one of the special cases where it *is* possible.)

Fig. 3-1 a_1 a_2 a_3 a_4 a_5 b M

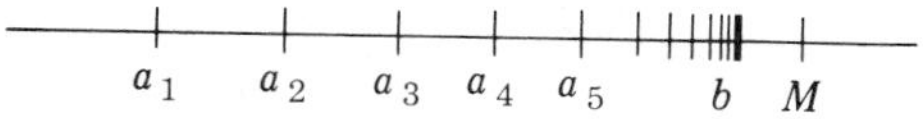

In this section, we present some *indirect* tests for convergence and divergence which apply only to series with positive (or at least nonnegative) terms. That is, we assume throughout this section that $u_n \geq 0$ for $n = 1, 2, \ldots$. Tests for series with terms which may be positive or negative will be discussed in the following sections.

Before establishing the next theorem we introduce the Axiom of Continuity, a property of the real number system which is useful throughout calculus and analysis.

Axiom C (Axiom of Continuity). Suppose that an infinite sequence $a_1, a_2, \ldots, a_n, \ldots$ has the properties (1) $a_{n+1} \geq a_n$ for all n, and (2) there is a number M such that $a_n \leq M$ for all n. Then there is a number $b \leq M$ such that

$$\lim_{n\to\infty} a_n = b \qquad \text{and} \qquad a_n \leq b$$

for all n.

Figure 3-1 shows the situation. The numbers a_n move steadily to the right, and yet they can never get beyond M. It is reasonable to have an axiom which assumes that there must be some number b (perhaps M itself) toward which the a_n cluster. Axiom C is usually stated in the form: **Every bounded, nondecreasing sequence of numbers tends to a limit.**

Theorem 8. *Suppose that* $u_n \geq 0$, $n = 1, 2, \ldots$ *and* $s_n = \Sigma_{k=1}^{n} u_k$ *is the nth partial sum. Then, either* (a), *there is a number M such that all the* $s_n \leq M$, *in which case the series* $\Sigma_{k=1}^{\infty} u_k$ *converges to a value* $s \leq M$, *or else* (b), $s_n \to +\infty$ *and the series diverges.*

PROOF. By subtraction, we have

$$u_n = s_n - s_{n-1} \geq 0,$$

and so the s_n form an increasing (or at least nondecreasing) sequence. If all $s_n \leq M$, then by the Axiom of Continuity, we conclude that $s_n \to s \leq M$. Thus part (a) of the theorem is established. If there is no such M, then for each number E, no matter how large, there must be an $s_n > E$; and all s_m with $m > n$ are greater than or equal to s_n. This is another way of saying $s_n \to +\infty$.

The next theorem is one of the most useful tests for deciding convergence and divergence of series.

Theorem 9 (Comparison Test). *Suppose that all* $u_n \geq 0$. (a) *If* $\Sigma_{n=1}^{\infty} a_n$ *is a convergent series and* $u_n \leq a_n$ *for all n, then* $\Sigma_{n=1}^{\infty} u_n$ *is convergent and*

$$\sum_{n=1}^{\infty} u_n \leq \sum_{n=1}^{\infty} a_n.$$

(b) *If* $\Sigma_{n=1}^{\infty} a_n$ *is a divergent series of nonnegative terms and* $u_n \geq a_n$ *for all* n, *then* $\Sigma_{n=1}^{\infty} u_n$ *diverges.*

Proof. We let

$$s_n = u_1 + u_2 + \cdots + u_n, \qquad S_n = a_1 + a_2 + \cdots + a_n$$

be the nth partial sums. The s_n, S_n are both nondecreasing sequences. In case (a), we let S be the limit of S_n and, since $s_n \leq S_n \leq S$ for every n, we apply Theorem 8 to conclude that s_n converges. In case (b), we have $S_n \to +\infty$ and $s_n \geq S_n$ for every n. Hence, $s_n \to +\infty$.

Remarks. In order to apply the Comparison Test, the reader must show either (a), that the terms u_n of the given series are $\leq a_n$ where $\Sigma_{n=1}^{\infty} a_n$ is a *known* convergent series or (b), that each $u_n \geq a_n$ where $\Sigma_{n=1}^{\infty} a_n$ is a *known* divergent series. In all other cases, no conclusion can be drawn.

For the Comparison Test to be useful, we must have at hand as large a number as possible of series (of positive terms) about whose convergence and divergence we are fully informed. Then, when confronted with a new series of positive terms, we shall have available a body of series for comparison purposes. So far, the only series which we have shown to be convergent are the geometric series with $r < 1$, and the only series which we have shown to be divergent are those in which u_n does not tend to zero. We now study the convergence and divergence of a few special types of series in order to obtain material which can be used for the Comparison Test.

Definition. If n is a positive integer, we define $n!$ (read n factorial) $= 1 \cdot 2 \cdots n$; it is convenient to define $0! = 1$.

For example, $5! = 1 \cdot 2 \cdot 3 \cdot 4 \cdot 5 = 120$, etc. We see that

$$(n+1)! = (n+1) \cdot n!, \qquad n \geq 0.$$

Example 1. Test the series

$$\sum_{n=1}^{\infty} \frac{1}{n!}$$

for convergence or divergence.

Solution. Writing the first few terms, we obtain

$$\frac{1}{1!} = \frac{1}{1}, \qquad \frac{1}{2!} = \frac{1}{1 \cdot 2}, \qquad \frac{1}{3!} = \frac{1}{1 \cdot 2 \cdot 3}, \qquad \frac{1}{4!} = \frac{1}{1 \cdot 2 \cdot 3 \cdot 4}.$$

Since each factor except 1 and 2 in $n!$ is larger than 2, we have the inequalities (which may be rigorously established by mathematical induction):

$$n! \geq 2^{n-1} \quad \text{and} \quad \frac{1}{n!} \leq \frac{1}{2^{n-1}}.$$

The series $\sum_{n=1}^{\infty} a_n$ with $a_n = 1/2^{n-1}$ is a geometric series with $r = \frac{1}{2}$, and is therefore convergent. Hence, by the Comparison Test, $\sum_{n=1}^{\infty} 1/n!$ converges.

REMARK. Since any *finite* number of terms at the beginning of a series does not affect convergence or divergence, the comparison between u_n and a_n in Theorem 9 is not required for all n. It is required for all *n except a finite number*.

The next theorem gives us an entire collection of series useful for comparison purposes.

Theorem 10 (The p-series). *The series*

$$\sum_{n=1}^{\infty} \frac{1}{n^p},$$

known as the p-series, is convergent if $p > 1$ and divergent if $p \leq 1$.

The proof of this theorem is deferred until later in the section. We note that it is not necessary for p to be an integer.

EXAMPLE 2. Test the series

$$\sum_{n=1}^{\infty} \frac{1}{n(n+1)}$$

for convergence or divergence.

SOLUTION. For each n we have

$$\frac{1}{n(n+1)} \leq \frac{1}{n^2}.$$

Since $\sum_{n=1}^{\infty} 1/n^2$ is a p-series with $p = 2$ and so converges, we are in a position to use the Comparison Test. Therefore, the series

$$\sum_{n=1}^{\infty} \frac{1}{n(n+1)}$$

converges.

EXAMPLE 3. Test the series

$$\sum_{n=1}^{\infty} \frac{1}{n+10}$$

for convergence or divergence.

SOLUTION 1. Writing out a few terms, we have

$$\tfrac{1}{11} + \tfrac{1}{12} + \tfrac{1}{13} + \tfrac{1}{14} + \cdots,$$

and we see that the series is just like $\Sigma_{n=1}^{\infty} 1/n$, except that the first ten terms are missing. According to the Remark before Theorem 10, we may compare the given series with the p-series for $p = 1$. The comparison establishes divergence.

SOLUTION 2. We have, for every $n \geq 1$,

$$n + 10 \leq 11n, \qquad \text{and so} \qquad \frac{1}{n+10} \geq \frac{1}{11n}.$$

The series

$$\sum_{n=1}^{\infty} \frac{1}{11n} = \frac{1}{11} \sum_{n=1}^{\infty} \frac{1}{n}$$

is divergent (p-series with $p = 1$) and, therefore, the given series diverges.

The next theorem yields another test which is used frequently in conjunction with the Comparison Test.

Theorem 11 (Integral Test). *Assume that f is a continuous, nonnegative, and nonincreasing function defined for all $x \geq 1$. That is, we suppose that*

$$f(x) \geq 0, \ (\textit{nonnegative})$$

and

$$f(x) \geq f(y) \qquad \textit{for } x \leq y \ (\textit{nonincreasing}).$$

Suppose that $\Sigma_{n=1}^{\infty} u_n$ is a series with

$$u_n = f(n) \qquad \textit{for each } n \geq 1.$$

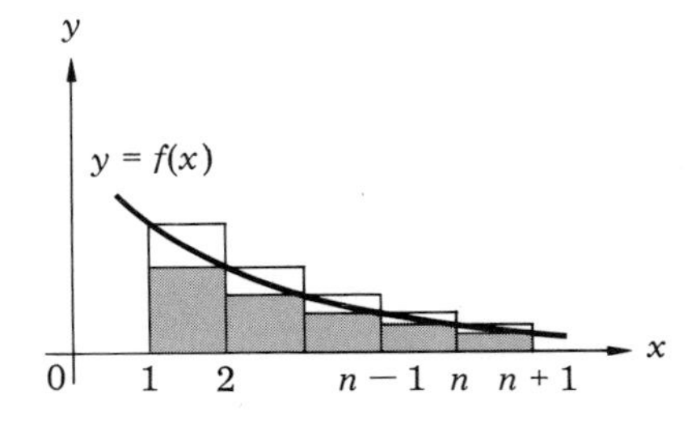

Fig. 3-2

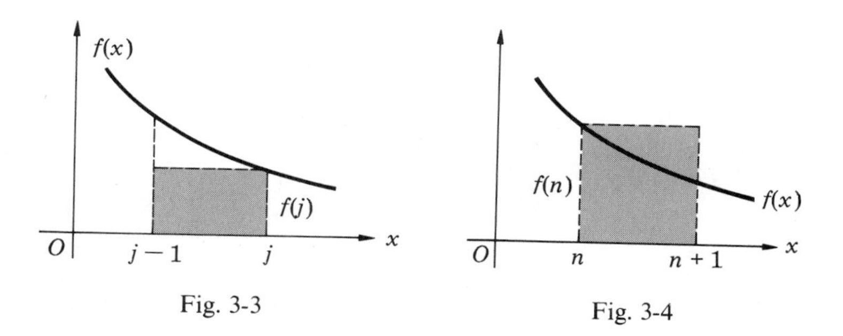

Fig. 3-3

Fig. 3-4

Then (a) $\Sigma_{n=1}^{\infty} u_n$ *is convergent if the improper integral* $\int_1^{\infty} f(x)\,dx$ *is convergent and, conversely* (b), *the improper integral converges if the series does.*

Proof. (See Fig. 3-2.) (a) Suppose first that the improper integral is convergent. Then, since $f(x) \geq f(j)$ for $x \leq j$, we see that

$$\int_{j-1}^{j} f(x)\,dx \geq f(j), \tag{1}$$

a fact verified by noting in Fig. 3-3 that $f(j)$ is the shaded area and the integral is the area under the curve. We define

$$a_j = \int_{j-1}^{j} f(x)\,dx,$$

and then

$$a_1 + \int_1^n f(x)\,dx = a_1 + a_2 + \cdots + a_n,$$

where we have set $a_1 = f(1) = u_1$. By hypothesis, $\int_1^{\infty} f(x)\,dx$ is finite, and so the series $\Sigma_{n=1}^{\infty} a_n$ is convergent. Since $f(j) = u_j$, the inequality $a_j \geq u_j$ is a restatement of (1), and now the comparison theorem applies to yield the result.

b) Suppose now that $\Sigma_{n=1}^{\infty} u_n$ converges. Let

$$v_n = \int_n^{n+1} f(x)\,dx \leq f(n) = u_n, \qquad n = 1, 2, \ldots,$$

the inequality holding since $f(x) \leq f(n)$ for $n \leq x$ (Fig. 3-4). Because $f(x) \geq 0$, we have each $v_n \geq 0$, and so $\Sigma_{n=1}^{\infty} v_n$ converges to some number S. That is,

$$\sum_{k=1}^{n} v_k = \int_1^{n+1} f(x)\,dx \leq S$$

for every n. Let ε be any positive number. There is an N such that

$$S - \varepsilon < \int_1^{n+1} f(x)\,dx \leq S \qquad \text{for all} \quad n \geq N.$$

Since $f(x) \geq 0$, we see that

$$S - \varepsilon < \int_1^{n+1} f(x)\,dx \leq \int_1^X f(x)\,dx \leq S \qquad \text{if} \quad X \geq N + 1.$$

Hence the improper integral converges.

We shall now employ the Integral Test to establish the convergence and divergence of the p-series.

PROOF OF THEOREM 10. We define the function $f(x) = 1/x^p$, which satisfies all the conditions of the integral test if $p > 0$. We have

$$\int_1^\infty \frac{dx}{x^p} = \lim_{t\to\infty} \int_1^t \frac{dx}{x^p} = \lim_{t\to\infty} \left(\frac{t^{1-p} - 1}{1 - p}\right) \qquad \text{for} \quad p \neq 1.$$

The limit exists for $p > 1$ and fails to exist for $p < 1$. As for the case $p = 1$, we have

$$\int_1^t \frac{dx}{x} = \log t,$$

which tends to ∞ as $t \to \infty$. Thus Theorem 10 is established.

EXAMPLE 4. Test the series

$$\sum_{n=1}^{\infty} \frac{1}{(n+1)\log(n+1)}$$

for convergence or divergence.

SOLUTION. Let

$$f(x) = \frac{1}{(x+1)\log(x+1)}$$

and note that all conditions for the Integral Test are fulfilled. We obtain

$$\int_1^t \frac{dx}{(x+1)\log(x+1)} = \int_2^{t+1} \frac{du}{u \log u} = \int_2^{t+1} \frac{d(\log u)}{\log u}$$
$$= \log[\log(t+1)] - \log(\log 2).$$

The expression on the right diverges as $t \to \infty$ and, therefore, the given series is divergent.

The particular p-series with $p = 1$, known as the **harmonic series**, is interesting, as it is on the borderline between convergence and divergence of the various p-series. It is an example of a series in which the general term u_n tends to zero while the series diverges. (See page 102.) We can prove divergence of the harmonic series without recourse to the Integral Test. To do so, we write

$$\tfrac{1}{1} + \tfrac{1}{2} + (\tfrac{1}{3} + \tfrac{1}{4}) + (\tfrac{1}{5} + \tfrac{1}{6} + \tfrac{1}{7} + \tfrac{1}{8})$$
$$+ (\tfrac{1}{9} + \tfrac{1}{10} + \tfrac{1}{11} + \tfrac{1}{12} + \tfrac{1}{13} + \tfrac{1}{14} + \tfrac{1}{15} + \tfrac{1}{16}) + (\text{next 16 terms}) + \cdots. \tag{2}$$

We have the obvious inequalities

$$\tfrac{1}{3} + \tfrac{1}{4} > \tfrac{1}{4} + \tfrac{1}{4} = \tfrac{1}{2},$$
$$\tfrac{1}{5} + \tfrac{1}{6} + \tfrac{1}{7} + \tfrac{1}{8} > \tfrac{1}{8} + \tfrac{1}{8} + \tfrac{1}{8} + \tfrac{1}{8} = \tfrac{1}{2},$$
$$\tfrac{1}{9} + \tfrac{1}{10} + \tfrac{1}{11} + \tfrac{1}{12} + \tfrac{1}{13} + \tfrac{1}{14} + \tfrac{1}{15} + \tfrac{1}{16}$$
$$> \tfrac{1}{16} + \tfrac{1}{16} + \tfrac{1}{16} + \tfrac{1}{16} + \tfrac{1}{16} + \tfrac{1}{16} + \tfrac{1}{16} + \tfrac{1}{16} = \tfrac{1}{2},$$

and so forth.

In other words, each set of terms in a set of parentheses in series (2) is larger than $\frac{1}{2}$. By taking a sufficient number of parentheses, we can make the partial sum s_n of the harmonic series as large as we please. Therefore the series diverges.

EXAMPLE 5. Show that the series

$$\sum_{n=1}^{\infty} \frac{1}{3n - 2}$$

diverges.

SOLUTION. We may use the Integral Test, or observe that

$$\sum_{n=1}^{\infty} \frac{1}{3n - 2} = \frac{1}{3} \sum_{n=1}^{\infty} \frac{1}{(n - \frac{2}{3})} \quad \text{and} \quad \frac{1}{n - \frac{2}{3}} \ge \frac{1}{n} \quad \text{for all} \quad n \ge 1.$$

The comparison test shows divergence. The divergence may also be shown directly by recombination of terms, as we did for the harmonic series.

PROBLEMS

In each of Problems 1 through 30, test for convergence or divergence.

1. $\displaystyle\sum_{n=1}^{\infty} \frac{1}{n\sqrt{n}}$
2. $\displaystyle\sum_{n=1}^{\infty} \frac{1}{\sqrt{n}}$
3. $\displaystyle\sum_{n=1}^{\infty} \frac{1}{(n + 1)(n + 2)}$
4. $\displaystyle\sum_{n=1}^{\infty} \frac{n + 1}{n\sqrt{n}}$
5. $\displaystyle\sum_{n=1}^{\infty} \frac{2n + 3}{n^2 + 3n + 2}$
6. $\displaystyle\sum_{n=1}^{\infty} \frac{1}{n \cdot 2^n}$
7. $\displaystyle\sum_{n=1}^{\infty} \frac{n - 1}{n^3}$
8. $\displaystyle\sum_{n=1}^{\infty} \frac{n^2 + 3n - 6}{n^4}$
9. $\displaystyle\sum_{n=1}^{\infty} \frac{1}{\sqrt{n(n + 1)}}$
10. $\displaystyle\sum_{n=1}^{\infty} \frac{1}{2n + 3}$

11. $\sum_{n=1}^{\infty} \frac{1}{n+100}$

12. $\sum_{n=1}^{\infty} \frac{2}{2^n+3}$

13. $\sum_{n=1}^{\infty} \frac{1}{n^2+2}$

14. $\sum_{n=1}^{\infty} \frac{2^n}{1000}$

15. $\sum_{n=1}^{\infty} \frac{2n+5}{n^3}$

16. $\sum_{n=1}^{\infty} \frac{n-1}{n^2}$

17. $\sum_{n=1}^{\infty} \frac{\log n}{n}$

18. $\sum_{n=1}^{\infty} \frac{1}{(n+1)[\log(n+1)]^2}$

19. $\sum_{n=1}^{\infty} \frac{\log(n+1)}{(n+1)^3}$

20. $\sum_{n=1}^{\infty} \frac{n!}{1\cdot 3\cdot 5\cdot 7\ldots(2n-1)}$

21. $\sum_{n=1}^{\infty} \frac{n}{e^n}$

22. $\sum_{n=1}^{\infty} \frac{1}{\sqrt{n^2+1}}$

23. $\sum_{n=1}^{\infty} \frac{n+1}{n\cdot 2^n}$

24. $\sum_{n=1}^{\infty} \frac{n^4}{e^n}$

25. $\sum_{n=1}^{\infty} \frac{n}{2^n}$

26. $\sum_{n=1}^{\infty} \frac{n^2}{2^n}$

27. $\sum_{n=1}^{\infty} \frac{\log n}{n^2}$

28. $\sum_{n=1}^{\infty} \frac{\log n}{n\sqrt{n}}$

29. $\sum_{n=1}^{\infty} \frac{n^4}{n!}$

30. $\sum_{n=1}^{\infty} \frac{(2n)!}{(3n)!}$

31. For what values of p does the series

$$\sum_{n=2}^{\infty} \frac{1}{n(\log n)^p}$$

converge?

32. For what values of q does the series

$$\sum_{n=1}^{\infty} \frac{\log n}{n^q}$$

converge?

33. For what values of p and q does the series

$$\sum_{n=1}^{\infty} \frac{(\log n)^p}{n^q}$$

converge?

34. State and prove as general an Integral Test as you can in which f is not nonincreasing. [*Hints:* The behavior of f for small values of x can be disregarded. Also, the precise behavior of f between two successive values of n may not be crucial.]

35. Let $f(x) = \sin^2 \pi x$. Show that the integral $\int_1^\infty \sin^2 \pi x\, dx$ diverges. However, $\Sigma_{n=1}^{\infty} f(n)$ converges. Explain.

36. Suppose that the series $\Sigma_{n=1}^{\infty} a_n$ is convergent with $a_n \geq 0$ for all n. Show that $\Sigma_{n=1}^{\infty} a_n^2$ is a convergent series.

37. Let $a_n \geq 0$ for all n and consider $r_n = \sqrt[n]{a_n}$. Show that if $\lim_{n\to\infty} r_n = r < 1$, then $\Sigma_{n=1}^{\infty} a_n$ converges. If $r > 1$, the series diverges. [*Hint:* show that for sufficirntly large n, a comparison with a geometric series can be made.]

*38. Given the polynomials

$$P(x) = a_n x^n + a_{n-1}x^{n-1} + \cdots + a_1 x + a_0, \quad a_n > 0,$$

$$Q(x) = b_m x^m + b_{m-1}x^{m-1} + \cdots + b_1 x + b_0, \quad b_m > 0.$$

Show that the infinite series

$$\sum_{r=r_0}^{\infty} \frac{P(r)}{Q(r)}$$

converges if $m > n + 1$ and r_0 is a sufficiently large integer.

4. Series of Positive and Negative Terms

In this section we establish three theorems which serve as important tests for the convergence and divergence of series whose terms are not necessarily positive.

Theorem 12. *If* $\Sigma_{n=1}^{\infty} |u_n|$ *converges, then* $\Sigma_{n=1}^{\infty} u_n$ *converges, and*

$$\left| \sum_{n=1}^{\infty} u_n \right| \leq \sum_{n=1}^{\infty} |u_n|.$$

PROOF. We define numbers $v_1, v_2, \ldots, v_n, \ldots$ by the relations

$$v_n = \begin{cases} u_n & \text{if} \quad u_n \quad \text{is nonnegative,} \\ 0 & \text{if} \quad u_n \quad \text{is negative.} \end{cases}$$

In other words, the series $\Sigma_{n=1}^{\infty} v_n$ consists of all the nonnegative entries in $\Sigma_{n=1}^{\infty} u_n$. Similarly, we define the sequence w_n by

$$w_n = \begin{cases} 0 & \text{if} \quad u_n \quad \text{is nonnegative,} \\ -u_n & \text{if} \quad u_n \quad \text{is negative.} \end{cases}$$

The w_n are all nonnegative, and we have

$$v_n + w_n = |u_n|, \qquad v_n - w_n = u_n \tag{1}$$

for each n. Since $v_n \leq |u_n|$ and $w_n \leq |u_n|$, and since, by hypothesis, $\Sigma_{n=1}^{\infty} |u_n|$ converges, we may apply the comparison test to conclude that the series

$$\sum_{n=1}^{\infty} v_n, \qquad \sum_{n=1}^{\infty} w_n$$

converge. Consequently, (see Theorem 7) the series

$$\sum_{n=1}^{\infty} v_n - \sum_{n=1}^{\infty} w_n = \sum_{n=1}^{\infty} (v_n - w_n) = \sum_{n=1}^{\infty} u_n$$

converges. Also,

$$\left|\sum_{n=1}^{\infty} u_n\right| = \left|\left(\sum_{n=1}^{\infty} v_n\right) - \left(\sum_{n=1}^{\infty} w_n\right)\right|$$

$$\leq \sum_{n=1}^{\infty} v_n + \sum_{n=1}^{\infty} w_n = \sum_{n=1}^{\infty} |u_n|.$$

Remark. Theorem 12 shows that if the series of absolute values $\Sigma_{n=1}^{\infty} |u_n|$ is convergent, then the series itself also converges. The converse is not necessarily true. We give an example below (page 115) in which $\Sigma_{n=1}^{\infty} u_n$ converges while $\Sigma_{n=1}^{\infty} |u_n|$ diverges.

Definitions. A series $\Sigma_{n=1}^{\infty} u_n$ which is such that $\Sigma_{n=1}^{\infty} |u_n|$ converges is said to be **absolutely convergent**. However, if $\Sigma_{n=1}^{\infty} u_n$ converges and $\Sigma_{n=1}^{\infty} |u_n|$ diverges, then the series $\Sigma_{n=1}^{\infty} u_n$ is said to be **conditionally convergent**.

The next theorem yields a test for series whose terms are alternately positive and negative. Since the hypotheses are rather stringent, the test can be used only under special circumstances.

Theorem 13 (Alternating Series Theorem). *Suppose that the numbers $u_1, u_2, \ldots, u_n, \ldots$ satisfy the hypotheses:*

i) *the u_n are alternately positive and negative,*
ii) *$|u_{n+1}| < |u_n|$ for every n, and*
iii) *$\lim_{n\to\infty} u_n = 0$.*

Then $\Sigma_{n=1}^{\infty} u_n$ is convergent. Furthermore, if the sum is denoted by s, then s lies between the partial sums s_n and s_{n+1} for each n.

Proof. Assume that u_1 is positive. (If it is not, we can consider the series beginning with u_2, since discarding a finite number of terms does not affect convergence.) Therefore, all u_k with odd subscripts are positive and all u_k with even subscripts are negative. We state this fact in the form

$$u_{2n-1} > 0, \qquad u_{2n} < 0$$

for each n. We now write

$$s_{2n} = (u_1 + u_2) + (u_3 + u_4) + (u_5 + u_6) + \cdots + (u_{2n-1} + u_{2n}).$$

Since $|u_{2k}| < u_{2k-1}$ for each k, we know that each quantity in the parentheses is positive. Hence s_{2n} increases for all n. On the other hand,

$$s_{2n} = u_1 + (u_2 + u_3) + (u_4 + u_5) + \cdots + (u_{2n-2} + u_{2n-1}) + u_{2n}.$$

Fig. 3-5 $s_2 \quad s_4 \quad \cdots \quad s_{2n} \quad \cdots \quad s \cdots s_{2n+1} \cdots \quad s_5 \quad s_3 \quad s_1$

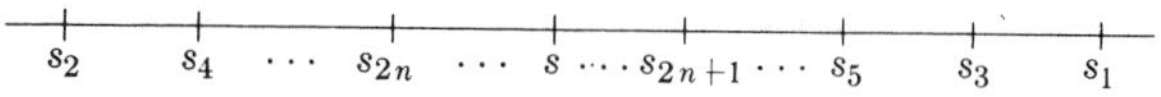

Each quantity in parentheses in the above expression is negative, and so is u_{2n}. Therefore $s_{2n} < u_1$ for all n. We conclude that s_{2n} is an increasing sequence bounded by the number u_1. It must tend to a limit (Axiom of Continuity).

We apply similar reasoning to s_{2n-1}. We have

$$s_{2n-1} + u_{2n} = s_{2n}$$

and, since $u_{2n} < 0$, we have $s_{2n-1} > s_{2n}$ for every n. Therefore for $n > 1$, s_{2n-1} is bounded from below by $s_2 = u_1 + u_2$. Also,

$$s_{2n+1} = s_{2n-1} + (u_{2n} + u_{2n+1}).$$

Since the quantity in parentheses on the right is negative, $s_{2n-1} > s_{2n+1}$; in other words, the partial sums with odd subscripts form a decreasing bounded sequence (Fig. 3-5). The limits approached by s_{2n} and s_{2n-1} must be the same, since by hypothesis (iii),

$$u_{2n} = s_{2n} - s_{2n-1} \to 0.$$

If s is the limit, we see that every even sum is less than or equal to s, while every odd sum is greater than or equal to s.

EXAMPLE 1. Test the series

$$\sum_{n=1}^{\infty} \frac{(-1)^{n+1}}{n}$$

for convergence or divergence. If it is convergent, determine whether it is conditionally convergent or absolutely convergent.

SOLUTION. We set $u_n = (-1)^{n+1}/n$ and observe that the three hypotheses of Theorem 13 hold; i.e., the terms alternate in sign, $1/(n+1) < 1/n$ for each n, and $\lim_{n\to\infty} (-1)^n/n = 0$. Therefore the given series converges. However, the series $\Sigma_{n=1}^{\infty} |u_n|$ is the harmonic series

$$\sum_{n=1}^{\infty} \frac{1}{n},$$

which is divergent. Therefore the original series is conditionally convergent.

The next test is one of the most useful for determining absolute convergence of series.

Theorem 14 (Ratio Test). *Suppose that in the series $\Sigma_{n=1}^{\infty} u_n$ every $u_n \neq 0$ and that*

$$\lim_{n\to\infty}\left|\frac{u_{n+1}}{u_n}\right| = \rho \qquad or \qquad \left|\frac{u_{n+1}}{u_n}\right| \to +\infty \text{ as } n\to\infty.$$

Then

i) *if* $\rho < 1$, *the series* $\Sigma_{n=1}^{\infty} u_n$ *converges absolutely;*
ii) *if* $\rho > 1$, *or if* $|u_{n+1}/u_n| \to +\infty$, *the series diverges;*
iii) *if* $\rho = 1$, *the test gives no information.*

PROOF. (i) Suppose that $\rho < 1$. Choose any ρ' such that $\rho < \rho' < 1$. Then, since

$$\lim_{n\to\infty}\left|\frac{u_{n+1}}{u_n}\right| = \rho,$$

there must be a sufficiently large N for which

$$\left|\frac{u_{n+1}}{u_n}\right| < \rho', \qquad \text{for all} \quad n \geq N.$$

Then we obtain

$$|u_{N+1}| < \rho'|u_N|, \qquad |u_{N+2}| < \rho'|u_{N+1}|, \qquad |u_{N+3}| < \rho'|u_{N+2}|, \qquad \text{etc.}$$

By substitution we find

$$|u_{N+2}| < \rho'^2|u_N|, \qquad |u_{N+3}| < \rho'^3|u_N|, \qquad |u_{N+4}| < \rho'^4|u_N|, \qquad \text{etc.}$$

and, in general,

$$|u_{N+k}| < (\rho')^k|u_N| \qquad \text{for } k = 1, 2, \ldots. \tag{2}$$

The series

$$\sum_{k=1}^{\infty} |u_N|(\rho')^k = |u_N| \sum_{k=1}^{\infty} (\rho')^k$$

is a geometric series with ratio less than 1 and hence convergent. From (2) and the Comparison Test, we conclude that

$$\sum_{k=1}^{\infty} |u_{N+k}| \tag{3}$$

converges. Since (3) differs from the series

$$\sum_{n=1}^{\infty} |u_n|$$

in only a finite number of terms (N, to be exact), statement (i) of the theorem is established.

ii) Suppose that $\rho > 1$ or $|u_{n+1}/u_n| \to +\infty$. There is an N such that

$$\frac{|u_{n+1}|}{|u_n|} > 1 \qquad \text{for all} \quad n \geq N.$$

By induction $|u_n| > |u_N|$ for all $n > N$. Therefore u_n does not tend to zero, and the series diverges (corollary to Theorem 6).

To establish (iii), we exhibit two cases in which $\rho = 1$, one of them corresponding to a divergent series, the other to a convergent series. The p series

$$\sum_{n=1}^{\infty} \frac{1}{n^p}$$

has $u_n = 1/n^p$. Therefore

$$\left|\frac{u_{n+1}}{u_n}\right| = \frac{n^p}{(n+1)^p} = 1\bigg/\left(1 + \frac{1}{n}\right)^p.$$

Since for any p

$$\lim_{n\to\infty} 1\bigg/\left(1 + \frac{1}{n}\right)^p = 1 = \rho,$$

we see that if $p > 1$ the series converges (and $\rho = 1$), while if $p \le 1$, the series diverges (and $\rho = 1$).

Remarks. A good working procedure for a reader is to try first the ratio test for convergence or divergence. If the limit ρ turns out to be 1, some other test must then be tried. The integral test is one possibility. (We observe that the integral test establishes convergence and divergence for the p series, while the ratio test fails.) When the terms have alternating signs, the Alternating Series Theorem is suggested. In addition, we may also try comparison theorems.

In the statement of Theorem 14, it may appear at first glance that all possible situations for ρ have been considered. That is not the case, since it may happen that

$$\left|\frac{u_{n+1}}{u_n}\right|$$

does not tend to any limit and does not tend to $+\infty$. In such circumstances, more sophisticated ratio tests are available—ones which are discussed in specialized texts on infinite series.

Example 2. Test for absolute convergence:

$$\sum_{n=1}^{\infty} \frac{2^n}{n!}.$$

Solution. Applying the ratio test, we have

$$u_{n+1} = \frac{2^{n+1}}{(n+1)!}, \qquad u_n = \frac{2^n}{n!}, \qquad \text{and} \qquad \left|\frac{u_{n+1}}{u_n}\right| = \frac{2^{n+1}}{(n+1)!}\cdot\frac{n!}{2^n} = \frac{2}{n+1}.$$

Therefore

$$\lim_{n\to\infty}\left|\frac{u_{n+1}}{u_n}\right| = 0 = \rho.$$

The series converges absolutely.

EXAMPLE 3. Test for absolute and conditional convergence.

$$\sum_{n=1}^{\infty}\frac{(-1)^n n}{2^n}.$$

SOLUTION. We try the ratio test:

$$u_{n+1} = (-1)^{n+1}\frac{n+1}{2^{n+1}}, \qquad u_n = \frac{(-1)^n n}{2^n},$$

and therefore

$$\left|\frac{u_{n+1}}{u_n}\right| = \frac{n+1}{2^{n+1}}\cdot\frac{2^n}{n} = \frac{1}{2}\left(\frac{n+1}{n}\right) = \frac{1}{2}\left(\frac{1+1/n}{1}\right).$$

Hence

$$\lim_{n\to\infty}\left|\frac{u_{n+1}}{u_n}\right| = \frac{1}{2} = \rho.$$

The series converges absolutely.

EXAMPLE 4. Test for absolute convergence:

$$\sum_{n=1}^{\infty}\frac{(2n)!}{n^{100}}.$$

SOLUTION. We try the ratio test:

$$u_n = \frac{(2n)!}{n^{100}} \qquad \text{and} \qquad u_{n+1} = \frac{[2(n+1)]!}{(n+1)^{100}}.$$

Therefore

$$\left|\frac{u_{n+1}}{u_n}\right| = \frac{(2n+2)!}{(n+1)^{100}}\cdot\frac{n^{100}}{(2n)!} = (2n+1)(2n+2)\left(\frac{n}{1+n}\right)^{100}$$

$$= (2n+1)(2n+2)\left(\frac{1}{1+1/n}\right)^{100}.$$

Hence

$$\lim_{n\to\infty}\left|\frac{u_{n+1}}{u_n}\right| = +\infty,$$

and the series is divergent.

Problems

In each of Problems 1 through 28, test for convergence or divergence. If the series is convergent, determine whether it is absolutely or conditionally convergent.

1. $\sum_{n=1}^{\infty} \frac{n!}{10^n}$

2. $\sum_{n=1}^{\infty} \frac{10^n}{n!}$

3. $\sum_{n=1}^{\infty} \frac{(-1)^{n-1}n!}{10^n}$

4. $\sum_{n=1}^{\infty} \frac{(-1)^{n-1}10^n}{n!}$

5. $\sum_{n=1}^{\infty} n\left(\frac{3}{4}\right)^n$

6. $\sum_{n=1}^{\infty} n^2\left(\frac{3}{4}\right)^n$

7. $\sum_{n=1}^{\infty} \frac{(-1)^n}{\sqrt{n}}$

8. $\sum_{n=1}^{\infty} \frac{(-1)^n}{n^p}, \quad 0 < p < 1$

9. $\sum_{n=1}^{\infty} \frac{(-1)^n}{n\sqrt{n}}$

10. $\sum_{n=1}^{\infty} \frac{(-1)^{n+1}(n-1)}{n^2+1}$

11. $\sum_{n=1}^{\infty} \frac{(-1)^n n^2}{2^n}$

12. $\sum_{n=1}^{\infty} \frac{(-1)^{n+1}(n-1)^2}{n^3}$

13. $\sum_{n=1}^{\infty} \frac{(-1)^{n-1}(4/3)^n}{n^2}$

14. $\sum_{n=1}^{\infty} \frac{(-1)^n(3/2)^2}{n^4}$

15. $\sum_{n=1}^{\infty} \frac{(-5)^{n-1}}{n \cdot n!}$

16. $\sum_{n=1}^{\infty} \frac{(-2)^{n-1}\cdot(n+1)}{(2n)!}$

17. $\sum_{n=1}^{\infty} \frac{(-1)^{n-1}n!}{1\cdot 3\cdot 5\cdots(2n-1)}$

18. $\sum_{n=1}^{\infty} \frac{(-1)^{n-1}(n!)^2\cdot 2^n}{(2n)!}$

19. $\sum_{n=1}^{\infty} \frac{(-1)^{n-1}(n+1)}{n\sqrt{n}}$

20. $\sum_{n=1}^{\infty} \frac{(n!)^2 5^n}{(2n)!}$

21. $\sum_{n=1}^{\infty} \frac{(-1)^n 2\cdot 4\cdot 6\cdots(2n)}{1\cdot 4\cdot 7\cdots(3n-2)}$

22. $\sum_{n=1}^{\infty} \frac{(-1)^{n+1}3^{n+1}}{2^{4n}}$

23. $\sum_{n=1}^{\infty} \frac{(-1)^{n-1}n}{n+1}$

24. $\sum_{n=1}^{\infty} \frac{(-1)^n(n-2)}{n^{7/4}}$

25. $\sum_{n=1}^{\infty} \frac{(-1)^n(6n^2-9n+4)}{n^3}$

26. $\sum_{n=1}^{\infty} \frac{(-1)^{n+1}}{(n+1)\log(n+1)}$

27. $\sum_{n=1}^{\infty} \frac{(-1)^{n+1}\log(n+1)}{n+1}$

28. $\sum_{n=1}^{\infty} \frac{(-1)^{n-1}\log n}{n^2}$

29. For what values of p does the series

$$\sum_{n=1}^{\infty} \frac{n!}{n^p}$$

converge?

30. For what values of p does the series

$$\sum_{n=1}^{\infty} \frac{n^p}{n!}$$

converge?

31. Given a series $\Sigma_{n=1}^{\infty} u_n$ in which, for each $n \geq 1$, u_{4n} and u_{4n+1} are positive while u_{4n+2} and u_{4n+3} are negative. After determining appropriate additional hypotheses on the $\{u_n\}$, state and prove a theorem analogous to the Alternating Series Theorem. Generalize your result if possible.

32. Given that $\Sigma_{n=1}^{\infty} u_n$ is a convergent series of positive terms, show that $\Sigma_{n=1}^{\infty} u_n^p$ is convergent for every $p > 1$.

33. Given that $\Sigma_{n=1}^{\infty} u_n$ is a divergent series of positive terms, show that $\Sigma_{n=1}^{\infty} u_n^p$ diverges for $0 < p < 1$.

*34. Consider the series $\Sigma_{n=1}^{\infty} (-1)^{n-1}/n$ and let A be any real number. Show that by rearranging the terms of the series, the sum will be a number in the interval $(A - 1, A + 1)$.

35. Suppose that ρ is a number less than 1 and that there is a positive integer N such that $|u_{n+1}/u_n| < \rho$ for all integers $n > N$. Prove that $\Sigma_{n=1}^{\infty} |u_n|$ converges even though $\lim_{n\to\infty} |u_{n+1}/u_n|$ may not exist.

36. Let a, b be two numbers larger than 1 with $a \neq b$. Show that if $u_n = (1/a^n)$ when n is odd and $u_n = (1/b^n)$ when n is even, then the series $\Sigma_{n=1}^{\infty} u_n$ converges, although $|u_{n+1}/u_n|$ does not tend to a limit as $n \to \infty$.

37. The **root test** states that a series $\Sigma_{n=1}^{\infty} a_n$ converges if the $\lim_{n\to\infty} \sqrt[n]{|a_n|}$ exists and has a value less than 1. The series diverges if the limit is larger than 1. Show that the series

$$\sum_{n=1}^{\infty} \frac{a^n}{2 + b^n}$$

converges if $0 < a < b$.

38. Let $\Sigma_{n=1}^{\infty} a_n$ be an absolutely convergent series. Show that

$$\sum_{n=1}^{\infty} \frac{\sqrt{|a_n|}}{n^p}$$

converges for $p > \frac{1}{2}$. What happens if $a_n = n^{-1}(\log(n + 1))^{-3/2}$ and $p = \frac{1}{2}$?

5. Power Series

A **power series** is a series of the form

$$c_0 + c_1(x - a) + c_2(x - a)^2 + \cdots + c_n(x - a)^n + \cdots,$$

in which a and the c_i, $i = 0, 1, 2, \ldots$, are constants. If a particular value is given to x, we then obtain an infinite series of numbers of the type we have been considering. The special case $a = 0$ occurs frequently, in which case the series becomes

$$c_0 + c_1x + c_2x^2 + c_3x^3 + \cdots + c_nx^n + \cdots.$$

Most often, we use the Σ-notation, writing

$$\sum_{n=0}^{\infty} c_n(x-a)^n \qquad \text{and} \qquad \sum_{n=0}^{\infty} c_nx^n.$$

If a power series converges for certain values of x, we may define a function of x by setting

$$f(x) = \sum_{n=0}^{\infty} c_n(x-a)^n \qquad \text{or} \qquad g(x) = \sum_{n=0}^{\infty} c_nx^n$$

for those values of x. We shall see that all the functions we have studied can be represented by convergent power series (with certain exceptions for the value a).

The Ratio Test may be used to determine when a power series converges. We begin with several examples.

EXAMPLE 1. Find the values of x for which the series

$$\sum_{n=1}^{\infty} \frac{1}{n}x^n$$

converges.

SOLUTION. We apply the Ratio Test, noting that

$$u_n = \frac{1}{n}x^n, \qquad u_{n+1} = \frac{1}{n+1}x^{n+1}.$$

Then

$$\left|\frac{u_{n+1}}{u_n}\right| = \frac{|x|^{n+1}}{n+1}\cdot\frac{n}{|x|^n} = |x|\frac{n}{n+1}.$$

It is important to observe that x remains *unaffected* as $n \to \infty$. Hence

$$\lim_{n\to\infty}\left|\frac{u_{n+1}}{u_n}\right| = \lim_{n\to\infty}|x|\frac{n}{n+1} = |x|\lim_{n\to\infty}\frac{1}{1+1/n} = |x|.$$

That is, $\rho = |x|$ in the Ratio Test.

We conclude: (a) the series converges if $|x| < 1$; (b) the series diverges if $|x| > 1$; (c) if $|x| = 1$, the Ratio Test gives no information. The last case corresponds to $x = \pm 1$, and we may try other methods for these two series, which are

$$\sum_{n=1}^{\infty}\frac{1}{n} \quad (\text{if } x = 1) \qquad \text{and} \qquad \sum_{n=1}^{\infty}\frac{(-1)^n}{n} \quad (\text{if } x = -1).$$

The first series above is the divergent harmonic series. The second series converges by the Alternating Series Theorem. Therefore the given series converges for $-1 \le x < 1$.

EXAMPLE 2. Find the values of x for which the series

$$\sum_{n=1}^{\infty} \frac{(-1)^n(x+1)^n}{2^n n^2}$$

converges.

SOLUTION. We apply the Ratio Test:

$$u_n = \frac{(-1)^n(x+1)^n}{2^n n^2}, \qquad u_{n+1} = \frac{(-1)^{n+1}(x+1)^{n+1}}{2^{n+1}(n+1)^2}.$$

Therefore

$$\left|\frac{u_{n+1}}{u_n}\right| = \frac{|x+1|^{n+1}}{2^{n+1}(n+1)^2} \cdot \frac{2^n n^2}{|x+1|^n} = \frac{1}{2}|x+1|\left(\frac{n}{n+1}\right)^2$$

and

$$\lim_{n\to\infty} \frac{|u_{n+1}|}{|u_n|} = \frac{1}{2}|x+1| \lim_{n\to\infty}\left(\frac{1}{1+1/n}\right)^2 = \frac{|x+1|}{2}.$$

According to the Ratio Test: (a) the series converges if $\frac{1}{2}|x+1| < 1$; (b) the series diverges if $\frac{1}{2}|x+1| > 1$; and (c) if $|x+1| = 2$, the test fails.

The inequality $|x+1| < 2$ may be written

$$-2 < x+1 < 2 \quad \Leftrightarrow \quad -3 < x < 1,$$

and the series converges in this interval, while it diverges for x outside this interval. The values $x = -3$ and $x = 1$ remain for consideration. The corresponding series are

$$\sum_{n=1}^{\infty} \frac{(-1)^n(-2)^n}{2^n n^2} = \sum_{n=1}^{\infty} \frac{1}{n^2} \qquad \text{and} \qquad \sum_{n=1}^{\infty} \frac{(-1)^n 2^n}{2^n n^2} = \sum_{n=1}^{\infty} \frac{(-1)^n}{n^2}.$$

Both series converge by the p series test. The original series converges for x in the interval $-3 \le x \le 1$.

EXAMPLE 3. Find the values of x for which the series

$$\sum_{n=0}^{\infty} \frac{(-1)^n x^n}{n!}$$

converges.

SOLUTION. We apply the Ratio Test:

$$u_n = \frac{(-1)^n x^n}{n!}, \qquad u_{n+1} = \frac{(-1)^{n+1} x^{n+1}}{(n+1)!},$$

and

$$\left|\frac{u_{n+1}}{u_n}\right| = \frac{|x|^{n+1}}{(n+1)!} \cdot \frac{n!}{|x|^n} = |x|\frac{1}{n+1}.$$

Hence $\rho = 0$, regardless of the value of $|x|$. The series converges for all values of x; that is, for $-\infty < x < \infty$.

EXAMPLE 4. Find the values of x for which the series

$$\sum_{n=0}^{\infty} \frac{(-1)^n n! x^n}{10^n}$$

converges.

SOLUTION. We apply the Ratio Test:

$$u_n = \frac{(-1)^n n! x^n}{10^n}, \qquad u_{n+1} = \frac{(-1)^{n+1}(n+1)! x^{n+1}}{10^{n+1}}$$

and

$$\left|\frac{u_{n+1}}{u_n}\right| = \frac{|x|^{n+1}(n+1)!}{10^{n+1}} \cdot \frac{10^n}{|x|^n n!} = |x| \cdot \frac{n+1}{10}.$$

Therefore, if $x \neq 0$, $|u_{n+1}/u_n| \to \infty$ and the series diverges. The series converges only for $x = 0$.

The convergence properties of the most general power series are illustrated in the examples above. However, the proof of the theorem which states this fact (given below in Theorem 16) will be proved in section 13. In all the examples above we see that it always happened that $|u_{n+1}/u_n|$ tended to a limit or to $+\infty$. The examples are deceptive, since there are cases in which $|u_{n+1}/u_n|$ may neither tend to a limit nor tend to $+\infty$.

Lemma. *If the series $\Sigma_{n=0}^{\infty} u_n$ converges, there is a number M such that $|u_n| \leq M$ for every n.*

PROOF. By Theorem 6 we know that $\lim_{n\to\infty} u_n = 0$. From the definition of a limit, there must be a number N such that

$$|u_n| < 1 \qquad \text{for all} \qquad n > N$$

(by taking $\varepsilon = 1$ in the definition of limit). We define M to be the largest of the numbers

$$|u_0|, \quad |u_1|, \quad |u_2|, \quad \ldots, \quad |u_N|, \quad 1,$$

and the result is established.

Theorem 15. *If the series* $\Sigma_{n=0}^{\infty} a_n x^n$ *converges for some* $x_1 \neq 0$, *then the series converges absolutely for all* x *for which* $|x| < |x_1|$, *and there is a number* M *such that*

$$|a_n x^n| \leq M \left|\frac{x}{x_1}\right|^n \qquad \textit{for all } n.$$

PROOF. Since the series $\Sigma_{n=0}^{\infty} a_n x_1^n$ converges, we know from the Lemma above that there is a number M such that

$$|a_n x_1^n| \leq M \qquad \text{for all } n.$$

Then

$$|a_n x^n| = \left|a_n x_1^n \frac{x^n}{x_1^n}\right| = |a_n x_1^n| \cdot \left|\frac{x}{x_1}\right|^n \leq M \left|\frac{x}{x_1}\right|^n.$$

The series

$$\sum_{n=0}^{\infty} M \left|\frac{x}{x_1}\right|^n$$

is a geometric series with ratio less than 1, and so convergent. Hence, by the Comparison Test, the series

$$\sum_{n=0}^{\infty} a_n x^n$$

converges absolutely.

REMARK. Theorem 15 may be established for series of the form

$$\sum_{n=0}^{\infty} a_n (x - a)^n$$

in a completely analogous manner.

Theorem 16. *Let* $\Sigma_{n=0}^{\infty} a_n (x - a)^n$ *be any given power series. Then either*

i) *the series converges only for* $x = a$;
ii) *the series converges for all values of* x; *or*
iii) *there is a number* $R > 0$ *such that the series converges for all* x *for which* $|x - a| < R$ *and diverges for all* x *for which* $|x - a| > R$.

The proof is given in section 13. The consequence (iii) in Theorem 16 states that there is an *interval of convergence* $-R < x - a < R$ or $a - R < x < a + R$. Nothing is stated about what happens when $x = a - R$ or $a + R$. These *endpoint* problems must be settled on a case-by-case basis. The alternatives (i) and (ii) correspond to $R = 0$ and $R = +\infty$, respectively.

PROBLEMS

In Problems 1 through 27, find the values of x for which the following power series converge. Include a discussion of the endpoints.

1. $\sum_{n=0}^{\infty} x^n$

2. $\sum_{n=0}^{\infty} (-1)^n x^n$

3. $\sum_{n=0}^{\infty} (2x)^n$

4. $\sum_{n=0}^{\infty} (\frac{1}{4}x)^n$

5. $\sum_{n=0}^{\infty} (-1)^n (n+1) x^n$

6. $\sum_{n=1}^{\infty} \frac{(x-1)^n}{3^n n^2}$

7. $\sum_{n=1}^{\infty} \frac{(x-1)^n}{2^n n^3}$

8. $\sum_{n=1}^{\infty} \frac{(-1)^{n+1}(x-2)^n}{n\sqrt{n}}$

9. $\sum_{n=1}^{\infty} \frac{(x+2)^n}{\sqrt{n}}$

10. $\sum_{n=0}^{\infty} \frac{(10x)^n}{n!}$

11. $\sum_{n=0}^{\infty} \frac{x^n}{(2n)!}$

12. $\sum_{n=0}^{\infty} \frac{n!(x+1)^n}{5^n}$

13. $\sum_{n=0}^{\infty} \frac{(-1)^n (3/2)^n x^n}{n+1}$

14. $\sum_{n=0}^{\infty} \frac{(2n)!x^n}{n!}$

15. $\sum_{n=1}^{\infty} n^2 (x-1)^n$

16. $\sum_{n=1}^{\infty} \frac{n(x+2)^n}{2^n}$

17. $\sum_{n=1}^{\infty} \frac{(-1)^{n-1}(x+4)^n}{3^n \cdot n^2}$

18. $\sum_{n=1}^{\infty} \frac{n!(x-3)^n}{1 \cdot 3 \cdot 5 \cdots (2n-1)}$

19. $\sum_{n=1}^{\infty} \frac{(-1)^{n+1}(n!)^2(x-2)^n}{2^n (2n)!}$

20. $\sum_{n=1}^{\infty} \frac{n!(x-1)^n}{4^n \cdot 1 \cdot 3 \cdot 5 \cdots (2n-1)}$

21. $\sum_{n=1}^{\infty} \frac{(-1)^{n-1} n! (3/2)^n x^n}{1 \cdot 3 \cdot 5 \cdots (2n-1)}$

22. $\sum_{n=0}^{\infty} \frac{(-1)^n 3^{n+1} x^n}{2^{3n}}$

23. $\sum_{n=1}^{\infty} \frac{(n-2)x^n}{n^2}$

24. $\sum_{n=0}^{\infty} \frac{(6n^2+3n+1)x^n}{2^n (n+1)^3}$

25. $\sum_{n=1}^{\infty} \frac{(-1)^{n-1} x^n}{(n+1)\log(n+1)}$

26. $\sum_{n=1}^{\infty} \frac{\log(n+1) 3^n (x-1)^n}{n+1}$

27. $\sum_{n=1}^{\infty} \frac{(-1)^{n-1}(\log n) 2^n x^n}{3^n n^2}$

28. Prove Theorem 15 for series of the form

$$\sum_{n=0}^{\infty} a_n (x-a)^n.$$

29. a) Find the interval of convergence of the series

$$\sum_{n=1}^{\infty} \frac{1 \cdot 3 \cdot 5 \cdots (2n-1)}{2 \cdot 4 \cdot 6 \cdots (2n)} x^n.$$

b) Show that the series in (a) is identical with the series

$$\sum_{n=1}^{\infty} \frac{(2n)!}{2^{2n}(n!)^2} x^n.$$

30. Find the interval of convergence of the series

$$\sum_{n=1}^{\infty} \frac{1 \cdot 3 \cdot 5 \cdots (2n-1)(x-2)^n}{2^n \cdot 1 \cdot 4 \cdot 7 \cdots (3n-2)}.$$

31. Find the interval of convergence of the **binomial series**

$$1 + \sum_{n=1}^{\infty} \frac{m(m-1)\cdots(m-n+1)}{n!} x^n; \qquad m \text{ fixed}.$$

*32. Suppose that the series $\Sigma_{n=0}^{\infty} a_n x^n$ has an interval of convergence $(-R, R)$. (a) Show that the series $\Sigma_{n=0}^{\infty} a_n x^{n+1}/(n+1)$ has the same interval of convergence.
b) Show that the series $\Sigma_{n=1}^{\infty} n a_n x^{n-1}$ has the same interval of convergence.

33. Given the series $\Sigma_{n=0}^{\infty} a_n x^n$. Suppose that $\lim_{n\to\infty} \sqrt{|a_n|} = r$. Show that the series converges for

$$-\frac{1}{r} < x < \frac{1}{r}.$$

34. Given the two series $\Sigma_{n=0}^{\infty} a_n x^n$ and $\Sigma_{n=0}^{\infty} b_n x^n$. Suppose there is an N such that $|a_n| \leq |b_n|$ for all $n \geq N$. Show that the interval of convergence of the first series is at least as large as the interval of convergence of the second series.

35. Suppose the series $\Sigma_{n=0}^{\infty} a_n x^n$ converges for $-R < x < R$. Show that the series $\Sigma_{n=0}^{\infty} a_n x^{kn}$ for some fixed positive integer k converges for $-\sqrt[k]{R} < x < \sqrt[k]{R}$.

6. Taylor's Series

Suppose that a power series

$$\sum_{n=0}^{\infty} a_n (x-a)^n$$

converges in some interval $-R < x - a < R$ $(R > 0)$. Then the sum of the series has a value for each x in this interval and so defines a function of x. We can therefore write

$$\begin{aligned} f(x) &= a_0 + a_1(x-a) + a_2(x-a)^2 + a_3(x-a)^3 + \cdots, \\ &a - R < x < a + R. \end{aligned} \tag{1}$$

We ask the question: What is the relationship between the coefficients a_1, $a_2, a_3, \ldots, a_n, \ldots$ and the function f?

We shall proceed naïvely, as if the right side of (1) were a polynomial. Setting $x = a$, we find at once that

$$f(a) = a_0.$$

We differentiate (1) (as if the right side were a polynomial) and get

$$f'(x) = a_1 + 2a_2(x-a) + 3a_3(x-a)^2 + 4a_4(x-a)^3 + \cdots.$$

For $x = a$, we find that

$$f'(a) = a_1.$$

We continue both differentiating and setting $x = a$, to obtain

$$f''(x) = 2a_2 + 3\cdot 2a_3(x-a) + 4\cdot 3a_4(x-a)^2 + 5\cdot 4a_5(x-a)^3 + \cdots,$$

$$f''(a) = 2a_2 \qquad \text{or} \qquad a_2 = \frac{f''(a)}{2!},$$

$$f'''(x) = 3\cdot 2a_3 + 4\cdot 3\cdot 2a_4(x-a) + 5\cdot 4\cdot 3a_5(x-a)^2 + 6\cdot 5\cdot 4a_6(x-a)^3 + \cdots,$$

$$f'''(a) = 3\cdot 2a_3 \qquad \text{or} \qquad a_3 = \frac{f'''(a)}{3!},$$

and so forth. The pattern is now clear. The general formula for the coefficients $a_0, a_1, \ldots, a_n, \ldots$ is

$$a_n = \frac{f^{(n)}(a)}{n!}.$$

In Section 8, we will show that all of the above steps are legitimate so long as the series is convergent in some interval. Substituting the formulas for the coefficients a_n into the power series, we obtain

$$f(x) = \sum_{n=0}^{\infty} \frac{f^{(n)}(a)}{n!}(x-a)^n. \tag{2}$$

Definition. The right side of Eq. (2) is called the **Taylor series for f about the point** a or the **expansion of f into a power series about** a.

For the special case $a = 0$, the Taylor series is

$$f(x) = \sum_{n=0}^{\infty} \frac{f^{(n)}(0)}{n!} x^n. \tag{3}$$

The right side of (3) is called the **Maclaurin series** *for f*.

EXAMPLE 1. Assuming that $f(x) = \sin x$ is given by its Maclaurin series, expand $\sin x$ into such a series.

SOLUTION. We have

$$f(x) = \sin x, \qquad f(0) = 0,$$

$$f'(x) = \cos x, \qquad f'(0) = 1,$$

$$f''(x) = -\sin x, \qquad f''(0) = 0,$$
$$f^{(3)}(x) = -\cos x, \qquad f^{(3)}(0) = -1,$$
$$f^{(4)}(x) = \sin x, \qquad f^{(4)}(0) = 0.$$

It is clear that $f^{(5)} = f'$, $f^6 = f''$, etc., so that the sequence 0, 1, 0, -1, 0, 1, 0, -1, ... repeats itself indefinitely. Therefore, from (3) we obtain

$$\sin x = x - \frac{x^3}{3!} + \frac{x^5}{5!} - \frac{x^7}{7!} + \frac{x^9}{9!} \cdots$$
$$\sin x = \sum_{k=0}^{\infty} \frac{(-1)^k x^{2k+1}}{(2k+1)!}. \tag{4}$$

Remark. It may be verified (by the Ratio Test, for example) that the series (4) converges for all values of x.

Example 2. Expand the function

$$f(x) = \frac{1}{x}$$

into a Taylor series about $x = 1$, assuming that such an expansion is valid.

Solution. We have

$$f(x) = x^{-1}, \qquad f(1) = 1,$$
$$f'(x) = (-1)x^{-2}, \qquad f'(1) = -1,$$
$$f''(x) = (-1)(-2)x^{-3}, \qquad f''(1) = (-1)^2 \cdot 2!,$$
$$f^{(3)}(x) = (-1)(-2)(-3)x^{-4}, \qquad f^{(3)}(1) = (-1)^3 \cdot 3!,$$
$$f^{(n)}(x) = (-1)(-2)\cdots(-n)x^{-n-1}, \qquad f^{(n)}(1) = (-1)^n \cdot n!$$

Therefore from (2) with $a = 1$, we obtain

$$f(x) = \frac{1}{x} = \sum_{n=0}^{\infty} (-1)^n (x-1)^n. \tag{5}$$

Remark. The series (5) converges for $|x - 1| < 1$ or $0 < x < 2$, as may be confirmed by the Ratio Test.

Examples 1 and 2 have meaning only if it is known that the functions are representable by means of power series. There are examples of functions for which it is possible to compute all the quantities $f^{(n)}(x)$ at a given value a, and yet the Taylor series about a will not represent the function. (See Problem 33 at the end of this Section.)

EXAMPLE 3. Compute the first six terms of the Maclaurin expansion of the function.

$$f(x) = \tan x,$$

assuming that such an expansion is valid.

SOLUTION. We have

$$\begin{aligned}
f(x) &= \tan x, & f(0) &= 0,\\
f'(x) &= \sec^2 x, & f'(0) &= 1,\\
f''(x) &= 2\sec^2 x \tan x, & f''(0) &= 0,\\
f^{(3)}(x) &= 2\sec^4 x + 4\sec^2 x \tan^2 x, & f^{(3)}(0) &= 2,\\
f^{(4)}(x) &= 8\tan x \sec^2 x(2 + 3\tan^2 x), & f^{(4)}(0) &= 0,\\
f^{(5)}(x) &= 48\tan^2 x \sec^4 x + 8\sec^2 x(2 + 3\tan^2 x)(\sec^2 x + 2\tan^2 x), & &\\
& & f^{(5)}(0) &= 16.
\end{aligned}$$

Therefore

$$f(x) = \tan x = x + \frac{x^3}{3} + \frac{2x^5}{15} + \cdots.$$

REMARK. Example 3 shows that the general pattern for the successive derivatives may not always be readily discernible. Examples 1 and 2, on the other hand, show how the general formula for the nth derivative may be arrived at simply.

PROBLEMS

In Problems 1 through 16, find the Taylor (Maclaurin if $a = 0$) series for each function f about the given value of a.

1. $f(x) = e^x, \quad a = 0$
2. $f(x) = \cos x, \quad a = 0$
3. $f(x) = \log(1 + x), \quad a = 0$
4. $f(x) = \log(1 + x), \quad a = 1$
5. $f(x) = (1 - x)^{-2}, \quad a = 0$
6. $f(x) = (1 - x)^{-1/2}, \quad a = 0$
7. $f(x) = (1 + x)^{1/2}, \quad a = 0$
8. $f(x) = e^x, \quad a = 1$
9. $f(x) = \log x, \quad a = 3$
10. $f(x) = \sin x, \quad a = \pi/4$
11. $f(x) = \cos x, \quad a = \pi/3$
12. $f(x) = \sin x, \quad a = 2\pi/3$
13. $f(x) = \sqrt{x}, \quad a = 4$
14. $f(x) = \sin(x + \frac{1}{2}), \quad a = 0$
15. $f(x) = \cos(x + \frac{1}{2}), \quad a = 0$
16. $f(x) = x^m, \quad a = 1$

In each of Problems 17 through 31, find the first few terms of the Taylor expansion about the given value of a. Carry out the process to include the term $(x-a)^n$ for the given integer n.

17. $f(x) = e^{-x^2}$, $a = 0$, $n = 4$
18. $f(x) = xe^x$, $a = 0$, $n = 4$
19. $f(x) = \dfrac{1}{1+x^2}$, $a = 0$, $n = 4$
20. $f(x) = \arctan x$, $a = 0$, $n = 5$
21. $f(x) = e^x \cos x$, $a = 0$, $n = 4$
22. $f(x) = \dfrac{1}{\sqrt{1-x^2}}$, $a = 0$, $n = 4$
23. $f(x) = \arcsin x$, $a = 0$, $n = 5$
24. $f(x) = \tan x$, $a = \pi/4$, $n = 5$
25. $f(x) = \log \sec x$, $a = 0$, $n = 6$
26. $f(x) = \sec x$, $a = 0$, $n = 4$
27. $f(x) = \csc x$, $a = \pi/6$, $n = 4$
28. $f(x) = \csc x$, $a = \pi/2$, $n = 4$
29. $f(x) = \sec x$, $a = \pi/3$, $n = 3$
30. $f(x) = \log \sin x$, $a = \pi/4$, $n = 4$
31. $f(x) = \tan x$, $a = 0$, $n = 7$ *Hint:* See Example 3; express $f^{(5)}(x)$ in terms of $\sec x$.]
32. a) Given the polynomial

$$f(x) = 3 + 2x - x^2 + 4x^3 - 2x^4, \tag{6}$$

show that f may be written in the form

$$f(x) = a_0 + a_1(x-1) + a_2(x-1)^2 + a_3(x-1)^3 + a_4(x-1)^4.$$

[*Hint:* Use the Taylor expansion (2) and (6) to get each a_i.]

*b) Given the same polynomial in two forms,

$$f(x) = \sum_{k=0}^{n} a_k(x-a)^k, \qquad f(x) = \sum_{k=0}^{n} b_k(x-b)^k,$$

express each b_i in terms of a, b, and the a_i.

*33. a) Given the function (see Fig. 3-6)

$$F(x) = \begin{cases} e^{-1/x^2}, & x \neq 0, \\ 0, & x = 0, \end{cases}$$

use l'Hôpital's Rule to show that

$$F'(0) = 0.$$

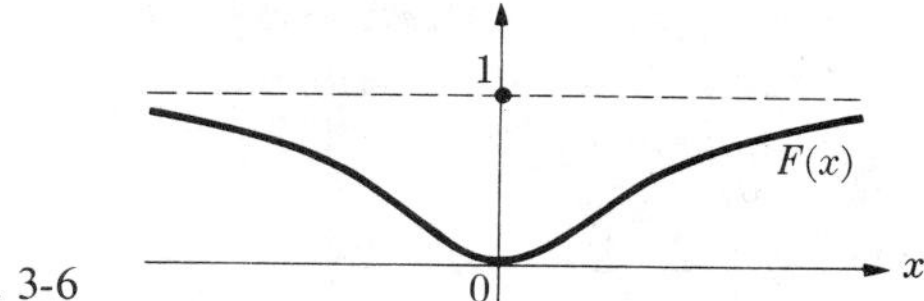

Fig. 3-6

b) Show that $F^{(n)}(0) = 0$ for every positive integer n.

c) What can be said about the Taylor series for F?

34. For any real number n and any positive number a show that the Maclaurin series for $(a + x)^n$ is given by

$$(a + x)^n = a^n + na^{n-1}x + \frac{n(n-1)}{2!}a^{n-2}x^2 + \cdots$$
$$+ \frac{n(n-1)\cdots(n-k+1)}{k!}a^{n-k}x^k + \cdots.$$

Set $a = 1$ and $n = \frac{1}{2}$ and find the Maclaurin series for $(1 + x)^{1/2}$. Find the interval of convergence. Also, find the Maclaurin expansion and the interval of convergence for $(1 + x)^{-1}$.

35. Find the Taylor expansion of $(3 + x)^{-1/2}$ about $a = 1$. [*Hint:* Write $3 + x = 4 + (x - 1)$ and use the result in Problem 34 with $a = 4$ and x replaced by $x - 1$.

7. Taylor's Theorem with Remainder

If a function f possesses only a finite number—say n—of derivatives, then it is clear that it is not possible to represent it by a Taylor series, since the coefficients $a_k = f^{(k)}(a)/k!$ cannot be computed beyond a_n. In such cases it is still possible to obtain a *finite version* of a Taylor expansion.

Suppose that $f(x)$ possesses n continuous derivatives in some interval about the value a. Then it is always possible to write

$$f(x) = f(a) + \frac{f'(a)}{1!}(x - a) + \frac{f^{(2)}(a)}{2!}(x - a)^2 + \cdots$$
$$+ \frac{f^{(n)}(a)}{n!}(x - a)^n + R_n \tag{1}$$

for x in this interval. The right side consists of a polynomial in x of degree n and a *remainder* R_n about which, as yet, we have no knowledge. In fact, the quantity R_n is defined by the formula (1). For example, in the unusual case that f and its n derivatives all vanish at $x = a$, the "remainder" R_n is just f itself. The content of **Taylor's Theorem** concerns useful information about the nature of R_n. This theorem is not only of great theoretical value but may also be used in approximations and numerical computations.

Theorem 17 (Taylor's Theorem with Derivative Form of Remainder). *Suppose that $f, f', f^{(2)}, \ldots, f^{(n)}, f^{(n+1)}$ are all continuous on some interval containing a and b. Then there is a number ξ between a and b such that*

$$f(b) = f(a) + \frac{f'(a)}{1!}(b-a) + \frac{f^{(2)}(a)}{2!}(b-a)^2 + \cdots$$

$$+ \frac{f^{(n)}(a)}{n!}(b-a)^n + \frac{f^{(n+1)}(\xi)(b-a)^{n+1}}{(n+1)!}.$$

That is, the remainder R_n is given by the formula

$$R_n = \frac{f^{(n+1)}(\xi)(b-a)^{n+1}}{(n+1)!}. \tag{2}$$

REMARKS. (i) We see that R_n depends on both b and a, and we write, in general, $R_n = R_n(a, b)$, a function of two variables.
ii) If we take the special case $n = 0$, we obtain

$$f(b) = f(a) + f'(\xi)(b-a),$$

which we recognize as the Theorem of the Mean. Thus this form of Taylor's Theorem is a direct generalization of the Theorem of the Mean.

PROOF OF THEOREM 17. The proof makes use of Rolle's Theorem. We create a function $\phi(x)$ which is zero at a and b and so, by Rolle's Theorem, there must be a number ξ between a and b where $\phi'(\xi) = 0$. The algebra is lengthy, and the reader should write out the details for the cases $n = 1, 2, 3$ in order to grasp the essence of the proof. We use the form (1) for $x = b$ and write

$$f(b) = f(a) + \frac{f'(a)(b-a)}{1!} + \frac{f^{(2)}(a)(b-a)^2}{2!} + \cdots$$

$$+ \frac{f^{(n)}(a)(b-a)^n}{n!} + R_n(a, b);$$

we wish to find $R_n(a, b)$. We define the function

$$\phi(x) = f(b) - f(x) - \frac{f(x)(b-x)}{1!} - \frac{f^{(2)}(x)(b-x)^2}{2!}$$

$$- \frac{f^{(3)}(x)(b-x)^3}{3!} - \cdots - \frac{f^{(n-1)}(x)(b-x)^{n-1}}{(n-1)!}$$

$$- \frac{f^{(n)}(x)(b-x)^n}{n!} - R_n(a, b)\frac{(b-x)^{n+1}}{(b-a)^{n+1}}.$$

The function ϕ was concocted in such a way that $\phi(a) = 0$ and $\phi(b) = 0$, facts which are easily checked by straight substitution. We compute the derivative $\phi'(x)$ (using the formula for the derivative of a product wherever necessary):

$$\phi'(x) = -f'(x) + f'(x) - \frac{f^{(2)}(x)(b-x)}{1!} + \frac{2f^{(2)}(x)(b-x)}{2!}$$

$$- \frac{f^{(3)}(x)(b-x)^2}{2!} + \frac{3f^{(3)}(x)(b-x)^2}{3!} - \frac{f^{(4)}(x)(b-x)^3}{3!} + \cdots$$

$$- \frac{f^{(n+1)}(x)(b-x)^n}{n!} + \frac{R_n(a,b)(n+1)(b-x)^n}{(b-a)^{n+1}}.$$

Amazingly, all the terms cancel except the last two, and we find

$$\phi'(x) = -\frac{f^{(n+1)}(x)(b-x)^n}{n!} + R_n(a,b)(n+1)\frac{(b-x)^n}{(b-a)^{n+1}}.$$

Using Rolle's Theorem, we know there must be a value ξ between a and b such that $\phi'(\xi) = 0$. Therefore we get

$$0 = -\frac{f^{(n+1)}(\xi)(b-\xi)^n}{n!} + R_n(a,b)(n+1)\frac{(b-\xi)^n}{(b-a)^{n+1}}$$

or, upon solving for $R_n(a,b)$, the formula (2) exactly.

Remarks. (i) If we know that $f(x)$ has continuous derivatives of all orders and if $R_n(a,b) \to 0$ as $n \to \infty$, then we can establish the validity of the Taylor series. ii) In any case, R_n is a measure of how much f differs from a certain polynomial of degree n. If R_n is small, then the polynomial may be used for an approximation to f.

When we use Taylor's Theorem in the computation of functions from the approximating polynomial, errors may arise from two sources: the error R_n, made above by neglecting the powers of $(b - a)$ beyond the nth; and the "round-off error" made by expressing each term in decimal form. If we wish to compute the value of some function $f(b)$ to an accuracy of four decimal places, it is essential to be able to say for certain that $f(b)$ is between some decimal fraction with four decimals -0.00005 and the same decimal fraction $+0.00005$. Time is saved by computing each term to two decimals more than are required. Frequently R_n is close to the value of the first term in the series omitted, and this fact can be used as a guide in choosing the number of terms. Although we do not know R_n exactly, we can often show that there are two numbers m and M with

$$m \le f^{(n+1)}(x) \le M \qquad \text{for } all\ x \text{ between } a \text{ and } b.$$

Then we get for $R_n(a,b)$ the inequality

$$\frac{m(b-a)^{n+1}}{(n+1)!} \le R_n(a,b) \le \frac{M(b-a)^{n+1}}{(n+1)!}.$$

Example 1. Compute $(1.1)^{1/5}$ to an accuracy of four decimal places.

SOLUTION. The key to the solution, using Taylor's Theorem, is the fact that we can set

$$f(x) = (1 + x)^{1/5}, \qquad a = 0, \qquad b = 0.1.$$

Then

$$\begin{aligned}
f(x) &= (1 + x)^{1/5}, & f(a) &= 1 &&= 1.0000\ 00\\
f'(x) &= \frac{1}{5}(1 + x)^{-4/5} & f'(a)(b - a) &= \frac{1}{5}(0.1)^1 &&= 0.0200\ 00\\
f''(x) &= -\frac{4}{25}(1 + x)^{-9/5}, & \frac{f''(a)(b - a)^2}{2!} &= -\frac{2}{25}(0.1)^2 &&= -0.0008\ 00\\
f'''(x) &= \frac{36}{125}(1 + x)^{-14/5}, & \frac{f'''(a)(b - a)^3}{3!} &= \frac{6}{125}(0.1)^3 &&= 0.0000\ 48
\end{aligned}$$

For all x between 0 and 1 we have

$$0 < (1 + x)^{-14/5} < 1.$$

Therefore we can estimate R_n for $n = 2$:

$$0 < R_2 = \frac{36}{125}(1 + \xi)^{-14/5}\frac{(b - a)^3}{3!} < \frac{6}{125}(0.1)^3 = 0.0000\ 48.$$

Adding the terms in the Taylor expansion through $n = 2$, we get

$$(1.1)^{1/5} = 1.0192, \text{ approximately;}$$

in fact, a more precise statement is

$$1.0192 < (1.1)^{1/5} < 1.0192\ 48.$$

REMARK. In Example 1, we could have selected $f(x) = x^{1/5}$ with $a = 1$, $b = 1.1$. The result is the same.

EXAMPLE 2. Compute $\sqrt[3]{7}$ to an accuracy of four decimal places.

SOLUTION. Set

$$f(x) = x^{1/3}, \qquad a = 8, \qquad b = 7.$$

Then $b - a = -1$ and

$$\begin{aligned}
f(x) &= x^{1/3}, & f(a) &= 2 &&= 2.0000\ 00\\
f'(x) &= \frac{1}{3}x^{-2/3}, & f'(a)(b - a) &= -\frac{1}{12} &&= -0.0833\ 33\\
f''(x) &= -\frac{2}{9}x^{-5/3}, & \frac{f''(a)(b - a)^2}{2!} &= -\frac{1}{9 \cdot 2^5} &&= -0.0034\ 72
\end{aligned}$$

$$f^{(3)}(x) = \frac{10}{27}x^{-8/3}, \qquad \frac{f'''(a)(b-a)^3}{3!} = -\frac{5}{81 \cdot 2^8} = -0.0002\ 41$$

$$f^{(4)}(x) = \frac{-80}{81}x^{-11/3}, \qquad \frac{f^{(4)}(a)(b-a)^4}{4!} = -\frac{5}{243 \cdot 2^{10}} = -0.0000\ 20$$

It would appear to be sufficient to use only the terms through $(b-a)^3$. However, by computing the sum of the decimal fractions given, we obtain 1.9129 54. But the next term is $-0.0000\ 20$ which, if included, would reduce the value to 1.9129 34. If we stopped with the $(b-a)^3$ term and rounded off, we would obtain 1.9130 whereas, if we keep the next term and round off, we obtain 1.9129. So the term in $(b-a)^4$ should be retained, and the remainder R_4 must be estimated. We have

$$f^{(5)}(x) = \frac{880}{243}x^{-14/3} = \frac{880x^{1/3}}{243x^5}.$$

Since we are concerned with the interval $7 < x < 8$, we see that $x^{1/3} < 2$ and $x^5 > (49)(343)$. Hence

$$0 < \frac{f^{(5)}(x)}{5!} < \frac{1760}{(243)(49)(343)(120)} < \frac{1}{270{,}000} < 0.000004.$$

Since $(b-a)^5 = -1$, we conclude that

$$-0.000004 < R_4 < 0$$

and that

$$\sqrt[3]{7} = 1.9129$$

to the required accuracy. Actually, if we merely keep an extra decimal in each term retained, we see that

$$\sqrt[3]{7} = 1.91293$$

to five decimals.

REMARKS. In Example 1 there was no round-off error, since each decimal fraction gave the exact value of the corresponding term. This was not true in Example 2, however. In general, the round-off error in each term may be as much as $\frac{1}{2}$ in the last decimal place retained. Round-off errors may tend to cancel each other if there is a large number of computations in a given problem.

The remainder $R_n(a, b)$ in Taylor's Theorem may be given in many forms. The next theorem, stated without proof, gives the remainder in the form of an integral.

Theorem 18 (Taylor's Theorem with Integral Form of Remainder). *Suppose that $f, f', f^{(2)}, \ldots, f^{(n)}, f^{(n+1)}$ are all continuous in some interval containing a and b. Then $f(x)$ may be written in the form*

$$f(b) = f(a) + \frac{f'(a)}{1!}(b-a) + \frac{f^{(2)}(a)(b-a)^2}{2!} + \cdots$$
$$+ \frac{f^{(n)}(a)(b-a)^n}{n!} + R_n$$

where

$$R_n = \frac{1}{n!}\int_a^b f^{(n+1)}(t)(b-t)^n\,dt.$$

EXAMPLE 3. Write $\log(1+x)$ as a polynomial of the third degree, and estimate the remainder R_n for $0 < x < \frac{1}{2}$.

SOLUTION. We select

$$f(x) = \log(1+x), \qquad a = 0, \qquad b = x.$$

Then we write successive derivatives,

$$f'(x) = \frac{1}{1+x}, \qquad f'(0) = 1,$$

$$f''(x) = -\frac{1}{(1+x)^2}, \qquad f''(0) = -1,$$

$$f^{(3)}(x) = \frac{2}{(1+x)^3}, \qquad f^{(3)}(0) = 2,$$

$$f^{(4)}(x) = -\frac{6}{(1+x)^4}.$$

Therefore

$$\log(1+x) = x - \frac{x^2}{2} + \frac{x^3}{3} + R_n,$$

with

$$R_n = -\frac{6}{6}\int_0^x \frac{1}{(1+t)^4}(x-t)^3\,dt.$$

A simple estimate replaces $(1+t)^{-4}$ by its smallest value 1, and we find

$$|R_n| < \int_0^{1/2}\left(\frac{1}{2} - t\right)^3 dt = \frac{1}{64}.$$

PROBLEMS

In each of Problems 1 through 20, compute the given quantities to the specified number of decimal places. Make sure of your accuracy by using Taylor's Theorem with Remainder. Use the fact that $2 < e < 4$ wherever necessary.

1. $e^{-0.2}$,	5 decimals	2. $e^{-0.4}$,	4 decimals
3. $e^{0.2}$,	5 decimals	4. $\sin(0.5)$,	5 decimals
5. $\cos(0.5)$,	5 decimals	6. $\tan(0.1)$,	3 decimals
7. $\log(1.2)$,	4 decimals	8. $\log(0.9)$,	5 decimals
9. e^{-1},	5 decimals	10. e,	5 decimals
11. $(1.08)^{1/4}$,	5 decimals	12. $(0.92)^{1/4}$,	5 decimals
13. $(0.91)^{1/3}$,	5 decimals	14. $(0.90)^{1/5}$,	5 decimals
15. $(30)^{1/5}$,	5 decimals	16. $(15)^{1/4}$,	5 decimals
17. $(0.8)^{1/5}$,	5 decimals	18. $(65)^{1/6}$,	5 decimals
19. $\log(0.8)$,	5 decimals	20. $\log(0.6)$,	3 decimals

Given that $1° = \pi/180$ radians $= 0.0174533$ radians and $5° = \pi/36$ radians $= 0.0872655$ radians, in each of Problems 21 through 23 compute to the number of decimal places required.

21. $\sin 1°$, 6 decimals 22. $\sin 5°$, 5 decimals 23. $\cos 5°$, 5 decimals

24. Find the largest interval about $x = 0$ in which the function $f(x) = \sin x$ may be approximated to four decimal places by $x - \frac{1}{6}x^3$, the first two terms in the Maclaurin expansion.

25. Given the function $f: x \to \cos^2 x$, find an upper and lower bound for $R_n(a, b)$ where $a = 0$, $b = \pi/4$, $n = 4$.

26. Verify the identity in x:

$$\frac{1}{1+x} = 1 - x + x^2 - \cdots + (-1)^{n-1}x^{n-1} + \frac{(-1)^n x^n}{1+x}.$$

Integrate between 0 and b to obtain the formula

$$\log(1 + b) = -\sum_{k=1}^{n} (-1)^k \frac{b^k}{k} + (-1)^n \int_0^b \frac{x^n}{1+x} dx.$$

Show that this is Taylor's formula with integral remainder. Obtain the estimate for R_n:

$$|R_n| \leq \frac{|b|^{n+1}}{1 - |b|} \qquad \text{for } |b| < 1.$$

27. Apply the method of Problem 26 to the function $f(x) = 1/(1 + x^2)$ and obtain Taylor's formula for $\arctan x$.

28. Let $\Sigma_{n=0}^{\infty} u_n$ be a convergent infinite series whose terms are alternately positive and negative and also such that $|u_{n+1}| < |u_n|$ for each n. It can be shown that the error in approximating the above series by $\Sigma_{n=0}^{N} u_n$ is always less than $|u_{N+1}|$. Apply this result to approximate $\cos(\pi/6)$ by its Taylor series with remainder and thereby find $\sqrt{3}$ to three decimal places.

8. Differentiation and Integration of Series

In Section 6 we developed the formula for the Taylor series of a function and, in so doing, we ignored the validity of the manipulations which were performed. Now we shall establish the theorems which verify the correctness of the results already obtained.

Theorem 19 (Validity of Term-by-Term Differentiation of Power Series). *If $R > 0$ and the series*

$$\sum_{n=0}^{\infty} a_n x^n \tag{1}$$

converges for $|x| < R$, then the series obtained from (1) *by term-by-term differentiation converges obsolutely for $|x| < R$.*

PROOF. Term-by-term differentiation of (1) yields

$$\sum_{n=1}^{\infty} n a_n x^{n-1}. \tag{2}$$

Choose any value x such that $|x| < R$ and choose x_1 so that $|x| < |x_1| < R$. According to the Lemma of Section 5 (page 123), there is a positive number M with the property that

$$|a_n x_1^n| \le M \qquad \text{for all } n.$$

We have the relation

$$|n a_n x^{n-1}| = \left| n a_n \frac{x^{n-1}}{x_1^n} \cdot x_1^n \right| \le n \frac{M}{x_1} \left| \frac{x}{x_1} \right|^{n-1},$$

and now we can apply the comparison test to the series (2). The series

$$\frac{M}{x_1} \sum_{n=1}^{\infty} n \left| \frac{x}{x_1} \right|^{n-1}$$

converges by the Ratio Test, since $\rho = |x/x_1| < 1$; hence, so does the series (2). Since x is any number in the interval $(-R, R)$, the interval of convergence of (2) is the same as that of (1).

Corollary. *Under the hypotheses of Theorem* 19, *the series* (1) *may be differentiated any number of times and each of the differentiated series converges for $|x| < R$.*

REMARKS. (i) The Corollary is obtained by induction, since each differentiated series has the same radius of convergence as the one before. (ii) The results of Theorem 19 and the Corollary are valid for a series of the form

$$\sum_{n=0}^{\infty} a_n (x - a)^n,$$

which converges for $|x - a| < R$ so long as $R > 0$. The proof is the same. (iii) The quantity R may be $+\infty$, in which case the series and its derived ones converge for all values of x.

Theorem 20. *If $R > 0$ and f is defined by*

$$f(x) = \sum_{n=0}^{\infty} a_n x^n \qquad \text{for } |x| < R,$$

then f is continuous for $|x| < R$.

PROOF. Let x_0 be any number such that $-R < x_0 < R$; we wish to show that f is continuous at x_0. In other words we must show that

$$f(x) \to f(x_0) \quad \text{as} \quad x \to x_0.$$

We have

$$|f(x) - f(x_0)| = \left| \sum_{n=0}^{\infty} a_n(x^n - x_0^n) \right| \le \sum_{n=0}^{\infty} |a_n||x^n - x_0^n|.$$

We apply the Theorem of the Mean to the function $g(x) = x^n$; that is, the relation

$$g(x) - g(x_0) = g'(\xi)(x - x_0), \qquad (\xi \text{ is between } x \text{ and } x_0),$$

applied to the function $g(x) = x^n$, is

$$x^n - x_0^n = n\xi_n^{n-1}(x - x_0), \qquad (\xi_n \text{ between } x \text{ and } x_0).$$

The subscript n has been put on ξ to identify the particular exponent of the function x^n. We may also write

$$|x^n - x_0^n| = n|\xi_n|^{n-1}|x - x_0|.$$

Thus we find

$$|f(x) - f(x_0)| \le \sum_{n=0}^{\infty} n|a_n||\xi_n|^{n-1}|x - x_0|.$$

So long as x is in the interval of convergence there is an x_1 such that $|\xi_n| < x_1$ for all n. We deduce that

$$|f(x) - f(x_0)| \le |x - x_0| \sum_{n=0}^{\infty} n|a_n|x_1^{n-1}.$$

Now we apply Theorem 19 to conclude that the series on the right converges; call its sum K. Then

$$|f(x) - f(x_0)| \le |x - x_0| \cdot K.$$

As x tends to x_0 the quantity on the right tends to zero, and so $f(x)$ tends to $f(x_0)$. That is, $f(x)$ is continuous at x_0.

Theorem 21 (Term-by-Term Integration of Power Series). *Suppose that $R > 0$ and*

$$f(x) = \sum_{n=0}^{\infty} a_n x^n \tag{3}$$

converges for $|x| < R$. We define

$$F(x) = \int_0^x f(t)\,dt.$$

Then the formula

$$F(x) = \sum_{n=0}^{\infty} a_n \frac{x^{n+1}}{n+1} \tag{4}$$

holds for $|x| < R$.

PROOF. Let x be any number such that $-R < x < R$. Choose x_1 so that $|x| < |x_1| < R$. We note that for any n

$$\int_0^x a_n t^n\,dt = \frac{a_n x^{n+1}}{n+1}.$$

Therefore

$$F(x) - \sum_{n=0}^{N} a_n \frac{x^{n+1}}{n+1} = \int_0^x \left[f(t) - \sum_{n=0}^{N} a_n t^n \right] dt. \tag{5}$$

But now

$$f(t) - \sum_{n=0}^{N} a_n t^n = \sum_{n=0}^{\infty} a_n t^n - \sum_{n=0}^{N} a_n t^n = \sum_{n=N+1}^{\infty} a_n t^n,$$

and for all t such that $-|x| < t < |x|$ and $|x| < |x_1| < R$,

$$\left| f(t) - \sum_{n=0}^{N} a_n t^n \right| \le \sum_{n=N+1}^{\infty} |a_n||x|^n. \tag{6}$$

Since the series (3) converges absolutely at x, the right side of (6)—being the remainder—tends to zero as $N \to \infty$. We conclude from (5) that

$$\left| F(x) - \sum_{n=0}^{N} a_n \frac{x^{n+1}}{n+1} \right| \le \left| \int_0^x \left(\sum_{n=N+1}^{\infty} |a_n||x|^n \right) dt \right| = \left(\sum_{n=N+1}^{\infty} |a_n||x| \right) \cdot |x|.$$

As N tends to ∞, the right side above tends to zero, and the left side above yields (4).

EXAMPLE 1. Assuming that the function $f(x) = \sin x$ is given by the series

$$\sin x = x - \frac{x^3}{3!} + \frac{x^5}{5!} - \cdots + \frac{(-1)^n x^{2n+1}}{(2n+1)!} + \cdots,$$

find the Maclaurin series for $\cos x$.

SOLUTION. Applying the Ratio Test to the series

$$\sum_{n=0}^{\infty} \frac{(-1)^n x^{2n+1}}{(2n+1)!},$$

we see that it converges for all values of x. We define

$$F(x) = \int_0^x \sin t\, dt = -\cos x + 1.$$

Integrating the above series for $\sin x$ term by term, we obtain

$$F(x) = 1 - \cos x = \sum_{n=0}^{\infty} \frac{(-1)^n x^{2n+2}}{(2n+2)!}$$

or

$$\cos x = 1 - \frac{x^2}{2!} + \frac{x^4}{4!} - \frac{x^6}{6!} + \cdots + \frac{(-1)^n x^{2n}}{(2n)!} + \cdots = \sum_{n=0}^{\infty} \frac{(-1)^n x^{2n}}{(2n)!}.$$

The next theorem relates the derivative of a function given by a series with the term-by-term differentiation of the series.

Theorem 22. *If* $R > 0$ *and*

$$f(x) = \sum_{n=0}^{\infty} a_n x^n \qquad \text{for } |x| < R, \tag{7}$$

then $f(x)$ has continuous derivatives of all orders for $|x| < R$ which are given there by series obtained by successive term-by-term differentiations of (7).

PROOF. The fact that

$$\sum_{n=1}^{\infty} n a_n x^{n-1} \tag{8}$$

converges for $|x| < R$ was shown in Theorem 19. We must show that $g(x) = \sum_{n=1}^{\infty} n a_n x^{n-1}$ is the derivative of f. Theorem 20 establishes the fact that g is continuous. Then, integrating the series (8) term by term we get, on the one hand,

$$a_0 + \int_0^x g(t)\, dt$$

and, on the other, the series for $f(x)$. That is,

$$f(x) = a_0 + \int_0^x g(t)\, dt.$$

The Fundamental Theorem of the Calculus then asserts that

$$f'(x) = g(x),$$

which is the result we wished to establish.

REMARK. The result of Theorem 22 holds equally well for functions f given by series of the form

$$f(x) = \sum_{n=0}^{\infty} a_n (x-a)^n,$$

which converge for $|x-a| < R$ with $R > 0$. The kth derivative of f is given by

$$\boxed{f^{(k)}(x) = \sum_{n=k}^{\infty} n(n-1)\cdots(n-k+1)(x-a)^{n-k}}$$

EXAMPLE 2. Assuming that the expansion of $f: x \to (4+x^2)^{-1}$ is given by

$$\frac{1}{4+x^2} = \sum_{n=0}^{\infty} \frac{(-1)^n}{2^{2n+2}} x^{2n},$$

valid for $|x| < 2$, obtain an expansion for $2x/(4+x^2)^2$.

SOLUTION. Differentiating term by term, we find

$$-\frac{2x}{(4+x^2)^2} = \sum_{n=1}^{\infty} \frac{(-1)^n 2n}{2^{2n+2}} x^{2n-1}.$$

We may replace n by $n+1$ in the infinite series if we change the lower limit from 1 to 0. We get

$$\frac{2x}{(4+x^2)^2} = \sum_{n=0}^{\infty} \frac{(-1)^n (2n+2)}{2^{2n+4}} x^{2n+1}.$$

We now make use of the above theorems and the fact that the function $1/(1+x)$ may be expanded in the simple geometric series

$$\frac{1}{1+x} = \sum_{n=0}^{\infty} (-1)^n x^n, \qquad |x| < 1,$$

to obtain additional series expansions.

Theorem 23. *For* $|x| < 1$, *we have the expansion*:

$$\log(1+x) = \sum_{n=1}^{\infty} \frac{(-1)^{n-1} x^n}{n}.$$

PROOF. Letting $F(x) = \log(1+x)$ and differentiating, we find

$$F'(x) = f(x) = \frac{1}{1+x} = \sum_{n=0}^{\infty} (-1)^n x^n, \qquad |x| < 1.$$

Now, by Theorem 21, we may integrate term by term and get

$$F(x) = \int_0^x f(t)\,dt = \sum_{n=0}^{\infty} \frac{(-1)^n x^{n+1}}{n+1},$$

which is equivalent to the statement of the theorem.

EXAMPLE 3. Find the Maclaurin series for

$$f(x) = \frac{1}{(1-x)^2}.$$

SOLUTION. If we define $F(x) = (1-x)^{-1}$, we see that $F'(x) = f(x)$. Now

$$F(x) = \sum_{n=0}^{\infty} x^n, \qquad |x| < 1$$

[which we can obtain from the expansion for $(1+x)^{-1}$ if we replace x by $-x$]. Therefore

$$f(x) = \frac{1}{(1-x)^2} = \sum_{n=1}^{\infty} n x^{n-1} = \sum_{k=0}^{\infty} (k+1)x^k, \qquad |x| < 1.$$

EXAMPLE 4. Find the Maclaurin expansion for $f(x) = (1+x^2)^{-1}$.

SOLUTION. The geometric series

$$\frac{1}{1+u} = \sum_{n=0}^{\infty} (-1)^n u^n, \qquad |u| < 1$$

after substitution of x^2 for u, becomes

$$\frac{1}{1+x^2} = \sum_{n=0}^{\infty} (-1)^n x^{2n},$$

valid for $x^2 < 1$. The inequality $x^2 < 1$ is equivalent to the inequality $|x| < 1$.

In the problems below, assume that the following formulas hold. They will be proved in the next section.

$$(1-x)^{-1} = \sum_{n=0}^{\infty} x^n; \qquad e^x = \sum_{n=0}^{\infty} \frac{x^n}{n!};$$

$$\sin x = \sum_{n=0}^{\infty} \frac{(-1)^n x^{2n+1}}{(2n+1)!}; \qquad \cos x = \sum_{n=0}^{\infty} \frac{(-1)^n x^{2n}}{(2n)!};$$

$$(1+x)^m = 1 + \sum_{n=1}^{\infty} \frac{m(m-1)\cdots(m-n+1)}{n!} x^n.$$

PROBLEMS

In Problems 1 and 2, assume the Maclaurin series for $\sin x$ and $\cos x$.

1. Find the Maclaurin series for

$$f(x) = \begin{cases} (\sin x)/x, & x \neq 0, \\ 1, & x = 0. \end{cases}$$

2. Find the Maclaurin series for

$$f(x) = \begin{cases} (1 - \cos x)/x, & x \neq 0, \\ 0, & x = 0. \end{cases}$$

3. Find the Maclaurin series for e^x, determine the interval of convergence, and then find the Maclaurin series for f given by

$$f(x) = \begin{cases} (e^x - 1)/x, & x \neq 0, \\ 1, & x = 0. \end{cases}$$

In Problems 4 through 8, use term-by-term differentiation and integration of known series as in Examples 2, 3, and 4 to determine Taylor or Maclaurin series for the function f. State the interval of convergence.

4. $f(x) = (1 + x)^{-2}$
5. $f(x) = \log \dfrac{1 + x}{1 - x}$
6. $f(x) = (1 - x)^{-3}$
7. $f(x) = (1 + x)^{-3}$
8. $f(x) = x \log(1 + x^2)$
9. Find the Maclaurin expansion for $(1 + x^2)^{-1}$ and then use term-by-term integration to get a Maclaurin expansion for $\arctan x$.
10. Find the Maclaurin expansion for $(1 - x^2)^{-1/2}$, and then find one for $\arcsin x$.
11. Find the Maclaurin expansion for $\cos 2x$ and then find one for $\sin^2 x$.
12. If $f(x) = e^{\alpha x}$, show that the Maclaurin series for $f^{(k)}(x)$ is α^k times the Maclaurin series for f.
13. Find the sum of the series*

$$\sum_{k=1}^{\infty} kx^{k-1} \qquad \text{for } |x| < 1.$$

14. Find the sum of the series

$$\sum_{k=1}^{\infty} \frac{x^k}{k(k+1)}.$$

15. Find the Maclaurin expansion for $\log(x + \sqrt{1 + x^2})$.

* That is, find an expression in closed form which is equal to the sum of the series for all x in the interval of convergence.

16. Find the sum of the series

$$\sum_{n=0}^{\infty} \frac{(n+2)(n+1)x^n}{n!}$$

and conclude that

$$e = \frac{1}{7} \sum_{n=0}^{\infty} \frac{(n+2)(n+1)}{n!}.$$

17. Find the first five terms of the Maclaurin expansion of

$$F(x) = \int_0^x e^{-t^2}\, dt.$$

Obtain an approximate value of $F(1)$. Can you approximate the error?

18. Show that for every positive integer p, we have

$$(1-x)^{-p-1} = \sum_{n=0}^{\infty} \binom{n+p}{p} x^n, \qquad |x| < 1,$$

where the symbol $\binom{n+p}{p}$ is an abbreviation of

$$\frac{(n+p)(n+p-1)\cdots(n+1)}{1\cdot 2\cdot 3\cdots p}.$$

Deduce the formula $3^{p+1} = \sum_{n=0}^{\infty} \binom{n+p}{p}(\frac{2}{3})^n$.

*19. Let

$$f(x) = \frac{1}{2} + \frac{x}{5} + \frac{x^2}{8} + \frac{x^3}{11} + \cdots.$$

a) Show that the expansion for $x/(1-x^3)$ is precisely $(x^2 f(x^3))'$.
b) Find an expression for $f(x)$.

20. Use the fact that $x^p = e^{p \log x}$ to express x^p as a power series in $\log x$:

$$x^p = 1 + \frac{p(\log x)}{1!} + \cdots + \frac{p^n(\log x)^n}{n!} + \cdots.$$

Find the interval of convergence. For what values of p is the above expansion valid?

21. Suppose that $f(x)$ has a Maclaurin series with a positive radius of convergence: $f(x) = \sum_{n=0}^{\infty} a_n x^n$. Show that if f is an even function of x, that is, if $f(-x) = f(x)$ for all x, then $a_n = 0$ for all odd integers n. If f is odd, i.e., if $f(-x) = -f(x)$, show that $a_n = 0$ for all even n.

22. Writing the identity

$$\frac{1}{1-x} = \frac{-1}{x\left(1 - \dfrac{1}{x}\right)},$$

we get the formal expansion

$$\frac{1}{1-x} = -\sum_{n=0}^{\infty} \frac{1}{x^{n+1}}.$$

For what values of x does this series converge? Under what conditions can the series be differentiated and integrated term by term?

23. Find an expansion for $\arctan x$ in powers of $1/x$. [*Hint:* Consider the expansion of $1/(1+x^2)$ as in Problem 22 and then integrate.]

9. Validity of Taylor Expansions and Computations with Series

The theorems of this section state that certain functions are truly represented by their Taylor series.

Theorem 24. *For any values of a and x, we have*

$$e^x = e^a \sum_{n=0}^{\infty} \frac{(x-a)^n}{n!}; \tag{1}$$

i.e., the Taylor series for e^x about $x = a$ converges to e^x for any a and x.

PROOF. For simplicity, set $a = 0$, the proof being analogous when $a \neq 0$. If we let $f(x) = e^x$, then $f^{(n)}(x) = e^x$ for all n. Now, using Taylor's Theorem with Derivative Form of the Remainder, we have

$$e^x = \sum_{k=0}^{n} \frac{x^k}{k!} + R_n, \qquad \text{where} \qquad R_n = \frac{e^{\xi}x^{n+1}}{(n+1)!},$$

with ξ between 0 and x. If x is positive, then $e^{\xi} < e^x$ while, if x is negative, then $e^{\xi} < e^0 = 1$. In either case,

$$R_n \leq C\frac{|x|^{n+1}}{(n+1)!}, \tag{2}$$

where C is the larger of 1 and e^x but is *independent of n*. If the right side of (2) is the general term of a series then, by the Ratio Test, that series is convergent for all x. The general term of any convergent series must tend to zero, and so

$$C\frac{|x|^{n+1}}{(n+1)!} \to 0 \qquad \text{as} \qquad n \to \infty \text{ for each } x.$$

We conclude that $R_n \to 0$ as $n \to \infty$, and so (1) is established.

Theorem 25. *The following functions are given by their Maclaurin series:*

$$\sin x = \sum_{n=0}^{\infty} \frac{(-1)^n x^{2n+1}}{(2n+1)!}, \qquad \cos x = \sum_{n=0}^{\infty} \frac{(-1)^n x^{2n}}{(2n)!}.$$

The proofs of these results follow the same outline as the proof of Theorem 24 and are left as exercises for the reader at the end of this section.

Theorem 26 (Binomial Theorem). *For each real number m, we have*

$$(1+x)^m = 1 + \sum_{n=1}^{\infty} \frac{m(m-1)(m-2)\cdots(m-n+1)}{n!} x^n \qquad \textit{for } |x| < 1.$$

PROOF. To show that the series on the right converges absolutely for $|x| < 1$, we apply the Ratio Test:

$$\left|\frac{u_{n+1}}{u_n}\right|$$

$$= \left|\frac{m(m-1)\cdots(m-n+1)(m-n)}{(n+1)!} \cdot \frac{n!}{m(m-1)\cdots(m-n+1)} \cdot \frac{x^{n+1}}{x^n}\right|$$

$$= \frac{|m-n|}{n+1}|x| = \frac{|1-m/n|}{1+1/n}|x|.$$

The quantity on the right tends to $|x|$ as $n \to \infty$, and so the series converges for $|x| < 1$. We define

$$f(x) = 1 + \sum_{n=1}^{\infty} \frac{m(m-1)\cdots(m-n+1)}{n!} x^n, \tag{3}$$

and we must show that $f(x) = (1+x)^m$ if $|x| < 1$. Employing Theorem 22, we get $f'(x)$ by term by term differentiation of the series for f. We have

$$f'(x) = m + \sum_{n=2}^{\infty} \frac{m(m-1)\cdots(m-n+1)}{(n-1)!} x^{n-1}. \tag{4}$$

Multiplying both sides of (4) by x, we get

$$xf'(x) = \sum_{n=1}^{\infty} n\frac{m(m-1)\cdots(m-n+1)}{n!} x^n. \tag{5}$$

We add (4) and (5) to obtain

$$(1+x)f'(x) = m\left\{1 + \sum_{n=1}^{\infty} \frac{m(m-1)\cdots(m-n+1)}{n!} x^n\right\} = mf(x).$$

In other words, the function f satisfies the relation

$$(1+x)f'(x) - mf(x) = 0.$$

We now use a trick to find $f(x)$. We multiply this equation by $(1+x)^{-m-1}$, getting

$$(1 + x)^{-m}f'(x) - m(1 + x)^{-m-1}f(x) = 0.$$

This last equation can be written

$$\frac{d}{dx}[(1 + x)^{-m}f(x)] = 0.$$

A function whose derivative is identically zero must be constant, so that

$$(1 + x)^{-m}f(x) = K,$$

where K is some constant. Setting $x = 0$ in (3), we see that $f(0) = 1$, and this fact yields $K = 1$. We conclude finally that

$$f(x) = (1 + x)^m \qquad \text{for } |x| < 1.$$

EXAMPLE 1. Write the first 5 terms of the series expansion for $(1 + x)^{3/2}$.

SOLUTION. We have $m = \frac{3}{2}$, and therefore

$$\begin{aligned}(1 + x)^{3/2} &= 1 + \frac{\frac{3}{2}}{1!}x + \frac{(\frac{3}{2})(\frac{1}{2})}{2!}x^2 + \frac{(\frac{3}{2})(\frac{1}{2})(-\frac{1}{2})}{3!}x^3 + \frac{(\frac{3}{2})(\frac{1}{2})(-\frac{1}{2})(-\frac{3}{2})}{4!}x^4 \\ &= 1 + \frac{3}{2}x + \frac{3}{8}x^2 - \frac{3}{2^3 \cdot 3!}x^3 + \frac{3^2}{2^4 \cdot 4!}x^4 - \cdots.\end{aligned}$$

EXAMPLE 2. Write the binomial series for $(1 + x)^7$.

SOLUTION. We have

$$(1 + x)^7 = 1 + \sum_{n=1}^{\infty} \frac{7(6) \cdots (7 - n + 1)}{n!}x^n.$$

We now observe that beginning with $n = 8$ all the terms have a zero in the numerator. Therefore,

$$\begin{aligned}(1 + x)^7 &= 1 + \sum_{n=1}^{7} \frac{7(6) \cdots (7 - n + 1)}{n!}x^n \\ &= 1 + 7x + \frac{7 \cdot 6}{2!}x^2 + \frac{7 \cdot 6 \cdot 5}{3!}x^3 + \cdots + \frac{7!}{7!}x^7.\end{aligned}$$

For m a positive integer, the binomial series always terminates after a finite number of terms.

EXAMPLE 3. Compute

$$\int_0^{0.5} e^{x^2}\,dx$$

to an accuracy of five decimal places.

SOLUTION. We have

$$e^u = \sum_{n=0}^{\infty} \frac{u^n}{n!} \quad \text{for all } u, \quad \text{and so} \quad e^{x^2} = \sum_{n=0}^{\infty} \frac{x^{2n}}{n!} \quad \text{for all } x.$$

By Theorem 21 we can integrate term by term to get

$$\int_0^x e^{t^2}\,dt = \sum_{n=0}^{\infty} \frac{x^{2n+1}}{n!(2n+1)}.$$

For $x = 0.5$, we now compute

$$\begin{aligned}
x &= 0.5 = \frac{1}{2} &&= 0.50000\ 00,\\
\frac{x^3}{1!\cdot 3} &= \frac{(0.5)^3}{3} = \frac{1}{24} &&= 0.04166\ 67^-,\\
\frac{x^5}{2!\cdot 5} &= \frac{(0.5)^5}{10} = \frac{1}{320} &&= 0.00312\ 50,\\
\frac{x^7}{3!\cdot 7} &= \frac{(0.5)^7}{42} = \frac{1}{5376} &&= 0.00018\ 60^+,\\
\frac{x^9}{4!\cdot 9} &= \frac{(0.5)^9}{216} = \frac{1}{110{,}592} &&= 0.00000\ 90^+,
\end{aligned}$$

$$\text{Sum of the right-hand column} = 0.54498\ 67.$$

If we wish to stop at this point we must estimate the error made by neglecting all the remaining terms. The remainder is

$$\begin{aligned}
\sum_{n=5}^{\infty} \frac{x^{2n+1}}{n!(2n+1)} &\le \frac{x^{11}}{5!\cdot 11}\left(1 + \frac{x^2}{6} + \frac{x^4}{6^2} + \frac{x^6}{6^3} + \cdots\right)\\
&\le \frac{x^{11}}{1320}\frac{1}{(1 - x^2/6)} = \frac{24}{23}\cdot\frac{1}{1320}\cdot\frac{1}{2048}\\
&\le 0.00000\ 04.
\end{aligned}$$

Therefore

$$\int_0^{0.5} e^{x^2}\,dx = \begin{cases} 0.54499 \text{ to an accuracy of 5 decimals,}\\ 0.544987 \text{ to an accuracy of 6 decimals.}\end{cases}$$

PROBLEMS

In each of Problems 1 through 7, write the beginning of the binomial series for the given expression to the required number of terms.

1. $(1 + x)^{-3/2}$, 5 terms
2. $(1 - x)^{1/2}$, 4 terms

3. $(1 + x^2)^{-2/3}$, 5 terms

4. $(1 - x^2)^{-1/2}$, 6 terms

5. $(1 + x^3)^7$, all terms

6. $(5 + x)^{1/2}$, 4 terms

7. $(3 + \sqrt{x})^{-3}$, 5 terms

In each of Problems 8 through 17, compute the value of the definite integral to the number of decimal places specified. Estimate the remainder.

8. $\int_0^1 \sin(x^2)\,dx$, 5 decimals

9. $\int_0^1 \cos(x^2)\,dx$, 5 decimals

10. $\int_0^1 e^{-x^2}\,dx$, 5 decimals

11. $\int_0^{0.5} \frac{dx}{1 + x^3}$, 5 decimals

12. $\int_0^1 \frac{\sin x}{x}\,dx$, 5 decimals

13. $\int_0^1 \frac{e^x - 1}{x}\,dx$, 5 decimals

14. $\int_0^{0.5} \frac{dx}{\sqrt{1 + x^3}}$, 5 decimals

15. $\int_0^{1/3} \frac{dx}{\sqrt[3]{1 + x^2}}$, 5 decimals

16. $\int_0^{1/3} \frac{dx}{\sqrt[3]{1 - x^2}}$, 5 decimals

17. $\int_0^{0.5} \frac{dx}{\sqrt{1 - x^3}}$, 5 decimals

18. Use the series for $\log \frac{1 + x}{1 - x}$ to find log 1.5 to 5 decimals of accuracy.

19. Same as Problem 18, to find log 2.

20. Prove that

$$\sin x = \sum_{n=0}^{\infty} \frac{(-1)^n x^{2n+1}}{(2n + 1)!}.$$

21. Prove that

$$\cos x = \sum_{n=0}^{\infty} \frac{(-1)^n x^{2n}}{(2n)!}.$$

22. If $f(x) = \sin x$ show, by induction, that $f^{(k)}(x) = \sin(x + \frac{1}{2}k\pi)$.

23. Use the result of Problem 22 to show that for all a and x the Taylor series for $\sin x$ is given by

$$\sin x = \sum_{n=0}^{\infty} \frac{\sin(a + \frac{1}{2}n\pi)}{n!}(x - a)^n.$$

24. If $f(x) = \cos x$ show, by induction, that $f^{(k)}(x) = \cos(x + \frac{1}{2}k\pi)$.

25. Use the result of Problem 24 to show that for all a and x the Taylor series for $\cos x$ is given by

$$\cos x = \sum_{n=0}^{\infty} \frac{\cos(a + \frac{1}{2}n\pi)}{n!}(x - a)^n.$$

26. Use the series for $\log(1 + x)/(1 - x)$ to obtain the formula

$$\log(y+1) = \log y + 2\left[\frac{1}{2y+1} + \frac{1}{3(2y+1)^3} + \frac{1}{5(2y+1)^5} + \cdots\right].$$

[*Hint:* Let $x = (2y+1)^{-1}$.]

27. Assuming that $\log 2$ and $\log 5$ are known, use the expansion in Problem 26 to estimate the error in finding $\log 11$ when three terms of the series are used.

10. Algebraic Operations with Series

In previous sections we discussed the question of term-by-term differentiation and integration of series. Now we turn to the question of multiplication and division of power series.

Suppose we are given two power series

$$\sum_{n=0}^{\infty} a_n x^n = a_0 + a_1 x + a_2 x^2 + \cdots + a_n x^n + \cdots,$$

$$\sum_{n=0}^{\infty} b_n x^n = b_0 + b_1 x + b_2 x^2 + \cdots + b_n x^n + \cdots.$$

Without considering questions of convergence, we multiply the two series by following the rules for multiplying two polynomials. We obtain the successive lines, each obtained by multiplying an element of the second series with all the terms of the first series:

$$\begin{array}{ll}
b_0: & a_0b_0 + a_1b_0x + a_2b_0x^2 + \cdots + a_nb_0x^n + \cdots, \\
b_1x: & \qquad a_0b_1x + a_1b_1x^2 + \cdots + a_{n-1}b_1x^n + a_nb_1x^{n+1} + \cdots, \\
b_2x^2: & \qquad\qquad a_0b_2x^2 + \cdots + a_{n-2}b_2x^n + a_{n-1}b_2x^{n+1} + a_nb_2x^{n+2} + \cdots, \\
\vdots & \\
b_nx^n: & \qquad\qquad\qquad a_0b_nx^n + a_1b_nx^{n+1} + a_2b_nx^{n+2} + \cdots.
\end{array}$$

Adding the columns, we obtain the power series

$$\begin{aligned}
a_0b_0 &+ (a_1b_0 + a_0b_1)x + (a_2b_0 + a_1b_1 + a_0b_2)x^2 \\
&+ (a_3b_0 + a_2b_1 + a_1b_2 + a_0b_3)x^3 + \cdots \\
&+ (a_nb_0 + a_{n-1}b_1 + \cdots + a_0b_n)x^n + \cdots.
\end{aligned}$$

The technique for computing the coefficients of any term is easy to determine. The subscripts of the a's decrease by one as the subscripts of the b's increase, the total always remaining the same.

Definition. Given the two series

$$\sum_{n=0}^{\infty} a_n x^n, \qquad \sum_{n=0}^{\infty} b_n x^n,$$

we define the **Cauchy Product** to be the series

$$c_0 + c_1 x + c_2 x^2 + \cdots + c_n x^n + \cdots$$

where

$$c_n = a_n b_0 + a_{n-1} b_1 + \cdots + a_0 b_n = \sum_{k=0}^{n} a_{n-k} b_k.$$

EXAMPLE 1. Given the Maclaurin series for e^x and $\cos x$, find the first seven terms of the Cauchy Product of these two series—i.e., the terms through x^6.

SOLUTION. We have

$$e^x = 1 + x + \frac{x^2}{2} + \frac{x^3}{6} + \frac{x^4}{24} + \frac{x^5}{120} + \frac{x^6}{720} + \cdots,$$

$$\cos x = 1 \qquad - \frac{x^2}{2} \qquad + \frac{x^4}{24} \qquad - \frac{x^6}{720} + \cdots.$$

We multiply term by term to obtain

$$1 + x + \frac{x^2}{2} + \frac{x^3}{6} + \frac{x^4}{24} + \frac{x^5}{120} + \frac{x^6}{720} + \cdots$$

$$- \frac{x^2}{2} - \frac{x^3}{2} - \frac{x^4}{4} - \frac{x^5}{12} - \frac{x^6}{48} + \cdots$$

$$+ \frac{x^4}{24} + \frac{x^5}{24} + \frac{x^6}{48} + \cdots$$

$$- \frac{x^6}{720} + \cdots$$

$$\text{Cauchy Product} = 1 + x - \frac{x^3}{3} - \frac{x^4}{6} - \frac{x^5}{30} + 0 \cdot x^6 + \cdots.$$

Theorem 27. *If*

$$f(x) = \sum_{n=0}^{\infty} a_n x^n, \qquad g(x) = \sum_{n=0}^{\infty} b_n x^n,$$

both converge for $|x| < R$, then the Cauchy Product of the two series converges to $f(x) \cdot g(x)$ for $|x| < R$.

This theorem will be proved in Section 14 (page 181).

On the basis of Theorem 27, we see that, in Example 1 above, the Cauchy Product is actually the Maclaurin expansion for the function $e^x \cos x$, valid for all x.

Now we apply the process of long division to two series as if they were polynomials. We write

$$
\begin{array}{r|l}
 & c_0 + c_1 x + \cdots \\
\hline
b_0 + b_1 x + b_2 x^2 + \cdots & a_0 + a_1 x + a_2 x^2 + \cdots \\
 & c_0 b_0 + c_0 b_1 x + c_0 b_2 x^2 + \cdots \\
\cline{2-2}
 & + (a_1 - c_0 b_1)x + (a_2 - c_0 b_2)x^2 + \cdots \\
 & \qquad c_1 b_0\, x + \qquad c_1 b_1\, x^2 + \cdots \\
\cline{2-2}
 & \qquad\qquad + (a_2 - c_0 b_2 - c_1 b_1)x^2 + \cdots \\
 & \qquad\qquad\qquad c_2 b_0\, x^2 + \cdots \\
\cline{2-2}
\end{array}
$$

In order for the division process to proceed, we must have (assuming $b_0 \neq 0$)

$$
\begin{aligned}
a_0 &= c_0 b_0 &\quad \text{or} \quad && c_0 &= a_0/b_0, \\
a_1 - c_0 b_1 &= c_1 b_0 &\quad \text{or} \quad && c_1 &= (a_1 - c_0 b_1)/b_0, \\
a_2 - c_0 b_2 - c_1 b_1 &= c_2 b_0 &\quad \text{or} \quad && c_2 &= (a_2 - c_0 b_2 - c_1 b_1)/b_0,
\end{aligned}
$$

etc. By induction it can be established that the Cauchy Product of the quotient series with the divisor series yields the series which is the dividend.

Theorem 28. *Under the hypotheses of Theorem 27, the quotient series converges to $f(x)/g(x)$ for $|x| < T$ for some $T > 0$ so long as $b_0 \neq 0$.*

We omit the proof.

EXAMPLE 2. By dividing the Maclaurin series for $\sin x$ by the one for $\cos x$, find the terms up to x^5 in the Maclaurin series for $\tan x$.

SOLUTION. We have

$$
\begin{array}{r|l}
 & x + \dfrac{x^3}{3} + \dfrac{2x^5}{15} + \cdots \\
\hline
1 - \dfrac{x^2}{2} + \dfrac{x^4}{24} - \cdots & x - \dfrac{x^3}{6} + \dfrac{x^5}{120} - \cdots \\
 & x - \dfrac{x^3}{2} + \dfrac{x^5}{24} - \cdots \\
\cline{2-2}
 & \quad + \dfrac{x^3}{3} - \dfrac{x^5}{30} + \cdots \\
 & \quad + \dfrac{x^3}{3} - \dfrac{x^5}{6} + \cdots \\
\cline{2-2}
 & \qquad\qquad \dfrac{2x^5}{15} + \cdots
\end{array}
$$

PROBLEMS

In each of Problems 1 through 20, find the Maclaurin series to and including the term in x^n.

1. $\dfrac{\sin x}{1 + x}$, $n = 5$

2. $\dfrac{\sin x}{1 - x}$, $n = 5$

3. $\dfrac{\cos x}{1+x}$, $n = 5$

4. $\dfrac{\cos x}{1-x}$, $n = 5$

5. $\sqrt{1+x}\log(1+x)$, $n = 5$

6. $\sqrt{1-x}\log(1-x)$, $n = 5$

7. $\dfrac{\log(1+x)}{\sqrt{1+x}}$, $n = 5$

8. $\dfrac{\log(1-x)}{\sqrt{1+x}}$, $n = 5$

9. $e^x \sec x$, $n = 5$

10. $e^{-x}\tan x$, $n = 3$

11. $\dfrac{e^x}{\sqrt{1+x^2}}$, $n = 5$

12. $\dfrac{e^{-x}}{\sqrt{1+x^2}}$, $n = 5$

13. $\dfrac{\arctan x}{1+x}$, $n = 5$

14. $\dfrac{1-x}{1-x^3} = \dfrac{1}{1+x+x^2}$, all n

15. $\dfrac{\arcsin x}{\cos x}$, $n = 5$

16. $\sec x$, $n = 6$

17. $(1+x)^{1/3}(1+x^2)^{4/3}$, $n = 4$

18. $\sin^2 x \cos x$, $n = 4$

19. $(1+x^2)^{3/2}(1-x^3)^{-1/2}$, $n = 4$

20. $\tan^2 x$, $n = 6$

21. Use series expansions to show that

$$e^{x+y} = e^x e^y.$$

22. Use series expansions to show that

$$\sin(x+y) = \sin x \cos y + \cos x \sin y.$$

23. Use series expansions to show that for all x, we have

$$\sin^2 x + \cos^2 x = 1.$$

24. Writing $A(x) = \frac{1}{2}(e^x + e^{-x})$ and $B(x) = \frac{1}{2}(e^x - e^{-x})$, use series expansions to show:

a) $A^2(x) - B^2(x) = 1$ b) $A'(x) = B(x)$ c) $B'(x) = A(x)$

11. Uniform Convergence. Sequences of Functions

In this section and in the remaining sections of this chapter we discuss more advanced topics in infinite series.

We recall that **an infinite sequence**

$$a_1, a_2, \ldots, a_n, \ldots$$

has the limit c if and only if for each $\varepsilon > 0$ there is a positive integer N (the size of N depending on ε) such that

$$|a_n - c| < \varepsilon \qquad \text{for all } n > N.$$

Frequently the dependence of N upon ε is indicated by writing $N = N(\varepsilon)$.

Consider a sequence of functions

$$f_1(x), f_2(x), \ldots, f_n(x), \ldots \tag{1}$$

with each function in the sequence defined for x in an interval $[a, b]$. For any particular value of x, sequence (1) is a sequence of numbers which may or may not have a limit. If the limit does exist for some values of x, the limiting values form a function of x which we may denote by f. We write

$$\lim_{n\to\infty} f_n(x) = f(x)$$

where it is understood that f is defined only for those x for which the sequence converges.

For example, the collection of functions $f_n(x) = x^n$, $n = 1, 2, \ldots$, which we may write

$$x, x^2, x^3, \ldots, x^n, \ldots,$$

is a sequence of functions which converges if x is in the half-open interval $(-1, 1]$ but does not converge otherwise. The student may easily verify that the limit function f is

$$f(x) = \begin{cases} 0 & \text{for } -1 < x < 1, \\ 1 & \text{for } x = 1. \end{cases}$$

Among the most important limiting processes studied in the calculus are those of differentiation and integration of elementary functions and those concerned with the convergence of infinite sequences and series. In the applications of mathematics to various branches of technology as well as in the development of mathematical theory, it frequently happens that two, three, or even more limiting processes have to be applied successively. Does it matter in which order these limiting processes are performed? The answer is emphatically yes! The computation of two successive limits in one order usually yields an answer different from that obtained by computing them in the reverse order. Therefore it is of the utmost importance to know exactly when reversing the order in which two limits are computed does not change the answer. Mathematical literature abounds with erroneous results caused by the invalid interchange of two successive limiting processes.

Topics concerning sequences of functions frequently involve several successive limiting processes. A basic tool in the study of the evaluation of such limiting processes is the notion of the uniform convergence of a sequence.

Definition. We say that a sequence

$$f_1(x), f_2(x), \ldots, f_n(x), \ldots$$

of functions **converges uniformly on the interval** I to the function $f(x)$ if and

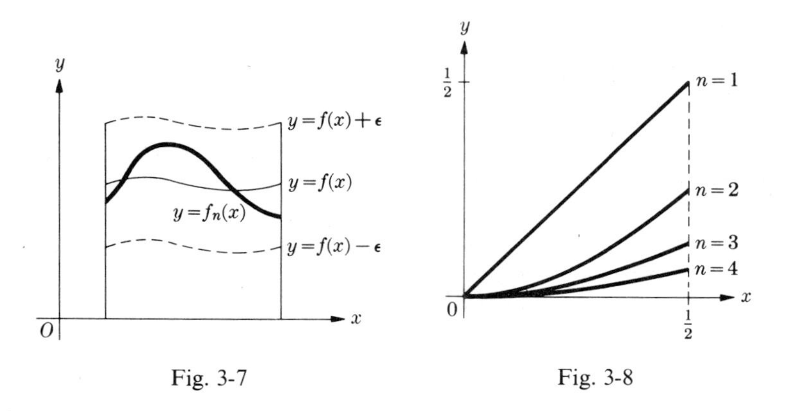

Fig. 3-7

Fig. 3-8

only if for each $\varepsilon > 0$ there is an integer N **independent of** x such that

$$|f_n(x) - f(x)| < \varepsilon \qquad \text{for all } x \text{ on } I \text{ and all } n > N. \tag{2}$$

The important fact which makes uniform convergence differ from ordinary convergence is that N does not depend on x, although naturally it depends on ε.

The geometric meaning of uniform convergence is illustrated in Fig. 3-7. Condition (2) in the definition states that if ε is any positive number, then the graph of $f_n(x)$ lies below the graph of $f(x) + \varepsilon$ and above the graph of $f(x) - \varepsilon$. The *uniformity* condition states that the graph of f_n must lie in this band of width 2ε not only for all n larger than N but also for all x in the *entire* length of the interval I.

An example of uniform convergence is given by the sequence of functions $f_n(x) = x^n$, $0 \le x \le \frac{1}{2}$, $n = 1, 2, \ldots$ (Fig. 3-8). We see that the limit function is zero and that the largest value of $f_n(x)$ occurs at $x = \frac{1}{2}$ with the value $1/2^n$. If we are given an $\varepsilon > 0$, we select N so that $1/2^N < \varepsilon$, and then $f_n(x)$ will be in the desired band for all $n > N$ and *all* x *in* the interval $[0, \frac{1}{2}]$. Note that the selection of N depended only on ε and not on the various values of x in the interval $[0, \frac{1}{2}]$.

It is important to observe that a sequence may converge uniformly on one interval and not on another. The same sequence, $f_n(x) = x^n$, $n = 1, 2, \ldots,$ does not converge uniformly on the interval $[0, 1]$, although it does converge at every point of this interval. (See Problem 15.) The limit function is discontinuous at $x = 1$.

It can happen that a sequence of continuous functions converges to a limit $f(x)$ for each x in an interval, that the limit function f is continuous, and that the convergence is not uniform. We shall show that such is the case for the sequence

$$f_n(x) = \frac{2nx}{1 + n^2x^2}, \qquad I = \{x : 0 \le x \le 1\}, \qquad n = 1, 2, \ldots$$

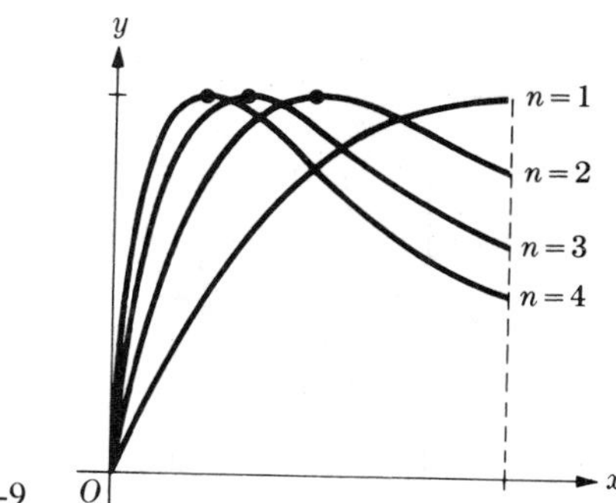

Fig. 3-9

The graphs for the first few values of n are shown in Fig. 3-9. If $x \neq 0$, we write

$$f_n(x) = \frac{2x/n}{x^2 + (1/n^2)},$$

and $\lim_{n\to\infty} f_n(x) = 0$ for $0 < x \leq 1$. Furthermore, $f_n(0) = 0$ for all n, and we conclude that $f_n(x) \to 0$ for all x on the closed interval I. The graph of $f_n(x)$ can be plotted accurately if we locate the maximum of this function. Taking the derivative,

$$f_n'(x) = \frac{2n(1 - n^2x^2)}{(1 + n^2x^2)^2},$$

we see that f_n has a maximum at $x = 1/n$ with the value $f_n(1/n) = 1$. Thus, as we proceed from right to left, every function in the sequence rises to the value 1 and then drops off to zero at $x = 0$. If we select $\varepsilon = \frac{1}{2}$, the condition of uniform convergence requires that

$$f(x) - \tfrac{1}{2} \leq f_n(x) \leq f(x) + \tfrac{1}{2}.$$

Since, in our example, $f(x) = 0$ for $0 \leq x \leq 1$, the above condition becomes $-\frac{1}{2} \leq f_n(x) \leq \frac{1}{2}$. But *every* function $f_n(x)$ has the value 1 somewhere in the interval, and so the sequence cannot converge uniformly.

Remark. In calculating the limit of $f_n(x)$, it is necessary to consider the case $x = 0$ separately since, for $x = 0$, the limit of the denominators $x^2 + (1/n^2)$ is zero.

In most cases it is not possible to verify directly from the definition whether or not a sequence converges uniformly. Therefore it is important to develop simple criteria which guarantee that a given sequence converges uniformly. We now derive a useful rule.

Theorem 29. *Suppose that* $\{f_n(x)\}$, $n = 1, 2, \ldots,$ *and* $f(x)$ *are continuous on the closed interval* $I = \{x : a \leq x \leq b\}$. *Then the sequence* $\{f_n(x)\}$ *converges* *uniformly to $f(x)$ on I if and only if the maximum value ε_n of $|f_n(x) - f(x)|$ converges to zero.*

PROOF. We must show (a), that if the convergence is uniform, then $\varepsilon_n \to 0$, and (b), that if $\varepsilon_n \to 0$, the convergence is uniform. To prove (a), we let

$$\varepsilon_n = \max|f_n(x) - f(x)|$$

for x on I. Then, since $f_n(x) - f(x)$ is continuous on I, for each n there is a value x_n such that $\varepsilon_n = |f_n(x_n) - f(x_n)|$. From the definition of uniform convergence, we know that for any $\varepsilon > 0$ there is an N such that

$$|f_n(x) - f(x)| < \varepsilon$$

for all $n > N$ and all x on I. Thus, for $n > N$, we must have

$$\varepsilon_n = |f_n(x_n) - f(x_n)| < \varepsilon.$$

Therefore $\varepsilon_n \to 0$ as $n \to \infty$.
b) Suppose that $\lim_{n\to\infty} \varepsilon_n = 0$. Let $\varepsilon > 0$ be given. There is an N such that $\varepsilon_n < \varepsilon$ for $n > N$. But then, since $\varepsilon_n = \max|f_n(x) - f(x)|$, we must have

$$0 \le |f_n(x) - f(x)| \le \varepsilon_n < \varepsilon$$

for all x on I and all $n > N$. The convergence is uniform.

EXAMPLE 1. Given the sequence

$$f_n(x) = \frac{n^2 x}{1 + n^3 x^2},$$

show that $f_n(x) \to 0$ for each x on $[0, 1]$, and determine whether or not the convergence is uniform.

SOLUTION. Since $f_n(0) = 0$ for every n, we have $f_n(0) \to 0$ as $n \to \infty$. For $0 < x \le 1$, we see that

$$\lim_{n\to\infty} f_n(x) = \lim_{n\to\infty} \frac{(x/n)}{x^2 + (1/n^3)} = \frac{0}{x^2} = 0.$$

The limit function $f(x)$ vanishes for $0 \le x \le 1$. Therefore

$$\varepsilon_n = \max|f_n(x) - f(x)| = \max f_n(x).$$

To find ε_n, we differentiate $f_n(x)$, getting

$$f_n'(x) = \frac{(1 + n^3 x^2)\cdot n^2 - n^2 x \cdot 2n^3 x}{(1 + n^3 x^2)^2} = \frac{n^2(1 - n^3 x^2)}{(1 + n^3 x^2)^2}.$$

Setting this derivative equal to zero, we obtain

$$x = x_n = n^{-3/2}; \qquad \varepsilon_n = f_n(x_n) = \frac{n^2 \cdot n^{-3/2}}{1 + n^3 \cdot n^{-3}} = \frac{\sqrt{n}}{2} \to \infty \quad \text{as } n \to \infty.$$

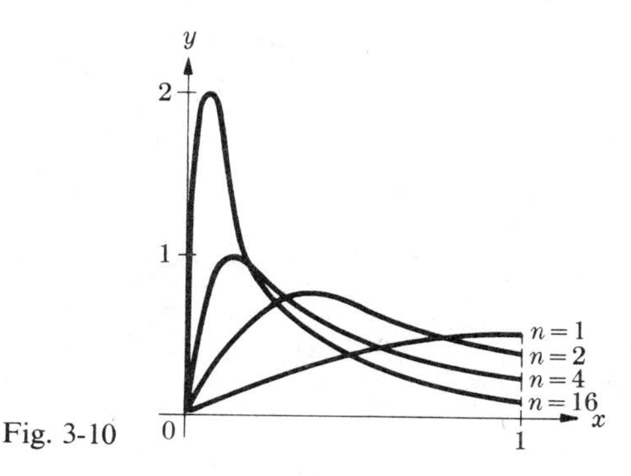

Fig. 3-10

Since ε_n does not tend to 0, the convergence is not uniform (Fig. 3-10).

Although Theorem 29 is useful in illustrating the idea of uniform convergence, it cannot be applied unless an explicit expression for the limit function $f(x)$ is known.

The next result shows that if a sequence of continuous functions converges uniformly on some interval, the limit function must be continuous on this interval.

Theorem 30. *Suppose that the sequence $\{f_n\}$, $n = 1, 2, \ldots$, converges uniformly to f on the interval I, with each f_n continuous on I. Then f is continuous on I.*

PROOF. Suppose that I is the open interval $I = \{x : a < x < b\}$. Let x_0 be any point of I and suppose that ε is any positive number. They, by the definition of uniform convergence, we can select N so large that

$$|f_n(x) - f(x)| < \frac{\varepsilon}{3} \qquad \text{for all } n > N \text{ and all } x \text{ on } I. \tag{3}$$

Since f_{N+1} is continuous on I, we apply the definition of continuity to state that for any $\varepsilon > 0$ there is a $\delta > 0$ such that

$$|f_{N+1}(x) - f_{N+1}(x_0)| < \frac{\varepsilon}{3} \qquad \text{for all } x \text{ in the interval } |x - x_0| < \delta. \tag{4}$$

At this point, we use a trick which expresses $f(x) - f(x_0)$ in a more complicated way. We write

$$f(x) - f(x_0) = f(x) - f_{N+1}(x) + f_{N+1}(x) - f_{N+1}(x_0) + f_{N+1}(x_0) - f(x_0)$$

and, using our knowledge of absolute values,

$$|f(x) - f(x_0)| \le |f(x) - f_{N+1}(x)| + |f_{N+1}(x) - f_{N+1}(x_0)| + |f_{N+1}(x_0) - f(x_0)|.$$

From (3), the first and third terms on the right are less than $\varepsilon/3$ while, according to (4), the second term on the right is less than $\varepsilon/3$. Hence

$$|f(x) - f(x_0)| \le \frac{\varepsilon}{3} + \frac{\varepsilon}{3} + \frac{\varepsilon}{3} = \varepsilon. \tag{5}$$

Since (5) holds for $|x - x_0| < \delta$, we conclude that f is continuous at each point x_0 on I.

If I includes an endpoint and if each f_n is continuous at this endpoint, the proof above shows that f is continuous at the endpoint also.

Suppose that $\{f_n\}$ is a uniformly convergent sequence of continuous functions on an interval I. If c is any point on I, we may form the new sequence $\{F_n\}$, where for each n,

$$F_n(x) = \int_c^x f_n(t)\,dt.$$

The next theorem establishes the convergence of such an integrated sequence.

Theorem 31. *Suppose that the sequence $\{f_n\}$ converges uniformly to f on the bounded interval I, and suppose that each f_n is continuous on I. Then the sequence $\{F_n(x)\}$ defined by*

$$F_n(x) = \int_c^x f_n(t)\,dt, \qquad n = 1, 2, \ldots$$

converges uniformly to $F(x) = \int_c^x f(t)\,dt$ on I.

Proof. First of all, by Theorem 2, f is continuous and so F may be defined.

Let L be the length of I. For any $\varepsilon > 0$ it follows from the uniform convergence that there is an N such that

$$|f_n(t) - f(t)| < \frac{\varepsilon}{L} \qquad \text{for all } n > N \text{ and all } t \text{ on } I.$$

We recall from elementary calculus the basic rule for estimating the size of an integral. If $f(x)$ is integrable on an interval $a \le x \le b$ and m and M are numbers such that

$$m \le f(x) \le M \qquad \text{for } a \le x \le b,$$

then

$$m(b-a) \le \int_a^b f(x)\,dx \le M(b-a).$$

It is also not difficult to show that

$$\left| \int_a^b f(x)\,dx \right| \le \int_a^b |f(x)|\,dx.$$

We use this last inequality to conclude that

$$|F_n(x) - F(x)| = \left| \int_c^x \{f_n(t) - f(t)\}\, dt \right| \leq \left| \int_c^x |f_n(t) - f(t)|\, dt \right|$$

$$\leq \frac{\varepsilon}{L}|x - c| \leq \varepsilon \qquad \text{for } n > N \text{ and all } x \text{ on } I.$$

Hence $\{F_n\}$ converges uniformly to F.

Example 2. Show that the sequence

$$F_n(x) = \frac{\log(1 + n^3x^2)}{n^2}, \qquad n = 1, 2, \ldots,$$

converges uniformly on the interval $I = \{x : 0 \leq x \leq 1\}$.

Solution. The sequence

$$F_n'(x) = \frac{2nx}{1 + n^3x^2} \equiv f_n(x)$$

is easily shown to converge to zero uniformly. In fact, we have

$$\lim_{n \to \infty} f_n(x) = 0 \qquad \text{for } 0 \leq x \leq 1,$$

and we can calculate ε_n in order to apply Theorem 29. The result is $\varepsilon_n = f_n(x_n) = 1/\sqrt{n} \to 0$ as $n \to \infty$. Therefore $\{f_n\}$ converges uniformly, and we now apply Theorem 31 to conclude that $\{F_n\}$ converges uniformly.

The next theorem allows us to draw conclusions about sequences which are differentiated term by term.

Theorem 32. *Suppose that $\{f_n\}$ is a sequence of functions each having a continuous derivative on a bounded interval I. Suppose that $f_n(x)$ converges to $f(x)$ on I and that the derived sequence $\{f_n'(x)\}$ converges uniformly to $g(x)$ on I. Then $g(x) = f'(x)$ on I.*

Proof. According to Theorem 30, g is continuous on I. Let c be any point on I. For each n we write

$$\int_c^x f_n'(t)\, dt = f_n(x) - f_n(c),$$

and we observe that the hypotheses of Theorem 31 are satisfied for this sequence of integrals. Therefore

$$\int_c^x g(t)\, dt = \lim_{n \to \infty} \int_c^x f_n'(t)\, dt = \lim_{n \to \infty} [f_n(x) - f_n(c)].$$

However, $\{f_n(x)\}$ converges to $f(x)$ for each x, and so

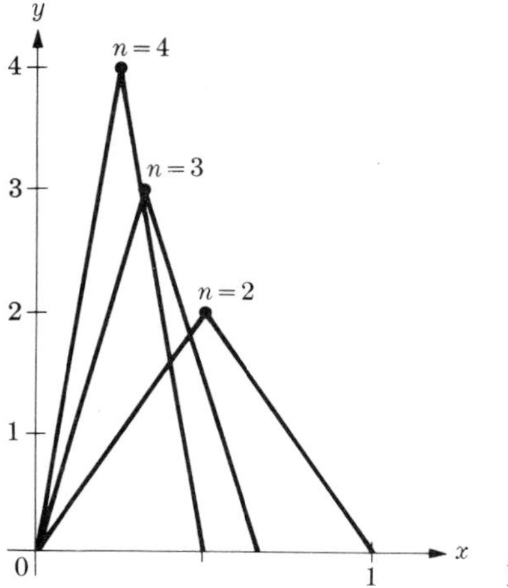

Fig. 3-11

$$\int_c^x g(t)\,dt = f(x) - f(c).$$

The left side may be differentiated with respect to x and, applying the Fundamental Theorem of Calculus, we find $g(x) = f'(x)$.

It is natural to ask whether or not the various hypotheses of the above theorems are absolutely essential. For example, suppose that $\{f_n\}$ is a sequence of continuous functions which converges on an interval I to a continuous function f. Is it always true that

$$\int_c^x f_n(x)\,dx \to \int_c^x f(x)\,dx \qquad \text{as } n \to \infty\,?$$

The answer is no! The hypothesis of uniform convergence (or some similar hypothesis) is required. To illustrate this point, we form the sequence

$$f_n(x) = \begin{cases} n^2 x, & 0 \le x \le \dfrac{1}{n}, \\ -n^2 x + 2n, & \dfrac{1}{n} \le x \le \dfrac{2}{n}, \\ 0, & \dfrac{2}{n} \le x \le 1. \end{cases}$$

The graph of f_n is shown in Fig. 3-11. As n increases, the triangle becomes narrower and taller. We see easily that

$$f_n(x) \to f(x) \equiv 0 \qquad \text{for all } x \text{ on } [0, 1].$$

On the other hand, $\int_0^1 f_n(x)\,dx = 1$ for every n, since the integral is exactly the area of the triangle. This area is the same for every function. But

$$\int_0^1 f(x)\,dx = 0.$$

Hence

$$1 = \int_0^1 f_n(x)\,dx \nrightarrow \int_0^1 f(x)\,dx = 0.$$

(See also Problem 16.)

Problems

In Problems 1 through 14, show that $f_n(x)$ converges to $f(x)$ for each x on I, and determine whether or not the convergence is uniform.

1. $f_n(x) = \dfrac{x}{1+nx}$, $f(x) = 0$, $I = [0, 1]$

2. $f_n(x) = \dfrac{\sin nx}{\sqrt{n}}$, $f(x) = 0$, $I = [0, 1]$

3. $f_n(x) = \dfrac{nx^2}{1+nx}$, $f(x) = x$, $I = [0, 1]$

4. $f_n(x) = \dfrac{nx}{1+n^2x^2}$, $f(x) = 0$, $I = [1, 2]$

5. $f_n(x) = \dfrac{n^2x}{1+n^3x^2}$, $f(x) = 0$, $I = [a, \infty)$, $a > 0$

6. $f_n(x) = nxe^{-nx^2}$, $f(x) = 0$, $I = [0, 1]$

7. $f_n(x) = \dfrac{1}{x} + \dfrac{1}{n}\sin\dfrac{1}{nx}$, $f(x) = \dfrac{1}{x}$, $I = (0, 1]$

8. $f_n(x) = \dfrac{\sin nx}{nx}$, $f(x) = 0$, $I = (0, \infty)$

9. $f_n(x) = (x^n - x^{n+1})$, $f(x) = 0$, $I = [0, 1]$

10. $f_n(x) = \sqrt{n}(x^n - x^{n+1})$, $f(x) = 0$, $I = [0, 1]$

11. $f_n(x) = \dfrac{1-x^{n+1}}{1-x}$, $f(x) = \dfrac{1}{1-x}$, $I = [-\frac{1}{2}, \frac{1}{2}]$

12. $f_n(x) = \dfrac{1-x^{n+1}}{1-x}$, $f(x) = \dfrac{1}{1-x}$, $I = (-1, 1)$

13. $f_n(x) = n^2x^n(1-x)$, $f(x) = 0$, $I = [0, 1]$

14. $f_n(x) = n^3x(1-x)^n$, $f(x) = 0$, $I = [0, 1]$

15. Show that the sequence $f_n(x) = x^n$ does not converge uniformly on $[0, 1]$.

16. Given the sequence $f_n(x) = (n+1)(n+2)x^n(1-x)$, show that $f_n(x) \to f(x) \equiv 0$ for x on $[0, 1]$. Decide whether or not

$$\int_0^1 f_n(x)\,dx \to \int_0^1 f(x)\,dx.$$

What can you conclude about the uniformity of the convergence of $f_n(x)$?

17. Given $f_n(x) = 2n^\alpha x/(1 + n^\beta x^2)$ with $\beta > \alpha \geq 0$. Find the values of α and β for which this sequence converges uniformly on $[0, 1]$.

18. a) If $f_n(x)$ and $g_n(x)$ converge uniformly on I, show that $f_n(x) + g_n(x)$ converges uniformly on I.
    b) If $f_n(x)$ converges uniformly on I and if c is a constant, show that $g_n(x) = cf_n(x)$ converges uniformly on I.

19. Show that the sequence

$$f_n(x) = x - \frac{x^n}{n}$$

converges uniformly on $[0, 1]$. Show that the sequence

$$f_n'(x) = 1 - x^{n-1}$$

does not converge uniformly on $[0, 1]$.

In each of Problems 20 through 23, a sequence $f_n(x)$ is given. Decide whether or not $f_n'(x)$ converges uniformly. Also decide whether or not

$$F_n(x) = \int_0^x f_n(t)\,dt$$

converges uniformly.

20. $f_n(x) = \dfrac{x}{1 + n^2 x}$, $I = [0, 1]$

21. $f_n(x) = \dfrac{n^2 x}{1 + n^3 x^2}$, $I = [0, 1]$

22. $f_n(x) = \dfrac{2 + nx^2}{2 + nx}$, $I = [0, 1]$

23. $f_n(x) = nxe^{-nx^2}$, $I = [0, 1]$

24. Show that the sequence $f_n(x) = (\sin x)^{1/n}$ converges, but not uniformly, on $I = \{x : 0 \leq x \leq \pi\}$.

25. Show that the sequence

$$f_n(x) = \left(\frac{\sin x}{x}\right)^{1/n}$$

converges, but not uniformly, on $I = \{x : 0 < x < \pi\}$.

26. Suppose that $f_n(x)$ and $g_n(x)$ are sequences of continuous functions which converge uniformly on a closed, bounded interval I, to the functions $f(x)$ and $g(x)$, respectively. Show that $h_n(x) = f_n(x)g_n(x)$ converges uniformly on I to $h(x) = f(x)g(x)$.

12. Uniform Convergence of Series

An infinite series

$$\sum_{k=1}^{\infty} a_k = a_1 + a_2 + \cdots + a_n + \cdots \tag{1}$$

has the sequence of partial sums

$$s_1, s_2, \ldots, s_n, \ldots \tag{2}$$

defined by

$$s_n = \sum_{k=1}^{n} a_k.$$

The convergence of the infinite series (1) is equivalent, by definition, to the convergence of the sequence of partial sums (2). The infinite series of functions

$$\sum_{k=1}^{\infty} u_k(x)$$

converges uniformly on an interval I if and only if the sequence $\{s_n(x)\}$ of partial sums converges uniformly on this interval. We observe that all the results of Section 11 on the uniform convergence of sequences translate at once into the corresponding results for series. In this way we obtain the next three theorems.

Theorem 30′. *Suppose that the series $\Sigma_{k=1}^{\infty} u_k(x)$ converges uniformly to $s(x)$ on the interval I. If each u_k is continuous on I, then s is continuous on I.*

Theorem 31′. *Suppose that the series $\Sigma_{k=1}^{\infty} u_k(x)$ converges uniformly to $s(x)$ on I and that each u_k is continuous on I. If we define $U_k(x) = \int_c^x u_k(t)\,dt$, $S(x) = \int_c^x s(t)\,dt$, then $\Sigma_{k=1}^{\infty} U_k(x)$ converges uniformly to $S(x)$ on I.*

Theorem 32′. *Suppose that $\Sigma_{k=1}^{\infty} u_k(x)$ converges to $s(x)$ on I and that*

$$\sum_{k=1}^{\infty} u_k'(x)$$

converges uniformly to $t(x)$ on I. If each u_k' is continuous on I, then $t(x) = s'(x)$.

The following theorem gives a simple and useful *indirect* test for uniform convergence of series. One of its virtues is that we may apply the test without finding the sum of the series—i.e., without obtaining the limit of the sequence of partial sums.

Theorem 33 (Weierstrass M-test). *Suppose that $|u_k(x)| \le M_k$ for all x on I where, for each k, M_k is a positive constant. If the series $\Sigma_{k=1}^{\infty} M_k$ converges, then the series $\Sigma_{k=1}^{\infty} u_k(x)$ converges uniformly on I.*

PROOF. For each x on I, the series $\Sigma_{k=1}^{\infty} |u_k(x)|$ *converges by the Comparison* Test (Theorem 9). Therefore $\Sigma_{k=1}^{\infty} u_k(x)$ converges, and we call the limit $s(x)$. We define

$$S = \sum_{j=1}^{\infty} M_j, \qquad S_n = \sum_{j=1}^{n} M_j,$$

and we note that $\lim_{n\to\infty} (S - S_n) = 0$. Since $s_n(x) = \Sigma_{j=1}^{n} u_j(x)$, we have

$$|s(x) - s_n(x)| = \left| \sum_{j=n+1}^{\infty} u_j(x) \right| \leq \sum_{j=n+1}^{\infty} |u_j(x)| \leq \sum_{j=n+1}^{\infty} M_j$$
$$= S - S_n \to 0.$$

Since the numbers $S - S_n$ are independent of x, the convergence is uniform.

Remarks. If the interval I is closed and if each u_n is continuous, then the theorem requires that M_n be either the *maximum* of $|u_n(x)|$ on I or some conveniently chosen number larger than this maximum. If I is open or half-open and $u_n(x)$ tends to a limit at each endpoint, then M_n must be larger than the maximum of the function extended to the closed interval. We shall see that the quantities M_n may frequently be found by inspection. It is important also that we remember many of the convergent series of positive constants studied in Sections 2, 3, and 4.

Example 1. Show that the series

$$\sum_{n=1}^{\infty} \frac{\cos nx}{n^2}$$

converges uniformly for all x.

Solution. We have

$$\left| \frac{\cos nx}{n^2} \right| = \frac{|\cos nx|}{n^2} \leq \frac{1}{n^2}.$$

Since $\Sigma_{n=1}^{\infty} 1/n^2$ is a known convergent series, the result follows.

The Weierstrass M-test may frequently be combined with the ratio test for convergence of constants to yield results on uniform convergence. We illustrate with an example.

Example 2. Given the series

$$\sum_{n=0}^{\infty} (n + 1)x^n,$$

determine an interval of the form $|x| \leq h$ on which the series is uniformly convergent.

Solution. If $|x| \leq h$, then $|(n + 1)x^n| \leq (n + 1)h^n = M_n$. We apply the Ratio Test to determine when $\Sigma_{n=1}^{\infty} M_n$ converges. We have

$$\frac{M_{n+1}}{M_n} = \frac{(n+2)h^{n+1}}{(n+1)h^n} = h\frac{1+(2/n)}{1+(1/n)} \to h \quad \text{as } n \to \infty.$$

Accordingly, the series $\Sigma_{n=1}^{\infty} M_n$ converges if $h < 1$ and does not converge if $h \geq 1$. The original series converges uniformly on any interval of the form $|x| \leq h$ with $h < 1$.

EXAMPLE 3. Discuss for uniform convergence the series

$$\sum_{n=1}^{\infty} \frac{(-1)^{n-1}x^n}{n^2}.$$

SOLUTION. If $|x| \leq h$, then

$$\left|\frac{(-1)^{n-1}x^n}{n^2}\right| \leq \frac{h^n}{n^2} = M_n.$$

The series $\Sigma_{n=1}^{\infty} M_n$ converges for $h \leq 1$. The given series converges uniformly on any interval of the form $|x| \leq h$ with $h \leq 1$.

EXAMPLE 4. Given the relation

$$\frac{1}{1-x^4} = \sum_{n=0}^{\infty} x^{4n}, \qquad -1 < x < 1,$$

show that

$$\frac{1}{(1-x^4)^2} = \sum_{n=1}^{\infty} nx^{4(n-1)}, \qquad -1 < x < 1.$$

SOLUTION. Setting $u_n(x) = x^{4n}$, we see that the series

$$\sum_{n=1}^{\infty} u_n'(x) = \sum_{n=1}^{\infty} 4nx^{4n-1}$$

converges uniformly for $|x| \leq h$ with $h < 1$. Therefore, applying Theorem 32′, we find that the sum of the derived series is

$$\frac{d}{dx}[(1-x^4)^{-1}] = 4x^3(1-x^4)^{-2} = \sum_{n=1}^{\infty} 4nx^{4n-1}.$$

Dividing by $4x^3$, we get the desired expansion.

PROBLEMS

In each of Problems 1 through 14, determine h such that the given series converges uniformly on I.

1. $\sum_{n=0}^{\infty} x^n, \quad I = \{x : |x| \leq h\}$

2. $\sum_{n=1}^{\infty} \frac{(-1)^n x^n}{n^2}$, $I = \{x : |x| \le h\}$

3. $\sum_{n=0}^{\infty} n^2 x^n$, $I = \{x : |x| \le h\}$

4. $\sum_{n=0}^{\infty} \frac{x^n}{n!}$, $I = \{x : |x| \le h\}$

5. $\sum_{n=0}^{\infty} \frac{(10x)^n}{n!}$, $I = \{x : |x| \le h\}$

6. $\sum_{n=0}^{\infty} \frac{(-1)^n 3^n x^n}{(n+1)2^n}$, $I = \{x : |x| \le h\}$

7. $\sum_{n=1}^{\infty} \frac{n!(x-3)^n}{1 \cdot 3 \cdot 5 \cdots (2n-1)}$, $I = \{x : |x-3| \le h\}$

8. $\sum_{n=1}^{\infty} \frac{(n!)^2 (x+1)^n}{(2n)!}$, $I = \{x : |x+1| \le h\}$

9. $\sum_{n=1}^{\infty} \frac{(-1)^n x^n}{(n+1)\log(n+1)}$, $I = \{x : |x| \le h\}$

10. $\sum_{n=1}^{\infty} \frac{(\log n) 2^n x^n}{3^n n^2}$, $I = \{x : |x| \le h\}$

11. $\sum_{n=1}^{\infty} \frac{(1 - x^{2n})^{1/2}}{3^n}$, $I = \{x : |x| \le h\}$

12. $\sum_{n=1}^{\infty} x(1-x)^n$, $I = \{x : |x| \le h\}$

13. $\sum_{n=1}^{\infty} \frac{x^2}{n(4 + nx^2)}$, $I = \{x : |x| \le h\}$

14. $\sum_{n=1}^{\infty} \frac{1}{n^x}$, $I = \{x : h \le x < \infty\}$

15. Given that $\Sigma_{n=1}^{\infty} |b_n|$ converges, show that $\Sigma_{n=1}^{\infty} b_n \sin nx$ converges uniformly for all x.

16. Given that $\Sigma_{n=1}^{\infty} n|b_n|$ converges and that $f(x) = \Sigma_{n=1}^{\infty} b_n \sin nx$, show that $f'(x) = \Sigma_{n=1}^{\infty} nb_n \cos nx$, and that both series converge uniformly for all x.

17. Find those values of h for which $\Sigma_{n=1}^{\infty} (x \log x)^n$ converges uniformly for $0 < x \le h$.

18. Find the values of h that $\Sigma_{n=1}^{\infty} (\sin x)^n$ converges uniformly for $|x| \le h$.

19. Show that $\Sigma_{n=0}^{\infty} (1+x)x^n$ converges uniformly for $-1 \le x \le 0$.

20. Given the series expansions

$$A(x, \lambda) = \sum_{n=0}^{\infty} \frac{(-1)^n \lambda^{2n+1} x^{2n+1}}{(2n+1)!},$$

$$B(x, \lambda) = \sum_{n=0}^{\infty} \frac{(-1)^n \lambda^{2n} x^{2n}}{(2n)!},$$

use the theorems of this section to show that

a) $dA/dx = \lambda B$, b) $dB/dx = -\lambda A$,

c) $d^2A/dx^2 + \lambda^2 A = 0$, d) $d^2B/dx^2 + \lambda^2 B = 0$.

Do not use the fact that $A(x, \lambda) = \sin \lambda x$, $B(x, \lambda) = \cos \lambda x$.

21. Show, by successive differentiation of the series $(1 - x)^{-1} = \Sigma_{n=0}^{\infty} x^n$ that

$$\frac{1}{(1-x)^k} = \sum_{n=0}^{\infty} \frac{(n+1)(n+2)\cdots(n+k-1)}{(k-1)!} x^n, \qquad k \geq 2$$

and that the convergence is uniform for $|x| \leq h$, with $h < 1$.

22. Let A be an $n \times n$ matrix, and consider the series

$$I + \frac{1}{1!}A + \frac{1}{2!}A^2 + \cdots + \frac{1}{k!}A^k + \cdots$$

a) Define convergence for such a series in terms of the limit of the matrices forming the partial sums.

b) Show that the above series converges if

$$A = \begin{pmatrix} 1 & 1 \\ 0 & 1 \end{pmatrix}.$$

13. Integration and Differentiation of Power Series

We discussed the elementary properties of power series in Sections 5 through 10. In this section we shall show how the notion of uniform convergence can be used to extend and amplify many of the earlier results.

Theorem 34. *Suppose that the power series $\Sigma_{n=0}^{\infty} a_n x^n$ converges for some $x_1 \neq 0$. Then* (a), *the series converges absolutely for all x with $|x| < |x_1|$ and* (b), *it converges uniformly on any interval $|x| \leq h$ with $h < |x_1|$.*

PROOF. We recognize part (a) as Theorem 15 (page 124). As for (b), the proof of Theorem 15 given on page 124 shows that the convergence is uniform (although we did not say so at the time). Referring to the proof there, we see that the series

$$\sum_{n=0}^{\infty} M\left|\frac{x}{x_1}\right|^n \leq \sum_{n=0}^{\infty} M\left(\frac{h}{|x_1|}\right)^n = \sum_{n=0}^{\infty} M_n.$$

The series $\Sigma_{n=0}^{\infty} M_n$ is a convergent geometric series, and the Weierstrass M-test may be applied to yield uniform convergence.

REMARK. Theorem 34 also holds for the series $\Sigma_{n=0}^{\infty} a_n(x - c)^n$ with $|x - c| \leq h$.

Corollary. *If $\Sigma_{n=0}^{\infty} a_n x_1^n$ diverges, then $\Sigma_{n=0}^{\infty} a_n x^n$ diverges for all x with $|x| > |x_1|$.*

PROOF. If $\Sigma_{n=0}^{\infty} a_n x_2^n$ were convergent for $|x_2| > |x_1|$, then we could apply the above theorem to conclude that $\Sigma_{n=0}^{\infty} a_n x_1^n$ converges, a contradiction.

The next theorem shows the pattern for the convergence properties of power series in general. The result, which we now prove was stated without proof on page 124.

Theorem 16. *Let $\Sigma_{n=0}^{\infty} a_n x^n$ be any power series. Then either*

i) *the series converges only for $x = 0$; or*
ii) *the series converges for all x; or*
iii) *there is a number R such that the series converges for all x with $|x| < R$ and diverges for all x with $|x| > R$.*

PROOF. We have already encountered examples of series where (i) and (ii) hold. The series $\Sigma_{n=0}^{\infty} n!x^n$ converges only for $x = 0$, and the series

$$\sum_{n=0}^{\infty} \frac{x^n}{n!}$$

converges for all x. Therefore we need consider only alternative (iii). Then there is an $x_1 \neq 0$ where the series converges and an X_1 where the series diverges. We let

$$r_1 = |x_1| \qquad \text{and} \qquad R_1 = |X_1|$$

and note that $R_1 \geq r_1 > 0$. If $R_1 = r_1$, we select this value for R and (iii) is established. So we suppose that $R_1 > r_1$. We define an increasing sequence $r_1, r_2, \ldots, r_n, \ldots$ and a decreasing sequence $R_1, R_2, \ldots, R_n, \ldots$ in the following way. If the series converges for $\frac{1}{2}(r_1 + R_1)$, we define

$$r_2 = \frac{r_1 + R_1}{2} \qquad \text{and} \qquad R_2 = R_1.$$

If the series diverges for $\frac{1}{2}(r_1 + R_1)$, we define

$$r_2 = r_1 \qquad \text{and } R_2 = \frac{r_1 + R_1}{2}.$$

We continue the process inductively. If the series converges for $\frac{1}{2}(r_k + R_k)$, we define

$$r_{k+1} = \frac{r_k + R_k}{2} \qquad \text{and} \qquad R_{k+1} = R_k.$$

If the series diverges for $\frac{1}{2}(r_k + R_k)$, we define

$$r_{k+1} = r_k \qquad \text{and} \qquad R_{k+1} = \frac{r_k + R_k}{2}.$$

Figure 3-12 shows a typical situation for the two sequences. From the

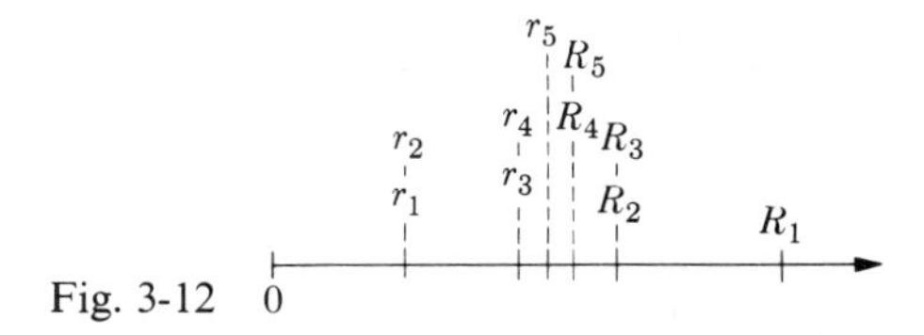

Fig. 3-12

manner of construction we see that

a) $r_k \leq r_{k+1} < R_{k+1} \leq R_k$, $\quad k = 1, 2, \ldots$
b) $R_k - r_k = (R_1 - r_1)/2^{k-1}$, $\quad k = 1, 2, \ldots$
c) The series converges for all x with $|x| < r_k$ and diverges for all x with $|x| > R_k$, $k = 1, 2, \ldots$ Now, from the Axiom of Continuity, (a) and (b) together imply that there is a number R such that $R_k \to R$ and $r_k \to R$. Part (c) shows that the series converges if $|x| < R$ and diverges if $|x| > R$.

Definitions. The number R in Theorem 7 is called the **radius of convergence** of the series. We define $R = 0$ when (i) holds and $R = +\infty$ when (ii) holds.

Corollary. *If a power series has a positive radius of convergence R, it converges uniformly on any interval $|x| \leq h$ where $h < R$. In fact, the series of absolute values $\Sigma_{n=0}^{\infty} |a_n x^n|$ also converges uniformly.*

PROOF. Given h, we select $x_1 = \frac{1}{2}(h + R)$ and observe that the series $\Sigma_{n=0}^{\infty} a_n x_1^n$ converges. Now we apply Theorem 34.

Theorems 19 and 21 on term-by-term differentiation and term-by-term integration of power series which were proved earlier may now be established as simple consequences of the uniform convergence properties of power series. We state the results, leaving as exercises for the reader those proofs which use the more advanced methods. (See Exercises 12 through 15.)

Theorem 35. *Suppose that $R > 0$ and that the power series $\Sigma_{n=0}^{\infty} a_n x^n$ converges for $|x| < R$. Then the series obtained by differentiating it term by term also converges for $|x| < R$.*

(The above theorem is proved on page 138.)

Theorem 36. *Suppose that $R > 0$ and that*

$$f(x) = \sum_{n=0}^{\infty} a_n (x - c)^n$$

converges for $|x - c| < R$. Then f is continuous and has continuous derivatives of all orders, which are given on $|x - c| < R$ by differentiating the series the appropriate number of times. Moreover, if

$$F(x) = \int_c^x f(t)\,dt,$$

then F is given on the same interval by the series obtained by term-by-term integration of the series for f.

The fact that f is continuous is the content of Theorem 20. However, since $u_n(x) = a_n(x - c)^n$ is continuous for each n, Theorem 30′ yields the continuity of f at once.

Corollary. *If $R > 0$ and*

$$f(x) = \sum_{n=0}^{\infty} a_n(x - c)^n \qquad \text{for } |x - c| < R,$$

then

$$a_n = \frac{f^{(n)}(c)}{n!}.$$

The reader may verify the formula for a_n by repeated differentiation of the series for f, after which the value $x = c$ is inserted.

Remark. In Theorems 35 and 36 nothing has been said about convergence at the endpoints. If a power series with radius of convergence $R > 0$ converges uniformly for $|x - c| \leq R$, then the *integrated* series will also converge uniformly for $|x - c| \leq R$ (Theorem 30′). On the other hand, the differentiated series may not converge at all at the endpoints. For example, the series

$$f(x) = \sum_{n=2}^{\infty} \frac{x^n}{n(n-1)}$$

converges uniformly for $|x| \leq 1$. But the series

$$f''(x) = \sum_{n=0}^{\infty} x^n, \qquad |x| < 1,$$

does not converge at either endpoint.

Once a power-series expansion for a function is known (say by a Taylor or Maclaurin expansion), the series expansions for related functions may be found by differentiation and integration. Further results may be obtained by using these in connection with the next theorem, on substitution.

Theorem 37 (Simple Substitutions). (a) *If $f(u) = \Sigma_{n=0}^{\infty} a_n(u - c_0)^n$ for $|u - c_0| < R$, and if $c_0 = bc + d$ with $b \neq 0$, then*

$$f(bx + d) = \sum_{n=0}^{\infty} a_n b^n (x - c)^n \qquad \text{for } |x - c| < \frac{R}{|b|}.$$

b) *If* $f(u) = \Sigma_{n=0}^{\infty} a_n u^n$ *for* $|u| < R$, *then for any positive integer* k,

$$f(x^k) = \sum_{n=0}^{\infty} a_n x^{kn} \qquad \textit{for } |x| < R^{1/k}.$$

PROOF. (a) If $u = bx + d$, then $u - c_0 = b(x - c)$ and

$$a_n(u - c_0)^n = a_n b^n (x - c)^n$$

for each n. Therefore the following inequalities are equivalent:

$$|u - c_0| < R \Leftrightarrow |b||x - c| < R \Leftrightarrow |x - c| < \frac{R}{|b|}.$$

The proof of part (b) is similar.

EXAMPLE 1. Use simple substitution to obtain the expansion

$$(1 - x^8)^{-1} = \sum_{n=0}^{\infty} x^{8n}, \qquad |x| < 1.$$

SOLUTION. We have

$$(1 - u)^{-1} = \sum_{n=0}^{\infty} u^n,$$

and we use Theorem 10(b) with $u = x^8$ to get the result. Note that since $R = 1$, then $R^{1/8} = 1$.

EXAMPLE 2. Find the Maclaurin expansion for Arctan (x^2).

SOLUTION. We have the expansion

$$\frac{1}{1 + u} = \sum_{n=0}^{\infty} (-1)^n u^n, \qquad |u| < 1.$$

Therefore, letting $u = v^2$, we get

$$\frac{1}{1 + v^2} = \sum_{n=0}^{\infty} (-1)^n v^{2n}, \qquad |v| < 1.$$

Now we integrate term by term, obtaining

$$\operatorname{Arctan} v = \sum_{n=0}^{\infty} \frac{(-1)^n v^{2n+1}}{2n + 1}, \qquad |v| < 1.$$

Setting $v = x^2$, we conclude that

$$\operatorname{Arctan}(x^2) = \sum_{n=0}^{\infty} \frac{(-1)^n x^{4n+2}}{2n + 1}, \qquad |x| < 1.$$

PROBLEMS

In each of Problems 1 through 10, find the Maclaurin expansion for the function f, using differentiation, integration, and simple substitution, whenever necessary.

1. $f(x) = \dfrac{1}{(1-x^2)^2}$
2. $f(x) = \log(1 + x^3)$
3. $f(x) = e^{(1+x^2)}$
4. $f(x) = \arctan(x^3)$
5. $f(x) = [1 + (x/3)^3]^{-3}$
6. $f(x) = \arcsin(x^2)$
7. $f(x) = \sin\left(\dfrac{3x^2}{\pi}\right)$
8. $f(x) = \arccos(x^3)$
9. $f(x) = \begin{cases} (e^{x^2} - 1)/x^2, & x \neq 0 \\ 1 \quad , & x = 0 \end{cases}$
10. $f(x) = (1 + x^2)^{-3/2}$
11. Write a complete proof of Theorem 34(b).
12. Use the results on uniform convergence to prove Theorem 35.
13. Use the results on uniform convergence to prove Theorem 36.
14. Write a complete proof of the Corollary to Theorem 36.
15. Prove Theorem 37(b).
16. Show that if a power series converges absolutely at an endpoint, then the series obtained by term-by-term integration converges at the same point.
17. Let k be any positive integer. Give an example of a power series for a function f such that all the series obtained by term-by-term differentiation k times converge at the endpoint, while the series obtained by differentiating $k + 1$ times diverges at the endpoint.

14. Double Sequences and Series

A **double sequence** is a function which has as its domain S some set of ordered pairs (i, j) of nonnegative integers and as its range a portion of the real number system. We may write $f: S \to R^1$ for such a function. A more natural way uses double subscripts for elements of the domain and a letter with these subscripts as an element of the range corresponding to the subscript. We may write f in the form

$$a_{ij}, \qquad i, j = 0, 1, 2, \ldots .$$

We obtain a simple way of looking at a double sequence by writing a rectangular array, as shown in Fig. 3-13. If all the rows and columns of such an array terminate, then the sequence is called **finite**; if the rows and columns continue indefinitely to the right and downward, the double sequence is called **infinite**.

$$\begin{array}{ccccc}
u_{11} & u_{12} & \cdots & u_{1n} & \cdots \\
u_{21} & u_{22} & \cdots & u_{2n} & \cdots \\
\vdots & \vdots & \cdots & \vdots & \cdots \\
u_{m1} & u_{m2} & \cdots & u_{mn} & \cdots \\
\vdots & \vdots & \cdots & \vdots & \cdots
\end{array}$$

Fig. 3-13

$$\begin{array}{cccc|cc}
a_{11} & a_{12} & \cdots & a_{1n} & a_{1,n+1} & \cdots \\
a_{21} & a_{22} & \cdots & a_{2n} & a_{2,n+1} & \cdots \\
\vdots & \vdots & \cdots & \vdots & & \cdots \\
a_{m1} & a_{m2} & \cdots & a_{mn} & a_{m,n+1} & \cdots \\
\hline
a_{m+1,1} & a_{m+1,2} & \cdots & \cdots & \cdots & \cdots
\end{array}$$

Fig. 3-14

Definition. We say that **the double sequence** $\{u_{mn}\}$, m, $n = 1, 2, \ldots$, **tends to a limit** L **as** $(m, n) \to \infty$ if and only if for each $\varepsilon > 0$ there is a positive integer N such that

$$|u_{mn} - L| < \varepsilon \qquad \text{whenever both } m > N \text{ and } n > N.$$

It can be shown that if such a number L exists, then it must be unique. Also, the customary theorems on limits which we established for ordinary sequences are easily extended to double sequences. We shall use the symbols

$$\lim_{(m,n)\to\infty} u_{mn} = L \qquad \text{and} \qquad u_{mn} \to L \text{ as } (m, n) \to \infty$$

for the double limit of a sequence.

Given a double sequence $\{a_{mn}\}$, m, $n = 1, 2, \ldots$, we define its **partial sum** s_{mn} by the formula

$$s_{mn} = \sum_{j=1}^{m} \sum_{k=1}^{n} a_{jk}. \tag{1}$$

Pictorially, s_{mn} denotes the sum of all the terms in the rectangular array indicated in Fig. 3-14.

Definition. The **sum of a double sequence** $\{a_{mn}\}$, m, $n = 1, 2, \ldots$, is defined as

$$\lim_{m,n\to\infty} s_{mn} \tag{2}$$

where s_{mn} is given by (1). The limit may or may not exist. We write the expression

$$\sum_{m,n=1}^{\infty} a_{mn}$$

instead of (2) and call this the **infinite double series* whose terms are** a_{mn}. If the limit in (2) exists, we say that the double series **converges**; otherwise we say it **diverges**.

* The definition given here is informal. A more formal definition is the following: *An* **infinite double series** *is an order pair* $(\{u_{mn}\}, \{s_{mn}\})$ *of infinite double sequences in which*

$$s_{mn} = \sum_{i=1}^{m} \sum_{j=1}^{n} u_{ij} \qquad \textit{for each } m \textit{ and } n.$$

The finite double series $(\{u_{mn}\}, \{s_{mn}\})$ *is denoted by* $\sum_{m,n=1}^{\infty} u_{mn}$. When no confusion can arise we also denote by $\sum_{m,n=1}^{\infty} u_{mn}$ the limit of the double sequence $\{s_{mn}\}$ when it exists.

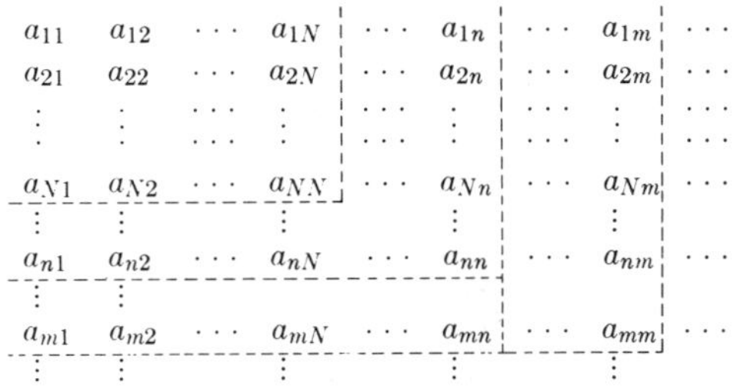

$$
\begin{array}{cccc|ccc|ccc|c}
a_{11} & a_{12} & \cdots & a_{1N} & \cdots & a_{1n} & & \cdots & a_{1m} & & \cdots \\
a_{21} & a_{22} & \cdots & a_{2N} & \cdots & a_{2n} & & \cdots & a_{2m} & & \cdots \\
\vdots & \vdots & \cdots & \vdots & \cdots & \vdots & & \cdots & \vdots & & \cdots \\
a_{N1} & a_{N2} & \cdots & a_{NN} & \cdots & a_{Nn} & & \cdots & a_{Nm} & & \cdots \\
\hline
\vdots & \vdots & & \vdots & & \vdots & & & \vdots & & \\
a_{n1} & a_{n2} & \cdots & a_{nN} & \cdots & a_{nn} & & \cdots & a_{nm} & & \cdots \\
\hline
\vdots & \vdots & & & & & & & \vdots & & \\
a_{m1} & a_{m2} & \cdots & a_{mN} & \cdots & a_{mn} & & \cdots & a_{mm} & & \cdots \\
\hline
\vdots & \vdots & & \vdots & & \vdots & & & \vdots & &
\end{array}
$$

Fig. 3-15

Many of the theorems on double series are direct extensions of those for single series. The next theorem follows immediately from the theory of limits.

Theorem 38. *Suppose that* $\Sigma_{m,n=1}^{\infty} a_{mn}$ *and* $\Sigma_{m,n=1}^{\infty} b_{mn}$ *are convergent double series and suppose that* c *and* d *are constants. Then* $\Sigma_{m,n=1}^{\infty}(ca_{mn} + db_{mn})$ *is convergent and*

$$\sum_{m,n=1}^{\infty} (ca_{mn} + db_{mn}) = c \sum_{m,n=1}^{\infty} a_{mn} + d \sum_{m,n=1}^{\infty} b_{mn}.$$

The following theorem on double series with positive terms is a direct analog of the corresponding theorem for single series. Note that part (ii) of the theorem is a comparison test.

Theorem 39. *Suppose that* $a_{mn} \geq 0$ *for all* m *and* n.

i) *If there is a number* M *such that the partial sums* $s_{mn} \leq M$, *then* $\Sigma_{m,n=1}^{\infty} a_{mn}$ *converges to a number* $s \leq M$, *and each* $s_{mn} \leq s$.
ii) **(Comparison Test).** *If* $0 \leq a_{mn} \leq A_{mn}$ *and* $\Sigma_{m,n=1}^{\infty} A_{mn}$ *converges, then* $\Sigma_{m,n=1}^{\infty} a_{mn}$ *does also, and* $\Sigma_{m,n=1}^{\infty} a_{mn} \leq \Sigma_{m,n=1}^{\infty} A_{mn}$.

PROOF. (i) It is clear that the *special* partial sums s_{nn} are nondecreasing and $s_{nn} \leq M$ for all n. By the Axiom of Continuity, we know there is a number $s \leq M$ such that $s_{nn} \to s$ and $s_{nn} \leq s$ for each n. Therefore, for any $\varepsilon > 0$, there is an N such that

$$s - \varepsilon \leq s_{nn} \leq s \qquad \text{for all } n > N.$$

Let m, n be any two numbers larger than N. For convenience, suppose that $m \geq n$. It follows that

$$s - \varepsilon \leq s_{nn} \leq s_{mn} \leq s_{mm} \leq s.$$

Figure 3-15 shows the various rectangular blocks of a_{ij} which make up the partial sums. Hence $s_{mn} \to s$ as $(m, n) \to \infty$ and each $s_{mn} \leq s$. Part (ii) follows directly from (i).

EXAMPLE 1. Show that the double series

$$\sum_{p,q=1}^{\infty} \frac{1}{p^2 q^2}$$

is convergent.

SOLUTION. We have

$$\sum_{p,q=1}^{m,n} \frac{1}{p^2 q^2} = s_{mn},$$

and it is easy to see that

$$s_{mn} = \left(\sum_{p=1}^{m} \frac{1}{p^2}\right)\left(\sum_{q=1}^{n} \frac{1}{q^2}\right).$$

Since $\Sigma_{k=1}^{\infty} 1/k^2 = M < \infty$, it follows that $s_{mn} < M^2$ for all m, n. Therefore the double series is convergent.

EXAMPLE 2. Show that the series

$$\sum_{m,n=1}^{\infty} \frac{1}{m^4 + n^4}$$

is convergent.

SOLUTION. Since, by Example 1, the double series

$$\frac{1}{2} \sum_{p,q=1}^{\infty} \frac{1}{p^2 q^2} = \sum_{p,q=1}^{\infty} \frac{1}{2p^2 q^2}$$

converges and since, for any numbers p and q, we have $2p^2q^2 \leq p^4 + q^4$, the comparison test may be used to yield the result.

For single series with positive and negative terms the notion of absolute convergence plays an important part. We now establish the basic theorems for absolute convergence of double series.

Theorem 40. *Suppose that* $\Sigma_{m,n=1}^{\infty} |a_{mn}|$ *converges. If we define*

$$b_{mn} = \begin{cases} a_{mn} & \text{whenever} \quad a_{mn} \geq 0, \\ 0 & \text{whenever} \quad a_{mn} < 0, \end{cases}$$

$$c_{mn} = \begin{cases} 0 & \text{whenever} \quad a_{mn} \geq 0, \\ -a_{mn} & \text{whenever} \quad a_{mn} < 0, \end{cases}$$

then $b_{mn} + c_{mn} = |a_{mn}|$, $b_{mn} - c_{mn} = a_{mn}$, *and* $\Sigma_{m,n=1}^{\infty} b_{mn}$, $\Sigma_{m,n=1}^{\infty} c_{mn}$ *converge. Calling the sum of these last two series B and C, respectively, we have*

$$\sum_{m,n=1}^{\infty} a_{mn} = B - C, \qquad \sum_{m,n=1}^{\infty} |a_{mn}| = B + C.$$

Furthermore,

$$\left| \sum_{m,n=1}^{\infty} a_{mn} \right| \leq \sum_{m,n=1}^{\infty} |a_{mn}|.$$

PROOF. Since $b_{mn} \geq 0$, $c_{mn} \geq 0$, and $b_{mn} + c_{mn} = |a_{mn}|$, the Comparison Test (Theorem 39 (ii)) shows that $\Sigma_{m,n=1}^{\infty} b_{mn}$ and $\Sigma_{m,n=1}^{\infty} c_{mn}$ converge. The remaining results follow from Theorem 38 and from the fact that

$$|B - C| \leq B + C,$$

when B and C are any nonnegative numbers.

Definition. A series $\Sigma_{m,n=1}^{\infty} a_{mn}$ is called **absolutely convergent** if the series $\Sigma_{m,n=1}^{\infty} |a_{mn}|$ converges.

Theorem 40 states that *a series which is absolutely convergent is itself convergent.*

The next theorem establishes the basic relation between double series, single series, and repeated or iterated series.

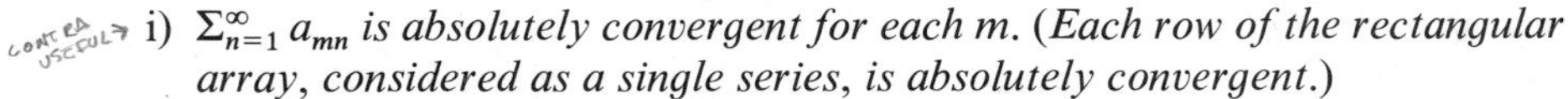

Theorem 41. *Suppose that $\Sigma_{m,n=1}^{\infty} a_{mn}$ is absolutely convergent. Then*

i) *$\Sigma_{n=1}^{\infty} a_{mn}$ is absolutely convergent for each m. (Each row of the rectangular array, considered as a single series, is absolutely convergent.)*

ii) *$\Sigma_{m=1}^{\infty} [\Sigma_{n=1}^{\infty} a_{mn}]$ is absolutely convergent. (The iterated sum, taking rows first, is absolutely convergent.)*

iii) *$\Sigma_{m=1}^{\infty} [\Sigma_{n=1}^{\infty} a_{mn}] = \Sigma_{m,n=1}^{\infty} a_{mn}$. (The iterated sum is equal to the double sum.) The same results hold if the roles of m and n are interchanged.*

iv) *$\Sigma_{p=2}^{\infty} [\Sigma_{m+n=p} a_{mn}]$ converges absolutely and equals the double sum.*

PROOF. Because of Theorem 40, it suffices to prove the result when $a_{mn} \geq 0$. Let

$$s = \sum_{m,n=1}^{\infty} a_{mn}.$$

Then it follows that $\Sigma_{n=1}^{N} a_{mn} \leq s$ for every N and every m. Thus (i) holds. Therefore we may write

$$A_m = \sum_{n=1}^{\infty} a_{mn} = \lim_{N \to \infty} \sum_{n=1}^{N} a_{mn}.$$

Also, since $s_{MN} \leq s$ always, we see that

$$\sum_{m=1}^{M} A_m = \lim_{N \to \infty} s_{MN} \leq s \qquad \text{for each } M.$$

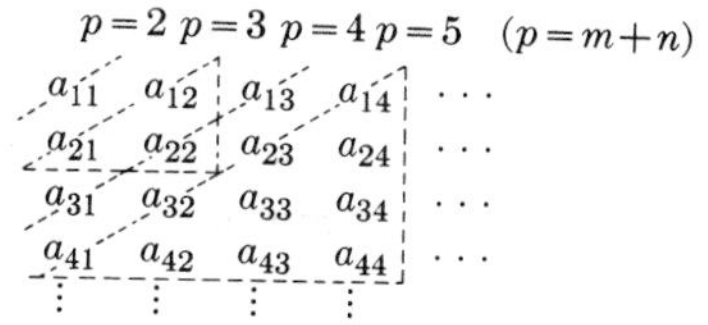

$$\begin{array}{ccccl} p=2 & p=3 & p=4 & p=5 & (p=m+n) \\ a_{11} & a_{12} & a_{13} & a_{14} & \cdots \\ a_{21} & a_{22} & a_{23} & a_{24} & \cdots \\ a_{31} & a_{32} & a_{33} & a_{34} & \cdots \\ a_{41} & a_{42} & a_{43} & a_{44} & \cdots \\ \vdots & \vdots & \vdots & \vdots & \end{array}$$

Fig. 3-16

We conclude that $\Sigma_{m=1}^{\infty} A_m$ is a convergent series, and so (ii) holds. Now let $\varepsilon > 0$ be given. There is an N_0 such that

$$s - \varepsilon < s_{MN} \le \sum_{m=1}^{M} A_m \le s \qquad \text{if } M > N_0 \text{ and } N > N_0.$$

This last statement establishes (iii).

To prove (iv), we first consider the meaning of the sum

$$\sum_{p=2}^{\infty} \sum_{m+n=p} a_{mn}.$$

The inner sum adds the elements of a diagonal, as shown in Fig. 3-16, and the outer sum adds all the diagonals. For any $\varepsilon > 0$, there is an N_0 such that $s - \varepsilon < s_{MN} \le s$ whenever $M > N_0$ and $N > N_0$. Suppose that $P = 2N_0 + 2$. Then the triangular set (Fig. 3-16) of all (m, n) such that

$$m + n = p \le P$$

contains the set of all (m, n) such that $m \le N_0 + 1$ and $n \le N_0 + 1$. This set is also contained in the set of all (m, n) such that $m \le P$ and $n \le P$. Therefore

$$s - \varepsilon < s_{N_0+1, N_0+1} \le \sum_{p=0}^{P} \left[\sum_{m+n=p} a_{mn} \right] \le s_{PP} \le s,$$

which implies statement (iv).

Theorem 42. *Suppose that $\Sigma_{n=0}^{\infty} a_{mn}$ converges absolutely for each m and that*

$$\sum_{m=1}^{\infty} \left[\sum_{n=1}^{\infty} |a_{mn}| \right] \tag{3}$$

converges. Then the double series converges absolutely. The same result holds with the roles of m and n interchanged.

PROOF. If s is the sum of the iterated series (3), we see immediately that

$$s_{MN} = \sum_{m=1}^{M} \left[\sum_{n=1}^{N} |a_{mn}| \right] \le \sum_{m=1}^{M} \left[\sum_{n=1}^{\infty} |a_{mn}| \right] \le s.$$

The result follows from Theorems 39 and 41.

REMARK. The above theorem states that *if each row of a double series is absolutely convergent and if the iterated series of absolute values converges, then the double series converges.*

A similar statement is true for columns.

EXAMPLE 3. Show that the double series

$$\sum_{m,n=1}^{\infty} \frac{1}{m^2(1+n^{3/2})}$$

converges.

SOLUTION. For fixed m, the series

$$\sum_{n=1}^{\infty} \frac{1}{m^2(1+n^{3/2})} < \sum_{n=1}^{\infty} \frac{1}{n^{3/2}},$$

which is a convergent p-series with $p = 3/2$. Denoting

$$A = \sum_{n=1}^{\infty} \frac{1}{n^{3/2}},$$

we see that

$$\sum_{m=1}^{\infty} \left[\sum_{n=1}^{\infty} \frac{1}{m^2(1+n^{3/2})} \right] \leq \sum_{m=1}^{\infty} \frac{1}{m^2} A,$$

which converges. The hypotheses of Theorem 42 are satisfied and the double series converges.

*****Theorem 43.** *Suppose that $\Sigma_{m=1}^{\infty} a_m$ and $\Sigma_{n=1}^{\infty} b_n$ are each absolutely convergent. Then the double series $\Sigma_{m,n=1}^{\infty} a_m b_n$ is absolutely convergent, and*

$$\sum_{m,n=1}^{\infty} a_m b_n = \left[\sum_{m=1}^{\infty} a_m \right] \left[\sum_{n=1}^{\infty} b_n \right]. \tag{4}$$

PROOF. Let

$$A = \sum_{m=1}^{\infty} |a_m|, \qquad B = \sum_{n=1}^{\infty} |b_n|,$$

$$s_{mn} = \sum_{j=1}^{m} \sum_{k=1}^{n} a_j b_k, \qquad \text{and} \qquad S_{mn} = \sum_{j=1}^{m} \sum_{k=1}^{n} |a_j||b_k|.$$

Then we see that $S_{mn} \leq AB$ for every m and n. Hence the double series is absolutely convergent. Also

$$s_{mn} = \left[\sum_{j=1}^{m} a_j \right] \left[\sum_{k=1}^{n} b_k \right]. \tag{5}$$

The formula (4) results from passing to the limit in (5).

$$\begin{array}{lllll}
a_0b_0 & + a_0b_1x & + \cdots + a_0b_nx^n & + \cdots \\
a_1b_0x & + a_1b_1x^2 & + \cdots + a_1b_nx^{n+1} & + \cdots \\
\vdots & \vdots & \vdots & \\
a_mb_0x^m & + a_mb_1x^{m+1} & + \cdots + a_mb_nx^{m+n} & + \cdots \\
\vdots & \vdots & \vdots &
\end{array}$$

Fig. 3-17

In section 10, page 152, we stated without proof the following theorem concerning the Cauchy product of two power series.

Theorem 27. *Suppose that*

$$f(x) = \sum_{m=0}^{\infty} a_m x^m \qquad \textit{for } |x| < R,$$

$$g(x) = \sum_{n=0}^{\infty} b_n x^n \qquad \textit{for } |x| < R,$$

where $R > 0$. *Then*

$$f(x)g(x) = \sum_{p=0}^{\infty} \left[\sum_{m+n=p} a_m b_n \right] x^p \qquad \textit{for } |x| < R.$$

PROOF. Since the series for $f(x)$ and $g(x)$ both converge absolutely for $|x| < R$ then, by Theorem 43, their product $\Sigma_{m,n=0}^{\infty} a_m b_n x^{m+n}$ converges absolutely as a double series (Fig. 3-17). The convergence of the Cauchy product series to $f(x)g(x)$ follows from Theorem 41 (iv).

REMARK. Theorem 27 applies equally well for power series of the form

$$\sum_{m=0}^{\infty} a_m(x-c)^m \qquad \text{and} \qquad \sum_{n=0}^{\infty} b_n(x-c)^n.$$

EXAMPLE 4. Use Theorem 27 to find the power-series expansion of all terms up to x^5 of the function $\sqrt{1+x}\cos x$.

SOLUTION. We have

$$(1+x)^{1/2} = 1 + \frac{1}{2}x - \frac{1}{8}x^2 + \frac{1}{16}x^3 - \frac{5}{128}x^4 + \frac{7}{256}x^5 - \cdots,$$

$$\cos x = 1 - \frac{x^2}{2!} + \frac{x^4}{4!} - \cdots.$$

Taking the Cauchy product, we find

$$\sqrt{1+x}\cos x \approx \sum_{p=0}^{5} \left[\sum_{m+n=p} a_m b_n \right] x^p$$

$$= 1 + \frac{1}{2}x - \frac{5}{8}x^2 - \frac{3}{16}x^3 + \frac{25}{384}x^4 + \frac{13}{768}x^5.$$

Double sequences and series in which the elements are functions are defined in the same way as are single sequences and series of functions.

Definitions. The sequence $s_{mn}(x)$ is said to **converge uniformly to** $s(x)$ for x on some interval I if and only if for each $\varepsilon > 0$ there is an N **independent of** x such that

$$|s_{mn}(x) - s(x)| \le \varepsilon \qquad \text{for all } m > N \text{ and } n > N.$$

The **uniform convergence of a double series** is equivalent to the uniform convergence of the sequence of its partial sums. The individual terms in a sequence or series may consist of functions of several variables. We define uniform convergence in the natural way; a sequence such as $\{s_{mn}(x_1, x_2, x_3)\}$ is said to **converge uniformly for** (x_1, x_2, x_3) **in some region** R of R^3 if the index N in the definition above does not depend on the location of the point (x_1, x_2, x_3) in R.

We can now extend the Weierstrass M-test to double series.

Theorem 44. *Suppose that* $|u_{mn}(x, y)| < M_{mn}$ *for all* (x, y) *in some region* R *of the plane. If the double series* $\Sigma_{m,n=0}^{\infty} M_{mn}$ *converges, then* $\Sigma_{m,n=0}^{\infty} u_{mn}(x, y)$ *converges uniformly on* R.

PROOF. By Theorem 39, $\Sigma_{m,n=0}^{\infty} u_{mn}(x, y)$ converges absolutely for each fixed (x, y) in R. Let $s(x, y)$ be the sum of the series and $s_{mn}(x, y)$ its partial sum. Denote

$$S = \sum_{m,n=0}^{\infty} M_{mn} \qquad \text{and} \qquad S_{mn} = \sum_{j,k=0}^{m,n} M_{jk}.$$

Let $\varepsilon > 0$ be given. Then there is an N such that

$$|S - S_{mn}| < \varepsilon \quad \text{for all} \quad (m, n) \qquad \text{with} \quad m > N, \qquad n > N.$$

Therefore, for each (x, y) in R,

$$|s(x, y) - s_{mn}(x, y)| \le |S - S_{mn}| < \varepsilon, \qquad \text{whenever } m > N \text{ and } n > N.$$

Since N was chosen without regard to (x, y), the convergence is uniform.

Theorems on the continuity of the uniform limit of double series of continuous functions read the same as for single series. Similarly the results on term-by-term integration and differentiation (partial differentiation for functions of several variables) are all quite analogous to those obtained in Section 1. (See Problems 24, 25, and 26.) We shall restrict ourselves to the statement of some results for double power series of the form

$$\sum_{m,n=0}^{\infty} a_{mn} x^m y^n. \tag{6}$$

Theorem 45. *If the double power series* (6) *converges absolutely for some* $x_0 \neq 0$ *and* $y_0 \neq 0$, *then the series* (6) *and the series of absolute values converge uniformly for* $|x| \leq |x_0|$ *and* $|y| \leq |y_0|$.

PROOF. Since $|a_{mn}x^m y^n| \leq |a_{mn}x_0^m y_0^n| \equiv M_{mn}$, we may apply Theorem 39, and the result follows.

Theorem* 46. *If the double power series* (6) *converges absolutely for* $x_0 \neq 0$ *and* $y_0 \neq 0$, *then all the series obtained by differentiating term by term with respect to x and y converge for all* (x, y) *in the rectangle*

$$R = \{(x, y) : |x| < |x_0|, |y| < |y_0|\}.$$

The convergence is uniform on any rectangle $R = \{(x, y) : |x| \leq h, |y| \leq k\}$, *where* $h < |x_0|$, $k < |y_0|$.

The proof parallels that for single power series.

Theorem* 47. *If the double power series* (6) *converges absolutely for* $x_0 \neq 0$, *and* $y_0 \neq 0$, *and if f is defined by the series, so that*

$$f(x, y) = \sum_{m,n=0}^{\infty} a_{mn}x^m y^n \qquad \text{for } |x| < |x_0|, \quad |y| < |y_0|, \tag{7}$$

then f is continuous and has partial derivatives of all orders which are given in the rectangle $R = \{(x, y) : |x| < |x_0|, |y| < |y_0|\}$ *by the appropriate series obtained by term-by-term differentiation.*

Corollary. *Under the assumptions of Theorem* 47, *we have*

$$a_{mn} = \frac{1}{m!n!} \frac{\partial^{m+n} f(0, 0)}{\partial x^m \partial y^n}. \tag{8}$$

REMARK. All the results on double power series are valid for series of the form $\Sigma a_{mn}(x - c)^m(y - d)^n$, with the usual modifications; e.g., the evaluation in (8) is at (c, d) instead of $(0, 0)$.

EXAMPLE 5. Find the first six nonvanishing terms of the double power series expansion of e^{xy} about the point $(1, 0)$. Assume that the series is convergent in a rectangle containing $(1, 0)$.

SOLUTION. We have $f(1, 0) = 1$. Evaluating all partial derivatives at $(1, 0)$, we find

* Partial derivatives are discussed in Chapter 4. Readers unfamiliar with partial differentiation may postpone or skip Theorems 46 and 47, the Corollary, and Example 5.

$$\frac{\partial f}{\partial x} = 0, \qquad \frac{\partial f}{\partial y} = 1, \qquad \frac{\partial^2 f}{\partial x^2} = 0, \qquad \frac{\partial^2 f}{\partial x \partial y} = 1, \qquad \frac{\partial^2 f}{\partial y^2} = 1,$$

$$\frac{\partial^3 f}{\partial x^3} = \frac{\partial^3 f}{\partial x^2 \partial y} = 0, \qquad \frac{\partial^3 f}{\partial x \partial y^2} = 2, \qquad \frac{\partial^3 f}{\partial y^3} = 1,$$

Therefore

$$f(x, y) = 1 + y + (x - 1)y + \frac{1}{2!}y^2 + (x - 1)y^2 + \frac{1}{3!}y^3 + \cdots$$

EXAMPLE 6. Using power-series expansions, estimate the error made in computing $(e^{0.2} - 1)^2$ from the terms of the series for $(e^x - 1)^2$ out to and including the terms in x^5.

SOLUTION. We use the Cauchy product of the series $e^x - 1$ with itself. We write

$$e^x - 1 = x + \frac{x^2}{2} + \frac{x^3}{6} + \frac{x^4}{24} + \frac{x^5}{120} + \cdots$$

$$e^x - 1 = x + \frac{x^2}{2} + \frac{x^3}{6} + \frac{x^4}{24} + \frac{x^5}{120} + \cdots$$

$$\begin{aligned} & x^2 + \frac{x^3}{2} + \frac{x^4}{6} + \frac{x^5}{24} + \frac{x^6}{120}\left(1 + \frac{x}{6} + \frac{x^2}{6 \cdot 7} + \cdots\right) \\ & + \frac{x^3}{2} + \frac{x^4}{4} + \frac{x^5}{12} + \frac{x^6}{48}\left(1 + \frac{x}{5} + \frac{x^2}{5 \cdot 6} + \cdots\right) \\ & + \frac{x^4}{6} + \frac{x^5}{12} + \frac{x^6}{36}\left(1 + \frac{x}{4} + \frac{x^2}{4 \cdot 5} + \cdots\right) \\ & + \frac{x^5}{24} + \frac{x^6}{48}\left(1 + \frac{x}{3} + \frac{x^2}{3 \cdot 4} + \cdots\right) \\ & + \frac{x^6}{120}\left(1 + \frac{x}{6} + \frac{x^2}{6 \cdot 7} + \cdots\right)\left(1 + \frac{x}{2} + \frac{x^2}{2 \cdot 3} + \cdots\right). \end{aligned}$$

Therefore

$$(e^x - 1)^2 = x^2 + x^3 + \frac{7x^4}{12} + \frac{x^5}{4} + \varepsilon$$

where, by replacing each series in parentheses by the geometric series with the same first two terms, we obtain

$$\begin{aligned} \varepsilon < {} & \frac{x^6}{120(1 - x/6)} + \frac{x^6}{48(1 - x/5)} + \frac{x^6}{36(1 - x/4)} + \frac{x^6}{48(1 - x/3)} \\ & + \frac{x^6}{120(1 - x/6)(1 - x/2)} \end{aligned}$$

$$< \frac{(0.2)^6}{12}\left(\frac{30}{290} + \frac{25}{96} + \frac{20}{19 \cdot 3} + \frac{15}{56} + \frac{30 \cdot 10}{10 \cdot 29 \cdot 9}\right)$$

$$< \frac{1}{12}(0.000064)(0.104 + 0.261 + 0.351 + 0.268 + 0.115)$$

$$< 59 \times 10^{-7}.$$

In an actual computation the rounding error would have to be added to obtain an estimate of the total error committed. Thus

$$(e^{0.2} - 1)^2 = 0.049013 \pm 0.000006.$$

Problems

1. Show that the double series

$$\sum_{m,n=0}^{\infty} \frac{1}{(m+n)!}$$

is convergent. [*Hint*. Show that $(m+n)! \geq m!n!$.]

2. Show that

$$\sum_{m,n=1}^{\infty} \frac{1}{mn}$$

is divergent.

3. Test for convergence:

$$\sum_{m,n=0}^{\infty} \frac{2^{m+n}}{m!n!}.$$

4. Show that

$$\sum_{m,n=1}^{\infty} \frac{1}{(m^2+n^2)^p}$$

converges if $p > 1$. [*Hint*. Show first that $m^2 + n^2 \geq 2mn$.]

5. Show that

$$\sum_{m,n=1}^{\infty} \frac{1}{m^2+n^2}$$

is divergent. *Hint*. Note that

$$\sum_{m,n=1}^{N} \frac{1}{(m^2+n^2)} > \sum_{m=1}^{N}\left[\sum_{n=1}^{m} \frac{1}{(m^2+n^2)}\right] \geq \sum_{m=1}^{N} \frac{1}{2m}.$$

6. Show that

$$\sum_{m,n=0}^{\infty} \frac{(m+n)^5}{m!n!}$$

is convergent. [*Hint*. Let

$$a_m = \sum_{n=0}^{\infty} \frac{(m+n)^5}{m!n!}$$
$$= \sum_{n=0}^{m} \frac{(m+n)^5}{m!n!} + \sum_{n=m+1}^{\infty} \frac{(m+n)^5}{m!n!}, \qquad m \geq 0,$$

and show that

$$a_m \leq e \cdot \frac{(2m)^5}{m!} + \frac{1}{m!} \sum_{n=0}^{\infty} \frac{(2n)^5}{n!} = A_m.$$

Then show that $\Sigma_{m=0}^{\infty} A_m$ converges and use Theorem 42.]

7. Test for convergence:

$$\sum_{\substack{m,n=1 \\ m \neq n}}^{\infty} \frac{1}{m^2 - n^2}.$$

In each of Problems 8 through 12, use Maclaurin expansions for the separate functions, and the formula for the Cauchy product to obtain the Maclaurin expansions for the given functions. Carry the process out to the term ax^n, where n is given.

8. $e^{2x} \sin 3x, \quad n = 4$
9. $(1 + x)^{-2} \cos x, \quad n = 5$
10. $(1 + x)^{-1/2} e^x, \quad n = 4$
11. $e^x \log(1 + x), \quad n = 5$
12. $(\cos x) \log(1 + x), \quad n = 5.$

In each of Problems 13 through 18, find the terms in the double (Maclaurin) series of the given function up to and including terms of degree three.

13. $e^x \cos y$
14. $\dfrac{1}{1 - x - 2y + x^2}$
15. $\dfrac{1}{e^x \cos y}$
16. $e^{-2x} \log(1 + y)$
17. $\cos xy$
18. $(1 + x + y)^{-1/2}$

In each of Problems 19 through 22, estimate the error made in computing each function from its series, as in Example 6 above.

19. $\dfrac{e^x}{1 - x}, \quad x = 0.2, \quad n = 4$
20. $\dfrac{\cos x}{1 - x^2}, \quad x = 0.2, \quad n = 4$
21. $\dfrac{e^{-x}}{1 + x}, \quad x = 0.2, \quad n = 4$
22. $\dfrac{\sin x}{1 - x^2}, \quad x = 0.2, \quad n = 5$
23. Prove Theorem 38.
24. State and prove a theorem on the limit of a uniformly convergent double series of continuous functions of n variables.
25. State and prove a theorem on the term-by-term differentiation of a uniformly convergent double series of functions of n variables.

26. State and prove a theorem on the term-by-term integration of a uniformly convergent double series of functions of three variables.
27. Write out the proof of Theorem 46.
28. Write out the proof of Theorem 47.
29. Prove the Corollary to Theorem 47.

15. Complex Functions. Complex Series

Although functions of one or more complex variables are defined formally in the same manner as are functions of real variables, we shall see that the implications of the definition are quite different in the complex case. Denoting the collection of all complex numbers $a + bi$ by $\mathbb{C}$, we define a **relation from** $\mathbb{C}$ **to** $\mathbb{C}$ as a collection of ordered pairs (z, w) in which z is a complex number and w is a complex number. A **relation from** $\mathbb{C}^n$ **to** $\mathbb{C}$ is a set of ordered pairs $[(z_1, z_2, \ldots, z_n), w]$ in which $(z_1, z_2, \ldots, z_n)$ is an ordered array of n complex numbers and w is a complex number. A relation from $\mathbb{C}$ to $\mathbb{C}$ is a **function on** $\mathbb{C}$ if and only if no two distinct pairs have the same first element. We use the usual functional notation and write $w = f(z)$ for a function of one variable. Functions of more than one variable are defined similarly, and we write $w = f(z_1, z_2, \ldots, z_n)$ for a function on $\mathbb{C}^n$. The **domain** and **range** of a function are defined precisely as in the case of a real variable.

For problems concerning a function of a real variable, we found it helpful to interpret $y = f(x)$ as a set of points (frequently a curve or an arc) in the plane of analytic geometry. If $w = f(z)$ is a complex-valued function of a complex variable z, then both the domain and the range are sets of complex numbers. An aid to visualization is obtained by drawing *two* complex planes side by side, denoting one of them the z-plane and the other the w-plane (Fig. 3-18). The domain of a function f is a set of points S in the z-plane, and the range is a set of points T in the w-plane. The function f assigns a value Q in T to each point P in S. The functions we shall consider will usually have domains which are either a region in the z-plane or the entire z-plane.

Definition. Suppose that f is a function on $\mathbb{C}$ and c and L are complex num-

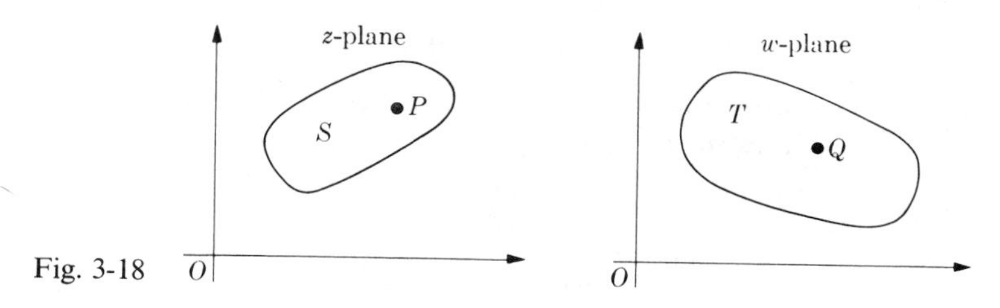

Fig. 3-18

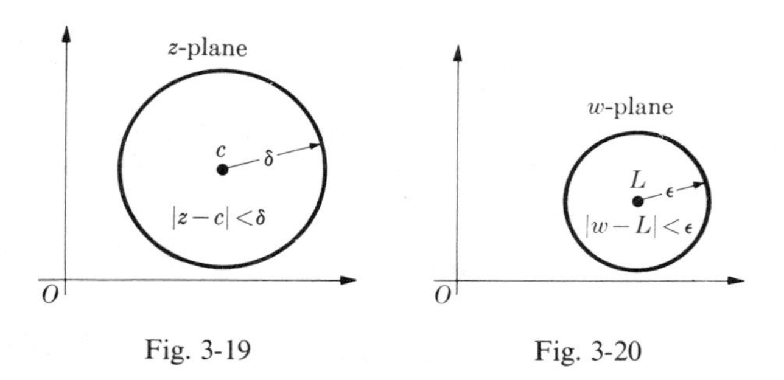

Fig. 3-19 Fig. 3-20

bers. We say that $f(z)$ **has the limit** L **as** z **tends to** c if and only if for each $\varepsilon > 0$ there is a $\delta > 0$ such that

$$|f(z) - L| < \varepsilon \qquad \text{whenever} \qquad 0 < |z - c| < \delta.$$

We also write $f(z) \to L$ as $z \to c$ and $\lim_{z \to c} f(z) = L$.

REMARKS. (i) Since absolute values are defined for complex numbers, the above definition makes sense when the function is complex-valued as well as when it is real-valued.
ii) The definition has a simple interpretation in terms of two complex planes. The set of points $|z - c| < \delta$ consists of all the points z which are not farther away from c than δ. That is, the set consists of all points in a circle of radius δ with center at c (Fig. 3-19). The inequality $0 < |z - c|$ means that z is not allowed to be equal to c itself. The points $w = f(z)$ which satisfy the inequality $|f(z) - L| < \varepsilon$ must lie in the circle of radius ε with center at L (Fig. 3-20).
iii) In order for the definition to make sense, the punctured disk $0 < |z - c| < \delta$ must lie in the domain S of f. If S consists of only part of this disk, then it is automatically understood that z is restricted to be a point of S.
iv) The definition of limit for a function of several complex variables is analogous to that for real variables. However, since a geometric interpretation is no longer readily available, we must lean more heavily on the analytic statements.
v) The usual theorems on limits, such as those for limits of sums, products, and quotients, have the identical statements and proofs given for real variables and need not be repeated.

If we set $z = x + iy$ and write $w = u + iv$, then a function $w = f(z)$ may be written

$$f(z) = u(x, y) + iv(x, y).$$

in which u and v are each a function of the *two real variables* x and y. In other words, one complex-valued function of one complex variable may be considered as two real-valued functions of two real variables. If a, b, M, N are real numbers, then we see that the statement

$$f(z) \to M + iN \qquad \text{as} \qquad z \to a + ib$$

is equivalent to

$$u(x, y) \to M, \qquad v(x, y) \to N \qquad \text{as} \qquad (x, y) \to (a, b).$$

Thus notions of limits and continuity for a complex function may be reduced to the corresponding statements for pairs of functions of two real variables. We say that f **is continuous at** z_0 if and only if

$$\lim_{z \to z_0} f(z) = f(z_0).$$

If u and v are continuous at (x_0, y_0), then f is continuous at $z_0 = x_0 + iy_0$, and conversely.

The processes of differentiation and integration for complex functions are substantially *different* from those for real-valued functions. There is a wealth of material in complex analysis, and the interested student will find many books devoted entirely to this subject.

We shall be concerned here mostly with sequences and series of complex numbers. A sequence

$$s_1, s_2, \ldots, s_n, \ldots$$

of complex numbers may be written in the form

$$r_1 + it_1, r_2 + it_2, \ldots, r_n + it_n, \ldots,$$

where r_k and t_k are the real and imaginary parts of s_k, respectively. We say that the sequence $\{s_n\}$ **is convergent** if and only if the two sequences $\{r_n\}$, $\{t_n\}$ of real numbers are convergent. The **infinite series**

$$\sum_{n=1}^{\infty} b_n \tag{1}$$

of complex numbers is convergent if and only if the sequence

$$s_n = \sum_{k=1}^{n} b_k$$

of partial sums is convergent. Many of the theorems established for sequences and series of real numbers carry over to complex sequences and series without change in statement or proof. A series of complex numbers such as (1) is said to be **absolutely convergent** if the real series

$$\sum_{n=1}^{\infty} |b_n|$$

is convergent. As in the case of Theorem 12, we can easily show that an absolutely convergent series of complex numbers is convergent.

A power series with complex terms is one of the form

$$\sum_{n=0}^{\infty} a_n z^n$$

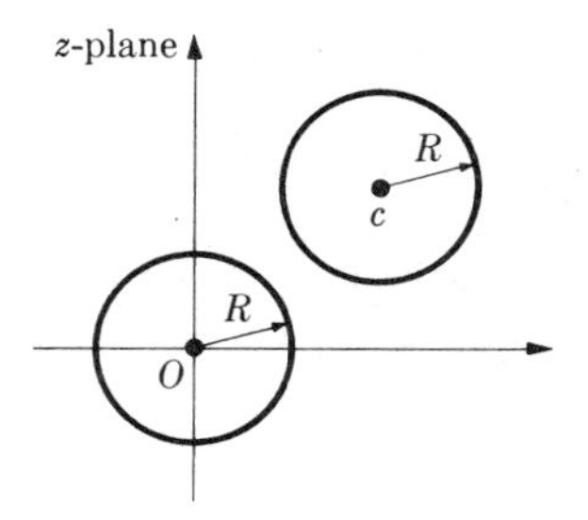

Fig. 3-21

in which the coefficients $a_0, a_1, \ldots$ are complex numbers. The Ratio Test (Theorem 14, page 115) is valid without change for complex numbers as well as for real numbers. We now state the basic theorem on the convergence of complex power series.

Theorem 48. *Let $\Sigma_{n=0}^{\infty} a_n z^n$ be any power series. Then either*

i) *the series converges only for $z = 0$; or*
ii) *the series converges for all z; or*
iii) *there is a number R such that the series converges for all z with $|z| < R$ and diverges for all z with $|z| > R$.*

The statement and proof of this theorem are the same as those of Theorem 16 in Section 13. However, we now see that the interpretation of the set $|z| < R$ of points of convergence is a disk of radius R with center at the origin (Fig. 3-21). We also see why the quantity R is called the *radius of convergence*. For series of the form $\Sigma_{n=0}^{\infty} a_n(z - c)^n$, the circle of convergence has radius R with its center at the number c (Fig. 3-21).

We may define the elementary functions of algebra, trigonometry, and calculus as functions of a complex variable. A certain amount of care is necessary, since a function such as $\sin \omega$, which has a perfectly good meaning in terms of angles when ω is real, has no geometric interpretation or definition when ω is complex. We solve the problem by defining functions in terms of their power-series expansions. We recall that e^x with x real has the Taylor expansion

$$e^x = \sum_{n=0}^{\infty} \frac{x^n}{n!}, \tag{2}$$

which is convergent for *all* x. We *define* e^z by the series

$$e^z = \sum_{n=0}^{\infty} \frac{z^n}{n!} \qquad \text{for all complex } z. \tag{3}$$

It is a simple exercise (Ratio Test) to show that the series (3) is absolutely convergent for all z. If z is real, then (3) becomes (2) and the definition is consistent. The next theorem gives the basic properties of e^z when z is complex.

Theorem 49. (a) $e^z \cdot e^w = e^{z+w}$ *for all complex* z *and* w.
b) $e^{x+yi} = e^x(\cos y + i \sin y)$ *for all real* x *and* y.
c) *If* $f(x) = e^{(a+bi)x}$, x *real, then* $f'(x) = (a + bi)e^{(a+bi)x}$.

PROOF. (a) We write

$$e^z \cdot e^w = \left(\sum_{m=0}^{\infty} \frac{z^m}{m!}\right)\left(\sum_{n=0}^{\infty} \frac{w^n}{n!}\right)$$

and apply Theorems 43 and 41 of Section 14. We obtain

$$e^z \cdot e^w = \sum_{p=0}^{\infty} \left[\frac{1}{p!} \sum_{m+n=p} \frac{(m+n)!}{m!n!} z^m w^n\right] = \sum_{p=0}^{\infty} \frac{(z+w)^p}{p!} = e^{z+w}.$$

b) From (a) we find

$$\begin{aligned} e^{x+yi} &= e^x \cdot e^{yi} = e^x \sum_{n=0}^{\infty} \frac{(yi)^n}{n!} \\ &= e^x \left\{\sum_{k=0}^{\infty} \frac{(-1)^k y^{2k}}{(2k)!} + i \sum_{k=0}^{\infty} \frac{(-1)^k y^{2k+1}}{(2k+1)!}\right\} \\ &= e^x(\cos y + i \sin y). \end{aligned}$$

To prove (c), we have

$$f(x) = e^{ax}(\cos bx + i \sin bx)$$

and

$$\begin{aligned} f'(x) &= e^{ax}(-b \sin bx + ib \cos bx) + ae^{ax}(\cos bx + i \sin bx) \\ &= (a + ib)e^{ax}(\cos bx + i \sin bx) = (a + ib)f(x). \end{aligned}$$

The trigonometric and hyperbolic functions of a complex variable are *defined* by the formulas

$$\sin z = \frac{e^{iz} - e^{-iz}}{2i}, \qquad \cos z = \frac{e^{iz} + e^{-iz}}{2},$$

$$\sinh z = \frac{e^z - e^{-z}}{2}, \qquad \cosh z = \frac{e^z + e^{-z}}{2},$$

for all complex z. The remaining trigonometric and hyperbolic functions are defined in the usual way. That is,

$$\tan z = \frac{\sin z}{\cos z}, \qquad \sec z = \frac{1}{\cos z},$$

and so forth.

The power series expansion for e^z and the definitions of the trigonometric functions may be used to get the familiar expansions

$$\sin z = \sum_{k=0}^{\infty} \frac{(-1)^k z^{2k+1}}{(2k+1)!}, \qquad \cos z = \sum_{k=0}^{\infty} \frac{(-1)^k z^{2k}}{(2k)!}, \tag{4}$$

valid for all complex z. Expansions for the remaining functions are obtained similarly.

Theorem 50. (a) *The addition theorems and double-angle formulas for* $\sin z$, $\cos z$, $\sinh z$ *and* $\cosh z$ *hold for all complex numbers.*
b) *We have for all* z:

$$\cos iz = \cosh z, \qquad \cosh iz = \cos z,$$
$$\sin iz = i \sinh z, \qquad \sinh iz = i \sin z,$$
$$\sin(x + iy) = \sin x \cosh y + i \cos x \sinh y,$$
$$\cos(x + iy) = \cos x \cosh y - i \sin x \sinh y.$$

PROOF. (a) We show that

$$\sin(z + w) = \sin z \cos w + \cos z \sin w.$$

From the definition of the trigonometric functions, we may write

$$\begin{aligned} \sin z \cos w + \cos z \sin w &= \frac{1}{4i}[(e^{iz} - e^{-iz})(e^{iw} + e^{-iw}) \\ &\quad + (e^{iz} + e^{-iz})(e^{iw} - e^{-iw})] \\ &= \frac{1}{4i}[e^{i(z+w)} + e^{i(z-w)} - e^{-i(z-w)} - e^{-i(z+w)} \\ &\quad + e^{i(z+w)} - e^{i(z-w)} + e^{-i(z-w)} - e^{-i(z+w)}] \\ &= \frac{1}{2i}[e^{i(z+w)} - e^{-i(z+w)}] = \sin(z + w). \end{aligned}$$

The remaining portions of the theorem are proved in a similar manner.

EXAMPLE 1. Write $\sin(1 + i)$ in the form $a + ib$.

SOLUTION. $\sin(1 + i) = \sin 1 \cdot \cosh 1 + i \cos 1 \cdot \sinh 1$. Therefore

$$a = \sin 1 \cosh 1, \qquad b = \cos 1 \sinh 1.$$

The reader may verify the following formulas:

$$\tan z = z + \frac{z^3}{3} + \frac{2z^5}{15} + \cdots, \tag{5}$$

$$\sec z = 1 + \frac{z^2}{2} + \frac{5z^4}{24} + \frac{61z^6}{720} + \cdots, \tag{6}$$

$$\tanh z = z - \frac{z^3}{3} + \frac{2z^5}{15} - \cdots, \tag{7}$$

$$\operatorname{sech} z = 1 - \frac{z^2}{2} + \frac{5z^4}{24} - \frac{61z^6}{720} + \cdots \tag{8}$$

EXAMPLE 2. Solve for $z : \cos z = 2$.

SOLUTION. We have

$$\cos z = \cos(x + iy) = \cos x \cosh y - i \sin x \sinh y = 2.$$

Therefore

$$\cos x \cosh y = 2 \qquad \text{and} \qquad \sin x \sinh y = 0.$$

Taking $\sin x \sinh y = 0$ first, we see that $\sinh y = 0$ if and only if $y = 0$. But then from the first equation $\cosh y = 1$ and $\cos x$ must be 2, which is impossible if x is *real.* So $y = 0$ is excluded. The equation $\sin x \sinh y = 0$ can also hold if $\sin x = 0$, which occurs if $x = \pm n\pi$. Then $\cos x = \pm 1$ and, ruling out the negative values, we get $x = \pm 2n\pi$ and $\cos x = +1$. The first equation then implies that $\cosh y = 2$ or $y = \operatorname{argcosh} 2 = \log(2 + \sqrt{3})$. The answer is

$$z = x + iy = \pm 2n\pi + i \log(2 + \sqrt{3}).$$

REMARK. Example 2 shows that the rules about the range of the various trigonometric functions, which we learned for the case when the domain is real, no longer hold when the domain is complex. For example, the functions $\sin z$ and $\cos z$ may have arbitrarily large values if z is complex.

When x is real, the function e^x is the inverse of the logarithm function. For complex exponential and logarithmic functions we must proceed quite differently. With the observation that

$$e^{2\pi i} = (\cos 2\pi + i \sin 2\pi) = 1,$$

we conclude

$$e^{z+2\pi i} = e^z \cdot e^{2\pi i} = e^z,$$

and so e^z *is a periodic function with period* $\mathbf{2\pi i}$. In attempting to define the logarithm as the inverse of $w = e^z$ we are stymied, because to each value of w there corresponds the infinite collection of values $z \pm 2n\pi i$, $n = 1, 2, \ldots$ The inverse relation of e^z is not a function. However, we can proceed by writing w in polar form. That is, if $w = r(\cos\theta + i \sin\theta)$, then

$$w = e^z = e^{x+iy} = e^x(\cos y + i \sin y) = r(\cos\theta + i \sin\theta).$$

The last equality yields $r = e^x$ and $y = \theta \pm 2n\pi$. This suggests defining the **principal inverse function** of e^z by

$$\operatorname{Log} w = \log r + i\theta \qquad \text{where} \qquad w = re^{i\theta}, \qquad -\pi < \theta \leq \pi.$$

Thus we find that

$$e^z = w \qquad \Leftrightarrow \qquad z = \operatorname{Log} w \pm 2n\pi i.$$

We observe that if w is real and positive, then $\theta = 0$ and $\operatorname{Log} w$ is just the ordinary natural logarithm of w.

EXAMPLE 3. Find the value of $\operatorname{Log}(-3)$.

SOLUTION. We have $-3 = 3e^{\pi i}$, since $e^{\pi i} = \cos\pi + i\sin\pi = -1$. Therefore $\operatorname{Log}(-3) = \log 3 + \pi i$.

REMARKS. The trick in Example 3 is to write the number -3 in the form $re^{i\theta}$ with $r \geq 0$ and with θ always in the interval $-\pi < \theta \leq \pi$. *Every* complex number can be so written. Note that the myth prevalent in elementary trigonometry courses concerning the nonexistence of logarithms of negative numbers evaporates as we enter the complex domain.

The inverses of the trigonometric and hyperbolic functions are also multiple-valued relations and so are not functions. The definitions of principal inverses of these functions are usually given in texts on complex function theory.

EXAMPLE 4. Express all the solutions of $\sinh w = z$ in terms of the Log function.

SOLUTION. From the equation $\sinh w = z$, we have

$$e^w - e^{-w} = 2z \qquad \text{or} \qquad e^{2w} - 2ze^w - 1 = 0.$$

This is a quadratic equation in e^w. Therefore

$$e^w = z \pm \sqrt{z^2 + 1}$$

and

$$w = \operatorname{Log}(z \pm \sqrt{z^2 + 1}) \pm 2n\pi i.$$

REMARK. Since z is complex, it is not clear what meaning should be attached to the expression $\sqrt{z^2 + 1}$. If $\zeta = \rho e^{i\phi}$, $-\pi < \phi \leq \pi$ is any complex number, the two square roots of ζ are

$$\sqrt{\rho}e^{i\phi/2} \qquad \text{and} \qquad -\sqrt{\rho}e^{i\phi/2}, \qquad \text{with} \quad -\pi < \phi \leq \pi.$$

The first of these numbers is in the right-hand portion of the z-plane (has positive real part), while the second is in the left-hand portion. We call the first one the *positive* square root of ζ and the second the negative square root (except when $\phi = \pi$ and both square roots are on the imaginary axis).

PROBLEMS

1. Show, by the Ratio Test, that

$$\sum_{n=0}^{\infty} \frac{z^n}{n!}$$

converges for all complex z.

2. Prove that $\sin z$ and $\cos z$ are given by their series expansions (4).

3. Prove that the functions $\sinh z$ and $\cosh z$ are given, respectively, by the series

$$\sum_{n=0}^{\infty} \frac{z^{2n+1}}{(2n+1)!}, \qquad \sum_{n=0}^{\infty} \frac{z^{2n}}{(2n)!},$$

which are valid for all z.

4. Prove that for all complex z, w:

$$\cos(z+w) = \cos z \cos w - \sin z \sin w.$$

5. Prove that for all complex z, w:

$$\sinh(z+w) = \sinh z \cosh w + \cosh z \sinh w.$$

6. Prove that for all complex z, w:

$$\cosh(z+w) = \cosh z \cosh w + \sinh z \sinh w.$$

7. Derive formulas for $\sin 2z$, $\cos 2z$, $\sinh 2z$, $\cosh 2z$, z complex.

8. Prove the validity of the formulas in Theorem 50(b).

9. Verify the formula for $\tan z$ in (5).

10. Verify the formula for $\sec z$ in (6).

11. Verify the formula for $\tanh z$ in (7).

12. Verify the formula for $\operatorname{sech} z$ in (8).

13. Write in the form $a + ib$:

    a) e^{1+i}, b) $\sin\left(\frac{4\pi}{3} + i\right)$, c) $\cosh\left(2 + \frac{i\pi}{3}\right)$

14. Write in the form $a + ib$:

    a) $\sinh(1 - i)$, b) $\tanh(2 + 3i)$, c) $e^{\pi + 2i}$

15. Write in the form $a + ib$:

    a) $\operatorname{Log}(-4)$, b) $\operatorname{Log}(1 + i)$, c) $\operatorname{Log}(-i)$

16. Write in the form $a + ib$:

    a) $e^{\sin[(\pi+i)/2]}$, b) $\operatorname{Log}(e^{(1+\pi i)/4})$, c) $\sin(\sqrt{1 - i})$

17. Write in the form $u(x, y) + iv(x, y)$:

    a) $\sin^2 z$, b) $\tan z$

18. Write in the form $u(x, y) + iv(x, y)$:

    a) e^{z^2}, b) $e^{(1/z)}$

19. Write in the form $u(x, y) + iv(x, y)$:

    a) $\text{Log}(3 + 3i)$, b) $e^{\sin(z^2)}$

20. Find the radius of convergence of the series

$$\text{Log}(1 + z) = z - \frac{z^2}{2} + \frac{z^3}{3} - \frac{z^4}{4} + \cdots (-1)^{n+1}\frac{z^n}{n} + \cdots$$

21. Find the Maclaurin expansion for $(1 + x)^{1/2}$, replace each x by z and find the radius of convergence of the resulting complex series.

22. Repeat Problem 21 for $(1 - x^2)^{-1/2}$.

23. Show that $\tanh w = z$ if and only if

$$w = \frac{1}{2}\text{Log}\frac{1 + z}{1 - z} \pm in\pi.$$

24. Show that $\tan w = z$ if and only if

$$w = \pm n\pi + \frac{i}{2}\text{Log}\frac{i + z}{i - z}.$$

25. Show that $\sin w = z$ if and only if

$$w = \frac{1}{i}\text{Log}(iz \pm \sqrt{1 - z^2}) \pm 2n\pi.$$

    Are there any complex numbers z for which there is no solution w? Note that $\text{Log}\, z$ is defined for all $z \neq 0$.

26. Show that $\cos w = z$ if and only if

$$w = \frac{1}{i}\text{Log}(z \pm \sqrt{z^2 - 1}) \pm 2n\pi.$$

    Are there any complex numbers z for which there is no solution w?

CHAPTER 4

Partial Derivatives. Applications

1. Limits and Continuity. Partial Derivatives

A function f is a mapping which takes each element of the domain D into an element of the range S. We write $f: D \to S$. If the domain D consists of ordered pairs of numbers, then we have a function on R^2 or a function of two variables. We employ the notation

$$z = f(x, y) \qquad \text{or} \qquad f: (x, y) \to z$$

to indicate a typical function on R^2 when the elements of the range are real numbers (elements of R^1).

We now give a precise definition of a function of two variables.

Definition. Consider a collection of ordered pairs (A, w) where the elements A are themselves ordered pairs of real numbers and the elements w are real numbers. If no two members of the collection have the same item A as a first element—i.e., if it can never happen that there are two members (A_1, w_1) and (A_1, w_2) with $w_1 \neq w_2$—then we call this collection a **function on $\mathbf{R}^2$** or a **function of two variables**. The totality of possible ordered pairs A is called the **domain** of the function. The totality of possible values for w is called the **range** of the function.

REMARK. The definition of a *function of three variables* is precisely the same as that for a function of two variables except that the elements A are ordered triples rather than ordered pairs. A *function of n variables* is defined by considering the elements A to be ordered n tuples of real numbers.

A function of three variables is frequently denoted by

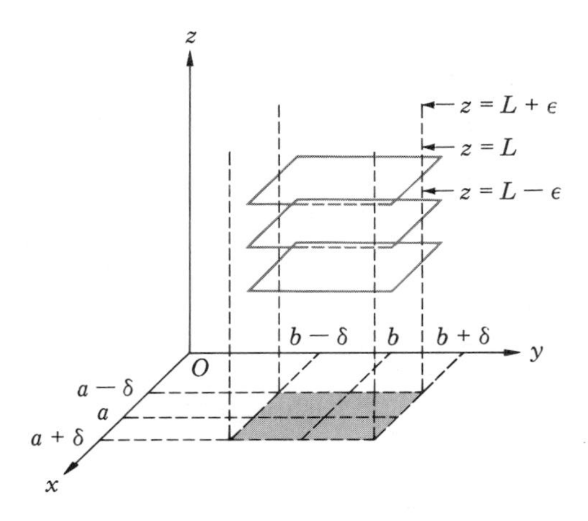

Fig. 4-1

$$w = G(x, y, z) \qquad \text{or} \qquad G : (x, y, z) \to w,$$

where (x, y, z) is an element of R^3 and w is an element of R^1. We can consider functions of four, five, or of any number of variables. If the exact number of variables is n, we usually write

$$y = f(x_1, x_2, \ldots, x_n)$$

in which $(x_1, x_2, \ldots, x_n)$ is an element of R^n and y is an element of R^1. The same letters f, F, g, G, ϕ, and so forth, are used for functions of any number of variables.

We are now ready to define a limit of a function of two variables. If f is such a function, we may write

$$z = f(x, y).$$

We wish to examine the behavior of f as the pair (x, y) tends to the pair (a, b); equivalently, we consider x tending to a and y tending to b.

Definition. We say that $f(x, y)$ **tends to the number** L **as** (x, y) **tends to** (a, b), and we write

$$f(x, y) \to L \qquad \text{as} \qquad (x, y) \to (a, b)$$

if and only if for each $\varepsilon > 0$ there is a $\delta > 0$ such that

$$|f(x, y) - L| < \varepsilon$$

whenever

$$|x - a| < \delta \qquad \text{and} \qquad |y - b| < \delta \qquad \text{and} \qquad (x, y) \neq (a, b).$$

A geometric interpretation of this definition is exhibited in Fig. 4-1. The definition asserts that whenever (x, y) are in the shaded square, as shown, then the function values, which we represent by z, must lie in the rectangular box of height 2ε between the values $L - \varepsilon$ and $L + \varepsilon$. This interpretation is an extension of the one usually given for functions of one variable.

REMARK. It is not necessary that (a, b) be in the domain of f. That is, a limit may exist with $f(a, b)$ being undefined.

Definition. We say that f **is continuous at** (a, b) if and only if

i) $f(a, b)$ is defined, and
ii) $f(x, y) \to f(a, b)$ as $(x, y) \to (a, b)$.

Definitions of limits and continuity for functions of three, four, and more variables are completely similar.

Functions of two or more variables do not have ordinary derivatives of the type we studied for functions of one variable. If f is a function of two variables, say x and y, then for each *fixed* value of y, f is a function of a single variable x. The derivative with respect to x (keeping y fixed) is then called the partial derivative with respect to x. For x fixed and y varying, we obtain a partial derivative with respect to y.

Definitions. We define **the partial derivatives of a function** f on R^2 by

$$f_x(x, y) = \lim_{h \to 0} \frac{f(x + h, y) - f(x, y)}{h},$$

$$f_y(x, y) = \lim_{h \to 0} \frac{f(x, y + h) - f(x, y)}{h},$$

where y is kept fixed in the first limit and x is kept fixed in the second. We use the symbols f_x and f_y for these partial derivatives.

If F is a function on R^3, we define the **partial derivatives** F_x, F_y and F_z by

$$F_x(x, y, z) = \lim_{h \to 0} \frac{F(x + h, y, z) - F(x, y, z)}{h}, \quad (y, z \text{ fixed})$$

$$F_y(x, y, z) = \lim_{h \to 0} \frac{F(x, y + h, z) - F(x, y, z)}{h}, \quad (x, z \text{ fixed})$$

$$F_z(x, y, z) = \lim_{h \to 0} \frac{F(x, y, z + h) - F(x, y, z)}{h}, \quad (x, y \text{ fixed}).$$

In other words, to find the partial derivative F_x of a function $F(x, y, z)$ of three variables, regard y and z as constants and find the **usual derivative** with respect to x; the derivatives F_y and F_z are found correspondingly. The same procedure applies for functions of any number of variables.

EXAMPLE 1. Given $f(x, y) = x^3 + 7x^2y + 8y^3 + 3x - 2y + 7$, find f_x and f_y.

SOLUTION. Keeping y fixed and differentiating with respect to x, we find that

$$f_x = 3x^2 + 14xy + 3.$$

Similarly, keeping x fixed, we get

$$f_y = 7x^2 + 24y^2 - 2.$$

EXAMPLE 2. Given $f(x, y, z) = x^3 + y^3 + z^3 + 3xyz$, find $f_x(x, y, z), f_y(x, y, z)$, and $f_z(x, y, z)$.

SOLUTION. We have

$$f_x(x, y, z) = 3x^2 + 3yz, f_y(x, y, z) = 3y^2 + 3xz, f_z(x, y, z) = 3z^2 + 3xy.$$

REMARKS ON NOTATION. The notation presented here is classical, although it may at times lead to confusion. A less ambiguous but less common notation uses numerical subscripts with a comma to indicate partial differentiation. For example, the symbols

$$f_{,1}(x, y) \qquad \text{and} \qquad f_{,2}(x, y)$$

stand for the partial derivatives with respect to x and y, respectively. These symbols have an advantage in that the subscript 1 represents the derivative with respect to the first variable, irrespective of the letter used. That is, if (r, s) are the independent variables with f the *same function* as the one above, then

$$f_{,1}(r, s) \qquad \text{and} \qquad f_{,2}(r, s)$$

have a clear meaning, while in classical notation f_r and f_s might sometimes be interpreted differently from f_x and f_y.

Another common symbol for partial differentiation is

$$\frac{\partial f}{\partial x} \qquad \text{(read: partial of } f \text{ with respect to } x\text{).}$$

This notation has the double disadvantage of using the letter x and of giving the impression (incorrectly) that the partial derivative is a fraction with ∂f and ∂x having independent meanings (which they do not). If we write $z = f(x, y)$, still another symbol for partial derivative is the expression

$$\frac{\partial z}{\partial x}.$$

Because of the multiplicity of symbols for partial derivatives used in texts on mathematics and various related branches of technology, it is important that the reader familiarize himself with *all* of them. For this reason we shall employ all the above symbols for partial derivative throughout the chapter.

EXAMPLE 3. Given $f(x, y) = e^{xy} \cos x \sin y$, find $f_x(x, y)$ and $f_y(x, y)$.

SOLUTION. We have

$$f_x(x, y) = e^{xy} \sin y(-\sin x) + \cos x \sin y e^{xy} \cdot y$$
$$= e^{xy} \sin y(y \cos x - \sin x),$$

and

$$f_y(x, y) = e^{xy} \cos x(x \sin y + \cos y).$$

EXAMPLE 4. Given that $z = x \arctan(y/x)$, find $\partial z/\partial x$ and $\partial z/\partial y$.

SOLUTION. We have

$$\frac{\partial z}{\partial x} = x \cdot \frac{-(y/x^2)}{1 + (y^2/x^2)} + \arctan(y/x) = \arctan(y/x) - \frac{xy}{x^2 + y^2},$$

and

$$\frac{\partial z}{\partial y} = x \frac{1/x}{1 + (y^2/x^2)} = \frac{x^2}{x^2 + y^2}.$$

EXAMPLE 5. Given $f(x, y) = x^2 - 3xy + 2x - 3y + 5$, find $f_{,1}(2, 3)$.

SOLUTION. We have

$$f_{,1}(x, y) = 2x - 3y + 2, \quad f_{,1}(2, 3) = -3.$$

PROBLEMS

In each of Problems 1 through 12, find f_x and f_y.

1. $f(x, y) = 2x^2 - 3xy + 4x$
2. $f(x, y) = x^2y^4 + 2xy - 6$
3. $f(x, y) = x^3 + y^3 + 3x^2y - 3xy^2 + 7$
4. $f\colon (x, y) \to \sqrt{x^2 + 1} + y^3$
5. $f(x, y) = \sqrt{x^2 + y^2}$
6. $f(x, y) = \dfrac{xy}{x^2 + y^2}$
7. $f(x, y) = \log(x^2 + y^2)$
8. $f(x, y) = \log \sqrt{x^2 + 3y^2}$
9. $f\colon (x, y) \to \arctan(y/x)$
10. $f(x, y) = \arcsin \dfrac{x}{1 + y}$
11. $f(x, y) = xye^{x^2 + y^2}$
12. $f\colon (x, y) \to \cos(xe^y)$

In each of Problems 13 through 18, find $f_{,1}$ and $f_{,2}$ at the values indicated.

13. $f(x, y) = x \arcsin(x - y), \quad x = 1, \quad y = 2$
14. $f(u, v) = e^{uv} \sec\left(\dfrac{u}{v}\right), \quad u = v = 3$
15. $f(x, z) = e^{\sin x} \tan xz, \quad x = \dfrac{\pi}{4}, \quad z = 1$

16. $f(t, u) = \dfrac{\cos 2tu}{t^2 + u^2}$, $t = 0$, $u = 1$

17. $f(y, x) = x^{xy}$, $x = y = 2$

18. $f(s, t) = s^t + t^s$, $s = t = 3$

In each of Problems 19 through 22, find $f_x(x, y, z)$, $f_y(x, y, z)$ and $f_z(x, y, z)$.

19. $f(x, y, z) = x^2y - 2x^2z + 3xyz - y^2z + 2xz^2$

20. $f(x, y, z) = \dfrac{xyz}{x^2 + y^2 + z^2}$

21. $f(x, y, z) = e^{xyz} \sin xy \cos 2xz$

22. $f(x, y, z) = x^2 + z^2 + y^3 + 2x - 3y + 4z$

In Problem 23 through 26 find in each case the indicated partial derivative.

23. $w = \log\left(\dfrac{xy}{x^2 + y^2}\right)$; $\dfrac{\partial w}{\partial x}, \dfrac{\partial w}{\partial y}$

24. $w = (r^2 + s^2 + t^2)\cos(rst)$; $\dfrac{\partial w}{\partial r}, \dfrac{\partial w}{\partial t}$

25. $w = e^{\sin(y/x)}$; $\dfrac{\partial w}{\partial y}, \dfrac{\partial w}{\partial x}$

26. $w = (\sec tu)\arcsin tv$; $\dfrac{\partial w}{\partial t}, \dfrac{\partial w}{\partial u}, \dfrac{\partial w}{\partial v}$

27. If $z = \log(y/x)$, show that

$$x\frac{\partial z}{\partial x} + y\frac{\partial z}{\partial y} = 0.$$

28. Let $P(x_1, x_2, \ldots, x_n) = a_1x_1^k + a_2x_2^k + \cdots + a_nx_n^k$, where $a_1, a_2, \ldots, a_n$ are numbers. Show that

$$x_1P_{,1} + x_2P_{,2} + \cdots + x_nP_{,n} = kP.$$

29. Let $Q(x, y, z) = \Sigma_{k=1}^n a_k x^{2n-2k}y^{k+1}z^{k-1}$, where the a_k are numbers and n is an integer greater than 1. Show that

$$xQ_x + yQ_y + zQ_z = 2nQ.$$

30. i) Given $f(x, y) = xy$. Show that f is continuous at $(0, 0)$ by finding a value of δ corresponding to each given ε such that $|xy - 0| < \varepsilon$ for all (x, y) for which $|x - 0| < \delta$ and $|y - 0| < \delta$.

    ii) For the same function f, show that it is continuous at (a, b) by finding a value of δ corresponding to each given ε such that $|xy - ab| < \varepsilon$ for all (x, y) for which $|x - a| < \delta$ and $|y - b| < \delta$. [*Hint:* Write $xy - ab = xy - ay + ay - ab$ and

$$|xy - ab| \le |xy - ay| + |ay - ab| \le |x - a| \cdot |y| + |a| \cdot |y - b|.]$$

31. Consider the function

$$g(x, y) = \begin{cases} \dfrac{xy}{x^2 + y^2} & \text{for } (x, y) \ne (0, 0), \\ 0 & \text{for } x = y = 0. \end{cases}$$

a) If x tends to zero and if we select $y = x^2$, then y also tends to zero. Show that under these conditions

$$g(x, y) \to 0 \qquad \text{as} \qquad x \to 0, \quad y \to 0.$$

b) If x tends to zero and we select $y = x$, show that

$$g(x, y) \to \tfrac{1}{2} \qquad \text{as} \qquad x \text{ and } y \text{ tend to } 0.$$

Conclude that g is not continuous at $(0, 0)$.

32. Use the method of Problem 31 to show that

$$h(x, y) = \begin{cases} \dfrac{x^2 - y^2}{x^2 + y^2} & \text{for } (x, y) \neq (0, 0), \\ 0 & \text{for } x = y = 0 \end{cases}$$

is not continuous at $(0, 0)$.

33. Show that the function

$$f(x, y) = \begin{cases} \dfrac{x^2 y^2}{x^2 + y^2} & \text{for } (x, y) \neq (0, 0), \\ 0 & \text{for } x = y = 0 \end{cases}$$

is continuous at $(0, 0)$. [*Hint:* $x^2 y^2/(x^2 + y^2) \leq x^2$.]

34. Show that the function $f(x, y) = \sqrt{|x|}\sqrt{|y|}$ is continuous at $(0, 0)$, but that f_x and f_y are not continuous at $(0, 0)$.

35. Let $f(x, y) = \sin(xy)$; $g(x, y) = \sin(x/y)$ for $y \neq 0$, $g(x, 0) = 0$. Is f continuous at $(0, 0)$? Is g continuous at $(0, 0)$?

36. Given the function

$$g(x, y) = \begin{cases} 0 & \text{for } \quad x = y = 0, \\ \dfrac{xy}{(x^2 + y^2)} & \text{otherwise.} \end{cases}$$

Show that g_x and g_y exist at $x = y = 0$. Use the result of Problem 31 to conclude that a function may have a partial derivative at a point and yet not be continuous there.

2. Implicit Differentiation

An equation involving x, y, and z establishes a relation among the variables. If we can solve for z in terms of x and y, then we may have one or more functions determined by the relation. For example, the equation

$$2x^2 + y^2 + z^2 - 16 = 0 \tag{1}$$

may be solved for z to give

$$z = \pm\sqrt{16 - 2x^2 - y^2}. \tag{2}$$

If one or more functions are determined by a relation, it is possible to compute partial derivatives implicitly in a way that is completely similar to the methods used for ordinary derivatives.

For example, in Equation (1) above, considering z as a function of x and y, we can compute $\partial z/\partial x$ directly from (1) without resorting to (2). We keep y fixed and in (1) differentiate implicitly with respect to x, getting

$$4x + 2z\frac{\partial z}{\partial x} = 0 \quad \text{and} \quad \frac{\partial z}{\partial x} = -\frac{2x}{z}.$$

Further examples exhibit the method.

EXAMPLE 1. Suppose that x, y, and z are variables and that z is a function of x and y which satisfies

$$x^3 + y^3 + z^3 + 3xyz = 5.$$

Find $\partial z/\partial x$ and $\partial z/\partial y$.

SOLUTION. Holding y constant and differentiating z with respect to x implicitly, we obtain

$$3x^2 + 3z^2\frac{\partial z}{\partial x} + 3xy\frac{\partial z}{\partial x} + 3yz = 0.$$

Therefore

$$\frac{\partial z}{\partial x} = -\frac{x^2 + yz}{xy + z^2}.$$

Holding x constant and differentiating with respect to y, we get

$$3y^2 + 3z^2\frac{\partial z}{\partial y} + 3xy\frac{\partial z}{\partial y} + 3xz = 0$$

and

$$\frac{\partial z}{\partial y} = -\frac{y^2 + xz}{xy + z^2}.$$

The same technique works with equations relating four or more variables, as the next example shows.

EXAMPLE 2. If r, s, t, and w are variables and if w is a function of r, s, and t which satisfies

$$e^{rt} - 2se^{w} + wt - 3w^2r = 5,$$

find $\partial w/\partial r$ and $\partial w/\partial t$.

SOLUTION. To find $\partial w/\partial r$, we keep s and t fixed and differentiate implicitly with respect to r. The result is

$$te^{rt} - 2se^{w}\frac{\partial w}{\partial r} + t\frac{\partial w}{\partial r} - 3w^2 - 6rw\frac{\partial w}{\partial r} = 0$$

or

$$\frac{\partial w}{\partial r} = \frac{3w^2 - te^{rt}}{t - 2se^{w} - 6rw}.$$

Similarly, with r and s fixed the result is

$$re^{rt} - 2se^{w}\frac{\partial w}{\partial t} + w + t\frac{\partial w}{\partial t} - 6rw\frac{\partial w}{\partial t} = 0$$

or

$$\frac{\partial w}{\partial t} = \frac{w + re^{rt}}{2se^{w} - t + 6rw}.$$

PROBLEMS

In each of Problems 1 through 12, assume that w is a function of all other variables. Find the partial derivatives as indicated in each case.

1. $3x^2 + 2y^2 + 6w^2 - x + y - 12 = 0;\quad \dfrac{\partial w}{\partial x}, \dfrac{\partial w}{\partial y}$

2. $x^2 + y^2 + w^2 + 3xy - 2xw + 3yw = 36;\quad \dfrac{\partial w}{\partial x}, \dfrac{\partial w}{\partial y}$

3. $x^2 - 2xy + 2xw + 3y^2 + w^2 = 21;\quad \dfrac{\partial w}{\partial x}, \dfrac{\partial w}{\partial y}$

4. $x^2y - x^2w - 2xy^2 - yw^2 + w^3 = 7;\quad \dfrac{\partial w}{\partial x}, \dfrac{\partial w}{\partial y}$

5. $w - (r^2 + s^2)\cos rw = 0;\quad \dfrac{\partial w}{\partial r}, \dfrac{\partial w}{\partial s}$

6. $w - e^{w\sin(y/x)} = 1;\quad \dfrac{\partial w}{\partial x}, \dfrac{\partial w}{\partial y}$

7. $e^{xyw}\sin xy\cos 2xw - 4 = 0;\quad \dfrac{\partial w}{\partial x}, \dfrac{\partial w}{\partial y}$

8. $w^2 - 3xw - \log\left(\dfrac{xy}{x^2 + y^2}\right) = 0;\quad \dfrac{\partial w}{\partial x}, \dfrac{\partial w}{\partial y}$

9. $xyz + x^2z + xzw - yzw + yz^2 - w^3 = 3;\quad \dfrac{\partial w}{\partial x}, \dfrac{\partial w}{\partial y}, \dfrac{\partial w}{\partial z}$

10. $r^2 + 3s^2 - 2t^2 + 6tw - 8w^2 + 12sw^3 = 4;\quad \dfrac{\partial w}{\partial r}, \dfrac{\partial w}{\partial s}, \dfrac{\partial w}{\partial t}$

11. $we^{xw} - ye^{yw} + e^{xy} = 1$; $\quad \dfrac{\partial w}{\partial x}, \dfrac{\partial w}{\partial y}$

12. $x^3 + 3x^2y + 2z^2t - 4zt^3 + 7xw - 8yw^2 + w^4 = 5$; $\quad \dfrac{\partial w}{\partial x}, \dfrac{\partial w}{\partial y}, \dfrac{\partial w}{\partial z}, \dfrac{\partial w}{\partial t}$

13. Given that $x^2 + y^2 - z^2 = 4$. Show that

$$\frac{\partial z}{\partial x} \cdot \frac{\partial x}{\partial y} \cdot \frac{\partial y}{\partial z} = -1.$$

*14. Suppose that the relation $f(x, y, z) = 0$ may be solved so that z is a function of x and y. Then we may compute $\partial z/\partial x$. If also we may compute $\partial x/\partial y$ and $\partial y/\partial z$, show that

$$\frac{\partial z}{\partial x} \cdot \frac{\partial x}{\partial y} \cdot \frac{\partial y}{\partial z} = -1.$$

15. The transformation of rectangular coordinates to spherical coordinates is given by

$$x = \rho \sin \phi \cos \theta, \qquad y = \rho \sin \phi \sin \theta, \qquad z = \rho \cos \phi.$$

Find

$$\frac{\partial x}{\partial \rho}, \frac{\partial x}{\partial \phi}, \frac{\partial x}{\partial \theta}, \frac{\partial y}{\partial \rho}, \frac{\partial y}{\partial \phi}, \frac{\partial y}{\partial \theta}, \frac{\partial z}{\partial \rho}, \frac{\partial z}{\partial \phi}, \frac{\partial z}{\partial \theta}.$$

16. In Problem 15 assume that ρ, ϕ, θ are functions of x, y, z and find

$$\frac{\partial \rho}{\partial x}, \frac{\partial \rho}{\partial y}, \frac{\partial \rho}{\partial z}, \frac{\partial \phi}{\partial x}, \frac{\partial \phi}{\partial y}, \frac{\partial \phi}{\partial z}, \frac{\partial \theta}{\partial x}, \frac{\partial \theta}{\partial y}, \frac{\partial \theta}{\partial z}.$$

in terms of x, y, and z.

17. An *affine transformation* in R^3 is one which changes a rectangular (x, y, z) coordinate system into a (u, v, w) coordinate system by the equations

$$u = a_{11}x + a_{12}y + a_{13}z$$
$$v = a_{21}x + a_{22}y + a_{23}z$$
$$w = a_{31}x + a_{32}y + a_{33}z$$

where all the a_{ij} are numbers. Find $\partial v/\partial y$ and $\partial y/\partial v$. How are these quantities related?

3. The Chain Rule

The Chain Rule is one of the most effective devices for calculating ordinary derivatives. In this section we show how to extend the Chain Rule for the computation of partial derivatives. The basis of the Rule in the case of functions of one variable is the Fundamental Lemma on Differentiation, which we now recall.

Theorem 1 (Fundamental Lemma on Differentiation in $\mathbf{R}^1$). *Suppose that F has a derivative at a value u so that $F'(u)$ exists. We define the function*

$$G(h) = \begin{cases} \dfrac{F(u+h) - F(u)}{h} - F'(u), & \text{if } h \neq 0, \\ 0, & \text{if } h = 0. \end{cases}$$

Then (a) *G is continuous at $h = 0$, and* (b) *the formula*

$$F(u+h) - F(u) = [F'(u) + G(h)]h \tag{1}$$

holds.

PROOF. From the definition of derivative, we know that

$$\lim_{h \to 0} \frac{F(u+h) - F(u)}{h} = F'(u)$$

or

$$\lim_{h \to 0} \left[\frac{F(u+h) - F(u)}{h} - F'(u) \right] = 0.$$

Hence $G(h) \to 0$ as $h \to 0$. Therefore G is continuous at 0, and (a) holds. To establish (b), we observe that for $h \neq 0$, the formula (1) is a restatement of the definition of G. For $h = 0$, both sides of (1) are zero.

The above theorem has a natural generalization for functions of two variables.

Theorem 2 (Fundamental Lemma on Differentiation in $\mathbf{R}^2$). *Suppose that f is a continuous function of two variables (say x and y) and that f_x and f_y are continuous at (x_0, y_0). Then there are two functions, $G_1(h, k)$ and $G_2(h, k)$ continuous at* $(0, 0)$ *with* $G_1(0, 0) = G_2(0, 0) = 0$, *such that*

$$\begin{aligned} f(x_0 + h, y_0 + k) - f(x_0, y_0) = {} & f_x(x_0, y_0)h + f_y(x_0, y_0)k \\ & + G_1(h, k)h + G_2(h, k)k. \end{aligned} \tag{2}$$

PROOF. The proof depends on writing the left side of (2) in a more complicated way:

$$\begin{aligned} f(x_0 + h, y_0 + k) - f(x_0, y_0) = {} & [f(x_0 + h, y_0 + k) - f(x_0 + h, y_0)] \\ & + [f(x_0 + h, y_0) - f(x_0, y_0)]. \end{aligned} \tag{3}$$

Figure 4-2 shows the points at which f is evaluated in (3) (h and k are taken to be positive in the figure). We apply the Theorem of the Mean to each of the quantities in brackets in (3) above. The result for the first quantity is

$$\frac{f(x_0 + h, y_0 + k) - f(x_0 + h, y_0)}{(y_0 + k) - y_0} = f_y(x_0 + h, \eta), \tag{4}$$

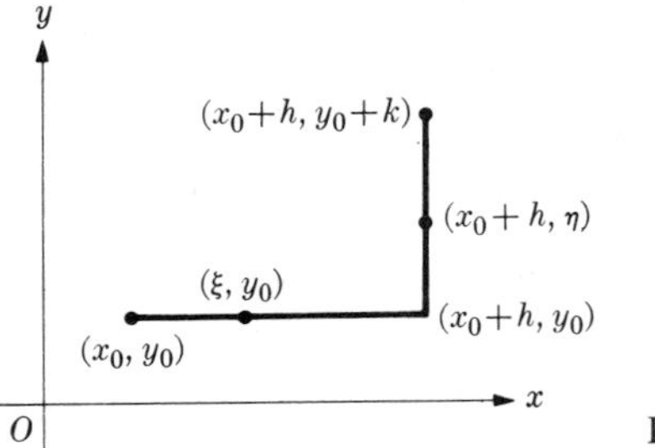

Fig. 4-2

and for the second quantity,

$$\frac{f(x_0+h, y_0) - f(x_0, y_0)}{(x_0+h) - x_0} = f_x(\xi, y_0), \tag{5}$$

where η is between y_0 and $y_0 + k$ and ξ is between x_0 and $x_0 + h$. Typical locations for ξ and η are shown in Fig. 4-2. Substituting (4) and (5) into the right side of (3), we find

$$f(x_0+h, y_0+k) - f(x_0, y_0) = f_x(\xi, y_0)h + f_y(x_0+h, \eta)k. \tag{6}$$

The quantities G_1 and G_2 are defined by*

$$G_1 = f_x(\xi, y_0) - f_x(x_0, y_0),$$
$$G_2 = f_y(x_0+h, \eta) - f_y(x_0, y_0).$$

Multiplying the expression for G_1 by h and that for G_2 by k and inserting the result in the right side of (6), we obtain the statement of the theorem. Since

$$\xi \to x_0 \quad \text{as } h \to 0 \qquad \text{and} \qquad \eta \to y_0 \quad \text{as } k \to 0,$$

it follows (since f_x and f_y are continuous) that G_1 and G_2 tend to zero as h and k tend to zero. Thus G_1 and G_2 are continuous at $(0, 0)$ if we define them as being equal to zero there.

Theorem 3 (Chain Rule). *Suppose that $z = f(x, y)$ is continuous and that $\partial f/\partial x$, $\partial f/\partial y$ are continuous. Assume that $x = x(r, s)$ and $y = y(r, s)$ are functions of r and s such that $\partial x/\partial r$, $\partial x/\partial s$, $\partial y/\partial r$, $\partial y/\partial s$ all exist. Then z is a function of r and s and the following formulas hold:*

$$\left.\begin{aligned} \frac{\partial z}{\partial r} &= \left(\frac{\partial f}{\partial x}\right)\left(\frac{\partial x}{\partial r}\right) + \left(\frac{\partial f}{\partial y}\right)\left(\frac{\partial y}{\partial r}\right) \\ \frac{\partial z}{\partial s} &= \left(\frac{\partial f}{\partial x}\right)\left(\frac{\partial x}{\partial s}\right) + \left(\frac{\partial f}{\partial y}\right)\left(\frac{\partial y}{\partial s}\right) \end{aligned}\right\} \tag{7}$$

Proof. The first formula will be established; the second is proved similarly.

* Of course, if $k = 0$, we define $G_2 = f_y(x_0 + h, y_0) - f_y(x_0, y_0)$, etc.

We use the Δ notation. A change Δr in r induces a change Δx in x and a change Δy in y. That is,

$$\Delta x = x(r + \Delta r, s) - x(r, s),$$
$$\Delta y = y(r + \Delta r, s) - y(r, s).$$

The function z has the partial derivative

$$\frac{\partial z}{\partial r} = \lim_{\Delta r \to 0} \frac{\Delta z}{\Delta r},$$

where Δz, the change in z due to the change Δr in r, is given by

$$\Delta z = f(x + \Delta x, y + \Delta y) - f(x, y),$$

which latter expression we also denote by Δf. The Fundamental Lemma on Differentiation in R^2 with $h = \Delta x$ and $k = \Delta y$ reads

$$\Delta f = \frac{\partial f}{\partial x}\Delta x + \frac{\partial f}{\partial y}\Delta y + G_1 \Delta x + G_2 \Delta y,$$

where we have changed notation by using $\partial f/\partial x$ in place of $f_x(x, y)$ and $\partial f/\partial y$ for $f_y(x, y)$. Dividing by Δr in the above equation for Δf, we get

$$\frac{\Delta z}{\Delta r} = \frac{\Delta f}{\Delta r} = \frac{\partial f}{\partial x}\frac{\Delta x}{\Delta r} + \frac{\partial f}{\partial y}\frac{\Delta y}{\Delta r} + G_1 \frac{\Delta x}{\Delta r} + G_2 \frac{\Delta y}{\Delta r}.$$

Letting Δr tend to zero and remembering that $G_1 \to 0$, $G_2 \to 0$, we obtain the first formula in (7), as desired.

REMARKS. (i) The formulas (7) may be written in various notations. Two common expressions are

$$\left.\begin{aligned} \frac{\partial z}{\partial r} &= \left(\frac{\partial z}{\partial x}\right)\left(\frac{\partial x}{\partial r}\right) + \left(\frac{\partial z}{\partial y}\right)\left(\frac{\partial y}{\partial r}\right) \\ \frac{\partial z}{\partial s} &= \left(\frac{\partial z}{\partial x}\right)\left(\frac{\partial x}{\partial s}\right) + \left(\frac{\partial z}{\partial y}\right)\left(\frac{\partial y}{\partial s}\right) \end{aligned}\right\} \tag{8}$$

and

$$\left.\begin{aligned} f_r &= f_x x_r + f_y y_r \\ f_s &= f_x x_s + f_y y_s \end{aligned}\right\}. \tag{9}$$

ii) To use the comma notation we define the function $g(r, s) = f[x(r, s), y(r, s)]$. The formulas expressing the Chain Rule are then

$$\left.\begin{aligned} g_{,1}(r, s) &= f_{,1}(x, y)x_{,1}(r, s) + f_{,2}(x, y)y_{,1}(r, s) \\ g_{,2}(r, s) &= f_{,1}(x, y)x_{,2}(r, s) + f_{,2}(x, y)y_{,2}(r, s) \end{aligned}\right\} \tag{10}$$

iii) For functions of one variable, the Chain Rule is easily remembered as the rule which allows us to think of derivatives as fractions. The formula

$$\frac{dy}{dx} = \frac{dy}{du} \cdot \frac{du}{dx}$$

is an example. The symbol du has a meaning of its own. To attempt to draw such an analogy with the Chain Rule for Partial Derivatives leads to disaster. Formulas (7) are the ones we usually employ in the applications. The parentheses around the individual terms are used to indicate the inseparable nature of each item. Actually, the forms (9) and (10) for the same formulas avoid the danger of erroneously treating partial derivatives as fractions.

EXAMPLE 1. Suppose that $z = x^3 + y^3$, $x = 2r + s$, $y = 3r - 2s$. Find $\partial z/\partial r$ and $\partial z/\partial s$.

SOLUTION. We can employ the Chain Rule and obtain

$$\frac{\partial z}{\partial x} = 3x^2, \qquad \frac{\partial z}{\partial y} = 3y^2,$$

$$\frac{\partial x}{\partial r} = 2, \qquad \frac{\partial x}{\partial s} = 1, \qquad \frac{\partial y}{\partial r} = 3, \qquad \frac{\partial y}{\partial s} = -2.$$

Therefore

$$\frac{\partial z}{\partial r} = (3x^2)(2) + (3y^2)(3) = 6x^2 + 9y^2 = 6(2r + s)^2 + 9(3r - 2s)^2,$$

$$\frac{\partial z}{\partial s} = (3x^2)(1) + (3y^2)(-2) = 3x^2 - 6y^2 = 3(2r + s)^2 - 6(3r - 2s)^2.$$

In Theorem 3 (the Chain Rule), the variables r and s are the **independent variables**; we denote the variables x and y **intermediate variables**. The formulas we derived extend easily to any number of independent variables and any number of intermediate variables. For example,

$$\text{if} \quad w = f(x, y, z) \quad \text{and if} \quad x = x(r, s), \quad y = y(r, s), \quad z = z(r, s),$$

then

$$\frac{\partial w}{\partial r} = \left(\frac{\partial f}{\partial x}\right)\left(\frac{\partial x}{\partial r}\right) + \left(\frac{\partial f}{\partial y}\right)\left(\frac{\partial y}{\partial r}\right) + \left(\frac{\partial f}{\partial z}\right)\left(\frac{\partial z}{\partial r}\right),$$

and there is a similar formula for $\partial w/\partial s$. The case of four intermediate variables and one independent variable—that is,

$$w = f(x, y, u, v), \quad x = x(t), \quad y = y(t), \quad u = u(t), \quad v = v(t)$$

—leads to the formula

$$\frac{dw}{dt} = \frac{\partial f}{\partial x}\frac{dx}{dt} + \frac{\partial f}{\partial y}\frac{dy}{dt} + \frac{\partial f}{\partial u}\frac{du}{dt} + \frac{\partial f}{\partial v}\frac{dv}{dt}.$$

The ordinary d (rather than the round ∂) is used for derivatives with respect to t, since w, x, y, u, and v are all functions of the *one* variable t.

REMARK. As an aid in remembering the Chain Rule, we note that *there are as many terms in the formula as there are intermediate variables.*

EXAMPLE 2. If $z = f(x, y) = 2x^2 + xy - y^2 + 2x - 3y + 5$, $x = 2s - t$, $y = s + t$, find $\partial z/\partial t$.

SOLUTION. We use the Chain Rule:

$$\frac{\partial f}{\partial x} = 4x + y + 2, \qquad \frac{\partial f}{\partial y} = x - 2y - 3, \qquad \frac{\partial x}{\partial t} = -1, \qquad \frac{\partial y}{\partial t} = 1.$$

Therefore

$$\begin{aligned}\frac{\partial z}{\partial t} = (4x + y + 2)(-1) + (x - 2y - 3)(1) &= -3x - 3y - 5 \\ &= -3(2s - t) - 3(s + t) - 5 \\ &= -9s - 5.\end{aligned}$$

EXAMPLE 3. Given $w = f(x, y, z) = x^2 + 3y^2 - 2z^2 + 4x - y + 3z - 1$, $x = t^2 - 2t + 1$, $y = 3t - 2$, $z = t^2 + 4t - 3$, find dw/dt when $t = 2$.

SOLUTION. Employing the Chain Rule, we find

$$\frac{\partial f}{\partial x} = 2x + 4, \qquad \frac{\partial f}{\partial y} = 6y - 1, \qquad \frac{\partial f}{\partial z} = -4z + 3,$$

$$\frac{dx}{dt} = 2t - 2, \qquad \frac{dy}{dt} = 3, \qquad \frac{dz}{dt} = 2t + 4.$$

Therefore

$$\frac{dw}{dt} = (2x + 4)(2t - 2) + (6y - 1)(3) + (-4z + 3)(2t + 4).$$

When $t = 2$, we have $x = 1$, $y = 4$, $z = 9$, and so

$$\frac{dw}{dt} = (6)(2) + (23)(3) + (-33)(8) = -183.$$

PROBLEMS

In each of Problems 1 through 12, use the Chain Rule to obtain the indicated partial derivatives.

1. $z = f(x, y) = x^2 + y^2$; $x = s - 2t$, $y = 2s + t$; $\dfrac{\partial z}{\partial s}, \dfrac{\partial z}{\partial t}$

2. $z = f(x, y) = x^2 - xy - y^2$; $x = s + t$, $y = -s + t$; $\dfrac{\partial z}{\partial s}, \dfrac{\partial z}{\partial t}$

3. $z = f(x, y) = x^2 + y^2$; $x = s^2 - t^2$, $y = 2st$; $\dfrac{\partial z}{\partial s}, \dfrac{\partial z}{\partial t}$

4. $z = f(x, y) = \dfrac{x}{x^2 + y^2}$; $x = s\cos t$, $y = s\sin t$; $\dfrac{\partial z}{\partial s}, \dfrac{\partial z}{\partial t}$

5. $z = f(x, y) = \dfrac{x}{\sqrt{x^2 + y^2}}$; $x = 2s - t$, $y = s + 2t$; $\dfrac{\partial z}{\partial s}, \dfrac{\partial z}{\partial t}$

6. $z = f(x, y) = e^x \cos y$; $x = s^2 - t^2$, $y = 2st$; $\dfrac{\partial z}{\partial s}, \dfrac{\partial z}{\partial t}$

7. $w = f(x, y, z) = x^2 + y^2 + z^2 + 3xy - 2xz + 4$; $x = 3s + t$, $y = 2s - t$, $z = s + 2t$; $\partial w/\partial s$, $\partial w/\partial t$

8. $w = f(x, y, z) = x^3 + 2y^3 + z^3$; $x = s^2 - t^2$, $y = s^2 + t^2$, $z = 2st$; $\partial w/\partial s$, $\partial w/\partial t$

9. $w = f(x, y, z) = x^2 - y^2 + 2z^2$; $x = r^2 + 1$, $y = r^2 - 2r + 1$, $z = r^2 - 2$; dw/dr

10. $z = f(x, y) = \dfrac{x - y}{1 + x^2 + y^2}$; $x = r + 3s - t$, $y = r - 2s + 3t$; $\partial z/\partial r$, $\partial z/\partial s$, $\partial z/\partial t$

11. $w = f(u) = u^3 + 2u^2 - 3u + 1$; $u = r^2 - s^2 + t^2$; $\dfrac{\partial w}{\partial r}, \dfrac{\partial w}{\partial s}, \dfrac{\partial w}{\partial t}$

12. $w = f(x, y, u, v) = x^2 + y^2 - u^2 - v^2 + 3x - 2y + u - v$; $x = 2r + s - t$, $y = r - 2s + t$, $u = 3r - 2s + t$, $v = r - s - t$; $\dfrac{\partial w}{\partial r}, \dfrac{\partial w}{\partial s}, \dfrac{\partial w}{\partial t}$

In Problems 13 through 20, use the Chain Rule to find the indicated derivatives at the values given.

13. $z = x^2 - y^2$; $x = r\cos\theta$, $y = r\sin\theta$; $\dfrac{\partial z}{\partial r}, \dfrac{\partial z}{\partial \theta}$ where $r = \sqrt{2}$, $\theta = \dfrac{\pi}{4}$

14. $w = x^2 + y^2 - z^2$; $x = 1 - t^2$, $y = 2t + 3$, $z = t^2 + t$, $\dfrac{dw}{dt}$ where $t = -1$

15. $w = xy + yz + zx$; $x = t\cos t$, $y = t\sin t$, $z = t$; $\dfrac{dw}{dt}$ where $t = \dfrac{\pi}{4}$

16. $z = u^3 + 2u - 3$; $u = s^2 + t^2 - 4$; $\dfrac{\partial z}{\partial s}, \dfrac{\partial z}{\partial t}$ where $s = 1$ and $t = 2$

17. $z = \dfrac{xy}{x^2 + y^2}$; $x = r\cos\theta$, $y = r\sin\theta$; $\dfrac{\partial z}{\partial r}, \dfrac{\partial z}{\partial \theta}$ where $r = 3$, $\theta = \dfrac{\pi}{6}$

18. $f = e^{xyz}\cos xyz$; $x = r^2 + t^2$, $y = t^2 + r$, $z = r + t$; $\dfrac{\partial f}{\partial x}, \dfrac{\partial f}{\partial y}, \dfrac{\partial f}{\partial z}$, where $r = 2$, $t = -2$

19. $w = x^3 + y^3 + z^3 - u^2 - v^2$; $x = r^2 + s^2 + t^2$, $y = r^2 + s^2 - t^2$, $z = r^2 - s^2 - t^2$, $u = r^2 + t^2$, $v = r^2 - s^2$; $\dfrac{\partial w}{\partial s}, \dfrac{\partial w}{\partial t}$ where $r = 1$, $s = 0$, $t = -1$

20. $w = x^4 - y^4 - z^4$; $x = 5r + 3s - 2t + u - v$, $y = 2r - 4s + t - u^2 + v^2$, $z = s^3 - 2t^2 + 3v^2$; $\dfrac{\partial w}{\partial s}, \dfrac{\partial w}{\partial v}$ where $r = 1$, $s = -1$, $t = 0$, $u = 3$, $v = -2$

21. Suppose that $w = f(x, y, z, t)$ and $z = g(x, y, t)$. Using the notation as given in (8), we may find $\partial w/\partial x$ by the formula

$$\frac{\partial w}{\partial x} = \frac{\partial w}{\partial x}\frac{\partial x}{\partial x} + \frac{\partial w}{\partial y}\frac{\partial y}{\partial x} + \frac{\partial w}{\partial z}\frac{\partial z}{\partial x} + \frac{\partial w}{\partial t}\frac{\partial t}{\partial x}.$$

Since x, y, and t are independent, we have

$$\frac{\partial x}{\partial x} = 1, \qquad \frac{\partial y}{\partial x} = 0, \qquad \text{and} \qquad \frac{\partial t}{\partial x} = 0,$$

and we find

$$\frac{\partial w}{\partial x} = \frac{\partial w}{\partial x} + \frac{\partial w}{\partial z}\frac{\partial z}{\partial x} \qquad \text{or} \qquad 0 = \frac{\partial w}{\partial z}\frac{\partial z}{\partial x}.$$

This formula is incorrect. (a) Verify this fact by substituting specific functions for f and g. (b) Using formulas (10), show that

$$\frac{\partial w}{\partial x} = f_{,1} + f_{,3}g_{,1}.$$

Verify the correctness of this result for the specific functions employed in part (a).

22. Suppose that $z = f(x, y, z)$ is continuous and that $\partial f/\partial x$, $\partial f/\partial y$, $\partial f/\partial z$ are continuous. Let $x = x(r, s)$, $y = y(r, s)$, $z = z(r, s)$ be functions all of whose first partial derivatives exist. Derive the Chain Rule formula for $\partial z/\partial s$.

23. State and prove the Fundamental Lemma on Differentiation (Theorem 2) for functions f of three variables (say x, y, and z).

24. Let $f = f(x, y, z)$ be a function with continuous partial derivatives and define $g(t) = f(tx, ty, tz)$. Show that $g'(t) = xf_{,1} + yf_{,2} + zf_{,3}$.

4. Applications of the Chain Rule

The Chain Rule may be employed profitably in many types of applications. These are best illustrated with examples, and we shall begin with two problems in related rates.

EXAMPLE 1. At a certain instant the altitude of a right circular cone is 30 cm and is increasing at the rate of 2 cm/sec. At the same instant, the radius of the base is 20 cm. and is increasing at the rate of 1 cm/sec. At what rate is the volume increasing at that instant? (See Fig. 4-3.)

SOLUTION. The volume V is given by $V = \frac{1}{3}\pi r^2 h$, with r and h functions of the

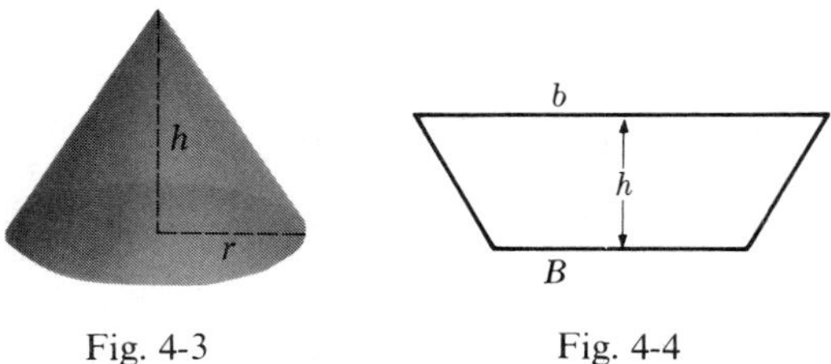

Fig. 4-3 Fig. 4-4

time t. We can apply the Chain Rule to obtain

$$\begin{aligned}\frac{dV}{dt} &= \frac{\partial V}{\partial r}\frac{dr}{dt} + \frac{\partial V}{\partial h}\frac{dh}{dt} \\ &= \frac{2}{3}\pi r h \frac{dr}{dt} + \frac{1}{3}\pi r^2 \frac{dh}{dt}.\end{aligned}$$

At the given instant,

$$\frac{dV}{dt} = \frac{2}{3}\pi(20)(30)(1) + \frac{1}{3}\pi(20)^2(2) = \frac{2000\pi}{3}\text{cm}^3/\text{sec}.$$

EXAMPLE 2. The base B of a trapezoid increases in length at the rate of 2 cm/sec and the base b decreases in length at the rate of 1 cm/sec. If the altitude h is increasing at the rate of 3 cm/sec, how rapidly is the area A changing when $B = 30$ cm., $b = 50$ cm., and $h = 10$ cm.? (See Fig. 4-4.)

SOLUTION. The area A is given by $A = \frac{1}{2}(B + b)h$, with B, b, and h functions of time. We apply the Chain Rule to get

$$\begin{aligned}\frac{dA}{dt} = \frac{\partial A}{\partial B}\frac{dB}{dt} + \frac{\partial A}{\partial b}\frac{db}{dt} + \frac{\partial A}{\partial h}\frac{dh}{dt} &= \frac{1}{2}h\frac{dB}{dt} + \frac{1}{2}h\frac{db}{dt} + \frac{1}{2}(B + b)\frac{dh}{dt} \\ &= (5)(2) + (5)(-1) + (40)(3) \\ &= 125 \text{ in}^2/\text{sec}.\end{aligned}$$

Note that since b is decreasing, db/dt is negative.

The next example shows that a clear understanding of the symbolism in partial differentiation is required in many applications.

EXAMPLE 3. Suppose that $z = f(x + at)$ and a is constant. Show that

$$\frac{\partial z}{\partial t} = a\frac{\partial z}{\partial x}.$$

SOLUTION. We observe that f is a function of *one argument* (in which, however, two variables occur in a particular combination). We let $u = x + at$ and, if

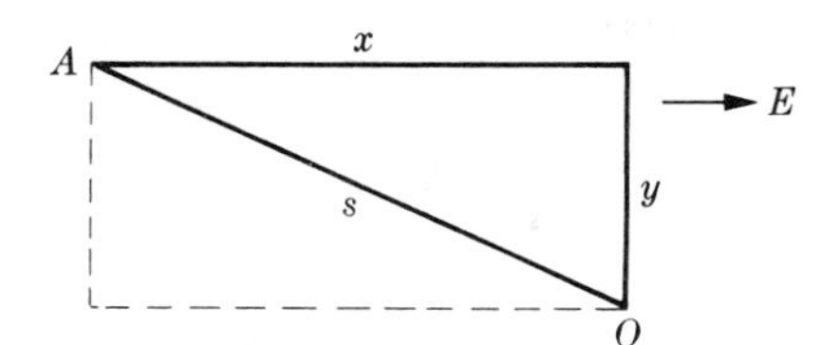

Fig. 4-5

we now write

$$z = f(u), \qquad u = x + at,$$

we recognize the applicability of the Chain Rule. Therefore

$$\frac{\partial z}{\partial x} = \frac{dz}{du}\frac{\partial u}{\partial x} = f'(u) \cdot 1,$$

$$\frac{\partial z}{\partial t} = \frac{dz}{du}\frac{\partial u}{\partial t} = f'(u) \cdot a.$$

We conclude that

$$\frac{\partial z}{\partial t} = a\frac{\partial z}{\partial x}.$$

EXAMPLE 4. An airplane is traveling directly east at 300 km/hr and is climbing at the rate of 600 meters/min. At a certain instant, the airplane is 12,000 meters above ground and 5 km directly west of an observer on the ground. How fast is the distance changing between the airplane and the observer at this instant?

SOLUTION. Referring to Fig. 4-5, with the observer at O and the airplane at A, we see that x, y, and s are functions of the time t. The distance s between the airplane and the observer is given by $s = (x^2 + y^2)^{1/2}$, and we wish to find ds/dt. Using the Chain Rule, we get

$$\frac{ds}{dt} = \frac{\partial s}{\partial x}\frac{dx}{dt} + \frac{\partial s}{\partial y}\frac{dy}{dt} = \frac{x}{\sqrt{x^2 + y^2}}\frac{dx}{dt} + \frac{y}{\sqrt{x^2 + y^2}}\frac{dy}{dt}.$$

From the given data we see that, at the instant in question,

$$y = 12{,}000,\ x = 5{,}000,\quad \frac{dx}{dt} = -83.3 \text{ m/sec.} \quad \text{and} \quad \frac{dy}{dt} = +10 \text{ m/sec.}$$

Therefore

$$\begin{aligned}\frac{ds}{dt} &= \frac{5{,}000}{13{,}000}(-83.3) + \frac{12{,}000}{13{,}000}(10)\\ &= -22.8 \text{ m/sec.}\end{aligned}$$

The negative sign indicates that the airplane is approaching the observer.

Problems

1. Find the rate at which the lateral area of the cone in Example 1 is increasing at the given instant.

2. At a certain instant a right circular cylinder has radius of base 10 cm. and altitude 15 cm. At this instant the radius is decreasing at the rate of 5 cm/sec and the altitude is increasing at the rate of 4 cm/sec. How rapidly is the volume changing at this moment?

3. A gas obeys the law $pv = RT$ ($R =$ const). At a certain instant while the gas is being compressed, $v = 15$ m^3, $p = 25$ kg/cm^2, v is decreasing at the rate of 3 m^3/min, and p is increasing at the rate of $6\frac{2}{3}$ kg/cm^2/min. Find dT/dt. (Answer in terms of R.)

4. (a) In Problem 2, find how rapidly the lateral surface area of the cylinder is changing at the same instant. (b) What would the result be for the area A consisting of the top and bottom of the cylinder as well as the lateral surface?

5. At a certain instant of time, the angle A of a triangle ABC is 60° and increasing at the rate of 5°/sec, the side AB is 10 cm. and increasing at the rate of 1 cm/sec, and side AC is 16 cm. and decreasing at the rate of $\frac{1}{2}$ cm/sec. Find the rate of change of side BC.

6. A point moves along the surface $z = x^2 + 2y^2 - 3x + y$ in such a way that $dx/dt = 3$ and $dy/dt = 2$. Find how z changes with time when $x = 1$, $y = 4$.

7. Water is leaking out of a conical tank at the rate of 0.5 m^3/min. The tank is also stretching in such a way that, while it remains conical, the distance across the top at the water surface is increasing at the rate of 0.2 m/min. How fast is the height h of water changing at the instant when $h = 10$ and the volume of water is 75 cu. meters?

8. A rectangular bin is changing in size in such a way that its length is increasing at the rate of 3 cm/sec, its width is decreasing at the rate of 2 cm/sec, and its height is increasing at the rate of 1 cm/sec. (a) How fast is the volume changing at the instant when the length is 15, the width is 10, and the height is 8? (b) How fast is the total surface area changing at the same instant?

9. a) Given $z = f(y/x)$. Find $\partial z/\partial x$ and $\partial z/\partial y$ in terms of $f'(y/x)$ and x and y. [*Hint:* Let $u = y/x$.] (b) Show that $x(\partial z/\partial x) + y(\partial z/\partial y) = 0$.

10. a) Given that $w = f(y - x - t, z - y + t)$. By letting $u = y - x - t$, $v = z - y + t$, find $\partial w/\partial x$, $\partial w/\partial y$, $\partial w/\partial z$, $\partial w/\partial t$ in terms of $f_{,1}$ and $f_{,2}$.
    b) Show that

$$\frac{\partial w}{\partial x} + 2\frac{\partial w}{\partial y} + \frac{\partial w}{\partial z} + \frac{\partial w}{\partial t} = 0.$$

11. Suppose that $z = f(x, y)$ and $x = r\cos\theta$, $y = r\sin\theta$. (a) Express $\partial z/\partial r$ and $\partial z/\partial\theta$ in terms of $\partial z/\partial x$ and $\partial z/\partial y$. (b) Show that

$$\left(\frac{\partial z}{\partial r}\right)^2 + \frac{1}{r^2}\left(\frac{\partial z}{\partial \theta}\right)^2 = \left(\frac{\partial z}{\partial x}\right)^2 + \left(\frac{\partial z}{\partial y}\right)^2.$$

12. Suppose that $z = f(x, y)$, $x = e^s \cos t$, $y = e^s \sin t$. Show that

$$\left(\frac{\partial z}{\partial s}\right)^2 + \left(\frac{\partial z}{\partial t}\right)^2 = e^{2s}\left[\left(\frac{\partial z}{\partial x}\right)^2 + \left(\frac{\partial z}{\partial y}\right)^2\right].$$

13. Suppose that $u = f(x + at, y + bt)$, with a and b constants. Show that

$$\frac{\partial u}{\partial t} = a\frac{\partial u}{\partial x} + b\frac{\partial u}{\partial y}.$$

14. Given that

$$f(x, y) = \frac{x + y}{x^2 - xy + y^2},$$

show that

$$xf_{,1} + yf_{,2} = -f.$$

15. Given that $f(x, y) = x^2 - y^2 + xy \log(y/x)$, show that $xf_{,1} + yf_{,2} = 2f$.

16. If in Example 4 a second observer is situated at a point O', 12 km west of the one at O, find the rate of change of the distance between A and O' at the same instant.

17. If in Example 2 one of the acute angles is held constant at 60°, find the rate of change of the perimeter of the trapezoid at the instant in question.

18. A rectangular bin (without a top) is changing in size in such a way that its total surface area always maintains the value of 100 sq. cm. The length and width are each increasing at the rate of 1 cm/sec. Find how fast the height and the volume are changing when the length and width are each 2 cm.

19. The number A of bacteria at time t (in hours) of a certain type follows a growth law given by

$$\frac{\partial A}{\partial t} = x\frac{\partial A}{\partial x} + 2y\frac{\partial A}{\partial y},$$

where x and y are two types of liquid (in cc) in which the culture is grown. Show that $A(x, y, t) = c(x^2 + y)e^{2t}$, where c is a constant, is a possible law of growth. If there are 4 cc of x-type liquid and 10 cc of y-type and if there are 1000 bacteria at time $t = 0$, find the number of bacteria after four hours.

5. Directional Derivatives. Gradient

The partial derivative of a function with respect to x may be considered as the derivative in the x direction; the partial derivative with respect to y is the derivative in the y direction. We now show how we may define the derivative in *any direction*. To see this, we consider a function $f(x, y)$ and a point $P(x, y)$ in the xy plane. A particular direction is singled out by specifying the angle θ which a line through P makes with the positive x axis (Fig. 4-6). We may also prescribe the direction by drawing the directed line segment $\overrightarrow{PP'}$ of *unit*

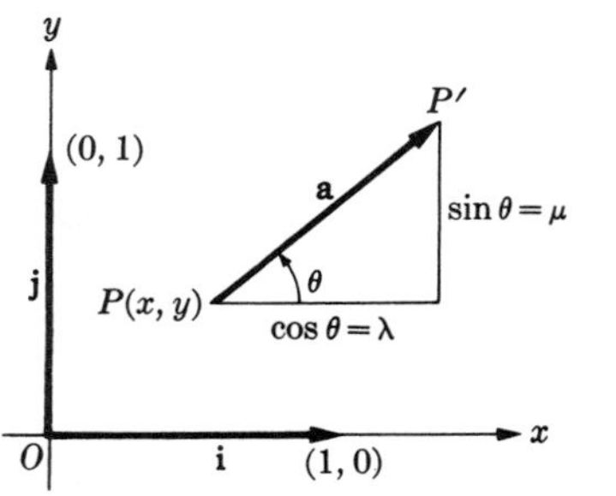

Fig. 4-6

length, as shown, and defining the vector **a** by the relation

$$\mathbf{a} = \lambda\mathbf{i} + \mu\mathbf{j},$$

with $\lambda = \cos\theta$, $\mu = \sin\theta$, and **i** and **j** the customary unit vectors. We note that **a** is a vector of unit length. The vector **a** determines the same direction as the angle θ.

Definition. Let f be a function of two variables. We define the **directional derivative** $D_\mathbf{a}f$ **of** f **in the direction of a** *by*

$$D_\mathrm{a}f(x, y) = \lim_{h\to 0}\frac{f(x + \lambda h, y + \mu h) - f(x, y)}{h}$$

whenever the limit exists.

REMARK. We note that, when $\theta = 0$, then $\lambda = 1$, $\mu = 0$, and the direction is the positive x direction. The directional derivative is exactly $\partial f/\partial x$. Similarly, if $\theta = \pi/2$, we have $\lambda = 0$, $\mu = 1$, and the directional derivative is $\partial f/\partial y$.

The working formula for directional derivatives is established in the next theorem.

Theorem 4. *If $f(x, y)$ and its partial derivatives are continuous and*

$$\mathbf{a} = (\cos\theta)\mathbf{i} + (\sin\theta)\mathbf{j},$$

then

$$\boxed{D_\mathrm{a}f(x, y) = f_x(x, y)\cos\theta + f_y(x, y)\sin\theta.}$$

PROOF. The proof uses the following artificial device. We define the function $g(s)$ by

$$g(s) = f(x + s\cos\theta, y + s\sin\theta),$$

in which we keep x, y, and θ fixed and allow s to vary. The Chain Rule now yields

$$g'(s) = f_x(x + s\cos\theta, y + s\sin\theta)\cos\theta + f_y(x + s\cos\theta, y + s\sin\theta)\sin\theta,$$

and we see that

$$g'(0) = f_x(x, y) \cos \theta + f_y(x, y) \sin \theta.$$

By definition we know that

$$g'(0) = \lim_{s \to 0} \frac{g(s) - g(0)}{s} = \lim_{s \to 0} \frac{f(x + s \cos \theta, y + s \sin \theta) - f(x, y)}{s},$$

and hence $g'(0)$ is precisely $D_{\mathbf{a}} f(x, y)$. The result is established.

EXAMPLE 1. Given $f(x, y) = x^2 + 2y^2 - 3x + 2y$, find the directional derivative of f in the direction $\theta = \pi/6$. What is the value of this derivative at the point $(2, -1)$?

SOLUTION. We compute

$$\partial f/\partial x = 2x - 3, \qquad \partial f/\partial y = 4y + 2.$$

Therefore

$$D_{\mathbf{a}} f = (2x - 3) \cdot \tfrac{1}{2}\sqrt{3} + (4y + 2) \cdot \tfrac{1}{2}.$$

In particular, when $x = 2$ and $y = -1$, we obtain

$$D_{\mathbf{a}} f = \tfrac{1}{2}\sqrt{3} - 1.$$

An alternate notation for directional derivative for functions of two variables is the symbol

$$d_\theta f(x, y),$$

in which θ is the angle the direction makes with the positive x axis. For a fixed value of x and y, the directional derivative is a function of θ. It is an ordinary problem in maxima and minima to find the value of θ which makes the directional derivative at a given point the largest or the smallest. The next example shows the method.

EXAMPLE 2. Given $f(x, y) = x^2 - xy - y^2$, find $d_\theta f(x, y)$ at the point $(2, -3)$. For what value of θ does $d_\theta(2, -3)$ take on its maximum value?

SOLUTION. We have

$$f_{,1}(x, y) = 2x - y, \qquad f_{,2}(x, y) = -x - 2y.$$

Therefore for $x = 2$, $y = -3$,

$$d_\theta f(2, -3) = 7 \cos \theta + 4 \sin \theta.$$

To find the maximum of this function of θ we differentiate the function

$$k(\theta) = 7 \cos \theta + 4 \sin \theta$$

and set the derivative equal to zero. We get

$$k'(\theta) = -7\sin\theta + 4\cos\theta = 0 \qquad \text{or} \qquad \tan\theta = \tfrac{4}{7}.$$

The result is $\cos\theta = \pm 7/\sqrt{65}$, $\sin\theta = \pm 4/\sqrt{65}$, and

$$\tan\theta = \tfrac{4}{7}, \qquad \theta = \begin{cases} 29°45' \text{ approximately}, \\ 209°45' \text{ approximately}. \end{cases}$$

It is clear by substitution that the first choice for θ makes $k(\theta)$ a maximum, while the second makes it a minimum.

The definition of directional derivative for functions of two variables has a natural extension to functions of three variables. In three dimensions, a direction is determined by a set of direction cosines λ, μ, ν or, equivalently, by a vector

$$\mathbf{a} = \lambda\mathbf{i} + \mu\mathbf{j} + \nu\mathbf{k}.$$

We recall that $\lambda^2 + \mu^2 + \nu^2 = 1$, and so $\mathbf{a}$ is a *unit* vector.

Definition. We define the **directional derivative** $D_{\mathbf{a}}f$ **of** $f(x, y, z)$ **in the direction of a** *by*

$$D_{\mathbf{a}}f(x, y, z) = \lim_{h\to 0} \frac{f(x + \lambda h, y + \mu h, z + \nu h) - f(x, y, z)}{h}$$

whenever the limit exists.

The proof of the next theorem is entirely analogous to (and employs the same device as) the proof of Theorem 4.

Theorem 5. *If* $f(x, y, z)$ *and its partial derivatives are continuous and*

$$\mathbf{a} = \lambda\mathbf{i} + \mu\mathbf{j} + \nu\mathbf{k}$$

is a unit vector, then

$$\boxed{D_{\mathbf{a}}f(x, y, z) = \lambda f_x(x, y, z) + \mu f_y(x, y, z) + \nu f_z(x, y, z).}$$

EXAMPLE 3. Find the directional derivative of

$$f(x, y, z) = x^2 + y^2 + z^2 - 3xy + 2xz - yz$$

at the point $(1, 2, -1)$.

SOLUTION. We have

$$f_x(x, y, z) = 2x - 3y + 2z; \quad f_y(x, y, z) = 2y - 3x - z;$$
$$f_z(x, y, z) = 2z + 2x - y.$$

Denoting the direction by $\mathbf{a} = \lambda\mathbf{i} + \mu\mathbf{j} + \nu\mathbf{k}$, we get

$$D_{\mathbf{a}}f(1, 2, -1) = -6\lambda + 2\mu - 2\nu.$$

EXAMPLE 4. Given the function

$$f(x, y, z) = xe^{yz} + ye^{xz} + ze^{xy},$$

find the directional derivative at $P(1, 0, 2)$ in the direction going from P to $P'(5, 3, 3)$.

SOLUTION. A set of direction numbers for the line through P and P' is 4, 3, 1. The corresponding direction cosines are $4/\sqrt{26}$, $3/\sqrt{26}$, $1/\sqrt{26}$, which we denote by λ, μ, ν. The direction $\mathbf{a}$ is given by

$$\mathbf{a} = \frac{4}{\sqrt{26}}\mathbf{i} + \frac{3}{\sqrt{26}}\mathbf{j} + \frac{1}{\sqrt{26}}\mathbf{k}.$$

We find that

$$\frac{\partial f}{\partial x}(x, y, z) = e^{yz} + yze^{xz} + zye^{xy} \quad \text{and} \quad \frac{\partial f}{\partial x}(1, 0, 2) = 1,$$

$$\frac{\partial f}{\partial y}(x, y, z) = xze^{yz} + e^{xz} + xze^{xy} \quad \text{and} \quad \frac{\partial f}{\partial y}(1, 0, 2) = 4 + e^2,$$

$$\frac{\partial f}{\partial z}(x, y, z) = xye^{yz} + xye^{xz} + e^{xy} \quad \text{and} \quad \frac{\partial f}{\partial z}(1, 0, 2) = 1.$$

Therefore

$$D_{\mathbf{a}} f(1, 0, 2) = \frac{4}{\sqrt{26}} + (4 + e^2)\frac{3}{\sqrt{26}} + \frac{1}{\sqrt{26}} = \frac{17 + 3e^2}{\sqrt{26}}.$$

As the next definition shows, the *gradient* of a function is a vector containing the partial derivatives of the function.

Definitions. (i) If $f(x, y)$ has partial derivatives, we define the **gradient vector**

$$\mathbf{grad}\, f(x, y) = f_x(x, y)\mathbf{i} + f_y(x, y)\mathbf{j}.$$

ii) If $g(x, y, z)$ has partial derivatives, we define

$$\mathbf{grad}\, g(x, y, z) = g_x(x, y, z)\mathbf{i} + g_y(x, y, z)\mathbf{j} + g_z(x, y, z)\mathbf{k}.$$

The symbol ∇, an inverted delta, is called "del" and is a common one used to denote the gradient. We will frequently write ∇f for $\mathbf{grad}\, f$.

If $\mathbf{b}$ and $\mathbf{c}$ are two vectors, we recall that the scalar product $\mathbf{b} \cdot \mathbf{c}$ of $\mathbf{b} = b_1\mathbf{i} + b_2\mathbf{j} + b_3\mathbf{k}$ and $\mathbf{c} = c_1\mathbf{i} + c_2\mathbf{j} + c_3\mathbf{k}$ is given by

$$\mathbf{b} \cdot \mathbf{c} = b_1c_1 + b_2c_2 + b_3c_3.$$

For two-dimensional vectors the result is the same, with $b_3 = c_3 = 0$.

We recognize that if $\mathbf{a}$ is a *unit* vector so that $\mathbf{a} = \lambda\mathbf{i} + \mu\mathbf{j} + \nu\mathbf{k}$, then we have the formula

$$D_{\mathbf{a}} f = \lambda f_x + \mu f_y + \nu f_z = \mathbf{a} \cdot \nabla f.$$

Looked at another way, the scalar product of $\mathbf{a}$ and ∇f is given by

$$\mathbf{a} \cdot \nabla f = |\mathbf{a}|\,|\nabla f| \cos\phi = D_{\mathbf{a}} f,$$

where ϕ is the angle between the vectors $\mathbf{a}$ and ∇f.

From the above formula we can conclude that *$D_{\mathbf{a}} f$ is a maximum when ϕ is zero*—i.e., when $\mathbf{a}$ *is in the direction of* $\mathbf{grad}\, f$.

EXAMPLE 5. Given the function

$$f(x, y, z) = x^3 + 2y^3 + z^3 - 4xyz,$$

find the maximum value of $D_{\mathbf{a}} f$ at the point $P = (-1, 1, 2)$.

SOLUTION. We have

$$\partial f/\partial x = 3x^2 - 4yz; \qquad \partial f/\partial y = 6y^2 - 4xz; \qquad \partial f/\partial z = 3z^2 - 4xy.$$

Therefore

$$\nabla f(-1, 1, 2) = -5\mathbf{i} + 14\mathbf{j} + 16\mathbf{k},$$

and a unit vector $\mathbf{a}$ in the direction of ∇f is

$$\mathbf{a} = -\frac{5}{3\sqrt{53}}\mathbf{i} + \frac{14}{3\sqrt{53}}\mathbf{j} + \frac{16}{3\sqrt{53}}\mathbf{k}.$$

The maximum value of $D_{\mathbf{a}} f$ is given by

$$D_{\mathbf{a}} f = -5\left(\frac{-5}{3\sqrt{53}}\right) + 14\left(\frac{14}{3\sqrt{53}}\right) + 16\left(\frac{16}{3\sqrt{53}}\right) = 3\sqrt{53}.$$

PROBLEMS

In Problems 1 through 6, find in each case $d_\theta f(x, y)$ at the given point.

1. $f(x, y) = x^2 + y^2$; $(3, 4)$
2. $f(x, y) = x^3 + y^3 - 3x^2y - 3xy^2$; $(1, -2)$
3. $f(x, y) = \arctan(y/x)$; $(4, 3)$
4. $f(x, y) = \sin(xy)$; $(2, \pi/4)$
5. $f(x, y) = e^x \cos y$; $(0, \pi/3)$
6. $f(x, y) = (\sin x)^{xy}$; $(\pi/2, 0)$

In each of Problems 7 through 10 find the value of $d_\theta f(x, y)$ at the given point. Also, find the value of θ which makes $d_\theta f$ a maximum at this point. Express your answer in terms of $\sin\theta$ and $\cos\theta$.

7. $f(x, y) = x^2 + y^2 - 2x + 3y$; $(2, -1)$
8. $f(x, y) = \arctan(x/y)$; $(3, 4)$
9. $f(x, y) = e^x \sin y$; $(0, \pi/6)$
10. $f(x, y) = (\sin y)^{xy}$; $(0, \pi/2)$

In each of Problems 11 through 14, find $D_{\mathbf{a}}f$ at the given point.

11. $f(x, y, z) = x^2 + xy - xz + y^2 - z^2$; $(2, 1, -2)$

12. $f(x, y, z) = x^2y + xze^y - xye^z$; $(-2, 3, 0)$

13. $f(x, y, z) = \cos xy + \sin xz$; $(0, 2, -1)$

14. $f(x, y, z) = \log(x + y + z) - xyz$; $(-1, 2, 1)$

In Problems 15 through 18, in each case find $D_{\mathbf{a}}f$ at the given point P when $\mathbf{a}$ is the given unit vector.

15. $f(x, y, z) = x^2 + 2xy - y^2 + xz + z^2$; $P(2, 1, 1)$; $\mathbf{a} = \frac{1}{3}\mathbf{i} - \frac{2}{3}\mathbf{j} + \frac{2}{3}\mathbf{k}$

16. $f(x, y, z) = x^2y + xye^z - 2xze^y$; $P(1, 2, 0)$; $\mathbf{a} = \frac{2}{7}\mathbf{i} - \frac{3}{7}\mathbf{j} + \frac{6}{7}\mathbf{k}$

17. $f(x, y, z) = \sin xz + \cos xy$; $P(0, -1, 2)$; $\mathbf{a} = \frac{1}{\sqrt{6}}\mathbf{i} - \frac{1}{\sqrt{6}}\mathbf{j} + \frac{2}{\sqrt{6}}\mathbf{k}$

18. $f(x, y, z) = \tan xyz + \sin xy - \cos xz$; $P(0, 1, 1)$;

$$\mathbf{a} = \frac{1}{\sqrt{26}}\mathbf{i} + \frac{3}{\sqrt{26}}\mathbf{j} + \frac{4}{\sqrt{26}}\mathbf{k}$$

19. The temperature at any point of a rectangular plate in the xy plane is given by the formula $T = 50(x^2 - y^2)$ (in degrees Celsius). Find $d_\theta T(4, 3)$, and find $\tan\theta$ when $d_\theta T(4, 3) = 0$. Find also the slope of the curve $T = \text{const}$ which passes through that point.

In each of Problems 20 through 23, find ∇f at the given point.

20. $f(x, y) = x^3 - 2x^2y + xy^2 - y^3$; $P(3, -2)$

21. $f(x, y) = \log(x^2 + y^2 + 1) + e^{2xy}$; $P(0, -2)$

22. $f(x, y, z) = \sin xy + \sin xz + \sin yz$; $P(1, 2, -1)$

23. $f(x, y, z) = xze^{xy} + yze^{xz} + xye^{yz}$; $P(-1, 2, 1)$

In each of Problems 24 through 27, find $D_{\mathbf{a}}f$ at the given point P where $\mathbf{a}$ is a unit vector in the direction $\overrightarrow{PP'}$. Also, find at P the value of $D_{\bar{\mathbf{a}}}f$ where $\bar{\mathbf{a}}$ is a unit vector such that $D_{\bar{\mathbf{a}}}f$ is a maximum.

24. $f(x, y, z) = x^2 + 3xy + y^2 + z^2$; $P(1, 0, 2)$; $P'(-1, 3, 4)$

25. $f(x, y, z) = e^x\cos y + e^y\sin z$; $P(2, 1, 0)$; $P'(-1, 2, 2)$

26. $f(x, y, z) = \log(x^2 + y^2) + e^z$; $P(0, 1, 0)$; $P'(-4, 2, 3)$

27. $f(x, y, z) = x\cos y + y\cos z + z\cos x$; $P(2, 1, 0)$; $P'(1, 4, 2)$

28. Prove Theorem 5.

29. Given $g(x, y, z) = 2x^2 + y^2 - z^2 + 2xy - 3x + 2y + z$, find the points, if any, such that $|\nabla g| = 0$.

30. Given $g(x, y, z) = x^3 - 2y^2 + z^2 - 2xz + 4y^2z$, find the points, if any, such that $|\nabla g| = 0$.

31. Suppose that $f(x, y, z)$ and $g(x, y, z)$ are given functions and that a and b are constants. Prove the formulas:

$$\nabla(af + bg) = a\nabla f + b\nabla g,$$

$$\nabla(fg) = f\nabla g + g\nabla f.$$

32. Suppose that the domain of $F = F(u)$ contains the range of $g = g(x, y, z)$. Prove that

$$\nabla(F(g)) = F'(u)\nabla g.$$

33. Suppose that $\mathbf{F}(x, y, z) = f_1(x, y, z)\mathbf{i} + f_2(x, y, z)\mathbf{j} + f_3(x, y, z)\mathbf{k}$ is a vector function. We define the vector $\mathbf{\nabla}$ by the symbolic formula

$$\mathbf{\nabla} = \frac{\partial}{\partial x}\mathbf{i} + \frac{\partial}{\partial y}\mathbf{j} + \frac{\partial}{\partial z}\mathbf{k}$$

and the cross product $\nabla \times \mathbf{F}$ is given by

$$\mathbf{\nabla} \times \mathbf{F} = \left(\frac{\partial f_3}{\partial y} - \frac{\partial f_2}{\partial z}\right)\mathbf{i} + \left(\frac{\partial f_1}{\partial z} - \frac{\partial f_3}{\partial x}\right)\mathbf{j} + \left(\frac{\partial f_2}{\partial x} - \frac{\partial f_1}{\partial y}\right)\mathbf{k}.$$

If $g(x, y, z)$ is a scalar function show that

$$\mathbf{\nabla} \times (g\mathbf{F}) = g\mathbf{\nabla} \times \mathbf{F} + (\mathbf{\nabla} g) \times \mathbf{F}.$$

34. Refer to Problem 33 for definitions and show that

$$\nabla \cdot (\mathbf{F} \times \mathbf{G}) = \mathbf{G} \cdot \mathbf{\nabla} \times \mathbf{F} - \mathbf{F} \cdot \mathbf{\nabla} \times \mathbf{G}.$$

35. Suppose that $f\colon R^3 \to R^1$ has the property that $\nabla f \equiv 0$. Show that f is a constant.

36. Let $\mathbf{F} = x^3\mathbf{i} + y^3\mathbf{j} + z^3\mathbf{k}$. Show that $\mathbf{\nabla} \times \mathbf{F} \equiv \mathbf{0}$; conclude that $\mathbf{\nabla} \times F = \mathbf{0}$ does not imply $\mathbf{F} =$ const. (See Problem 35.) [*Hint:* See the proof of Theorem 2.]

6. Geometric Interpretation of Partial Derivatives. Tangent Planes

From the geometric point of view, a function of one variable represents a curve in the plane. The derivative at a point on the curve is the slope of the line tangent to the curve at this point. A function of two variables $z = f(x, y)$ represents a surface in three-dimensional space. If (x_0, y_0) are the coordinates of a point in the xy plane, then $P(x_0, y_0, z_0)$, with $z_0 = f(x_0, y_0)$, is a point on the surface. Consider the vertical plane $y = y_0$, as shown in Fig. 4-7. This plane cuts the surface $z = f(x, y)$ in a curve C_1 which contains the point P. From the definition of partial derivative we see that

$$f_x(x_0, y_0)$$

is the slope of the line tangent to the curve C_1 at the point P. This line is labeled L_1 in Fig. 4-7 and, of course, is in the plane $y = y_0$. In a completely analogous manner we construct a plane $x = x_0$ intersecting the surface with

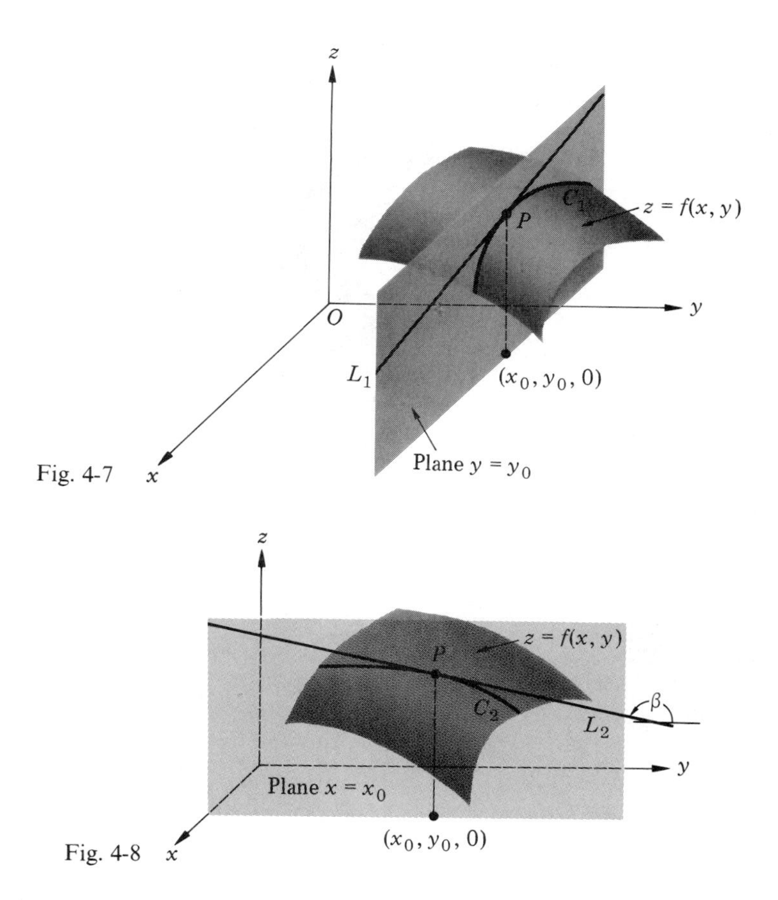

Fig. 4-7

Fig. 4-8

equation $z = f(x, y)$ in a curve C_2. The partial derivative

$$f_y(x_0, y_0)$$

is the slope of the line tangent to C_2 at the point P. The line is denoted L_2 in Fig. 4-8, and its slope is the tangent of the angle β, as shown.

The directional derivative $d_\theta f(x_0, y_0)$ has a similar interpretation. We construct the vertical plane through $(x_0, y_0, 0)$ which makes an angle θ with the positive x direction. Such a plane is shown in Fig. 4-9. The curve C_3 is the intersection of this plane with the surface, and the line L_3 is the line tangent to C_3 at P. Then $d_\theta f(x_0, y_0)$ is the slope of L_3.

According to Theorem 2, we may write the formula

$$\begin{aligned} f(x, y) - f(x_0, y_0) = f_x(x_0, y_0)(x - x_0) + f_y(x_0, y_0)(y - y_0) \\ + G_1 \cdot (x - x_0) + G_2 \cdot (y - y_0), \end{aligned}$$

in which we take

$$x - x_0 = h \qquad \text{and} \qquad y - y_0 = k.$$

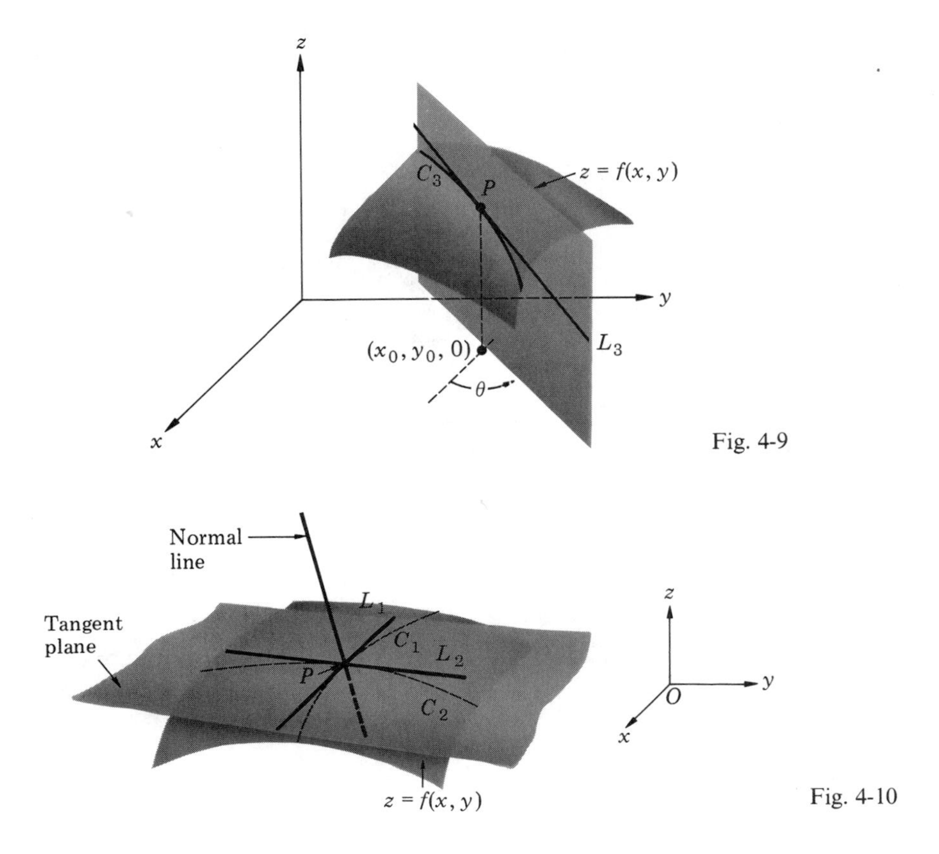

Fig. 4-9

Fig. 4-10

Since G_1 and G_2 tend to zero as $(x, y) \to (x_0, y_0)$, the next definition has an intuitive geometric meaning.

Definition. The plane whose equation is

$$z - z_0 = m_1(x - x_0) + m_2(y - y_0)$$

where

$$z_0 = f(x_0, y_0), \qquad m_1 = f_x(x_0, y_0), \qquad m_2 = f_y(x_0, y_0)$$

is called the **tangent plane** to the surface $z = f(x, y)$ at (x_0, y_0).

REMARKS. (i) According to the geometric interpretation of partial derivative which we gave, it is easy to verify that the tangent plane contains the lines L_1 and L_2, tangents to C_1 and C_2, respectively. (See Fig. 4-10.) (ii) From the definition of tangent plane we observe at once that

$$f_x(x_0, y_0), \qquad f_y(x_0, y_0), \qquad -1$$

is a set of attitude numbers for the plane. (Attitude numbers were defined on page 19.) The equation of the tangent plane is conveniently expressed in vector notation. Define

$$\mathbf{n} = m_1\mathbf{i} + m_2\mathbf{j} - \mathbf{k} \qquad \text{where} \qquad m_1 = f_x(x_0, y_0),\ m_2 = f_y(x_0, y_0).$$

Also, let $\mathbf{v} = x\mathbf{i} + y\mathbf{j} + z\mathbf{k}$ and $\mathbf{v}_0 = x_0\mathbf{i} + y_0\mathbf{j} + z_0\mathbf{k}$. Then the definition of the tangent plane takes the form

$$\boxed{\mathbf{n} \cdot (\mathbf{v} - \mathbf{v}_0) = 0.}$$

Geometrically, the plane is traced by the heads of the directed line segments having base at the origin which are representatives of the vector $\mathbf{v}$. The geometric importance of the vector $\mathbf{n}$ is given in the next definition.

Definition. The line with equations

$$\boxed{\frac{x - x_0}{m_1} = \frac{y - y_0}{m_2} = \frac{z - z_0}{-1},}$$

with $z_0 = f(x_0, y_0)$, $m_1 = f_x(x_0, y_0)$, $m_2 = f_y(x_0, y_0)$, is called the **normal line** to the surface at the point $P(x_0, y_0, z_0)$. Clearly, the normal line is perpendicular to the tangent plane. (See Fig. 4-10.)

REMARK. In addition to $f_x(x_0, y_0)$ we shall use the symbols

$$f_{,1}(x_0, y_0), \qquad \frac{\partial f(x_0, y_0)}{\partial x}, \qquad \text{and} \qquad \left.\frac{\partial z}{\partial x}\right|_{\substack{x=x_0\\ y=y_0}}$$

for m_1, and analogous notations for m_2.

EXAMPLE 1. Find the equation of the tangent plane and the equations of the normal line to the surface

$$z = x^2 + xy - y^2$$

at the point where $x = 2$, $y = -1$.

SOLUTION. We have $z_0 = f(2, -1) = 1$; $f_{,1}(x, y) = 2x + y$, $f_{,2}(x, y) = x - 2y$. Therefore $m_1 = 3$, $m_2 = 4$, and the desired equation for the tangent plane is

$$z - 1 = 3(x - 2) + 4(y + 1).$$

The equations of the normal line are

$$\frac{x - 2}{3} = \frac{y + 1}{4} = \frac{z - 1}{-1}.$$

If the equation of the surface is given in implicit form it is possible to use the methods of Section 2 to find $\partial z/\partial x$ and $\partial z/\partial y$ at the desired point. The next example shows how we obtain the equations of the tangent plane and normal line under such circumstances.

EXAMPLE 2. Find the equation of the tangent plane and the equations of the normal line at $(3, -1, 2)$ to the graph of

$$xy + yz + xz - 1 = 0.$$

SOLUTION. Holding y constant and differentiating with respect to x, we get

$$y + y\frac{\partial z}{\partial x} + z + x\frac{\partial z}{\partial x} = 0 \qquad \text{and} \qquad \frac{\partial z}{\partial x} = -\frac{y+z}{y+x} = -\frac{1}{2} = m_1.$$

Similarly, holding x constant, we obtain

$$x + y\frac{\partial z}{\partial y} + z + x\frac{\partial z}{\partial y} = 0, \qquad \frac{\partial z}{\partial y} = -\frac{x+z}{x+y} = -\frac{5}{2} = m_2.$$

The equation of the tangent plane is

$$z - 2 = -\tfrac{1}{2}(x-3) - \tfrac{5}{2}(y+1) \qquad \Leftrightarrow \qquad x + 5y + 2z - 2 = 0.$$

The normal line has equations

$$\frac{x-3}{-1/2} = \frac{y+1}{-5/2} = \frac{z-2}{-1} \qquad \Leftrightarrow \qquad \frac{x-3}{1} = \frac{y+1}{5} = \frac{z-2}{2}.$$

The methods of the calculus of functions of several variables enable us to establish purely geometric facts, as the next example shows.

EXAMPLE 3. Show that any line normal to the sphere

$$x^2 + y^2 + z^2 = a^2$$

always passes through the center of the sphere.

SOLUTION. Since the sphere has center at $(0, 0, 0)$, we must show that the equations of every normal line are satisfied for $x = y = z = 0$. Let (x_0, y_0, z_0) be a point on the sphere. Differentiating implicitly, we find

$$\frac{\partial z}{\partial x} = -\frac{x}{z}, \qquad \frac{\partial z}{\partial y} = -\frac{y}{z}.$$

The normal line to the sphere at (x_0, y_0, z_0) has equations

$$\frac{x - x_0}{-x_0/z_0} = \frac{y - y_0}{-y_0/z_0} = \frac{z - z_0}{-1}. \tag{1}$$

Letting $x = y = z = 0$, we get an identity for (1); hence the line passes through the origin.

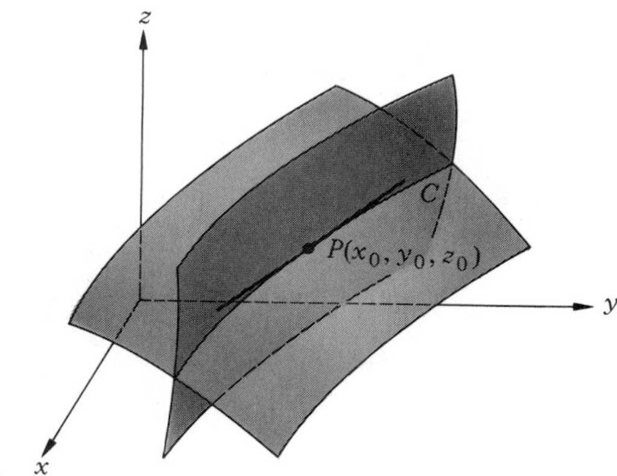

Fig. 4-11

Let the equations

$$z = f(x, y) \qquad \text{and} \qquad z = g(x, y)$$

represent surfaces which intersect in a curve C. The **tangent line** to the intersection at a point P on C is by definition the line of intersection of the tangent planes to f and g at P (Fig. 4-11).

We can use vector algebra in the following way to find the equations of this tangent line. If P has coordinates (x_0, y_0, z_0), then

$$\mathbf{u} = f_x(x_0, y_0)\mathbf{i} + f_y(x_0, y_0)\mathbf{j} + (-1)\mathbf{k}$$

is the vector perpendicular to the plane tangent to f at P. Similarly,

$$\mathbf{v} = g_x(x_0, y_0)\mathbf{i} + g_y(x_0, y_0)\mathbf{j} + (-1)\mathbf{k}$$

is the vector perpendicular to the plane tangent to g at P. The line of intersection of these tangent planes is perpendicular to both $\mathbf{u}$ and $\mathbf{v}$. We recall from the study of vectors that, if $\mathbf{u}$ and $\mathbf{v}$ are nonparallel vectors, the vector $\mathbf{u} \times \mathbf{v}$ is perpendicular to both $\mathbf{u}$ and $\mathbf{v}$. Defining the vector $\mathbf{w} = a\mathbf{i} + b\mathbf{j} + c\mathbf{k}$ by the relation

$$\mathbf{w} = \mathbf{u} \times \mathbf{v},$$

we see that the equations of the line of intersection of the two tangent planes are

$$\frac{x - x_0}{a} = \frac{y - y_0}{b} = \frac{z - z_0}{c}.$$

The next example shows how the method works.

EXAMPLE 4. Find the equations of the line tangent to the intersection of the surfaces

$$z = f(x, y) = x^2 + 2y^2, \qquad z = g(x, y) = 2x^2 - 3y^2 + 1$$

at the point (2, 1, 6).

SOLUTION. We have

$$f_x = 2x, \quad f_y = 4y; \qquad g_x = 4x, \quad g_y = -6y.$$

Therefore

$$\mathbf{u} = 4\mathbf{i} + 4\mathbf{j} - \mathbf{k}; \qquad \mathbf{v} = 8\mathbf{i} - 6\mathbf{j} - \mathbf{k};$$

$$\mathbf{u} \times \mathbf{v} = -10\mathbf{i} - 4\mathbf{j} - 56\mathbf{k}.$$

The desired equations are

$$\frac{x-2}{10} = \frac{y-1}{4} = \frac{z-6}{56}.$$

PROBLEMS

In Problems 1 through 14, find in each case the equation of the tangent plane and the equations of the normal line to the given surface at the given point.

1. $z = x^2 + 2y^2$; $(2, -1, 6)$
2. $z = 3x^2 - y^2 - 2$; $(-1, 2, -3)$
3. $z = xy$; $(2, -1, -2)$
4. $z = x^2y^2$; $(-2, 2, 16)$
5. $z = e^x \sin y$; $(1, \pi/2, e)$
6. $z = e^{2x} \cos 3y$; $(1, \pi/3, -e^2)$
7. $z = \log\sqrt{x^2 + y^2}$; $(-3, 4, \log 5)$
8. $x^2 + 2y^2 + 3z^2 = 6$; $(1, 1, -1)$
9. $x^2 + 2y^2 - 3z^2 = 3$; $(2, 1, -1)$
10. $x^2 + 3y^2 - z^2 = 0$; $(2, -2, 4)$
11. $x^2 + z^2 = 25$; $(4, -2, -3)$
12. $xy + yz + xz = 1$; $(2, 3, -1)$
13. $x^{1/2} + y^{1/2} + z^{1/2} = 6$; $(4, 1, 9)$
14. $y^{1/2} + z^{1/2} = 7$; $(3, 16, 9)$
15. Show that the equation of the plane tangent at (x_1, y_1, z_1) to the surface (an **ellipsoid** unless $A = B = C$ in which case it is a sphere)

$$\frac{x^2}{A^2} + \frac{y^2}{B^2} + \frac{z^2}{C^2} = 1 \quad \text{is} \quad \frac{x_1x}{A^2} + \frac{y_1y}{B^2} + \frac{z_1z}{C^2} = 1.$$

16. Show that every plane tangent to the surface (a **cone**)

$$x^2 + y^2 = z^2$$

passes through the origin.

17. Show that every line normal to the cone

$$z^2 = 3x^2 + 3y^2$$

intersects the z axis.

18. Show that the sum of the squares of the intercepts of any plane tangent to the

surface

$$x^{2/3} + y^{2/3} + z^{2/3} = a^{2/3}$$

is constant.

In Problems 19 through 22, find the equations of the line tangent to the intersection of the two surfaces at the given point.

19. $z = x^2 + y^2, \quad z = 2x + 4y + 20; \quad (4, -2, 20)$

20. $z = \sqrt{x^2 + y^2}, \quad z = 2x - 3y - 13; \quad (3, -4, 5)$

21. $z = x^2, \quad z = 25 - y^2; \quad (4, -3, 16)$

22. $z = \sqrt{25 - 9x^2}, \quad z = e^{xy} + 3; \quad (1, 0, 4)$

23. Find the point or points on the surface

$$z = x^2 - 2y^2 + 3y - 6$$

where the tangent plane is parallel to the plane $2x + 3y + z = 5$.

*24. Consider the surface (called a **hyperboloid**) $x^2 + y^2 - z^2 = 1$. Show that at each point parts of this surface lie on both sides of the tangent plane.

25. Given the surface

$$x^\alpha + y^\alpha + z^\alpha = a^\alpha$$

where α ($\neq 1$) and a are positive constants. Let $\bar{x}$, $\bar{y}$, $\bar{z}$ be the intercepts of any tangent plane to the surface with the coordinate axes. Show that

$$\bar{x}^{\alpha/(1-\alpha)} + \bar{y}^{\alpha/(1-\alpha)} + \bar{z}^{\alpha/(1-\alpha)} = \text{const.}$$

Find the value of the constant

26. Find the point or points (if any) on the surface

$$z = x^2 + 2xy - y^2 + 3x - 2y - 4$$

where the tangent plane is parallel to the xy-plane.

27. Find the point or points (if any) on the surface

$$x^2 + 2xy - y^2 + 3z^2 - 2x + 2y - 6z - 2 = 0$$

where the tangent plane is parallel to the yz-plane.

7. The Total Differential. Approximation

The differential of a function of one variable is a function of two variables selected in a special way. If $y = f(x)$, then the quantity df, called the **differential** of f, is defined by the relation

$$df = f'(x)h,$$

where h and x are independent variables.

Let f be a function of several variables; the next definition is the appropriate one for generalizing the notion of differential to such functions.

Definitions. The **total differential** of $f(x, y)$ is the function df of four variables x, y, h, k given by the formula

$$df(x, y, h, k) = f_x(x, y)h + f_y(x, y)k.$$

If F is a function of three variables—say x, y, and z—we define the **total differential** as the function of six variables x, y, z, h, k, l given by

$$dF(x, y, z, h, k, l) = F_x(x, y, z)h + F_y(x, y, z)k + F_z(x, y, z)l.$$

A quantity associated with the total differential is the *difference* of the values of a function at two nearby points. As is customary, we use Δ notation and, for functions of two variables, we define the quantity Δf by the formula

$$\Delta f \equiv \Delta f(x, y, h, k) = f(x + h, y + k) - f(x, y).$$

Here Δf is a function of four variables, as is df. If f is a function of x, y, and z, then Δf is a function of six variables defined by

$$\Delta f \equiv \Delta f(x, y, z, h, k, l) = f(x + h, y + k, z + l) - f(x, y, z).$$

EXAMPLE 1. Given the function

$$f(x, y) = x^2 + xy - 2y^2 - 3x + 2y + 4,$$

find $df(a, b, h, k)$ and $\Delta f(a, b, h, k)$ with $a = 3$, $b = 1$.

SOLUTION. We have

$$f(3, 1) = 7; \qquad f_{,1}(x, y) = 2x + y - 3; \qquad f_{,2}(x, y) = x - 4y + 2;$$

$$f_{,1}(3, 1) = 4; \qquad f_{,2}(3, 1) = 1.$$

Also

$$f(3 + h, 1 + k) = (3 + h)^2 + (3 + h)(1 + k) - 2(1 + k)^2 - 3(3 + h)$$
$$+ 2(1 + k) + 4 = h^2 + hk - 2k^2 + 4h + k + 7.$$

Therefore

$$df(3, 1, h, k) = 4h + k; \qquad \Delta f(3, 1, h, k) = 4h + k + h^2 + hk - 2k^2.$$

The close relationship between df and Δf is exhibited in Theorem 2 on page 207. Equation (1) of that theorem may be written

$$\Delta f(x_0, y_0, h, k) = df(x_0, y_0, h, k) + G_1(h, k)h + G_2(h, k)k.$$

The conclusion of the theorem implies that

$$\frac{\Delta f - df}{|h| + |k|} \to 0 \qquad \text{as} \qquad h, k \to 0,$$

since both G_1 and G_2 tend to zero with h and k. In many problems Δf is difficult to calculate, while df is easy. If h and k are both "small," we can use df as an approximation to Δf. The next example shows the technique.

EXAMPLE 2. Find, approximately, the value of $\sqrt{(5.98)^2 + (8.01)^2}$.

SOLUTION. We consider the function

$$z = f(x, y) = \sqrt{x^2 + y^2},$$

and we wish to find $f(5.98, 8.01)$. We see at once that $f(6, 8) = 10$; hence we may write

$$f(5.98, 8.01) = f(6, 8) + \Delta f,$$

where Δf is defined as above with $x_0 = 6$, $y_0 = 8$, $h = -0.02$, $k = 0.01$. The approximation consists of replacing Δf by df. We have

$$\frac{\partial z}{\partial x} = \frac{x}{\sqrt{x^2 + y^2}}, \qquad \frac{\partial z}{\partial y} = \frac{y}{\sqrt{x^2 + y^2}},$$

and so

$$df(6, 8, -0.02, 0.01) = \tfrac{6}{10}(-0.02) + \tfrac{8}{10}(0.01) = -0.004.$$

We conclude that $\sqrt{(5.98)^2 + (8.01)^2} = 10 - 0.004 = 9.996$, approximately.

As in the case of functions of one variable, the symbolism for the total differential may be used as an aid in differentiation. We let $z = f(x, y)$ and employ the symbols

$$dz \text{ for } df, \quad dx \text{ for } h, \quad \text{and} \quad dy \text{ for } k.$$

As in the case of one variable, there is a certain ambiguity, since dz has a precise definition as the total differential, while dx and dy are used as independent variables. The next theorem shows how the Chain Rule comes to our rescue and removes all difficulties when dx and dy are in turn functions of other variables (i.e., dx and dy are what we call intermediate variables).

Theorem 6. *Suppose that $z = f(x, y)$, and x and y are functions of some other variables. Then**

$$dz = \frac{\partial z}{\partial x}dx + \frac{\partial z}{\partial y}dy.$$

The result for $w = F(x, y, z)$ is similar. That is, the formula

* Of course, $\partial z/\partial x = f_x(x, y)$; $\partial w/\partial x = F_x(x, y, z)$, etc.

$$dw = \frac{\partial w}{\partial x} dx + \frac{\partial w}{\partial y} dy + \frac{\partial w}{\partial z} dz$$

holds when x, y, and z are either independent or intermediate variables.

PROOF. We establish the result for $z = f(x, y)$ with x and y functions of two variables, say r and s. The proof in all other cases is analogous. We write

$$x = x(r, s), \qquad y = y(r, s),$$

and then

$$z = f(x, y) = f[x(r, s), y(r, s)] \equiv g(r, s).$$

The definition of total differential yields

$$dz = g_r(r, s)h + g_s(r, s)k,$$

$$dx = x_r(r, s)h + x_s(r, s)k,$$

$$dy = y_r(r, s)h + y_s(r, s)k.$$

According to the Chain Rule, we have

$$g_r(r, s) = \frac{\partial f}{\partial x}\frac{\partial x}{\partial r} + \frac{\partial f}{\partial y}\frac{\partial y}{\partial r} = f_x(x, y)x_r(r, s) + f_y(x, y)y_r(r, s);$$

$$g_s(r, s) = \frac{\partial f}{\partial x}\frac{\partial x}{\partial s} + \frac{\partial f}{\partial y}\frac{\partial y}{\partial s} = f_x(x, y)x_s(r, s) + f_y(x, y)y_s(r, s).$$

Substituting the above expressions for g_r and g_s into that for dz, we obtain

$$dz = \frac{\partial f}{\partial x}[x_r(r, s)h + x_s(r, s)k] + \frac{\partial f}{\partial y}[y_r(r, s)h + y_s(r, s)k]$$

or

$$dz = \frac{\partial f}{\partial x} dx + \frac{\partial f}{\partial y} dy.$$

We recognize this last formula as the statement of the theorem.

EXAMPLE 3. Given

$$z = e^x \cos y + e^y \sin x, \qquad x = r^2 - t^2, \qquad y = 2rt,$$

find $dz(r, t, h, k)$ in two ways and verify that the results coincide.

SOLUTION. We have, by one method,

$$dz = \frac{\partial z}{\partial x} dx + \frac{\partial z}{\partial y} dy = (e^x \cos y + e^y \cos x)\, dx + (-e^x \sin y + e^y \sin x)\, dy;$$

$$dx = 2rh - 2tk;$$

$$dy = 2th + 2rk.$$

Therefore

$$\begin{aligned} dz &= (e^x \cos y + e^y \cos x)(2rh - 2tk) + (-e^x \sin y + e^y \sin x)(2th + 2rk) \\ &= 2[(e^x \cos y + e^y \cos x)r + (e^y \sin x - e^x \sin y)t]h \\ &\quad + 2[(-e^y \cos x - e^x \cos y)t + (e^y \sin x - e^x \sin y)r]k. \end{aligned} \tag{1}$$

On the other hand, the second method gives

$$dz = \frac{\partial z}{\partial r}h + \frac{\partial z}{\partial t}k, \tag{2}$$

and using the Chain Rule, we find

$$\frac{\partial z}{\partial r} = \frac{\partial z}{\partial x}\frac{\partial x}{\partial r} + \frac{\partial z}{\partial y}\frac{\partial y}{\partial r}, \qquad \frac{\partial z}{\partial t} = \frac{\partial z}{\partial x}\frac{\partial x}{\partial t} + \frac{\partial z}{\partial y}\frac{\partial y}{\partial t}.$$

We compute the various quantities in the two formulas above and find that

$$\frac{\partial z}{\partial r} = (e^x \cos y + e^y \cos x)(2r) + (e^y \sin x - e^x \sin y)(2t),$$

$$\frac{\partial z}{\partial t} = (e^x \cos y + e^y \cos x)(-2t) + (e^y \sin x - e^x \sin y)(2r).$$

Substituting these expressions in (2), we get (1) precisely.

Problems

In each of Problems 1 through 6, find $df(x, y, h, k)$ and $\Delta f(x, y, h, k)$ for the given values of x, y, h, and k.

1. $f(x, y) = x^2 - xy + 2y^2$; $x = 2$, $y = -1$, $h = -0.01$, $k = 0.02$
2. $f(x, y) = 2x^2 + 3xy - y^2$; $x = 1$, $y = 2$, $h = 0.02$, $k = -0.01$
3. $f(x, y) = \sin xy + \cos(x + y)$; $x = \pi/6$, $y = 0$, $h = 2\pi$, $k = 3\pi$
4. $f(x, y) = e^{xy} \sin(x + y)$; $x = \pi/4$, $y = 0$, $h = -\pi/2$, $k = 4\pi$
5. $f(x, y) = x^3 - 3xy + y^3$; $x = -2$, $y = 1$, $h = -0.03$, $k = -0.02$
6. $f(x, y) = x^2y - 2xy^2 + 3x$; $x = 1$, $y = 1$, $h = 0.02$, $k = 0.01$

In each of Problems 7 through 10, find $df(x, y, z, h, k, l)$ and $\Delta f(x, y, z, h, k, l)$ for the given values of x, y, z, h, k, and l.

7. $f(x, y, z) = x^2 - 2y^2 + z^2 - xz$; $(x, y, z) = (2, -1, 3)$;
   $(h, k, l) = (0.01, -0.02, 0.03)$
8. $f(x, y, z) = xy - xz + yz + 2x - 3y + 1$; $(x, y, z) = (2, 0, -3)$;
   $(h, k, l) = (0.1, -0.2, 0.1)$
9. $f(x, y, z) = x^2y - xyz + z^3$; $(x, y, z) = (1, 2, -1)$;
   $(h, k, l) = (-0.02, 0.01, 0.02)$

10. $f(x, y, z) = \sin(x + y) - \cos(x - z) + \sin(y + 2z)$;
$(x, y, z) = (\pi/3, \pi/6, 0)$; $(h, k, l) = (\pi/4, \pi/2, 2\pi)$

11. We define the **approximate percentage error** of a function f by the formula
$$\text{Approximate percentage error} = 100\left(\frac{df}{f}\right).$$
Find the approximate percentage error if $f(x, y, z) = 3x^3y^7z^4$.

12. Find the approximate percentage error (see Problem 11) if $f = cx^my^nz^p$ ($c =$ const).

13. A box has square ends, 11.98 cm. on each side, and has a length of 30.03 cm. Find its approximate volume, using differentials.

14. Use differentials to find the approximate value of
$$\sqrt{(5.02)^2 + (11.97)^2}.$$

15. The legs of a right triangle are measured and found to be 6.0 and 8.0 cm., with a possible error of 0.1 cm. Find approximately the maximum possible value of the error in computing the hypotenuse. What is the maximum approximate percentage error? (See Problem 11.)

16. Find in degrees the maximum possible approximate error in the computed value of the smaller acute angle in the triangle of Problem 15.

17. The diameter and height of a right circular cylinder are found by measurement to be 8.0 and 12.5 cm., respectively, with possible errors of 0.05 cm. in each measurement. Find the maximum possible approximate error in the computed volume.

18. A right circular cone is measured, and the radius of the base is 12.0 cm. with the height 16.0 cm. If the possible error in each measurement is 0.06, find the maximum possible error in the computed volume. What is the maximum possible approximate error in the lateral surface area?

19. By measurement, a triangle is found to have two sides of length 50 cm. and 70 cm.; the angle between them is 30°. If there are possible errors of $\frac{1}{2}\%$ in the measurements of the sides and $\frac{1}{2}$ degree in that of the angle, find the maximum approximate percentage error in the measurement of the area. (See Problem 11.)

20. Use differentials to find the approximate value of
$$\sqrt{(3.02)^2 + (1.99)^2 + (5.97)^2}.$$

21. Use differentials to find the approximate value of
$$[(3.01)^2 + (3.98)^2 + (6.02)^2 + 5(1.97)^2]^{-1/2}.$$

In each of Problems 22 through 26, find dz in two ways (as in Example 3,) in terms of the independent variables.

22. $z = x^2 + xy - y^2$; $x = r^2 + 2s^2$, $y = rs + 2$

23. $z = 2x^2 + 3xy + y^2$; $x = t^3 + 2t - 1$, $y = t^2 + t - 3$

24. $z = x^3 + y^3 - x^2y$; $x = r + 2s - t$, $y = r - 3s + 2t$

25. $z = u^2 + 2v^2 - x^2 + 3y^2$; $u = r^2 - s^2$, $v = r^2 + s^2$, $x = 2rs$, $y = 2r/s$
26. $z = u^3 + v^3 + w^3$; $u = r^2 + s^2 + t^2$, $v = r^2 - s^2 + t^2$, $w = r^2 + s^2 - t^2$
27. Using the formulas in the proof of Theorem 2, find explicit expressions for the functions G_1 and G_2 if

    $z = x^2 + 2y^2 + 6xy.$

    Conclude that $(\Delta z - dz)/(|h| + |k|)$ tends to zero as $h, k \to 0$.
28. Same as Problem 27 for $z = 3x^3 + 2y^3 + 2xy$.

*29. Let $z = f(x_1, x_2, \ldots, x_n)$ be a function of n variables. Define the **total differential** df by the formula

$$\boxed{df = f_{x_1}h_1 + f_{x_2}h_2 + \cdots + f_{x_n}h_n}$$

where $h_1, h_2, \ldots, h_n$ are independent variables. Assume each x_i is a function of the m variables $y_1, y_2, \ldots, y_m$. State and prove the analog of Theorem 6.

30. Use the formula in Problem 29 to find approximately the value of

$$[(3.01)^2 + (2.97)^2 + (5.02)^2 + (3.99)^2 + (7.01)^2 + (6.02)^2]^{1/2}.$$

31. The period T of a simple pendulum is given by $T = 2\pi(l/g)^{1/2}$ where l is the length and g is the gravitational constant. If l is measured to be 20 cm with an error of 0.2, if g is 980 with an error of 7, and if π is computed as 3.14 with an error of 0.002, use differentials to find an approximate value of T.

8. Applications of the Total Differential

Once we clearly understand the concept of function we are able to use the notation of the total differential to obtain a number of useful differentiation formulas.

One of the simplest formulas, which we now develop, uses the fact that *the total differential of a constant is zero*. Suppose that x and y are related by some equation such as

$$f(x, y) = 0.$$

If it turns out that y is a function of x, we can compute the derivative dy/dx by implicit methods in the usual way. However, we may also use an alternate procedure. Since $f = 0$, the differential df also vanishes. Therefore we can write

$$df = \frac{\partial f}{\partial x}dx + \frac{\partial f}{\partial y}dy = 0$$

or

$$\frac{dy}{dx} = -\frac{\partial f/\partial x}{\partial f/\partial y} \qquad \left(\text{if } \frac{\partial f}{\partial y} \neq 0\right). \tag{1}$$

EXAMPLE 1. Use the methods of partial differentiation to compute dy/dx if

$$x^4 + 3x^2y^2 - y^4 + 2x - 3y = 5.$$

SOLUTION. Setting

$$f(x, y) = x^4 + 3x^2y^2 - y^4 + 2x - 3y - 5 = 0,$$

we find

$$f_x = 4x^3 + 6xy^2 + 2, \qquad f_y = 6x^2y - 4y^3 - 3.$$

Therefore, using (1) above,

$$\frac{dy}{dx} = -\frac{4x^3 + 6xy^2 + 2}{6x^2y - 4y^3 - 3}.$$

Of course, the same result is obtained by the customary process of implicit differentiation.

The above method for ordinary differentiation may be extended to yield partial derivatives. Suppose x, y, and z are connected by a relation of the form

$$F(x, y, z) = 0,$$

and we imagine that z is a function of x and y. That is, we make the assumption that it is possible to solve for z in terms of x and y even though we have no intention of doing so; in fact, we may find it exceptionally difficult (if not downright impossible) to perform the necessary steps. If z is a function of x and y, then we have the formula for the total differential:

$$dz = \frac{\partial z}{\partial x}dx + \frac{\partial z}{\partial y}dy. \tag{2}$$

On the other hand, since $F = 0$, the differential dF is also. Therefore

$$dF = F_x\,dx + F_y\,dy + F_z\,dz = 0.$$

Solving this last equation for dz, we get

$$dz = \left(-\frac{F_x}{F_z}\right)dx + \left(-\frac{F_y}{F_z}\right)dy, \tag{3}$$

assuming that $F_z \neq 0$. Comparing Eqs. (2) and (3), it is possible to prove (although we shall not do so) that

$$\frac{\partial z}{\partial x} = -\frac{F_x}{F_z} \qquad \text{and} \qquad \frac{\partial z}{\partial y} = -\frac{F_y}{F_z}. \tag{4}$$

We exhibit the utility of formulas (4) in the next example.

EXAMPLE 2. Use formulas (4) to find $\partial z/\partial x$ and $\partial z/\partial y$ if

$$e^{xy}\cos z + e^{-xz}\sin y + e^{yz}\cos x = 0.$$

SOLUTION. We set

$$F(x, y, z) = e^{xy}\cos z + e^{-xz}\sin y + e^{yz}\cos x,$$

and compute

$$F_x = ye^{xy}\cos z - ze^{-xz}\sin y - e^{yz}\sin x,$$

$$F_y = xe^{xy}\cos z + e^{-xz}\cos y + ze^{yz}\cos x,$$

$$F_z = -e^{xz}\sin z - xe^{-xz}\sin y + ye^{yz}\cos x.$$

Therefore

$$\frac{\partial z}{\partial x} = -\frac{ye^{xy}\cos z - ze^{-xz}\sin y - e^{yz}\sin x}{-e^{xy}\sin z - xe^{-xz}\sin y + ye^{yz}\cos x},$$

$$\frac{\partial z}{\partial y} = -\frac{xe^{xy}\cos z + e^{-xz}\cos y + ze^{yz}\cos x}{-e^{xy}\sin z - xe^{-xz}\sin y + ye^{yz}\cos x}.$$

REMARKS. (i) Note that we also could have found the derivatives by the implicit methods described in Section 2. (ii) Formulas similar to (4) may be established for a single relation with any number of variables. For example, if we are given $G(x, y, u, v, w) = 0$ and we assume w is a function of the remaining variables with $G_w \neq 0$, then

$$\frac{\partial w}{\partial x} = -\frac{G_x}{G_w}, \qquad \frac{\partial w}{\partial y} = -\frac{G_y}{G_w}, \qquad \frac{\partial w}{\partial u} = -\frac{G_u}{G_w}, \qquad \frac{\partial w}{\partial v} = -\frac{G_v}{G_w}.$$

A more complicated application of differentials is exhibited in the derivation of the next set of formulas. Suppose, for example, that x, y, u, v are related by *two* equations, so that

$$F(x, y, u, v) = 0 \qquad \text{and} \qquad G(x, y, u, v) = 0.$$

If we could solve one of them for, say u, and substitute in the other, we would get a single equation for x, y, v. Then, solving for v, we would find that v is a function of x and y. Similarly, we might find u as a function of x and y. Of course, all this work is purely fictitious, since we have no intention of carrying out such a process. In fact, it may be impossible. The main point is that we know that under appropriate circumstances the process is *theoretically feasible*. (This fact is discussed in Chapter 9.) Therefore whenever $u = u(x, y)$ and $v = v(x, y)$ it makes sense to write the symbols

$$\frac{\partial u}{\partial x}, \quad \frac{\partial u}{\partial y}, \quad \frac{\partial v}{\partial x}, \quad \frac{\partial v}{\partial y}. \tag{5}$$

Furthermore, the selection of u and v in terms of x and y is arbitrary. We could equally well attempt to solve for v and x in terms of u and y or for any two of the variables in terms of the remaining two variables.

The problem we pose is one of determining the quantities in (5) without actually finding the functions $u(x, y)$ and $v(x, y)$. We use the total differential. Since $F = 0$ and $G = 0$, so are dF and dG. We have

$$dF = F_x\,dx + F_y\,dy + F_u\,du + F_v\,dv = 0,$$

$$dG = G_x\,dx + G_y\,dy + G_u\,du + G_v\,dv = 0.$$

We write these equations,

$$F_u\,du + F_v\,dv = -F_x\,dx - F_y\,dy, \qquad G_u\,du + G_v\,dv = -G_x\,dx - G_y\,dy,$$

and consider du and dv as unknowns with everything else known. Solving two equations in two unknowns is easy. We obtain

$$du = \frac{G_xF_v - G_vF_x}{F_uG_v - F_vG_u}dx + \frac{G_yF_v - G_vF_y}{F_uG_v - F_vG_u}dy, \tag{6}$$

$$dv = \frac{G_uF_x - G_xF_u}{F_uG_v - F_vG_u}dx + \frac{G_uF_y - G_yF_u}{F_uG_v - F_vG_u}dy. \tag{7}$$

(We suppose, of course, that $F_uG_v - F_vG_u \neq 0$.) On the other hand, we know that if $u(x, y)$, $v = v(x, y)$, then

$$du = \frac{\partial u}{\partial x}dx + \frac{\partial u}{\partial y}dy, \tag{8}$$

$$dv = \frac{\partial v}{\partial x}dx + \frac{\partial v}{\partial y}dy. \tag{9}$$

Therefore, comparing (6) with (8) and (7) with (9), we find

$$\frac{\partial u}{\partial x} = \frac{G_xF_v - G_vF_x}{F_uG_v - F_vG_u}, \tag{10}$$

and similar formulas for $\partial u/\partial y$, $\partial v/\partial x$, and $\partial v/\partial y$. If F and G are specific functions, the right side of (10) is computable.

EXAMPLE 3. Given the relations for x, y, u, v:

$$u^2 - uv - v^2 + x^2 + y^2 - xy = 0,$$

$$uv - x^2 + y^2 = 0,$$

and assuming that $u = u(x, y)$, $v = v(x, y)$, find $\partial u/\partial x$, $\partial u/\partial y$, $\partial v/\partial x$, and $\partial v/\partial y$.

SOLUTION. We could find F_x, $F_y, \ldots, G_u$, G_v and then substitute for the coefficients in (6) and (7) to obtain the result. Instead we make use of the fact that we can treat differentials both as independent variables and as total differentials, with no fear of difficulty (because of the Chain Rule).

Taking such differentials in each of the given equations, we get

$$2u\,du - u\,dv - v\,du - 2v\,dv + 2x\,dx + 2y\,dy - x\,dy - y\,dx = 0,$$

$$v\,du + u\,dv - 2x\,dx + 2y\,dy = 0.$$

Solving the two equations simultaneously for du and dv in terms of dx and dy, we obtain

$$du = \frac{uy + 4xv}{2(u^2 + v^2)}dx + \frac{ux - 4y(u + v)}{2(u^2 + v^2)}dy,$$

$$dv = \frac{4xu - yv}{2(u^2 + v^2)}dx + \frac{4y(v - u) - xv}{2(u^2 + v^2)}dy.$$

From these equations and equations (8) and (9), we read off the results. For example,

$$\frac{\partial u}{\partial y} = \frac{ux - 4y(u + v)}{2(u^2 + v^2)},$$

and there are corresponding expressions for $\partial u/\partial x$, $\partial v/\partial x$, $\partial v/\partial y$.

Problems

In each of Problems 1 through 7, find the derivative dy/dx by the methods of partial differentiation.

1. $x^2 + 3xy - 4y^2 + 2x - 6y + 7 = 0$
2. $x^3 + 3x^2y - 4xy^2 + y^3 - x^2 + 2y - 1 = 0$
3. $\log(1 + x^2 + y^2) + e^{xy} = 5$
4. $x^4 - 3x^2y^2 + y^4 - x^2y + 2xy^2 = 3$
5. $e^{xy} + \sin xy + 1 = 0$
6. $xe^y + ye^x + \sin(x + y) - 2 = 0$
7. $\arctan(y/x) + (x^2 + y^2)^{3/2} = 2$

In each of Problems 8 through 12, assume that w is a function of the remaining variables. Find the partial derivatives as indicated by the method of Example 2.

8. $x^2 + y^2 + w^2 - 3xyw - 4 = 0; \quad \dfrac{\partial w}{\partial y}$
9. $x^3 + 3x^2w - y^2w + 2yw^2 - 3w + 2x = 8; \quad \dfrac{\partial w}{\partial x}$
10. $e^{xy} + e^{yw} - e^{xw} + xyw = 4; \quad \dfrac{\partial w}{\partial y}$
11. $\sin(xyw) + x^2 + y^2 + w^2 = 3; \quad \dfrac{\partial w}{\partial x}$

12. $(w^2 - y^2)(w^2 + x^2)(x^2 - y^2) = 1; \qquad \dfrac{\partial w}{\partial y}$

In Problems 13 and 14, use the methods of this section to find the partial derivatives as indicated.

13. $x^2 + y^2 - z^2 - w^2 + 3xy - 2xz + 4xw - 3zw + 2x - 3y = 0; \qquad \dfrac{\partial w}{\partial y}$

14. $x^2y^2z^2w^2 + x^2z^2w^4 - y^4w^4 + x^6w^2 - 2y^3w^3 = 8; \qquad \dfrac{\partial w}{\partial z}$

If $F(x, y, z) = 0$ and $G(x, y, z) = 0$, then we may consider z and y as functions of the single variable x; that is, $z = z(x)$, $y = y(x)$. Using differentials, we obtain

$$F_x\,dx + F_y\,dy + F_z\,dz = 0, \qquad G_x\,dx + G_y\,dy + G_z\,dz = 0,$$

and so we can get the ordinary derivatives dz/dx and dy/dx. Use this method in Problems 15 through 18 to obtain these derivatives.

15. $z = x^2 + y^2, \qquad y^2 = 4x + 2z$

16. $x^2 - y^2 + z^2 = 7, \qquad 2x + 3y + 4z = 15$

17. $2x^2 + 3y^2 + 4z^2 = 12, \qquad x = yz$

18. $xyz = 5, \qquad x^2 + y^2 - z^2 = 16$

In each of Problems 19 through 23, find $\partial u/\partial x$, $\partial u/\partial y$, $\partial v/\partial x$, and $\partial v/\partial y$ by the method of differentials.

19. $x = u^2 - v^2, \qquad y = 2uv$

20. $x = u + v, \qquad y = uv$

21. $u + v - x^2 = 0, \qquad u^2 - v^2 - y = 0$

22. $u^3 + xv^2 - xy = 0, \qquad u^2y + v^3 + x^2 - y^2 = 0$

23. $u^2 + v^2 + x^2 - y^2 = 4, \qquad u^2 - v^2 - x^2 - y^2 = 1$

24. Given that $F(x, y, z, u) = 0$, show that if each of the partial derivatives in the expressions below actually exists, then the two formulas are valid:

$$\frac{\partial x}{\partial y}\cdot\frac{\partial y}{\partial z}\cdot\frac{\partial z}{\partial x} = -1; \qquad \frac{\partial x}{\partial y}\cdot\frac{\partial y}{\partial z}\cdot\frac{\partial z}{\partial u}\cdot\frac{\partial u}{\partial x} = 1.$$

25. Given that $x = f(u, v)$, $y = g(u, v)$, find $\partial u/\partial x$, $\partial u/\partial y$, $\partial v/\partial x$, $\partial v/\partial y$ in terms of u and v and the derivatives of f and g.

26. Given that $F(u, v, x, y, z) = 0$ and $G(u, v, x, y, z) = 0$, assume that $u = u(x, y, z)$ and $v = v(x, y, z)$ and find formulas for $\partial u/\partial x$, $\partial v/\partial x$, $\ldots$, $\partial v/\partial z$ in terms of the derivatives of F and G.

27. Given $F(x_1, x_2, \ldots, x_n, u_1, u_2) = 0$ and $G(x_1, x_2, \ldots, x_n, u_1, u_2) = 0$. Find formulas for

$$\frac{\partial u_1}{\partial x_i}, \frac{\partial u_2}{\partial x_i}, \qquad i = 1, 2, \ldots, n,$$

assuming that u_1 and u_2 are functions of $x_1, x_2, \ldots, x_n$.

28. Given the equations $F_i(x_1, x_2, \ldots, x_n, u_1, u_2, \ldots, u_k) = 0$ where $i = 1, 2, \ldots, k$. Assuming that $u_i = u_i(x_1, x_2, \ldots, x_n)$ where $i = 1, 2, \ldots, k$, find formulas for

$$\frac{\partial u_i}{\partial x_j}, \qquad i = 1, 2, \ldots, k, \quad j = 1, 2, \ldots, n.$$

29. Given $F(x, y, z, u, v, w) = 0$, $G(x, y, z, u, v, w) = 0$, and $H(x, y, z, u, v, w) = 0$. Use the total differentials dF, dG, and dH to derive formulas for

$$\frac{\partial u}{\partial x}, \frac{\partial u}{\partial y}, \ldots, \frac{\partial w}{\partial z}$$

similar to formula (10) in the case of two functions.

Answer: Let

$$D = \begin{vmatrix} F_u & F_v & F_w \\ G_u & G_v & G_w \\ H_u & H_v & H_w \end{vmatrix} \quad \text{and} \quad D_1 = \begin{vmatrix} F_x & F_v & F_w \\ G_x & G_v & G_w \\ H_x & H_v & H_w \end{vmatrix}.$$

Then $\partial u/\partial x = -(D_1/D)$ if the determinant D is not zero. Similar formulas hold for the remaining partial derivatives.

30. Given

$$x^2 - y^2 + z^2 + 2u - v^2 + w^2 - 3 = 0,$$
$$x^2 - z^2 + 2u^2 - w^3 - 1 = 0,$$
$$y^2 + 2z^2 - u^2 + 2v - 3w^3 - 4 = 0.$$

Use the results in Problem 29 to find $\partial v/\partial y$.

9. Second and Higher Derivatives

If f is a function of two variables—say x and y—then f_x and f_y are also functions of the same two variables. When we differentiate f_x and f_y, we obtain second partial derivatives. The **second partial derivatives** of f are defined by the formulas

$$f_{xx}(x, y) = \lim_{h \to 0} \frac{f_x(x + h, y) - f_x(x, y)}{h},$$

$$f_{xy}(x, y) = \lim_{k \to 0} \frac{f_x(x, y + k) - f_x(x, y)}{k}.$$

The first derivatives of f_y are defined by similar expressions. We observe that if f is a function of two variables, there are four second partial derivatives.

There is a multiplicity of notations for partial derivatives which at times may lead to confusion. For example, if we write $z = f(x, y)$, then the following five symbols all have the same meaning:

$$f_{xx}; \qquad \frac{\partial^2 z}{\partial x^2}; \qquad \frac{\partial^2 f}{\partial x^2}; \qquad f_{,1,1}; \qquad z_{xx}.$$

For other partial derivatives we have the variety of expressions:

$$f_{xy} = f_{,1,2} = \frac{\partial}{\partial y}\left(\frac{\partial z}{\partial x}\right) = \frac{\partial^2 z}{\partial y\,\partial x} = \frac{\partial^2 f}{\partial y\,\partial x} = z_{xy},$$

$$f_{yx} = f_{,2,1} = \frac{\partial}{\partial x}\left(\frac{\partial z}{\partial y}\right) = \frac{\partial^2 z}{\partial x\,\partial y} = \frac{\partial^2 f}{\partial x\,\partial y} = z_{yx},$$

$$f_{xyx} = \frac{\partial}{\partial x}\left(\frac{\partial^2 z}{\partial y\,\partial x}\right) = \frac{\partial^3 z}{\partial x\,\partial y\,\partial x} = \frac{\partial^3 f}{\partial x\,\partial y\,\partial x} = f_{,1,2,1} = z_{xyx},$$

and so forth. Note that, in the subscript notation, symbols such as f_{xyy} or z_{xyy} mean that the order of partial differentiation is taken from left to right—that is, first with respect to x and then twice with respect to y. On the other hand, the symbol

$$\frac{\partial^3 z}{\partial x\,\partial y\,\partial y}$$

asserts that we first take two derivatives with respect to y and then one with respect to x. The denominator symbol and the subscript symbol are the reverse of each other.

EXAMPLE 1. Given $z = x^3 + 3x^2y - 2x^2y^2 - y^4 + 3xy$, find

$$\frac{\partial z}{\partial x}, \qquad \frac{\partial z}{\partial y}, \qquad \frac{\partial^2 z}{\partial x^2}, \qquad \frac{\partial^2 z}{\partial x\,\partial y}, \qquad \frac{\partial^2 z}{\partial y\,\partial x}, \qquad \frac{\partial^2 z}{\partial y^2}.$$

SOLUTION. We have

$$\frac{\partial z}{\partial x} = 3x^2 + 6xy - 4xy^2 + 3y; \qquad \frac{\partial z}{\partial y} = 3x^2 - 4x^2y - 4y^3 + 3x;$$

$$\frac{\partial^2 z}{\partial x^2} = 6x + 6y - 4y^2; \qquad \frac{\partial^2 z}{\partial y^2} = -4x^2 - 12y^2;$$

$$\frac{\partial^2 z}{\partial y\,\partial x} = 6x - 8xy + 3; \qquad \frac{\partial^2 z}{\partial x\,\partial y} = 6x - 8xy + 3.$$

In the example above, it is not accidental that $\partial^2 z/\partial y\,\partial x = \partial^2 z/\partial x\,\partial y$, as the next theorem shows.

Theorem 7. *Assume that $f(x, y), f_x, f_y, f_{xy}$, and f_{yx} are all continuous at (x_0, y_0). Then*

$$f_{xy}(x_0, y_0) = f_{yx}(x_0, y_0).$$

In comma notation, the formula reads

$$f_{,1,2}(x_0, y_0) = f_{,2,1}(x_0, y_0).$$

(The order of partial differentiation may be reversed without affecting the result.)

PROOF. The result is obtained by use of a quantity we call the **double difference**, denoted by $\Delta_2 f$, and defined by the formula

$$\Delta_2 f = [f(x_0 + h, y_0 + h) - f(x_0 + h, y_0)] - [f(x_0, y_0 + h) - f(x_0, y_0)]. \tag{1}$$

We shall show that, as h tends to zero, the quantity $\Delta_2 f/h^2$ tends to $f_{xy}(x_0, y_0)$. On the other hand, we shall also show that the same quantity tends to $f_{yx}(x_0, y_0)$. The principal tool is the repeated application of the Theorem of the Mean. (See Chapter 3, page 91.) We may write $\Delta_2 f$ in a more transparent way by defining

$$\phi(s) = f(x_0 + s, y_0 + h) - f(x_0 + s, y_0), \tag{2}$$

$$\psi(t) = f(x_0 + h, y_0 + t) - f(x_0, y_0 + t). \tag{3}$$

(The quantities x_0, y_0, h are considered fixed in the definition of ϕ and ψ.) Then straight substitution in (1) shows that

$$\Delta_2 f = \phi(h) - \phi(0) \tag{4}$$

and

$$\Delta_2 f = \psi(h) - \psi(0). \tag{5}$$

We apply the Theorem of the Mean in (4) and (5), getting two expressions for $\Delta_2 f$. They are

$$\Delta_2 f = \phi'(s_1) \cdot h \quad \text{with} \quad 0 < s_1 < h,$$

$$\Delta_2 f = \psi'(t_1) \cdot h \quad \text{with} \quad 0 < t_1 < h.$$

The derivatives $\phi'(s_1)$ and $\psi'(t_1)$ are easily computed from (2) and (3). We obtain

$$\phi'(s_1) = f_x(x_0 + s_1, y_0 + h) - f_x(x_0 + s_1, y_0),$$

$$\psi'(t_1) = f_y(x_0 + h, y_0 + t_1) - f_y(x_0, y_0 + t_1),$$

and the two expressions for $\Delta_2 f$ yield

$$\frac{1}{h}\Delta_2 f = [f_x(x_0 + s_1, y_0 + h) - f_x(x_0 + s_1, y_0)], \tag{6}$$

$$\frac{1}{h}\Delta_2 f = [f_y(x_0 + h, y_0 + t_1) - f_y(x_0, y_0 + t_1)]. \tag{7}$$

We may apply the Theorem of the Mean to the expression on the right in (6) with respect to $y_0 + h$ and y_0. We get

$$\frac{1}{h}\Delta_2 f = f_{xy}(x_0 + s_1, y_0 + t_2)h \qquad \text{with} \qquad 0 < t_2 < h. \tag{8}$$

Similarly, the Theorem of the Mean may be applied in (7) to the expression on the right with respect to $x_0 + h$ and x_0. The result is

$$\frac{1}{h}\Delta_2 f = f_{yx}(x_0 + s_2, y_0 + t_1)h \qquad \text{with} \qquad 0 < s_2 < h. \tag{9}$$

Dividing by h in (8) and (9), we find that

$$\frac{1}{h^2}\Delta_2 f = f_{xy}(x_0 + s_1, y_0 + t_2) = f_{yx}(x_0 + s_2, y_0 + t_1).$$

Letting h tend to zero and noticing that s_1, s_2, t_1, and t_2 all tend to zero with h, we obtain the result. The argument is similar if his negative.

Corollary 1. *Suppose that f is a function of any number of variables and s and t are any two of them. Then under appropriate hypotheses on the continuity of the partial derivatives (as in Theorem 7), we have*

$$f_{st} = f_{ts}.$$

For example, if the function is $f(x, y, s, t, u, v)$, then

$$f_{xt} = f_{tx}, \qquad f_{yu} = f_{uy}, \qquad f_{yv} = f_{vy}, \qquad etc.$$

The proof of the corollary is identical with the proof of the theorem.

Corollary 2. *For derivatives of the third, fourth, or any order, it does not matter in what order the differentiations with respect to the various variables are performed. For instance, assuming that all fourth-order partial derivatives are continuous, we have*

$$\frac{\partial^4 z}{\partial x\,\partial x\,\partial y\,\partial y} = \frac{\partial^4 z}{\partial x\,\partial y\,\partial x\,\partial y} = \frac{\partial^4 z}{\partial x\,\partial y\,\partial y\,\partial x}$$
$$= \frac{\partial^4 z}{\partial y\,\partial x\,\partial x\,\partial y} = \frac{\partial^4 z}{\partial y\,\partial x\,\partial y\,\partial x} = \frac{\partial^4 z}{\partial y\,\partial y\,\partial x\,\partial x}.$$

REMARKS. (i) It is true that there are functions for which f_{xy} is not equal to f_{yx}. Of course, the hypotheses of Theorem 7 are violated for such functions. (ii) All the functions we have considered thus far and all the functions we shall consider from now on will always satisfy the hypotheses of Theorem 7. Therefore the order of differentiation will be reversible throughout. (See, however, Problem 33 at the end of this Section.)

EXAMPLE 2. Given $u = e^x \cos y + e^y \sin z$, find all first partial derivatives and verify that

$$\frac{\partial^2 u}{\partial x\,\partial y} = \frac{\partial^2 u}{\partial y\,\partial x}, \qquad \frac{\partial^2 u}{\partial x\,\partial z} = \frac{\partial^2 u}{\partial z\,\partial x}, \qquad \frac{\partial^2 u}{\partial y\,\partial z} = \frac{\partial^2 u}{\partial z\,\partial y}.$$

SOLUTION. We have

$$\frac{\partial u}{\partial x} = e^x \cos y; \qquad \frac{\partial u}{\partial y} = -e^x \sin y + e^y \sin z; \qquad \frac{\partial u}{\partial z} = e^y \cos z.$$

Therefore

$$\frac{\partial^2 u}{\partial y\,\partial x} = -e^x \sin y = \frac{\partial^2 u}{\partial x\,\partial y},$$

$$\frac{\partial^2 u}{\partial z\,\partial x} = 0 \qquad = \frac{\partial^2 u}{\partial x\,\partial z},$$

$$\frac{\partial^2 u}{\partial z\,\partial y} = e^y \cos z \quad = \frac{\partial^2 u}{\partial y\,\partial z}.$$

EXAMPLE 3. Suppose that $u = F(x, y, z)$ and $z = f(x, y)$. Obtain a formula for $\partial^2 u/\partial x^2$ in terms of the derivatives of F (that is, F_x, F_y, F_z, F_{xx}, etc.) and the derivatives of f (or, equivalently, z). That is, in the expression for F we consider x, y, z *intermediate variables*, while in the expression for f we consider x and y *independent variables*.

SOLUTION. We apply the Chain Rule to F to obtain $\partial u/\partial x$ with x and y as independent variables. We get

$$\frac{\partial u}{\partial x} = F_x \frac{\partial x}{\partial x} + F_y \frac{\partial y}{\partial x} + F_z \frac{\partial z}{\partial x}.$$

Since x and y are independent, $\partial y/\partial x = 0$; also, $\partial x/\partial x = 1$. Therefore

$$\frac{\partial u}{\partial x} = F_x + F_z \frac{\partial z}{\partial x}.$$

In order to differentiate a second time, we must recognize that F_x and F_z are again functions of *all three* intermediate variables. We find that

$$\frac{\partial^2 u}{\partial x^2} = F_{xx}\frac{\partial x}{\partial x} + F_{xy}\frac{\partial y}{\partial x} + F_{xz}\frac{\partial z}{\partial x} + \frac{\partial z}{\partial x}\left(F_{zx}\frac{\partial x}{\partial x} + F_{zy}\frac{\partial y}{\partial x} + F_{zz}\frac{\partial z}{\partial x}\right) + F_z \frac{\partial^2 z}{\partial x^2}.$$

The result is

$$\frac{\partial^2 u}{\partial x^2} = F_{xx} + 2F_{xz}\frac{\partial z}{\partial x} + F_{zz}\left(\frac{\partial z}{\partial x}\right)^2 + F_z \frac{\partial^2 z}{\partial x^2}.$$

Problems

In each of Problems 1 through 10, verify that $f_{xy} = f_{yx}$

1. $f(x, y) = x^3 + 7x^2y - 2xy^2 + y^2$
2. $f(x, y) = x^4 + 4x^2y^2 - 2y^4 + 4x^2y - 3xy^2$
3. $f(x, y) = x^5 + 2y^4 - 3x^3y^2 + x^2y^3 - xy^3 + 2x$
4. $f(x, y, z) = x^2 + 2xyz + y^2 + z^2 + 3xz - 2xy$
5. $f: (x, y) \to e^{xy} \sin x + e^{xy} \cos y$
6. $f(r, s) = e^{r/s} \cos(rs) + e^r \sin s$
7. $f: (x, y) \to \arctan\left(\dfrac{xy}{x + y}\right)$
8. $f(x, y, z) = \log \dfrac{1 + x}{1 + y} - e^{xyz}$
9. $f: (x, y, z) \to (x^2 + y^2 - z^2)^{1/2}$
10. $f: (x, y, z, t) \to x^3 - 2y^3 + z^3 + 2t^3 - 3xyzt$

In each of Problems 11 through 15, verify that $u_{xy} = u_{yx}$ and $u_{xz} = u_{zx}$.

11. $u = \log\sqrt{x^2 + y^2 + z^2}$
12. $u = \log(x + \sqrt{y^2 + z^2})$
13. $u = x^3 + y^3 + z^3 - 3xyz$
14. $u = e^{xy} + e^{2xz} - e^{3yz}$
15. $u = e^{xy}/\sqrt{x^2 + z^2}$
16. Given that $u = 1/\sqrt{x^2 + y^2 + z^2}$, verify that

$$\frac{\partial^2 u}{\partial x^2} + \frac{\partial^2 u}{\partial y^2} + \frac{\partial^2 u}{\partial z^2} = 0.$$

17. Given that $u = xe^x \cos y$, verify that

$$\frac{\partial^4 u}{\partial x^4} + 2\frac{\partial^4 u}{\partial x^2\,\partial y^2} + \frac{\partial^4 u}{\partial y^4} = 0.$$

In each of Problems 18 through 21, r and s are independent variables. Find $\partial^2 z/\partial r^2$ by (a) the Chain Rule and by (b) finding z in terms of r and s first.

18. $z = x^2 - xy - y^2, \quad x = r + s, \quad y = s - r$
19. $z = x^2 - y^2, \quad x = r\cos s, \quad y = r\sin s$
20. $z = x^3 - y^3, \quad x = 2r - s, \quad y = s + 2r$
21. $z = x^2 - 2xy - y^2, \quad x = r^2 - s^2, \quad y = 2rs$
22. Given that $u = F(x, y, z)$ and $z = f(x, y)$, find $\partial^2 u/\partial y\,\partial x$ with all variables satisfying the conditions in Example 3.

23. Given that $u = F(x, y, z)$ and $z = f(x, y)$, find $\partial^2 u/\partial y^2$ with all variables satisfying the conditions in Example 3.

24. If $u = F(x, y)$, $y = f(x)$, find d^2u/dx^2.

25. Given that $u = f(x + 2y) + g(x - 2y)$, show that

$$u_{xx} - \tfrac{1}{4}u_{yy} = 0.$$

26. Given that $u = F(x, y)$, $x = r\cos\theta$, $y = r\sin\theta$, find $\partial^2 u/\partial r^2$, r and θ being independent variables.

27. Given $u = F(x, y)$, $x = f(r, s)$, $y = g(r, s)$, find $\partial^2 u/\partial r\,\partial s$, r and s being independent variables.

28. If $F(x, y) = 0$, find d^2y/dx^2 in terms of partial derivatives of F.

29. Given that $u = F(x, y)$, $x = e^s \cos t$, $y = e^s \sin t$, use the Chain Rule to show that

$$\frac{\partial^2 u}{\partial s^2} + \frac{\partial^2 u}{\partial t^2} = e^{2s}\left(\frac{\partial^2 u}{\partial x^2} + \frac{\partial^2 u}{\partial y^2}\right).$$

where s and t are independent variables and x and y are intermediate variables.

30. Given $V = F(x, y)$, $x = \tfrac{1}{2}r(e^s + e^{-s})$, $y = \tfrac{1}{2}r(e^s - e^{-s})$, show that

$$V_{xx} - V_{yy} = V_{rr} + \frac{1}{r}V_r - \frac{1}{r^2}V_{ss}.$$

*31. If $u = f(x - ut)$, show that $u_t + uu_x = 0$.

32. Assume that $f(x, y)$ has all continuous partial derivatives to the nth order. Use mathematical induction to prove that every nth partial derivative is independent of the order in which it is taken.

*33. Consider the function

$$f(x, y) = \begin{cases} xy\dfrac{x^2 - y^2}{x^2 + y^2}, & x^2 + y^2 > 0, \\ 0, & x = y = 0. \end{cases}$$

a) Show that $f_x(0, 0) = 0$ and $f_y(0, 0) = 0$.
b) Show, by appealing to the definition of derivative, that

$$f_{xy}(0, 0) = \lim_{k \to 0} \frac{f_x(0, k) - f_x(0, 0)}{k} = -1,$$

$$f_{yx}(0, 0) = \lim_{h \to 0} \frac{f_y(h, 0) - f_y(0, 0)}{h} = 1.$$

c) Show that the result of (b) follows because f_{xy} and f_{yx} are not continuous at $x = y = 0$.

34. a) Show that $f(x_1, x_2, \ldots, x_n) = [x_1^2 + x_2^2 + \cdots + x_n^2]^{(2-n)/2}$ for $n > 2$ is a solution of the equation

$$\frac{\partial^2 f}{\partial x_1^2} + \cdots + \frac{\partial^2 f}{\partial x_n^2} = 0$$

except when $x_1 = x_2 = \cdots = x_n = 0$.

b) Show that $g(x_1, x_2, \ldots, x_n) = [x_1^2 + \cdots + x_{n-1}^2 - x_n^2]^{(2-n)/2}$ for $n > 2$ is a solution of the equation

$$\frac{\partial^2 f}{\partial x_1^2} + \cdots + \frac{\partial^2 f}{\partial x_{n-1}^2} - \frac{\partial^2 f}{\partial x_n^2} = 0$$

except when $(x_1, \ldots, x_n)$ is on the cone $x_1^2 + x_2^2 + \cdots + x_{n-1}^2 = x_n^2$.

35. Suppose that $u = F(x, y)$, $x = f(s, t)$, $y = g(s, t)$. Show that

$$\frac{\partial^2 u}{\partial s\,\partial t} = F_{xx}\frac{\partial x}{\partial s}\frac{\partial x}{\partial t} + F_{xy}\left(\frac{\partial x}{\partial s}\frac{\partial y}{\partial t} + \frac{\partial x}{\partial t}\frac{\partial y}{\partial s}\right) + F_{yy}\frac{\partial y}{\partial s}\frac{\partial y}{\partial t} + F_x\frac{\partial^2 x}{\partial s\,\partial t} + F_y\frac{\partial^2 y}{\partial s\,\partial t}.$$

Also find an expression for $\partial^2 u/\partial s^2$.

10. Taylor's Theorem with Remainder

Taylors' theorem for functions of one variable was established in Chapter 3 on page 132. There we found that if $F(x)$ has $n + 1$ derivatives in an interval containing a value x_0, then we can obtain the expansion

$$F(x) = F(x_0) + F'(x_0)(x - x_0) + \cdots + \frac{F^{(n)}(x_0)(x - x_0)^n}{n!} + R_n, \qquad (1)$$

where the remainder R_n is given by the formula

$$R_n = \frac{F^{(n+1)}(\xi)(x - x_0)^{n+1}}{(n + 1)!},$$

with ξ some number between x_0 and x.

Taylor's theorem in several variables is a generalization of the expansion (1). We carry out the procedure for a function of two variables $f(x, y)$, the process for functions of more variables being completely analogous. Consider the function

$$\phi(t) = f(x + \lambda t, y + \mu t),$$

in which the quantities x, y, λ, and μ are temporarily kept constant. Then ϕ is a function of the single variable t, and we may compute its derivative. Using the Chain Rule, we obtain

$$\phi'(t) = f_x(x + \lambda t, y + \mu t)\lambda + f_y(x + \lambda t, y + \mu t)\mu.$$

We will simplify matters further by omitting the arguments in f. We write

$$\phi'(t) = f_x\lambda + f_y\mu.$$

It is important to compute second, third, fourth, etc., derivatives of ϕ. We do so by applying the Chain Rule repeatedly. The results are

$$\phi^{(2)}(t) = \lambda^2 f_{xx} + 2\lambda\mu f_{xy} + \mu^2 f_{yy},$$
$$\phi^{(3)}(t) = \lambda^3 f_{xxx} + 3\lambda^2\mu f_{xxy} + 3\lambda\mu^2 f_{xyy} + \mu^2 f_{yyy},$$
$$\phi^{(4)}(t) = \lambda^4 f_{xxxx} + 4\lambda^3\mu f_{xxxy} + 6\lambda^2\mu^2 f_{xxyy} + 4\lambda\mu^3 f_{xyyy} + \mu^4 f_{yyyy}.$$

Examining the pattern in each of the above derivatives, we see that the coefficients in $\phi^{(2)}$ are formed by the **symbolic** expression

$$\left(\lambda\frac{\partial}{\partial x} + \mu\frac{\partial}{\partial y}\right)^2 f,$$

provided that the numerical exponent applied to a partial derivative is interpreted as *repeated differentiation instead of multiplication.* Using this new symbolism we can easily write any derivative of ϕ. The kth derivative is

$$\phi^{(k)}(t) = \left(\lambda\frac{\partial}{\partial x} + \mu\frac{\partial}{\partial y}\right)^k f. \tag{2}$$

For instance, with $k = 7$ we obtain

$$\phi^{(7)}(t) = \lambda^7 f_{xxxxxxx} + 7\lambda^6\mu f_{xxxxxxy} + \frac{7\cdot 6}{1\cdot 2}\lambda^5\mu^2 f_{xxxxxyy} + \cdots + \mu^7 f_{yyyyyyy}.$$

Of course all derivatives are evaluated at $(x + \lambda t, y + \mu t)$. The formula (2) may be established by mathematical induction.

Before stating Taylor's theorem for functions of two variables, we introduce still more symbols. The quantity

$$\sum_{1\le r+s\le p} (\quad)$$

means that the sum of the terms in parentheses is taken over all possible combinations of r and s which add up to a number between 1 and p. Neither r nor s is allowed to be negative. For example, if $p = 3$ the combinations are

$$\begin{array}{lll} (r=0, s=1), & (r=0, s=2), & (r=0, s=3), \\ (r=1, s=0), & (r=1, s=1), & (r=1, s=2), \\ (r=2, s=0), & (r=2, s=1), & \\ (r=3, s=0). & & \end{array}$$

The terms in the sum are shown schematically in Fig. 4-12 as the circled lattice points.

The symbol

$$\sum_{r+s=p} (\quad)$$

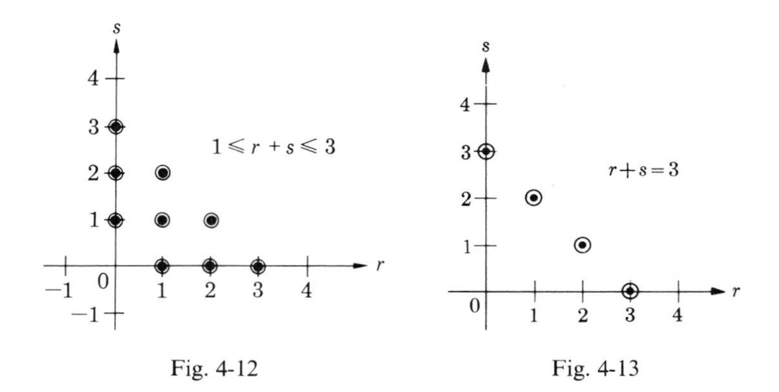

Fig. 4-12 Fig. 4-13

means that the sum is taken over all possible nonnegative combinations of r and s which add up to p exactly. For instance, if $p = 3$, then the combinations are

$$(r = 0, s = 3), \qquad (r = 1, s = 2), \qquad (r = 2, s = 1), \qquad (r = 3, s = 0).$$

The terms in this sum are indicated schematically in Fig. 4-13.

Theorem 8 (Taylor's Theorem). *Suppose that f is a function of two variables and that f and all of its partial derivatives of order up to $p + 1$ are continuous in a neighborhood of the point (a, b). Then we have the expansion*

$$f(x, y) = f(a, b) + \sum_{1 \le r+s \le p} \frac{\partial^{r+s} f(a, b)}{\partial x^r \, \partial y^s} \frac{(x-a)^r}{r!} \cdot \frac{(y-b)^s}{s!} + R_p, \tag{3}$$

where the remainder R_p is given by the formula

$$R_p = \sum_{r+s=p+1} \frac{\partial^{r+s} f(\xi, \eta)}{\partial x^r \, \partial y^s} \frac{(x-a)^r}{r!} \frac{(y-b)^s}{s!},$$

with the value (ξ, η) situated on the line segment joining the points (a, b) and (x, y). (See Fig. 4-14.)

Proof. We let $d = \sqrt{(x-a)^2 + (y-b)^2}$ and define

$$\lambda = \frac{x-a}{d}, \qquad \mu = \frac{y-b}{d}.$$

The function

$$\phi(t) = f(a + \lambda t, b + \mu t), \qquad 0 \le t \le d$$

may be differentiated according to the rules described at the beginning of the section. Taylor's theorem for $\phi(t)$ (a function of *one* variable) taken about $t = 0$ and evaluated at d yields

$$\phi(d) = \phi(0) + \phi'(0)d + \phi''(0)\frac{d^2}{2!} + \cdots + \frac{\phi^{(p)}(0)d^p}{p!} + \frac{\phi^{(p+1)}(\tau)d^{p+1}}{(p+1)!}. \tag{4}$$

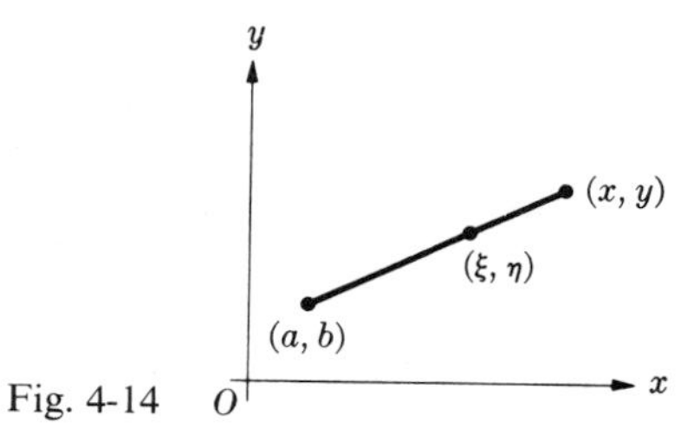

Fig. 4-14

The kth derivative of ϕ evaluated at 0 is given symbolically by

$$\phi^{(k)}(0) = \left(\lambda \frac{\partial}{\partial x} + \mu \frac{\partial}{\partial y}\right)^k f(a, b).$$

We now recall the binomial formula:

$$(A + B)^k = A^k + \frac{k}{1} A^{k-1} B + \frac{k(k-1)}{1 \cdot 2} A^{k-2} B^2 + \cdots + B^k$$

$$= \sum_{q=0}^{k} \frac{k!}{q!(k-q)!} A^{k-q} B^q.$$

Applying the binomial formula to the symbolic expression for $\phi^{(k)}(0)$, we obtain

$$\phi^{(k)}(0) = \sum_{q=0}^{k} \frac{k!}{q!(k-q)!} \frac{\partial^k f(a, b)}{\partial x^{k-q} \partial y^q} \lambda^{k-q} \mu^q. \tag{5}$$

Noting that $\phi(d) = f(a + \lambda d, b + \mu d) = f(x, y)$ and that $\phi(0) = f(a, b)$, we find, upon substitution of (5) into (4):

$$f(x, y) = f(a, b) + \frac{\partial f(a, b)}{\partial x}(x - a) + \frac{\partial f(a, b)}{\partial y}(y - b)$$

$$+ \frac{\partial^2 f(a, b)}{\partial x^2} \frac{(x - a)^2}{2!} + \frac{\partial^2 f(a, b)}{\partial x \partial y}(x - a)(y - b)$$

$$+ \frac{\partial^2 f(a, b)}{\partial y^2} \frac{(y - b)^2}{2!} + \cdots,$$

which is precisely the formula in the statement of the theorem. The remainder term shows that when $0 < \tau < d$, then

$$\xi = a + \frac{(x - a)}{d} \tau, \qquad \eta = b + \frac{(y - b)}{d} \tau,$$

which places (ξ, η) on the line segment joining (a, b) and (x, y).

REMARKS. (i) **Taylor's formula** [as (3) is usually called] is also often written in the form

$$f(x, y) = f(a, b) + \sum_{q=1}^{p} \frac{1}{q!} \left[\sum_{r=0}^{q} \frac{q!}{(q-r)!r!} \frac{\partial^q f(a, b)}{\partial x^{q-r} \partial y^r} (x - a)^{q-r} (y - b)^r \right] + R_p.$$

ii) For some functions f, if we let p tend to infinity we may find that $R_p \to 0$. We thereby obtain a representation of f as an infinite series in x and y. Such a series is called a **double series**, and we say that f is **expanded about the point** (a, b). For functions of three variables we obtain a triple sum, the Taylor formula in this case being

$$f(x, y, z) = f(a, b, z) + \sum_{1 \le r+s+t \le p} \frac{\partial^{r+s+t} f(a, b, c)}{\partial x^r \, \partial y^s \, \partial z^t} \frac{(x-a)^r}{r!} \frac{(y-b)^s}{s!} \frac{(z-c)^t}{t!} + R_p,$$

with

$$R_p = \sum_{r+s+t=p+1} \frac{\partial^{p+1} f(\xi, \eta, \zeta)}{\partial x^r \, \partial y^s \, \partial z^t} \frac{(x-a)^r}{r!} \frac{(y-b)^s}{s!} \frac{(z-c)^t}{t!},$$

where (ξ, η, ζ) is on the line segment joining (a, b, c) and (x, y, z).

EXAMPLE 1. Expand x^2y about the point $(1, -2)$ out to and including the terms of the second degree. Find R_2.

SOLUTION. Setting $f(x, y) = x^2y$, we obtain

$$f_x = 2xy, \quad f_y = x^2, \quad f_{xx} = 2y, \quad f_{xy} = 2x, \quad f_{yy} = 0,$$
$$f_{xxx} = 0, \quad f_{xxy} = 2, \quad f_{xyy} = f_{yyy} = 0.$$

Noting that $f(1, -2) = -2$, we find

$$x^2y = -2 - 4(x-1) + (y+2) + \frac{1}{2!}[-4(x-1)^2 + 4(x-1)(y+2)] + R_2,$$

with

$$R_2 = \frac{1}{2!} 2(x-1)^2(y+2) = (x-1)^2(y+2).$$

EXAMPLE 2. Given $f(x, y, z) = e^x \cos y + e^y \cos z + e^z \cos x$. Define

$$\phi(t) = f(x + \lambda t, y + \mu t, z + \nu t).$$

Find $\phi'(0)$ and $\phi^{(2)}(0)$ in terms of x, y, z, λ, μ, ν.

SOLUTION. We have

$$\phi'(0) = f_x(x, y, z)\lambda + f_y(x, y, z)\mu + f_z(x, y, z)\nu.$$

Computing the derivatives, we obtain

$$\phi'(0) = (e^x \cos y - e^z \sin x)\lambda + (e^y \cos z - e^x \sin y)\mu + (e^z \cos x - e^y \sin z)\nu.$$

The formula for $\phi^{(2)}(0)$ is

$$\begin{aligned}\phi^{(2)}(0) &= \left(\lambda\frac{\partial}{\partial x} + \mu\frac{\partial}{\partial y} + v\frac{\partial}{\partial z}\right)^2 f \\ &= \lambda^2(e^x \cos y - e^z \cos x) + \mu^2(e^y \cos z - e^x \cos y) \\ &\quad + v^2(e^z \cos x - e^y \cos z) + 2\lambda\mu(-e^x \sin y) \\ &\quad + 2\lambda v(-e^z \sin x) + 2\mu v(-e^y \sin z).\end{aligned}$$

Problems

1. Expand $x^3 + xy^2$ about the point $(2, 1)$.

2. Expand $x^4 + x^2y^2 - y^4$ about the point $(1, 1)$ out to terms of the second degree. Find the form of R_2.

3. Find the expansion of $\sin(x + y)$ about $(0, 0)$ out to and including the terms of the third degree in (x, y). Compare the result with that which you get by writing $\sin u \approx u - \frac{1}{6}u^3$ and setting $u = x + y$.

4. Find the expansion of $\cos(x + y)$ about $(0, 0)$ out to and including terms of the fourth degree in (x, y). Compare the result with that which you get by writing $\cos u \approx 1 - \frac{1}{2}u^2 + \frac{1}{24}u^4$ and setting $u = x + y$.

5. Find the expansion of e^{x+y} about $(0, 0)$ out to and including the terms of the third degree in (x, y). Compare the result with that which you get by setting $e^u \approx 1 + u + \frac{1}{2}u^2 + \frac{1}{6}u^3$, and then setting $u = x + y$. Next compare the result with that obtained by multiplying the series for e^x by that for e^y and keeping terms up to and including the third degree.

6. Find the expansion of $\sin x \sin y$ about $(0, 0)$ out to and including the terms of the fourth degree in (x, y). Compare the result with that which you get by multiplying the series for $\sin x$ and $\sin y$.

7. Do the same as Problem 6 for $\cos x \cos y$.

8. Expand $e^x \arctan y$ about $(1, 1)$ out to and including the terms of the second degree in $(x - 1)$ and $(y - 1)$.

9. Expand $x^2 + 2xy + yz + z^2$ about $(1, 1, 0)$.

10. Expand $x^3 + x^2y - yz^2 + z^3$ about $(1, 0, 1)$ out to and including the terms of the second degree in $(x - 1)$, y, and $(z - 1)$.

11. If $f(x, y) = x^2 + 4xy + y^2 - 6x$ and $\phi(t) = f(x + \lambda t, y + \mu t)$, find $\phi^{(2)}(0)$ when $x = -1$ and $y = 2$. Is $\phi^{(2)}(0) > 0$ when $(x, y) = (-1, 2)$ if λ and μ are related so that
$$\lambda^2 + \mu^2 = 1?$$

12. If $f(x, y) = x^3 + 3xy^2 - 3x^2 - 3y^2 + 4$ and $\phi(t) = f(x + \lambda t, y + \mu t)$, find $\phi^{(2)}(0)$ for $(x, y) = (2, 0)$. Show that $\phi^{(2)}(0) > 0$ for all λ and μ such that $\lambda^2 + \mu^2 = 1$.

*13. a) Write the appropriate expansion formula using binomial coefficients for
$$(A + B + C)^k,$$
with k a positive integer. (b) If $f(x, y, z)$ and $\phi(t) = f(x + \lambda t, y + \mu t, z + vt)$ are

sufficiently differentiable, show the relationship between the symbolic expression

$$\left(\lambda\frac{\partial}{\partial x}+\mu\frac{\partial}{\partial y}+\nu\frac{\partial}{\partial z}\right)^k f(x+\lambda t, y+\mu t, z+\nu t) \qquad \text{and} \qquad \phi^{(k)}(t).$$

14. Write Taylor's formula for a function $f(x,y,u,v)$ of four variables expanded about the point a, b, c, d. How many second-derivative terms are there? Third-derivative terms?

*15. Use mathematical induction to establish formula (2) on page 251.

16. Referring to the symbol $\Sigma_{1\le r+s+t\le p}$, find the number of terms when $p=3$; when $p=4$. Exhibit the corresponding lattices with three-dimensional diagrams.

17. Write Taylor's formula for a function $f(x_1,x_2,\ldots,x_k)$ of k variables. Give an expression for R_p, the remainder.

18. Write out a proof of Taylor's Theorem (Theorem 8) for a function $f(x,y,z)$ of three variables.

*19. a) Given the positive integer p, how many solutions are there of the equation

$$r_1+r_2+\cdots+r_k=p,$$

where $r_1, r_2, \ldots, r_k$ are positive integers? (b) Same problem with $r_1, r_2, \ldots, r_k$ nonnegative integers. (c) How many different partial derivatives of order p are there for a function $f(x_1,x_2,\ldots,x_k)$?

20. Expand the function $f(x,y)=\sin x\cos y$ about the values $x=\pi/6$ and $y=\pi/3$ to terms of order 2 and write an approximate formula for $\sin 29^\circ\cos 61^\circ$. Estimate the remainder term. Evaluate the third derivatives at the point $(\pi/6,\pi/3)$.

21. Let $f(x,y,z)$ be a polynomial in (x,y,z). That is,

$$f(x,y,z)=\sum_{i+j+k=1}^{n} a_{ijk}x^iy^jz^k.$$

Show that if all partial derivatives of f up to and including order n vanish at some point (x_0,y_0,z_0), then $f\equiv 0$.

11. Maxima and Minima

One of the principal applications of differentiation of functions of one variable occurs in the study of maxima and minima. In the calculus of functions of one variable we learn various tests using first and second derivatives which enable us to determine relative maxima and minima of functions of a single variable. These tests are useful for graphing functions, for solving problems involving related rates, and for attacking a variety of geometrical and physical problems.

The study of maxima and minima for functions of two, three, or more variables has its basis in the Extreme Value theorem, which we state without proof.

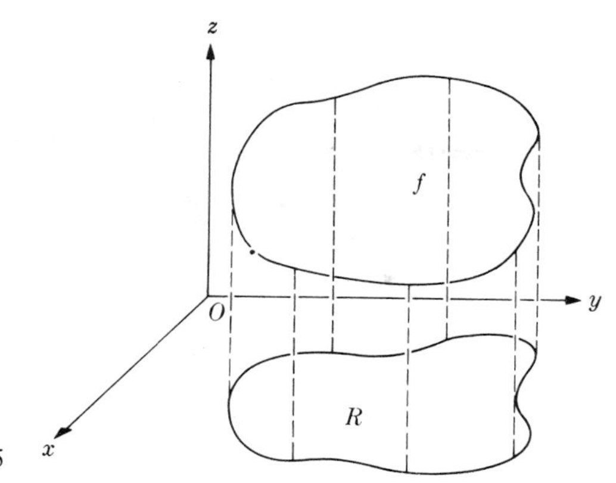

Fig. 4-15

Theorem 9 (Extreme Value Theorem). *Let R be a region in the xy plane with the boundary curve of R considered as part of R also (Fig. 4-15). If f is a function of two variables defined and continuous on R, then there is (at least) one point in R where f takes on a maximum value and there is (at least) one point in R where f takes on a minimum value.*

REMARKS. (i) Analogous theorems may be stated for functions of three, four, or more variables. (ii) The maximum and minimum may occur on the boundary of R. Thus the region R *must contain its boundary* in order to guarantee the validity of the result.

Definition. A function $f(x, y)$ is said to have a **relative maximum** at (x_0, y_0) if there is some region containing (x_0, y_0) in its interior such that

$$f(x, y) \leq f(x_0, y_0)$$

for all (x, y) in this region. (See Fig. 4-16.) More precisely, there must be some positive number δ (which may be "small") such that the above inequality

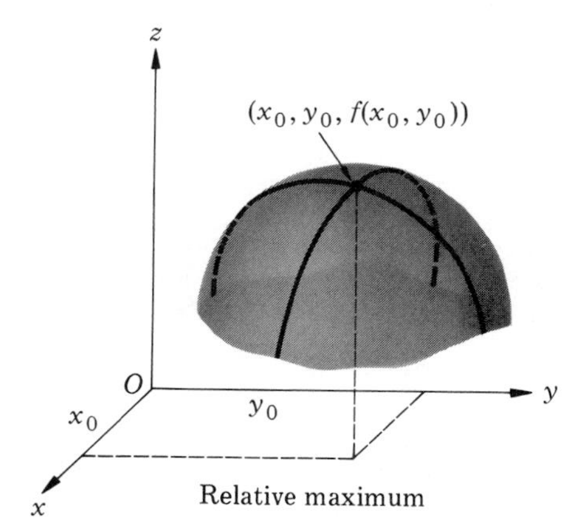

Fig. 4-16

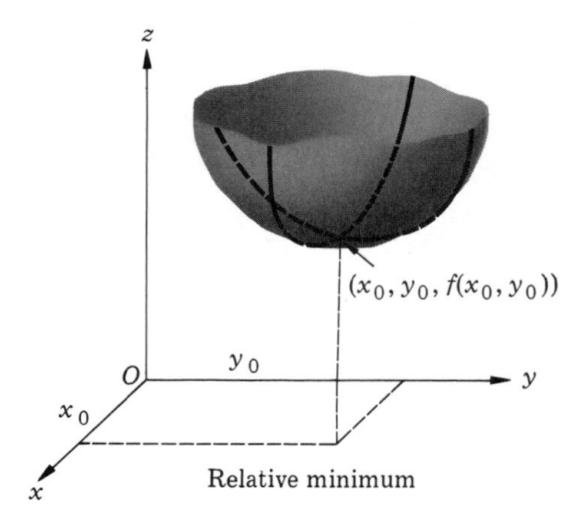

Fig. 4-17

holds for all (x, y) in the square

$$|x - x_0| < \delta, \qquad |y - y_0| < \delta.$$

A similar definition holds for **relative minimum** when the inequality $f(x, y) \geq f(x_0, y_0)$ is satisfied in a square about (x_0, y_0). (See Fig. 4-17.) The above definitions are easily extended to functions of three, four, or more variables.

Theorem 10. *Suppose that $f(x, y)$ is defined in a region R containing (x_0, y_0) in its interior. Suppose that $f_x(x_0, y_0)$ and $f_y(x_0, y_0)$ are defined and that*

$$f(x, y) \leq f(x_0, y_0)$$

for all (x, y) in R; that is, $f(x_0, y_0)$ is a relative maximum. Then

$$f_x(x_0, y_0) = f_y(x_0, y_0) = 0.$$

PROOF. We show that $f_x(x_0, y_0) = 0$, the proof for f_y being analogous. By definition,

$$f_x(x_0, y_0) = \lim_{h \to 0} \frac{f(x_0 + h, y_0) - f(x_0, y_0)}{h}.$$

By hypothesis,

$$f(x_0 + h, y_0) - f(x_0, y_0) \leq 0.$$

for all h sufficiently small so that $(x_0 + h, y_0)$ is in R. If h is positive, then

$$\frac{f(x_0 + h, y_0) - f(x_0, y_0)}{h} \leq 0,$$

and as $h \to 0$ we conclude that $f_{,1}$ must be nonpositive. On the other hand, if $h < 0$, then

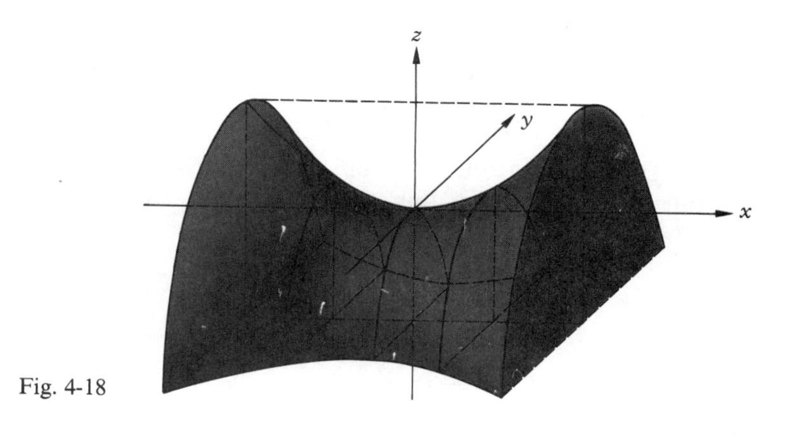

Fig. 4-18

$$\frac{f(x_0 + h, y_0) - f(x_0, y_0)}{h} \geq 0,$$

since division of both sides of an inequality by a negative number reverses its direction. Letting $h \to 0$, we conclude that f_x is nonnegative. A quantity which is both nonnegative and nonpositive vanishes.

Corollary. *The conclusion of Theorem* 10 *holds at a relative minimum.*

Definition. A value (x_0, y_0) at which both f_x and f_y vanish is called a **critical point** of f.

DISCUSSION. The conditions that f_x and f_y vanish at a point are *necessary* conditions for a relative maximum or a relative minimum. It is easy to find a function for which f_x and f_y vanish at a point, with the function having neither a relative maximum nor a relative minimum at that point. A critical point at which f is neither a maximum nor a minimum is a "**saddle point**." A simple example of a function which has such a point is given by

$$f(x, y) = x^2 - y^2.$$

We see that $f_x = 2x$, $f_y = -2y$, and $(0, 0)$ is a critical point. However, as Fig. 4-18 shows, the function is "saddle-shaped" in a neighborhood of $(0, 0)$.

While we shall develop a test which, under certain conditions, guarantees that a function has a maximum or minimum at a critical point, it is sometimes possible to make this decision from the nature of the problem itself. We exhibit such an example.

EXAMPLE 1. In three-dimensional space find the point on the plane

$$S = \{(x, y, z) : 2x + 3y - z = 1\}$$

which is closest to the origin.

SOLUTION. The function $d = \sqrt{x^2 + y^2 + z^2}$ represents a distance function which is defined at each point on the plane S. The minimum of the function

$$f(x, y, z) = x^2 + y^2 + z^2$$

occurs at the same point as the minimum of d, and f is simpler to handle. We substitute for z from the equation of the plane, and so we must minimize

$$\begin{aligned} f(x, y) &= x^2 + y^2 + (1 - 2x - 3y)^2 \\ &= 5x^2 + 10y^2 + 12xy - 4x - 6y + 1. \end{aligned}$$

A critical point must be a solution of the equations

$$f_x = 10x + 12y - 4 = 0, \qquad f_y = 20y + 12x - 6 = 0.$$

Solving these equations simultaneously, we find

$$x = \tfrac{1}{7}, \qquad y = \tfrac{3}{14}.$$

From the geometric character of the problem we know that $(\frac{1}{7}, \frac{3}{14})$ corresponds to a minimum. The point on S corresponding to $x = \frac{1}{7}$, $y = \frac{3}{14}$ is found by substitution in the equation for S. The answer is $(\frac{1}{7}, \frac{3}{14}, -\frac{1}{14})$.

The basic criterion for finding maxima and minima for functions of two variables is the so-called Second Derivative Test, which we now establish.

Theorem 11 (Second Derivative Test). *Suppose that f and its partial derivatives up to and including those of the third order are continuous near the point (a, b), and suppose that*

$$f_x(a, b) = f_y(a, b) = 0;$$

that is, (a, b) is a critical point. Then we have

i) *a local minimum if*

$$f_{xx}(a, b) f_{yy}(a, b) - f_{xy}^2(a, b) > 0 \qquad \text{and} \qquad f_{xx}(a, b) > 0;$$

ii) *a local maximum if*

$$f_{xx}(a, b) f_{yy}(a, b) - f_{xy}^2(a, b) > 0 \qquad \text{and} \qquad f_{xx}(a, b) < 0;$$

iii) *a saddle point if*

$$f_{xx}(a, b) f_{yy}(a, b) - f_{xy}^2(a, b) < 0;$$

iv) *no information if*

$$f_{xx}(a, b) f_{yy}(a, b) - f_{xy}^2(a, b) = 0.$$

PROOF. For convenience, we define

$$A = f_{xx}(a, b), \qquad B = f_{xy}(a, b), \qquad C = f_{yy}(a, b).$$

From the Taylor expansion (Theorem 8, page 252) of $f(x, y)$ about the point (a, b), we find

$$f(x,y) = f(a,b) + \tfrac{1}{2}[A(x-a)^2 + 2B(x-a)(y-b) + C(y-b)^2] + R_2. \quad (1)$$

The first derivative terms are absent because (a,b) is a critical point. The term R_2 is given by

$$R_2 = \frac{1}{6}\left[\frac{\partial^3 f(\xi,\eta)}{\partial x^3}(x-a)^3 + 3\frac{\partial^3 f(\xi,\eta)}{\partial x^2\,\partial y}(x-a)^2(y-b) + 3\frac{\partial^3 f(\xi,\eta)}{\partial x\,\partial y^2}(x-a)(y-b)^2 + \frac{\partial^3 f(\xi,\eta)}{\partial y^3}(y-b)^3\right].$$

We define $r = \sqrt{(x-a)^2 + (y-b)^2}$ and the quantities λ, μ by the formulas

$$\lambda = \frac{x-a}{r}, \qquad \mu = \frac{y-b}{r}.$$

Note that for any x, y, a, b the relation $\lambda^2 + \mu^2 = 1$ prevails. The Taylor expansion (1) now becomes

$$f(x,y) - f(a,b) = \tfrac{1}{2}r^2(A\lambda^2 + 2B\lambda\mu + C\mu^2 + r\rho), \tag{2}$$

where

$$\rho = \frac{1}{3}\left(\frac{\partial^3 f}{\partial x^3}\lambda^3 + 3\frac{\partial^3 f}{\partial x^2\,\partial y}\lambda^2\mu + 3\frac{\partial^3 f}{\partial x\,\partial y^2}\lambda\mu^2 + \frac{\partial^3 f}{\partial y^3}\mu^3\right)_{\substack{x=\xi\\ y=\eta}}.$$

The quantity ρ is bounded since, by hypothesis, f has continuous third derivatives. The behavior of $f(x,y) - f(a,b)$ is determined completely by the size of $r\rho$ and the size of the quadratic expression

$$A\lambda^2 + 2B\lambda\mu + C\mu^2, \tag{3}$$

with $\lambda^2 + \mu^2 = 1$. If

$$B^2 - AC < 0 \qquad \text{and} \qquad A > 0,$$

then there are no real roots to (3) and it has a positive minimum value. (Call it m.) Now, selecting r so small that $r\rho$ is negligible compared with m, we deduce that the right side of (2) is always positive if (x, y) is sufficiently close to (a, b). Hence

$$f(x,y) - f(a,b) > 0$$

and f is a minimum at (a, b). We have just established part (i) of the theorem. By the same argument, if

$$B^2 - AC < 0 \qquad \text{and} \qquad A < 0,$$

then (3) is always negative and

$$f(x,y) - f(a,b) < 0$$

for (x, y) near (a, b). Thus the statement of (ii) follows. Part (iii) results when $B^2 - AC > 0$, in which case the expression (3) (and therefore (2)) is sometimes positive and sometimes negative. Then f can have neither a maximum nor a minimum at (a, b) and the surface $z = f(x, y)$ can be shown to be saddle-shaped near (a, b). Part (iv) is provided for completeness.

EXAMPLE 2. Test for relative maxima and minima the function f defined by

$$f(x, y) = x^3 + 3xy^2 - 3x^2 - 3y^2 + 4.$$

SOLUTION. We have

$$f_x = 3x^2 + 3y^2 - 6x$$

and

$$f_y = 6xy - 6y.$$

We set these equations equal to zero and solve simultaneously. Writing

$$x^2 + y^2 - 2x = 0, \qquad y(x - 1) = 0,$$

we see that the second equation vanishes only when $y = 0$ or $x = 1$. If $y = 0$, the first equation gives $x = 0$ or 2; if $x = 1$, the first equation gives $y = \pm 1$. The critical points are

$$(0, 0), \qquad (2, 0), \qquad (1, 1), \qquad (1, -1).$$

To apply the Second Derivative Test, we compute

$$A = f_{xx} = 6x - 6, \qquad B = f_{xy} = 6y, \qquad C = f_{yy} = 6x - 6.$$

At $(0, 0)$: $AC - B^2 > 0$ and $A < 0$, a maximum.
At $(2, 0)$: $AC - B^2 > 0$ and $A > 0$, a minimum.
At $(1, 1)$: $AC - B^2 < 0$, saddle point.
At $(1, -1)$: $AC - B^2 < 0$, saddle point.

EXAMPLE 3. Find the dimensions of the rectangular box, open at the top, which has maximum volume if the surface area is 12.

SOLUTION. Let V be the volume of the box; let (x, y) be the horizontal directions and z the height. Then $V = xyz$, and the surface area is given by

$$xy + 2xz + 2yz = 12. \tag{4}$$

Solving this equation for z and substituting its value in the expression for V, we get

$$V = \frac{xy(12 - xy)}{2(x + y)} = \frac{12xy - x^2y^2}{2(x + y)}.$$

The domains for x, y, z are restricted by the inequalities

$$x > 0, \qquad y > 0, \qquad xy < 12.$$

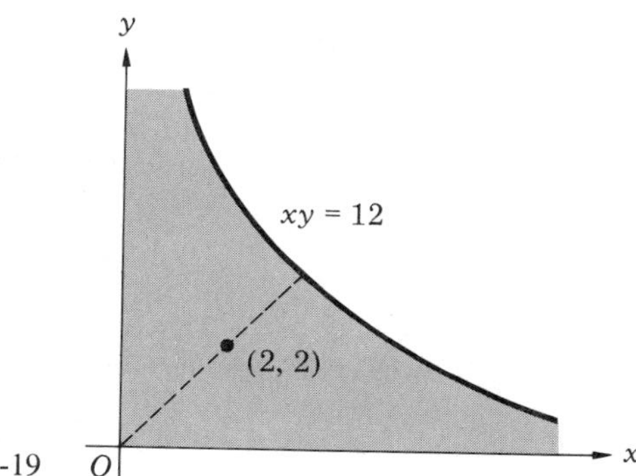

Fig. 4-19

In other words, x and y must lie in the shaded region shown in Fig. 4-19. To find the critical points, we compute

$$V_x = \frac{y^2(12 - x^2 - 2xy)}{2(x+y)^2}, \qquad V_y = \frac{x^2(12 - y^2 - 2xy)}{2(x+y)^2}$$

and set these expressions equal to zero. We obtain (excluding $x = y = 0$)

$$x^2 + 2xy = 12, \qquad y^2 + 2xy = 12.$$

Subtracting, we find that $x = \pm y$. If $x = y$, then the positive solution is $x = y = 2$. We reject $x = -y$, since both quantities must be positive. From formula (4) for surface area we conclude that $z = 1$ when $x = y = 2$. From geometrical considerations, we conclude that these are the dimensions which give a maximum volume.

REMARKS. (i) The determination of maxima and minima hinges on our ability to solve the two simultaneous equations in two unknowns resulting when we set $f_x = 0$ and $f_y = 0$. In Example 1 these equations are linear and so quite easy to solve. In Examples 2 and 3, however, the equations are nonlinear, and there are no routine methods for solving nonlinear simultaneous equations. Elementary courses in algebra usually avoid such topics, and the reader is left to his own devices. The only general rule we can state is: try to solve one of the equations for one of the unknowns in terms of the other. Substitute this value in the second equation and try to find all solutions of the second equation. Otherwise use trickery and guesswork. In actual practice, systems of nonlinear equations may be solved by a variety of numerical techniques. The use of a computer is particularly valuable in such problems. (ii) Definitions of critical point, relative maximum and minimum, etc., for functions of three, four, and more variables are simple extensions of the two-variable case. If $f(x, y, z)$ has first derivatives, then a point where

$$f_x = 0, \qquad f_y = 0, \qquad f_z = 0$$

is a **critical point**. We obtain such points by solving simultaneously three equations in three unknowns. To obtain the critical points for functions of n variables, we set all n first derivatives equal to zero and solve simultaneously the n equations in n unknowns. (iii) Extensions of the Second-Derivative Test (Theorem 11) for functions of three or more variables may be developed in a similar manner.

PROBLEMS

In each of Problems 1 through 13, test the functions f for relative maxima and minima.

1. $f(x, y) = x^2 + 2y^2 - 4x + 4y - 3$
2. $f(x, y) = x^2 - y^2 + 2x - 4y - 2$
3. $f(x, y) = x^2 + 2xy + 3y^2 + 2x + 10y + 9$
4. $f(x, y) = x^2 - 3xy + y^2 + 13x - 12y + 13$
5. $f(x, y) = y^3 + x^2 - 6xy + 3x + 6y - 7$
6. $f(x, y) = x^3 + y^2 + 2xy + 4x - 3y - 5$
7. $f(x, y) = 3x^2y + x^2 - 6x - 3y - 2$
8. $f(x, y) = xy + 4/x + 2/y$
9. $f(x, y) = \sin x + \sin y + \sin(x + y)$
10. $f(x, y) = x^3 - 6xy + y^3$
11. $f(x, y) = 8^{2/3} - x^{2/3} - y^{2/3}$
12. $f(x, y) = e^x \cos y$
13. $f(x, y) = e^{-x} \sin^2 y$

In each of Problems 14 through 17, find the critical points.

14. $f(x, y, z) = x^2 + 2y^2 + z^2 - 6x + 3y - 2z - 5$
15. $f(x, y, z) = x^2 + y^2 - 2z^2 + 3x + y - z - 2$
16. $f(x, y, z) = x^2 + y^2 + z^2 + 2xy - 3xz + 2yz - x + 3y - 2z - 5$
17. $f(x, y, z, t) = x^2 + y^2 + z^2 - t^2 - 2xy + 4xz + 3xt - 2yt + 4x - 5y - 3$
18. In three-dimensional space find the minimum distance from the origin to the plane
$$S = \{(x, y, z) : 3x + 4y + 2z = 6\}.$$
19. In the plane find the minimum distance from the point $(-1, -3)$ to the line
$$l = \{(x, y) : x + 3y = 7\}.$$
20. In three-dimensional space find the minimum distance from the point $(-1, 3, 2)$ to the plane
$$S = \{(x, y, z) : x + 3y - 2z = 8\}.$$

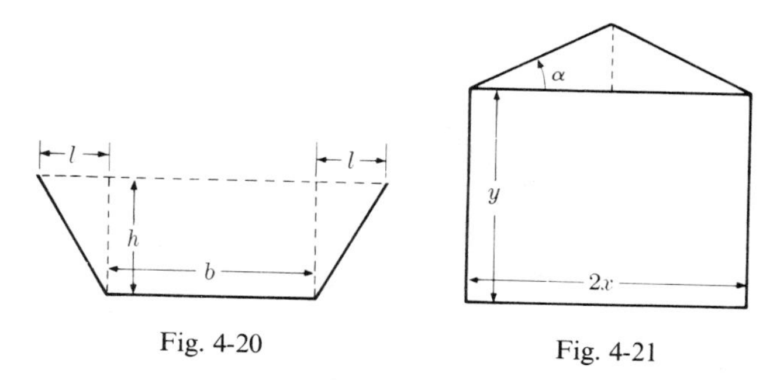

Fig. 4-20

Fig. 4-21

21. In three-dimensional space find the minimum distance from the origin to the surface (a cone)

$$C = \{(x, y, z) : z^2 = (x-1)^2 + (y-2)^2\}.$$

22. For a package to go by parcel post, the sum of the length and girth (perimeter of cross-section) must not exceed 100 cm. Find the dimensions of the package of largest volume which can be sent; assume the package has the shape of a rectangular box.

23. Find the dimensions of the rectangular parallelepiped of maximum volume with edges parallel to the axes which can be inscribed in the surface (an ellipsoid)

$$E = \left\{(x, y, z) : \frac{x^2}{9} + \frac{y^2}{4} + \frac{z^2}{16} = 1\right\}.$$

24. Find the shape of the closed rectangular box of largest volume with a surface area of 16 sq. cm.

25. The base of an open rectangular box costs half as much per square cm as the sides. Find the dimensions of the box of largest volume which can be made for D dollars.

*26. The cross-section of a trough is an isosceles trapezoid (see Fig. 4-20). If the trough is made by bending up the sides of a strip of metal 18 cm wide, what should the dimensions be in order for the area of the cross-section to be a maximum? Choose h and l as independent variables.

*27. A pentagon is composed of a rectangle surmounted by an isosceles triangle (see Fig. 4-21). If the pentagon has a given perimeter P, find the dimensions for maximum area. Choose variables as indicated in Fig. 4-21.

28. A curve C in three-space is given by

$$C = \{(x, y, z) : x^2 - xy + y^2 - z^2 = 1 \quad \text{and} \quad x^2 + y^2 = 1\}.$$

Find the point or points on C closest to the origin.

29. a) Find a function $f(x, y)$ with $f_{xx}f_{yy} - f_{xy}^2 = 0$ at a point (a, b) and such that f has a maximum at (a, b).
    b) Same as part (a) except that f has a minimum at (a, b).
    c) Same as part (a) except that f has neither a maximum nor a minimum at (a, b).

*30. State and prove a Second-Derivative Test such as Theorem 11 for functions $f(x, y, z)$ of three variables.

*31. Let $f(x_1, x_2, \ldots, x_n)$ be a function with the property that $f_{x_i} = 0$ at a point $P(a_1, a_2, \ldots, a_n)$ for $i = 1, 2, \ldots, n$. Show that if at the point P we have $f_{x_i x_j} = 0$, $i \neq j$, $i, j = 1, 2, \ldots, n$ and $f_{x_i x_i} < 0$, $i = 1, 2, \ldots, n$, then f has a relative maximum at P.

32. Show that of all triangles having a given perimeter L, the equilateral triangle has the largest area. [*Hint:* Use two sides and the included angle as variables.]

33. Suppose that a rectangular parallelepiped is inscribed in a sphere of radius R. That is, all eight vertices lie on the sphere. Show that the inscribed parallelepiped of largest volume is a cube.

12. Maxima and Minima by the Method of Lagrange Multipliers

In Example 2 of Section 11 (page 262), we solved the problem of finding the relative maxima and minima of the function

$$f(x, y) = x^3 + 3xy^2 - 3x^2 - 3y^2 + 4. \tag{1}$$

In Example 1 of the same section (page 259), we solved the problem of finding the minimum of the function

$$f(x, y, z) = x^2 + y^2 + z^2, \tag{2}$$

subject to the condition that (x, y, z) is on the plane

$$S = \{(x, y, z) : 2x + 3y - z - 1 = 0\}. \tag{3}$$

The problem of finding the critical points of (1) is quite different from that of finding those of (2) because, in the latter case, the additional condition (3) is attached. This distinction leads to the following definitions.

Definitions. The problem of finding maxima and minima of a function of several variables [such as (1) above] without added conditions is called a problem in **free maxima and minima**. When a condition such as (3) is imposed on a function such as (2) above, the problem of determining the maximum and minimum of that function is called a problem in **constrained maxima and minima**. The added condition (3) is called a **side condition**.

Problems in maxima and minima may have one or more side conditions. When side conditions occur, they are crucial. For example, the minimum of the function f given by (2) without a side condition is clearly zero.

While the problem of minimizing (2) with the side condition (3) has already been solved, we shall do it again by a different and important method. This method, due to Lagrange, changes a problem in constrainted maxima and minima to a problem in free maxima and minima.

We first introduce a new variable, traditionally denoted by λ, and form the function

$$F(x, y, z, \lambda) = (x^2 + y^2 + z^2) + \lambda(2x + 3y - z - 1).$$

The problem of finding the critical points of (2) with side condition (3) can be shown to be equivalent (under rather general circumstances) to that of finding the critical points of F considered as a function of the *four* variables x, y, z, λ. (See Theorem 12 at the end of this section.) We proceed by computing F_x, F_y, F_z, and F_λ and setting each of these expressions equal to zero. We obtain

$$F_x = 2x + 2\lambda = 0,$$

$$F_y = 2y + 3\lambda = 0,$$

$$F_z = 2z - \lambda = 0,$$

$$F_\lambda = 2x + 3y - z - 1 = 0.$$

Note that the equation $F_\lambda = 0$ is precisely the side condition (3). That is, any solution to the problem will automatically satisfy the side condition. We solve these equations simultaneously by writing

$$x = -\lambda, \qquad y = -\tfrac{3}{2}\lambda, \qquad z = \tfrac{1}{2}\lambda,$$

$$2(-\lambda) + 3(-\tfrac{3}{2}\lambda) - (\tfrac{1}{2}\lambda) - 1 = 0,$$

and we get $\lambda = -\frac{1}{7}$, $x = \frac{1}{7}$, $y = \frac{3}{14}$, $z = -\frac{1}{14}$. The solution satisfies $F_\lambda = 0$ and so is on the plane (2).

The general method, known as the **method of Lagrange multipliers**, may be stated as follows: In order to find the critical points of a function

$$f(x, y, z)$$

subject to the side condition

$$\phi(x, y, z) = 0,$$

form the function

$$F(x, y, z, \lambda) = f(x, y, z) + \lambda\phi(x, y, z)$$

and find the critical points of F considered as a function of the four variables x, y, z, λ.

The method is quite general in that several "multipliers" may be introduced if there are several side conditions. To find the critical points of

$$f(x, y, z),$$

subject to the conditions

$$\phi_1(x, y, z) = 0 \qquad \text{and} \qquad \phi_2(x, y, z) = 0, \tag{4}$$

form the function

$$F(x, y, z, \lambda_1, \lambda_2) = f(x, y, z) + \lambda_1 \phi_1(x, y, z) + \lambda_2 \phi_2(x, y, z)$$

and find the critical points of F as a function of the five variables x, y, z, λ_1, and λ_2.

We exhibit the method by working three examples.

EXAMPLE 1. Find the minimum of the function

$$f(x, y) = x^2 + 2y^2 + 2xy + 2x + 3y,$$

subject to the condition that x and y satisfy the equation

$$x^2 - y = 1.$$

SOLUTION. We form the function

$$F(x, y, \lambda) = (x^2 + 2y^2 + 2xy + 2x + 3y) + \lambda(x^2 - y - 1).$$

Then

$$F_x = 2x + 2y + 2 + 2x\lambda = 0,$$

$$F_y = 4y + 2x + 3 - \lambda = 0,$$

$$F_\lambda = x^2 - y - 1 = 0.$$

Substituting $y = x^2 - 1$ in the first two equations, we get

$$x + x^2 - 1 + 1 + \lambda x = 0, \qquad 4x^2 - 4 + 2x + 3 = \lambda.$$

Solving, we obtain

$$x = 0, \qquad y = -1, \qquad \lambda = -1,$$

or

$$x = -\tfrac{3}{4}, \qquad y = -\tfrac{7}{16}, \qquad \lambda = -\tfrac{1}{4}.$$

Evaluating f at these points, we find that a lower value occurs when $x = -\frac{3}{4}$, $y = -\frac{7}{16}$. From geometrical considerations we conclude that f is a minimum at this value.

REMARKS. We could have solved this problem as a simple maximum and minimum problem by substituting $y = x^2 - 1$ in the equation for f and finding the critical points of the resulting function of the single variable x. However, in some problems the side condition may be so complicated that we cannot easily solve for one of the variables in terms of the others, although it may be possible to do so theoretically. It is in such cases that the power of the method of Lagrange multipliers becomes apparent. The system of equations obtained by setting the first derivatives equal to zero may be solvable even though the side condition alone may not be. The next example illustrates this point.

EXAMPLE 2. Find the critical values of

$$f(x, y) = x^2 + y^2, \tag{5}$$

subject to the condition that

$$x^3 + y^3 - 6xy = 0. \tag{6}$$

SOLUTION. We form the function

$$F(x, y, \lambda) = x^2 + y^2 + \lambda(x^3 + y^3 - 6xy)$$

and obtain the derivatives

$$F_x = 2x + 3x^2\lambda - 6y\lambda = 0,$$
$$F_y = 2y + 3y^2\lambda - 6x\lambda = 0,$$
$$F_\lambda = x^3 + y^3 - 6xy = 0.$$

Solving simultaneously, we find from the first two equations that

$$\lambda = \frac{-2x}{3x^2 - 6y}, \qquad \lambda = \frac{-2y}{3y^2 - 6x}, \qquad \text{and} \qquad x(3y^2 - 6x) = y(3x^2 - 6y).$$

The equations

$$x^2y - xy^2 + 2x^2 - 2y^2 = 0, \qquad x^3 + y^3 - 6xy = 0$$

may be solved simultaneoulsy by a trick. Factoring the first equation, we see that

$$(x - y)(2x + 2y + xy) = 0$$

and $x = y$ is a solution. When $x = y$, the second equation yields

$$2x^3 - 6x^2 = 0, \qquad x = 0, 3.$$

We discard the **complex valued** solutions obtained by setting $2x + 2y + xy = 0$. The values $x = 0$, $y = 0$ clearly yield a minimum, while from geometric considerations (Fig. 4-22), the point $x = 3$, $y = 3$ corresponds to a relative

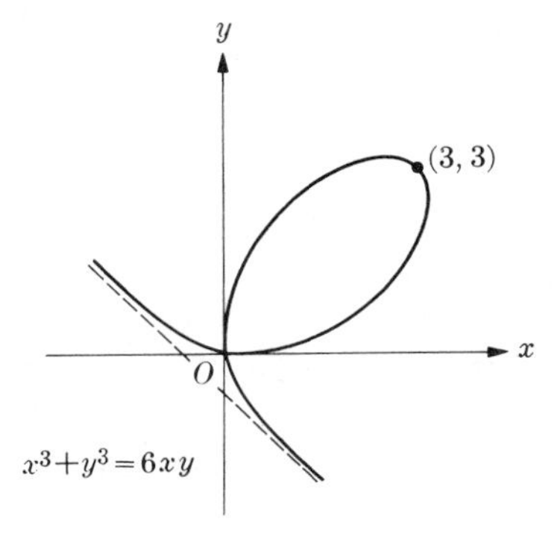

Fig. 4-22

maximum. There is no true maximum of f, since $x^2 + y^2$ (the square of the distance from the origin to the curve) grows without bound if either x or y does.

Note that it is not easy to solve Eq. (6) for either x or y and substitute in (5) to get a function of one variable. Therefore the methods of one-dimensional calculus are not readily usable in this problem.

The next example illustrates the technique of Lagrange multipliers when there are two side conditions.

EXAMPLE 3. Find the minimum of the function

$$f(x, y, z, t) = x^2 + 2y^2 + z^2 + t^2,$$

subject to the conditions

$$x + 3y - z + t = 2, \tag{7}$$

$$2x - y + z + 2t = 4. \tag{8}$$

SOLUTION. We form the function

$$F(x, y, z, t, \lambda_1, \lambda_2) = (x^2 + 2y^2 + z^2 + t^2) + \lambda_1(x + 3y - z + t - 2) + \lambda_2(2x - y + z + 2t - 4).$$

We have

$$F_x = 2x + \lambda_1 + 2\lambda_2 = 0, \qquad F_t = 2t + \lambda_1 + 2\lambda_2 = 0,$$

$$F_y = 4y + 3\lambda_1 - \lambda_2 = 0, \qquad F_{\lambda_1} = x + 3y - z + t - 2 = 0,$$

$$F_z = 2z - \lambda_1 + \lambda_2 = 0, \qquad F_{\lambda_2} = 2x - y + z + 2t - 4 = 0.$$

Solving these six linear equations in six unknowns is tedious but routine. We obtain

$$x = \tfrac{67}{69}, \qquad y = \tfrac{6}{69}, \qquad z = \tfrac{14}{69}, \qquad t = \tfrac{67}{69}.$$

The corresponding values of λ_1 and λ_2 are: $\lambda_1 = -26/69$, $\lambda_2 = -54/69$.

The validity of the method of Lagrange multipliers hinges on the ability to solve an equation for a side condition such as

$$\phi(x, y, z) = 0 \tag{9}$$

for one of the unknowns in terms of the other two. Theorems which state when such a process can be performed (theoretically, that is, not actually) are called *implicit function theorems* and are established in Chapter 9.

We now show that the method of Lagrange multipliers is valid.

Theorem 12 (Lagrange Multiplier Method). *Suppose that $f(x, y, z)$ subject to the side condition*

$$\phi(x, y, z) = 0 \tag{10}$$

has a critical point at (x_0, y_0, z_0). *If* $\phi_z(x_0, y_0, z_0) \neq 0$ *so that* (10) *can be solved for* z *in terms of* x *and* y, *then* (x_0, y_0, z_0) *is a critical point of*

$$F(x, y, z, \lambda) = f(x, y, z) + \lambda\phi(x, y, z).$$

PROOF. We solve (10) for z in terms of x and y and write the function $z = g(x, y)$. Next, we set

$$H(x, y) = f[x, y, g(x, y)],$$

and we note that H has a critical point at (x_0, y_0). Therefore

$$H_x = f_x + f_z g_x = 0, \qquad H_y = f_y + f_z g_y = 0. \tag{11}$$

But, by differentiating (11) implicitly, we obtain

$$\frac{\partial z}{\partial x} = g_x = -\frac{\phi_x}{\phi_z}, \qquad \frac{\partial z}{\partial y} = g_y = -\frac{\phi_y}{\phi_z}, \tag{12}$$

[since $\phi_z \neq 0$ near (x_0, y_0, z_0)]. Substituting (12) into (11), we find

$$f_x - \frac{f_z}{\phi_z}\phi_x = 0 \qquad \text{and} \qquad f_y - \frac{f_z}{\phi_z}\phi_y = 0.$$

We add to these equations the obvious identity

$$f_z - \frac{f_z}{\phi_z}\phi_z = 0,$$

and then we set $\lambda_0 = -f_z(x_0, y_0, z_0)/\phi_z(x_0, y_0, z_0)$. In this way we obtain at (x_0, y_0, z_0) the equations

$$f_x + \lambda_0\phi_x = 0, \qquad f_y + \lambda_0\phi_y = 0, \qquad f_z + \lambda_0\phi_z = 0, \qquad \phi = 0,$$

which are just the equations satisfied at a critical point of $F = f + \lambda\phi$.

The proof when there are more side conditions is similar but somewhat more complicated.

PROBLEMS

Solve the following problems by the method of Lagrange multipliers.

1. Find the minimum of $f(x, y, z) = x^2 + y^2 + z^2$ subject to the condition that $x + 3y - 2z = 4$.

2. Find the minimum of $f(x, y, z) = 3x^2 + 2y^2 + 4z^2$ subject to the condition that $2x + 4y - 6z + 5 = 0$.

3. Find the minimum of $f(x, y, z) = x^2 + y^2 + z^2$ subject to the condition that $ax + by + cz = d$.

4. Find the minimum of $f(x, y, z) = ax^2 + by^2 + cz^2$ subject to the condition that $dx + ey + gz + h = 0$ (a, b, c positive).

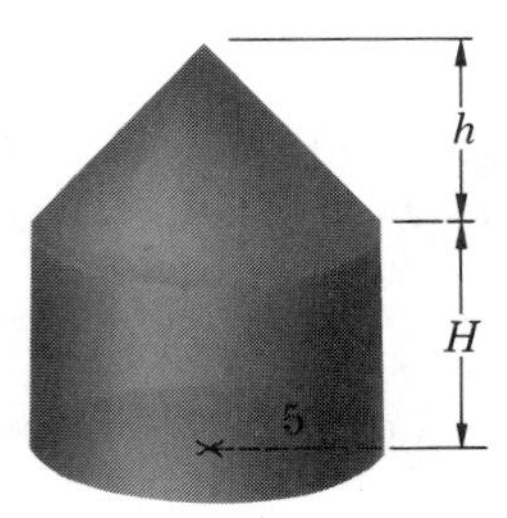

Fig. 4-23

5. Find the minimum of $f(x, y, z) = x^2 + y^2 + z^2$ if (x, y, z) is on the line of intersection of the planes

$$x + 2y + z - 1 = 0, \qquad 2x - y - 3z - 4 = 0.$$

6. Find the minimum of $f(x, y, z) = 2x^2 + y^2 + 3z^2$ if (x, y, z) is on the line of intersection of the planes

$$2x + y - 3z = 4, \qquad x - y + 2z = 6.$$

7. Find the points on the curve $x^2 + 2xy + 2y^2 = 100$ which are closest to the origin.

8. Find the relative maxima and minima of the function

$$f(x, y, z) = x^3 + y^3 + z^3$$

where (x, y, z) is on the plane $x + y + z = 4$.

9. Find the dimensions of the rectangular box, open at the top, which has maximum volume if the surface area is 12. (Compare with Example 3, page 262.)

10. A tent is made in the form of a cylinder surmounted by a cone (Fig. 4-23). If the cylinder has radius 5 and the total surface area is 100, find the height H of the cylinder and the height h of the cone which make the volume a maximum.

11. A container is made of a right circular cylinder with radius 5 and with a conical cap at each end. If the volume is given, find the height H of the cylinder and the height h of each of the conical caps which together make the total surface area as small as possible.

12. Find the minimum of the function

$$f(x, y, z, t) = x^2 + y^2 + z^2 + t^2$$

subject to the condition $3x + 2y - 4z + t = 2$.

13. Find the minimum of the function

$$f(x, y, z, t) = x^2 + y^2 + z^2 + t^2$$

subject to the conditions

$$x + y - z + 2t = 2, \qquad 2x - y + z + 3t = 3.$$

14. Find the minimum of the function

$$f(x, y, z, t) = 2x^2 + y^2 + z^2 + 2t^2$$

subject to the conditions

$$x + y + z + 2t = 1, \qquad 2x + y - z + 4t = 2,$$
$$x - y + z - t = 4.$$

15. Find the points on the curve $x^4 + y^4 + 3xy = 2$ which are closest to the origin; find those which are farthest from the origin.

16. Find three critical points of the function $x^4 + y^4 + z^4 + 3xyz$ subject to the condition that (x, y, z) is on the plane $x + y + z = 3$. Can you identify these points?

17. Work Exercise 22 of Section 11 by the method of Lagrange multipliers.

18. Find the dimensions of the rectangular parallelepiped of maximum volume with edges parallel to the axes which can be inscribed in the ellipsoid

$$E = \left\{(x, y, z) : \frac{x^2}{a^2} + \frac{y^2}{b^2} + \frac{z^2}{c^2} = 1\right\}.$$

19. If the base of an open rectangular box costs three times as much per square cm as the sides, find the dimensions of the box of largest volume which can be made for D dollars.

20. Find and identify the critical points of the function

$$f(x, y, z) = 2x^2 + y^2 + z^2$$

subject to the condition that (x, y, z) is on the surface $x^2yz = 1$.

21. Find the critical points of the function $f(x, y, z) = x^a y^b z^c$ if

$$x + y + z = A,$$

where a, b, c, A are given positive numbers.

22. Let $b_1, b_2, \ldots, b_k$ be k positive numbers. Find the maximum of

$$f(x_1, x_2, \ldots, x_k) = b_1x_1 + b_2x_2 + \cdots + b_kx_k,$$

subject to the condition

$$x_1^2 + x_2^2 + \cdots + x_k^2 = 1.$$

23. Find the maximum of the function

$$f(x_1, x_2, \ldots, x_k) = x_1^2 \cdot x_2^2 \cdot \cdots \cdot x_k^2$$

subject to the condition

$$x_1^2 + x_2^2 + \cdots + x_k^2 = 1.$$

*24. If $a_1, a_2, \ldots, a_k$ are positive numbers, prove that

$$(a_1 \cdot a_2 \cdot \cdots \cdot a_k)^{1/k} \le \frac{a_1 + a_2 + \cdots + a_k}{k}. \qquad (*)$$

[*Hint:* Define $x_1, x_2, \ldots, x_k$ by the relations $x_i^2 = \alpha a_i$, $i = 1, 2, \ldots, k$ where $\alpha = 1/\Sigma_{i=1}^k a_i$. Then $(*)$ is equivalent to the inequality

$$(x_1^2 \cdot x_2^2 \cdot \dots \cdot x_k^2)^{1/k} \le 1/k,$$

with the side condition

$$\sum_{i=1}^{k} x_i^2 = 1.$$

Now use Lagrange multipliers (see Problem 23).]

25. State and prove a theorem on the validity of the Lagrange multiplier method for obtaining the critical points of the function

$$f(x_1, x_2, \dots, x_k)$$

subject to the side condition $\phi(x_1, x_2, \dots, x_k) = 0$.

26. Given the quadratic function

$$f(x_1, x_2, \dots, x_n) = \sum_{i,j=1}^{n} a_{ij} x_i x_j$$

where a_{ij}, $i, j = 1, 2, \dots, n$ are numbers such that $a_{ij} = a_{ji}$. Write the n equations which must be satisfied by a maximum of f on the unit sphere in $R^n : x_1^2 + x_2^2 + \dots + x_n^2 = 1$. The n values of λ obtained when these equations are solved are called **eigenvalues**. If $n = 2$ and $f(x_1, x_2) = 2x_1^2 + 3x_1x_2 + 4x_2^2$, find the two eigenvalues.

27. A manufacturer makes three types of automobile tires labeled A, B, and C. Let x, y, and z be the number of tires made each day of types A, B, and C, respectively. The profit on each tire of type A is \$2, on type B it is \$3, and on type C it is \$5. The total number of tires which can be produced each day is subject to the constraint $2x^2 + y^2 + 3z^2 = 500$. Find how many tires of each type should be produced in order to maximize profit.

13. Exact Differentials

In Section 7 we saw that the total differential of a function $f(x, y)$ is given by

$$df = \frac{\partial f}{\partial x} dx + \frac{\partial f}{\partial y} dy. \tag{1}$$

The quantity df is a function of four variables, since $\partial f/\partial x$ and $\partial f/\partial y$ are functions of x and y and dx and dy are additional independent variables. Expressions of the form

$$P(x, y)\, dx + Q(x, y)\, dy$$

occur frequently in problems in engineering and physics. It is natural to ask when such an expression is the total differential of a function f. For example, if we are given

$$(3x^2 + 2y)\, dx + (2x - 3y^2)\, dy,$$

we may guess (correctly) that the function $f(x, y) = x^3 + 2xy - y^3$ has the above expression as its total differential, df. On the other hand, if we are given

$$(2x^2 - 3y)\,dx + (2x - y^3)\,dy, \tag{2}$$

then it can be shown that *there is no function f* whose total differential is the expression (2).

Definition. If there is a function $f(x, y)$ such that

$$df = P(x, y)\,dx + Q(x, y)\,dy$$

for all (x, y) in some region and for all values of dx and dy, we say that

$$P(x, y)\,dx + Q(x, y)\,dy$$

is an **exact differential**. If there is a function $F(x, y, z)$ such that

$$dF = P(x, y, z)\,dx + Q(x, y, z)\,dy + R(x, y, z)\,dz$$

for all (x, y, z) in some region and for all values of dx, dy, and dz, we say that $P\,dx + Q\,dy + R\,dz$ is an **exact differential**. For functions with any number of variables the extension is immediate.

The next theorem gives a precise criterion for determining when a differential expression is an exact differential.

Theorem 13. *Suppose that $P(x, y)$, $Q(x, y)$, $\partial P/\partial y$, $\partial Q/\partial x$ are continuous in a rectangle S. Then the expression*

$$P(x, y)\,dx + Q(x, y)\,dy \tag{3}$$

is an exact differential for (x, y) in the region S if and only if

$$\boxed{\frac{\partial P}{\partial y} = \frac{\partial Q}{\partial x} \qquad \text{for all } (x, y) \text{ in } S.} \tag{4}$$

Proof. The theorem has two parts: we must show (a), that if (3) is an exact differential, then (4) holds; and (b), that if (4) holds, then the expression (3) is an exact differential.

To establish (a) we start with the assumption that there is a function f such that

$$df = P\,dx + Q\,dy,$$

and so $\partial f/\partial x = P(x, y)$ and $\partial f/\partial y = Q(x, y)$. We differentiate and obtain

$$\frac{\partial^2 f}{\partial y\,\partial x} = \frac{\partial P}{\partial y} \qquad \text{and} \qquad \frac{\partial^2 f}{\partial x\,\partial y} = \frac{\partial Q}{\partial x}.$$

Now Theorem 7, page 244, which states that the order of differentiation is immaterial, may be invoked to conclude that (4) holds.

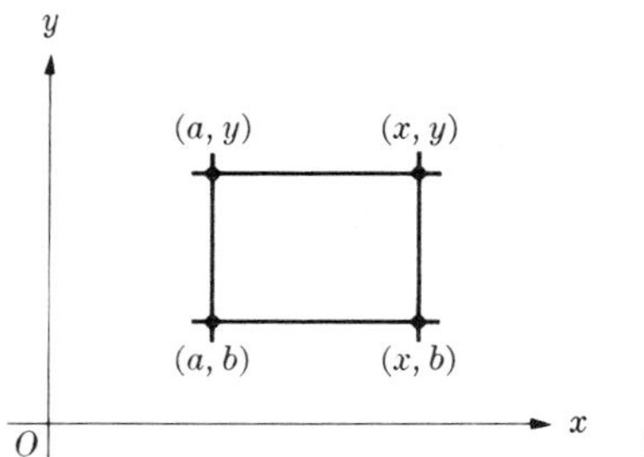

Fig. 4-24

To prove (b) we assume that (4) holds, and we must construct a function f such that df is equal to the differential expression (3). That is, we must find a function f such that

$$\frac{\partial f}{\partial x} = P(x, y) \qquad \text{and} \qquad \frac{\partial f}{\partial y} = Q(x, y). \tag{5}$$

Let (a, b) be a point of S; suppose we try to solve these two partial differential equations for the function f. We integrate the first with respect to x, getting

$$f(x, y) = C(y) + \int_a^x P(\xi, y)\, d\xi$$

where, instead of a "constant" of integration, we get a function of the remaining variable. Letting $x = a$, we find that $f(a, y) = C(y)$, and we can write

$$f(x, y) = f(a, y) + \int_a^x P(\xi, y)\, d\xi. \tag{6}$$

Setting $x = a$ in the second equation of (5) and integrating with respect to y, we obtain

$$f(a, y) = C_1 + \int_b^y Q(a, \eta)\, d\eta;$$

letting $y = b$, we see that $C_1 = f(a, b)$ and we conclude that

$$f(a, y) = f(a, b) + \int_b^y Q(a, \eta)\, d\eta.$$

Substitution of this expression for $f(a, y)$ into the equation (6) yields (Fig. 4-24)

$$f(x, y) = f(a, b) + \int_b^y Q(a, \eta)\, d\eta + \int_a^x P(\xi, y)\, d\xi. \tag{7}$$

We may repeat the entire process by integrating with respect to y first and with respect to x second. The three equations are

$$f(x, y) = f(x, b) + \int_b^y Q(x, \eta)\, d\eta,$$

$$f(x, b) = f(a, b) + \int_a^x P(\xi, b)\, d\xi,$$

and

$$f(x, y) = f(a, b) + \int_a^x P(\xi, b)\, d\xi + \int_b^y Q(x, \eta)\, d\eta. \tag{8}$$

The two expressions for f given by (7) and (8) will be identical if and only if (after subtraction) the equation

$$\int_a^x [P(\xi, y) - P(\xi, b)]\, d\xi = \int_b^y [Q(x, \eta) - Q(a, \eta)]\, d\eta \tag{9}$$

holds. To establish (9), we start with the observation that

$$P(\xi, y) - P(\xi, b) = \int_b^y \frac{\partial P(\xi, \eta)}{\partial y}\, d\eta = \int_b^y \frac{\partial Q(\xi, \eta)}{\partial x}\, d\eta,$$

where, for the first time, we have used the hypothesis that $\partial P/\partial y = \partial Q/\partial x$. Therefore, upon integration,

$$\int_a^x [P(\xi, y) - P(\xi, b)]\, d\xi = \int_a^x \left[\int_b^y \frac{\partial Q(\xi, \eta)}{\partial x}\, d\eta\right] d\xi.$$

It will be shown in Chapter 5, Section 3, that the order of integrations in the term on the right may be interchanged, so that

$$\begin{aligned} \int_a^x [P(\xi, y) - P(\xi, b)]\, d\xi &= \int_b^y \left[\int_a^x \frac{\partial Q(\xi, \eta)}{\partial x}\, d\xi\right] d\eta \\ &= \int_b^y [Q(x, \eta) - Q(a, \eta)]\, d\eta. \end{aligned}$$

But this equality is (9) precisely; the theorem is established when we observe that as a result of (7) or (8), the relations

$$\frac{\partial f}{\partial x} = P \quad \text{and} \quad \frac{\partial f}{\partial y} = Q$$

hold.

The proof of Theorem 13 contains in it the method for finding the function f when it exists. Examples illustrate the technique.

Example 1. Show that

$$(3x^2 + 6y)\, dx + (3y^2 + 6x)\, dy$$

is an exact differential, and find the function f of which it is the total differential.

Solution. Setting $P = 3x^2 + 6y$, $Q = 3y^2 + 6x$, we obtain

$$Q_x = 6, \qquad P_y = 6,$$

so that $P\,dx + Q\,dy$ is an exact differential. We write (as in the proof of the theorem)

$$f_x = 3x^2 + 6y$$

and integrate to get

$$f = x^3 + 6xy + C(y).$$

We differentiate with respect to y. We find

$$f_y = 6x + C'(y),$$

and this expression must be equal to Q. Therefore

$$6x + C'(y) = 3y^2 + 6x \qquad \text{or} \qquad C'(y) = 3y^2, \qquad C(y) = y^3 + C_1.$$

Thus

$$f(x, y) = x^3 + 6xy + y^3 + C_1.$$

A constant of integration will always appear in the integration of exact differentials.

Example 2. Show that

$$(e^x \cos y - e^y \sin x)\,dx + (e^y \cos x - e^x \sin y)\,dy$$

is an exact differential, and find the function f of which it is the differential.

Solution. Setting $P = e^x \cos y - e^y \sin x$, $Q = e^y \cos x - e^x \sin y$, we have

$$\frac{\partial P}{\partial y} = -e^x \sin y - e^y \sin x = \frac{\partial Q}{\partial x},$$

and the differential is exact. Integrating $f_x = P$, we get

$$f(x, y) = e^x \cos y + e^y \cos x + C(y). \tag{10}$$

Differentiating (10) with respect to y, we find

$$f_y = -e^x \sin y + e^y \cos x + C'(y) = Q = e^y \cos x - e^x \sin y.$$

Therefore, $C'(y) = 0$ and C is a constant. The function f is given by

$$f(x, y) = e^x \cos y + e^y \cos x + C.$$

The next theorem is an extension of Theorem 13 to functions of three variables.

Theorem 14. *Suppose that $P(x, y, z)$, $Q(x, y, z)$, $R(x, y, z)$ are continuous on some rectangular parallelepiped S. Then*

$$P(x, y, z)\,dx + Q(x, y, z)\,dy + R(x, y, z)\,dz$$

is an exact differential on S if and only if

$$\boxed{\frac{\partial P}{\partial y} = \frac{\partial Q}{\partial x}, \qquad \frac{\partial R}{\partial x} = \frac{\partial P}{\partial z}, \qquad \frac{\partial Q}{\partial z} = \frac{\partial R}{\partial y}.}$$

It is assumed that all the above partial derivatives are continuous functions of (x, y, z) on S.

The proof of this theorem follows the lines (and uses the proof) of Theorem 13.

The next example shows how to integrate an exact differential in three variables.

EXAMPLE 3. Determine whether or not

$$(3x^2 - 4xy + z^2 + yz - 2)\,dx + (xz - 6y^2 - 2x^2)\,dy + (9z^2 + 2xz + xy + 6z)\,dz$$

is an exact differential and, if so, find the function f of which it is the total differential.

SOLUTION. Setting P, Q, R equal to the coefficients of dx, dy, and dz, respectively, we obtain

$$P_y = -4x + z = Q_x,$$

$$P_z = 2z + y = R_x,$$

$$Q_z = x = R_y.$$

Therefore $P\,dx + Q\,dy + R\,dz$ is an exact differential, and we proceed to find f. Writing $f_x = P$, we integrate to get

$$f(x, y, z) = x^3 - 2x^2y + xz^2 + xyz - 2x + C(y, z).$$

We differentiate with respect to y:

$$f_y(x, y, z) = -2x^2 + xz + C_y(y, z) = Q = xz - 6y^2 - 2x^2.$$

Hence

$$C_y(y, z) = -6y^2$$

and, upon integration with respect to y,

$$C(y, z) = -2y^3 + C_1(z).$$

We may write

$$f(x, y, z) = x^3 - 2x^2y + xz^2 + xyz - 2x - 2y^3 + C_1(z),$$

and we wish to find $C_1(z)$. We differentiate f with respect to z:

$$f_z = 2xz + xy + C_1'(z) = R = 9z^2 + 2xz + xy + 6z.$$

We obtain

$$C_1'(z) = 9z^2 + 6z \qquad \text{and} \qquad C_1(z) = 3z^3 + 3z^2 + C_2.$$

Therefore

$$f(x, y, z) = x^3 - 2y^3 + 3z^3 - 2x^2y + xz^2 + xyz + 3z^2 - 2x + C_2.$$

PROBLEMS

In each of Problems 1 through 18, determine which of the differentials are exact. In case a differential is exact, find the functions of which it is the total differential.

1. $(x^3 + 3x^2y)\,dx + (x^3 + y^3)\,dy$
2. $(2x + 3y)\,dx + (3x + 2y)\,dy$
3. $\left(2y - \dfrac{1}{x}\right) dx + \left(2x + \dfrac{1}{y}\right) dy$
4. $(x^2 + 2xy)\,dx + (y^3 - x^2)\,dy$
5. $x^2 \sin y\,dx + x^2 \cos y\,dy$
6. $\dfrac{x^2 + y^2}{2y^2}\,dx - \dfrac{x^3}{3y^3}\,dy$
7. $2xe^{x^2} \sin y\,dx + e^{x^2} \cos y\,dy$
8. $(ye^{xy} + 3x^2)\,dx + (xe^{xy} - \cos y)\,dy$
9. $\dfrac{x\,dy - y\,dx}{x^2 + y^2}, \quad x > 0$
10. $(2x \log y)\,dx + \dfrac{x^2}{y}\,dy, \quad y > 0$
11. $(x + \cos x \tan y)\,dx + (y + \tan x \cos y)\,dy$
12. $\dfrac{1}{y} e^{2x/y}\,dx - \dfrac{1}{y^3} e^{2x/y}(y + 2x)\,dy$
13. $(3x^2 \log y - x^3)\,dx + \dfrac{3x^2}{y}\,dy$
14. $\dfrac{x\,dx}{\sqrt{x^2 + y^2}} + \left(\dfrac{y}{\sqrt{x^2 + y^2}} - 2\right) dy$
15. $(2x - y + 3z)\,dx + (3y + 2z - x)\,dy + (2x + 3y - z)\,dz$
16. $(2xy + z^2)\,dx + (2yz + x^2)\,dy + (2xz + y^2)\,dz$
17. $(e^x \sin y \cos z)\,dx + (e^x \cos y \cos z)\,dy - (e^x \sin y \sin z)\,dz$
18. $\left(\dfrac{1}{y^2} - \dfrac{y}{x^2z} - \dfrac{z}{x^2y}\right) dx + \left(\dfrac{1}{xz} - \dfrac{x}{y^2z} - \dfrac{z}{xy^2}\right) dy + \left(\dfrac{1}{xy} - \dfrac{x}{yz^2} - \dfrac{y}{xz^2}\right) dz$

*19. a) Given the differential expression

$$P\,dx + Q\,dy + R\,dz + S\,dt,$$

where P, Q, R, S are functions of x, y, z, t, state a theorem which is a plausible generalization of Theorem 13 in order to decide when the above expression is exact. (b) Use the result of part (a) to show that the following expression is exact. Find the function f of which it is the total differential.

$$(3x^2 + 2z + 3)\,dx + (2y - t - 2)\,dy + (3z^2 + 2x)\,dz + (4 - 3t^2 - y)\,dt.$$

*20. Prove Theorem 14.

*21. If $\omega = P(x, y, z)\,dx + Q(x, y, z)\,dy + R(x, y, z)\,dz$, define

$$d\omega = \left(\frac{\partial P}{\partial y} - \frac{\partial Q}{\partial x}\right) dx\,dy + \left(\frac{\partial P}{\partial z} - \frac{\partial R}{\partial x}\right) dx\,dz + \left(\frac{\partial Q}{\partial z} - \frac{\partial R}{\partial y}\right) dy\,dz.$$

Verify that if ω is an exact differential so that $\omega = df$, then Theorem 14 states that $d(df) = 0$. Now by analogy state a formula for $d\omega$ if $\omega = \Sigma_{i=1}^{k} P_i(x_1, x_2, \ldots, x_k)\,dx_i$. Under what conditions is $d\omega = 0$?

22. Let $f(x)$, $g(y)$ be arbitrary integrable functions. Show that

$$\left[yf(x) + \int_a^y g(\eta)\,d\eta\right] dx + \left[\int_b^x f(\xi)\,d\xi + xg(y)\right] dy$$

is an exact differential. If $f(x)$, $g(y)$, $h(z)$ are integrable, generalize the above result to functions of three variables.

It may happen that the expression $P(x, y)\,dx + Q(x, y)\,dy$ is not exact but that a function $I(x, y)$ can be found so that $I(x, y)[P(x, y)\,dx + Q(x, y)\,dy]$ is an exact differential. The function I is called an **integrating factor**. In each of Problems 23 through 25, show that the function I is an integrating factor. Then find the function f which yields the total differential.

23. $(xy + x^2 + 1)\,dx + x^2\,dy, \quad I = 1/x$

24. $x\,dy - y\,dx - 2x^2 \log y\,dy, \quad I = 1/x^2$

25. $(2 - xy)y\,dx + (2 + xy)x\,dy, \quad I = 1/x^2y^2$

In the left side of the differential equation of the form $P(x, y)\,dx + Q(x, y)\,dy = 0$ is an exact differential and if $f(x, y)$ is the function which yields the total differential, then $f(x, y) = c$ where c is any constant is the **general solution** of the equation. In each of Problems 26 through 28, solve the differential equation by first finding an integrating factor and then determining the function f.

26. $ye^{-x/y}\,dx - (xe^{-x/y} + y^3)\,dy = 0$

27. $x\,dx + y\,dy + (x^2 + y^2)(y\,dx - x\,dy) = 0$

28. $x\,dy - (y + x^3e^{2x})\,dx = 0$

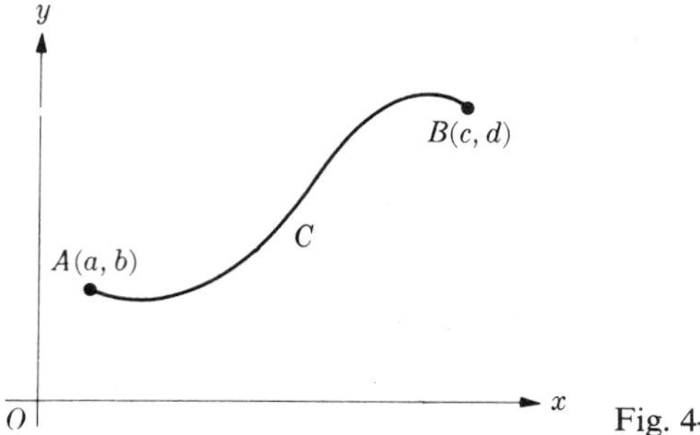

Fig. 4-25

14. Definition of a Line Integral

Let C be an arc in the plane extending from the point $A(a, b)$ to the point $B(c, d)$, as shown in Fig. 4-25. Suppose that $f(x, y)$ is a continuous function defined in a region which contains the arc C in its interior. We make a decomposition of the arc C by introducing $n - 1$ points between A and B along C. We label these points $P_1, P_2, \ldots, P_{n-2}, P_{n-1}$, and set $A = P_0$, $B = P_n$. Denote the coordinates of the point P_i by (x_i, y_i), $i = 0, 1, 2, \ldots, n$. (See Fig. 4-26.) Between each two successive points of the subdivision we select a point on the curve. Call these points $Q_1, Q_2, \ldots, Q_n$, and denote the coordinates of Q_i by (ξ_i, η_i), $i = 1, 2, \ldots, n$. This selection may be made in any manner whatsoever so long as Q_i is on the part of C between P_{i-1} and P_i (Fig. 4-27).

We form the sum

$$f(\xi_1, \eta_1)(x_1 - x_0) + f(\xi_2, \eta_2)(x_2 - x_1) + \cdots + f(\xi_n, \eta_n)(x_n - x_{n-1}),$$

or, written more compactly,

$$\sum_{i=1}^{n} f(\xi_i, \eta_i)(x_i - x_{i-1}) \equiv \sum_{i=1}^{n} f(\xi_i, \eta_i)\,\Delta_i x. \tag{1}$$

We define the **norm of the subdivision** $P_0, P_1, P_2, \ldots, P_n$ of the curve C to be the maximum distance between any two successive points of the subdivision. We denote the norm by $\|\Delta\|$.

Definition. Suppose there is a number L with the following property: for each $\varepsilon > 0$ there is a $\delta > 0$ such that

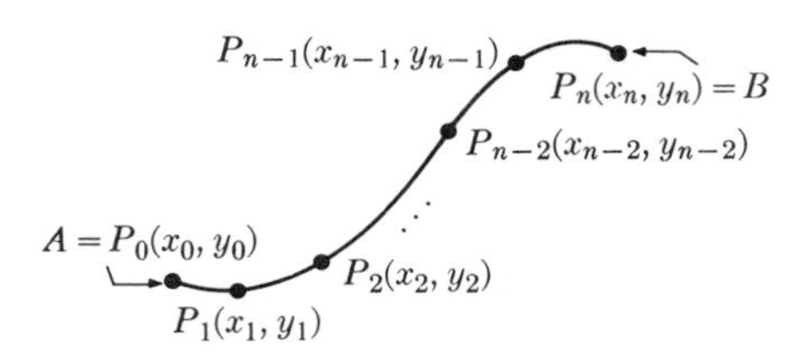

Fig. 4-26

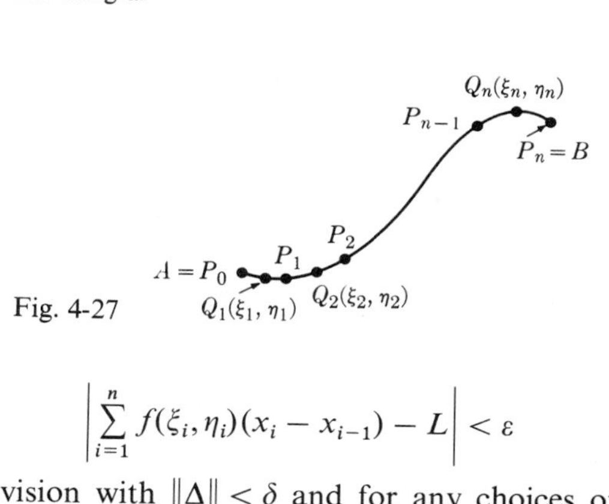

Fig. 4-27

$$\left| \sum_{i=1}^{n} f(\xi_i, \eta_i)(x_i - x_{i-1}) - L \right| < \varepsilon$$

for every subdivision with $\|\Delta\| < \delta$ and for any choices of the (ξ_i, η_i) as described above. Then we say that **the line integral of f with respect to x along the curve C exists and its value is L.** There are various symbols for this line integral such as

$$\int_C f(x, y)\, dx \qquad \text{and} \qquad (C) \int_A^B f(x, y)\, dx. \tag{2}$$

Note that the value of the integral will depend, in general, not only on f and the points A and B but also on the particular arc C selected. The number L is unique if it exists at all.

The expression (1) is one of several types of sums which are commonly formed in line integrations. We also introduce the sum

$$\sum_{i=1}^{n} f(\xi_i, \eta_i)(y_i - y_{i-1})$$

in which the points (ξ_i, η_i) are selected as before. The limit, if it exists (as $\|\Delta\| \to 0$), is the line integral

$$(C) \int_A^B f(x, y)\, dy, \tag{3}$$

and will generally have a value different from (2).

If the arc C happens to be a segment of the x axis, then the line integral $\int_C f(x, y)\, dx$ reduces to an ordinary integral. To see this we note that in the approximating sums all the $\eta_i = 0$. Therefore we have

$$(C) \int_A^B f(x, y)\, dx = \int_a^c f(x, 0)\, dx.$$

On the other hand, when C is a segment of the x axis, the integral $\int_C f(x, y)\, dy$ always vanishes, since in each approximating sum $y_i - y_{i-1} = 0$ for every i.

Simple properties of line integrals, analogous to those for ordinary integrals, may be derived directly from the definition. For example, if the arc C is traversed in the opposite direction, the line integral changes sign. That is,

$$(C)\int_A^B f(x,y)\,dx = -(C)\int_B^A f(x,y)\,dx.$$

If C_1 is an arc extending from A_1 to A_2 and C_2 is an arc extending from A_2 to A_3, then

$$(C_1)\int_{A_1}^{A_2} f(x,y)\,dx + (C_2)\int_{A_2}^{A_3} f(x,y)\,dx = (C_1 \cup C_2)\int_{A_1}^{A_3} f(x,y)\,dx, \tag{4}$$

where the symbol $C_1 \cup C_2$ has the obvious meaning. As in the case of ordinary integrals, line integrals satisfy the additive property:

$$\int_C [f(x,y) + g(x,y)]\,dx = \int_C f(x,y)\,dx + \int_C g(x,y)\,dx.$$

Statements similar to those above hold for integrals of the type $\int_C f(x,y)\,dy$.

There is one more type of line integral which we can define. If the arc C and the function f are as before and if s denotes arc length along C measured from the point A to the point B, we can define **the line integral with respect to the arc length** s. We use the symbol

$$(C)\int_A^B f(x,y)\,ds$$

for this line integral. If C is given in the form $y = g(x)$, we use the relation $ds = [1 + (g'(x))^2]^{1/2}\,dx$ to define:

$$(C)\int_A^B f(x,y)\,ds = (C)\int_A^B f[x, g(x)]\sqrt{1 + (g'(x))^2}\,dx,$$

in which the right-hand side has already been defined. If the curve C is in the form $x = h(y)$, we may write

$$(C)\int_A^B f(x,y)\,ds = (C)\int_A^B f[h(y), y]\sqrt{1 + (h'(y))^2}\,dy.$$

If C is in neither the form $y = g(x)$ nor the form $x = h(y)$, it may be broken up into a sum of arcs, each one of which does have the appropriate functional behavior. Then the integrals over each piece may be calculated and the results added.

For ordinary integrals there is a simple theorem to the effect that if a function f is continuous on an interval $[a, b]$, then it is integrable there. It can be shown that if $f(x,y)$ is continuous and if the arc C is rectifiable (has finite length), then the line integrals exist. We shall consider throughout only functions and arcs which are sufficiently smooth so that the line integrals always exist. It is worth remarking that if C consists of a collection of smooth arcs joined together (Fig. 4-28), then because of (4) the line integral along C exists as the sum of the line integrals taken along each of the pieces.

Line integrals in three dimensions may be defined similarly to the way they were defined in the plane. An arc C joining the points A and B in three-space may be given either **parametrically** by three equations,

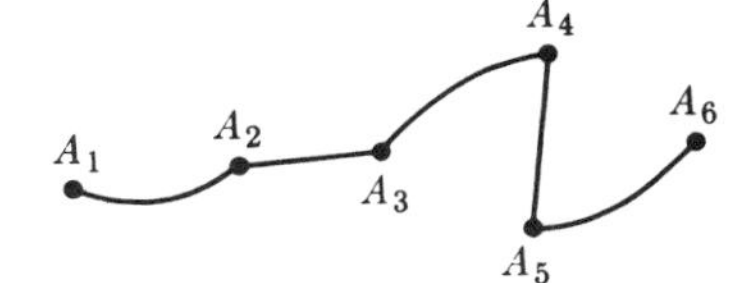

Fig. 4-28

$$x = x(t), \qquad y = y(t), \qquad z = z(t), \qquad t_0 \le t \le t_1,$$

or, in some cases, **nonparametrically** by two equations,

$$y = g_1(x), \qquad z = g_2(x).$$

If $f(x, y, z)$ is a function defined along C, then a subdivision of the arc C leads to a sum of the form

$$\sum_{i=1}^{n} f(\xi_i, \eta_i, \zeta_i)(x_i - x_{i-1})$$

which, in turn, is an approximation to the line integral

$$\int_C f(x, y, z)\, dx.$$

Line integrals such as $\int_C f(x, y, z)\, dy$, $\int_C f(x, y, z)\, dz$ and $\int_C f(x, y, z)\, ds$ are defined similarly.

15. Calculation of Line Integrals

In the study of integration of functions of one variable, the definition of integral is fairly worthless as a tool for finding the value of any specific integral. The methods actually employed for performing integration use the properties of integrals, special formulas for antiderivatives, and other devices.

The situation with line integrals is similar. In the last section we defined various types of line integrals, and now we shall exhibit methods for calculating the value of these integrals when the curve C and the function f are given specifically. It is an interesting fact that *all such integrals may be reduced to ordinary integrations of the type we have already studied.* Once the reduction is made, the problem becomes routine and all the formulas we learned for evaluation of integrals and tables of integrals may be used.

The next theorem, stated without proof, establishes the rule for reducing a line integration to an ordinary integration of a function of a single variable.

Theorem 15. *Let C be a (directed) rectifiable arc given in the form*

$$C = \{(x, y) : x = x(t), \qquad y = y(t), \qquad t_0 \le y \le t_1\}, \tag{1}$$

so that the point $A(a, b)$ corresponds to t_0, and $B(c, d)$ corresponds to t_1. Suppose $f(x, y)$ is a continuous function along C, and $x'(t)$, $y'(t)$ are continuous. Then

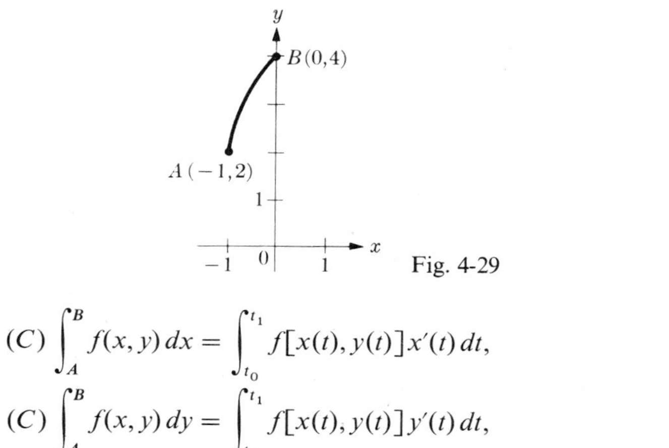

Fig. 4-29

$$(C)\int_A^B f(x,y)\,dx = \int_{t_0}^{t_1} f[x(t),y(t)]x'(t)\,dt,$$

$$(C)\int_A^B f(x,y)\,dy = \int_{t_0}^{t_1} f[x(t),y(t)]y'(t)\,dt,$$

$$(C)\int_A^B f(x,y)\,ds = \int_{t_0}^{t_1} f[x(t),y(t)]\sqrt{(x'(t))^2+(y'(t))^2}\,dt.$$

A similar theorem is valid for line integrals in three-space.

Corollary. *If the arc C is in the form $y = g(x)$, then*

$$(C)\int_A^B f(x,y)\,dx = \int_a^c f[x,g(x)]\,dx.$$

For, if $y = g(x)$, then x may be used as a parameter in place of t in (1) and the corollary is a restatement of the theorem. Similar statements may be made if C is given by an equation of the type $x = h(y)$.

EXAMPLE 1. Evaluate the integrals

$$\int_C (x^2 - y^2)\,dx - \int_C 2xy\,dy$$

where C is the arc (Fig. 4-29):

$$C = \{(x,y) : x = t^2 - 1, \quad y = t^2 + t + 2, \quad 0 \le t \le 1\}.$$

SOLUTION. According to Theorem 15, we compute

$$x'(t) = 2t, \qquad y'(t) = 2t + 1$$

and make the appropriate substitutions. We get

$$\int_C (x^2 - y^2)\,dx = \int_0^1 [(t^2-1)^2 - (t^2+t+2)^2]\cdot 2t\,dt,$$

$$-\int_C 2xy\,dy = -2\int_0^1 (t^2-1)(t^2+t+2)(2t+1)\,dt.$$

Multiplying out the integrands, we find

$$\int_C (x^2 - y^2)\,dx = 2\int_0^1 (-2t^3 - 7t^2 - 4t - 3)t\,dt,$$

$$-2\int_C xy\,dy = -2\int_0^1 (t^4 + t^3 + t^2 - t - 2)(2t + 1)\,dt.$$

The integration is now routine, and the final result is

$$\int_C [(x^2 - y^2)\,dx - 2xy\,dy] = -2\int_0^1 (2t^5 + 5t^4 + 10t^3 + 3t^2 - 2t - 2)\,dt$$

$$= -\frac{11}{3}.$$

EXAMPLE 2. Evaluate the integral

$$\int_C (x^2 - 3xy + y^3)\,dx$$

where C is the arc

$$C = \{(x, y) : y = 2x^2,\quad 0 \le x \le 2\}.$$

SOLUTION. We have

$$\int_C (x^2 - 3xy + y^3)\,dx = \int_0^2 [x^2 - 3x(2x^2) + (2x^2)^3]\,dx$$

$$= \left[\frac{x^3}{3} - \frac{3}{2}x^4 + \frac{8}{7}x^7\right]_0^2 = \frac{2624}{21}.$$

EXAMPLE 3. Evaluate

$$\int_C y\,ds$$

where C is the arc

$$C = \{(x, y) : y = \sqrt{x},\quad 0 \le x \le 6\}.$$

SOLUTION. We have

$$ds = \sqrt{1 + \left(\frac{dy}{dx}\right)^2}\,dx = \frac{1}{2}\sqrt{\frac{1 + 4x}{x}}\,dx,$$

and therefore

$$\int_C y\,ds = \frac{1}{2}\int_0^6 \sqrt{x}\sqrt{\frac{1 + 4x}{x}}\,dx = \frac{1}{8}\int_0^6 \sqrt{1 + 4x}\,d(1 + 4x)$$

$$= \left[\frac{1}{8}\cdot\frac{2}{3}(1 + 4x)^{3/2}\right]_0^6 = \frac{31}{3}.$$

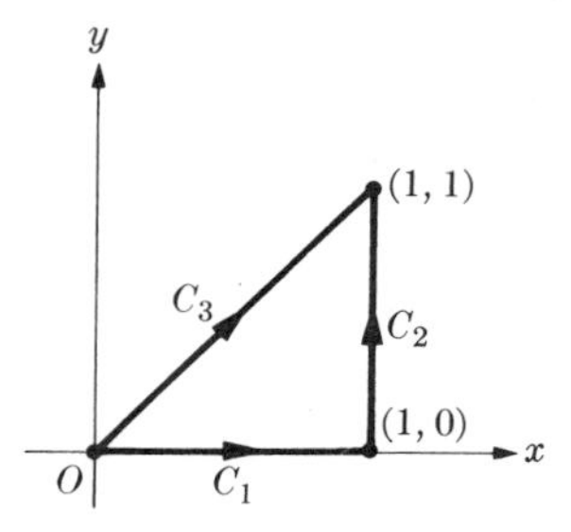

Fig. 4-30

The next example shows how we evaluate integrals when the arc C consists of several pieces.

EXAMPLE 4. Evaluate

$$\int_C [(x + 2y)\,dx + (x^2 - y^2)\,dy],$$

where C is the line segment C_1 from $(0,0)$ to $(1,0)$ followed by the line segment C_2 from $(1,0)$ to $(1,1)$ (Fig. 4-30).

SOLUTION. Along C_1 we have $x = x$, $y = 0$, $0 \leq x \leq 1$, so $dy = 0$, and

$$\int_{C_1} [(x + 2y)\,dx + (x^2 - y^2)\,dy] = \int_0^1 x\,dx = \frac{1}{2}.$$

Along C_2 we have $x = 1$, $y = y$, and so $dx = 0$. We obtain

$$\int_{C_2} [(x + 2y)\,dx + (x^2 - y^2)\,dy] = \int_0^1 (1 - y^2)\,dy = \frac{2}{3}.$$

Therefore

$$\int_C [(x + 2y)\,dx + (x^2 - y^2)\,dy] = \frac{1}{2} + \frac{2}{3} = \frac{7}{6}.$$

EXAMPLE 5. Evaluate the integral of Example 4 where the arc C is now the line segment C_3 from $(0,0)$ to $(1,1)$. (See Fig. 4-30.)

SOLUTION. Along C_3 we have $y = x$, and so $dy = dx$. Therefore

$$\int_{C_3} [(x + 2y)\,dx + (x^2 - y^2)\,dy] = \int_0^1 3x\,dx = \frac{3}{2}.$$

The next example illustrates the method for evaluation of line integrals in three space.

EXAMPLE 6. Evaluate the integral

$$\int_C [(x^2 + y^2 - z^2)\,dx + yz\,dy + (x - y)\,dz]$$

where C is the arc

$$C = \{(x, y, z) : x = t^2 + 2, \quad y = 2t - 1, \quad z = 2t^2 - t, \quad 0 \le t \le 1\}. \tag{2}$$

SOLUTION. We substitute for x, y, z from (2) and insert the values $dx = 2t\,dt$, $dy = 2\,dt$, $dz = (4t - 1)\,dt$, to obtain

$$\int_0^1 \{[(t^2 + 2)^2 + (2t - 1)^2 - (2t^2 - t)^2]2t\,dt + (2t - 1)(2t^2 - t)2\,dt + (t^2 - 2t + 3)(4t - 1)\,dt\}.$$

Upon multiplying out all the terms and performing the resulting routine integration we get the value 263/30.

PROBLEMS

In each of Problems 1 through 10, evaluate $\int_C (P\,dx + Q\,dy)$ and draw a sketch of the arc C.

1. $\int_C [(x + y)\,dx + (x - y)\,dy]$ where C is the line segment from $(0, 0)$ to $(2, 1)$.
2. $\int_C [(x + y)\,dx + (x - y)\,dy]$ where C consists of the line segment from $(0, 0)$ to $(2, 0)$ followed by that from $(2, 0)$ to $(2, 1)$.
3. $\int_C [(x^2 - 2y)\,dx + (2x + y^2)\,dy]$ where C is the arc of $y^2 = 4x - 1$ going from $(\frac{1}{2}, -1)$ to $(\frac{5}{4}, 2)$.
4. $\int_C [(x^2 - 2y)\,dx + (2x + y^2)\,dy]$ where C is the line segment going from $(\frac{1}{2}, -1)$ to $(\frac{5}{4}, 2)$.
5. $\int_C [y\,dx + (x^2 + y^2)\,dy]$ where C is the arc of the circle $y = \sqrt{4 - x^2}$ from $(-2, 0)$ to $(0, 2)$.
6. $\int_C [y\,dx + (x^2 + y^2)\,dy]$ where C consists of the line segment from $(-2, 0)$ to $(0, 0)$ followed by that from $(0, 0)$ to $(0, 2)$.
7. $$\int_C \left(\frac{x^2}{\sqrt{x^2 - y^2}}\,dx + \frac{2y}{4x^2 + y^2}\,dy\right)$$

   where C is the arc $y = \frac{1}{2}x^2$ from $(0, 0)$ to $(2, 2)$.
8. $$\int_C \left(\frac{x^2}{\sqrt{x^2 - y^2}}\,dx + \frac{2y}{4x^2 + y^2}\,dy\right)$$

   where C consists of the line segment from $(0, 0)$ to $(2, 0)$, followed by the line segment from $(2, 0)$ to $(2, 2)$.
9. $$\int_C \left(\frac{-y}{x\sqrt{x^2 - y^2}}\,dx + \frac{1}{\sqrt{x^2 - y^2}}\,dy\right)$$

   where C is the arc of $x^2 - y^2 = 9$ from $(3, 0)$ to $(5, 4)$.

10. Same integral as in Problem 9, where C consists of the line segment from $(3,0)$ to $(5,0)$, followed by the line segment from $(5,0)$ to $(5,4)$.

11. Calculate $\int_C \sqrt{x+(3y)^{5/3}}\,ds$ where C is the arc $y = \frac{1}{3}x^3$ going from $(0,0)$ to $(3,9)$.

12. Calculate $\int_C \sqrt{x+3y}\,ds$ where C is the straight line segment going from $(0,0)$ to $(3,9)$.

13. Calculate $\int_C y^2 \sin^3 x \sqrt{1+\cos^2 x}\,ds$ where C is the arc $y = \sin x$ going from $(0,0)$ to $(\pi/2, 1)$.

14. Calculate $\int_C (2x^2 + 3y^2 - xy)\,ds$ where C is the arc

$$\left.\begin{aligned} x &= 3\cos t \\ y &= 3\sin t \end{aligned}\right\}, \qquad 0 \le t \le \frac{\pi}{4}.$$

15. Calculate $\int_C x^2\,ds$ where C is the arc $x = 2y^{3/2}$ going from $(2,1)$ to $(16,4)$.

16. Calculate $\int_C [(x^2+y^2)\,dx + (x^2-y^2)\,dy]$ where C is the arc

$$C = \{(x,y) : x = t^2 + 3, \quad y = t - 1, \quad 1 \le t \le 2\}.$$

17. Calculate $\int_C [\sin x\,dy + \cos y\,dx]$ where C is the arc

$$C = \{(x,y) : x = t^2 + 3, \quad y = 2t^2 - 1, \quad 0 \le t \le 2\}.$$

18. Calculate $\int_C [(x-y)\,dx + (y-z)\,dy + (z-x)\,dz]$ where C is the line segment extending from $(1,-1,2)$ to $(2,3,1)$.

19. Calculate $\int_C [(x^2-y^2)\,dx + 2xz\,dy + (xy-yz)\,dz]$ where C is the line segment

$$C = \{(x,y,z) : x = 2t - 1, \quad y = t + 1, \quad z = t - 2, \quad 0 \le t \le 3\}.$$

20. Calculate $\int_C [(x-y+z)\,dx + (y+z-x)\,dy + (z+x-y)\,dz]$ where C consists of straight line segments connecting the points $(1,-1,2)$, $(2,-1,2)$, $(2,3,2)$, and $(2,3,1)$, in that order.

21. Calculate

$$\int_C \frac{x\,dx + y\,dy + z\,dz}{x^2+y^2+z^2}$$

where C is the arc $x = 2t$, $y = 2t+1$, $z = t^2 + t$, joining the points $(0,1,0)$ and $(2,3,2)$.

22. Same as Problem 21, where C is the straight line segment joining $(0,1,0)$ and $(2,3,2)$.

23. Evaluate

$$\int_C \frac{y\,dx + x\,dy}{\sqrt{x^2+y^2}}$$

where C is the *closed curve*

$$C = \{(x,y) : x = \cos t, \quad y = \sin t, \quad -\pi \le t \le \pi\}.$$

24. Evaluate

$$\int_C \frac{-y\,dx + x\,dy}{\sqrt{x^2+y^2}}$$

where C is the same curve as in Problem 23.

25. Write out a proof of the formula

$$(C)\int_A^B (f+g)\,dx = (C)\int_A^B f\,dx + (C)\int_A^B g\,dx.$$

26. Suppose $|f(x,y)| \le M$ for all points (x,y) on an arc C. If C is of length L, establish the inequality

$$\left|(C)\int_A^B f(x,y)\,ds\right| \le ML.$$

27. Suppose a closed curve C consists of the segment $L = \{(x,y): a \le x \le b, y = 0\}$ and the arc $K = \{(x,y): y = f(x),\quad a \le x \le b\}$ where the endpoints of K are at $(a,0)$ and $(b,0)$ and where $f(x) \ge 0$. Show that the area enclosed by C is given by $(C)\int x\,dy$ where the integral is taken counterclockwise. Extend the result to a general closed curve C.

28. Let $f(x,y)$ have continuous first derivatives in a region containing a smooth arc C with endpoints $A(x_0,y_0)$, $B(x_1,y_1)$. Show that

$$(C)\int_A^B [f_x\,dx + f_y\,dy] = f(x_1,y_1) - f(x_0,y_0).$$

[*Hint*: Assume C is given parametrically by $C = \{(x,y): x = x(t),\quad y = y(t),\ t_0 \le t \le t_1\}$. Then define $\phi(t) = f[x(t),y(t)]$ and use the Chain Rule.]

29. Suppose that $\mathbf{f}$ is a vector function in the plane. That is, $\mathbf{f}(x,y) = P(x,y)\mathbf{i} + Q(x,y)\mathbf{j}$. Define the vector differential $d\mathbf{v} = (dx)\mathbf{i} + (dy)\mathbf{j}$. Then it is natural to define

$$(C)\int \mathbf{f}\cdot d\mathbf{v} = (C)\int [P\,dx + Q\,dy].$$

If $g(x,y)$ is a scalar function which has two derivatives, show that

$$(C)\int \nabla g\cdot d\mathbf{v} = 0$$

if (C) is a smooth closed curve. [*Hint*: Recall the theorem on exact differentials or use the result in Problem 28.]

16. Path-Independent Line Integrals

In general, the value of a line integral depends on the integrand, on the two endpoints, and on the arc connecting these endpoints. However, there are special circumstances when the value of a line integral depends solely on the integrand and endpoints but *not* on the arc on which the integration is performed. When such conditions hold, we say that the integral is **independent of the path**. The next theorem establishes the connection between path-independent integrals and exact differentials. (See Section 13.)

Theorem 16. *Suppose that $P(x,y)\,dx + Q(x,y)\,dy$ is an exact differential. That is, there is a function $f(x,y)$ with*

$$df = P\,dx + Q\,dy.$$

Let C be an arc given parametrically by

$$x = x(t), \qquad y = y(t), \qquad t_0 \le t \le t_1$$

where $x'(t)$, $y'(t)$ are continuous. Then

$$\int_C (P\,dx + Q\,dy) = f[x(t_1), y(t_1)] - f[x(t_0), y(t_0)].$$

That is, the integral depends only on the endpoints and not on the arc C joining them.

PROOF. We define the function $F(t)$ by

$$F(t) = f[x(t), y(t)], \qquad t_0 \le t \le t_1.$$

We use the Chain Rule to calculate the derivative:

$$F'(t) = f_x x'(t) + f_y y'(t)$$

so that

$$F'(t) = P[x(t), y(t)]x'(t) + Q[x(t), y(t)]y'(t). \tag{1}$$

Integrating both sides of (1) with respect to t and employing Theorem 15, we conclude that

$$F(t_1) - F(t_0) = \int_C (P\,dx + Q\,dy).$$

The result follows when we note that

$$F(t_1) = f[x(t_1), y(t_1)] \qquad \text{and} \qquad F(t_0) = f[x(t_0), y(t_0)].$$

Corollary. *If $P\,dx + Q\,dy + R\,dz$ is an exact differential, then*

$$\int_C (P\,dx + Q\,dy + R\,dz) = f[x(t_1), y(t_1), z(t_1)] - f[x(t_0), y(t_0), z(t_0)]$$

where $df = P\,dx + Q\,dy + R\,dz$ and the curve C is given by

$$C = \{(x, y, z) : x = x(t), \quad y = y(t), \quad z = z(t), \quad t_0 \le t \le t_1\}.$$

EXAMPLE 1. Show that the integrand of

$$\int_C [(2x + 3y)\,dx + (3x - 2y)\,dy]$$

is an exact differential and find the value of the integral over any arc C going from the point $(1, 3)$ to the point $(-2, 5)$.

SOLUTION. Setting $P = 2x + 3y$, $Q = 3x - 2y$, we have

$$\frac{\partial P}{\partial y} = 3 = \frac{\partial Q}{\partial x}.$$

By Theorem 13, the integrand is an exact differential. Using the methods of Section 13 for integrating exact differentials, we find that

$$f(x, y) = x^2 + 3xy - y^2 + C_1.$$

Therefore

$$\int_C [(2x + 3y)\,dx + (3x - 2y)\,dy] = f(-2, 5) - f(1, 3) = -52.$$

Notice that in the evaluation process the constant C_1 disappears.

EXAMPLE 2. Show that the integrand of

$$\int_C [(3x^2 + 6xy)\,dx + (3x^2 - 3y^2)\,dy]$$

is an exact differential, and find the value of the integral over any arc C going from the point $(1, 1)$ to the point $(2, 3)$.

SOLUTION. Setting $P = 3x^2 + 6xy$, $Q = 3x^2 - 3y^2$, we have

$$\frac{\partial P}{\partial y} = 6x = \frac{\partial Q}{\partial x}.$$

Instead of finding the function f with the property that $df = P\,dx + Q\,dy$, we may pick *any* simple path joining $(1, 1)$ and $(2, 3)$ and evaluate the integral along that path. We select the horizontal path C_1 from $(1, 1)$ to $(2, 1)$, followed by the vertical path from $(2, 1)$ to $(2, 3)$. The result is

$$\int_{C_1} (3x^2 + 6x)\,dx + \int_{C_2} (12 - 3y^2)\,dy = [x^3 + 3x^2]_1^2 + [12y - y^3]_1^3 = 14.$$

PROBLEMS

In each of Problems 1 through 11, show that the integrand is an exact differential and evaluate the integral.

1. $\int_C [(x^2 + 2y)\,dx + (2y + 2x)\,dy]$ where C is any arc from $(2, 1)$ to $(4, 2)$.
2. $\int_C [(3x^2 + 4xy - 2y^2)\,dx + (2x^2 - 4xy - 3y^2)\,dy]$ where C is any arc from $(1, 1)$ to $(3, 2)$.
3. $\int_C (e^x \cos y\,dx - e^x \sin y\,dy)$ where C is any arc from $(1, 0)$ to $(0, 1)$.
4. $$\int_C \left[\left(\frac{2xy^2}{1 + x^2} + 3\right) dx + (2y \log(1 + x^2) - 2)\,dy\right]$$
   where C is any arc from $(0, 2)$ to $(5, 1)$.

5. $\displaystyle\int_C \left[\frac{y^2}{(x^2+y^2)^{3/2}}dx - \frac{xy}{(x^2+y^2)^{3/2}}dy\right]$
where C is any arc from $(4, 3)$ to $(-3, 4)$ which does not pass through the origin.

6. $\displaystyle\int_C \left[\frac{x}{\sqrt{1+x^2+y^2}}dx + \frac{y}{\sqrt{1+x^2+y^2}}dy\right]$
where C is any arc from $(-2, -2)$ to $(4, 1)$.

7. $\int_C \{[ye^{xy}(\cos xy - \sin xy) + \cos x]\,dx + [xe^{xy}(\cos xy - \sin xy) + \sin y]\,dy\}$
where C is any arc from $(0, 0)$ to $(3, -2)$.

8. $\int_C [(2x - 2y + z + 2)\,dx + (2y - 2x - 1)\,dy + (-2z + x)\,dz]$
where C is any arc from $(1, 0, 2)$ to $(3, -1, 4)$.

9. $\int_C [(2x + y - z)\,dx + (-2y + x + 2z + 3)\,dy + (4z - x + 2y - 2)\,dz]$
where C is any arc from $(0, 2, -1)$ to $(1, -2, 4)$.

10. $\int_C [(3x^2 - 3yz + 2xz)\,dx + (3y^2 - 3xz + z^2)\,dy + (3z^2 - 3xy + x^2 + 2yz)\,dz]$
where C is any are from $(-1, 2, 3)$ to $(3, 2, -1)$.

11. $\int_C [(yze^{xyz}\cos x - e^{xyz}\sin x + y\cos xy + z\sin xz)\,dx$
$\quad + (xze^{xyz}\cos x + x\cos xy)\,dy + (xye^{xyz}\cos x + x\sin xz)\,dz]$
where C is any arc from $(0, 0, 0)$ to $(-1, -2, -3)$.

12. Find the value of the integral in Problem 5 where C is a circle to which the origin is exterior. Show that the result is independent of the size of the circle selected.

13. Let α_i, β_i, $i = 1, 2, \ldots, k$ be constants, and suppose that $P_i\,dx + Q_i\,dy$, $i = 1, 2, \ldots, k$ are exact differentials. Show that

$$\sum_{i=1}^{k} [(\alpha_i P_i)\,dx + (\beta_i Q_i)\,y]$$

is exact if $\alpha_i = \beta_i$, $i = 1, 2, \ldots, k$. Is the condition $\alpha_i = \beta_i$ necessary?

14. Suppose that $P_i\,dx + Q_i\,dy$, $i = 1, 2$ are exact differentials. Prove that $P_1P_2\,dx + Q_1Q_2\,dy$ is exact if

$$(P_2 - Q_2)\frac{\partial P_1}{\partial y} = (Q_1 - P_1)\frac{\partial Q_2}{\partial x}.$$

Is this condition necessary? Extend the result to the product of two exact differentials in three variables.

*15. Let $P_1(x_1, x_2, x_3, x_4)$, $P_2(x_1, x_2, x_3, x_4)$, $\ldots$, $P_4(x_1, x_2, x_3, x_4)$ be four functions of four variables such that

$$\frac{\partial P_i}{\partial x_j} = \frac{\partial P_j}{\partial x_i}, \qquad i, j = 1, 2, 3, 4.$$

State and prove a theorem similar to Theorem 15. Note that it is first necessary to define a line integral in R^4.

CHAPTER 5

Multiple Integration

1. Definition of the Double Integral

Let F be a region of area A situated in the xy plane. We shall always assume that a region includes its boundary curve. Such regions are sometimes called **closed regions** in analogy with closed intervals on the real line—that is, ones which include their endpoints. We subdivide the xy plane into rectangles by drawing lines parallel to the coordinate axes. These lines may or may not be equally spaced (Fig. 5-1). Starting in some convenient place (such as the upper left-hand corner of F), we systematically number all the rectangles *lying entirely within F*. Suppose there are n such and we label them r_1,

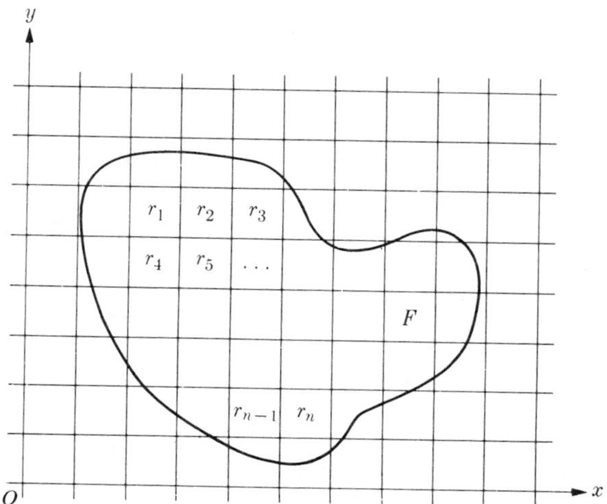

Fig. 5-1

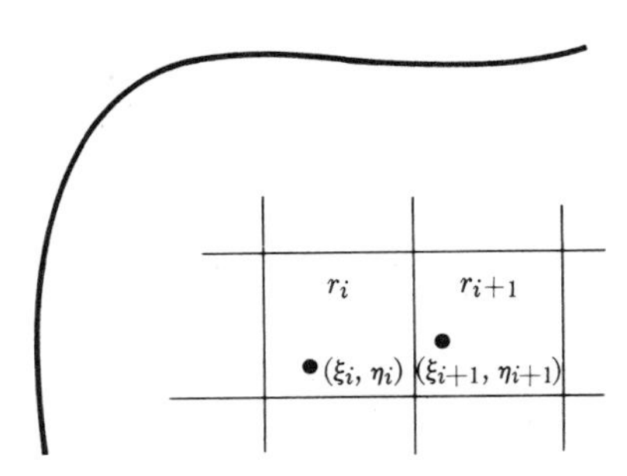

Fig. 5-2

$r_2, \ldots, r_n$. We use the symbols $A(r_1), A(r_2), \ldots, A(r_n)$ for the areas of these rectangles. The collection of n rectangles $\{r_1, r_2, \ldots, r_n\}$ is called a **subdivision** Δ of F. The **norm of the subdivision**, denoted by $\|\Delta\|$, is the length of the diagonal of the largest rectangle in the subdivision Δ.

Suppose that $f(x, y)$ is a function defined for all (x, y) in the region F. The definition of the *double integral of f over the region F* is similar to the definition of the integral for functions of one variable. Select arbitrarily a point in each of the rectangles of the subdivision Δ, denoting the coordinates of the point in the rectangle r_i by (ξ_i, η_i). (See Fig. 5-2.) Now form the sum

$$f(\xi_1, \eta_1)A(r_1) + f(\xi_2, \eta_2)A(r_2) + \cdots + f(\xi_n, \eta_n)A(r_n)$$

or, more compactly,

$$\sum_{i=1}^{n} f(\xi_i, \eta_i)A(r_i). \tag{1}$$

This sum is an approximation to the double integral we shall define. Sums such as (1) may be formed for subdivisions with any positive norm and with the ith point (ξ_i, η_i) chosen in any way whatsoever in the rectangle r_i.

Definition. We say that **a number L is the limit of sums of type** (1) and write

$$\lim_{\|\Delta\| \to 0} \sum_{i=1}^{n} f(\xi_i, \eta_i)A(r_i) = L$$

if the number L has the property: for each $\varepsilon > 0$ there is a $\delta > 0$ such that

$$\left| \sum_{i=1}^{n} f(\xi_i, \eta_i)A(r_i) - L \right| < \varepsilon$$

for every subdivision Δ with $\|\Delta\| < \delta$ and for all possible choices of the points (ξ_i, η_i) in the rectangles r_i.

It can be shown that if the number L exists, then it must be unique.

Definition. If f is defined on a region F and the number L defined above exists, we say that **f is integrable over F** and write

$$L = \iint_F f(x, y)\, dA.$$

We also call the expression above the **double integral of f over F.**

The double integral has a geometric interpretation in terms of the volume of a region in three-space. The definition of volume depends on (i), the definition of the volume of a cube—namely, length times width times height, and (ii), a limiting process.

Let S be a region in three-space. We divide all of space into cubes by constructing planes parallel to the coordinate planes at a distance apart of $1/2^n$ units, with n some positive integer. In such a network, the cubes are of three kinds: type (1), those cubes entirely within S; type (2), those cubes partly in S and partly outside S; and type (3), those cubes entirely outside S (Fig. 5-3). We define

$$V_n^-(S) = \frac{1}{8^n} \text{ times the number of cubes of type (1),}$$

$$V_n^+(S) = V_n^-(S) + \frac{1}{8^n} \text{ times the number of cubes of type (2).}$$

Intuitively we expect that, however the volume of S is defined, the number $V_n^-(S)$ would be smaller than the volume, while the number $V_n^+(S)$ would be larger. It can be shown that, as n increases, $V_n^-(S)$ gets larger or at least does not decrease, while $V_n^+(S)$ gets smaller or at least is nonincreasing. Clearly,

$$V_n^-(S) \leq V_n^+(S),$$

always. Since bounded increasing sequences and bounded decreasing sequences tend to limits, the following definitions are appropriate.

Definitions. The **inner volume** of a region S, denoted $V^-(S)$, is $\lim_{n\to\infty} V_n^-(S)$. The **outer volume**, denoted $V^+(S)$, is $\lim_{n\to\infty} V_n^+(S)$. A set of points S in three-space has a volume whenever

$$V^-(S) = V^+(S).$$

This common value is denoted by $V(S)$ and is called **the volume of S.**

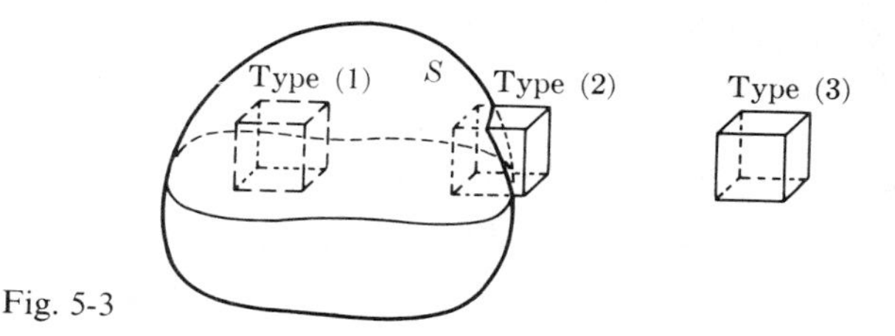

Fig. 5-3

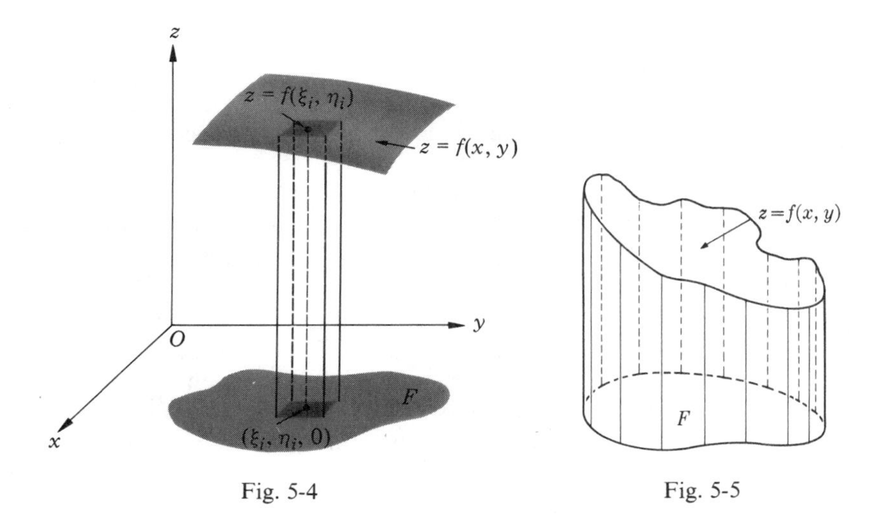

Fig. 5-4

Fig. 5-5

REMARK. It is not difficult to construct point sets for which $V^-(S) \neq V^+(S)$. For example, take S to be all points (x, y, z) such that x, y, and z are rational and

$$0 \leq x \leq 1, \qquad 0 \leq y \leq 1, \qquad 0 \leq z \leq 1.$$

The reader can verify that $V_n^-(S) = 0$ for every n, while $V_n^+(S) = 1$ for every n.

If S_1 and S_2 are two regions with no points in common, it can be shown, as expected, that $V(S_1 \cup S_2) = V(S_1) + V(S_2)$. Also, the subdivision of all of space into cubes is not vital. Rectangular parallelepipeds would do equally well, with the formula for the volume of a rectangular parallelepiped taken as length times width times height.

The volume of a region is intimately connected with the double integral in the same way that the area of a region is connected with the single integral. We now exhibit this connection.

Suppose that $f(x, y)$ is a positive function defined for (x, y) in some region F (Fig. 5-4). An item in the sum (1) approximating the double integral is

$$f(\xi_i, \eta_i)A(r_i),$$

which we recognize as the volume of the rectangular column of height $f(\xi_i, \eta_i)$ and area of base $A(r_i)$ (Fig. 5-4). The sum of the volumes of such columns is an approximation to the volume of the cylindrical region bounded by the surface $z = f(x, y)$, the plane figure F, and lines parallel to the z axis through the boundary of F (Fig. 5-5). It can be shown that, with appropriate hypotheses on the function f, the double integral

$$\iint_F f(x, y)\, dA$$

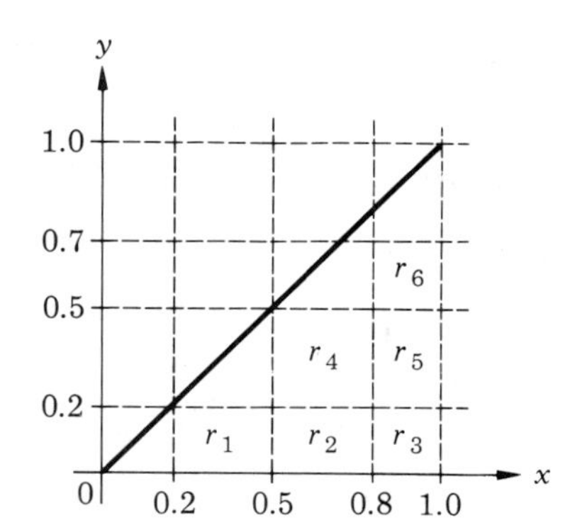

Fig. 5-6

measures the "volume under the surface" in the same way that a single integral of a positive function f

$$\int_a^b f(x)\,dx$$

measures the area under the curve.

The precise result is given in the next theorem which we state without proof.

Theorem 1. *If $f(x, y)$ is continuous for (x, y) in a closed region F, then f is integrable over F. Furthermore, if $f(x, y) > 0$ for (x, y) in F, then*

$$V(S) = \iint_F f(x, y)\,dA,$$

where $V(S)$ is the volume of the region defined by

$$S = \{(x, y, z) : (x, y) \text{ in } F \text{ and } 0 \leq z \leq f(x, y)\}.$$

We discuss methods for evaluating double integrals in Section 3.

EXAMPLE. Given $f(x, y) = 1 + xy$ and the region F bounded by the lines $y = 0$, $y = x$, and $x = 1$ (Fig. 5-6), let Δ be the subdivision formed by the lines $x = 0, 0.2, 0.5, 0.8, 1$ and $y = 0, 0.2, 0.5, 0.7, 1$. Find the value of the approximating sum

$$\sum_{i=1}^{n} f(\xi_i, \eta_i) A(r_i)$$

to the double integral

$$\iint_F f(x, y)\,dA$$

if the points (ξ_i, η_i) are selected at the centers of the rectangles.

SOLUTION. Referring to Fig. 5-6, we see that there are 6 rectangles in the subdivision which we label $r_1, r_2, \ldots, r_6$, as shown. We compute:

$$\begin{aligned}
A(r_1) &= 0.06, \qquad f(0.35,\ 0.1) = 1.035\\
A(r_2) &= 0.06, \qquad f(0.65,\ 0.1) = 1.065\\
A(r_3) &= 0.04, \qquad f(0.9,\ 0.1) = 1.090\\
A(r_4) &= 0.09, \qquad f(0.65, 0.35) = 1.2275\\
A(r_5) &= 0.06, \qquad f(0.9,\ 0.35) = 1.315\\
A(r_6) &= 0.04, \qquad f(0.9,\ 0.6) = 1.540
\end{aligned}$$

Multiplying and adding, we find that

$$\sum_{i=1}^{6} f(\xi_i, \eta_i) A(r_i) = 0.420575 \quad \text{(Answer)}.$$

PROBLEMS

In each of Problems 1 through 10, calculate the sum $\Sigma_{i=1}^{n} f(\xi_i, \eta_i) A(r_i)$ for the subdivision Δ of the region F formed by the given lines and with the points (ξ_i, η_i) selected as directed in each case.

1. $f(x, y) = x^2 + 2y^2$; F is the rectangle $0 \le x \le 1$, $0 \le y \le 1$. The subdivision Δ is: $x = 0, 0.4, 0.8, 1$; $y = 0, 0.3, 0.7, 1$. For each i the point (ξ_i, η_i) is taken at the center of the rectangle r_i.

2. Same as Problem 1, with (ξ_i, η_i) taken at the point of r_i which is closest to the origin.

3. $f(x, y) = 1 + x^2 - y^2$; F is the triangular region formed by the lines $y = 0$, $y = x$, $x = 2$. The subdivision Δ is: $x = 0, 0.5, 1, 1.6, 2$; $y = 0, 0.6, 1, 1.5, 2$. For each i, the point (ξ_i, η_i) is taken at the center of the rectangle r_i.

4. Same as Problem 3, with (ξ_i, η_i) taken at the point of r_i which is closest to the origin.

5. Same as Problem 3, with (ξ_i, η_i) selected on the lower edge of r_i, midway between the vertical subdivision lines.

6. $f(x, y) = x^2 - 2xy + 3x - 2y$; F is the trapezoid bounded by the lines $x = 0$, $x = 2$, $y = 0$, $y = x + 1$. The subdivision Δ is: $x = 0, 0.4, 1, 1.5, 2$; $y = 0, 0.6, 1, 1.4, 1.8, 2, 3$. For each i the point (ξ_i, η_i) is taken at the center of the rectangle r_i.

7. Same as Problem 6, with (ξ_i, η_i) taken at the point of r_i farthest from the origin.

8. Same as Problem 6, with (ξ_i, η_i) taken at the point of r_i closest to the origin.

9. Let

$$f(x, y) = \frac{x - y}{1 + x + y};$$

F is the region bounded by the line $y = 0$ and the curve $y = 2x - x^2$. The subdivision Δ is: $x = 0, 0.5, 1.0, 1.5, 2$; $y = 0, 0.2, 0.4, 0.6, 0.8, 1$. For each i, the point (ξ_i, η_i) is taken at the center of the rectangle r_i.

10. Same as Problem 9 with the point (ξ_i, η_i) taken at the point of r_i closest to the origin.

2. Properties of the Double Integral

In analogy with the properties of the definite integral of functions of one variable, we state several basic properties of the double integral. The simplest properties are given in the two following theorems.

Theorem 2. *If c is any number and f is integrable over a closed region F, then cf is integrable and*

$$\iint_F cf(x, y)\, dA = c \iint_F f(x, y)\, dA.$$

Theorem 3. *If f and g are integrable over a closed region F, then*

$$\iint_F [f(x, y) + g(x, y)]\, dA = \iint_F f(x, y)\, dA + \iint_F g(x, y)\, dA.$$

The result holds for the sum of any finite number of integrable functions. The proofs of Theorems 2 and 3 are obtained directly from the definition.

Theorem 4. *Suppose that f is integrable over a closed region F and*

$$m \le f(x, y) \le M \text{ for all } (x, y) \text{ in } F.$$

Then, if $A(F)$ denotes the area of F, we have

$$mA(F) \le \iint_F f(x, y)\, dA \le MA(F).$$

The proof of Theorem 4 follows exactly the same pattern as does the proof in the one-variable case. First it must be shown that the above inequalities hold for the sums

$$mA(F) \le \sum_{i=1}^{n} f(\xi_i, \eta_i) A(r_i) \le MA(F).$$

Hence the inequalities hold in the limit.

Theorem 5. *If f and g are integrable over F and $f(x, y) \leq g(x, y)$ for all (x, y) in F, then*

$$\iint_F f(x, y)\, dA \leq \iint_F g(x, y)\, dA.$$

The proof is established by the same argument used in the one-variable case. According to Theorem 4

$$\iint_F [g(x, y) - f(x, y)]\, dA \geq m(b - a) \geq 0$$

since $g(x, y) - f(x, y) \geq 0$. Then Theorem 3 yields the result.

Theorem 6. *If the closed region F is decomposed into (non-overlapping) regions F_1 and F_2 and if f is continuous over F, then*

$$\iint_F f(x, y)\, dA = \iint_{F_1} f(x, y)\, dA + \iint_{F_2} f(x, y)\, dA.$$

The proof depends on the definition of double integral and on the basic theorems on limits.

Problems

In Problems 1 through 7, use Theorem 4 to find in each case estimates for the largest and smallest values the given double integrals can possibly have.

1. $\iint_F xy\, dA$ where F is the region bounded by the lines $x = 0$, $y = 0$, $x = 2$, $y = x + 3$.
2. $\iint_F (x^2 + y^2)\, dA$ where F is the region bounded by the lines $x = -2$, $x = 3$, $y = x + 2$, $y = -2$.
3. $\iint_F (1 + 2x^2 + y^2)\, dA$ where is the region bounded by the lines $x = -3$, $x = 3$, $y = 4$, $y = -4$.
4. $\iint_F y^4\, dA$ where F is the region bounded by the line $y = 0$ and the curve $y = 2x - x^2$.
5. $\iint_F (x - y)\, dA$ where F is the region enclosed in the circle $x^2 + y^2 = 9$.
6. $\iint_F [1/(1 + x^2 + y^2)]\, dA$ where F is the region enclosed in the ellipse $4x^2 + 9y^2 = 36$.
7. $\iint_F \sqrt{1 + x^2 + y^2}\, dA$ where F is the region bounded by the curves $y = 3x - x^2$ and $y = x^2 - 3x$.
8. Write out a proof of Theorem 3.
9. Write out a proof of Theorem 4.
10. Write out a proof of Theorem 5.
11. Write out a proof of Theorem 6.

12. Let $F_1, F_2, \ldots, F_n$ be n nonoverlapping regions. State and prove a generalization of Theorem 6.

13. Let $f(x, y)$ be continuous on a closed region F.

    a) Show that $\left|\iint\limits_F f(x,y)\,dA\right| \le \iint\limits_F |f(x,y)|\,dA$.

    b) If $g(x, y)$ is also continuous on F, show that

$$\left|\iint\limits_F [f(x,y)+g(x,y)]\,dA\right| \le \iint\limits_F [|f(x,y)|+|g(x,y)|]\,dA.$$

14. Use the definition of double integral to show that the volume of a right circular cylinder with the height h and radius of base r is $\pi r^2 h$.

3. Evaluation of Double Integrals. Iterated Integrals

The definition of the double integral is useless as a tool for evaluation in any particular case. Of course, it may happen that the function $f(x, y)$ and the region F are particularly simple, so that the limit of the sum

$$\sum_{i=1}^{n} f(\xi_i, \eta_i) A(r_i)$$

can be found directly. However, such limits cannot generally be found. As in the case of ordinary integrals and line integrals, it is important to develop simple and routine methods for determining the value of a given double integral. In this section we show how the evaluation of a double integral may be performed by successive evaluations of single integrals. In other words, we reduce the problem to one we have already studied extensively. The reader will recall that in Chapter 4 the evaluation of line integrals was reduced to known techniques for single integrals in a similar way.

Let F be the rectangle with sides $x = a$, $x = b$, $y = c$, $y = d$, as shown in Fig. 5-7. Suppose that $f(x, y)$ is continuous for (x, y) in F. We form the ordinary integral with respect to x,

$$\int_a^b f(x, y)\,dx,$$

in which we keep y fixed when performing the integration. Of course, the

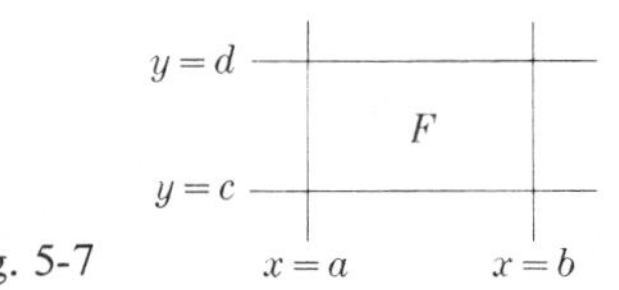

Fig. 5-7

value of the above integral will depend on the value of y used, and so we write

$$A(y) = \int_a^b f(x, y)\, dx.$$

The function $A(y)$ is defined for $c \le y \le d$ and, in fact, it can be shown that if $f(x, y)$ is continuous on F, then $A(y)$ is continuous on $[c, d]$. The integral of $A(y)$ may be computed, and we write

$$\int_c^d A(y)\, dy = \int_c^d \left[\int_a^b f(x, y)\, dx \right] dy. \tag{1}$$

We could start the other way around by fixing x and forming the integral

$$B(x) = \int_c^d f(x, y)\, dy.$$

Then

$$\int_a^b B(x)\, dx = \int_a^b \left[\int_c^d f(x, y)\, dy \right] dx. \tag{2}$$

Note that the integrals are computed *successively*; in (1) we first integrate with respect to x (keeping y constant) and then with respect to y; in (2) we first integrate with respect to y (keeping x constant) and then with respect to x.

Definition. The integrals

$$\int_c^d \left[\int_a^b f(x, y)\, dx \right] dy, \qquad \int_a^b \left[\int_c^d f(x, y)\, dy \right] dx$$

are called the **iterated integrals of** f. The terms **repeated** integrals and **successive** integrals are also used.

NOTATION. The brackets in iterated integrals are unwieldy, and we will write

$$\int_c^d \int_a^b f(x, y)\, dx\, dy \qquad \text{to mean} \qquad \int_c^d \left[\int_a^b f(x, y)\, dx \right] dy,$$

$$\int_a^b \int_c^d f(x, y)\, dy\, dx \qquad \text{to mean} \qquad \int_a^b \left[\int_c^d f(x, y)\, dy \right] dx.$$

Iterated integrals are computed in the usual way, as the next example shows.

EXAMPLE 1. Evaluate

$$\int_1^4 \int_{-2}^3 (x^2 - 2xy^2 + y^3)\, dx\, dy.$$

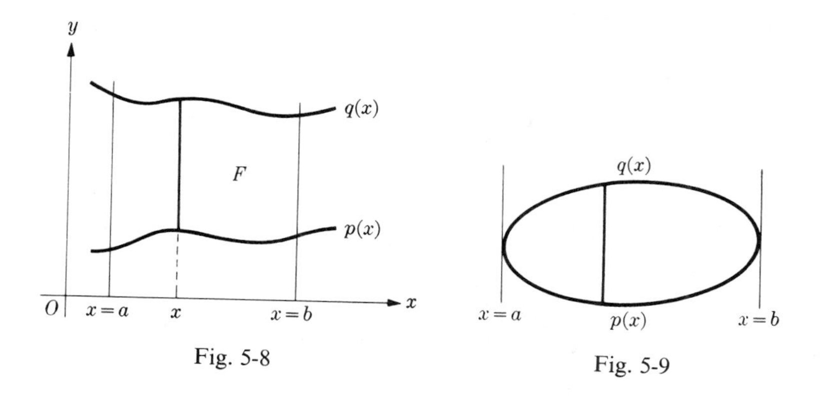

Fig. 5-8

Fig. 5-9

SOLUTION. Keeping y fixed, we have

$$\int_{-2}^{3} (x^2 - 2xy^2 + y^3)\,dx = \left[\frac{1}{3}x^3 - x^2y^2 + y^3x\right]_{-2}^{3}$$

$$= 9 - 9y^2 + 3y^3 - \left(-\frac{8}{3} - 4y^2 - 2y^3\right)$$

$$= \frac{35}{3} - 5y^2 + 5y^3.$$

Therefore

$$\int_1^4 \int_{-2}^{3} (x^2 - 2xy^2 + y^3)\,dx\,dy = \int_1^4 \left(\frac{35}{3} - 5y^2 + 5y^3\right) dy$$

$$= \left[\frac{35}{3}y - \frac{5}{3}y^3 + \frac{5}{4}y^4\right]_1^4 = \frac{995}{4}.$$

Iterated integrals may be defined over regions F which have curved boundaries. This situation is more complicated than the one we just discussed. Consider a region F such as that shown in Fig. 5-8, in which the boundary consists of the lines $x = a$, $x = b$, and the graphs of the functions $p(x)$ and $q(x)$ with $p(x) \leq q(x)$ for $a \leq x \leq b$. We may define

$$\int_a^b \int_{p(x)}^{q(x)} f(x, y)\,dy\,dx,$$

in which we first integrate (for fixed x) from the lower curve to the upper curve, i.e., along a typical line as shown in Fig. 5-8; then we integrate with respect to x over all such typical segments from a to b.

More generally, iterated integrals may be defined over a region F such as the one shown in Fig. 5-9. Integrating first with respect to y, we have

$$\int_a^b \int_{p(x)}^{q(x)} f(x, y)\,dy\,dx.$$

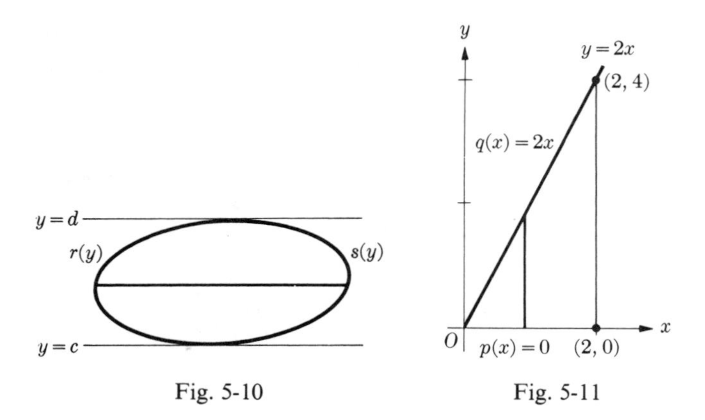

Fig. 5-10 Fig. 5-11

On the other hand, the integral taken first with respect to x requires that we represent F as shown in Fig. 5-10. Then we have

$$\int_c^d \int_{r(y)}^{s(y)} f(x, y)\, dx\, dy.$$

EXAMPLE 2. Given the function $f(x, y) = xy$ and the triangular region F bounded by the lines $y = 0$, $y = 2x$, $x = 2$ (Fig. 5-11), find the value of both iterated integrals.

SOLUTION. Referring to Fig. 5-11, we see that for

$$\int_a^b \int_{p(x)}^{q(x)} xy\, dy\, dx,$$

we have $p(x) = 0$, $q(x) = 2x$, $a = 0$, $b = 2$. Therefore

$$\int_0^2 \int_0^{2x} xy\, dy\, dx = \int_0^2 \left[\frac{1}{2}xy^2\right]_0^{2x} dx$$
$$= \int_0^2 2x^3\, dx = \left[\frac{1}{2}x^4\right]_0^2 = 8.$$

Integrating with respect to x first (Fig. 5-12), we have

$$\int_c^d \int_{r(y)}^{s(y)} xy\, dx\, dy \qquad \text{with} \qquad r(y) = \frac{1}{2}y, \quad s(y) = 2, \quad c = 0, \quad d = 4.$$

Therefore

$$\int_0^4 \int_{y/2}^2 xy\, dx\, dy = \int_0^4 \left[\frac{1}{2}x^2 y\right]_{y/2}^2 dy$$
$$= \int_0^4 \left(2y - \frac{1}{8}y^3\right) dy = \left[y^2 - \frac{1}{32}y^4\right]_0^4 = 8.$$

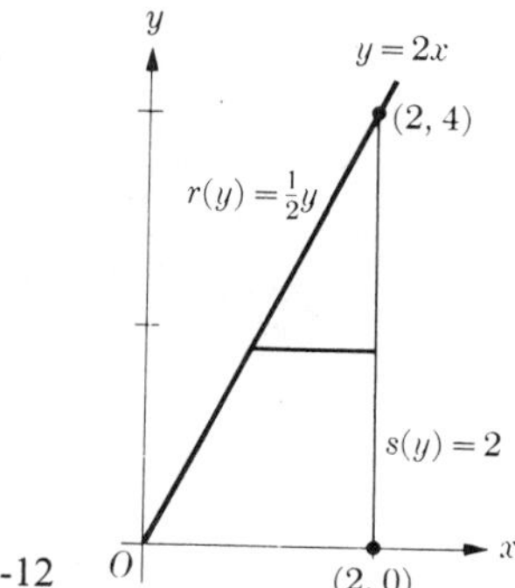

Fig. 5-12

It is not accidental that the two integrals in Example 2 have the same value. The next theorem describes the general situation.

Theorem 7. *Suppose that F is a region given by*

$$F = \{(x, y) : a \leq x \leq b, \quad p(x) \leq y \leq q(x)\},$$

where p and q are continuous and $p(x) \leq q(x)$ *for* $a \leq x \leq b$. *Suppose that* $f(x, y)$ *is continuous on F. Then*

$$\iint_F f(x, y)\, dA = \int_a^b \int_{p(x)}^{q(x)} f(x, y)\, dy\, dx.$$

The corresponding result holds if the closed region F has the representation

$$F = \{(x, y) : c \leq y \leq d, \quad r(y) \leq x \leq s(y)\}$$

where $r(y) \leq s(y)$ *for* $c \leq y \leq d$. *In such a case,*

$$\iint_F f(x, y)\, dA = \int_c^d \int_{r(y)}^{s(y)} f(x, y)\, dx\, dy.$$

In other words, both iterated integrals, when computable, are equal to the double integral and therefore equal to each other.

PARTIAL PROOF. We shall discuss the first result, the second being similar. Suppose first that $f(x, y)$ is positive. A plane $x = \text{const}$ intersects the surface $z = f(x, y)$ in a curve (Fig. 5-13). The area under this curve in the $x = \text{const}$ plane is shown as a shaded region. Denoting the area of this region by $A(x)$, we have the formula

$$A(x) = \int_{p(x)}^{q(x)} f(x, y)\, dy.$$

It can be shown that $A(x)$ is continuous. Furthermore, it can also be shown that if $A(x)$ is integrated between $x = a$ and $x = b$, the volume V under the surface $f(x, y)$ is swept out. The double integral yields the volume under the

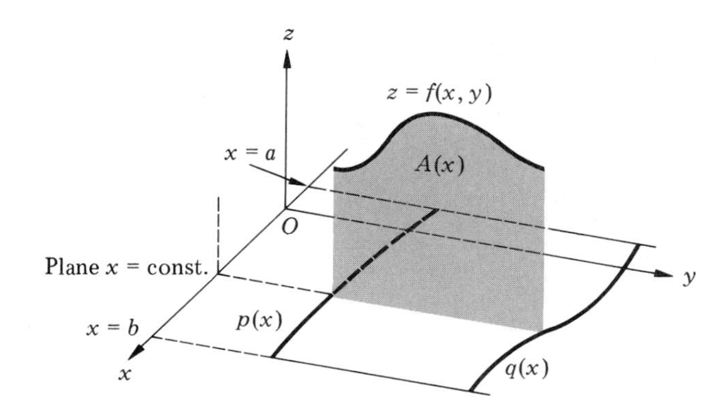

Fig. 5-13

surface, and so we write

$$V = \iint_F f(x, y)\, dA.$$

On the other hand, we obtain the volume by integrating $A(x)$; that is,

$$V = \int_a^b A(x)\, dx = \int_a^b \int_{p(x)}^{q(x)} f(x, y)\, dy\, dx.$$

If $f(x, y)$ is not positive but is bounded from below by the plane $z = c$, then subtraction of the volume of the cylinder of height c and cross-section F leads to the same result.

REMARKS. We have considered two ways of expressing a region F in the xy plane. They are

$$F = \{(x, y) : a \leq x \leq b, \quad p(x) \leq y \leq q(x)\} \tag{3}$$

and

$$F = \{(x, y) : c \leq y \leq d, \quad r(y) \leq x \leq s(y)\}. \tag{4}$$

It frequently happens that a region F is expressible more simply in one of the above forms than in the other. In doubtful cases, a sketch of F may show which is simpler and, therefore, which of the iterated integrals is evaluated more easily.

A region F may not be expressible in either the form (3) or the form (4). In such cases, F may sometimes be subdivided into a number of regions, each having one of the two forms. The integrations are then performed for each subregion and the results added. Figure 5-14 gives examples of how the subdivision process might take place.

EXAMPLE 3. Evaluate $\iint_F x^2 y^2\, dA$ where F is the figure bounded by the lines $y = 1$, $y = 2$, $x = 0$, and $x = y$ (Fig. 5-15).

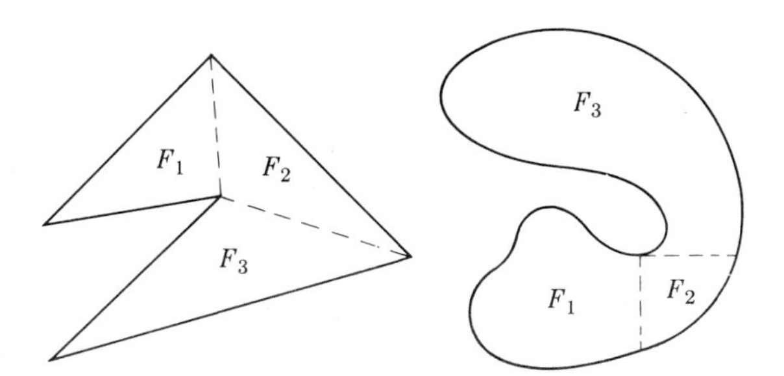

Fig. 5-14

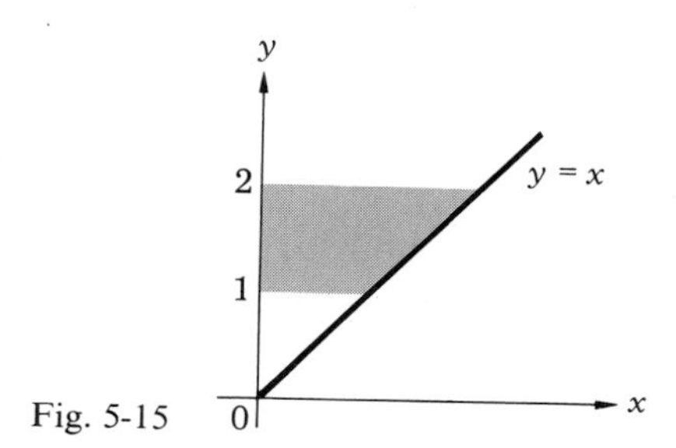

Fig. 5-15

SOLUTION. The region F is the set

$$F = \{(x, y) : 1 \le y \le 2, \quad 0 \le x \le y\}.$$

We use Theorem 7 and evaluate the iterated integral, to find

$$\iint_F x^2y^2\,dA = \int_1^2 \int_0^y x^2y^2\,dx\,dy = \int_1^2 \left[\frac{1}{3}x^3y^2\right]_0^y dy$$

$$= \frac{1}{3}\int_1^2 y^5\,dy = \frac{7}{2}.$$

Note that in the above example the iterated integral in the other order is a little more difficult, since the curves $p(x)$, $q(x)$ are

$$p(x) = \begin{Bmatrix} 1 \text{ for } 0 \le x \le 1 \\ x \text{ for } 1 \le x \le 2 \end{Bmatrix}, \qquad q(x) = 2, \quad 0 \le x \le 2.$$

The evaluation would have to take place in two parts, so that

$$\iint_F x^2y^2\,dA = \int_0^1 \int_1^2 x^2y^2\,dy\,dx + \int_1^2 \int_x^2 x^2y^2\,dy\,dx.$$

EXAMPLE 4. Evaluate

$$\int_0^2 \int_0^{x^2/2} \frac{x}{\sqrt{1 + x^2 + y^2}}\,dy\,dx.$$

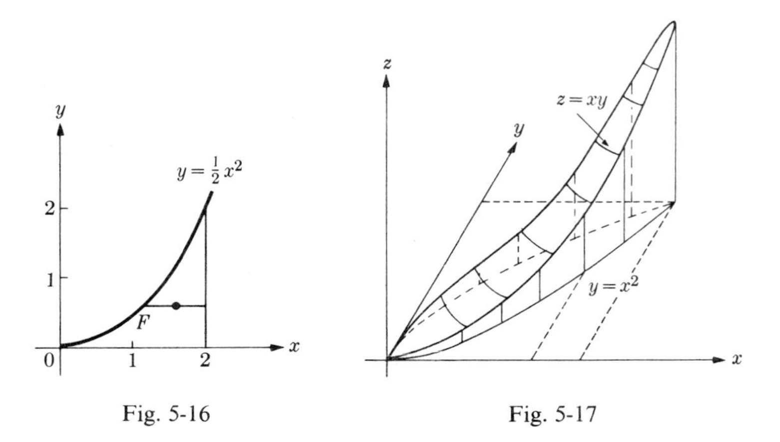

Fig. 5-16 Fig. 5-17

SOLUTION. Carrying out the integration first with respect to y is possible but difficult and leads to a complicated integral for x. Therefore we shall try to express the integral as an iterated integral in the opposite order and use Theorem 7. We construct the region F as shown in Fig. 5-16. The region is expressed by

$$F = \{(x, y) : 0 \le x \le 2 \quad \text{and} \quad 0 \le y \le \tfrac{1}{2}x^2\}.$$

However, it is also expressed by

$$F = \{(x, y) : 0 \le y \le 2 \quad \text{and} \quad \sqrt{2y} \le x \le 2\}.$$

Therefore, integrating with respect to x first, we have

$$\begin{aligned}\int_0^2 \int_0^{x^2/2} \frac{x}{\sqrt{1 + x^2 + y^2}}\, dy\, dx &= \iint_F \frac{x}{\sqrt{1 + x^2 + y^2}}\, dA = \int_0^2 \int_{\sqrt{2y}}^2 \frac{x}{\sqrt{1 + x^2 + y^2}}\, dx\, dy \\ &= \int_0^2 [\sqrt{1 + x^2 + y^2}]_{\sqrt{2y}}^2\, dy = \int_0^2 [\sqrt{5 + y^2} - (1 + y)]\, dy \\ &= [\tfrac{5}{2}\log(y + \sqrt{y^2 + 5}) + \tfrac{1}{2}y\sqrt{y^2 + 5} - y - \tfrac{1}{2}y^2]_0^2 \\ &= \tfrac{5}{2}\log 5 + 3 - 4 - \tfrac{5}{2}\log\sqrt{5} = -1 + \tfrac{5}{4}\log 5.\end{aligned}$$

The next example shows how the volume of a region in R^3 may be found by iterated integration.

EXAMPLE 5. Let S be the region bounded by the surface $z = xy$, the cylinders $y = x^2$ and $y^2 = x$, and the plane $z = 0$. Find the volume $V(S)$.

SOLUTION. The region S is shown in Fig. 5-17. It consists of all points "under" the surface $z = xy$, bounded by the cylinders, and "above" the xy plane.

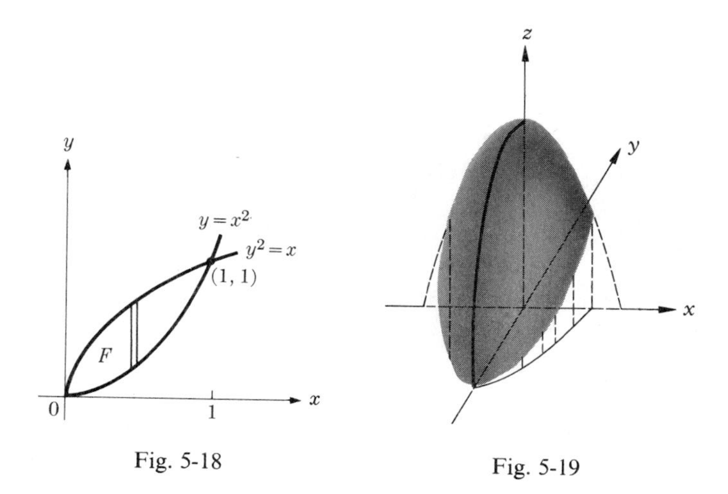

Fig. 5-18

Fig. 5-19

The region F in the xy plane is bounded by the curves $y = x^2$, $y^2 = x$ and is shown in Fig. 5-18. Therefore

$$V(S) = \iint_F xy\, dA = \int_0^1 \int_{x^2}^{\sqrt{x}} xy\, dy\, dx = \int_0^1 \left[\frac{xy^2}{2}\right]_{x^2}^{\sqrt{x}} dx$$

$$= \frac{1}{2}\int_0^1 (x^2 - x^5)\, dx = \frac{1}{12}.$$

If a region S is bounded by two surfaces of the form $z = f(x, y)$ and $z = g(x, y)$ with $f(x, y) \le g(x, y)$, then the volume between the surfaces may be found as a double integral, and that integral in turn may be evaluated by iterated integrals. The region F over which the integration is performed is found by the projection onto the xy plane of the curve of intersection of the two surfaces. To find this projection we merely set

$$f(x, y) = g(x, y)$$

and trace this curve in the xy plane. The next example shows the method.

EXAMPLE 6. Find the volume bounded by the surfaces

$$z = x^2$$

and

$$z = 4 - x^2 - y^2.$$

SOLUTION. A portion of the region S (the part corresponding to $y \le 0$) is shown in Fig. 5-19. We set

$$x^2 = 4 - x^2 - y^2$$

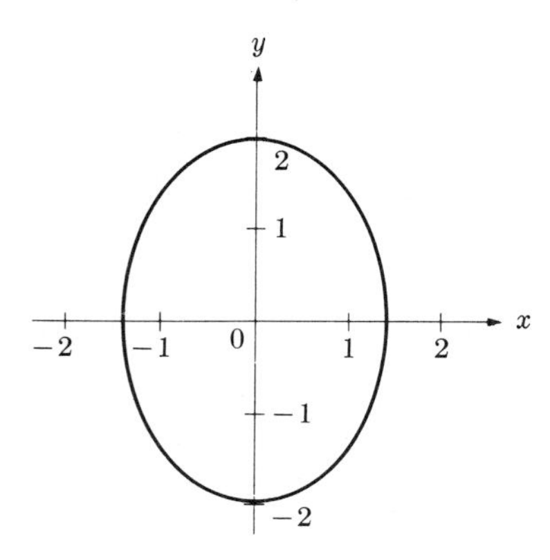

Fig. 5-20

and find that the region F in the xy plane is the elliptical disk (Fig. 5-20)

$$F = \left\{(x, y) : \frac{x^2}{2} + \frac{y^2}{4} \leq 1\right\}.$$

Note that the surface $z = 4 - x^2 - y^2 \equiv g(x, y)$ is above the surface $z = x^2 \equiv f(x, y)$ for (x, y) inside the above ellipse. Therefore

$$\begin{aligned}
V(S) &= \iint_F (4 - y^2 - x^2 - x^2)\, dA \\
&= \int_{-2}^{2} \int_{-\sqrt{4-y^2}/\sqrt{2}}^{+\sqrt{4-y^2}/\sqrt{2}} (4 - y^2 - 2x^2)\, dx\, dy \\
&= \int_{-2}^{2} \frac{2\sqrt{2}}{3}(4 - y^2)^{3/2}\, dy = \frac{4\sqrt{2}}{3} \int_0^2 (4 - y^2)^{3/2}\, dy \\
&= \frac{64\sqrt{2}}{3} \int_0^{\pi/2} \cos^4 \theta\, d\theta = \frac{16\sqrt{2}}{3} \int_0^{\pi/2} (1 + 2\cos 2\theta + \cos^2 2\theta)\, d\theta \\
&= \frac{8\pi\sqrt{2}}{3} + \left[\frac{16\sqrt{2}}{3} \sin 2\theta\right]_0^{\pi/2} + \frac{8\sqrt{2}}{3} \int_0^{\pi/2} (1 + \cos 4\theta)\, d\theta = 4\pi\sqrt{2}.
\end{aligned}$$

Problems

In Problems 1 through 10, evaluate the iterated integrals as indicated. Sketch the region F in the xy plane over which the integration is taken.

1. $\displaystyle\int_1^4 \int_2^5 (x^2 - y^2 + xy - 3)\, dx\, dy$
2. $\displaystyle\int_0^2 \int_{-3}^{2} (x^3 + 2x^2 y - y^3 + xy)\, dy\, dx$
3. $\displaystyle\int_1^4 \int_{\sqrt{x}}^{x^2} (x^2 + 2xy - 3y^2)\, dy\, dx$

4. $\int_0^1 \int_{x^3}^{x^2} (x^2 - xy)\,dy\,dx$

5. $\int_2^3 \int_{1+y}^{\sqrt{y}} (x^2 y + xy^2)\,dx\,dy$

6. $\int_{-2}^{2} \int_{-\sqrt{4-x^2}}^{+\sqrt{4-x^2}} y\,dy\,dx$

7. $\int_{-3}^{3} \int_{-\sqrt{18-2y^2}}^{+\sqrt{18-2y^2}} x\,dx\,dy$

8. $\int_{-3}^{3} \int_{x^2}^{18-x^2} xy^3\,dy\,dx$

9. $\int_0^2 \int_{x^2}^{2x^2} x \cos y\,dy\,dx$

10. $\int_1^2 \int_{x^3}^{4x^3} \frac{1}{y}\,dy\,dx$

In Problems 11 through 17, evaluate the double integrals as indicated. Sketch the region F.

11. $\iint_F (x^2 + y^2)\,dA; \quad F = \{(x, y) : y^2 \le x \le 4, \quad 0 \le y \le 2\}$

12. $\iint_F x \cos y\,dA; \quad F = \{(x, y) : 0 \le x \le \sqrt{\pi/2}, \quad 0 \le y \le x^2\}$

13. $\iint_F \frac{x}{x^2 + y^2}\,dA; \quad F = \{(x, y) : 1 \le x \le \sqrt{3}, \quad 0 \le y \le x\}$

14. $\iint_F \log y\,dA; \quad F = \{(x, y) : 2 \le x \le 3, \quad 1 \le y \le x - 1\}$

15. $\iint_F \frac{x}{\sqrt{1 - y^2}}\,dA; \quad F = \{(x, y) : 0 \le x \le \frac{1}{2}, \quad x \le y \le \frac{1}{2}\}$

16. $\iint_F \frac{x}{\sqrt{x^2 + y^2}}\,dA; \quad F = \{(x, y) : 1 \le x \le 2, \quad 1 \le y \le x\}$

17. $\iint_F \frac{1}{y^2} e^{x/\sqrt{y}}\,dA; \quad F = \{(x, y) : 1 \le x \le \sqrt{2}, \quad x^2 \le y \le 2\}$

In each of Problems 18 through 22, (a) sketch the domain F over which the integration is performed; (b) write the equivalent iterated integral in the reverse order; (c) evaluate the integral obtained in (b). Describe F using set notation (in both orders).

18. $\int_1^2 \int_1^x \frac{x^2}{y^2}\,dy\,dx$

19. $\int_{-2}^{2} \int_{-\sqrt{4-x^2}}^{+\sqrt{4-x^2}} xy\,dy\,dx$

20. $\int_0^a \int_0^{\sqrt{a^2-x^2}} (a^2 - y^2)^{3/2}\,dy\,dx$

21. $\int_0^1 \int_y^1 \sqrt{1 + x^2}\,dx\,dy$

22. $\int_0^1 \int_{\sqrt{x}}^1 \sqrt{1 + y^3}\,dy\,dx$

In each of Problems 23 through 32, find the volume $V(S)$ of the region S described. Sketch the domain F of integration and describe it using set notation.

23. S is bounded by the surfaces $z = 0$, $z = x$, and $y^2 = 2 - x$.
24. S is bounded by the planes $z = 0$, $y = 0$, $y = x$, $x + y = 2$, and $x + y + z = 3$.
25. S is bounded by the surfaces $x = 0$, $z = 0$, $y^2 = 4 - x$, and $z = y + 2$.
26. S is bounded by the surfaces $x^2 + z^2 = 4$, $y = 0$, and $x + y + z = 3$.
27. S is bounded by the surfaces $y^2 = z$, $y = z^3$, $z = x$, and $y^2 = 2 - x$.
28. S is bounded by the coordinate planes and the surface $x^{1/2} + y^{1/2} + z^{1/2} = a^{1/2}$.
29. S is bounded by the surfaces $y = x^2$ and $z^2 = 4 - y$.
30. S is bounded by the surfaces $y^2 = x$, $x + y = 2$, $x + z = 0$, and $z = x + 1$.
31. S is bounded by the surfaces $x^2 = y + z$, $y = 0$, $z = 0$, and $x = 2$.
32. S is bounded by the surfaces $y^2 + z^2 = 2x$ and $y = x - \frac{3}{2}$.

*33. Let $F = \{(x, y) : 0 \le x \le 1, \quad 0 \le y \le 1\}$. We define

$$f(x, y) = \begin{cases} 1 & \text{if } y \text{ is rational,} \\ 0 & \text{otherwise.} \end{cases}$$

Show that $\iint_F f(x, y)\, dA$ does not exist.

*34. Let $F = \{(x, y) : 0 \le x \le 1, \quad 0 \le y \le 1\}$. We define

$$f(x, y) = \begin{cases} 0 & \text{if } x = 1/n, \quad n = 1, 2, 3, \ldots, \\ 1 & \text{otherwise.} \end{cases}$$

Show that $\iint_F f(x, y)\, dA$ exists and has the value 1.

35. Let $F = \{(x, y) : a \le x \le b, c \le y \le d\}$ and suppose $f(x, y) = g(x) \cdot h(y)$. Show that

$$\iint_F f(x, y)\, dA = \left(\int_a^b g(x)\, dx\right)\left(\int_c^d h(y)\, dy\right).$$

*36. Suppose that $f(x, y)$ is continuous on a closed, bounded region F. Assume that

$$\iint_F f(x, y)\phi(x, y)\, dA = 0$$

for *all* functions $\phi(x, y)$ which are continuous on F. Show that $f(x, y) \equiv 0$ on F. [*Hint:* Assume there is a point in F where f is positive, choose ϕ carefully, and then use Theorem 4 with $m > 0$ thereby reaching a contradiction.]

37. Find the volume of the ellipsoid

$$F = \left\{(x, y, z) : \frac{x^2}{a^2} + \frac{y^2}{b^2} + \frac{z^2}{c^2} - 1 = 0\right\}.$$

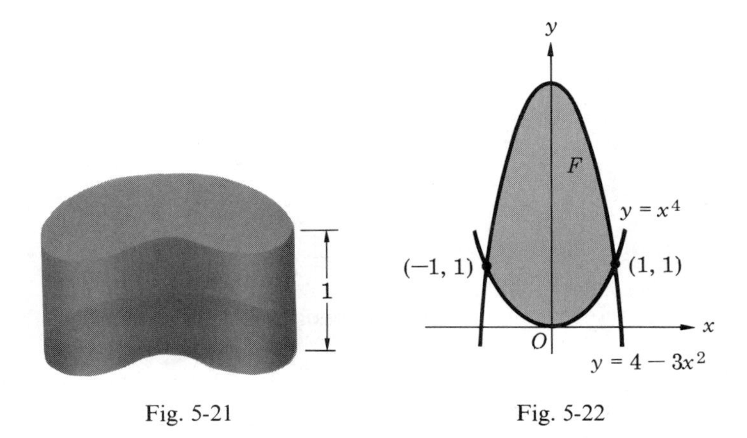

Fig. 5-21

Fig. 5-22

4. Area, Density, and Mass

The double integral of a nonnegative function $z = f(x, y)$ taken over a region F may be intepreted as a volume. The value of such an integral is the volume of the cylinder having generators parallel to the z axis and situated between the surface $z = f(x, y)$ and the region F in the xy plane.

If we select for the surface f the particularly simple function $z = 1$, then the volume V is given by the formula

$$V = \iint_F 1 \, dA.$$

On the other hand, the volume of a right cylinder of cross section F and height 1 is

$$V = A(F) \cdot 1.$$

(See Fig. 5-21.) Therefore

$$A(F) = \iint_F dA.$$

We see that *the double integral of the function* 1 *taken over F is precisely the area of F.* By Theorem 7, we conclude that the iterated integral of the function 1 also yields the area of F.

EXAMPLE 1. Use iterated integration to find the area of the region F given by

$$F = \{(x, y) : -1 \leq x \leq 1, \quad x^4 \leq y \leq 4 - 3x^2\}.$$

SOLUTION. The region F is shown in Fig. 5-22. One of the iterated integrals for the area is

$$A(F) = \int_{-1}^{1} \int_{x^4}^{4-3x^2} dy\,dx,$$

and its evaluation gives

$$A(F) = \int_{-1}^{1} [y]_{x^4}^{4-3x^2}\,dx = \int_{-1}^{1} (4 - 3x^2 - x^4)\,dx$$

$$= \left[4x - x^3 - \frac{1}{5}x^5\right]_{-1}^{1} = \frac{28}{5}.$$

Note that the iterated integral in the other direction is more difficult to evaluate.

If a flat object is made of an extremely thin uniform material, then the mass of the object is just a multiple of the area of the plane region on which the object rests. (The multiple depends on the units used.) If a thin object (resting on the xy plane) is made of a nonuniform material, then the mass of the object may be expressed in terms of the density $\rho(x, y)$ of the material at any point. We assume that the material is uniform in the z direction. Letting F denote the region occupied by the object, we decompose F into rectangles $r_1, r_2, \ldots, r_n$ in the usual way. Then an approximation to the mass of the ith rectangle is given by

$$\rho(\xi_i, \eta_i)A(r_i),$$

where $A(r_i)$ is the area of r_i and (ξ_i, η_i) is a point in r_i. The total mass of F is approximated by

$$\sum_{i=1}^{n} \rho(\xi_i, \eta_i)A(r_i),$$

and when we proceed to the limit in the customary manner, the mass $M(F)$ is

$$M(F) = \iint_F \rho(x, y)\,dA$$

In other words, the double integral is a useful device for finding the mass of a thin object with variable density.

Example 2. A thin object occupies the region

$$F = \{(x, y) : y^2 \leq x \leq 4 - y^2, \quad -\sqrt{2} \leq y \leq \sqrt{2}\}.$$

The density is given by $\rho(x, y) = 1 + 2x + y$. Find the total mass.

Solution. We have

$$M(F) = \iint_F (1 + 2x + y)\,dA.$$

Sketching the region F (Fig. 5-23), we obtain for $M(F)$ the iterated integral

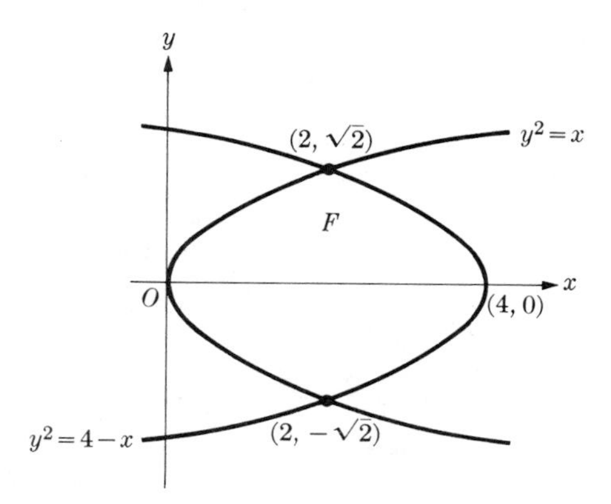

Fig. 5-23

$$\begin{aligned} M(F) &= \int_{-\sqrt{2}}^{\sqrt{2}} \int_{y^2}^{4-y^2} (1 + 2x + y)\,dx\,dy \\ &= \int_{-\sqrt{2}}^{\sqrt{2}} [x + x^2 + xy]_{y^2}^{4-y^2}\,dy \\ &= \int_{-\sqrt{2}}^{\sqrt{2}} (20 + 4y - 10y^2 - 2y^3)\,dy \\ &= \left[20y + 2y^2 - \frac{10}{3}y^3 - \frac{1}{2}y^4\right]_{-\sqrt{2}}^{\sqrt{2}} = \frac{80}{3}\sqrt{2}. \end{aligned}$$

PROBLEMS

In each of Problems 1 through 5 use iterated integration to find the area of the given region F. Subdivide F and do each part separately whenever necessary.

1. $F = \{(x, y) : 0 \le x \le 1, \quad x^3 \le y \le \sqrt{x}\}$.
2. $F = \{(x, y) : \frac{1}{4}y^2 \le x \le y, \quad 1 \le y \le 4\}$.
3. F is determined by the inequalities
$$xy \le 4, \qquad y \le x, \qquad 27y \ge 4x^2.$$
4. F consists of all (x, y) which satisfy the inequalities
$$y^2 \le x, \qquad y^2 \le 6 - x, \qquad y \le x - 2.$$
5. F consists of all (x, y) which satisfy the inequalities
$$x^2 + y^2 \le 9, \qquad y \le x + 3, \qquad x + y \le 0.$$

In each of Problems 6 through 14, find the mass of the given region F. Draw a sketch of F.

6. $F = \{(x, y) : x^2 + y^2 \le 64\}; \quad \rho = x^2 + y^2$.
7. $F = \{(x, y) : 0 \le x \le 1, \quad x^2 \le y \le \sqrt{x}\}; \quad \rho = 3y$.

8. $F = \{(x, y) : -1 \le x \le 2, \quad x^2 \le y \le x + 2\}; \quad \rho = x^2 y.$

9. $F = \{(x, y) : 0 \le x \le 1, \quad x^3 \le y \le \sqrt{x}\}; \quad \rho = 2x.$

10. $F = \left\{(x, y) : 1 \le x \le 4, \quad \dfrac{4}{x} \le y \le 5 - x\right\}; \quad \rho = 4y.$

11. $F = \{(x, y) : y^2 \le x \le y + 2, \quad -1 \le y \le 2\}; \quad \rho = x^2 y^2.$

12. F is the interior of the triangle with vertices at $(0, 0)$, $(a, 0)$, (b, c), $a > b > 0$, $c > 0$; $\rho = 2x$.

13. $F = \{(x, y) : -a \le x \le a, \quad 0 \le y \le \sqrt{a^2 - x^2}\}; \quad \rho = 3y.$

14. F is the interior of the rectangle with vertices at $(0, 0)$, $(a, 0)$, (a, b), $(0, b)$; $\rho = 3x/(1 + x^2 y^2)$.

15. Suppose the density $\rho(x, y)$ of a region F satisfies the inequalities $m_1 \le \rho(x, y) \le m_2$. Show that $M(F)$, the mass of F, is between the limits $m_1 A \le M(F) \le m_2 A$, where A is the area of F.

*16. Suppose that $\rho(x, y)$, the density, is continuous on a region F which has positive area. Show that if $\iint_F \rho(x, y)\, dA = 0$, then $\rho(x, y) \equiv 0$ on F.

*17. Suppose that the density $\rho(x, y)$ of a region F in the xy plane is of the form $\rho(x, y) = \rho_1(x)\rho_2(y)$. Let $M(F)$ be the total mass of F. Show that

$$M(F) \le \tfrac{1}{2}(M_1(F) + M_2(F))$$

where M_1 is the mass of F with density ρ_1^2 and M_2 is the mass of F with density ρ_2^2.

5. Evaluation of Double Integrals by Polar Coordinates

The polar coordinates (r, θ) of a point in the plane are related to the rectangular coordinates (x, y) of the same point by the equations

$$x = r\cos\theta, \qquad y = r\sin\theta, \qquad r \ge 0. \tag{1}$$

In the calculus of one variable we saw that certain problems concerned with finding areas by integration are solved more easily in polar coordinates than in rectangular coordinates. The same situation prevails in problems involving double integration.

Instead of considering (1) as a means of representing a point in two different coordinate systems, we interpret the equations as a mapping between the xy plane and the $r\theta$ plane. We draw the $r\theta$ plane as shown in Fig. 5-24, treating $r = 0$ and $\theta = 0$ as perpendicular straight lines. A rectangle G in the $r\theta$ plane bounded by the lines $r = r_1$, $r = r_2$, and $\theta = \theta_1$, $\theta = \theta_2$ (θ in radians) with $2\pi > \theta_2 > \theta_1 \ge 0$, $r_2 > r_1 > 0$ has an image F in the xy plane bounded by two circular arcs and two rays. For the area, denoted $A_{x,y}$, of F we have

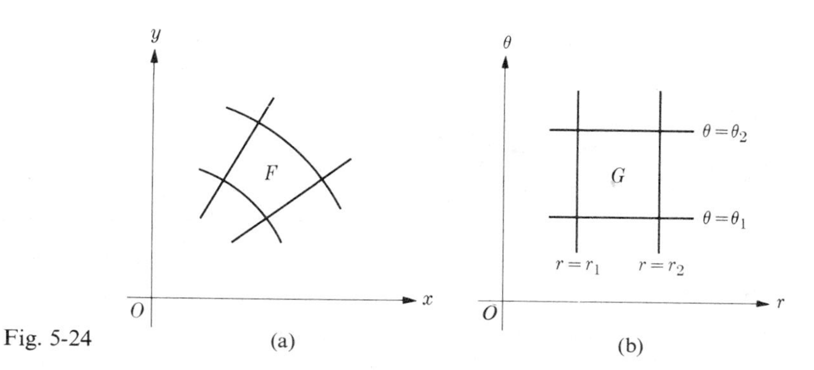

Fig. 5-24

$$A_{x,y}(F) = \tfrac{1}{2}(r_2^2 - r_1^2)(\theta_2 - \theta_1).$$

This area may be written as an iterated integral. A simple calculation shows that

$$A_{x,y}(F) = \int_{\theta_1}^{\theta_2} \left[\int_{r_1}^{r_2} r\,dr \right] d\theta.$$

Because double integrals and iterated integrals are equivalent for evaluation purposes, we can also write

$$A_{x,y}(F) = \iint_G r\,dA_{r,\theta}, \tag{2}$$

where $dA_{r,\theta}$ is an element of area in the $r\theta$ plane, r and θ being treated as rectangular coordinates. That is, $dA_{r,\theta} = dr\,d\theta$.

More generally, it can be shown that if G is *any region* in the $r\theta$ plane and F is its image under the transformation (1), then the area of F may be found by formula (2). Thus, areas of regions may be determined by expressing the double integral in polar coordinates as in (2) and then evaluating the double integral by iterated integrals in the usual way.

EXAMPLE 1. A region F above the x axis is bounded on the left by the line $y = -x$, and on the right by the curve

$$C = \{(x, y) : x^2 + y^2 = 3\sqrt{x^2 + y^2} - 3x\},$$

as shown in Fig. 5-25. Find its area.

SOLUTION. We employ polar coordinates to describe the region. The curve C is the cardioid $r = 3(1 - \cos\theta)$, and the line $y = -x$ is the ray $\theta = 3\pi/4$. The region F in Fig. 5-25 is the image under the mapping (1) of the set G in the (r, θ) plane (Fig. 5-26), given by

$$G = \{(r, \theta) : 0 \le r \le 3(1 - \cos\theta), \quad 0 \le \theta \le 3\pi/4\}.$$

Therefore, for the area $A(F)$ we obtain

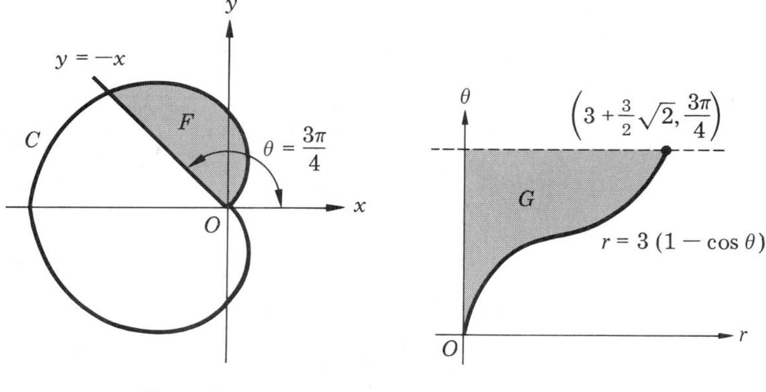

Fig. 5-25

Fig. 5-26

$$A(F) = \iint_F dA_{x,y} = \iint_G r\,dA_{r,\theta} = \int_0^{3\pi/4}\int_0^{3(1-\cos\theta)} r\,dr\,d\theta$$
$$= \int_0^{3\pi/4} \frac{1}{2}\Big[r^2\Big]_0^{3(1-\cos\theta)} d\theta = \frac{9}{2}\int_0^{3\pi/4}(1-\cos\theta)^2\,d\theta.$$

To perform the integration we multiply out and find that

$$A(F) = \frac{9}{2}\int_0^{3\pi/4}(1 - 2\cos\theta + \cos^2\theta)\,d\theta$$
$$= \frac{9}{2}\left[\theta - 2\sin\theta + \frac{1}{2}\theta + \frac{1}{4}\sin 2\theta\right]_0^{3\pi/4}$$
$$= \frac{9}{8}\left(\frac{9}{2}\pi - 4\sqrt{2} - 1\right).$$

The transformation of regions from the xy plane to the $r\theta$ plane is useful because general double integrals as well as areas may be evaluated by means of polar coordinates. The theoretical basis for the method is the Fundamental Lemma on Integration which we state without proof.

Theorem 8 (Fundamental Lemma on Integration). *Assume that f and g are continuous on some region F. Then for each $\varepsilon > 0$ there is a $\delta > 0$ such that*

$$\left|\sum_{i=1}^{n} f_i g_i A(F_i) - \iint_F f(x,y)g(x,y)\,dA\right| < \varepsilon$$

for every subdivision $F_1, F_2, \ldots, F_n$ of F with norm less than δ and any numbers $f_1, f_2, \ldots, f_n, g_1, g_2, \ldots, g_n$ where each f_i and each g_i is any number between the minimum and maximum values of f and g, respectively, on F_i.

We sketch the proof of the next theorem which is based on the Fundamental Lemma of Integration.

Theorem 9. *Suppose F and G are regions related according to the mapping $x = r\cos\theta$, $y = r\sin\theta$, and $f(x, y)$ is continuous on F. Then the function $g(r, \theta) = f(r\cos\theta, r\sin\theta)$ is defined and continuous on G and*

$$\iint_F f(x, y)\, dA_{x,y} = \iint_G g(r, \theta) r\, dA_{r,\theta}.$$

Sketch of proof. Consider a subdivision of G into "figures" $G_1, \ldots, G_n$. (See the discussion of volume in Section 1.) Let (r_i, θ_i) be in G_i for each i, and let (ξ_i, η_i) and F_i be the respective images of (r_i, θ_i) and G_i. Then $(F_1, \ldots, F_n)$ is a subdivision of F. From the expression for area in the xy plane as an integral, we obtain

$$A_{x,y}(F_i) = \iint_{G_i} r\, dA_{r,\theta},$$

Using Theorem 4 concerning bounds for integrals, we obtain

$$\iint_{G_i} r\, dA_{r,\theta} = \bar{r}_i A_{r,\theta}(G_i),$$

where $\bar{r}_i$ is between the minimum and maximum of r on G_i. Thus

$$\sum_{i=1}^{n} f(\xi_i, \eta_i) A_{x,y}(F_i) = \sum_{i=1}^{n} g(r_i, \theta_i) \cdot \bar{r}_i \cdot A_{r,\theta}(G_i).$$

The theorem follows by letting the norms of the subdivisions tend to zero, using the fundamental lemma to evaluate the limit of the sum on the right.

In terms of iterated integrals, the result in Theorem 8 yields the useful formula

$$\boxed{\iint_F f(x, y)\, dA = \iint_G g(r, \theta) r\, dr\, d\theta.}$$

Of course, the integral on the right is equal to the iterated integral in the reverse order. In any specific case the order of integration will usually depend on the determination of the limits of integration. One way may be easier than the other. We give an example.

Example 2. Use polar coordinates to evaluate

$$\iint_F \sqrt{x^2 + y^2}\, dA_{x,y},$$

where F is the region inside the circle $x^2 + y^2 = 2x$.

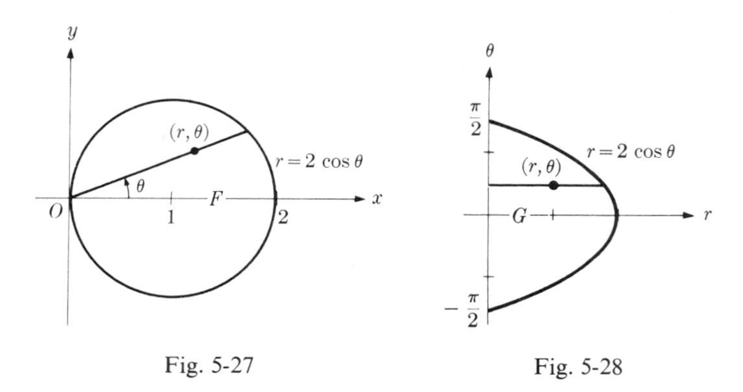

Fig. 5-27

Fig. 5-28

SOLUTION. F (see Fig. 5-27) is the image of the set G (Fig. 5-28) given by

$$G = \left\{(r, \theta): -\frac{\pi}{2} \leq \theta \leq \frac{\pi}{2}, \quad 0 \leq r \leq 2\cos\theta\right\}.$$

Therefore

$$\iint_F \sqrt{x^2 + y^2}\, dA_{x,y} = \iint_G r \cdot r\, dA_{r,\theta} = \int_{-\pi/2}^{\pi/2} \int_0^{2\cos\theta} r^2\, dr\, d\theta$$

$$= \int_{-\pi/2}^{\pi/2} \frac{8}{3} \cos^3\theta\, d\theta$$

$$= \frac{16}{3} \int_0^{\pi/2} (1 - \sin^2\theta) \cos\theta\, d\theta = \frac{32}{9}.$$

Although the construction of the region G in the $r\theta$ plane is helpful in understanding the transformation (1), it is not necessary for determining the limits in the iterated integrals. The limits of integration in polar coordinates may be found by using rectangular and polar coordinates in the same plane and using a sketch of the region F to read off the limits for r and θ.

Double integrals are useful for finding volumes bounded by surfaces. Cylindrical coordinates (r, θ, z) are a natural extension to three-space of polar coordinates in the plane. The z direction is selected as in rectangular coordinates, as shown in Fig. 5-29. If a closed surface in space is expressed in cylindrical coordinates, we may find the volume enclosed by this surface by evaluating a double integral in polar coordinates. An example illustrates the method.

EXAMPLE 3. A region S is bounded by the surfaces $x^2 + y^2 - 2x = 0$, $4z = x^2 + y^2$, $z^2 = x^2 + y^2$. Use cylindrical coordinates to find the volume $V(S)$.

SOLUTION. In cylindrical coordinates, the surface (a paraboloid) $4z = x^2 + y^2$ has equation $4z = r^2$; the cylinder $x^2 + y^2 - 2x = 0$ has equation $r = 2\cos\theta$;

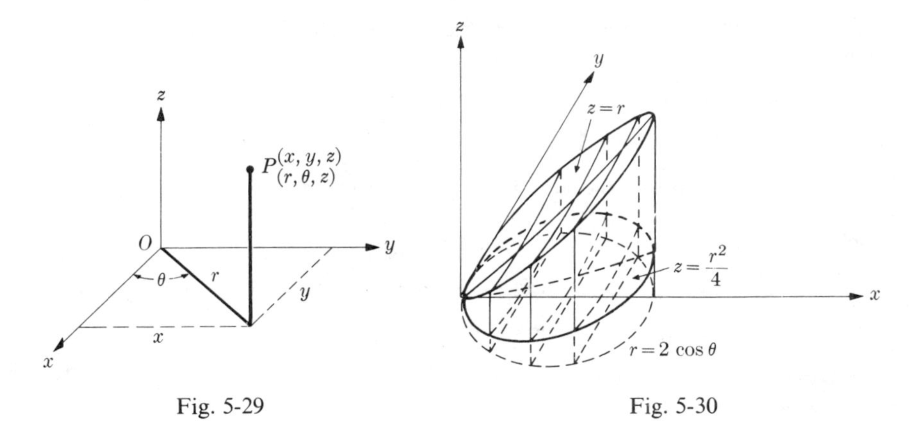

Fig. 5-29

Fig. 5-30

and the cone $z^2 = x^2 + y^2$ has equation $z^2 = r^2$. The region is shown in Fig. 5-30, and we note that the projection of S on the xy plane is precisely the plane region F of Example 2. We obtain

$$\begin{aligned} V(S) &= \iint_F \left[\sqrt{x^2+y^2} - \frac{x^2+y^2}{4}\right] dA_{x,y} \\ &= \iint_G \left(r - \frac{1}{4}r^2\right) r\, dA_{r,\theta} \\ &= \int_{-\pi/2}^{\pi/2}\int_0^{2\cos\theta} \left(r^2 - \frac{1}{4}r^3\right) dr\, d\theta \\ &= \int_{-\pi/2}^{\pi/2} \left(\frac{8}{3}\cos^3\theta - \cos^4\theta\right) d\theta \\ &= \frac{32}{9} - \frac{1}{2}\int_0^{\pi/2}\left(1 + 2\cos 2\theta + \frac{1+\cos 4\theta}{2}\right) d\theta = \frac{32}{9} - \frac{3\pi}{8}. \end{aligned}$$

Problems

In each of Problems 1 through 7, evaluate the given integral by first expressing it as a double integral and then changing to polar coordinates.

1. $\displaystyle\int_0^2 \int_0^{\sqrt{4-y^2}} \sqrt{x^2+y^2}\, dx\, dy$

2. $\displaystyle\int_{-2}^2 \int_{-\sqrt{4-x^2}}^{\sqrt{4-x^2}} e^{-(x^2+y^2)}\, dy\, dx$

3. $\displaystyle\int_{-\sqrt{\pi}}^{\sqrt{\pi}} \int_{-\sqrt{\pi-y^2}}^{\sqrt{\pi-y^2}} \sin(x^2+y^2)\, dx\, dy$

4. $\displaystyle\int_0^4 \int_{-\sqrt{4x-x^2}}^{\sqrt{4x-x^2}} \sqrt{x^2+y^2}\, dy\, dx$

5. $\displaystyle\int_{-2}^2 \int_{2-\sqrt{4-x^2}}^{2+\sqrt{4-x^2}} \sqrt{16-x^2-y^2}\, dy\, dx$

6. $\displaystyle\int_0^2 \int_0^x (x^2+y^2)\, dy\, dx$

7. $\displaystyle\int_0^1 \int_y^{\sqrt{y}} (x^2+y^2)^{-1/2}\, dx\, dy$

In each of Problems 8 through 10, use polar coordinates to find the area of the region given.

8. The region inside the circle $x^2 + y^2 - 8y = 0$ and outside the circle $x^2 + y^2 = 9$.

9. The region $F = \{(x, y) : \frac{1}{4}y^2 \leq x \leq 2y, \quad 0 \leq y \leq 8\}$.

10. The region interior to the curve $(x^2 + y^2)^3 = 16x^2$.

In each of Problems 11 through 26, find the volume of S.

11. S is the set bounded by the surfaces $z = 0$, $2z = x^2 + y^2$, and $x^2 + y^2 = 4$.

12. S is the set bounded by the cone $z^2 = x^2 + y^2$ and the cylinder $x^2 + y^2 = 4$.

13. S is the set cut from a sphere of radius 4 by a cylinder of radius 2 whose axis is a diameter of the sphere.

14. S is the set above the cone $z^2 = x^2 + y^2$ and inside the sphere
$$x^2 + y^2 + z^2 = a^2.$$

15. S is the set bounded by the cone $z^2 = x^2 + y^2$ and the cylinder
$$x^2 + y^2 - 2y = 0.$$

16. S is the set bounded by the sphere $x^2 + y^2 + z^2 = 4$ and the cylinder
$$x^2 + y^2 = 2x.$$

17. S is the set bounded by the cone $z^2 = x^2 + y^2$ and the paraboloid
$$3z = x^2 + y^2.$$

18. S is bounded by the surfaces $z = 0$, $2z = x^2 + y^2$, and $2y = x^2 + y^2$.

19. S is bounded by the cylinder $x^2 + y^2 = 4$ and the surface (a hyperboloid)
$$x^2 + y^2 - z^2 = 1.$$

20. S is bounded by the cone $z^2 = x^2 + y^2$ and the cylinder $r = 1 + \cos\theta$.

21. S is bounded by the surfaces $z = x$ and $2z = x^2 + y^2$.

22. S is bounded by the surfaces $z = 0$, $z = x^2 + y^2$, and $r = 2(1 + \cos\theta)$.

23. S is inside the sphere $x^2 + y^2 + z^2 = a^2$ and inside the cylinder erected on one loop of the curve $z = 0$, $r = a\cos 2\theta$.

24. S is inside the sphere $x^2 + y^2 + z^2 = 4$ and inside the cylinder erected on one loop of the curve $z = 0$, $r^2 = 4\cos 2\theta$.

25. S is bounded by the surfaces $z^2 = x^2 + y^2$, $y = 0$, $y = x$, and $x = a$.

*26. S is bounded by the surfaces $z^2 = x^2 + y^2$ and $x - 2z + 2 = 0$.

27. A wedge is cut from a spherical ball of radius C by two planes which intersect on a diameter of the ball. If the angle between the planes is $\pi/3$, find the volume of the wedge. What is the volume if the angle is ϕ?

28. A *torus* is generated by revolving a circular disk of radius a about an axis outside

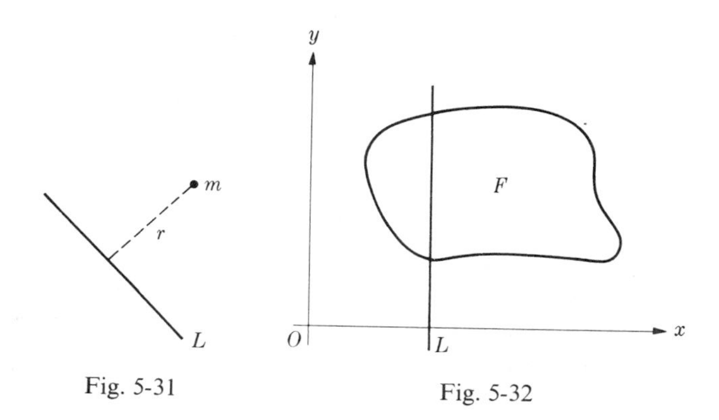

Fig. 5-31

Fig. 5-32

the disk. If the distance of the axis from the center of the disk is b, use polar coordinates to find the volume of the torus.

29. a) Suppose that $g(r, \theta) = g(r, -\theta)$ for all r, θ. Let F be a region which is symmetric with respect to the x axis. Let F_0 be the portion of F above the x axis. Show that

$$\iint_F g(r, \theta) r\, dr\, d\theta = 2 \iint_{F_0} g(r, \theta) r\, dr\, d\theta.$$

b) If $g(r, \theta) = -g(r, -\theta)$, show that $\iint_F g(r, \theta) r\, dr\, d\theta = 0$.

6. Moment of Inertia and Center of Mass

Consider the idealized situation in which an object of mass m occupies a single point. Let L be a line which we designate as an axis.

Definition. The **moment of inertia of a particle of mass m about the axis L** is mr^2, where r is the perpendicular distance of the object from the axis (Fig. 5-31). If we have a system of particles $m_1, m_2, \ldots, m_n$ at perpendicular distances, respectively, of $r_1, r_2, \ldots, r_n$ from the axis L, then the **moment of inertia of the system**, I, is given by

$$I = m_1 r_1^2 + m_2 r_2^2 + \cdots + m_n r_n^2 = \sum_{i=1}^{n} m_i r_i^2.$$

Let F be an object made of thin material occupying a region in the xy plane (Fig. 5-32). We wish to define the moment of inertia of F about an axis L. The axis L may be any line in three-dimensional space. We proceed as in the definition of integration. First we make a subdivision of the plane into rectangles or squares. We designate the rectangles either wholly or partly in F by $F_1, F_2, \ldots, F_n$. Since the object F may be of irregular shape and of variable density, the mass of the subregions may not be calculable exactly. We select a point in each subregion F_i and denote its coordinates (ξ_i, η_i). We assume that the entire mass of F_i, denoted $m(F_i)$, is concentrated at the point (ξ_i, η_i). Letting r_i be the perpendicular distance of (ξ_i, η_i) from the line L, we form the sum

$$\sum_{i=1}^{n} m(F_i) r_i^2.$$

Definition. If the above sums tend to a limit (called I) as the norms of the subdivisions tend to zero, and if this limit is independent of the manner in which the (ξ_i, η_i) are selected within the F_i, then we say that I is the **moment of inertia of the mass distribution about the axis L.**

The above definition of moment of inertia leads in a natural way to the next theorem.

Theorem 10. *Given a mass distribution occupying a region F in the xy plane and having a continuous density $\rho(x, y)$. Then the moment of inertia about the y axis (denoted by I_1) is given by*

$$I_1 = \iint_F x^2 \rho(x, y)\, dA.$$

Similarly, the moments of inertia about the x axis and the z axis are, respectively,

$$I_2 = \iint_F y^2 \rho(x, y)\, dA, \qquad I_3 = \iint_F (x^2 + y^2) \rho(x, y)\, dA.$$

The proof depends on the Fundamental Lemma on Integration. (See Problem 29 at the end of this section.)

Corollary. *The moments of inertia of F about the lines*

$$L_1 = \{(x, y, z) : x = a, \quad z = 0\}, \qquad L_2 = \{(x, y, z) : y = b, \quad z = 0\},$$

$$L_3 = \{(x, y, z) : x = a, \quad y = b\}$$

are, respectively,

$$I_1^a = \iint_F (x - a)^2 \rho(x, y)\, dA,$$

$$I_2^b = \iint_F (y - b)^2 \rho(x, y)\, dA,$$

$$I_3^{a,b} = \iint_F [(x - a)^2 + (y - b)^2] \rho(x, y)\, dA.$$

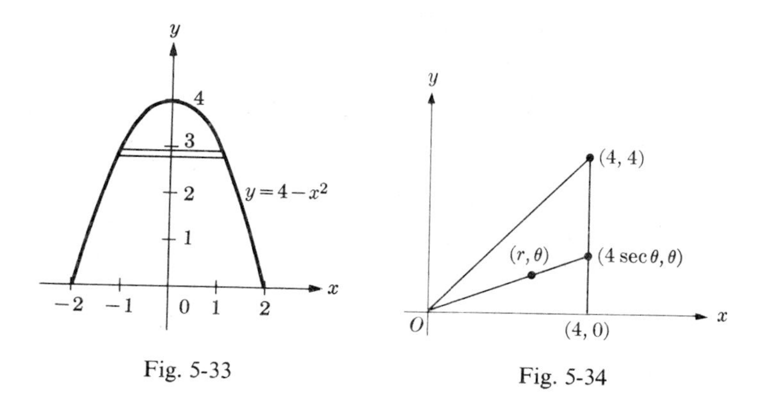

Fig. 5-33

Fig. 5-34

EXAMPLE 1. Find the moment of inertia about the x axis of the homogeneous plate bounded by the line $y = 0$ and the curve $y = 4 - x^2$ (Fig. 5-33).

SOLUTION. According to Theorem 10, we have

$$\begin{aligned} I_2 &= \iint_F y^2 \rho \, dA = \rho \iint_F y^2 \, dy \, dx \\ &= \rho \int_{-2}^{2} \int_0^{4-x^2} y^2 \, dy \, dx = \frac{\rho}{3} \int_{-2}^{2} (4 - x^2)^3 \, dx \\ &= \frac{\rho}{3} \int_{-2}^{2} (64 - 48x^2 + 12x^4 - x^6) \, dx = \frac{4096\rho}{105}. \end{aligned}$$

EXAMPLE 2. Find the moment of inertia about the z axis of the homogeneous triangular plate bounded by the lines $y = 0$, $y = x$, and $x = 4$.

SOLUTION 1. We have

$$\begin{aligned} I_3 &= \iint_F (x^2 + y^2) \rho \, dA = \rho \int_0^4 \int_0^x (x^2 + y^2) \, dy \, dx \\ &= \rho \int_0^4 \left[x^2 y + \frac{1}{3} y^3 \right]_0^x dx = \frac{4\rho}{3} \int_0^4 x^3 \, dx = \frac{256\rho}{3}. \end{aligned}$$

SOLUTION 2. We may introduce polar coordinates as shown in Fig. 5-34. Then

$$\begin{aligned} I_3 &= \iint_F (x^2 + y^2) \rho \, dA_{x,y} = \rho \iint_G r^2 \cdot r \, dr \, d\theta = \rho \int_0^{\pi/4} \int_0^{4 \sec \theta} r^3 \, dr \, d\theta \\ &= 64\rho \int_0^{\pi/4} \sec^4 \theta \, d\theta = 64\rho \left[\tan \theta + \frac{1}{3} \tan^3 \theta \right]_0^{\pi/4} = \frac{256\rho}{3}. \end{aligned}$$

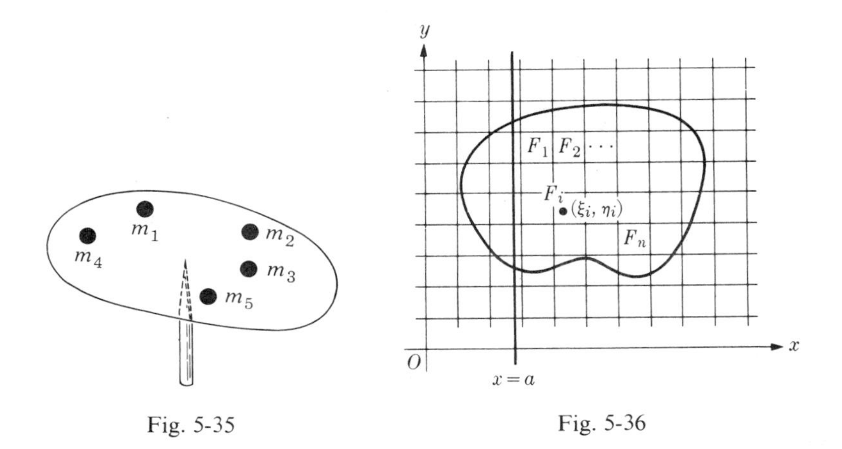

Fig. 5-35

Fig. 5-36

The moment of inertia about the z axis of two-dimensional objects in the xy plane is called the **polar moment of inertia**. Since the combination $x^2 + y^2 = r^2$ is always present in calculating polar moments, a change to polar coordinates is frequently advantageous.

Suppose that a number of masses, say five, are located at various points in the xy plane. We wish to find the center of mass of this system. From the point of view of mechanics, we imagine the masses supported by a weightless tray and assume that each mass occupies a single point. The center of mass is the point at which the tray will balance when supported by a sharp nail (Fig. 5-35). To calculate the center of mass we make use of the *moment of a mass m* with respect to one of the coordinate axes. If particles of masses $m_1, m_2, \ldots, m_n$ are situated at the points $(x_1, y_1), (x_2, y_2), \ldots, (x_n, y_n)$, respectively, then the **algebraic moment** (sometimes called **first moment** or simply **moment**) of this system **about the** y **axis** is defined by the quantity

$$m_1x_1 + m_2x_2 + \cdots + m_nx_n = \sum_{i=1}^{n} m_ix_i.$$

Its **algebraic moment about the** x **axis** is

$$\sum_{i=1}^{n} m_iy_i.$$

More generally, the algebraic moments about the line $x = a$ and about the line $y = b$ are, respectively,

$$\sum_{i=1}^{n} m_i(x_i - a) \qquad \text{and} \qquad \sum_{i=1}^{n} m_i(y_i - b).$$

We now define the moment of a thin object occupying a region F in the xy plane.

Definition. Assume that a thin mass occupies a region F in the xy plane. Let $F_1, F_2, \ldots, F_n$ be a subdivision of F as shown in Fig. 5-36. Choose a point

(ξ_i, η_i) in each F_i and replace the mass in F_i by a particle of mass $m(F_i)$ located at (ξ_i, η_i). The n idealized masses have moment

$$\sum_{i=1}^{n} (\xi_i - a) m(F_i)$$

about the line $x = a$. If the above sums tend to a limit M_1 as the norms of the subdivisions tend to zero and for any choices of the points (ξ_i, η_i) in F_i, then we define the limit M_1 as the **moment of the mass distribution about the line** $x = a$. An analogous definition for the limit M_2 of sums of the form

$$\sum_{i=1}^{n} (\eta_i - b) m(F_i)$$

yields the first moment about the line $y = b$.

The definition of first moment and the Fundamental Lemma on Integration yield the next theorem.

Theorem 11. *If a distribution of mass over a region F in the xy plane has a continuous density $\rho(x, y)$, then the moments M_1 and M_2 of F about the lines $x = a$ and $y = b$ are given by the formulas*

$$M_1 = \iint_F (x - a)\rho(x, y)\, dA, \qquad M_2 = \iint_F (y - b)\rho(x, y)\, dA.$$

Corollary. *Given a distribution of mass over a region F in the xy plane as in Theorem 11, then there are unique values of a and b (denoted $\bar{x}$ and $\bar{y}$, respectively) such that $M_1 = M_2 = 0$. In fact, the values of $\bar{x}$ and $\bar{y}$ are given by*

$$\boxed{\bar{x} = \frac{\iint_F x\rho(x, y)\, dA}{m(F)}, \qquad \bar{y} = \frac{\iint_F y\rho(x, y)\, dA}{m(F)},}$$

where

$$m(F) = \iint_F \rho(x, y)\, dA.$$

PROOF. If we set $M_1 = 0$, we get

$$0 = \iint_F (x - a)\rho(x, y)\, dA = \iint_F x\rho(x, y)\, dA - a \iint_F \rho(x, y)\, dA.$$

Since $m(F) = \iint_F \rho(x, y)\, dA$, we find for the value of a:

$$a = \frac{\iint_F x\rho(x, y)\, dA}{m(F)} = \bar{x}.$$

The value $\bar{y}$ is found similarly.

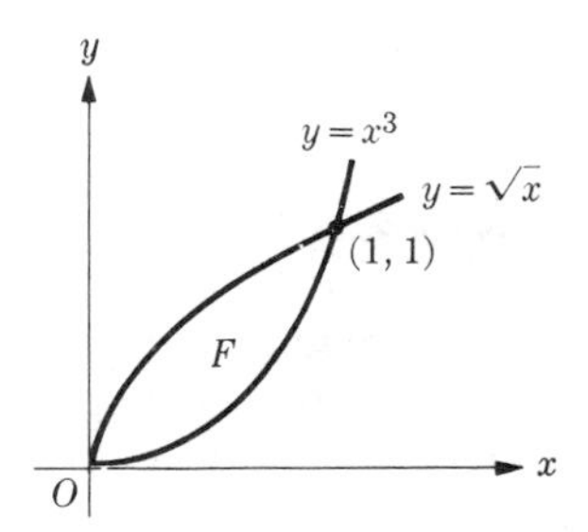

Fig. 5-37

Definition. The point $(\bar{x}, \bar{y})$ is called the **center of mass** of the distribution over F.

EXAMPLE 3. Find the center of mass of the region

$$F = \{(x, y) : 0 \leq x \leq 1, \quad x^3 \leq y \leq \sqrt{x}\}$$

if the density of F is given by $\rho = 3x$.

SOLUTION. (See Fig. 5-37.) For the first moments, we have

$$\begin{aligned}
M_1 &= \iint_F x\rho \, dA = 3 \int_0^1 \int_{x^3}^{\sqrt{x}} x^2 \, dy \, dx \\
&= 3 \int_0^1 x^2 [y]_{x^3}^{\sqrt{x}} \, dx = 3 \int_0^1 (x^{5/2} - x^5) \, dx = \frac{5}{14}, \\
M_2 &= \iint_F y\rho \, dA = 3 \int_0^1 \int_{x^3}^{\sqrt{x}} yx \, dy \, dx = \frac{3}{2} \int_0^1 x [y^2]_{x^3}^{\sqrt{x}} \, dx \\
&= \frac{3}{2} \int_0^1 (x^2 - x^7) \, dx = \frac{5}{16}, \\
m(F) &= \iint_F \rho \, dA = 3 \int_0^1 \int_{x^3}^{\sqrt{x}} x \, dy \, dx = 3 \int_0^1 x [y]_{x^3}^{\sqrt{x}} \, dx \\
&= 3 \int_0^1 (x^{3/2} - x^4) \, dx = \frac{3}{5}.
\end{aligned}$$

Therefore

$$\bar{x} = \frac{M_1}{m(F)} = \frac{5}{14} \cdot \frac{5}{3} = \frac{25}{42}; \qquad \bar{y} = \frac{M_2}{m(F)} = \frac{5}{16} \cdot \frac{5}{3} = \frac{25}{48}.$$

EXAMPLE 4. Find the center of mass of a plate in the form of a circular sector of radius a and central angle 2α if its thickness is proportional to its distance from the center of the circle from which the sector is taken.

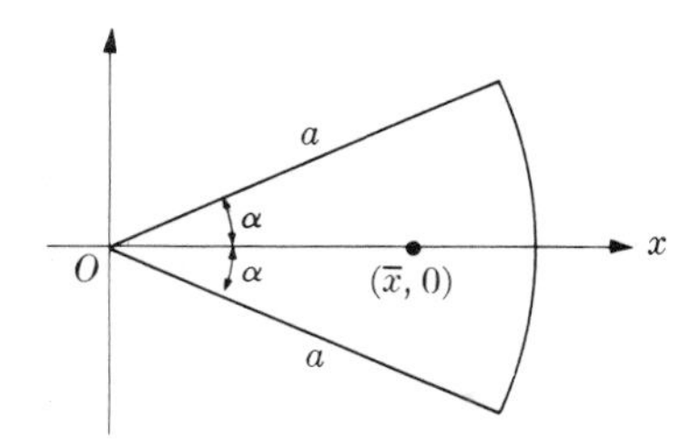

Fig. 5-38

Solution. Select the sector so that the vertex is at the origin and the x axis bisects the region (Fig. 5-38). Then the density is given by $\rho = kr$, where k is a proportionality constant. By symmetry we have $\bar{y} = 0$. Using polar coordinates, we obtain

$$M_1 = \iint_F xkr\,dA_{x,y} = k\int_{-\alpha}^{\alpha}\int_0^a r^2 \cos\theta r\,dr\,d\theta$$

$$= \frac{ka^4}{4}\int_{-\alpha}^{\alpha} \cos\theta\,d\theta = \frac{1}{2}ka^4 \sin\alpha,$$

$$m(F) = k\iint_F r\,dA_{x,y} = k\int_{-\alpha}^{\alpha}\int_0^a r^2\,dr\,d\theta = \frac{2}{3}ka^3\alpha.$$

Therefore

$$\bar{x} = \frac{M_1}{m(F)} = \frac{3a\sin\alpha}{4\alpha}.$$

Problems

In each of Problems 1 through 7, find the moment of inertia about the given axis of the plate F whose density is given.

1. F is the square with vertices $(0,0)$, $(a,0)$, (a,a), $(0,a)$, $\rho = \text{constant}$; y axis.

2. F is the triangle with vertices $(0,0)$, $(a,0)$, (b,c), with $a > 0$, $c > 0$, $\rho = \text{constant}$; x axis.

3. $F = \{(x,y): 0 \le x \le 1,\quad x^2 \le y \le \sqrt{x}\}$, $\quad \rho = \text{constant}$; $\quad y$ axis.

4. $F = \left\{(x,y): 1 \le x \le 4,\quad \frac{4}{x} \le y \le 5 - x\right\}$, $\quad \rho = ky$; $\quad x$ axis.

5. $F = \{(x,y): -a \le x \le a,\quad 0 \le y \le \sqrt{a^2 - x^2}\}$, $\quad \rho = ky$; $\quad x$ axis.

6. $F = \left\{(x,y): 1 \le x \le 4,\quad \frac{4}{x} \le y \le 5 - x\right\}$, $\quad \rho = \text{constant}$; $\quad x$ axis.

7. $F = \{(x,y): 0 \le x \le \pi,\quad 0 \le y \le \sin x\}$, $\quad \rho = \text{constant}$; $\quad y$ axis.

In each of Problems 8 through 17, find the moment of inertia about the given axis of the plate F whose density is given.

8. $F = \{(x,y) : x^2 + y^2 \le a^2\}$, $\rho = k\sqrt{x^2+y^2}$; z axis.

9. $F = \{(x,y) : -1 \le x \le 2, \quad x^2 \le y \le x+2\}$, $\rho =$ constant; x axis.

10. F is the interior of the square with vertices $(0,0)$, $(a,0)$, (a,a), $(0,a)$, $\rho =$ constant; z axis.

11. $F = \{(x,y) : 0 \le x \le 1, \quad x^2 \le y \le \sqrt{x}\}$, $\rho = ky$; y axis.

12. $F = \{(x,y) : -1 \le x \le 2, \quad x^2 \le y \le x+2\}$, $\rho =$ constant; axis is line $y = 4$.

13. $F = \{(x,y) : x^2 + y^2 \le a^2\}$, $\rho = k\sqrt{x^2+y^2}$; x axis.

14. F is bounded by the closed curve $r = 2a\cos\theta$, $\rho = kr$; z axis.

15. F is bounded by one loop of $r^2 = a^2\cos 2\theta$, $\rho =$ constant; z axis.

16. F is bounded by one loop of $r^2 = a^2\cos 2\theta$, $\rho =$ constant; x axis.

17. F is the region in the first quadrant inside the circle $r = 1$, and bounded by $r = 1$, $\theta = r$, and $\theta = \pi/2$; $\lambda =$ constant; z axis.

In each of Problems 18 through 28, find the center of mass of the plate F described.

18. $F = \left\{(x,y) : 1 \le x \le 4, \quad \frac{4}{x} \le y \le 5 - x\right\}$, $\rho = ky$.

19. $F = \{(x,y) : y^2 \le x \le y+2, \quad -1 \le y \le 2\}$, $\rho = kx$.

20. F is the interior of the triangle with vertices at $(0,0)$, $(a,0)$, (b,c), with $0 < b < a$, $0 < c$, $\rho = kx$.

21. $F = \{(x,y) : 0 \le x \le 1, \quad x^2 \le y \le \sqrt{x}\}$, $\rho = ky$.

22. $F = \{(x,y) : -1 \le x \le 2, \quad x^2 \le y \le x+2\}$, $\rho =$ constant.

23. F is the square with vertices at $(0,0)$, $(a,0)$, (a,a), $(0,a)$, $\rho = k(x^2+y^2)$.

24. F is the triangle with vertices at $(0,0)$, $(1,0)$, $(1,1)$, $\rho = kr^2$.

25. F is bounded by the cardioid $r = 2(1+\cos\theta)$, $\rho =$ constant.

26. F is bounded by one loop of the curve $r = 2\cos 2\theta$, $\rho =$ constant.

27. F is bounded by $3x^2 + 4y^2 = 48$ and $(x-2)^2 + y^2 = 1$, $\rho =$ constant.

28. F is bounded by one loop of the curve $r^2 = a^2\cos 2\theta$, $\rho =$ constant.

29. The **Theorem of the Mean** for double integrals states that *if f is integrable over a region F of area $A(F)$* and if $m \le f(x,y) \le M$ for all (x,y) on F, then there is a number $\bar{m}$ between m and M such that

$$\iint_F f(x,y)\,dA = \bar{m}A(F).$$

Use the Fundamental Lemma on Integration and the Theorem of the Mean to establish Theorem 10. Use the idea of the proof of Theorem 9.

30. Show that if a mass distribution F lies between the lines $x = a$ and $x = b$, then $a \le \bar{x} \le b$. Similarly, if F lies between the lines $y = c$ and $y = d$, then $c \le \bar{y} \le d$.

31. Let $F_1, F_2, \ldots, F_n$ be regions no two of which have any points in common, and let $(\bar{x}_1, \bar{y}_1), (\bar{x}_2, \bar{y}_2), \ldots, (\bar{x}_n, \bar{y}_n)$ be their respective centers of mass. Denote the mass of F_i by m_i. If F is the region containing all the points in every F_i, show that the center of mass $(\bar{x}, \bar{y})$ of F is given by

$$\bar{x} = \frac{m_1\bar{x}_1 + m_2\bar{x}_2 + \cdots + m_n\bar{x}_n}{m_1 + m_2 + \cdots + m_n},$$

$$\bar{y} = \frac{m_1\bar{y}_1 + m_2\bar{y}_2 + \cdots + m_n\bar{y}_n}{m_1 + m_2 + \cdots + m_n}.$$

32. Show that if F is symmetric with respect to the x axis and $\rho(x, -y) = \rho(x, y)$ for all (x, y) on F, then $\bar{y} = 0$. A similar result holds for symmetry with respect to the y axis.

33. Find the moment of inertia about the z axis of a ring of uniform density in the xy plane bounded by the circles $x^2 + y^2 = r_1^2$ and $x^2 + y^2 = r_2^2$ with $r_1 < r_2$. Find the result of the solid disk with the same density having the same moment of inertia about the z axis.

34. Let F be a region in the xy plane with mass $m(F)$. Show that in the notation of this section

$$I_1^a = I_1 - 2aM_1 + a^2 m(F)$$

where M_1 is the first moment of F about the y axis. Also, show that

$$I_3^{(a,b)} = I_3 - 2aM_1 - 2bM_2 + (a^2 + b^2)m(F)$$

where M_2 is the first moment of F with respect to the x axis.

7. Surface Area

To define surface area we employ a procedure similar to that used for defining area in the plane. First, we define surface area in the simplest case, and second, we employ a limiting process for the definition of surface area of a general curved surface.

Suppose two planes Γ_1 and Γ_2 intersect at an angle ϕ (Fig. 5-39). From

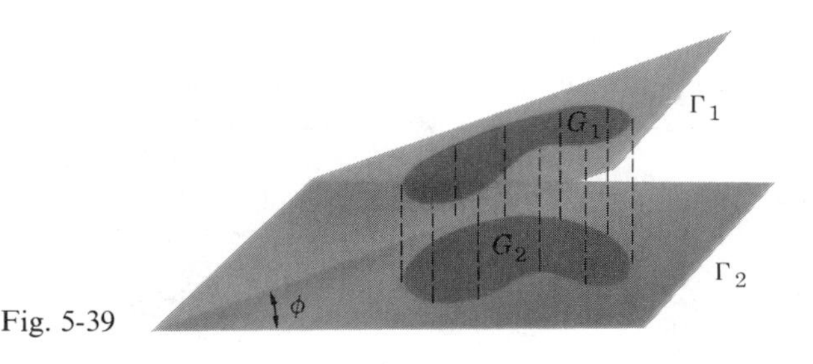

Fig. 5-39

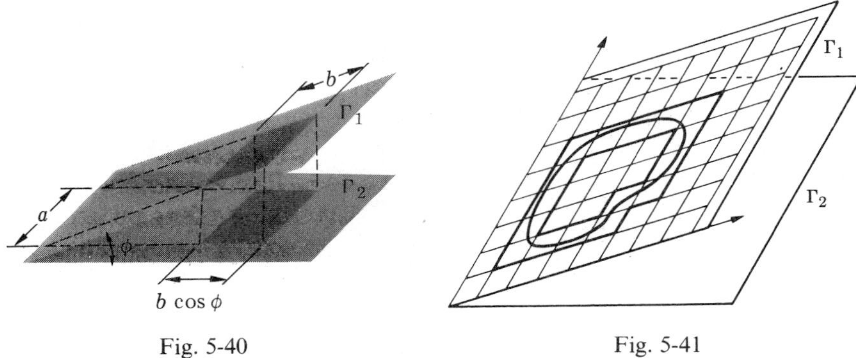

Fig. 5-40 Fig. 5-41

each point of G_1, a region in the plane Γ_1, we drop a perpendicular to the plane Γ_2. The set of points of intersection of these perpendiculars with Γ_2 forms a region which we denote G_2. The set G_2 is called the **projection of G_1 on Γ_2**. We shall now determine the relationship between the area $A(G_1)$ of G_1 and the area $A(G_2)$ of G_2. If G_1 is a rectangle—the simplest possible case—the problem may be solved by elementary geometry. For convenience, we select the rectangle in Γ_1 so that one side is parallel to the line of intersection of the two planes (Fig. 5-40). Let the lengths of the sides of the rectangle be a and b, as shown. The projection of the rectangle in Γ_1 onto Γ_2 is a rectangle, as the reader may easily verify. The lengths of the sides of the rectangle in Γ_2 are a and $b\cos\phi$. The area $A_1 = ab$ for the rectangle in Γ_1 and the area $A_2 = ab\cos\phi$ for the rectangle in Γ_2. They satisfy the relation

$$A_2 = A_1 \cos\phi. \tag{1}$$

Equation (1) is the basis of the next useful result.

Lemma. *Let G_1 be a region in a plane Γ_1 and let G_2 be the projection of G_1 onto a plane Γ_2. Then*

$$A(G_2) = A(G_1)\cos\phi, \tag{2}$$

where ϕ is the angle between the planes Γ_1 and Γ_2.

This lemma is proved by subdividing the plane Γ_1 into a network of rectangles and observing that these rectangles project onto rectangles in Γ_2 with areas related by equation (1). Since the areas of G_1 and G_2 are obtained as limits of sums of the areas of rectangles, formula (2) holds in the limit (Fig. 5-41). We observe that if the planes are parallel, then ϕ is zero, the region and its projection are congruent, and the areas are equal. If the planes are perpendicular, then $\phi = \pi/2$, the projection of G_1 degenerates into a line segment, and $A(G_2)$ vanishes. Thus formula (2) is valid for all angles ϕ such that $0 \le \phi \le \pi/2$.

Suppose we have a surface S represented by an equation

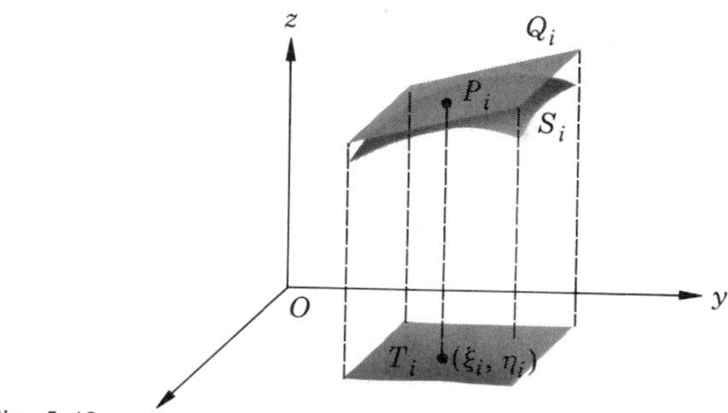

Fig. 5-42

$$z = f(x, y)$$

for (x, y) on some region F in the xy plane. We shall consider only functions f which have continuous first partial derivatives for all (x, y) on F.

To define the area of the surface S we begin by subdividing the xy plane into a rectangular mesh. Suppose T_i, a rectangle of the subdivision, is completely contained in F. Select a point (ξ_i, η_i) in T_i. This selection may be made in any manner whatsoever. The point $P_i(\xi_i, \eta_i, \zeta_i)$, with $\zeta_i = f(\xi_i, \eta_i)$, is on the surface S. Construct the plane tangent to the surface S at P_i (Fig. 5-42). Planes parallel to the z axis and through the edges of T_i cut out a portion (denoted S_i) of the surface, and they cut out a quadrilateral, denoted Q_i, from the tangent plane. The projection of Q_i on the xy plane is T_i. If the definition of surface area is to satisfy our intuition, then the area of S_i must be close to the area of Q_i whenever the subdivision in the xy plane is sufficiently fine. However, Q_i is a plane region, and its area can be found exactly. We recall from Chapter 4 (page 226) that we can determine the equation of a plane tangent to a surface $z = f(x, y)$ at a given point on the surface. Such a determination is possible because the quantities

$$f_x(\xi_i, \eta_i), \qquad f_y(\xi_i, \eta_i), \qquad -1$$

form a set of attitude numbers for the tangent plane at the point (ξ_i, η_i, ζ_i) where $\zeta_i = f(\xi_i, \eta_i)$.

On page 22 we showed that the formula for the angle between two planes is

$$\cos\phi = \frac{|a_1b_1 + a_2b_2 + a_3b_3|}{\sqrt{a_1^2 + a_2^2 + a_3^2}\sqrt{b_1^2 + b_2^2 + b_3^2}},$$

where a_1, a_2, a_3 and b_1, b_2, b_3 are sets of attitude numbers of the two planes. We now find the cosine of the angle between the plane tangent to the surface and the xy plane. Letting ϕ denote the angle between the tangent plane and the xy plane and recalling that the xy plane has attitude numbers 0, 0, -1, we get

$$\cos\phi = \frac{0 \cdot f_{,1} + 0 \cdot f_{,2} + 1 \cdot 1}{\sqrt{1 + f_{,1}^2 + f_{,2}^2}} = (1 + f_x^2 + f_y^2)^{-1/2}.$$

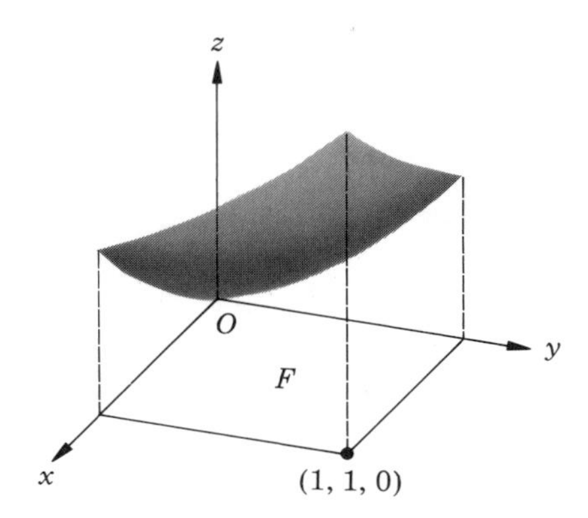

Fig. 5-43

According to the above lemma, we have

$$A(T_i) = A(Q_i)\cos\phi$$

or

$$A(Q_i) = A(T_i)\sqrt{1 + f_x^2(\xi_i, \eta_i) + f_y^2(\xi_i, \eta_i)}.$$

We add all expressions of the above type for rectangles T_i which are in F. We obtain the sum

$$\sum_{i=1}^{n} A(Q_i) = \sum_{i=1}^{n} A(T_i)\sqrt{1 + f_x^2(\xi_i, \eta_i) + f_y^2(\xi_i, \eta_i)}, \tag{3}$$

and we expect that this sum is a good approximation to the (as yet undefined) surface area if the norm of the rectangular subdivision in the xy plane is sufficiently small.

Definition. If the limit of the sums (3) exists as the norms of the subdivisions tend to zero and for arbitrary selections of the values (ξ_i, η_i) in T_i, then we say that the surface $z = f(x, y)$ has **surface area**. The **value of the surface area** $A(S)$ of S is the limit of the sum (3).

Theorem 12. *The sums in* (3) *tend to*

$$\boxed{A(S) = \iint_F \sqrt{1 + [f_x(x, y)]^2 + [f_y(x, y)]^2}\, dA}$$

whenever the first derivatives f_x and f_y are continuous on F.

This theorem is an immediate consequence of Theorem 1 and the fact that $\sqrt{1 + f_x^2 + f_y^2}$ is continuous if f_x and f_y are. The integration formula of the theorem may be used to calculate surface area, as the next examples show.

Example 1. Find the area of the surface $z = \frac{2}{3}(x^{3/2} + y^{3/2})$ situated above the square $F = \{(x, y) : 0 \le x \le 1, 0 \le y \le 1\}$ (Fig. 5-43).

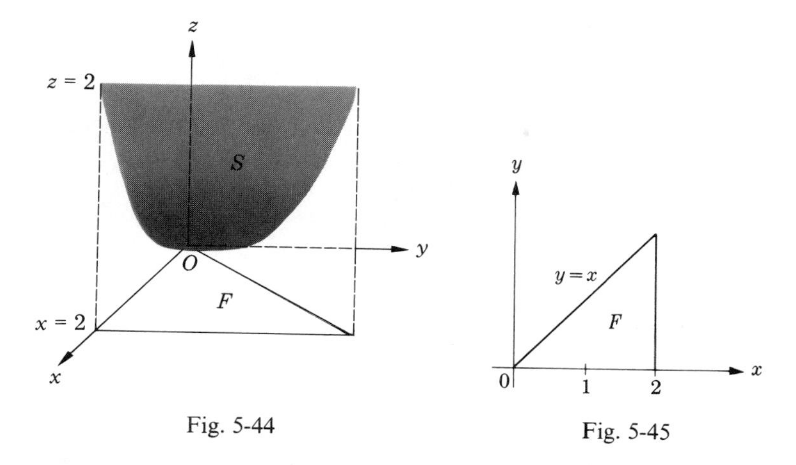

Fig. 5-44 Fig. 5-45

SOLUTION. Setting $z = f(x, y)$, we have $f_x = x^{1/2}, f_y = y^{1/2}$, and

$$A(S) = \iint_F (1 + x + y)^{1/2}\, dA$$

$$= \int_0^1 \int_0^1 (1 + x + y)^{1/2}\, dy\, dx.$$

Therefore

$$A(S) = \int_0^1 \frac{2}{3}[(1 + x + y)^{3/2}]_0^1\, dx = \frac{2}{3}\int_0^1 [(2 + x)^{3/2} - (1 + x)^{3/2}]\, dx$$

$$= \frac{4}{15}[(2 + x)^{5/2} - (1 + x)^{5/2}]_0^1 = \frac{4}{15}(1 + 9\sqrt{3} - 8\sqrt{2}).$$

EXAMPLE 2. Find the area of the part of the cylinder $z = \frac{1}{2}x^2$ cut out by the planes $y = 0$, $y = x$, and $x = 2$.

SOLUTION. See Figs. 5-44 and 5-45, which show the surface S and the projection F. We have $\partial z/\partial x = x$, $\partial z/\partial y = 0$. Therefore

$$A(S) = \iint_F \sqrt{1 + x^2}\, dA = \int_0^2 \int_0^x \sqrt{1 + x^2}\, dy\, dx$$

$$= \int_0^2 x\sqrt{1 + x^2}\, dx = \frac{1}{3}[(1 + x^2)^{3/2}]_0^2$$

$$= \frac{1}{3}(5\sqrt{5} - 1).$$

The next example shows that it is sometimes useful to use polar coordinates for the evaluation of the double integral.

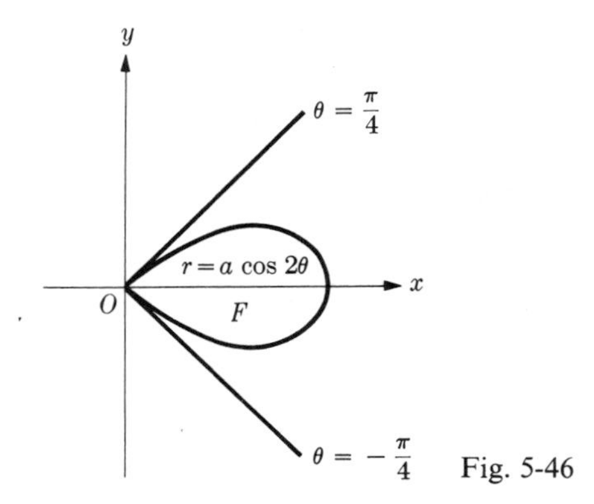

Fig. 5-46

EXAMPLE 3. Find the surface area of the part of the sphere $x^2 + y^2 + z^2 = a^2$ cut out by the vertical cylinder erected on one loop of the curve whose equation in polar coordinates is $r = a\cos 2\theta$.

SOLUTION. (See Fig. 5-46.) The surface consists of two parts, one above and one below the xy plane, symmetrically placed. The area of the upper half will be found. We have

$$z = \sqrt{a^2 - x^2 - y^2}, \qquad \frac{\partial z}{\partial x} = \frac{-x}{\sqrt{a^2 - x^2 - y^2}}, \qquad \frac{\partial z}{\partial y} = \frac{-y}{\sqrt{a^2 - x^2 - y^2}}.$$

Therefore (Fig. 5-46), we obtain

$$\begin{aligned} A(S) &= \iint_F \frac{a}{\sqrt{a^2 - x^2 - y^2}}\, dA_{x,y} \\ &= \int_{-\pi/4}^{\pi/4} \int_0^{a\cos 2\theta} \frac{ar}{\sqrt{a^2 - r^2}}\, dr\, d\theta. \end{aligned}$$

This integral is an improper integral, but it can be shown to be convergent. Taking this fact for granted, we get

$$\begin{aligned} A(S) &= 2a \int_0^{\pi/4} [-\sqrt{a^2 - r^2}]_0^{a\cos 2\theta}\, d\theta \\ &= 2a^2 \int_0^{\pi/4} (1 - \sin 2\theta)\, d\theta = \frac{1}{2}a^2(\pi - 2). \end{aligned}$$

The total surface area is $a^2(\pi - 2)$.

If the given surface is of the form $y = f(x, z)$ or $x = f(y, z)$, we get similar formulas for the surface area. These are

$$A(S) = \iint_F \sqrt{1 + \left(\frac{\partial z}{\partial x}\right)^2 + \left(\frac{\partial z}{\partial y}\right)^2}\, dA_{x,y} \qquad \text{if} \quad z = f(x, y),$$

$$A(S)=\iint_F \sqrt{1+\left(\frac{\partial y}{\partial x}\right)^2+\left(\frac{\partial y}{\partial z}\right)^2}\,dA_{x,z} \qquad \text{if} \quad y=f(x,z),$$

$$A(S)=\iint_F \sqrt{1+\left(\frac{\partial x}{\partial y}\right)^2+\left(\frac{\partial x}{\partial z}\right)^2}\,dA_{y,z} \qquad \text{if} \quad x=f(y,z).$$

Problems

In each of Problems 1 through 18, find the area of the surface described.

1. The portion of the surface $z=\frac{2}{3}(x^{3/2}+y^{3/2})$ situated above the triangle
$$F=\{(x,y):0\le x\le y,\quad 0\le y\le 1\}.$$
2. The portion of the plane $x/a+y/b+z/c=1$ in the first octant ($a>0,\quad b>0,\ c>0$).
3. The part of the cylinder $x^2+z^2=a^2$ inside the cylinder $x^2+y^2=a^2$.
4. The part of the cylinder $x^2+z^2=a^2$ above the square $|x|\le\frac{1}{2}a,\quad |y|\le\frac{1}{2}a$.
5. The part of the cone $z^2=x^2+y^2$ inside the cylinder $x^2+y^2=2x$.
6. The part of the cone $z^2=x^2+y^2$ above the figure bounded by one loop of the curve $r^2=4\cos 2\theta$.
7. The part of the cone $x^2=y^2+z^2$ between the cylinder $y^2=z$ and the plane $y=z-2$.
8. The part of the cone $y^2=x^2+z^2$ cut off by the plane $2y=(x+2)\sqrt{2}$.
9. The part of the cone $x^2=y^2+z^2$ inside the sphere $x^2+y^2+z^2=2z$.
10. The part of the surface $z=xy$ inside the cylinder $x^2+y^2=a^2$.
11. The part of the surface $4z=x^2-y^2$ above the region bounded by the curve $r^2=4\cos\theta$.
12. The part of the surface of a sphere of radius $2a$ inside a cylinder of radius a if the center of the sphere is on the surface of the cylinder.
13. The part of the surface of a sphere of radius a, center at the origin, inside the cylinder erected on one loop of the curve $r=a\cos 3\theta$.
14. The part of the sphere $x^2+y^2+z^2=4z$ inside the paraboloid $x^2+y^2=z$.
15. The part of the cylinder $y^2+z^2=2z$ cut off by the cone $x^2=y^2+z^2$.
16. The part of the cylinder $x^2+y^2=2ax$ inside the sphere $x^2+y^2+z^2=4a^2$.
17. The part of the cylinder $y^2+z^2=4a^2$ above the xy plane and bounded by the planes $y=0,\quad x=a$, and $y=x$.
18. The part of the paraboloid $y^2+z^2=4ax$ cut off by the cylinder $y^2=ax$ and the plane $x=3a$; outside the cylinder $y^2=ax$.

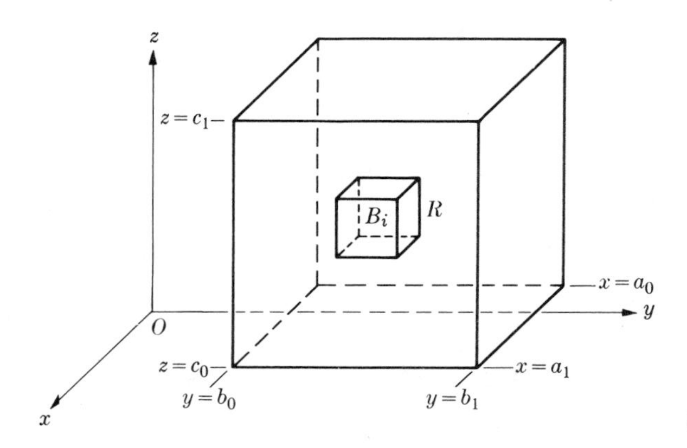

Fig. 5-47

19. (a) Use elementary geometry (and trigonometry) to establish equation (2) for an arbitrary triangle. (b) Use the result of (a) to establish equation (2) for an arbitrary polygon.

20. If $r = \sqrt{x^2 + y^2}$, $\theta = \arctan(y/x)$, and $z = f(x, y)$, establish the formula for surface area in polar coordinates:

$$A(S) = \iint_F \sqrt{1 + f_r^2 + \frac{1}{r^2} f_\theta^2}\, r\, dr\, d\theta.$$

Use the polar coordinates formula (Problem 20) to find the surface area in Problems 21 through 24.

21. The area of the surface of the paraboloid $z = x^2 + y^2$ which is inside the cylinder $x^2 + y^2 = 4$.

22. The portion of the cone $x^2 + y^2 = z^2$ inside the cylinder $(x^2 + y^2)^2 = 2xy$.

23. The portion of the cone $x^2 + y^2 = z^2$ inside the cylinder $x^2 + y^2 = 1$.

8. The Triple Integral

The definition of the triple integral parallels that of the double integral. In the simplest case, we consider a rectangular box R bounded by the six planes $x = a_0, x = a_1, y = b_0, y = b_1, z = c_0, z = c_1$ (Fig. 5-47). Let $f(x, y, z)$ be a function of three variables defined for (x, y, z) in R. We subdivide the entire three-dimensional space into rectangular boxes by constructing planes parallel to the coordinate planes. Let $B_1, B_2, \ldots, B_n$ be those boxes of the subdivision which contain points of R. Denote by $V(B_i)$ the volume of the ith box, B_i. We select a point $P_i(\xi_i, \eta_i, \zeta_i)$ in B_i; this selection may be made in any manner whatsoever. The sum

$$\sum_{i=1}^{n} f(\xi_i, \eta_i, \zeta_i) V(B_i)$$

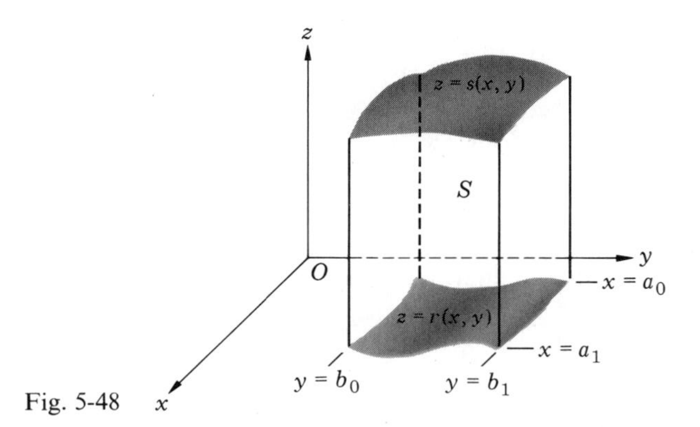

Fig. 5-48

is an approximation to the triple integral. The **norm of the subdivision** is the length of the longest diagonal of the boxes $B_1, B_2, \ldots, B_n$. If the above sums tend to a limit as the norms of the subdivisions tend to zero and for any choices of the points P_i, we call this limit the **triple integral of** f **over** R. The expression

$$\iiint_R f(x, y, z)\, dV$$

is used to represent this limit.

Just as the double integral is equal to a twice-iterated integral, so the triple integral has the same value as a threefold iterated integral. In the case of the rectangular box R, we obtain

$$\iiint_R f(x, y, z)\, dV = \int_{a_0}^{a_1} \left\{ \int_{b_0}^{b_1} \left[\int_{c_0}^{c_1} f(x, y, z)\, dz \right] dy \right\} dx.$$

Suppose a region S is bounded by the planes $x = a_0$, $x = a_1$, $y = b_0$, $y = b_1$, and by the surfaces $z = r(x, y)$, $z = s(x, y)$, as shown in Fig. 5-48. The triple integral may be defined in the same way as for a rectangular box R, and once again it is equal to the iterated integral. We have

$$\iiint_S f(x, y, z)\, dV = \int_{a_0}^{a_1} \left\{ \int_{b_0}^{b_1} \left[\int_{r(x,y)}^{s(x,y)} f(x, y, z)\, dz \right] dy \right\} dx.$$

We state without proof the following theorem, which applies in the general case.

Theorem 14. *Suppose that S is a region defined by the inequalities*

$$S = \{(x, y, z) : a \le x \le b, \quad p(x) \le y \le q(x), \quad r(x, y) \le z \le s(x, y)\},$$

where the functions p, q, r, and s are continuous. If f is a continuous function on S, then

$$\iiint_S f(x, y, z)\, dV = \int_a^b \left\{ \int_{p(x)}^{q(x)} \left[\int_{r(x,y)}^{s(x,y)} f(x, y, z)\, dz \right] dy \right\} dx.$$

The iterated integrations are performed in turn by holding all variables constant except the one being integrated. Brackets and braces in multiple integrals will be omitted unless there is danger of confusion.

EXAMPLE 1. Evaluate the iterated integral

$$\int_0^3 \int_0^{6-2z} \int_0^{4-(2/3)y-(4/3)z} yz\, dx\, dy\, dz.$$

SOLUTION. We have

$$\begin{aligned}
&\int_0^3 \int_0^{6-2z} \int_0^{4-(2/3)y-(4/3)z} yz\, dx\, dy\, dz \\
&= \int_0^3 \int_0^{6-2z} [xyz]_0^{4-(2/3)y-(4/3)z}\, dy\, dz \\
&= \int_0^3 \int_0^{6-2z} yz\left(4 - \frac{2}{3}y - \frac{4}{3}z\right) dy\, dz \\
&= \int_0^3 \left[2zy^2 - \frac{2}{9}zy^3 - \frac{2}{3}y^2z^2\right]_0^{6-2z} dz \\
&= \int_0^3 \left[\left(2z - \frac{2}{3}z^2\right)(6 - 2z)^2 - \frac{2}{9}z(6 - 2z)^3\right] dz \\
&= \frac{1}{9}\int_0^3 z(6 - 2z)^3\, dz.
\end{aligned}$$

The integration may be performed by the substitution $u = 6 - 2z$. The result is $54/5$.

The determination of the limits of integration is the principal difficulty in reducing a triple integral to an iterated integral. The reader who works a large number of problems will develop good powers of visualization of three-dimensional figures. There is no simple mechanical technique for determining the limits of integration in the wide variety of problems we encounter. The next examples illustrate the process.

EXAMPLE 2. Evaluate

$$\iiint_S x\, dV,$$

where S is the region bounded by the surfaces $y = x^2$, $y = x + 2$, $4z = x^2 + y^2$, and $z = x + 3$.

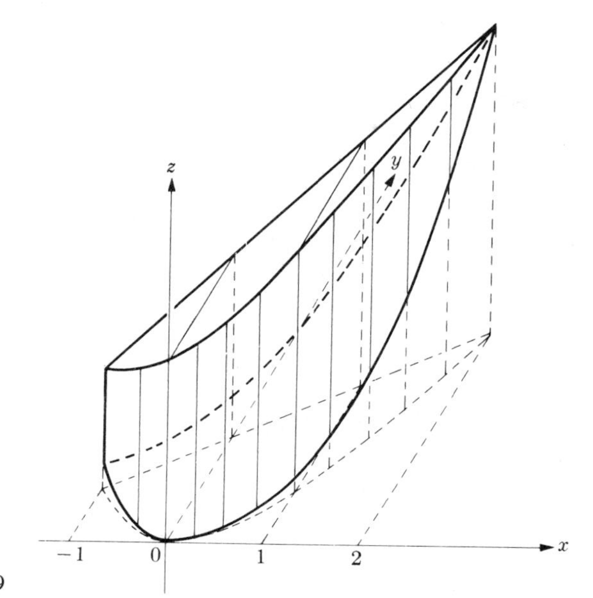

Fig. 5-49

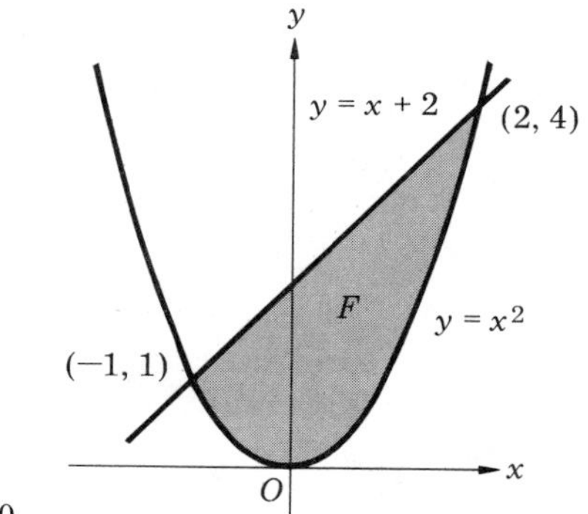

Fig. 5-50

SOLUTION. To transform the triple integral into an iterated integral, we must determine the limits of integration. The region S is sketched in Fig. 5-49. The projection of S on the xy plane is the region F bounded by the curves $y = x^2$ and $y = x + 2$, as shown in Fig. 5-50. From this projection, the region rises with vertical walls, bounded from below by the paraboloid $z = \frac{1}{4}(x^2 + y^2)$ and above by the plane $z = x + 3$. Since F is described by the inequalities

$$F = \{(x, y) : -1 \leq x \leq 2, \quad x^2 \leq y \leq x + 2\},$$

we have

$$S = \{(x, y, z) : -1 \leq x \leq 2, \quad x^2 \leq y \leq x + 2, \quad \tfrac{1}{4}(x^2 + y^2) \leq z \leq x + 3\}.$$

Therefore

$$\begin{aligned}\iiint_S x\,dV &= \int_{-1}^{2}\int_{x^2}^{x+2}\int_{(x^2+y^2)/4}^{x+3} x\,dz\,dy\,dx\\ &= \int_{-1}^{2}\int_{x^2}^{x+2}\left[x^2+3x-\frac{1}{4}(x^3+xy^2)\right]dy\,dx\\ &= \int_{-1}^{2}\left\{\left(3x+x^2-\frac{1}{4}x^3\right)(2+x-x^2)-\frac{x}{12}[(2+x)^3-x^6]\right\}dx\\ &= \frac{837}{160}.\end{aligned}$$

In the case of double integrals there are two possible orders of integration, one of them often being easier to calculate than the other. In the case of triple integrals there are six possible orders of integration. It becomes a matter of practice and trial and error to find which order is the most convenient.

The limits of integration may sometimes be found by projecting the region on one of the coordinate planes and then finding the equations of the "bottom" and "top" surfaces. This method was used in Example 2. If part of the boundary is a cylinder perpendicular to one of the coordinate planes, that fact can be used to determine the limits of integration.

EXAMPLE 3. Express the integral

$$I = \iiint_S f(x, y, z)\,dV$$

as an iterated integral in six different ways when S is the region bounded by the surfaces

$$z = 0, \qquad z = x, \qquad \text{and} \qquad y^2 = 4 - 2x.$$

SOLUTION. The region S is shown in Fig. 5-51. The projection of S on the xy plane is the two-dimensional region F_{xy} bounded by $x = 0$ and $y^2 = 4 - 2x$, as shown in Fig. 5-52. Therefore the integral may be written

$$\begin{aligned}I &= \int_{0}^{2}\int_{-\sqrt{4-2x}}^{+\sqrt{4-2x}}\int_{0}^{x} f(x, y, z)\,dz\,dy\,dx\\ &= \int_{-2}^{2}\int_{0}^{2-(1/2)y^2}\int_{0}^{x} f(x, y, z)\,dz\,dx\,dy.\end{aligned}$$

The projection of S on the xz plane is the triangular region bounded by the curves $z = 0$, $z = x$, and $x = 2$, as shown in Fig. 5-53. The iterated integral

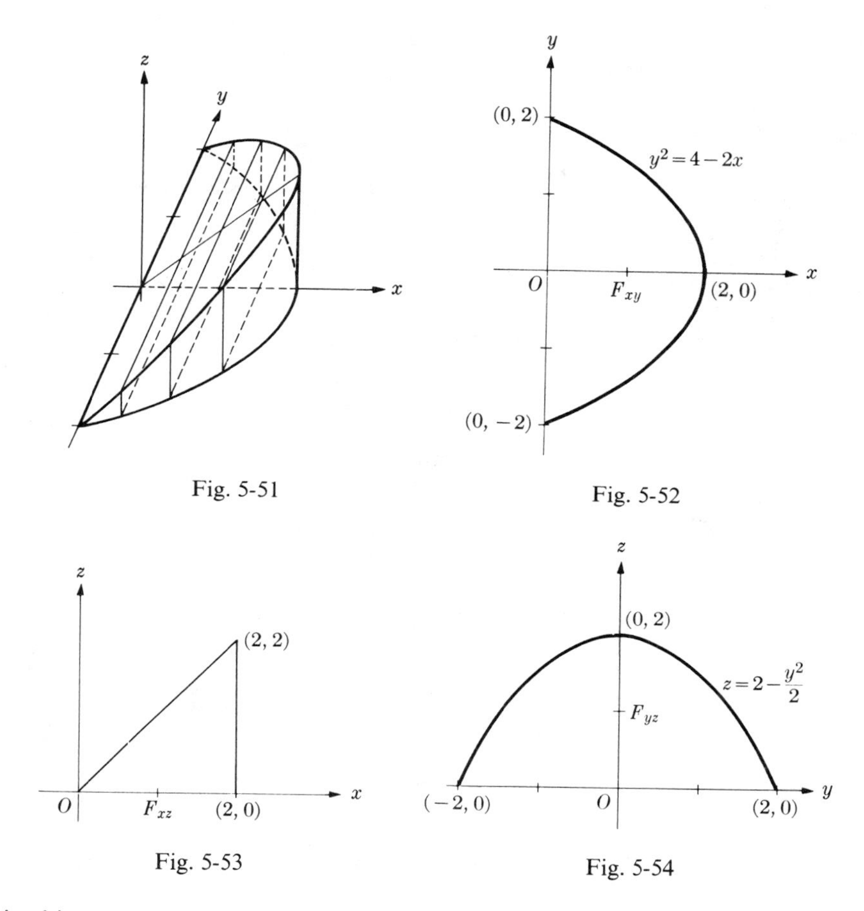

Fig. 5-51

Fig. 5-52

Fig. 5-53

Fig. 5-54

in this case becomes

$$I = \int_0^2 \int_0^x \int_{-\sqrt{4-2x}}^{+\sqrt{4-2x}} f(x, y, z)\, dy\, dz\, dx$$
$$= \int_0^2 \int_z^2 \int_{-\sqrt{4-2x}}^{+\sqrt{4-2x}} f(x, y, z)\, dy\, dx\, dz.$$

The projection of S on the yz plane is the plane region bounded by $z = 0$ and $z = 2 - \frac{1}{2}y^2$ (Fig. 5-54). Then I takes the form

$$I = \int_{-2}^2 \int_0^{2-(1/2)y^2} \int_z^{2-(1/2)y^2} f(x, y, z)\, dx\, dz\, dy$$
$$= \int_0^2 \int_{-\sqrt{4-2z}}^{+\sqrt{4-2z}} \int_z^{2-(1/2)y^2} f(x, y, z)\, dx\, dy\, dz.$$

PROBLEMS

In each of Problems 1 through 6, find the value of the interated integral. Express each region of integration in set notation.

1. $\displaystyle\int_0^1\int_0^x\int_0^{x-y} x\,dz\,dy\,dx$

2. $\displaystyle\int_{-1}^1\int_0^{1-y^2}\int_{-\sqrt{x}}^{\sqrt{x}} 2y^2\sqrt{x}\,dz\,dx\,dy$

3. $\displaystyle\int_0^1\int_{y^2}^{\sqrt{y}}\int_0^{y+z} xy\,dx\,dz\,dy$

4. $\displaystyle\int_0^4\int_0^{\sqrt{16-x^2}}\int_0^{\sqrt{16-x^2-y^2}} (x+y+z)\,dz\,dy\,dx$

5. $\displaystyle\int_0^2\int_0^{\sqrt{4-z^2}}\int_0^{2-z} z\,dx\,dy\,dz$

6. $\displaystyle\int_0^1\int_0^x\int_0^y \frac{1+\sqrt[3]{z}}{\sqrt{z}}\,dz\,dy\,dx$

In Problems 7 through 17, evaluate

$$\iiint_S f(x,y,z)\,dV$$

where S is bounded by the given surfaces and f is the given function. In each case, express S in set notation.

7. $z=0,\quad y=0,\quad y=x,\quad x+y=2,\quad x+y+z=3;\quad f(x,y,z)=x$
8. $x=0,\quad x=\sqrt{a^2-y^2-z^2};\quad f(x,y,z)=x$
9. $z=0,\quad x^2+z=1,\quad y^2+z=1,\quad f(x,y,z)=z^2$
10. $x^2+z^2=a^2,\quad y^2+z^2=a^2,\quad f(x,y,z)=x^2+y^2$
11. $x=0,\quad y=0,\quad z=0,\quad (x/a)+(y/b)+(z/c)=1,\quad (a,b,c>0);\quad f(x,y,z)=z$
12. $y=z^2,\quad y^2=z,\quad x=0,\quad x=y-z^2;\quad f(x,y,z)=y+z^2$
13. $x=0,\quad y=0,\quad z=0,\quad x^{1/2}+y^{1/2}+z^{1/2}=a^{1/2};\quad f(x,y,z)=z$
14. $x=0,\quad y=0,\quad z=0,\quad y^2=4-z,\quad x=y+2;\quad f(x,y,z)=x^2;\quad y\geq 0$
15. $z=x^2+y^2,\quad z=27-2x^2-2y^2;\quad f(x,y,z)=1$
16. $z^2=4ax,\quad x^2+y^2=2ax;\quad f(x,y,z)=1$

*17. $y^2+z^2=4ax,\quad y^2=ax,\quad x=3a;\quad f(x,y,z)=x^2$

In Problems 18 through 22, express each iterated integral as a triple integral by describing the set S over which the integration is performed. Sketch the set S and then express the iterated integral in two orders differing from the original. Do not evaluate the integrals.

18. $\displaystyle\int_0^1\int_0^x\int_0^{x-y} x\,dz\,dy\,dx$

19. $\displaystyle\int_{-1}^1\int_0^{1-y^2}\int_{-\sqrt{x}}^{\sqrt{x}} 2y^2\sqrt{x}\,dz\,dx\,dy$

20. $\displaystyle\int_0^1\int_{y^2}^{\sqrt{y}}\int_0^{y+z} xy\,dx\,dz\,dy$

21. $\displaystyle\int_{-2}^2\int_0^{4-y^2}\int_0^{y+2} (y^2+z^2)\,dz\,dx\,dy$

22. $\displaystyle\int_0^1\int_{x^2}^{\sqrt{x}}\int_0^{y-x^2} f(x,y,z)\,dz\,dy\,dx$

23. Express the integral of $f(x,y,z)$ over the region S bounded by the surface $z = \sqrt{16-x^2-y^2}$ and the plane $z=2$ in 6 ways.

24. Express the integral

$$\int_0^2\int_0^z\int_0^x (x^2+y^2+z^2)\,dy\,dx\,dz$$

in 5 additional ways.

25. Let S be the solid tetrahedron with vertices at $(0,0,0)$, $(1,0,0)$, $(0,1,0)$, and $(0,0,1)$. Find the value of $\iiint_S f(x,y,z)\,dV$ where $f(x,y,z) = (x^2+2xz+y^2)$.

26. Let S be the solid pyramid with vertices $(0,0,0)$, $(1,0,0)$, $(1,1,0)$, $(0,1,0)$, and $(0,0,1)$. Find the value of $\iiint_S f(x,y,z)\,dV$ where $f(x,y,z) = xyz + 2yz$.

9. Mass of a Region in R^3. Triple Integrals in Cylindrical and Spherical Coordinates

From the definition of triple integral we see that **if $f(x,y,z) \equiv 1$, then the triple integral taken over a region S is precisely the volume $V(S)$.** More generally, if an object occupies a region S, and if the density at any point is given by $\delta(x,y,z)$, then the total mass, $m(S)$, is given by the triple integral

$$m(S) = \iiint_S \delta(x,y,z)\,dV.$$

NOTATION. For the remainder of this chapter the symbol δ will be used for density. The quantity ρ, which we previously used for density, will denote one of the variables in spherical coordinates.

EXAMPLE 1. The region S in the first octant is bounded by the surfaces

$$z = 4 - x^2 - y^2, \qquad z = 0, \qquad x + y = 2, \qquad x = 0, \qquad y = 0.$$

The density is given by $\delta(x,y,z) = 2z$. Find the total mass.

SOLUTION. We have (Fig. 5-55)

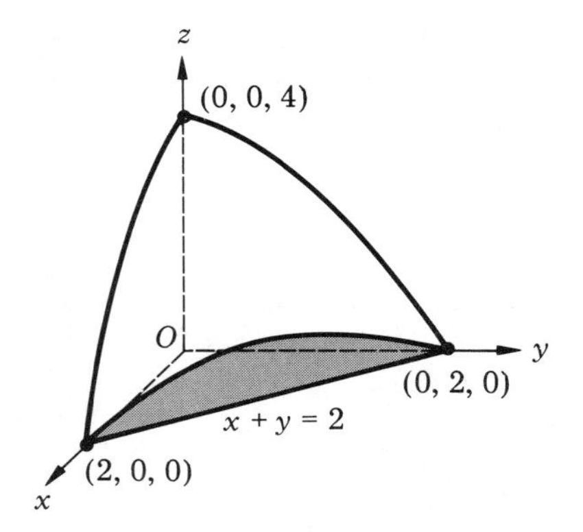

Fig. 5-55

$$\begin{aligned} m(S) &= \iiint_S 2z\,dV = 2\int_0^2\int_0^{2-x}\int_0^{4-x^2-y^2} z\,dz\,dy\,dx \\ &= \int_0^2\int_0^{2-x} (4-x^2-y^2)^2\,dy\,dx \\ &= \int_0^2\int_0^{2-x} (16+x^4+y^4-8x^2-8y^2+2x^2y^2)\,dy\,dx \\ &= \int_0^2 \left[(4-x^2)^2y - \frac{2}{3}(4-x^2)y^3 + \frac{1}{5}y^5\right]_0^{2-x} dx \\ &= \int_0^2 \left[2(4-x^2)^2 - x(4-x^2)^2 - \frac{2}{3}(4-x^2)(2-x)^3 \right. \\ &\qquad \left. + \frac{1}{5}(2-x)^5\right] dx. \end{aligned}$$

The above integral, a polynomial in x, can be evaluated. The answer is 704/45.

We found that certain double integrals are easy to evaluate if a polar coordinate system is used. Similarly, there are triple integrals which, although difficult to evaluate in rectangular coordinates, are simple integrations when transformed into other systems. The most useful transformations are those to cylindrical and spherical coordinates. (See Chapter 1, page 32 ff.)

Cylindrical coordinates consist of polar coordinates in the plane and a z coordinate as in a rectangular system. The transformation from rectangular to cylindrical coordinates is

$$x = r\cos\theta, \qquad y = r\sin\theta, \qquad z = z. \tag{1}$$

A region S in (x, y, z) space corresponds to a region U in (r, θ, z) space. The volume of S, $V(S)$, may be found in terms of a triple integral in (r, θ, z)-space by the formula (Fig. 5-56)

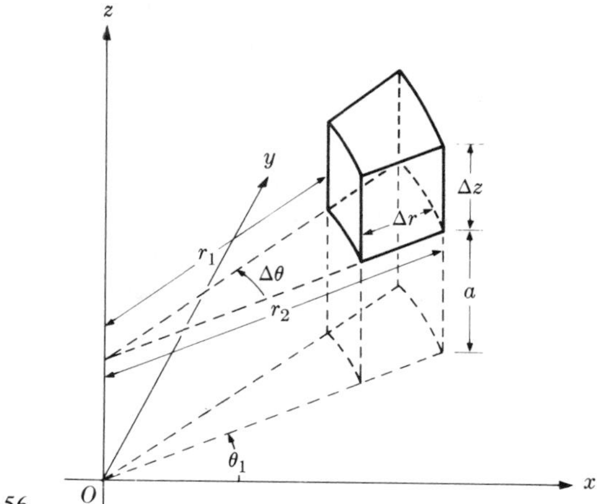

Fig. 5-56

$$V(S) = \iiint\limits_{U} r\,dV_{r\theta z}.$$

This formula is a natural extension of the formula relating area in rectangular and in polar coordinates. (See page 318 ff.)

More generally, if $f(x, y, z)$ is a continuous function, and if we define

$$g(r, \theta, z) = f(r\cos\theta, r\sin\theta, z).$$

then we have the following relationship between triple integrals:

$$\iiint\limits_{S} f(x, y, z)\,dV_{xyz} = \iiint\limits_{U} g(r, \theta, z) r\,dV_{r\theta z}.$$

A triple integral in cylindrical coordinates may be evaluated by iterated integrals. We write

$$\iiint g(r, \theta, z) r\,dV_{r\theta z} = \iiint g(r, \theta, z) r\,dr\,d\theta\,dz,$$

and, as before, there are five other orders of integration possible. Once again the major problem is the determination of the limits of integration. For this purpose it is helpful to superimpose cylindrical coordinates on a rectangular system, sketch the surface, and read off the limits of integration. The next example shows the method.

EXAMPLE 2. Find the mass of the region bounded by the cylinder $x^2 + y^2 = ax$ and the cone $z^2 = x^2 + y^2$ if the density is $\delta = k\sqrt{x^2 + y^2}$ (Fig. 5-57).

SOLUTION. We change to cylindrical coordinates. The cylinder is $r = a\cos\theta$ and the cone is $z^2 = r^2$. The density is kr. The region S corresponds to the

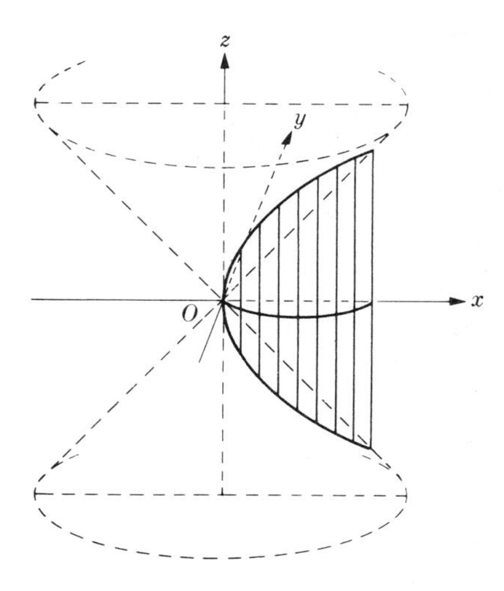

Fig. 5-57

region U given by

$$U = \left\{(r, \theta, z) : 0 \leq r \leq a\cos\theta, \quad -\frac{\pi}{2} \leq \theta \leq \frac{\pi}{2}, \quad -r \leq z \leq r\right\}.$$

Therefore

$$m(S) = \iiint_S k\sqrt{x^2 + y^2}\, dV_{xyz} = k \iiint_U r \cdot r \cdot dV_{r\theta z}$$

$$= k \int_{-\pi/2}^{\pi/2} \int_0^{a\cos\theta} \int_{-r}^{r} r^2\, dz\, dr\, d\theta$$

$$= 2k \int_{-\pi/2}^{\pi/2} \int_0^{a\cos\theta} r^3\, dr\, d\theta = \frac{1}{2}ka^4 \int_{-\pi/2}^{\pi/2} \cos^4\theta\, d\theta$$

$$= \frac{1}{8}ka^4 \int_{-\pi/2}^{\pi/2} \left(1 + 2\cos 2\theta + \frac{1 + \cos 4\theta}{2}\right) d\theta = \frac{3k\pi a^4}{16}.$$

The transformation from rectangular to spherical coordinates is given by the equations (see page 33)

$$x = \rho\cos\theta\sin\phi, \qquad y = \rho\sin\theta\sin\phi, \qquad z = \rho\cos\phi. \tag{2}$$

The region U given by

$$U = \{(\rho, \theta, \phi) : \rho_1 \leq \rho \leq \rho_2, \quad \theta_1 \leq \theta \leq \theta_2, \quad \phi_1 \leq \phi \leq \phi_2\}$$

corresponds to a rectangular box in (ρ, θ, ϕ)-space. We wish to find the volume of the region S in (x, y, z)-space which corresponds to U under the transformation (2). Referring to Fig. 5-58, we see that S is a region such as $ABCDA'B'C'D'$ between the spheres $\rho = \rho_1$ and $\rho = \rho_2$, between the planes $\theta = \theta_1$ and $\theta = \theta_2$, and between the cones $\phi = \phi_1(OADD'A')$ and $\phi =$

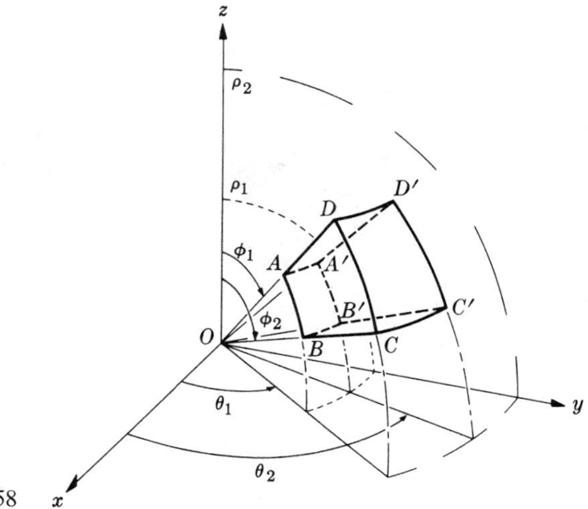

Fig. 5-58

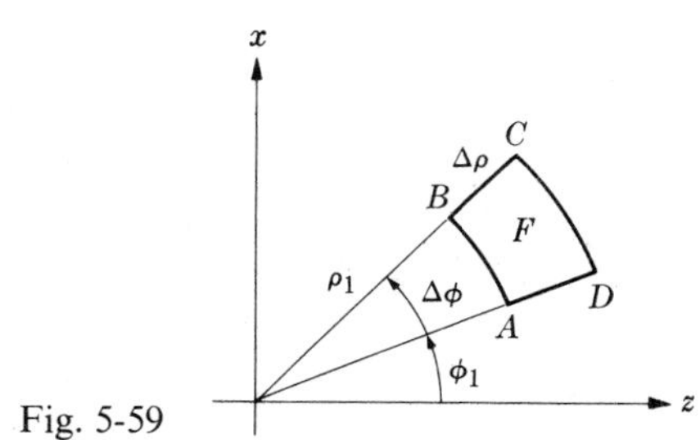

Fig. 5-59

$\phi_2(OBCC'B')$. The region S is obtained by sweeping the plane region $F(ABCD$ shown in Fig. 5-59) through an angle $\Delta\theta = \theta_2 - \theta_1$. The formula for determining the volume when an area is swept through an angle about an axis is given by

$$V(S) = \Delta\theta \iint_F x\, dA_{zx}.$$

Since ρ, ϕ are polar coordinates in the zx plane (Fig. 5-59), we have

$$x = \rho \sin\phi \qquad \text{and} \qquad dA_{zx} = \rho\, dA_{\rho\phi} = \rho\, d\rho\, d\phi.$$

Therefore

$$\begin{aligned} V(S) &= \Delta\theta \int_{\phi_1}^{\phi_2} \int_{\rho_1}^{\rho_2} \rho^2 \sin\phi\, d\rho\, d\phi \\ &= \int_{\theta_1}^{\theta_2} \int_{\phi_1}^{\phi_2} \int_{\rho_1}^{\rho_2} \rho^2 \sin\phi\, d\rho\, d\phi\, d\theta \\ &= \iiint_U \rho^2 \sin\phi\, dV_{\rho\theta\phi}. \end{aligned}$$

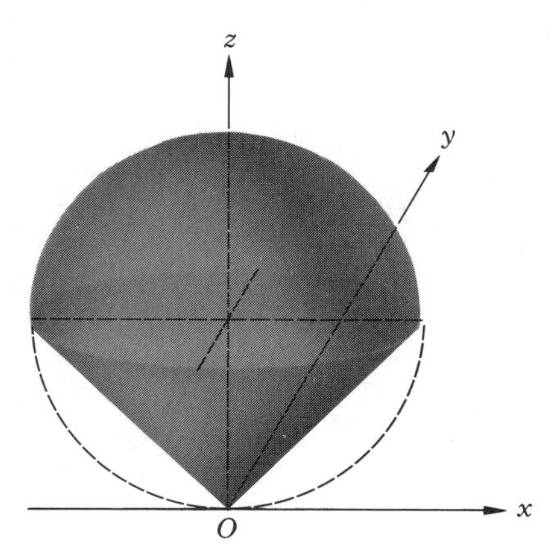

Fig. 5-60

More generally, if $f(x, y, z)$ is continuous on a region S and if

$$g(\rho, \theta, \phi) = f(\rho \cos\theta \sin\phi, \rho \sin\theta \sin\phi, \rho \cos\phi),$$

then the triple integral of f may be transformed according to the formula

$$\iiint_S f(x, y, z)\, dV_{xyz} = \iiint_U g(\rho, \theta, \phi)\rho^2 \sin\phi\, dV_{\rho\theta\phi}.$$

Once again, the triple integral is evaluated by iterated integrations. The next example illustrates the process.

EXAMPLE 3. Find the volume above the cone $z^2 = x^2 + y^2$ and inside the sphere $x^2 + y^2 + z^2 = 2az$ (Fig. 5-60).

SOLUTION. In spherical coordinates the cone and sphere have the equations

$$\phi = \frac{\pi}{4} \qquad \text{and} \qquad \rho = 2a\cos\phi,$$

respectively. Therefore

$$\begin{aligned} V(S) &= \iiint_S dV_{xyz} = \iiint_U \rho^2 \sin\phi\, dV_{\rho\theta\phi} \\ &= \int_0^{\pi/4} \int_0^{2a\cos\phi} \int_0^{2\pi} \rho^2 \sin\phi\, d\theta\, d\rho\, d\phi \\ &= 2\pi \int_0^{\pi/4} \int_0^{2a\cos\phi} \rho^2 \sin\phi\, d\rho\, d\phi = \frac{16a^3\pi}{3} \int_0^{\pi/4} \cos^3\phi \sin\phi\, d\phi \\ &= \frac{4a^3\pi}{3}[-\cos^4\phi]_0^{\pi/4} = \pi a^3. \end{aligned}$$

PROBLEMS

In each of Problems 1 through 16, find the mass of the region having the given density δ and bounded by the surfaces whose equations are given.

1. $z^2 = x^2 + y^2$, $x^2 + y^2 + z^2 = a^2$, above the cone, $\delta = \text{const}$.
2. The rectangular parallelepiped bounded by $x = -a$, $x = a$, $y = -b$, $y = b$, $z = -c$, $z = c$, $\delta = k(x^2 + y^2 + z^2)$.
3. $x^2 + y^2 + z^2 = a^2$, $x^2 + y^2 + z^2 = b^2$, $a < b$, $\delta = k\sqrt{x^2 + y^2 + z^2}$.
4. The rectangular parallelepiped bounded by $x = 0$, $x = 2a$, $y = 0$, $y = 2b$, $z = 0$, $z = 2c$, $\delta = k(x^2 + y^2 + z^2)$.
5. $x^2 + y^2 = a^2$, $x^2 + y^2 + z^2 = 4a^2$; $\delta = kz^2$; outside the cylinder.
6. The tetrahedron bounded by the coordinate planes and $x + y + z = 1$; $\delta = kxyz$.
7. $x^2 + y^2 = 2ax$, $x^2 + y^2 + z^2 = 4a^2$; $\delta = k(x^2 + y^2)$.
8. $z^2 = 25(x^2 + y^2)$, $z = x^2 + y^2 + 4$; $\delta = \text{const}$; above the paraboloid.
9. $z^2 = x^2 + y^2$, $x^2 + y^2 + z^2 = 2az$; above the cone; $\delta = kz$.
10. Interior of $x^2 + y^2 + z^2 = a^2$; $\delta = k(x^2 + y^2 + z^2)^n$, n a positive number.
11. $x^2 + y^2 = az$, $x^2 + y^2 + z^2 = 2az$; above the paraboloid; $\delta = \text{const}$.
12. $2z = x^2 + y^2$, $z = 2x$; $\delta = k\sqrt{x^2 + y^2}$.
13. $x^2 + y^2 + z^2 = a^2$, $r^2 = a^2 \cos 2\theta$ (cylindrical coordinates); $\delta = \text{const}$.
14. $x^2 + y^2 + z^2 = 4az$, $z = 3a$, above the plane; $\delta = k\sqrt{x^2 + y^2 + z^2}$.
15. $z^2 = x^2 + y^2$, $(x^2 + y^2)^2 = a^2(x^2 - y^2)$; $\delta = k\sqrt{x^2 + y^2}$.

*16. $z^2 = x^2 + y^2$, $x^2 + y^2 + z^2 = 2ax$; $\delta = \text{const}$; above the cone.

10. Moment of Inertia. Center of Mass

The definition of moment of inertia of a solid body is similar to the definition of moment of inertia of a plane region (page 326). The following definition is basic for the material of this section.

Definition. Suppose that a body occupies a region S and that L is any line in three-space. We make a subdivision of space into rectangular boxes and let $S_1, S_2, \ldots, S_n$ be those boxes which contain points of S. For each i, select any point $P_i(\xi_i, \eta_i, \zeta_i)$ in S_i. If the sums

$$\sum_{i=1}^{n} r_i^2 m(S_i) \qquad (r_i = \text{distance of } P_i \text{ from } L)$$

tend to a limit I as the norms of the subdivisions tend to zero, and for any choices of the P_i, then I is called the **moment of inertia of the solid S about L.**

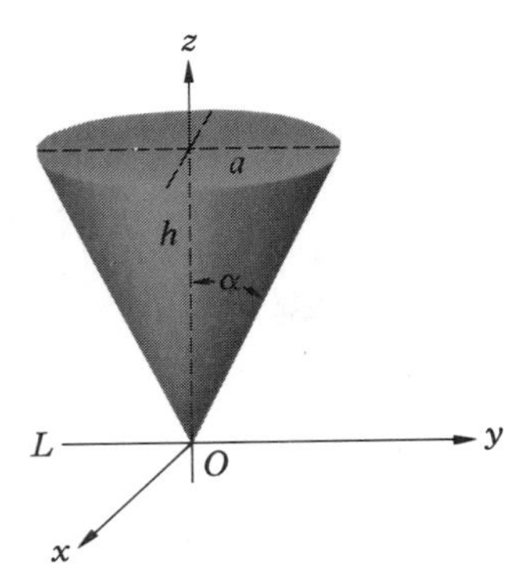

Fig. 5-61

It can be shown that if S has continuous density $\delta(x, y, z)$, then the moments of inertia I_x, I_y, and I_z about the x, y, and z axes, respectively, are given by the triple integrals

$$I_x = \iiint_S (y^2 + z^2)\delta(x, y, z)\, dV, \qquad I_y = \iiint_S (x^2 + z^2)\delta(x, y, z)\, dV,$$

$$I_z = \iiint_S (x^2 + y^2)\delta(x, y, z)\, dV.$$

If a region S has a density $\delta(x, y, z)$ and a mass $m(S)$, the point $(\bar{x}, \bar{y}, \bar{z})$, defined by the formulas

$$\bar{x} = \frac{\iiint_S x\delta(x, y, z)\, dV}{m(S)}, \quad \bar{y} = \frac{\iiint_S y\delta(x, y, z)\, dV}{m(S)}, \quad \bar{z} = \frac{\iiint_S z\delta(x, y, z)\, dV}{m(S)}$$

is called the **center of mass of** S.

In determining the center of mass, it is helpful to take into account all available symmetries. The following rules are useful:

a) *If S is symmetric in the xy plane and $\delta(x, y, -z) = \delta(x, y, z)$, then $\bar{z} = 0$. A similar result holds for other coordinate planes.*
b) *If S is symmetric in the x axis and $\delta(x, -y, -z) = \delta(x, y, z)$, then $\bar{y} = \bar{z} = 0$. A similar result holds for the other axes.*

EXAMPLE 1. Find the moment of inertia of the interior of a homogeneous cone of base radius a and altitude h about a line through the vertex and perpendicular to the axis.

SOLUTION. We take the vertex at the origin, the z axis as the axis of the cone, and the x axis as the line L about which the moment of inertia is to be computed (Fig. 5-61).

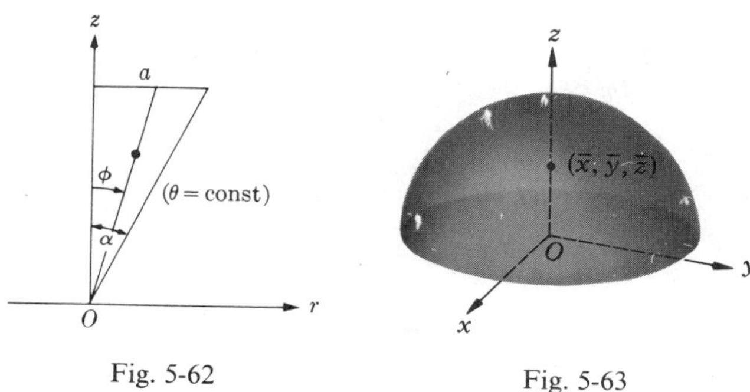

Fig. 5-62 Fig. 5-63

Let $\alpha = \arctan(a/h)$ be half the angle opening of the cone. In spherical coordinates we get (see Fig. 5-62)

$$I = \iiint_S (y^2 + z^2)\, dV_{xyz}$$
$$= \int_0^{\alpha} \int_0^{h\sec\phi} \int_0^{2\pi} \rho^2(\sin^2\phi \sin^2\theta + \cos^2\phi)\rho^2 \sin\phi \, d\theta\, d\rho\, d\phi.$$

Since

$$\int_0^{2\pi} \sin^2\theta\, d\theta = \pi,$$

we obtain

$$I = \pi \int_0^{\alpha} \int_0^{h\sec\phi} \rho^4(1 + \cos^2\phi)\sin\phi\, d\rho\, d\phi$$
$$= \frac{\pi h^5}{5} \int_0^{\alpha} [(\cos\phi)^{-5} + (\cos\phi)^{-3}]\sin\phi\, d\phi$$
$$= \frac{\pi h^5}{5}\left[\frac{\sec^4\alpha - 1}{4} + \frac{\sec^2\alpha - 1}{2}\right] = \frac{\pi h a^2}{20}(4h^2 + a^2),$$

since $\tan\alpha = a/h$.

EXAMPLE 2. Find the center of mass of the interior of a hemisphere of radius a which has density proportional to the distance from the center of the sphere.

SOLUTION. We select the hemisphere so that the plane section is in the xy plane and the z axis in axis of symmetry (Fig. 5-63). Then $\bar{x} = \bar{y} = 0$. Changing to spherical coordinates, we have

$$m(S) = \int_0^{2\pi} \int_0^{\pi/2} \int_0^{a} k\rho \cdot \rho^2 \sin\phi\, d\rho\, d\phi\, d\theta = \frac{1}{2}\pi k a^4$$

and from this we obtain

$$\bar{z} = \frac{\displaystyle\int_0^{2\pi}\int_0^{\pi/2}\int_0^a \rho\cos\phi\cdot k\rho\cdot\rho^2\sin\phi\,d\rho\,d\phi\,d\theta}{\frac{1}{2}\pi k a^4}$$

$$= \frac{2}{\pi a^4}\cdot\frac{1}{5}a^5\cdot 2\pi\int_0^{\pi/2}\cos\phi\sin\phi\,d\phi$$

$$= \frac{4}{5}a\left[\frac{\sin^2\phi}{2}\right]_0^{\pi/2} = \frac{2a}{5}.$$

The center of mass is at $(0, 0, 2a/5)$.

PROBLEMS

In each of Problems 1 through 15, find the moment of inertia about the given axis of the region having the specified density δ and bounded by the surfaces as described.

1. A cube of side a; $\delta = \text{const}$; about an edge.
2. A cube of side a; $\delta = \text{const}$; about a line parallel to an edge, at distance 2 from it, and in a plane of one of the faces.
3. Bounded by $x = 0$, $y = 0$, $z = 0$, $x + z = a$, $y = z$; $\delta = kx$; about the x axis.
4. $x^2 + y^2 = a^2$; $x^2 + z^2 = a^2$; $\delta = \text{const}$; about the z axis.
5. $z = x$, $y^2 = 4 - 2z$, $x = 0$; $\delta = \text{const}$; about the z axis.
6. $x = 0$, $y = 0$, $z^2 = 1 - x - y$; $\delta = \text{const}$; about the z axis.
7. $z^2 = y^2(1 - x^2)$; $y = 1$; $\delta = \text{const}$; about the x axis.
8. $x^2 + y^2 = a^2$, $x^2 + y^2 = b^2$, $z = 0$, $z = h$; $\delta = k\sqrt{x^2 + y^2}$; about the x axis ($a < b$).
9. $x^2 + y^2 + z^2 = a^2$, $x^2 + y^2 + z^2 = b^2$; $\delta = k\sqrt{x^2 + y^2 + z^2}$; about the z axis ($a < b$).
10. $\rho = 4$, $\rho = 5$, $z = 1$, $z = 3\,(1 \le z \le 3)$; $\delta = \text{const}$; about the z axis.
11. $z^2 = x^2 + y^2$, $(x^2 + y^2)^2 = a^2(x^2 - y^2)$; $\delta = k\sqrt{x^2 + y^2}$; about the z axis.
12. $z = 0$, $z = \sqrt{a^2 - x^2 - y^2}$; $\delta = kz$; about the x axis.
13. $x^2/a^2 + y^2/b^2 + z^2/c^2 = 1$; $\delta = \text{const}$; about the z axis. (Divide the integral into two parts.)
14. $(r - b)^2 + z^2 = a^2\,(0 < a < b)$; $\delta = kz$; about the z axis. Assume $z \ge 0$.
15. $\rho = b$, $\rho = c$, $r = a$ $(a < b < c)$; outside $r = a$, between $\rho = b$ and $\rho = c$; $\delta = \text{const}$; about the z axis.

In each of Problems 16 through 32, find the center of mass of the region having the given density and bounded by the surfaces as described. Describe each region in set notation.

16. $x = 0, \quad y = 0, \quad z = 0, \quad x/a + y/b + z/c = 1; \quad \delta = \text{const.}$

17. $x = 0, \quad y = 0, \quad z = 0, \quad x + z = a, \quad y = z; \quad \delta = kx.$

18. $x^2 + y^2 = a^2, \quad x^2 + z^2 = a^2, \quad \delta = \text{const};$ the portion where $x \geq 0$.

19. $z = x, \quad z = -x, \quad y^2 = 4 - 2x; \quad \delta = \text{const.}$

20. $z^2 = y^2(1 - x^2), \quad y = 1; \quad \delta = \text{const.}$

21. $z = 0, \quad x^2 + z = 1, \quad y^2 + z = 1; \quad \delta = \text{const.}$

22. $x = 0, \quad y = 0, \quad z = 0, \quad x^{1/2} + y^{1/2} + z^{1/2} = a^{1/2}; \quad \delta = \text{const.}$

23. $y^2 + z^2 = 4ax, \quad y^2 = ax, \quad x = 3a; \quad \delta = \text{const.}$ (Inside $y^2 = ax$.)

24. $z^2 = 4ax, \quad x^2 + y^2 = 2ax; \quad \delta = \text{const.}$

25. $z^2 = x^2 + y^2, \quad x^2 + y^2 + z^2 = a^2,$ above the cone; $\delta = \text{const.}$

26. $z^2 = x^2 + y^2, \quad x^2 + y^2 = 2ax; \quad \delta = k(x^2 + y^2).$

27. $z^2 = x^2 + y^2, \quad x^2 + y^2 + z^2 = 2az,$ above the cone; $\delta = kz.$

28. $x^2 + y^2 = az, \quad x^2 + y^2 + z^2 = 2az,$ above the paraboloid; $\delta = \text{const.}$

29. $x^2 + y^2 + z^2 = 4az, \quad z = 3a,$ above the plane; $\delta = k\sqrt{x^2 + y^2 + z^2}.$

30. $\rho = 4, \quad \rho = 5, \quad z = 1, \quad z = 3\ (1 \leq z \leq 3); \quad \delta = \text{const.}$

*31. $z^2 = x^2 + y^2, \quad (x^2 + y^2)^2 = a^2(x^2 - y^2); \quad \delta = k\sqrt{x^2 + y^2};$ the part for which $x \geq 0$.

*32. $z^2 = x^2 + y^2, \quad x^2 + y^2 + z^2 = 2ax,$ above the cone; $\delta = \text{const.}$

CHAPTER 6

Fourier Series

1. Fourier Series

In the study of infinite series, the functions

$$1, x, x^2, \ldots, x^n, \ldots$$

play a central role. Most of the elementary functions of algebra, trigonometry, and calculus may be expanded in series which are sums of powers of x, that is, in power series. The coefficients in such a Taylor or Maclaurin series are the successive derivatives of the given function evaluated at a point.

The simple function $f(x) = |x|$ cannot be expanded in a Maclaurin series. Since f does not have a derivative at $x = 0$ (Fig. 6-1), there is no way to compute the coefficient of x^n, $n \geq 1$, in such an expansion.

The collection of functions

$$1, \cos x, \cos 2x, \ldots, \cos nx, \ldots,$$

$$\sin x, \sin 2x, \ldots, \sin nx, \ldots$$

all have period 2π. We consider a function f which is periodic with period 2π and try to represent it in a series of the form

$$f(x) = \frac{a_0}{2} + \sum_{n=1}^{\infty} (a_n \cos nx + b_n \sin nx), \tag{1}$$

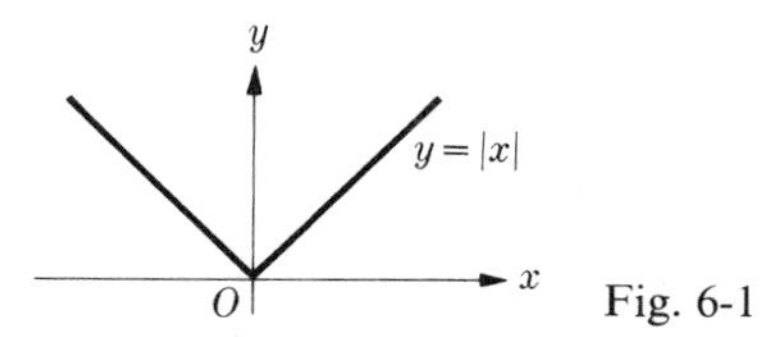

Fig. 6-1

where all the a_n and b_n, $n = 0, 1, 2, \ldots$, are constants. Suppose the above series (1) converges uniformly to $f(x)$ on the interval $-\pi \le x \le \pi$. Then since $u_n(x) = a_n \cos nx + b_n \sin nx$ is continuous, we know (Theorem 30′ in Chapter 3) that f is continuous on $-\pi \le x \le \pi$. Furthermore, we may integrate term by term and perform various manipulations with uniformly convergent series. Proceeding formally for the moment, we let m be a *fixed* integer and multiply the series (1) by $\cos mx$. We get

$$f(x)\cos mx = \frac{a_0}{2}\cos mx + \sum_{n=1}^{\infty} (a_n \cos nx \cos mx + b_n \sin nx \cos mx).$$

Now we integrate this series on the interval $[-\pi, \pi]$, obtaining

$$\begin{aligned} \int_{-\pi}^{\pi} f(x)\cos mx\,dx &= \frac{a_0}{2}\int_{-\pi}^{\pi} \cos mx\,dx \\ &\quad + \sum_{n=1}^{\infty}\left[a_n \int_{-\pi}^{\pi} \cos nx \cos mx\,dx + b_n \int_{-\pi}^{\pi} \sin nx \cos mx\,dx\right]. \end{aligned} \tag{2}$$

All the integrals on the right in the above expression may be calculated by elementary means. The reader can easily verify that

$$\int_{-\pi}^{\pi} \cos nx\,dx = \int_{-\pi}^{\pi} \sin nx\,dx = 0 \qquad \text{for } n = 1, 2, \ldots.$$

Also, using trigonometric relations such as

$$\cos mx \cos nx = \tfrac{1}{2}[\cos(m+n)x + \cos(m-n)x],$$

we find it a simple matter to verify that

$$\int_{-\pi}^{\pi} \cos mx \cos nx\,dx = \int_{-\pi}^{\pi} \sin mx \sin nx\,dx = \begin{cases} \pi & \text{if } m = n, \\ 0 & \text{if } m \ne n, \end{cases}$$

and

$$\int_{-\pi}^{\pi} \cos mx \sin nx\,dx = 0$$

for $m, n = 1, 2, \ldots$. In Eq. (2), all the integrals on the right have the value zero except the term in which $n = m$. We conclude (on the basis of the formal manipulations) that

$$\int_{-\pi}^{\pi} f(x)\cos mx\,dx = \pi a_m, \qquad m = 0, 1, 2, \ldots.$$

Next, multiplying the series (1) by $\sin mx$ with m fixed and then integrating term by term, we obtain the corresponding formulas for b_m. They are

$$\int_{-\pi}^{\pi} f(x)\sin mx\,dx = \pi b_m, \qquad m = 1, 2, \ldots.$$

The above development leads to the following theorem:

Theorem 1. *Suppose that f is continuous for all x and periodic with period 2π. Suppose that the series*

$$f(x) = \frac{a_0}{2} + \sum_{n=1}^{\infty} (a_n \cos nx + b_n \sin nx) \tag{3}$$

converges uniformly to $f(x)$ for all x. Then

$$\boxed{\begin{aligned} a_n &= \frac{1}{\pi}\int_{-\pi}^{\pi} f(t)\cos nt\,dt, \qquad n = 0, 1, 2, \ldots, \\ b_n &= \frac{1}{\pi}\int_{-\pi}^{\pi} f(t)\sin nt\,dt, \qquad n = 1, 2, 3, \ldots. \end{aligned}} \tag{4}$$

PROOF. Let

$$s_p(x) \equiv \frac{a_0}{2} + \sum_{k=1}^{p} (a_k \cos kx + b_k \sin kx)$$

be the pth partial sum of the series in (3). Since the sequence $s_p(x)$ converges uniformly to $f(x)$, it follows that $s_p(x)\cos nx$ converges uniformly to $f(x)\cos nx$ for each fixed n. In fact,

$$|s_p(x)\cos nx - f(x)\cos nx| = |s_p(x) - f(x)|\,|\cos nx| \le |s_p(x) - f(x)|,$$

and the last expression on the right tends to zero uniformly as $p \to \infty$. Similarly, $s_p(x)\sin nx$ converges uniformly to $f(x)\sin nx$. Therefore, the series for $s_p(x)\cos nx$ and for $s_p(x)\sin nx$ may be integrated term by term (Theorem 31 in Chapter 3). Performing this process as outlined in the discussion preceding the theorem, we obtain the formulas for a_n and b_n.

Definitions. The series (3) is called the **Fourier series** of the function f, and the numbers a_n and b_n as given by (4), are called the **Fourier coefficients of** f.

The fact that Theorem 1 is valid only for functions which are periodic with period 2π may be considered an unsatisfactory feature of a Fourier series. Functions such as e^x, $\log(1 + x)$, etc., which have Taylor expansions, are not periodic at all. Later we shall see how to extend the study of Fourier series so that this problem may be partially overcome. (See page 372.)

A more serious objection to Theorem 1 is the requirement that the series (3) converge uniformly. If f is any continuous function, the coefficients a_n and b_n may be calculated by (4). It turns out that there are continuous functions f such that the resulting Fourier series does not converge to the function. On the other hand, there are *discontinuous* functions f such that the series *does* converge to the function. *If f is discontinuous, the convergence cannot be uniform* because of Theorem 30 of Chapter 3. In order to obtain a useful convergence theorem—one which includes convergence for discontinuous functions—we introduce the class of functions described in the next paragraph.

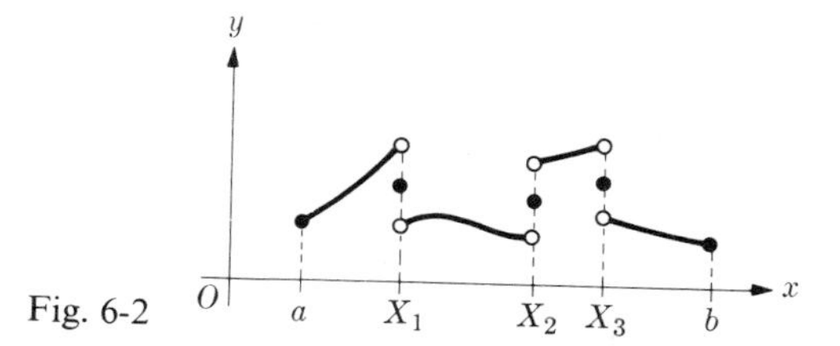

Fig. 6-2

Definitions. A function f is said to be **piecewise continuous** on an interval $[a, b]$ if and only if there is a finite subdivision

$$a = X_0 < X_1 < \cdots < X_{k-1} < X_k = b$$

such that the function f is continuous on each subinterval (X_{i-1}, X_i). Furthermore, the one-sided limits of f at each of the subdivision points X_i must exist (Fig. 6-2). At each subdivision point the function is discontinuous, and we call the difference $f(X_i +) - f(X_i -)$ the **jump** of f at X_i.

The function f may have any value at each X_i. In other words, a piecewise continuous function on $[a, b]$ is one which is continuous except at a finite number of points where it has jumps.

A function f is **piecewise smooth** on $[a, b] \Leftrightarrow f$ is piecewise continuous and f' is piecewise continuous, with the jumps of f' occurring at $X_0, X_1, \ldots, X_k$. The value of a piecewise continuous function at one of the points of discontinuity plays an important part in Fourier analysis. We say that a piecewise continuous function f is **normalized** if and only if its value at X_i is given by

$$f(X_i) = \tfrac{1}{2}[f(X_i -) + f(X_i +)], \qquad i = 1, 2, \ldots, k-1.$$

That is, the function value is halfway between the limit values from the left and right (Fig. 6-2). We say that a function f is **smooth** on $[a, b]$ if and only if f and f' are continuous throughout $[a, b]$.

The integral of a function f, thus far defined only for continuous functions, may easily be defined for piecewise continuous functions. *If f is continuous on $[a, b]$ except at $X_1, X_2, \ldots, X_{k-1}$ where it has jumps, we define*

$$\int_a^b f(x)\,dx = \sum_{i=1}^{k} \int_{X_{i-1}}^{X_i} f(x)\,dx. \tag{5}$$

The values of the integrals on the right in (5) are not influenced by the value of f at any X_i. Therefore, normalizing a piecesise continuous function by changing its values at the $\{X_i\}$ does not affect its integral. Since any finite linear combination of piecewise continuous functions is piecewise continuous, the formula

$$\int_a^b [cf(x) + dg(x)]\,dx = c\int_a^b f(x)\,dx + d\int_a^b g(x)\,dx$$

still holds. Furthermore, for piecewise continuous f,

$$m(b-a) \leq \int_a^b f(x)\,dx \leq M(b-a) \qquad \text{when } m \leq f(x) \leq M,$$

and

$$\int_a^b f(x)\,dx = \int_a^c f(x)\,dx + \int_c^b f(x)\,dx,$$

and so forth. In addition, if f is piecewise continuous on $[a, b]$, and if F is given by

$$F(x) = \int_a^x f(t)\,dt,$$

then F and F' are continuous except at $X_1, X_2, \ldots, X_{k-1}$. Moreover, F is continuous on $[a, b]$. To see this, we note that

$$|F(x_2) - F(x_1)| = \left|\int_{x_1}^{x_2} f(t)\,dt\right| \leq M|x_2 - x_1| \qquad \text{if } |f(x)| \leq M,$$

and so F satisfies the definition of continuity. Also, F is piecewise smooth on $[a, b]$.

The reader may verify that the product fg of a piecewise continuous function f and a continuous function g is piecewise continuous. Therefore $f(x) \cos nx$ and $f(x) \sin nx$ are piecewise continuous on $[-\pi, \pi]$ whenever f is. The coefficients a_n and b_n may be defined according to (4) for any piecewise continuous f, once the integral is defined for such functions. We next state a remarkable convergence theorem which, while not the most general one, is sufficiently broad for most applications.

Theorem 2. *Suppose that f is piecewise smooth, normalized, and periodic with period 2π. Then its Fourier series converges to $f(x)$ for each x. Furthermore, if f is smooth on the interval $I = \{x : c \leq x \leq d\}$, then the convergence is uniform on I.*

We shall prove the first part of this theorem in Section 4. (See page 378). For the second part, see Theorem 5, page 382.

Suppose we are given a piecewise smooth function f on the interval $[-\pi, \pi]$ and we wish to expand it in a Fourier series. First we define the **periodic extension** of f as the function $f_0(x)$ defined for all x by the relation

$$f_0(x) = \begin{cases} f(x) & \text{for } -\pi < x < \pi, \\ f_0(x - 2\pi) & \text{for all other } x. \end{cases}$$

Then we *normalize* f_0, if necessary (Fig. 6-3). No matter how smooth the function f may be, the periodic extension f_0 will introduce a discontinuity at π and $-\pi$ whenever $f(-\pi) \neq f(\pi)$. The study of Fourier series would have limited value if we were restricted from the beginning to functions

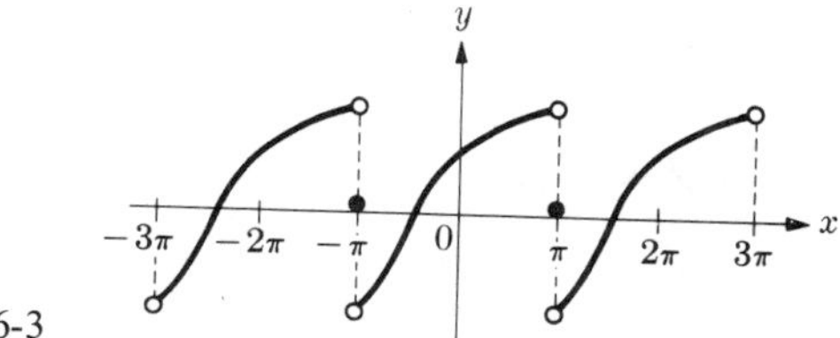

Fig. 6-3

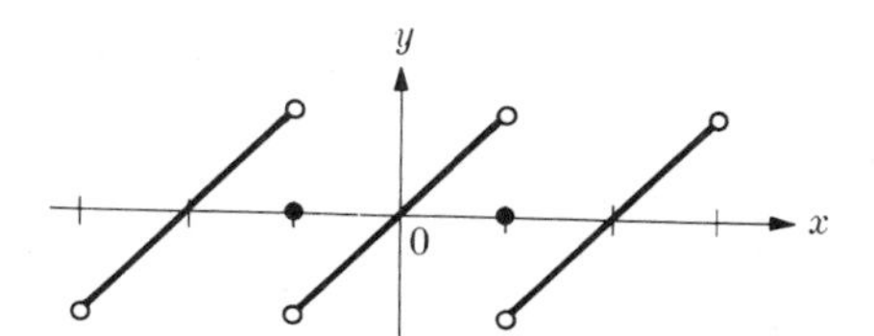

Fig. 6-4

which are smooth. The fact that the basic theory enables us to handle functions with jumps means that we may start with any integrable function defined on $[-\pi, \pi]$, form its periodic extension, and apply the theory. We illustrate this point with examples.

EXAMPLE 1. Find the Fourier series for the function

$$f(x) = x \qquad \text{on } (-\pi, \pi).$$

SOLUTION. We form the periodic extension of f and normalize it (Fig. 6-4). The normalization yields $f_0(-\pi) = f_0(\pi) = 0$. We compute a_n and b_n, using the formulas (4) for the coefficients:

$$a_n = \frac{1}{\pi}\int_{-\pi}^{\pi} x \cos nx\, dx, \qquad b_n = \frac{1}{\pi}\int_{-\pi}^{\pi} x \sin nx\, dx.$$

Integrating by parts, we find

$$a_n = \frac{1}{\pi}\left[\frac{x \sin nx}{n}\right]_{-\pi}^{\pi} - \frac{1}{n\pi}\int_{-\pi}^{\pi} \sin nx\, dx = 0, \qquad n = 1, 2, \ldots,$$

$$b_n = \frac{1}{\pi}\left[-\frac{x \cos nx}{n}\right]_{-\pi}^{\pi} + \frac{1}{n\pi}\int_{-\pi}^{\pi} \cos nx\, dx = -\frac{2}{n}\cos n\pi = (-1)^{n-1}\frac{2}{n}.$$

Since $a_0 = 0$, the Fourier series for $f(x) = x$ is

$$x = 2\left[\sin x - \frac{\sin 2x}{2} + \frac{\sin 3x}{3} - \frac{\sin 4x}{4} + \cdots\right], \qquad -\pi < x < \pi.$$

For $x = \pm\pi$, the series converges not to f but to the normalized value of f_0, the extension of f. In this case we notice that at $-\pi$ and π all the terms in the series vanish, and we verify that the series converges to $f_0(-\pi) = f_0(\pi) = 0$. A graph of the first few terms of the series is shown in Fig. 6-5.

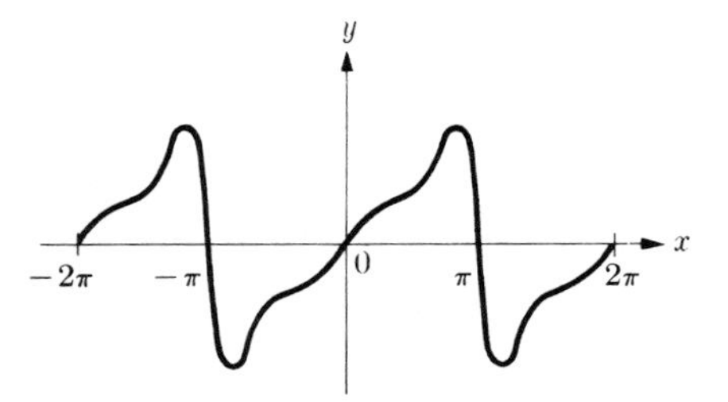

Fig. 6-5

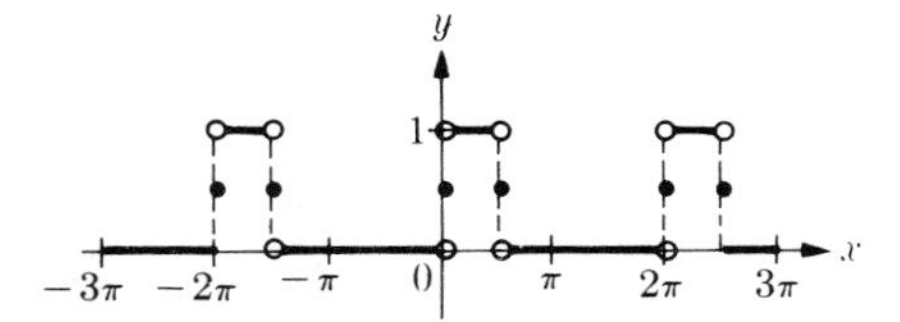

Fig. 6-6

EXAMPLE 2. Find the Fourier series of the function

$$f(x) = \begin{cases} 0, & -\pi \le x \le 0, \\ 1, & 0 < x < \dfrac{\pi}{2}, \\ 0, & \dfrac{\pi}{2} \le x \le \pi. \end{cases}$$

SOLUTION. We first extend f periodically and normalize it. The result is shown in Fig. 6-6. The solid dots in the figure show the normalized values at the jumps. We compute the coefficients:

$$a_0 = \frac{1}{\pi}\int_{-\pi}^{\pi} f(x)\,dx = \frac{1}{\pi}\int_0^{\pi/2} 1\,dx = \frac{1}{2},$$

$$a_n = \frac{1}{\pi}\int_{-\pi}^{\pi} f(x)\cos nx\,dx = \frac{1}{\pi}\int_0^{\pi/2} \cos nx\,dx = \frac{1}{n\pi}\sin nx\bigg]_0^{\pi/2}.$$

Therefore

$$a_n = \begin{cases} 0 & \text{if } n \text{ is even} \\ \dfrac{(-1)^k}{(2k+1)\pi} & \text{if } n = 2k+1, \quad k = 0, 1, 2, \ldots. \end{cases}$$

To find b_n, we write

$$b_n = \frac{1}{\pi}\int_0^{\pi/2} \sin nx\,dx = -\frac{\cos nx}{n\pi}\bigg]_0^{\pi/2} = \begin{cases} \dfrac{1}{(2k+1)\pi} & \text{if } n = 2k+1, \\ \dfrac{1-(-1)^k}{2k\pi} & \text{if } n = 2k. \end{cases}$$

The values of $\sin nx$ and $\cos nx$ at odd and even multiples of $\pi/2$ occur often in Fourier series; the reader should study carefully the evaluations above to be sure he understands how they are obtained. The desired Fourier series for f is

$$f(x) = \frac{1}{4} + \frac{1}{\pi}\left[\cos x + \sin x + \sin 2x - \frac{\cos 3x}{3} + \frac{\sin 3x}{3} + \frac{\cos 5x}{5} + \frac{\sin 5x}{5} + \frac{\sin 6x}{3} + \cdots\right].$$

Note that at $x = 0, \pi/2, 2\pi, 5\pi/2, \ldots$, the series must converge to the value $1/2$.

In computing the Fourier coefficients we may frequently save a great deal of labor by using certain properties of even and odd functions. We state that a function f is **even** if

$$f(-x) = f(x)$$

for all x. A function g is **odd** if

$$g(-x) = -g(x)$$

for all x. Using the definition of integral, we observe that if f is even and g is odd then, for any value a,

$$\int_{-a}^{a} f(x)\,dx = 2\int_{0}^{a} f(x)\,dx, \qquad \int_{-a}^{a} g(x)\,dx = 0.$$

The product of two even functions is even, the product of two odd functions is even, and the product of an even and an odd function is odd. For every positive integer n, the function $\cos nx$ is even and the function $\sin nx$ odd. Observe in Example 1 that $f(x) = x$ is *odd*; then clearly $f(x)\cos nx$ is odd for every n; we can thus conclude without any computation at all that $a_n = 0$, for $n = 0, 1, 2, \ldots$. In Example 2, since the function f is neither even nor odd, no simplification can be made on this basis. The preceding discussion is now stated in the form of a theorem.

Theorem 3. *Suppose that f is periodic with period 2π and is piecewise continuous. Then*

a) *if f is odd on $(-\pi, \pi)$, we have $a_n = 0$ for $n = 0, 1, 2, \ldots$, and*

$$b_n = \frac{2}{\pi}\int_0^{\pi} f(x)\sin nx\,dx;$$

b) *if f is even on $(-\pi, \pi)$, we have $b_n = 0$ for $n = 1, 2, \ldots$, and*

$$a_n = \frac{2}{\pi}\int_0^{\pi} f(x)\cos nx\,dx.$$

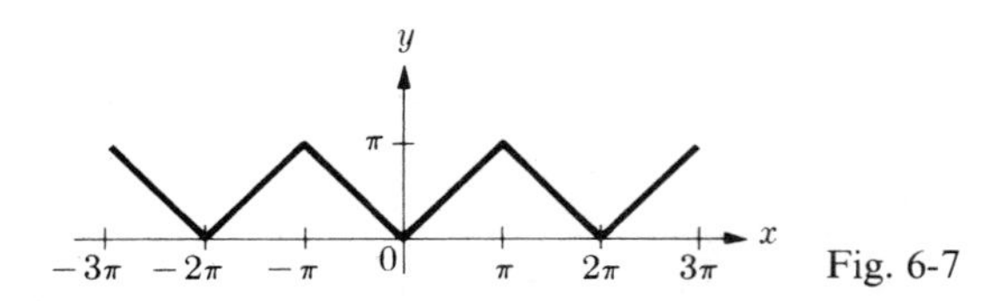

Fig. 6-7

EXAMPLE 3. Find the Fourier series of the function

$$f(x) = |x|, \qquad -\pi < x < \pi.$$

SOLUTION. We form the periodic extension of f as shown in Fig. 6-7. Note that the extended function happens to be continuous; normalization is therefore unnecessary. Because f is even, we conclude at once that $b_n = 0$ for all n, that $a_0 = \pi$, and that

$$a_n = \frac{2}{\pi}\int_0^\pi f(x)\cos nx\,dx.$$

Since $f(x) = x$ for $x \geq 0$ we obtain, upon integrating by parts,

$$\begin{aligned} a_n &= \frac{2}{\pi}\left[\frac{x\sin nx}{n}\right]_0^\pi - \frac{2}{n\pi}\int_0^\pi \sin nx\,dx = \left.\frac{2}{n^2\pi}\cos nx\right]_0^\pi \\ &= \frac{2}{n^2\pi}[\cos n\pi - 1] = \frac{2}{n^2\pi}[(-1)^n - 1] \\ &= \left\{\begin{array}{ll} \dfrac{-4}{(2k+1)^2\pi} & \text{for } n = 2k+1 \\ 0 & \text{for } n = 2k \end{array}\right\}, \qquad k = 0, 1, 2, \ldots. \end{aligned}$$

The desired series is

$$f(x) = \frac{\pi}{2} - \frac{4}{\pi}\left[\frac{\cos x}{1^2} + \frac{\cos 3x}{3^2} + \cdots + \frac{\cos(2k+1)x}{(2k+1)^2} + \cdots\right].$$

Setting $x = 0$ and noting that $f(0) = 0$, we obtain the remarkable formula

$$\pi^2 = 8(1 + \tfrac{1}{9} + \tfrac{1}{25} + \tfrac{1}{49} + \cdots).$$

PROBLEMS

In each of Problems 1 through 15, find the Fourier series for the given function. Draw a graph of the periodic, normalized extension f_0 on the interval $[-3\pi, 3\pi]$.

1. $f(x) = \begin{cases} \pi/4 & \text{for } 0 < x < \pi \\ -\pi/4 & \text{for } -\pi < x < 0 \end{cases}$

2. $f(x) = \begin{cases} 0 & \text{for } -\pi < x < 0 \\ 1 & \text{for } 0 < x < \pi \end{cases}$

3. $f(x) = x^2 \quad \text{for } -\pi \leq x \leq \pi$

4. $f(x) = \begin{cases} 0 & \text{for } -\pi < x < \pi/2 \\ 1 & \text{for } \pi/2 < x < \pi \end{cases}$

5. $f(x) = \begin{cases} -1 & \text{for } -\pi < x < -\pi/2 \\ 0 & \text{for } -\pi/2 < x < \pi/2 \\ 1 & \text{for } \pi/2 < x < \pi \end{cases}$

6. $f(x) = \begin{cases} 0 & \text{for } -\pi < x < 0 \\ x & \text{for } 0 \le x < \pi \end{cases}$

7. $f(x) = \begin{cases} 1 & \text{for } -\pi < x < -\pi/2 \\ 0 & \text{for } -\pi/2 < x < \pi/2 \\ 1 & \text{for } \pi/2 < x < \pi \end{cases}$

8. $f(x) = |\sin x| \quad \text{for } -\pi \le x \le \pi$

9. $f(x) = |\cos x| \quad \text{for } -\pi \le x \le \pi$

10. $f(x) = x^3 \quad \text{for } -\pi < x < \pi$

11. $f(x) = e^x \quad \text{for } -\pi < x < \pi$

12. $f(x) = \begin{cases} 0 & \text{for } -\pi < x < 0 \\ \sin x & \text{for } 0 < x < \pi \end{cases}$

13. $f(x) = \sin^2 x \quad \text{for } -\pi < x < \pi$

14. $f(x) = x \sin x \quad \text{for } -\pi < x < \pi$

15. $f(x) = \begin{cases} -\pi & \text{for } -\pi < x < 0 \\ x & \text{for } 0 < x < \pi \end{cases}$

16. Verify the formulas

$$\int_{-\pi}^{\pi} \cos mx \cos nx \, dx = \int_{-\pi}^{\pi} \sin mx \sin nx \, dx = \begin{cases} \pi & \text{if } m = n, \\ 0 & \text{if } m \neq n. \end{cases}$$

Also show that $\int_{-\pi}^{\pi} \cos mx \sin nx \, dx = 0 \quad$ for $m, n = 1, 2, \ldots$

17. For the series in Problem 15, show that for $x = 0$, we get the formula

$$\frac{\pi^2}{8} = \frac{1}{1^2} + \frac{1}{3^2} + \frac{1}{5^2} + \cdots$$

18. Find an expansion for π^2 in Example 3 by evaluation of the series at $x = \pi/4$. Can the error after n terms be estimated?

19. Given that $f(x) = x + x^2$, $-\pi < x < \pi$, find the Fourier expansion of f. Show that

$$\frac{\pi^2}{6} = \sum_{n=1}^{\infty} \frac{1}{n^2}.$$

20. Using the series expansion in Problem 1, show that

a) $\frac{\pi}{4} = 1 - \frac{1}{3} + \frac{1}{5} - \frac{1}{7} + \cdots$

b) $\frac{\pi}{3} = 1 + \frac{1}{5} - \frac{1}{7} - \frac{1}{11} + \frac{1}{13} + \frac{1}{17} - \cdots$

c) $\frac{\sqrt{3}}{6}\pi = 1 - \frac{1}{5} + \frac{1}{7} - \frac{1}{11} + \frac{1}{13} - \frac{1}{17} + \cdots$

21. Given the function $f(x) = x$ for $0 < x < \frac{\pi}{2}$. Extend the definition of f to the interval $-\pi < x < \pi$ in four different ways and compute the Fourier Series for these four functions.

22. Show that the derivative of an even function is an odd function and that the derivative of an odd function is an even function.

2. Half-Range Expansions

Suppose that a function f is defined on the interval $[0, \pi]$ and we wish to expand it in a Fourier series. Since the coefficients a_n and b_n involve integrals from $-\pi$ to π, we must somehow extend the definition of f to the interval $(-\pi, \pi)$. We can do this in many ways at our own convenience. One way is to extend f so that it is an *even* function on the interval $(-\pi, \pi)$ (Fig. 6-8). Since an even function has $b_n = 0$ for $n = 1, 2, \ldots$, the Fourier series has only cosine terms. We call such a series a **cosine series** and, as the original function is represented on $(0, \pi)$, the expansion is called a **half-range expansion**.

A function f defined on $(0, \pi)$ may also be extended to the interval $(-\pi, \pi)$ as an *odd* function. Figure 6-9 shows such an extension, and we notice that discontinuities are introduced at $-\pi$, 0, and π, unless $f(-\pi) = f(0) = f(\pi) = 0$. We are not disturbed by this fact, since the convergence theorem for Fourier series is valid for piecewise continuous functions. Since the Fourier series of an odd function has $a_n = 0$, $n = 0, 1, 2, \ldots$, the resulting series is called a **sine series**. We illustrate the process for obtaining half-range expansions with two examples.

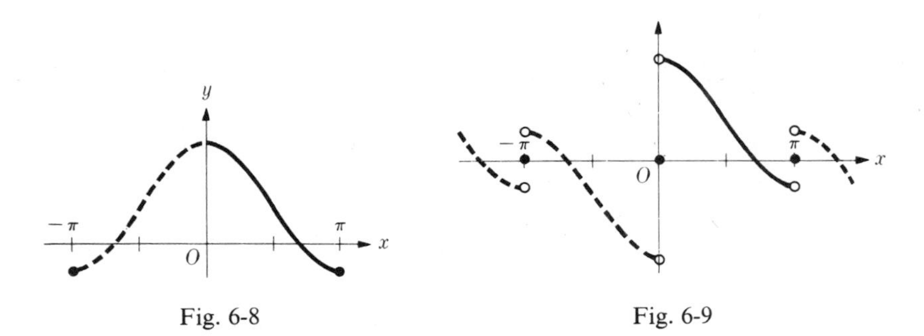

Fig. 6-8

Fig. 6-9

EXAMPLE 1. Given the function

$$f(x) = \begin{cases} 0 & \text{for } 0 < x < \pi/2, \\ 1 & \text{for } \pi/2 < x < \pi, \end{cases}$$

expand f in a cosine series and draw a graph of the extended function on the interval $[-3\pi, 3\pi]$.

SOLUTION. The graph of the even function is shown in Fig. 6-10. Since f is extended to be even, we have $b_n = 0$, $n = 1, 2, \ldots$, and

$$a_n = \frac{2}{\pi}\int_0^{\pi} f(x)\cos nx\,dx, \qquad n = 0, 1, 2, \ldots. \tag{1}$$

A simple calculation shows that

$$a_0 = \frac{2}{\pi}\int_{\pi/2}^{\pi} dx = 1 \qquad \text{and} \qquad a_n = \frac{2}{\pi}\int_{\pi/2}^{\pi} \cos nx\,dx = \left[\frac{2}{n\pi}\sin nx\right]_{\pi/2}^{\pi}$$

with

$$a_n = \left\{\begin{array}{ll} 0 & \text{if } n \text{ is even} \\ -\dfrac{2(-1)^{k-1}}{(2k-1)\pi} & \text{if } n = 2k-1 \end{array}\right\}, \qquad k = 1, 2, \ldots.$$

Therefore

$$f(x) = \frac{1}{2} - \frac{2}{\pi}\left[\frac{\cos x}{1} - \frac{\cos 3x}{3} + \frac{\cos 5x}{5} - \cdots\right], \qquad 0 < x < \pi.$$

REMARK. Setting $x = 0$ in this result, we obtain the interesting formula

$$\frac{\pi}{4} = 1 - \frac{1}{3} + \frac{1}{5} - \frac{1}{7} + \cdots.$$

(See also Problem 20 of Section 1.)

EXAMPLE 2. Expand the function f of Example 1 in a sine series and draw a graph of the extended function on the interval $[-3\pi, 3\pi]$.

SOLUTION. The graph of the odd function is shown in Fig. 6-11. We have $a_n = 0$, $n = 0, 1, 2, \ldots$, and

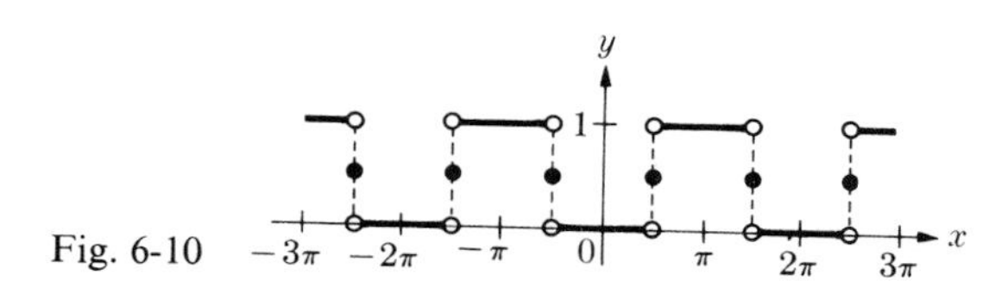

Fig. 6-10

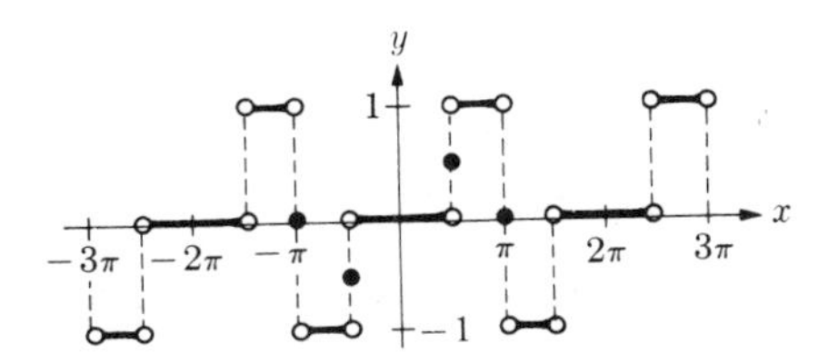

Fig. 6-11

$$b_n = \frac{2}{\pi}\int_0^{\pi} f(x)\sin nx\,dx, \qquad n = 1, 2, \ldots. \tag{2}$$

Therefore

$$b_n = \frac{2}{\pi}\int_{\pi/2}^{\pi} \sin nx\,dx = \left[-\frac{2}{n\pi}\cos nx\right]_{\pi/2}^{\pi} = \begin{cases} \dfrac{2}{n\pi} & \text{if } n \text{ is odd,} \\ \dfrac{2}{2k\pi}[(-1)^k - 1] & \text{if } n = 2k. \end{cases}$$

We conclude that

$$f(x) = \frac{2}{\pi}\left[\frac{\sin x}{1} - \frac{2\sin 2x}{2} + \frac{\sin 3x}{3} + \frac{0\cdot\sin 4x}{4} + \frac{\sin 5x}{5} - \frac{2\sin 6x}{6} + \frac{\sin 7x}{7} + \frac{0\cdot\sin 8x}{8} + \cdots\right], \qquad 0 < x < \pi.$$

REMARKS. Equation (1) shows that a cosine series may be obtained by using only the definition of f on $(0, \pi)$. The extension of f as an even function is a mental convenience which is used to set $b_n = 0$. The evaluation of a_n as in (1) does not use the extended function at all. Similarly, the evaluation of b_n in a sine series shows that the extended function plays no part except to help us set $a_n = 0$ and to permit us to use formula (2). Of course, for both cosine and sine series the expansion represents the original function on the interval $(0, \pi)$ only.

PROBLEMS

In each of Problems 1 through 8, expand each function in a cosine series on $(0, \pi)$ and draw the graph of the extended function on $[-3\pi, 3\pi]$.

1. $f(x) = \begin{cases} 1 & \text{for } 0 < x < \pi/2 \\ 0 & \text{for } \pi/2 < x < \pi \end{cases}$

2. $f(x) = \begin{cases} x & \text{for } 0 < x < \pi/2 \\ \pi - x & \text{for } \pi/2 < x < \pi \end{cases}$

3. $f(x) = \sin x$ for $0 \le x \le \pi$

4. $f(x) = |\cos x|$ for $0 \le x \le \pi$

5. $f(x) = x$ for $0 \le x \le \pi$

6. $f(x) = x^2$ for $0 \le x \le \pi$

7. $f(x) = x^3$ for $0 \le x \le \pi$

8. $f(x) = e^x$ for $0 \le x \le \pi$

In Problems 9 through 16, expand each function in a sine series on $(0, \pi)$ and draw the graph of the extended function on $[-3\pi, 3\pi]$.

9. f is the function in Problem 1.

10. f is the function in Problem 2.

11. $f(x) = \cos x \quad \text{for } 0 < x < \pi$

12. $f(x) = |\cos x| \quad \text{for } 0 < x < \pi$

13. $f(x) = x \quad \text{for } 0 < x < \pi$

14. $f(x) = x^2 \quad \text{for } 0 < x < \pi$

15. $f(x) = x^3 \quad \text{for } 0 < x < \pi$

16. $f(x) = e^{ax} \quad \text{for } 0 < x < \pi$, a constant.

17. Given the polynomial $f(x) = c_0 + c_1 x + c_2 x^2 + \cdots + c_n x^n$. Under what conditions will f be an even function? When will f be an odd function?

3. Expansions on Other Intervals

If f is a piecewise smooth function defined on the interval $[c - \pi, c + \pi]$, we form the periodic extension f_0 precisely as we did before. According to Theorem 2, the Fourier coefficients of the extended function, which we denote $\{a_n^0\}$ and $\{b_n^0\}$, are given by the formulas

$$a_n^0 = \frac{1}{\pi}\int_{-\pi}^{\pi} f_0(x)\cos nx\,dx, \qquad b_n^0 = \frac{1}{\pi}\int_{-\pi}^{\pi} f_0(x)\sin nx\,dx. \tag{1}$$

But, since f_0, $\cos nx$, and $\sin nx$ all have period 2π, we may replace the interval $[-\pi, \pi]$ in (1) by any other interval of length 2π. In particular, we may replace $[-\pi, \pi]$ by the interval $[c - \pi, c + \pi]$, on which f_0 coincides with f. Denoting by $\{a_n\}$ and $\{b_n\}$ the quantities

$$a_n = \frac{1}{\pi}\int_{c-\pi}^{c+\pi} f(x)\cos nx\,dx, \qquad b_n = \frac{1}{\pi}\int_{c-\pi}^{c+\pi} f(x)\sin nx\,dx, \tag{2}$$

we see that $a_n = a_n^0$ and $b_n = b_n^0$. The Fourier series formed with the coefficients in (2) yield an expansion for f which is valid in the interval

$$(c - \pi) < x < (c + \pi).$$

EXAMPLE 1. If $f(x) = x$ for $0 < x < 2\pi$, expand f in a Fourier series on the interval $(0, 2\pi)$. Draw a graph of f_0, the normalized periodic extension of f, on the interval $[-2\pi, 4\pi]$.

SOLUTION. The graph of f_0 is drawn in Fig. 6-12. We compute

$$a_0 = \frac{1}{\pi}\int_0^{2\pi} x\,dx = 2\pi,$$

$$a_n = \frac{1}{\pi}\int_0^{2\pi} x\cos nx\,dx = \frac{1}{\pi}\left[\frac{x\sin nx}{n} + \frac{\cos nx}{n^2}\right]_0^{2\pi} = 0,$$

$$b_n = \frac{1}{\pi}\int_0^{2\pi} x\sin nx\,dx = \frac{1}{\pi}\left[-\frac{x\cos nx}{n} + \frac{\sin nx}{n^2}\right]_0^{2\pi} = -\frac{2}{\pi}.$$

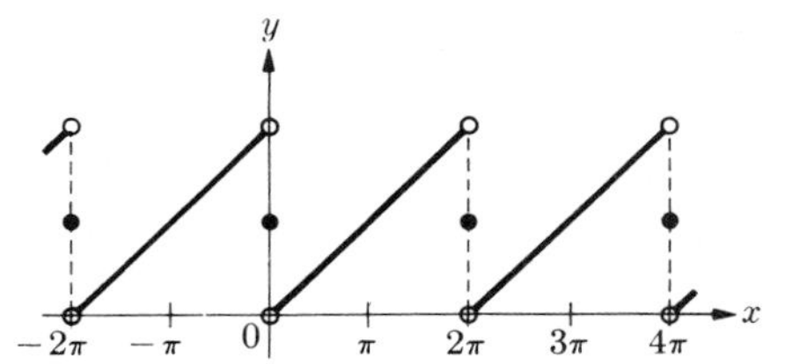

Fig. 6-12

Therefore

$$x = \pi - 2 \sum_{n=1}^{\infty} \frac{\sin nx}{n} \qquad \text{for } 0 < x < 2\pi.$$

REMARK. Of course, we could make the computations on the interval $[-\pi, \pi]$ using the extended function f_0. However, in that case in order to perform the integrations, the integrals which determine $\{a_n^0\}$ and $\{b_n^0\}$ would have to be separated into two parts, even though the final result would be the same.

A function f which is piecewise smooth and normalized on an interval $[-L, L]$ can be expanded in a **modified Fourier series**. If f is given on $[-L, L]$, we introduce a change of variable and define y and $g(y)$ by the formulas

$$y = \frac{\pi x}{L}, \qquad f(x) = f\left(\frac{Ly}{\pi}\right) = g(y) = g\left(\frac{\pi x}{L}\right).$$

This transformation maps $[-L, L]$ onto $[-\pi, \pi]$. We see that g is piecewise smooth and normalized on $[-\pi, \pi]$ and so can be expanded in a Fourier series

$$g(y) = \frac{a_0}{2} + \sum_{n=1}^{\infty} (a_n \cos ny + b_n \sin ny), \qquad -\pi < y < \pi, \tag{3}$$

with

$$a_n = \frac{1}{\pi}\int_{-\pi}^{\pi} g(y)\cos ny\, dy, \qquad b_n = \frac{1}{\pi}\int_{-\pi}^{\pi} g(y)\sin ny\, dy. \tag{4}$$

Using the change of variable $y = \pi x/L$, $dy = (\pi/L)\,dx$, we write (4) in the form

$$a_n = \frac{1}{L}\int_{-L}^{L} f(x)\cos\frac{n\pi x}{L}\, dx, \qquad b_n = \frac{1}{L}\int_{-L}^{L} f(x)\sin\frac{n\pi x}{L}\, dx. \tag{5}$$

The series (3) becomes

$$f(x) = \frac{a_0}{2} + \sum_{n=1}^{\infty}\left(a_n \cos\frac{n\pi x}{L} + b_n \sin\frac{n\pi x}{L}\right), \qquad -L < x < L. \tag{6}$$

The modified Fourier series (6), with formulas (5) for the coefficients, shows that a function f defined on any interval of *finite* length may be expanded in a Fourier series of the type we have been discussing. Hence such functions

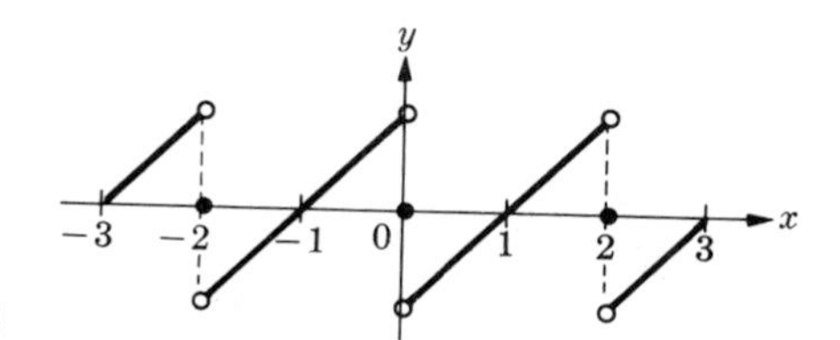

Fig. 6-13

as e^x, $1/(1 + x^2)$, etc., have Fourier expansions on intervals of any finite size. However, we recall that there are some power series such as the one for e^x which converge for $-\infty < x < \infty$. The corresponding result does not exist for Fourier series. There are no Fourier series which are valid on an infinite interval if the function is nonperiodic. The situation is not completely hopeless because an extension of the idea of Fourier series can be employed. This special representation, called the **Fourier Integral**, exhibits many properties for nonperiodic functions which Fourier series exhibit for periodic ones. Furthermore, the Fourier Integral is a useful tool in both theoretical and applied investigations of differential equations, as are Fourier series.

Theorem 3, for the expansions of even and odd functions, is applicable to modified Fourier series on an interval $(0, L)$. For example, a function f defined on $(0, L)$ has the sine series

$$f(x) = \sum_{n=1}^{\infty} b_n \sin\frac{n\pi x}{L}, \qquad b_n = \frac{2}{L}\int_0^L f(x)\sin\frac{n\pi x}{L}\,dx.$$

Half-range expansions may be carried out on any interval of length L, either by making a change of variable or by forming the periodic extension and computing all the formulas for the interval $(0, L)$.

EXAMPLE 2. Given

$$f(x) = \begin{cases} x + 1 & \text{for } -1 < x < 0, \\ x - 1 & \text{for } 0 < x < 1, \end{cases}$$

expand f into a Fourier series on $(-1, 1)$ and draw the graph of its periodic normalized extension on $[-3, 3]$.

SOLUTION. The graph of the periodic extension f_0 is drawn in Fig. 6-13. The function f_0 is odd, so all $a_n = 0$. Then

$$b_n = 2\int_0^1 (x - 1)\sin n\pi x\,dx$$

$$= 2\left[-\frac{(x-1)\cos n\pi x}{n\pi} + \frac{\sin n\pi x}{n^2\pi^2}\right]_0^1 = -\frac{2}{n\pi}.$$

We obtain

$$f(x) = -\frac{2}{\pi}\sum_{n=1}^{\infty}\frac{\sin n\pi x}{n}, \qquad -1 < x < 1.$$

PROBLEMS

In Problems 1 through 11, expand each function as indicated. Draw the appropriate extension f_0 on an interval of length three full periods.

1. $f(x) = \begin{cases} 1 & \text{for } 0 < x < \pi, \\ -1 & \text{for } \pi < x < 2\pi, \end{cases}$ full series on $[0, 2\pi]$

2. $f(x) = \begin{cases} 1 & \text{for } 0 < x < \pi/2, \\ 0 & \text{for } \pi/2 < x < 2\pi, \end{cases}$ full series on $[0, 2\pi]$

3. $f(x) = \sin(x/2)$, full series on $[0, 2\pi]$

4. $f(x) = x - \pi$, full series on $[\pi, 3\pi]$

5. $f(x) = \begin{cases} 0 & \text{for } -2 < x < 0, \\ x & \text{for } 0 \le x < 2, \end{cases}$ full series on $[-2, 2]$

6. $f(x) = \sin x$, full series on $[-\pi/2, \pi/2]$

7. $f(x) = 1 - |x|$, full series on $[-1, 1]$

8. $f(x) = \begin{cases} 1 & \text{for } 0 < x < \pi, \\ 0 & \text{for } \pi < x < 2\pi, \end{cases}$ cosine series on $[0, 2\pi]$

9. $f(x) = x^2$ for $0 < x < 2$, sine series on $[0, 2]$

10. $f(x) = x^2$ for $0 < x < 1$, cosine series on $[0, 1]$

11. $f(x) = 1 - 2x$ for $0 < x < 1$, sine series on $[0, 1]$

12. Denote the Fourier series of a function f by the symbol $\mathbf{F}$, and that of a function g by the symbol $\mathbf{G}$. Show that if c is a constant, the Fourier series of the function cf is $c\mathbf{F}$. If f and g are periodic with the same period, show that the Fourier series of $f + g$ is $\mathbf{F} + \mathbf{G}$.

13. By combining the series in Problems 1 and 2, and taking into account the results of Problem 12, find the Fourier series of

$$f(x) = \begin{cases} 2 & \text{for } 0 < x < \pi/2, \\ 1 & \text{for } \pi/2 < x < \pi, \\ -1 & \text{for } \pi < x < 2\pi, \end{cases} \qquad \text{full series on } [0, 2\pi].$$

14. Combine the results of Problems 2 and 3, and take the results of Problem 12 into account, to get an expansion for

$$f(x) = \begin{cases} 3 - 2\sin(x/2) & \text{for } 0 < x < \pi/2, \\ -2\sin(x/2) & \text{for } \pi/2 < x < 2\pi, \end{cases} \qquad \text{full series on } [0, 2\pi].$$

In each of Problems 15 through 18 expand the given function in a modified Fourier series on the interval $-L < x < L$.

15. $f(x) = e^x$.

16. $f(x) = x^3$

17. $f(x) = \frac{1}{2}(e^x - e^{-x})$, i.e. $f(x) = \sinh x$.

18. $f(x) = \cos^3 x$.

4. Convergence Theorems. Differentiation and Integration of Fourier Series

We now turn to the proof of Theorem 2, in Section 1 on the convergence of a Fourier series to the function it represents. Before repeating the statement of the theorem and deriving the main result, we prove four simple lemmas.

Lemma 1. *For every positive integer n and for all x we have the identity*

$$\frac{\sin(n+\frac{1}{2})x}{2\sin\frac{1}{2}x} = \frac{1}{2} + \sum_{k=1}^{n} \cos kx, \qquad x \neq 0.$$

PROOF. This is the type of trigonometric identity which occurs frequently in elementary trigonometry courses. It is sufficient to show that

$$\sin(n+\tfrac{1}{2})x = \sin\tfrac{1}{2}x + \sum_{k=1}^{n} 2\sin\tfrac{1}{2}x\cos kx.$$

We employ the formula

$$2\cos A\sin B = \sin(A+B) - \sin(A-B),$$

with $A = kx$, $B = \frac{1}{2}x$, to get

$$\sin\tfrac{1}{2}x + \sum_{k=1}^{n} 2\sin\tfrac{1}{2}x\cos kx = \sin\tfrac{1}{2}x + \sum_{k=1}^{n}\left[\sin(k+\tfrac{1}{2})x - \sin(k-\tfrac{1}{2})x\right].$$

All the terms on the right "telescope" except for the last which is $\sin(n+\frac{1}{2})x$, and the result follows.

Definition. We call the quantity

$$D_n(x) = \frac{\sin(n+\frac{1}{2})x}{2\sin\frac{1}{2}x} \equiv \frac{1}{2} + \sum_{k=1}^{n}\cos kx$$

the **Dirichlet kernal.** (It is an expression which occurs frequently in the study of Fourier series and plays a central role in many proofs.) Three properties of $D_n(x)$ which we shall use are the following:

i) $D_n(x)$ is an even function of x. This fact is readily seen when we note that $\cos kx$ is even for every k;
ii) for every n, we have
$$\frac{1}{\pi}\int_0^{\pi} D_n(x)\,dx = \frac{1}{2}. \tag{1}$$
This result is obtained directly by integrating $\frac{1}{2} + \Sigma_{k=1}^{n}\cos kx$ from 0 to π.
iii) $D_n(x)$ is periodic with period 2π. This result is evident from the formula defining $D_n(x)$.

Lemma 2. *For all x such that $0 < x \le \pi/2$, we have*

$$1 < \frac{x}{\sin x} \le \frac{\pi}{2}, \qquad \textit{Jordan's Inequality.}$$

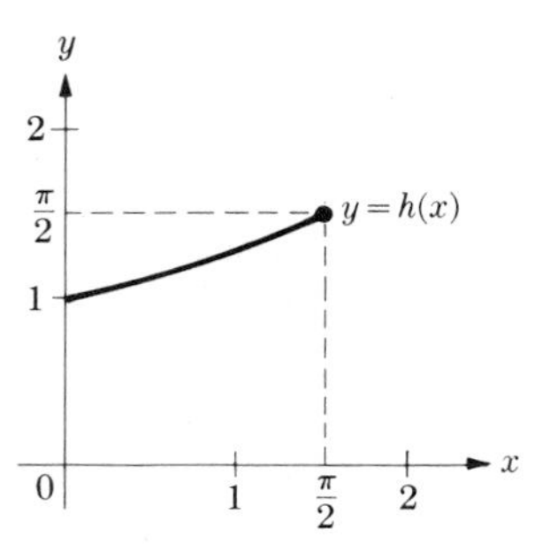

Fig. 6-14

Proof. We note that $h(x) = x/\sin x$ tends to 1 when $x \to 0$ and has the value $\pi/2$ when $x = \pi/2$. If we can show that h is an increasing function in this interval, the result is established (Fig. 6-14). It suffices to show that $h'(x) \geq 0$ on $(0, \pi/2)$. We leave this verification for the reader.

Let f be a piecewise smooth, periodic function with period 2π, and form the Fourier series

$$\frac{a_0}{2} + \sum_{k=1}^{\infty} (a_k \cos kx + b_k \sin kx), \tag{2}$$

where

$$a_k = \frac{1}{\pi}\int_{-\pi}^{\pi} f(t)\cos kt\,dt, \qquad b_k = \frac{1}{\pi}\int_{-\pi}^{\pi} f(t)\sin kt\,dt. \tag{3}$$

The partial sum $s_n(x)$ is the sum of the first n terms in (2).

The next lemma establishes a basic formula for $s_n(x)$.

Lemma 3. *If $s_n(x)$ is the* n*th partial sum of the Fourier series of f, and if $D_n(x)$* is the Dirichlet kernel, then

$$\begin{aligned} s_n(x) - f(x) &= \frac{1}{\pi}\int_0^{\pi} [f(x+u) - f(x+)]D_n(u)\,du \\ &\quad + \frac{1}{\pi}\int_0^{\pi} [f(x-u) - f(x-)]D_n(u)\,du. \end{aligned} \tag{4}$$

Proof. Since

$$s_n(x) = \frac{1}{2} + \sum_{k=1}^{n} (a_k \cos kx + b_k \sin kx),$$

we may insert the formulas for a_k and b_k from (3). We obtain

$$s_n(x) = \frac{1}{\pi}\int_{-\pi}^{\pi} f(t)\left[\frac{1}{2} + \sum_{k=1}^{n} (\cos kt \cos kx + \sin kt \sin kx)\right] dt.$$

From trigonometry we know that

$$\cos(kt - kx) = \cos kt \cos kx + \sin kt \sin kx,$$

and so

$$s_n(x) = \frac{1}{\pi}\int_{-\pi}^{\pi} f(t)\left[\frac{1}{2} + \sum_{k=1}^{n} \cos k(t - x)\right] dt. \tag{5}$$

Now, in (5), we hold x fixed and change the variable of integration to u by setting $t = x + u$. Then (5) becomes

$$s_n(x) = \frac{1}{\pi}\int_{-\pi-x}^{\pi-x} f(x + u)\left[\frac{1}{2} + \sum_{k=1}^{n} \cos ku\right] du.$$

Since both f and the quantity in the bracket are periodic with period 2π, we may adjust the interval of integration to $[-\pi, \pi]$. We also recognize the Dirichlet kernel in the bracket. Therefore

$$\begin{aligned} s_n(x) &= \frac{1}{\pi}\int_{-\pi}^{\pi} f(x + u)D_n(u)\,du \\ &= \frac{1}{\pi}\left[\int_{-\pi}^{0} f(x + v)D_n(v)\,dv + \int_{0}^{\pi} f(x + u)D_n(u)\,du\right]. \end{aligned}$$

Since $D_n(v)$ is even, we may replace v by $-u$ in the first integral on the right above and get

$$s_n(x) = \frac{1}{\pi}\int_{0}^{\pi} [f(x + u) + f(x - u)]D_n(u)\,du. \tag{6}$$

We have almost finished. We write $f(x)$ in a tricky form (a simple device used frequently in Fourier series),

$$f(x) = 2\cdot\frac{1}{\pi}\int_{0}^{\pi} f(x)D_n(u)\,du,$$

which we can do because of (1). Note that $f(x)$ is constant so far as the integration is concerned. Since $f(x) = \frac{1}{2}[f(x+) + f(x-)]$, we have

$$f(x) = \frac{1}{\pi}\int_{0}^{\pi} [f(x+) + f(x-)]D_n(u)\,du.$$

Subtracting this expression from (6), we obtain (4) precisely.

Lemma 4 (Bessel's Inequality). *Suppose that f is piecewise smooth and periodic with period 2π. Let*

$$\frac{a_0}{2} + \sum_{k=1}^{\infty} (a_k \cos kx + b_k \sin kx) \tag{7}$$

be the Fourier series of f. Then we have

$$\frac{a_0^2}{2} + \sum_{k=1}^{n} (a_k^2 + b_k^2) \le \frac{1}{\pi}\int_{-\pi}^{\pi} f^2(x)\,dx, \qquad n = 1, 2, \ldots. \tag{8}$$

PROOF. We have the identities

$$\begin{aligned}
\tfrac{1}{2}a_0^2 &= \frac{1}{\pi}\int_{-\pi}^{\pi} \frac{a_0}{2} f(t)\,dt,\\
a_k^2 &= \frac{1}{\pi}\int_{-\pi}^{\pi} a_k f(t)\cos kt\,dt,\\
b_k^2 &= \frac{1}{\pi}\int_{-\pi}^{\pi} b_k f(t)\sin kt\,dt, \qquad k = 1, 2, \ldots.
\end{aligned} \tag{9}$$

Denoting the nth partial sum of (7) by $s_n(x)$, and recalling the orthogonality relations of the functions $\{\sin nx, \cos nx\}$ on $[-\pi, \pi]$, we obtain the formulas

$$\begin{aligned}
\tfrac{1}{2}a_0^2 &= \frac{1}{\pi}\int_{-\pi}^{\pi} \tfrac{1}{2}a_0 s_n(t)\,dt,\\
a_k^2 &= \frac{1}{\pi}\int_{-\pi}^{\pi} a_k s_n(t)\cos kt\,dt,\\
b_k^2 &= \frac{1}{\pi}\int_{-\pi}^{\pi} b_k s_n(t)\sin kt\,dt;
\end{aligned} \tag{10}$$

and these formulas are valid for every positive integer n. Adding the expressions for a_0^2, a_k^2, b_k^2 and summing from 1 to n (first according to (9) and then according to (10)), we find

$$\tfrac{1}{2}a_0^2 + \sum_{k=1}^{n} (a_k^2 + b_k^2) = \frac{1}{\pi}\int_{-\pi}^{\pi} f(t)s_n(t)\,dt = \frac{1}{\pi}\int_{-\pi}^{\pi} s_n^2(t)\,dt. \tag{11}$$

Furthermore, we see that

$$\begin{aligned}
\int_{-\pi}^{\pi} [f(t) - s_n(t)]^2\,dt &= \int_{-\pi}^{\pi} f^2(t)\,dt - 2\int_{-\pi}^{\pi} f(t)s_n(t)\,dt + \int_{-\pi}^{\pi} s_n^2(t)\,dt\\
&= \int_{-\pi}^{\pi} f^2(t)\,dt - \int_{-\pi}^{\pi} f(t)s_n(t)\,dt.
\end{aligned}$$

The fact that $\int_{-\pi}^{\pi} [f(t) - s_n(t)]^2\,dt$ is nonnegative, combined with the first equality in (11), yields

$$\frac{1}{\pi}\int_{-\pi}^{\pi} f^2(t)\,dt - \left(\tfrac{1}{2}a_0^2 + \sum_{k=1}^{n} (a_k^2 + b_k^2)\right) \geq 0,$$

which is Bessel's Inequality.

We are ready to prove the convergence theorem of Section 1, which we restate.

Theorem 2. *Suppose that f is piecewise smooth, normalized, and periodic with period 2π. Then its Fourier series converges to $f(x)$ for each x.*

PROOF. To show that the Fourier series converges at a fixed value of x, we shall prove that the two integrals on the right in (4) on page 376 tend to zero as n tends to infinity. Writing

$$T_n(x) = \frac{1}{\pi}\int_0^\pi [f(x-u) - f(x-)]D_n(u)\,du,$$

for the second integral in (4), we employ Lemma 1 to obtain

$$\begin{aligned} T_n(x) &= \frac{1}{\pi}\int_0^\pi \frac{f(x-u)-f(x-)}{2\sin\frac{1}{2}u}\sin(n+\tfrac{1}{2})u\,du \\ &= \frac{1}{\pi}\int_0^\pi \frac{f(x-u)-f(x-)}{2\sin\frac{1}{2}u}[\sin nu\cos\tfrac{1}{2}u + \cos nu\sin\tfrac{1}{2}u]\,du. \end{aligned}$$

We define three functions

$$\phi(x,u) = \frac{f(x-u)-f(x-)}{2\sin\frac{1}{2}u},$$

$$\phi_1(x,u) = \phi(x,u)\sin\tfrac{1}{2}u, \qquad \phi_2(x,u) = \phi(x,u)\cos\tfrac{1}{2}u.$$

The functions ϕ, ϕ_1, and ϕ_2 are all piecewise smooth whenever f is, except possibly at $u = 0$. Applying L'Hôpital's Rule to ϕ at $u = 0$, we find that

$$\phi(x, 0+) = -f'(x-)$$

and conclude that ϕ, ϕ_1, and ϕ_2 are piecewise continuous everywhere. Therefore we find that

$$T_n(x) = \frac{1}{\pi}\int_0^\pi [\phi_1(x,u)\cos nu + \phi_2(x,u)\sin nu]\,du.$$

For *fixed* x this expression for $T_n(x)$ is the sum of the nth Fourier cosine and sine coefficient for $\frac{1}{2}\phi_1$ and $\frac{1}{2}\phi_2$, respectively. The conditions for the Bessel Inequality (Lemma 4, above) apply, and we conclude that

$$T_n(x) \to 0 \qquad \text{as } n \to \infty.$$

The proof for the first integral in (4) is identical.

If a function f is given by a power series with a positive radius of convergence, we saw (in Section 12 of Chapter 3) that f possesses derivatives of all orders. These derivatives are obtained by differentiating the power series for f the appropriate number of times, and the various derived series will have the same radius of convergence that f does. Moreover, if $F'(x) = f(x)$, then F is represented by integrating the series for f term by term. The situation is completely different in the case of Fourier series. We showed that any piecewise smooth function is representable by its Fourier series. Therefore it is intuitively clear that such a series cannot be differentiated at will. While it is true that if f is piecewise smooth, then f' is piecewise continuous, we observe that if f is not continuous everywhere, then it is not necessarily true that

$$f(\beta) - f(\alpha) = \int_{\alpha}^{\beta} f'(x)\, dx,$$

that is, the Fundamental Theorem of Calculus does not hold.

In general, if f is piecewise smooth but not continuous, the series obtained by term-by-term differentiation of the Fourier series for f does not converge to f'; in fact, it ordinarily does not converge at all.

An example to illustrate this point is given by

$$f(x) = \begin{cases} -1 & \text{for} \quad -\pi < x < 0, \\ 1 & \text{for} \quad 0 < x < \pi, \end{cases}$$

with the Fourier series

$$f(x) = \frac{4}{\pi}\left(\sin x - \frac{\sin 3x}{3} + \frac{\sin 5x}{5} - \cdots\right).$$

The derivative series is

$$\frac{4}{\pi}(\cos x - \cos 3x + \cos 5x - \cos 7x + \cdots),$$

which can be shown to diverge.

Since integration is a "smoothing" process while differentiation is a "scrambling" process, we can expect the behavior for term-by-term integration of Fourier series to be quite different from that for term-by-term differentiation. The main results in this direction are established in the following lemma and in the next two theorems.

Lemma 5. *Suppose that f is periodic with period 2π and piecewise smooth on every finite interval. Then its Fourier coefficients satisfy the bounds*

$$|a_n| \le \frac{C}{n}, \qquad |b_n| \le \frac{C}{n}, \qquad n = 1, 2, \ldots,$$

where C is a constant which depends on f but not on n.

PROOF. If the jumps of f occur at $X_1, X_2, \ldots, X_{L-1}$, we have

$$a_n = \frac{1}{\pi}\int_{-\pi}^{\pi} f(t)\cos nt\, dt = \frac{1}{\pi}\sum_{i=1}^{L}\int_{X_{i-1}}^{X_i} f(t)\cos nt\, dt,$$

where we set $-\pi = X_0$ and $\pi = X_L$. We integrate by parts in each of the integrals and obtain

$$\frac{1}{n}\sum_{i=1}^{L}\int_{X_{i-1}}^{X_i} f(t)\cos nt\, dt = \sum_{i=1}^{L} \left.\frac{f(t)\sin nt}{\pi n}\right]_{X_{i-1}}^{X_i} - \sum_{i=1}^{L}\frac{1}{\pi}\int_{X_{i-1}}^{X_i}\frac{f'(t)\sin nt}{n}\, dt.$$

We use the fact that $|f(x)| \leq M$ and $|f'(x)| \leq N$ to estimate the terms on the right. We find

$$\left| \frac{1}{\pi} \sum_{i=1}^{L} \int_{X_{i-1}}^{X_i} f(t) \cos nt \, dt \right| \leq \frac{2LM}{\pi n} + \frac{2N}{n}.$$

Now choose $C = \frac{2}{\pi} LM + 2N$ and the result follows.

Theorem 4. *Suppose that F is continuous for all x and periodic with period 2π, and suppose that F' is piecewise smooth. Then*

i) *the series obtained by differentiating the Fourier series for F converges at each point x to $\frac{1}{2}[F'(x+) + F'(x-)]$;*
ii) *the Fourier coefficients A_n, B_n of F satisfy the inequalities*

$$|A_n| \leq \frac{C}{n^2}, \qquad |B_n| \leq \frac{C}{n^2}, \qquad n = 1, 2, \ldots, \tag{12}$$

where C is a constant which depends on F but not on n;
iii) *the Fourier series for F converges uniformly to $F(x)$ for all x.*

Proof. Let the jumps of $F'(x)$ occur at $X_0, X_1, \ldots, X_L$. If we define

$$G(x) = \int_{-\pi}^{x} F'(t) \, dt,$$

we see that $G(x)$ is continuous and that $G'(x) - F'(x) = 0$ on each (X_{i-1}, X_i). By the Theorem of the Mean, the function $G(x) - F(x)$ is constant on each (X_{i-1}, X_i), and therefore for all x, since G and F are both continuous. Denote the Fourier coefficients of F' by a_n, b_n. We have, for every positive integer n,

$$A_n = \frac{1}{\pi} \int_{-\pi}^{\pi} F(x) \cos nx \, dx = \frac{1}{\pi} \sum_{i=1}^{L} \int_{X_{i-1}}^{X_i} F(x) \cos nx \, dx$$

and, upon integration by parts,

$$A_n = \frac{1}{\pi} \left[\sum_{i=1}^{k} \frac{F(X_i) \sin nX_i - F(X_{i-1}) \sin nX_{i-1}}{n} + \frac{1}{n} \int_{X_{i-1}}^{X_i} F'(x) \sin nx \, dx \right].$$

The terms in the first sum "telescope" except for the first and last ones, and the periodicity of F and $\sin nx$ cancels these two. Therefore

$$A_n = \frac{1}{n\pi} \int_{-\pi}^{\pi} F'(x) \sin nx \, dx = \frac{b_n}{n}. \tag{13}$$

Similarly,

$$B_n = -\frac{a_n}{n}. \tag{14}$$

The bounds in (12) follow from these formulas for A_n and B_n, taken in conjunction with the result of Lemma 5. Since F' is piecewise smooth, its

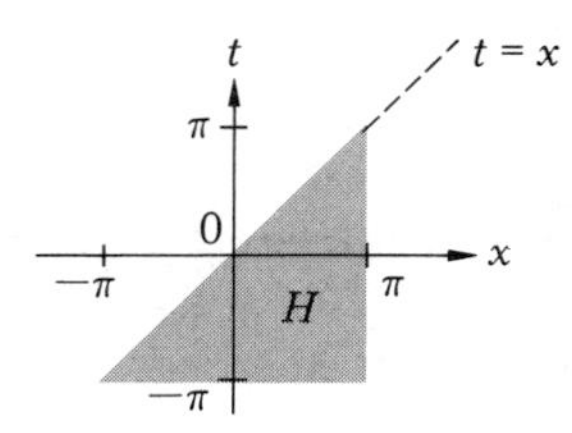

Fig. 6-15

series converges to F' for each x. Also, since F is piecewise smooth, its series converges for each x. The bounds given by (12) allow us to use the comparison test and conclude that the Fourier series for F converges uniformly. From (13) and (14), we see that the series for F' is obtained by differentiating the series for F term by term.

Theorem 5. *Suppose that f is periodic with period 2π and is piecewise smooth on any finite interval. Suppose that*

$$\pi a_0 = \int_{-\pi}^{\pi} f(x)\,dx = 0,$$

and define

$$F(x) = \int_{-\pi}^{x} f(t)\,dt.$$

Then

i) *the Fourier series for F is obtained by integrating that for f term by term, except for the constant term A_0 which is given by*

$$A_0 = -\frac{1}{\pi}\int_{-\pi}^{\pi} xf(x)\,dx;$$

ii) *F and its Fourier coefficients satisfy the conditions of Theorem* 4.

PROOF. If F is to be periodic, the condition $a_0 = 0$ is required. To find A_0, we have (Fig. 6-15)

$$\begin{aligned} A_0 &= \frac{1}{\pi}\int_{-\pi}^{\pi}\int_{-\pi}^{x} f(t)\,dt\,dx = \frac{1}{\pi}\iint_H f(t)\,dA_{xt} \\ &= \frac{1}{\pi}\int_{-\pi}^{\pi} f(t)\left[\int_t^{\pi} dx\right] dt = \frac{1}{\pi}\int_{-\pi}^{\pi} (\pi - t)f(t)\,dt \\ &= -\frac{1}{\pi}\int_{-\pi}^{\pi} tf(t)\,dt, \end{aligned}$$

since $\int_{-\pi}^{\pi} f(t)\,dt = 0$. All other statements in the theorem follow directly from Theorem 4.

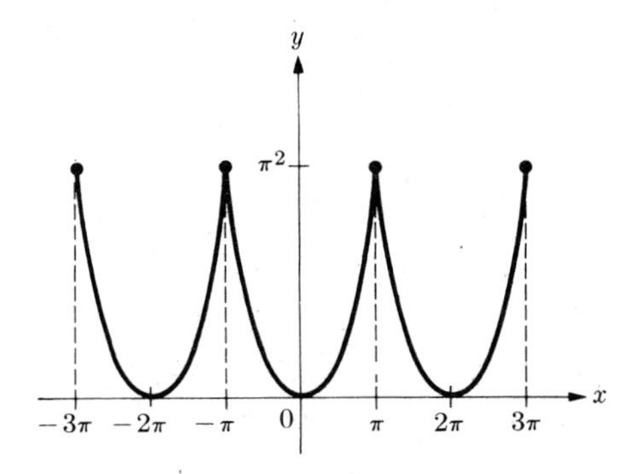

Fig. 6-16

REMARK. It is interesting that Theorem 5 does not require *uniform* convergence of the derivative series $F'(x) = f(x)$. In general, theorems involving term-by-term integration demand fewer hypotheses than those involving term-by-term differentiation.

EXAMPLE. Use the result of Example 1 in Section 1, which establishes the expansion

$$f(x) = x = 2\sum_{n=1}^{\infty} \frac{(-1)^{n-1}\sin nx}{n}, \qquad -\pi < x < \pi,$$

to obtain the Fourier series for $F(x) = x^2$ on $[-\pi, \pi]$.

SOLUTION. The periodic extension F_0 of F satisfies the hypotheses of Theorem 4, and we have

$$F_0'(x) = 4\sum_{n=1}^{\infty} \frac{(-1)^{n-1}\sin nx}{n}.$$

Therefore

$$F_0(x) = \frac{A_0}{2} + 4\sum_{n=1}^{\infty} \frac{(-1)^n \cos nx}{n^2},$$

with

$$A_0 = \frac{1}{\pi}\int_{-\pi}^{\pi} F(x)\,dx = \frac{2\pi^2}{3}.$$

REMARK. The graph of F_0 is shown in Fig. 6-16. We see that F_0 is continuous but has a corner at each of the points $(\pi \pm 2n\pi, \pi^2)$. These corners correspond to the jumps in the periodic extension f_0 of f, the graph of which is shown in Fig. 6-4.

PROBLEMS

1. If $h(x) = x/(\sin x)$, show that $h'(x) \geq 0$ for $0 < x \leq \pi$, thereby completing the proof of Lemma 2.

2. Using the results of the example on page 383, find the series for f, where

$$f(x) = \frac{\pi^2 x - x^3}{3} \qquad \text{on } [-\pi, \pi],$$

and show that

$$\sum_{n=1}^{\infty} \frac{1}{n^6} \le \frac{\pi^6}{945}.$$

Draw the graph of f_0.

$$\left[\textit{Answer:}\quad f(x) = 4 \sum_{n=1}^{\infty} \frac{(-1)^{n-1} \sin nx}{n^3}.\right]$$

3. Using Theorem 5, find the series for $f(x) = |x|$.

$$\left[\textit{Answer:}\ \left(|x| = \frac{\pi}{2} - \frac{4}{\pi}\ \frac{\cos x}{1^2} + \frac{\cos 3x}{3^2} + \frac{\cos 5x}{5^2} + \cdots\right).\right]$$

4. Using the method and result of Problem 3, find the Fourier series for f if

$$f(x) = \begin{cases} \dfrac{x^2 - \pi x}{2} & \text{for} \quad 0 \le x \le \pi, \\[2ex] \dfrac{-x^2 - \pi x}{2} & \text{for } -\pi \le x \le 0. \end{cases}$$

Show that

$$\frac{1}{1^6} + \frac{1}{3^6} + \frac{1}{5^6} + \cdots \le \frac{\pi^6}{960}.$$

Draw the graph of f_0.

$$\left[\textit{Answer:}\quad f(x) = -\frac{4}{\pi}\left(\frac{\sin x}{1^3} + \frac{\sin 3x}{3^3} + \frac{\sin 5x}{5^3} + \cdots\right).\right]$$

5. Using the result of Problem 2, find the series for

$$f(x) = \frac{(\pi^2 - x^2)^2}{12}.$$

Draw the graph of f_0.

$$\left[\textit{Answer:}\quad f(x) = \frac{2\pi^4}{45} + 4 \sum_{n=1}^{\infty} \frac{(-1)^{n-1} \cos nx}{n^4}.\right]$$

6. Find the Fourier series for f and F, given

$$f(x) = |\sin x|, \qquad -\pi \le x \le \pi,$$

$$F(x) = \begin{cases} -(1 - \cos x) - \dfrac{2x}{\pi} & \text{for } -\pi \le x \le 0, \\[2ex] (1 - \cos x) - \dfrac{2x}{\pi} & \text{for} \quad 0 \le x \le \pi. \end{cases}$$

Draw the graph of F_0.

7. Find the Fourier series for the function

$$f(x) = \begin{cases} -\pi - x & \text{for } -\pi \le x \le -\dfrac{\pi}{2}, \\ x & \text{for } -\dfrac{\pi}{2} \le x \le \dfrac{\pi}{2}, \\ \pi - x & \text{for } \dfrac{\pi}{2} \le x \le \pi. \end{cases}$$

8. Using the results of Problem 4 above and the expansion for $f(x) = x$, as in the example on page 383, find the Fourier series for the function

$$g(x) = \begin{cases} -\dfrac{x^2}{2} & \text{for } -\pi < x \le 0, \\ \dfrac{x^2}{2} & \text{for } \quad 0 \le x < \pi. \end{cases}$$

Note that $g'(x) = |x|$ on $(-\pi, \pi)$. Is the series obtained by term-by-term differentiation of that for g identical with that for $|x|$ given in Problem 3 above? Explain.

9. Given the *trigonometric polynomial*

$$f(x) = \sum_{k=0}^{m} (c_k \cos kx + d_k \sin kx).$$

Find the Fourier series of this function on $(-\pi, \pi)$.

*10. Suppose that f possesses continuous derivatives of all orders for $-\infty < x < \infty$ and is periodic with period 2π. What can be said about the ratios

$$\frac{a_n}{n^k}, \quad \frac{b_n}{n^k} \quad \text{as } n \to \infty,$$

where a_n, b_n, are the Fourier coefficients of f, and k is a positive integer?

11. The *Dirichlet conjugate kernel* is defined by

$$\bar{D}_n(u) = \sum_{k=1}^{n} \sin ku.$$

Show that

$$\bar{D}_n(u) = \frac{\cos \frac{1}{2}u - \cos (n + \frac{1}{2})u}{2 \sin \frac{1}{2}u}.$$

5. The Complex Form of Fourier Series

We consider complex-valued functions of a real variable of the form

$$f(x) = f_1(x) + if_2(x), \tag{1}$$

where $i = \sqrt{-1}$ and f_1, f_2 are real-valued, piecewise continuous functions on an interval $[a, b]$. The definitions of piecewise smooth, normalized, and periodic extension carry over directly to complex-valued functions. For any function f given by (1), we have

$$f'(x) = f_1'(x) + if_2'(x), \qquad \int_a^b f(x)\,dx = \int_a^b f_1(x)\,dx + i\int_a^b f_2(x)\,dx.$$

The definitions of evenness, oddness, and half-range expansions for complex functions are the same as for real-valued functions. The theorems for complex functions on convergence, integration, and differentiation of series are all established in the same way as for real functions.

We now introduce expansions of complex-valued functions in terms of certain exponential functions. There are problems for which such expansions are more convenient that the Fourier expansions discussed above. Using the formulas

$$\cos nx = \frac{e^{inx} + e^{-inx}}{2}, \qquad \sin nx = \frac{-i(e^{inx} - e^{-inx})}{2},$$

we obtain at once

$$a_n \cos nx + b_n \sin nx = c_n e^{inx} + c_{-n} e^{-inx},$$

where

$$c_n = \frac{a_n - ib_n}{2}, \qquad c_{-n} = \frac{a_n + ib_n}{2}, \qquad n = 1, 2, \ldots. \tag{2}$$

If, in addition, we set

$$c_0 = \frac{a_0}{2}, \tag{3}$$

we see that if f is represented by the convergent Fourier series

$$f(x) = \frac{a_0}{2} + \sum_{n=1}^{\infty} (a_n \cos nx + b_n \sin nx), \tag{4}$$

then f is also represented by the convergent series

$$f(x) = c_0 + \sum_{n=1}^{\infty} (c_n e^{inx} + c_{-n} e^{-inx}), \tag{5}$$

where c_n and c_{-n} are given in terms of a_n and b_n by (2) and (3). Moreover, if the series (5) converges to $f(x)$ with the terms grouped as indicated, then the series (4) does also, provided the a_n and b_n are related to c_n and c_{-n} by (2) and (3); that is,

$$a_0 = 2c_0, \qquad a_n = c_n + c_{-n}, \qquad b_n = i(c_n - c_{-n}).$$

Formally, the series (5) can be written in the form

$$f(x) = \sum_{n=-\infty}^{\infty} c_n e^{inx} = \sum_{n=1}^{\infty} c_n e^{inx} + \sum_{n=1}^{\infty} c_{-n} e^{-inx} + c_0$$

but, unless the two series $\Sigma_{n=1}^{\infty}|c_n|$ and $\Sigma_{n=1}^{\infty}|c_{-n}|$ are convergent, the rearrangement is not always valid.

For any complex number α, we denote its *complex conjugate* by $\bar{\alpha}$. Two complex functions f and g defined on an interval $[a, b]$ are **orthogonal** on $[a, b]$ if

$$\int_a^b f\bar{g}\,dx = 0.$$

We now show that the collection

$$\phi_0 = 1, \qquad \phi_1 = e^{ix}, \ldots, \qquad \phi_n = e^{inx}, \ldots,$$
$$\phi_{-1} = e^{-ix}, \ldots, \qquad \phi_{-n} = e^{-inx}, \ldots$$

forms an orthogonal set on $[-\pi, \pi]$. That is, we have

$$\int_{-\pi}^{\pi} e^{imx}(\overline{e^{inx}})\,dx = \int_{-\pi}^{\pi} e^{i(m-n)x}\,dx$$
$$= \left[\frac{e^{i(m-n)x}}{i(m-n)}\right]_{-\pi}^{\pi} = 0 \qquad \text{if } m \neq n,$$

and

$$\int_{-\pi}^{\pi} e^{imx} \cdot e^{-imx}\,dx = \int_{-\pi}^{\pi} 1 \cdot dx = 2\pi.$$

Since $c_n = \frac{1}{2}(a_n - ib_n)$, we find for $n > 0$

$$c_n = \frac{1}{2\pi}\int_{-\pi}^{\pi} f(x)(\cos nx - i\sin nx)\,dx = \frac{1}{2\pi}\int_{-\pi}^{\pi} f(x)e^{-inx}\,dx.$$

Also,

$$c_{-n} = \frac{1}{2\pi}\int_{-\pi}^{\pi} f(x)(\cos nx + i\sin nx)\,dx = \frac{1}{2\pi}\int_{-\pi}^{\pi} f(x)e^{inx}\,dx$$

and

$$c_0 = \frac{1}{2\pi}\int_{-\pi}^{\pi} f(x)\,dx = \frac{1}{2\pi}\int_{-\pi}^{\pi} f(x)e^{i\cdot 0\cdot x}\,dx.$$

Therefore the formula

$$c_k = \frac{1}{2\pi}\int_{-\pi}^{\pi} f(x)e^{-ikx}\,dx \tag{6}$$

holds for all integers k, positive, negative, or zero.

We have established the following result.

Theorem 6. *The set*

$$\left\{\frac{1}{\sqrt{2\pi}}e^{inx}\right\}, \qquad n = 0, \pm 1, \pm 2, \ldots,$$

is an orthogonal set on the interval $[-\pi, \pi]$. *If* f *satisfies the hypotheses of Theorem* 2 *and if the* c_k *are given by* (6), *then the series* (5) *converges to* $f(x)$ *for each* x. *Theorems* 4 *and* 5 *on the differentiation and integration of series remain valid.*

The effect on complex Fourier series of evenness or oddness of a function f is analogous to the effect in the real case. The next theorem describes the various possibilities.

Theorem 7. *Suppose that* f *satisfies the hypotheses of Theorem* 2 *and that the* c_n *are given by* (6). *Then*

i) f *is real if and only if* $c_{-n} = \bar{c}_n$ *for all* n;
ii) f *is even if and only if* $c_{-n} = c_n$ *for all* n;
iii) f *is even and real if and only if* $c_{-n} = c_n = \bar{c}_n$ *for all* n;
iv) f *is odd if and only if* $c_{-n} = -c_n$ *for all* n;
v) f *is odd and real if and only if* $c_{-n} = -c_n = \bar{c}_n$ *for all* n.

The series (5) is called the **complex Fourier series of** f.

EXAMPLE. If $f(x) = x$ for $-\pi < x < \pi$, find the complex Fourier series for f.

SOLUTION. For any integer $n \neq 0$, we have

$$c_n = \frac{1}{2\pi}\int_{-\pi}^{\pi} xe^{-inx}\,dx = \frac{1}{2\pi}\left[\frac{xe^{-inx}}{-in}\right]_{-\pi}^{\pi} + \frac{1}{2n\pi i}\int_{-\pi}^{\pi} e^{inx}\,dx$$

$$= -\frac{\pi}{2n\pi i}(e^{-in\pi} + e^{in\pi}) + \frac{1}{2n\pi i}\left[\frac{e^{-inx}}{-in}\right]_{-\pi}^{\pi}.$$

Since e^z is periodic with period $2\pi i$, we find

$$e^{-in\pi} = e^{in\pi} = \cos n\pi + i\sin n\pi = (-1)^n.$$

Therefore

$$c_n = \frac{(-1)^n}{-in} = \frac{i}{n}(-1)^n \qquad \text{and} \qquad c_0 = \frac{1}{2\pi}\int_{-\pi}^{\pi} x\,dx = 0.$$

Hence

$$x = \sum_{\substack{n=-\infty \\ n\neq 0}}^{\infty} \frac{i}{n}(-1)^n e^{inx}, \qquad -\pi < x < \pi.$$

Note that the c_n satisfy the conditions of Theorem 7(v) for an odd real function.

PROBLEMS

Find the complex Fourier series for each of the following functions.

1. $f(x) = x^2, \quad -\pi < x < \pi$

2. $f(x) = \frac{1}{3}(x^3 - \pi^2 x), \quad -\pi < x < \pi$

3. $f(x) = \begin{cases} -1 & \text{for } -\pi < x < 0 \\ 1 & \text{for } 0 < x < \pi \end{cases}$

4. $f(x) = |x| \quad \text{for } -\pi < x < \pi$

5. $f(x) = e^{ax}$, a real, $\quad -\pi < x < \pi$

6. $f(x) = \sin ax, \quad 0 < a < 1, \quad -\pi < x < \pi$

7. $f(x) = \frac{1}{2}(e^{ax} - e^{-ax}), \quad 0 < a < 1, \quad -\pi < x < \pi$

8. $f(x) = \cos ax, \quad 0 < a < 1, \quad -\pi < x < \pi$

9. $f(x) = 1 + ix, \quad -\pi < x < \pi$

10. $f(x) = 2x + ix^2, \quad -\pi < x < \pi$

11. Let f be given by its complex Fourier series.
    (a) Show that f is even if and only if $c_{-n} = c_n$ for all n.
    (b) Show that f is odd if and only if $c_{-n} = -c_n$ for all n.
    (c) Show that f is even and real if and only if $c_{-n} = c_n = \bar{c}_n$ for all n.

12. (a) Show that the derivative of an even complex-valued function is odd.
    (b) Show that the derivative of an odd complex-valued function is even.

CHAPTER 7

Implicit Function Theorems. Jacobians

1. Implicit Function Theorems

An equation such as

$$x^6 + 2y^8 + 7x^2y^2 - 8x + 2y = 0 \tag{1}$$

represents a *relation* between x and y. A pair of numbers which satisfies this equation corresponds to a point in the xy-plane. In general, the totality of points in R^2 which satisfy an equation of the form

$$F(x, y) = 0 \tag{2}$$

is called the **graph** of the equation. In the study of analytic geometry in the plane, the work with graphs consists mostly of the study of arcs or simple smooth curves. However, it is not at all obvious what the graph might be of an equation such as (1) above. Even equations which are quite simple in appearance sometimes have unusual graphs. For example, the equation $x^2 + 2y^2 + 9 = 0$ is in the form (2) but has no graph, since the sum of positive quantities can never be zero. The equation $(x - 1)^2 + (y + 2)^2 = 0$ is satisfied only when $x = 1$, $y = -2$, and the graph is a single point. The equation

$$\sin x + \sec y = 0, \tag{3}$$

also in the form of (2), has an interesting graph. Since $|\sin x| \leq 1$ and $|\sec y| \geq 1$, the equation can hold only when $\sin x = 1$ and $\sec y = -1$ or when $\sin x = -1$ and $\sec y = 1$. The graph consists of the isolated points

$$(\pi/2 \pm 2m\pi, \pi \pm 2n\pi)$$

and

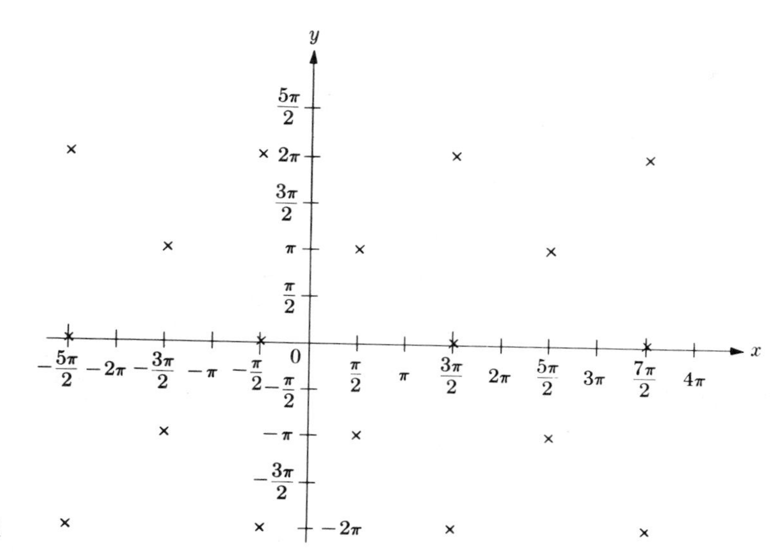

Fig. 7-1

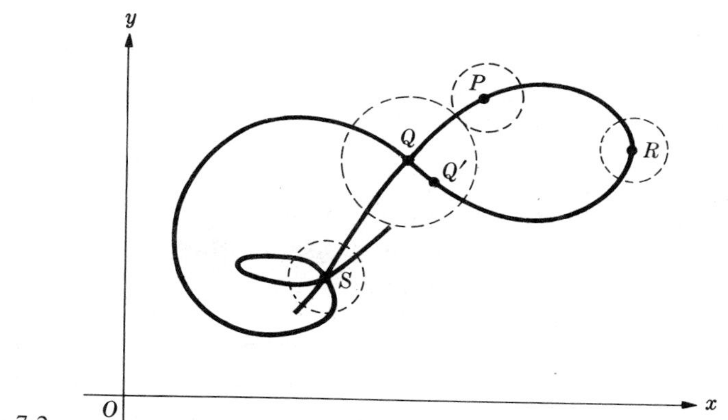

Fig. 7-2

$$(3\pi/2 \pm 2m\pi, \pm 2n\pi), \qquad m, n = 0, 1, 2, \ldots$$

as shown in Fig. 7-1.

If a function is given explicitly, say $y = f(x)$, then we calculate the derivative according to the usual rules. However, if the function is expressed in the form $F(x, y) = 0$, for example

$$x^5 + 2xy^4 + 3x - 2y^7 - 4 = 0, \tag{4}$$

then we must use the method of implicit differentiation to compute dy/dx. However, as we saw above, equation (4) may not represent y as a function of x, and in such cases dy/dx may not have a meaning.

Even under the most favorable circumstances, a relation of the form $F(x, y) = 0$ may have a complicated graph, such as the one shown in Fig. 7-2.

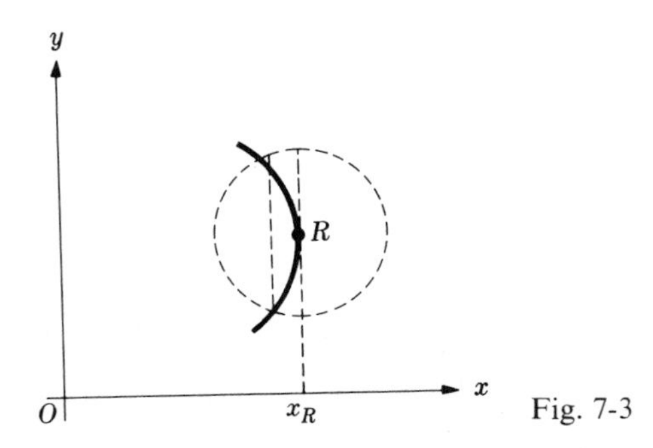

Fig. 7-3

We are interested in studying the behavior of the graph near such particular points as P, Q, R, or S. At each of these points, the graph has a special character, which we now examine. The next definition will be helpful in the discussion which follows.

Definition. By the **local behavior** of a graph, we mean the behavior of the graph in a neighborhood (small disk) of a particular point P. The term **local**, which is a technical one in mathematics, implies not only that the property we are describing persists no matter how small the neighborhood of P is, but also that it may be destroyed if a sufficiently large neighborhood is selected. It may also be destroyed if a different point is selected, no matter how close.

For example, in Fig. 7-2 we note that the slope of the graph is positive in a small neighborhood of P. This property concerns the local behavior, in that a selection of a smaller neighborhood will not change the correctness of the statement, but if the neighborhood selected is sufficiently large the statement is false. The graph in the neighborhood of R also has a special character. The tangent line at R is vertical, and therefore in a neighborhood of R we may think of x as a function of y. However, y is not a function of x since, for each value x_0 ($<x_R$) nearby, there correspond two values of y (Fig. 7-3). Thus we can describe the local behavior near R by saying that x is a function of y but that y is not a function of x. The statement is false if the neighborhood is sufficiently large. At the point Q the graph intersects itself (Fig. 7-2) and, in a neighborhood of Q, y is not representable as a function of x nor is x representable as a function of y. This statement concerns local behavior because the selection of another point Q', no matter how close to Q, changes the assertion. In a small neighborhood of Q', y is representable as a function of x and x is representable as a function of y.

If we are given a specific relation in the form

$$F(x, y) = 0, \tag{5}$$

we are interested in determining when this equation gives y as a function of x (or x as a function of y). We may also ask when we can solve for one of

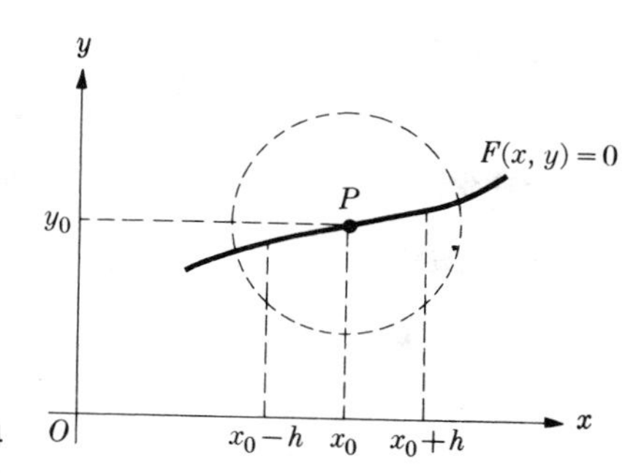

Fig. 7-4

the variables in terms of the other. In a simple case such as

$$x^2 + 3xy - 2x + 5y - 7 = 0,$$

we merely use elementary algebra to get

$$y = \frac{7 + 2x - x^2}{3x + 5},$$

and y is expressed as a function of x. However, even in cases where we can perform the algebra, certain questions arise. The equation

$$\frac{x^2}{4} + \frac{y^2}{9} - 1 = 0$$

may be solved for y to give

$$y = \pm\tfrac{3}{2}\sqrt{4 - x^2},$$

but still y is not represented as a function of x. If we write

$$f_1 : x \to +\tfrac{3}{2}\sqrt{4 - x^2} \qquad \text{and} \qquad f_2 : x \to -\tfrac{3}{2}\sqrt{4 - x^2},$$

then the entire graph (an ellipse) is described by these two functions.

Starting with the relation (5) and a point $P(x_0, y_0)$ on the graph, we suppose that in a neighborhood of P the relation represents y as a function of x. In other words, we assume that the local behavior of the relation allows us to write (Fig. 7-4)

$$y = f(x) \tag{6}$$

for $x_0 - h < x < x_0 + h$, where h is some positive number. If we substitute (6) into (5), then

$$F(x, f(x)) = 0$$

is an identity for $|x - x_0| < h$. Assuming that all quantities are smooth, we may use the chain rule to get

$$F_x[x, f(x)] + F_y[x, f(x)] \cdot f'(x) = 0.$$

If $F_y \neq 0$, we obtain the formula

$$f'(x) = -\frac{F_x}{F_y}.$$

Implicit function theorems are those which state conditions under which a relation (5) *may be put in the form* (6), *at least in the nieghborhood of some point P on the graph.* These theorems also decide when the differentiation formula (7) is valid. Before establishing the most important implicit function theorems, we recall several facts which were studied earlier. The first is the Intermediate Value Theorem.

Intermediate Value Theorem. *Suppose that f is continuous on an interval $[a, b]$ and that $f(a) = A$, $f(b) = B$. If C is any number between A and B, there is a number c between a and b such that $f(c) = C$.*

In other words, a continuous function must take on all intermediate values between any two values it assumes.

A second fact, used many times but never stated as a theorem, concerns the local behavior of a continuous function. Briefly, if a continuous function is positive for some value, it must be positive for all points in some sufficiently small region about this value. The next theorem establishes this result for functions in any number of variables.

Theorem 1. *Suppose $f(x_1, x_2, \ldots, x_n)$ is continuous at a point $(x_1^0, x_2^0, \ldots, x_n^0)$, and suppose $f(x_1^0, x_2^0, \ldots, x_n^0) > 0$. Then there is a positive number h such that $f(x_1, x_2, \ldots, x_n)$ is positive for all $(x_1, x_2, \ldots, x_n)$ in the neighborhood*

$$|x_1 - x_1^0| < h, \qquad |x_2 - x_2^0| < h, \ldots, \qquad |x_n - x_n^0| < h. \tag{8}$$

Proof. The domain (8) is an n-dimensional rectangular box (called a hypercube) of the type used in the definition of limit for functions of several variables. (See Chapter 4, page 198, for $n = 2$.) Let

$$\varepsilon = f(x_1^0, x_2^0, \ldots, x_n^0)$$

and apply the definition of continuity. For any $\varepsilon > 0$ there is a $\delta > 0$ such that

$$|f(x_1, x_2, \ldots, x_n) - f(x_1^0, x_2^0, \ldots, x_n^0)| < \varepsilon \tag{9}$$

whenever

$$|x_1 - x_1^0| < \delta, \qquad |x_2 - x_2^0| < \delta, \ldots, \qquad |x_n - x_n^0| < \delta.$$

Setting $\delta = h$, we see that (9) is the same as

$$0 = f(x_1^0, x_2^0, \ldots, x_n^0) - \varepsilon < f(x_1, x_2, \ldots, x_n) < f(x_1^0, x_2^0, \ldots, x_n^0) + \varepsilon$$

so long as $(x_1, x_2, \ldots, x_n)$ is in the hypercube (8).

A third fact we shall use in the proof of the first implicit function theorem is the Fundamental Lemma on Differentiation. For functions of two variables, the result is given by the formula

$$F(x_0 + h, y_0 + k) - F(x_0, y_0) = F_x(x_0, y_0)h + F_y(x_0, y_0)k + G_1(h,k)h + G_2(h,k)k, \tag{10}$$

where G_1 and G_2 tend to zero as h, k tend to zero, and

$$G_1(0,0) = G_2(0,0) = 0.$$

(See Chapter 4, page 207.)

We now prove the first implicit function theorem.

Theorem 2. *Suppose that F, F_x, and F_y are continuous near (x_0, y_0) and suppose that*

$$F(x_0, y_0) = 0,$$

$$F_y(x_0, y_0) \neq 0.$$

a) *Then there are two positive numbers h and k which determine a rectangle R about (x_0, y_0),*

$$R = \{(x,y) : |x - x_0| < h, |y - y_0| < k\},$$

such that for each x with $|x - x_0| < h$ there is a ***unique*** *number y with $|y - y_0| < k$ which satisfies the equation $F(x, y) = 0$. That is, y is a function of x, and we may write $y = f(x)$. The domain of f is*

$$|x - x_0| < h$$

and the range is in $|y - y_0| < k$.

b) *The function f determined in* (a) *and its derivative f' are continuous for $|x - x_0| < h$. Furthermore,*

$$F_y[x, f(x)] \neq 0$$

and

$$f'(x) = -\frac{F_x[x, f(x)]}{F_y[x, f(x)]}$$

for $|x - x_0| < h$.

Proof. We may assume that $F_y(x_0, y_0) > 0$; otherwise we simply replace F by $-F$ and repeat the argument. Since F_y is continuous, then from Theorem 1 there must be a number k such that $F_y(x, y) > 0$ in the square $S = \{(x, y) : |x - x_0| \leq k, |y - y_0| \leq k\}$ (Fig. 7-5). If we fix a value x in $|x - x_0| < k$, then $F(x, y)$, considered as a function of y alone, has a positive slope and therefore is *an increasing function of y in S*. In particular, $F(x_0, y)$ is an increasing function of y. Since by hypothesis

$$F(x_0, y_0) = 0,$$

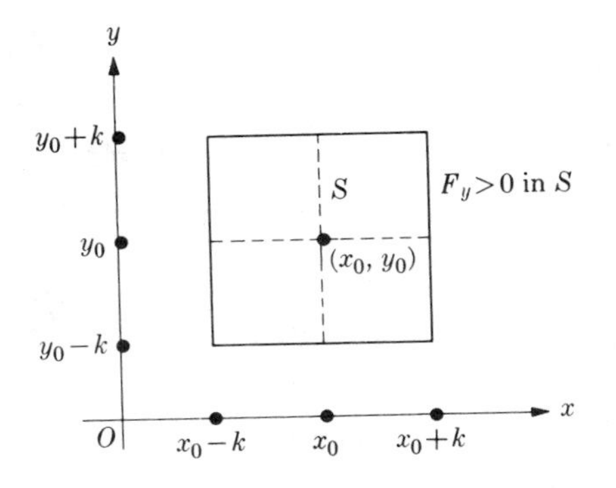

Fig. 7-5

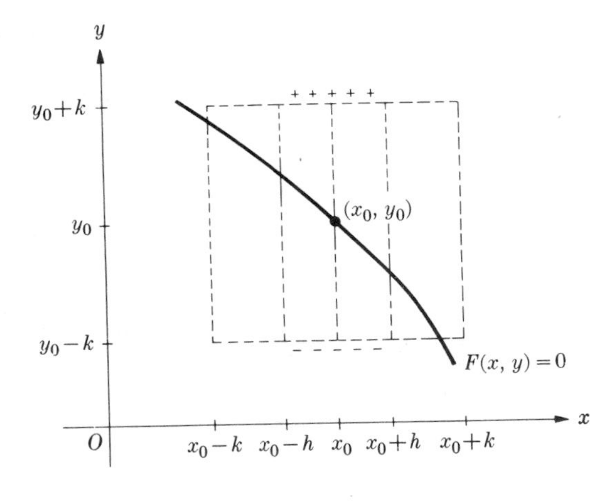

Fig. 7-6

it follows that

$$F(x_0, y_0 + k) > 0 \qquad \text{and} \qquad F(x_0, y_0 - k) < 0.$$

We now apply Theorem 1 to each of the functions

$$F(x, y_0 + k) \qquad \text{and} \qquad F(x, y_0 - k)$$

and conclude that there is an interval $|x - x_0| < h$ in which

$$F(x, y_0 + k) > 0 \qquad \text{and} \qquad F(x, y_0 - k) < 0.$$

We now return to the solution of the equation $F(x, y) = 0$. Fix x in the interval $x_0 - h < x < x_0 + h$ and concentrate on the rectangle

$$R = \{(x, y) : |x - x_0| < h, |y - y_0| < k\},$$

as shown in Fig. 7-6. Since $F(x, y)$ as a function of y is negative for $y = y_0 - k$ and positive for $y = y_0 + k$ then, according to the Intermediate Value Theorem, there is a value y such that $F(x, y) = 0$. Also, since $F_y > 0$, there cannot be more than one such value. The existence of the function $y = f(x)$ for $|x - x_0| < h$ is thus established.

To show that f is continuous at x_0, we must prove that for any $\varepsilon > 0$

$$|f(x) - f(x_0)| < \varepsilon,$$

provided that x is sufficiently close to x_0. The values of f are restricted by the choice of the square S—that is, by the size of k. If we select a k' smaller than k and go through the entire process described in the proof, we will obtain the same function f, but it may perhaps be defined on a smaller interval $|x - x_0| < h'$. Selecting $k' = \varepsilon$, we see that the choice of $h' = \delta$ in the definition of continuity yields the result. At any other point x_1 with

$$|x_1 - x_0| < h,$$

we establish the result by constructing the square S and the rectangle R with (x_1, y_1) as center, where $y_1 = f(x_1)$.

To establish part (b), we employ the Fundamental Lemma on Differentiation, Eq. (10). Writing $\Delta f = f(x + \Delta x) - f(x)$ and noting that $G_1(0,0) = G_2(0,0) = 0$, we have

$$\begin{aligned} 0 &= F[x + \Delta x, f(x + \Delta x)] - F[x, f(x)] \\ &= F_x[x, f(x)]\,\Delta x + F_y[x, f(x)]\,\Delta f + G_1(\Delta x, \Delta f)\,\Delta x + G_2(\Delta x, \Delta f)\,\Delta f. \end{aligned}$$

Therefore

$$F_x\,\Delta x + F_y\,\Delta f = -G_1\,\Delta x - G_2\,\Delta f$$

and

$$\frac{\Delta f}{\Delta x} = -\frac{F_x + G_1}{F_y + G_2}. \tag{11}$$

Since f is continuous, Δf tends to zero as Δx does. Consequently $G_1(\Delta x, \Delta f)$ and $G_2(\Delta x, \Delta f)$ tend to zero with Δx. The left side of (11) is the difference quotient of f and so tends to $f'(x)$ as the right side tends to $-F_x/F_y$. This establishes part (b). Furthermore, f' is continuous because of the hypotheses that F_x and F_y are.

REMARKS. (i) A geometric intepretation of the theorem is indicated in Fig. 7-7. The theorem states that there exists a rectangle

$$|x - x_0| < h, \qquad |y - y_0| < k$$

such that the part of the graph of $F(x, y) = 0$ which is in this rectangle lies along an arc of the form $y = f(x)$, where f is differentiable. Of course, the line $x = x_0$ may intersect the total graph at several points, as shown in the figure. (ii) The theorem is purely *local* in that nothing is determined about how far the domain of f can be extended. However, the result remains valid no matter how small a rectangle about (x_0, y_0) is selected. (iii) We have here an example of an *existence theorem*, in that the proof does not provide us with a method for finding the particular function f. A proof which enables

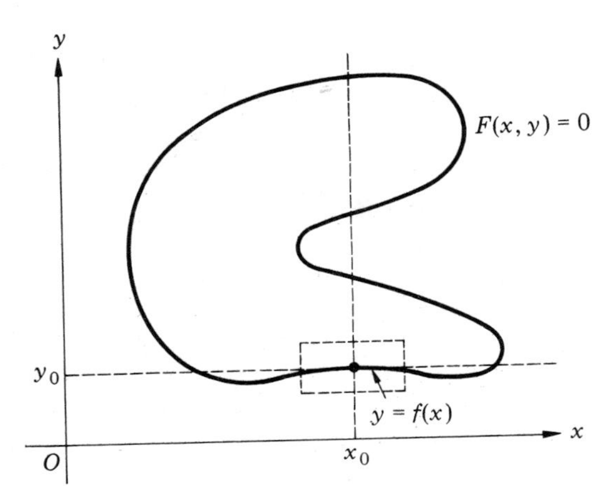

Fig. 7-7

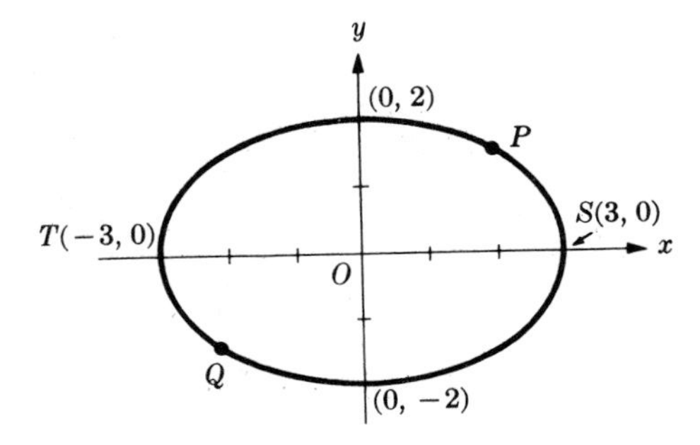

Fig. 7-8

us to determine the actual answer either numerically or analytically is called a *constructive proof*. Constructive proofs, although more desirable than existence proofs, are frequently more difficult. (iv) A corresponding result holds if the variables are interchanged. If $F_x(x_0, y_0) \neq 0$, then the Implicit Function Theorem allows us to write $x = g(y)$, at least locally.

Example 1. Apply Theorem 2 to the equation

$$\frac{x^2}{9} + \frac{y^2}{4} - 1 = 0,$$

and find the function f of the theorem whenever possible.

Solution. The graph of the equation is shown in Fig. 7-8. We see that if P is a point as shown, then $F_y = \frac{1}{2}y > 0$ and y is a function of x. In fact,

$$y = +\tfrac{2}{3}\sqrt{9 - x^2}.$$

At a point Q we have $F_y < 0$ and

$$y = -\tfrac{2}{3}\sqrt{9 - x^2}.$$

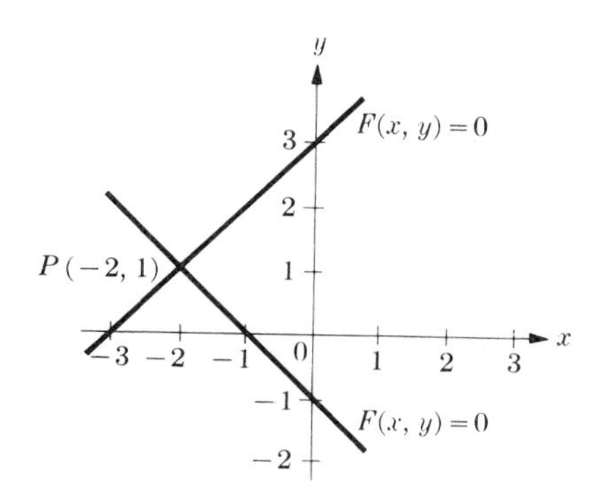

Fig. 7-9

However, $F_y = 0$ at points such as S or T, and the theorem fails. On the other hand, we observe that $F_x = \frac{2}{9}x > 0$ at S, and so we can write $x = g(y)$ in a neighborhood of S. We compute easily, $g(y) = \frac{3}{2}\sqrt{4 - y^2}$ near S.

EXAMPLE 2. Show that the entire graph of the equation

$$F(x, y) \equiv y^3 + 3x^2y - x^3 + 2x + 3y = 0$$

is a function which is defined for all x.

SOLUTION. We have

$$F_y = 3y^2 + 3x^2 + 3 > 0 \qquad \text{for all } (x, y).$$

Therefore, for each x, $F(x, y)$ is increasing in y for all y. Moreover,

$$F(x, y) \to -\infty$$

as $y \to -\infty$ and $F \to +\infty$ as $y \to +\infty$ for each fixed x. It follows that for each x there is a unique y such that $F(x, y) = 0$. Since the hypotheses of Theorem 2 are satisfied at any point (x_0, y_0) where $F(x_0, y_0) = 0$, the function f determined by the theorem is defined, continuous, and differentiable for all x.

EXAMPLE 3. Discuss the validity of Theorem 2 with regard to the function

$$F(x, y) \equiv x^2 - y^2 + 4x + 2y + 3 = 0.$$

SOLUTION. We have $F_x = 2x + 4$, $F_y = -2y + 2$. At any point on the graph with $y \neq 1$, we may solve for y as a function of x, since $F_y \neq 0$. Also, whenever $x \neq -2$ we may solve for x as a function of y. However, at the point $(-2, 1)$ which is on the graph, we have $F_x = F_y = 0$, and Theorem 2 fails (Fig. 7-9). The formula on page 394 gives

$$\frac{dy}{dx} = -\frac{F_x}{F_y} = \frac{x + 2}{1 - y},$$

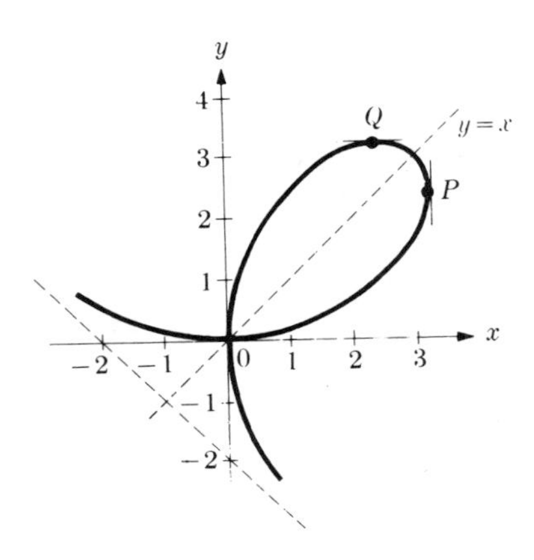

Fig. 7-10

which is indeterminate at $(-2, 1)$. The point $P(-2, 1)$ is a double point of the graph.

EXAMPLE 4. Locate the points on the graph

$$F(x, y) \equiv x^3 + y^3 - 6xy = 0$$

for which Theorem 2 is not applicable.

SOLUTION. This curve is the **folium of Descartes** (Fig. 7-10), which was discussed on page 269. A simple computation yields

$$F_x = 3x^2 - 6y, \qquad F_y = 3y^2 - 6x,$$

and we see that both F_x and F_y vanish at the origin, which is a double point. We cannot obtain y as a function of x at any point where $3y^2 - 6x = 0$. Substituting $x = \frac{1}{2}y^2$ into the equation of the curve, we find

$$x = 2\sqrt[3]{4}, \qquad y = 2\sqrt[3]{2}.$$

This point is designated P in Fig. 7-10. Similarly, setting $F_x = 0$ and solving, we find that we cannot express x as a function of y in a neighborhood of $Q(2\sqrt[3]{2}, 2\sqrt[3]{4})$. Theorem 2 is applicable at all other points on the curve.

Theorem 2 is valid in any number of variables, and the proof introduces no new difficulties. We state the result for reference.

Theorem 2′. *Suppose that $F(x_1, x_2, \ldots, x_n, y)$ and F_y are continuous near the point $(x_1^0, x_2^0, \ldots, x_n^0, y_0)$; suppose also that, for $i = 1, 2, \ldots, n$, each F_{x_i} is continuous near $(x_1^0, x_2^0, \ldots, x_n^0, y_0)$. Let*

$$F(x_1^0, x_2^0, \ldots, x_n^0, y_0) = 0, \qquad F_y(x_1^0, x_2^0, \ldots, x_n^0, y_0) \neq 0.$$

Then there are positive numbers h and k such that

a) *for each $(x_1, x_2, \ldots, x_n)$ in the hypercube $|x_i - x_i^0| < h$, $i = 1, 2, \ldots, n$,* *there is a* ***unique*** *number* y *with* $|y - y_0| < k$ *satisfying* $F(x_1, x_2, \ldots, x_n, y) = 0$;

b) *if we define* f *by* $f(x_1, x_2, \ldots, x_n) = y$, *then* f *and all first derivatives* f_{x_i}, $i = 1, 2, \ldots, n$ *are continuous and*

$$F_y[x_1, x_2, \ldots, x_n, f(x_1, x_2, \ldots, x_n)] \neq 0,$$

$$f_{x_i} = -\frac{F_{x_i}[x_1, x_2, \ldots, x_n, f(x_1, x_2, \ldots, x_n)]}{F_y[x_1, x_2, \ldots, x_n, f(x_1, x_2, \ldots, x_n)]}, \qquad i = 1, 2, \ldots, n,$$

for $|x - x_i| < h$, $i = 1, 2, \ldots, n$.

EXAMPLE 5. Using Theorem 2′, can we conclude that the part of the graph of the equation

$$F(x, y, z) = 3x^2 + 2y^2 + z^2 + 2xy + 2xz + 2yz - 9 = 0$$

within some box $|x - 2| < h$, $|y + 1| < h$, $|z + 1| < k$ lies along a surface $z = f(x, y)$ in which f is defined and differentiable for all x, y with $|x - 2| < h$ and $|y + 1| < h$? Solve for z in terms of x and y and discuss.

SOLUTION. Here $x_0 = 2$, $y_0 = -1$, $z_0 = -1$. [In Theorem 2′ we have (x_1, x_2, y) instead of (x, y, z).] Then $F_z = 2z + 2x + 2y$, $F_z(2, -1, -1) = 0$. Consequently, the hypotheses of the theorem *are not satisfied*. Solving for z, we obtain

$$\begin{gathered} z^2 + 2(x + y)z + (3x^2 + 2y^2 + 2xy - 9) = 0, \\ z = -(x + y) \pm \sqrt{9 - 2x^2 - y^2}. \end{gathered} \tag{12}$$

The domain of the two functions in (12) is the elliptical region

$$\{(x, y) : 0 \leq 2x^2 + y^2 \leq 9\},$$

and we note that the point $(2, -1)$ is on the boundary of this region. Thus there is no box satisfying the conditions. From our knowledge of analytic geometry we also observe that the plane tangent to the surface

$$F(x, y, z) = 0$$

at $(2, -1, -1)$ is parallel to the z axis.

PROBLEMS

In each of Problems 1 through 10, use Theorem 2 to show that the equation $F(x, y) = 0$ may be represented in the form $y = f(x)$ in a neighborhood of the given point (x_0, y_0). Draw a graph and compute $f'(x_0)$ in each case

1. $F(x, y) \equiv x + y + x \sin y = 0$; $(x_0, y_0) = (0, 0)$
2. $F(x, y) \equiv y^2 - 2xy + 5x^2 - 16 = 0$; $(x_0, y_0) = (1, 1 - 2\sqrt{3})$
3. $F(x, y) \equiv y^3 + y - x^2 = 0$; $(x_0, y_0) = (0, 0)$

4. $F(x, y) \equiv xe^y - y + 1 = 0; \quad (x_0, y_0) = (-1, 0)$
5. $F(x, y) \equiv x^{2/3} + y^{2/3} - 4 = 0; \quad (x_0, y_0) = (1, 3\sqrt{3})$
6. $F(x, y) \equiv (x^2 + y^2)^2 - 8(x^2 - y^2) = 0; \quad (x_0, y_0) = (\sqrt{3}, 1)$
7. $F(x, y) \equiv xy + \log(xy) - 1 = 0; \quad (x_0, y_0) = (1, 1)$
8. $F(x, y) \equiv x \cos xy; \quad (x_0, y_0) = (1, \pi/2)$
9. $F(x, y) \equiv x^5 + y^5 + xy + 4 = 0; \quad (x_0, y_0) = (2, -2)$
10. $F(x, y) \equiv 2 \sin x + \cos y - 1 = 0; \quad (x_0, y_0) = (\pi/6, 3\pi/2)$

In each of Problems 11 through 16, use Theorem 2′ to show that the equation $F(x, y, z) = 0$ may be represented in the form $z = f(x, y)$ in a neighborhood of the given point (x_0, y_0, z_0). Find $f_x(x_0, y_0)$ and $f_y(x_0, y_0)$.

11. $F(x, y, z) \equiv x^3 + y^3 + z^3 - 3xyz - 4 = 0; \quad (x_0, y_0, z_0) = (1, 1, 2)$
12. $F(x, y, z) \equiv x^4 + y^4 + z^4 - 18 = 0; \quad (x_0, y_0, z_0) = (1, 1, 2)$
13. $F(x, y, z) \equiv e^z - z^2 - x^2 - y^2 = 0; \quad (x_0, y_0, z_0) = (1, 0, 0)$
14. $F(x, y, z) \equiv z^3 - z - xy \sin z = 0; \quad (x_0, y_0, z_0) = (0, 0, 0)$
15. $F(x, y, z) \equiv x + y + z + \cos xyz = 0; \quad (x_0, y_0, z_0) = (0, 0, -1)$
16. $F(x, y, z) \equiv x + y + z - e^{xyz} = 0; \quad (x_0, y_0, z_0) = (0, \frac{1}{2}, \frac{1}{2})$
17. Do there exist numbers $h > 0$ and $k > 0$ such that all the points satisfying $y^2 - x^3 = 0$ and $|x| < h$, $|y| < k$
    a) lie along an arc $y = f(x)$ where $f(x)$ is defined and smooth for all x with $|x| < h$?
    b) lie along an arc $x = g(y)$ where g has the same properties for $|y| < k$?

In Problems 18 through 24, in each case plot and discuss the entire graph, indicate the different functions defined implicitly by the equation, and find any points on the graph where $F_y(x_0, y_0) = 0$. At any such points check to see whether $F_x(x_0, y_0) = 0$.

18. $F(x, y) \equiv (x - 3)^4 - (y + 2)^2 = 0$
19. $F(x, y) \equiv xe^y - 2y + 2 = 0$
20. $F(x, y) \equiv 2x^2 + y^2 - x^3 = 0$
21. $F(x, y) \equiv y^3 + x^2y - x^2 = 0$
22. $F(x, y) \equiv y^3 - 3y - x^2 = 0$
23. $F(x, y) \equiv (x^2 + y^2)^2 - 8(x^2 - y^2) = 0$ (lemniscate)
24. $F(x, y) \equiv e^{2xy} - \log[1/(1 + y^2)] = 0$

In Problems 25 and 26, show that $F_z(x_0, y_0, z_0) = 0$ and determine whether or not it is possible to express z as a function of x and y in a neighborhood of (x_0, y_0, z_0). Does the same situation prevail when y and z are interchanged; i.e., can we solve for y in terms of x and z in a neighborhood of (x_0, y_0, z_0)?

25. $F(x, y, z) \equiv 5x^2 + 3y^2 + z^2 - 4xy - 4xz + 2yz - 9 = 0$;
$(x_0, y_0, z_0) = (-1, 2, -4)$

26. $F(x, y, z) \equiv 2x^2 + 3y^2 + z^2 + 4xy + 2xz + 4yz - 7 = 0$;
$(x_0, y_0, z_0) = (4, -3, 2)$

27. Write out the proof of part (a) of Theorem 2′.

28. Write out the proof of part (b) of Theorem 2′.

29. Find an example of a relation $F(x, y, z) = 0$ such that $F_x(x_0, y_0, z_0) = F_y(x_0, y_0, z_0) = F_z(x_0, y_0, z_0) = 0$, and yet we are able to solve for z in terms of x and y in a neighborhood of (x_0, y_0, z_0).

30. Repeat Problem 29, for a relation $F(x_1, x_2, \ldots, x_n, y) = 0$.

2. Implicit Function Theorems for Systems

Suppose we have the system of two equations

$$\begin{aligned} x^2 + 2xy - 3xu + 4yv &= 0, \\ 4xy + x^3u - 8yv + 2 &= 0 \end{aligned} \tag{1}$$

in the four variables x, y, u, and v. It is possible in this case to express u and v as functions of x and y. Solving the first equation for u, we get

$$u = \frac{x^2 + 2xy + 4yv}{3x}, \tag{2}$$

and we label this equation $u = \phi(x, y, v)$. Next we substitute u from (2) into the second equation in (1) to obtain

$$4xy + x^3\left(\frac{x^2 + 2xy + 4yv}{3x}\right) - 8yv + 2 = 0.$$

This last equation may be solved for v in terms of x and y. The result is (after a little algebra)

$$v = \frac{x^4 + 2x^3y + 12xy + 6}{24y - 4x^2y}. \tag{3}$$

We now substitute v from (3) into (2) and find

$$u = \frac{(x^2 + 2xy)(6y - x^2) + (x^4 + 2x^3y + 12xy + 6)y}{3x(6y - x^2)}, \tag{4}$$

which expresses u as a function of x and y. In other words, starting with the system (1), we obtain a system $u = f(x, y)$ and $v = g(x, y)$ which, in this particular case, consists of the equations (4) and (3).

Suppose that we have a general system of two equations in four unknowns which we write

$$F(x, y, u, v) = 0 \qquad \text{and} \qquad G(x, y, u, v) = 0. \tag{5}$$

Proceeding in the most elementary manner, as in the example above, we imagine that we can solve the equation $F = 0$ for u in terms of x, y, and v. We write this solution

$$u = \phi(x, y, v).$$

Then, substituting this value of u in the equation $G = 0$, we get a single equation involving x, y, and v only. According to the Implicit Function Theorem (which we suppose it is possible to use), we extract from the relation connecting x, y, and v an explicit function

$$v = g(x, y).$$

Now, continuing to parallel the method of the above example, we substitute for v into ϕ to obtain

$$u = \phi[x, y, g(x, y)],$$

which we write

$$u = f(x, y).$$

In this way, we have expressed u in terms of x and y, and we have expressed v in terms of x and y. We observe that all these results are *local* in character, and thus the equations

$$u = f(x, y) \qquad \text{and} \qquad v = g(x, y) \tag{6}$$

are valid only in the neighborhood of some point.

Taking (6) into account, we can now write the equations $F = 0$ and $G = 0$ in the form

$$F[x, y, f(x, y), g(x, y)] = 0, \qquad G[x, y, f(x, y), g(x, y)] = 0. \tag{7}$$

Assuming that all functions are smooth and that all operations of differentiation are legitimate, we apply the chain rule in (5) to compute partial derivatives with respect to x. The result is (since x and y are the independent variables and u and v are the dependent variables)

$$F_x + F_u u_x + F_v v_x = 0 \qquad \text{and} \qquad G_x + G_u u_x + G_v v_x = 0. \tag{8}$$

In (8) we treat u_x and v_x as unknowns and the remaining quantities as known. Then the formula for solving two equations in two unknowns gives

$$u_x = -\frac{\begin{vmatrix} F_x & F_v \\ G_x & G_v \end{vmatrix}}{\begin{vmatrix} F_u & F_v \\ G_u & G_v \end{vmatrix}}, \qquad v_x = -\frac{\begin{vmatrix} F_u & F_x \\ G_u & G_x \end{vmatrix}}{\begin{vmatrix} F_u & F_v \\ G_u & G_v \end{vmatrix}}, \tag{9}$$

provided that the denominator $F_u G_v - F_v G_u$ is not zero. The derivatives u_y and v_y are obtained by differentiating the equations in (5) with respect to y.

The above development paid no attention either to the feasibility of carrying out the process or to the legitimacy of the various steps. At this time we are more interested in determining when the process leading to equations (6) and (9) is *possible* than we are in deciding when the actual steps can be performed, either analytically or numerically. The next theorem shows the circumstances under which all the various steps are mathematically correct.

Theorem 3A. *Suppose that $F(x, y, u, v)$ and $G(x, y, u, v)$ are continuous and have continuous first derivatives near a point (x_0, y_0, u_0, v_0). Also, suppose that*

$$F(x_0, y_0, {}_0, v_0) = 0, \qquad G(x_0, y_0, u_0, v_0) = 0$$

and

$$D_0 = \begin{vmatrix} F_u(x_0, y_0, u_0, v_0) & F_v(x_0, y_0, u_0, v_0) \\ G_u(x_0, y_0, u_0, v_0) & G_v(x_0, y_0, u_0, v_0) \end{vmatrix} \neq 0.$$

Then there are positive numbers h, k_1, and k_2 such that

a) *for each (x, y) with $|x - x_0| < h$, $|y - y_0| < h$, there is a* ***unique*** *solution (u, v) of the equations*

$$F(x, y, u, v) = 0, \qquad G(x, y, u, v) = 0$$

with $|u - u_0| < k_1$ and $|v - v_0| < k_2$. We denote these solutions

$$u = f(x, y), \qquad v = g(x, y).$$

b) *The functions f and g are continuous with continuous first derivatives, and the following formulas hold:*

$$f_x = -\frac{1}{D}\begin{vmatrix} F_x & F_v \\ G_x & G_v \end{vmatrix}, \qquad g_x = -\frac{1}{D}\begin{vmatrix} F_u & F_x \\ G_u & G_x \end{vmatrix},$$

$$f_y = -\frac{1}{D}\begin{vmatrix} F_y & F_v \\ G_y & G_v \end{vmatrix}, \qquad g_y = -\frac{1}{D}\begin{vmatrix} F_u & F_y \\ G_u & G_y \end{vmatrix},$$

where $D = F_u G_v - F_v G_u$.

PROOF. The proof consists of successive applications of the Implicit Function Theorem of Section 1. In this way the original formal description of the appropriate steps required for the elimination process can be made legitimate. Since $D_0 \neq 0$, it follows that G_u and G_v cannot both be zero at (x_0, y_0, u_0, v_0). Suppose that $G_v \neq 0$; the proof is similar if $G_u \neq 0$. Then, from Theorem 2′, we conclude that there are numbers m and r such that the entire portion of the graph of $G(x, y, u, v) = 0$, for which

$$|x - x_0| < m, \qquad |y - y_0| < m,$$

$$|u - u_0| < m, \qquad \text{and} \qquad |v - v_0| < r$$

may be represented in the form

$$v = H(x, y, u),$$

where H is continuous and differentiable; moreover

$$H_u(x, y, u) = -\frac{G_u[x, y, u, H(x, y, u)]}{G_v[x, y, u, H(x, y, u)]}. \tag{10}$$

(In applying Theorem 2′, we replace h by m, k by r, (x_1, x_2, x_3) by (x, y, u), and the variable y by the variable v. We also use the label H instead of f for the explicitly obtained function.)

Now we define

$$K(x, y, u) = F[x, y, u, H(x, y, u)]; \tag{11}$$

the function K is defined for the cube

$$|x - x_0| < m', \qquad |y - y_0| < m', \qquad |u - u_0| < m',$$

for some m' with $0 < m' \leq m$. Using the Chain rule to differentiate (11) with respect to u, we find that

$$K_u = F_u + F_v H_u = F_u + F_v\left(-\frac{G_u}{G_v}\right),$$

in which (10) has been used. Using the values at the point $P_0(x_0, y_0, u_0, v_0)$, we have

$$K_u(x_0, y_0, u_0) = \frac{F_u G_v - F_v G_u}{G_v(x_0, y_0, u_0, v_0)} = \frac{D_0}{G_v(P_0)} \neq 0. \tag{12}$$

Because of (12), we can apply the Implicit Function Theorem to the equation $K(x, y, u) = 0$. We conclude that there are positive numbers m'' and r'' with $m'' \leq m'$, $r'' \leq m'$ such that the part of the graph of $K(x, y, u) = 0$, for which

$$|x - x_0| < m'', \qquad |y - y_0| < m'' \qquad \text{and} \qquad |u - u_0| < r''$$

is representable in the form

$$u = f(x, y), \qquad \text{with} \qquad |x - x_0| < m'', \qquad |y - y_0| < m''.$$

Furthermore, f and its first derivatives are continuous. If we now define

$$v = g(x, y) \equiv H[x, y, f(x, y)] \qquad \text{for} \qquad |x - x_0| < m'', \qquad |y - y_0| < m'',$$

we see that g and its first derivatives are continuous.

The validity of the formulas for $f_x, \ldots, g_y$ follows at once from the Chain rule, as described at the beginning of the section.

Theorem 3A may be generalized to a pair of functions in any number of variables. If we are given

$$F(x_1, x_2, \ldots, x_n, u, v) = 0, \qquad G(x_1, x_2, \ldots, x_n, u, v) = 0,$$

and a point on the graph of F and G at which

$$D_0 = \begin{vmatrix} F_u(x_1^0, x_2^0, \dots, x_n^0, u^0, v^0) & F_v(x_1^0, x_2^0, \dots, x_n^0, u^0, v^0) \\ G_u(x_1^0, x_2^0, \dots, x_n^0, u^0, v^0) & G_v(x_1^0, x_2^0, \dots, x_n^0, u^0, v^0) \end{vmatrix} \neq 0,$$

then, under hypotheses analogous to those given in the above theorem, we may solve for u and v, obtaining

$$u = f(x_1, x_2, \dots, x_n), \qquad v = g(x_1, x_2, \dots, x_n) \tag{13}$$

in a neighborhood of $(x_1^0, x_2^0, \dots, x_n^0, u^0, v^0)$. Furthermore,

$$f_{x_i} = -\frac{\begin{vmatrix} F_{x_i} & F_v \\ G_{x_i} & G_v \end{vmatrix}}{\begin{vmatrix} F_u & F_v \\ G_u & G_v \end{vmatrix}}, \qquad g_{x_i} = -\frac{\begin{vmatrix} F_u & F_{x_i} \\ G_u & G_{x_i} \end{vmatrix}}{\begin{vmatrix} F_u & F_v \\ G_u & G_v \end{vmatrix}}. \tag{14}$$

EXAMPLE 1. Show that the graph of the equations

$$F(x, y, u, v) = x^2 - y^2 - u^3 + v^2 + 4 = 0,$$

$$G(x, y, u, v) = 2xy + y^2 - 2u^2 + 3v^4 + 8 = 0,$$

is representable in the form $u = f(x, y)$, $v = g(x, y)$ in a neighborhood of the point

$$P_0 = \{(x, y, u, v) : x = 2, y = -1, u = 2, v = 1\}.$$

Find the derivatives u_x, u_y, v_x, v_y at P_0.

SOLUTION. We have

$$F_u = -3u^2, \qquad F_v = 2v, \qquad G_u = -4u, \qquad G_v = 12v^3,$$

$$F_x = 2x, \qquad F_y = -2y, \qquad G_x = 2y, \qquad G_y = 2x + 2y.$$

At the point in question, we find

$$D_0 = F_u G_v - F_v G_u |_{P_0} = -128.$$

Since F and G are polynomial expressions (and hence smooth), and since $D_0 \neq 0$, Theorem 3A is applicable. We conclude that u and v are expressible as functions of x and y. A computation yields

$$u_x = \frac{1}{128}\begin{vmatrix} 4 & 2 \\ -2 & 12 \end{vmatrix} = \frac{13}{32}, \qquad v_x = \frac{7}{16}, \qquad u_y = \frac{5}{32}, \qquad v_y = -\frac{1}{16}.$$

EXAMPLE 2. Show that there is a box of the form $|x - 1| < h$, $|y + 1| < k$, $|z - 2| < k$ (h and $k > 0$) such that the part of the graph of the equations

$$z - 2x - 2y - 2 = 0 \qquad \text{and} \qquad z - x^2 - y^2 = 0$$

in that box lies along the curve determined by a pair of equations of the form

$y = f(x)$ and $z = g(x)$. Show that the functions f, g, f', g' are continuous for $|x-1| < h$. Also, find f and g explicitly. Does a similar box exist about the point $(-1, 1, 2)$?

SOLUTION. Let $F(x, y, z) = z - 2x - 2y - 2$, $G(x, y, z) = z - x^2 - y^2$. Then the Implicit Function Theorem for a general pair of equations is applicable with $x_1 = x$, $u = y$, $v = z$. Since

$$F_y = -2, \qquad F_z = 1, \qquad G_y = -2y, \qquad G_z = 1,$$

at the point $(1, -1, 2)$, we have

$$F_y = -2, \qquad F_z = 1, \qquad G_y = 2, \qquad G_z = 1,$$

and

$$F_uG_v - F_vG_u = F_yG_z - F_zG_y = -4 \neq 0.$$

Therefore such a box exists. To find the explicit functions, we substitute z from $F = 0$ into $G = 0$ and find

$$z = 2x + 2y + 2, \qquad x^2 + y^2 - 2x - 2y - 2 = 0$$

or

$$z = 2x + 2y + 2, \qquad (x-1)^2 + (y-1)^2 = 4.$$

Therefore

$$y = 1 \pm \sqrt{4 - (x-1)^2}$$

and

$$z = 2x + 2y + 2 \qquad \text{for } |x-1| \leq 2.$$

Since $y = -1$ when $x = 1$, we must take the branch $1 - \sqrt{4 - (x-1)^2}$. Hence

$$f(x) = 1 - \sqrt{4 - (x-1)^2},$$
$$g(x) = 2x + 4 - 2\sqrt{4 - (x-1)^2}$$

for $|x-1| \leq 2 \Leftrightarrow -1 \leq x \leq 3$. There is no box about the point $(-1, 1, 2)$, since x can never fall below -1 anywhere on the graph.

The implicit function theorems may be extended to cover the situation where we have any number of variables and any number of equations, so long as there are fewer equations than variables. The next theorem, which states the result in full generality, may be established by using an induction argument, of which Theorem 3A is the first step in the inductive process. We omit the proof.

Theorem 3. *Suppose that $F_1, F_2, \ldots, F_k$ are functions of the $n + k$ variables $x_1, x_2, \ldots, x_n, u_1, u_2, \ldots, u_k$, and suppose that each F_i and all its first derivatives are continuous in a neighborhood of a point*

$$P_0(x_1^0, x_2^0, \ldots, x_n^0, u_1^0, u_2^0, \ldots, u_k^0).$$

The k equations

$$F_i(x_1^0, x_2^0, \ldots, x_n^0, u_1^0, u_2^0, \ldots, u_k^0) = 0, \qquad i = 1, 2, \ldots, k,$$

are assumed to hold, and we suppose the determinant

$$D = \begin{vmatrix} \dfrac{\partial F_1}{\partial u_1} & \dfrac{\partial F_1}{\partial u_2} & \cdots & \dfrac{\partial F_1}{\partial u_k} \\ \dfrac{\partial F_2}{\partial u_1} & \dfrac{\partial F_2}{\partial u_2} & \cdots & \dfrac{\partial F_2}{\partial u_k} \\ \vdots & & & \\ \dfrac{\partial F_k}{\partial u_1} & \dfrac{\partial F_k}{\partial u_2} & \cdots & \dfrac{\partial F_k}{\partial u_k} \end{vmatrix} \tag{15}$$

evaluated at the point P_0 is different from zero. Then there are numbers h and r such that:

a) *for each $(x_1, x_2, \ldots, x_n)$ in the hypercube*

$$|x_i - x_i^0| < h, \qquad i = 1, 2, \ldots, n,$$

there is a unique solution $(u_1, u_2, \ldots, u_k)$ of the equations

$$F_i(x_1, x_2, \ldots, x_n, u_1, u_2, \ldots, u_k) = 0, \qquad i = 1, 2, \ldots, k,$$

for which $|u_j - u_j^0| < r, j = 1, 2, \ldots, k$. The solution is defined by

$$u_j = f_j(x_1, x_2, \ldots, x_n), \qquad j = 1, 2, \ldots, k.$$

b) *The functions f_j and all their first derivatives are continuous and the determinant D does not vanish for points P in the hypercube described in* (a). *The derivatives $\partial f_j / \partial x_k$ are obtained by applying Cramer's Rule for solving a linear system of equations to each of the n systems*

$$\sum_{j=1}^{k} \frac{\partial F_i}{\partial u_j} \frac{\partial f_j}{\partial x_p} + \frac{\partial F_i}{\partial x_p} = 0, \qquad \begin{matrix} i = 1, 2, \ldots, k, \\ p = 1, 2, \ldots, n. \end{matrix} \tag{16}$$

The determinant in (15) is called a **Jacobian determinant** or, simply, a **Jacobian**. Customary notations for the one in (15) are

$$J\left(\frac{F_1, F_2, \ldots, F_k}{u_1, u_2, \ldots, u_k}\right) \quad \text{and} \quad \frac{\partial(F_1, F_2, \ldots, F_k)}{\partial(u_1, u_2, \ldots, u_k)}.$$

EXAMPLE 3. Given the three equations

$$x_1^2 + 2x_2^2 - 3u_1^2 + 4u_1u_2 - u_2^2 + u_3^3 = 0,$$
$$x_1 + 3x_2 - 4x_1x_2 + 4u_1^2 - 2u_2^2 + u_3^2 = 0,$$
$$x_1^3 - x_2^3 + 4u_1^2 + 2u_2 - 3u_3^2 = 0,$$

assume that the conditions of Theorem 3 are valid, and use (16) to compute $\partial u_1/\partial x_1$, $\partial u_2/\partial x_1$, $\partial u_3/\partial x_1$.

SOLUTION. Differentiating each of the above equations implicitly with respect to u_1, we find

$$(-6u_1 + 4u_2)\frac{\partial u_1}{\partial x_1} + (4u_1 - 2u_2)\frac{\partial u_2}{\partial x_1} + 3u_3^2\frac{\partial u_3}{\partial x_1} + 2x_1 = 0,$$

$$8u_1\frac{\partial u_1}{\partial x_1} - 4u_2\frac{\partial u_2}{\partial x_1} + 2u_3\frac{\partial u_3}{\partial x_1} + 1 - 4x_2 = 0,$$

$$8u_1\frac{\partial u_1}{\partial x_1} + 2\frac{\partial u_2}{\partial x_1} - 6u_3\frac{\partial u_3}{\partial x_1} + 3x_1^2 = 0.$$

This linear system of three equations in the three unknowns $\partial u_1/\partial x_1$, $\partial u_2/\partial x_1$, $\partial u_3/\partial x_1$, is easily solved by Cramer's Rule. The details are left to the reader.

PROBLEMS

In Problems 1 through 6, use Theorem 3 to verify that there is a box

$$|x - x_0| < h, \qquad |y - y_0| < k, \qquad |z - z_0| < k$$

such that all the points (x, y, z) in that box which satisfy

$$F(x, y, z) = 0 \qquad \text{and} \qquad G(x, y, z) = 0$$

lie on the graph of equations of the form $y = f(x)$, $z = g(x)$, where f and g are smooth for $|x - x_0| < h$. In Problems 1 and 2, find in symmetric form the equations of the line. In Problems 3 through 6, find in symmetric form the equations of the tangent line at (x_0, y_0, z_0). Let P_0 denote (x_0, y_0, z_0).

1. $F(x, y, z) = 2x + y - z - 2$, $G(x, y, z) = x + 2y + z - 1$, $P_0 = (2, -1, 1)$.
2. $F = 3x + 2y - z - 8$, $G = x + y + 2z - 1$, $P_0 = (1, 2, -1)$.
3. $F = x^2 + 2y^2 - z^2 - 2$, $G = 2x - y + z - 1$, $P_0 = (2, 1, -2)$.
4. $F = 2x^2 + y^2 - z^2 + 3$, $G = 3x + 2y + z - 10$, $P_0 = (1, 2, 3)$.
5. $F = x^3 + y^3 + z^3 - 3xyz - 14$, $G = x^2 + y^2 + z^2 - 6$, $P_0 = (2, -1, 1)$.
6. $F = x^2 - xy + 2y^2 - 4xz + 2z^2 - 10$, $G = xyz - 6$, $P_0 = (2, 3, 1)$.

In Problems 7 through 10, show that there is a box

$$|x - x_0| < h, \qquad |y - y_0| < h, \qquad |u - u_0| < k, \qquad |v - v_0| < k$$

such that all the points (x, y, u, v) in that box which satisfy the equations

$$F(x, y, u, v) = 0 \qquad \text{and} \qquad G(x, y, u, v) = 0$$

lie along the graph of the equations $u = f(x, y)$, $v = g(x, y)$, where f and g are smooth for $|x - x_0| < h$ and $|y - y_0| < h$. Find the values of f_x, f_y, g_x, and g_y at (x_0, y_0). Let P_0 denote (x_0, y_0, u_0, v_0).

7. $F = 2x - 3y + u - v, \quad G = x + 2y + u + 2v, \quad P_0 = (0, 0, 0, 0)$.

8. $F = 2x - y + 2u - v, \quad G = 3x + 2y + u + v, \quad P_0 = (0, 0, 0, 0)$.

9. $F = x - 2y + u + v - 8. \quad G = x^2 - 2y^2 - u^2 + v^2 - 4, \quad P_0 = (3, -1, 2, 1)$.

10. $F = x^2 - y^2 + uv - v^2 + 3, \quad G = x + y^2 + u^2 + uv - 2, \quad P_0 = (2, 1, -1, 2)$.

11. Show that there is a box as in Problems 1 through 6 about each point, except $(2, -3, 6)$ and $(-10, -15, 30)$ of the graph of

$$z^2 - \tfrac{9}{2}x^2 - 2y^2 = 0 \qquad \text{and} \qquad x + y + z - 5 = 0.$$

Solve explicitly for y and z in terms of x.

12. For what values of (x_0, y_0, u_0, v_0) does there exist a box as in Problems 7 through 10 if

$$F = -x + u^2 - v^2, \qquad G = -y + 2uv?$$

Find explicit functions $f(x, y)$ and $g(x, y)$ which are smooth near

$$x_0 = 3, \qquad y_0 = -4$$

and which are such that

$$f(x_0, y_0) = -2 \qquad \text{and} \qquad g(x_0, y_0) = 1.$$

13. Complete Example 3 and determine $\partial u_1/\partial x_1, \partial u_2/\partial x_1, \partial u_3/\partial x_1$.

14. For the equations in Example 3, determine

$$\frac{\partial u_1}{\partial x_2}, \qquad \frac{\partial u_2}{\partial x_2}, \qquad \frac{\partial u_3}{\partial x_2}.$$

15. Given that

$$u = f(x, y), \qquad x = \phi(s, t),$$

$$v = g(x, y), \qquad y = \psi(s, t),$$

show, by using the Chain rule for partial derivatives that

$$J\left(\frac{u, v}{x, y}\right) \cdot J\left(\frac{x, y}{s, t}\right) = J\left(\frac{u, v}{s, t}\right).$$

16. Given that

$$u^5 + v^5 + x^5 + 2y = 0, \qquad u^3 + v^3 + y^3 + 2x = 0,$$

find, under the appropriate hypotheses, the quantities u_x, u_y, u_{xx}.

17. Given that

$$F(x, y, u, v) = 0, \qquad G(x, y, u, v) = 0,$$

and that the hypotheses of Theorem 3 are satisfied. State conditions under which the relation

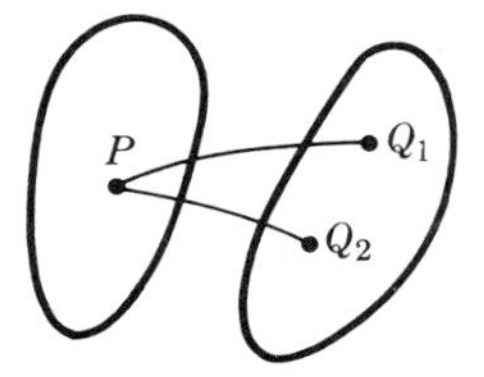

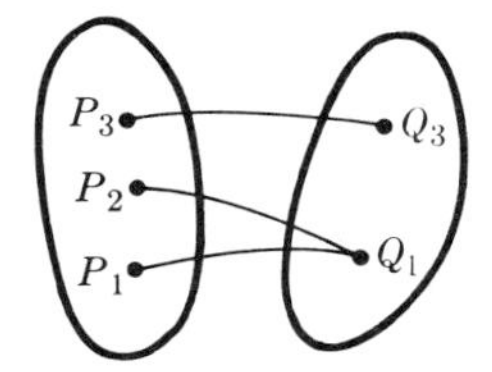

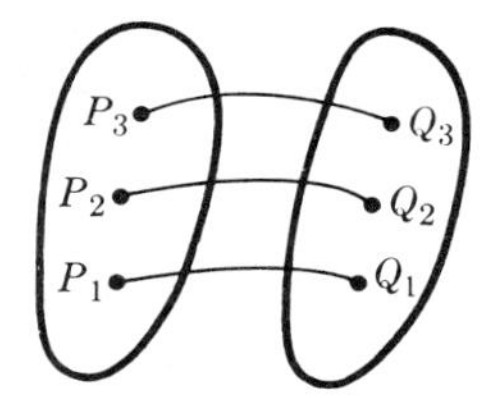

Fig. 7-11

$$\frac{\partial u}{\partial x}\frac{\partial y}{\partial u}+\frac{\partial v}{\partial x}\frac{\partial y}{\partial v}=0$$

is true.

18. State and prove an extension of Theorem 3A to a system of three equations $F(x_1, x_2, x_3, u_1, u_2, u_3) = 0$, $G(x_1, x_2, x_3, u_1, u_2, u_3) = 0$, $H(x_1, x_2, x_3, u_1, u_2, u_3) = 0$.

19. Extend the result of Problem 15 to show that

$$J\left(\frac{u_1, \ldots, u_k}{x_1, \ldots, x_k}\right)\cdot J\left(\frac{x_1, \ldots, x_k}{t_1, \ldots, t_k}\right)=J\left(\frac{u_1, \ldots, u_k}{t_1, \ldots, t_k}\right)$$

where $u_i = f_i(x_1, \ldots, x_k)$, $x_i = \varphi_i(t_1, \ldots, t_k)$, $i = 1, 2, \ldots, k$.

3. Transformations and Jacobians

Let M and N be two nonempty sets. A **relation** from M to N is a set of ordered pairs (P, Q) in which $P \in M$ and $Q \in N$. The **domain** of the relation consists of all the elements P which occur in any of the pairs, and its **range** consists of all the elements Q which occur (Fig. 7-11). If the relation is such that no two of the pairs have the same first element, the relation is called a **transformation**. We shall use capital letters such as T, U, V to denote transformations. For example, a *function* of any number of variables, say 3, is a particular case of a transformation in which the domain M is a set of ordered number triples (i.e., a set in R^3) and the range N is in R^1. However, transformations are more general in that we allow M and N to be arbitrary sets.

If T is a transformation and P is in its domain, we denote the unique corresponding element Q by $T(P)$ and call Q the **image of** P **under** T; we also often say that "T carries P into Q (or $T(P)$)." If E is a subset of the domain of T, we call the totality of images $T(P)$ of points P in E the **image of** E **under** T and denote it by $T(E)$ (Fig. 7-12).

If U is a relation (which may, in particular, be a transformation) from M to N, we often write

$$U: M \to N.$$

The **inverse relation** of a given relation $U: M \to N$ is defined as the set

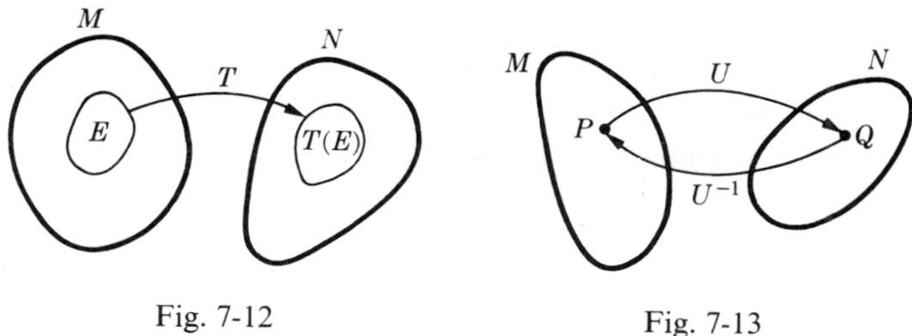

Fig. 7-12 Fig. 7-13

of all pairs (Q, P) for which the pair $(P, Q) \in U$. We denote the inverse of U by U^{-1}. As in the case of functions on R^1, the inverse of a transformation is not necessarily a transformation. If the transformation U is 1-1 (Fig. 7-11(c))—that is, if no two pairs have the same *second* element (as well as not having the same first element),—then U^{-1} is also a transformation. If U is any relation, $(U^{-1})^{-1} = U$ and the domain of U^{-1} is the range of U and the range of U^{-1} is the domain of U. If U is one-to-one and U carries P into Q, then U^{-1} carries Q back into P (Fig. 7-13).

In linear algebra we study linear transformations almost exclusively. However, many of the transformations we studied in elementary calculus are nonlinear. For example, the transformation from rectangular to polar coordinates in the plane

$$r = \sqrt{x^2 + y^2}, \qquad \theta = \arctan \frac{y}{x},$$

is a nonlinear transformation. Similarly, the transformations from rectangular to cylindrical coordinates and from rectangular to spherical coordinates are nonlinear. Also, it is frequently helpful to make nonlinear changes of variables in both single and multiple integrals. In this section we shall establish some general properties of nonlinear transformations and later we shall show how these results may be applied in the study of multiple integration.

Let D be a domain in the xy plane and T a transformation from D to a set of elements in another plane, which we denote the uv plane. We write

$$T: u = f(x, y), \qquad v = g(x, y), \qquad (x, y) \text{ in } D.$$

Definition. We say that the transformation T is **continuously differentiable** in D if and only if f and g are continuous and the first derivatives of f and g are continuous throughout D.

If T is a transformation from a domain in n dimensional space to n dimensional space, we write

$$T: \quad u_1 = f_1(x_1, x_2, \ldots, x_n), \qquad u_2 = f_2(x_1, x_2, \ldots, x_n), \ldots,$$
$$u_n = f_n(x_1, x_2, \ldots, x_n).$$

Just as the inverse relation of a function need not be a function, so the inverse relation of a transformation need not be a transformation. *The in-*

verse of a transformation T is also a transformation if and only if T establishes a one-to-one relation between its domain and its range. That is, if for each element Q in the range of T, there is only one P in its domain such that $T(P) = Q$, then T has an inverse, and we denote this inverse transformation by T^{-1}.

The implicit function theorems are the basis for the following inversion theorem for transformations.

Theorem 4 (Inverse Transformation Theorem). *Suppose*

$$T\colon\quad u_1 = F_1(x_1, x_2), \qquad u_2 = F_2(x_1, x_2), \qquad (x_1, x_2) \text{ in } D, \tag{1}$$

is a continuously differentiable transformation for (x_1, x_2) interior to D. Let $P_0 = (x_1^0, x_2^0)$ be a point in D and let $Q_0 = (u_1^0, u_2^0)$ be the image of P_0 under T. We define J_0, the Jacobian

$$J\left(\frac{F_1, F_2}{x_1, x_2}\right)$$

evaluated at P_0, and we suppose that

$$J_0 \neq 0. \tag{2}$$

Then there exist positive numbers h and k such that whenever $Q = (u_1, u_2)$ is within h of (u_1^0, u_2^0), that is $|u_1 - u_1^0| < h$, $|u_2 - u_2^0| < h$, there is one and only one point P which satisfies $T(P) = Q$ with $|x_1 - x_1^0| < k$, $|x_2 - x_2^0| < k$. If we denote the inverse transformation by

$$x_1 = f_1(u_1, u_2), \qquad x_2 = f_2(u_1, u_2), \tag{3}$$

then f_1 and f_2 are continuously differentiable for all Q for which

$$|u_1 - u_1^0| < h, \qquad |u_2 - u_2^0| < h.$$

Proof. If we rewrite equations (1) in the form

$$F(u_1, u_2, x_1, x_2) \equiv u_1 - F_1(x_1, x_2) = 0,$$

$$G(u_1, u_2, x_1, x_2) \equiv u_2 - F_2(x_1, x_2) = 0,$$

we see that for F and G all the conditions of Theorem 3 are fulfilled. Therefore we may solve for x_1 and x_2 in terms of u_1 and u_2, which is statement (3) precisely. The condition $J_0 \neq 0$ is identical with the condition $D_0 \neq 0$ in Theorem 3.

Remarks. (i) A schematic diagram of Theorem 4 is shown in Fig. 7-14. The transformation T takes the domain D into some region E in the u_1u_2-plane. The inverse mapping is defined in a square of side $2h$ about Q_0.
ii) Theorem 4 may be extended to n equations of the form

$$u_i = F_i(x_1, x_2, \ldots, x_n), \qquad i = 1, 2, \ldots, n.$$

The Inverse Transformation Theorem states that we may solve (at least

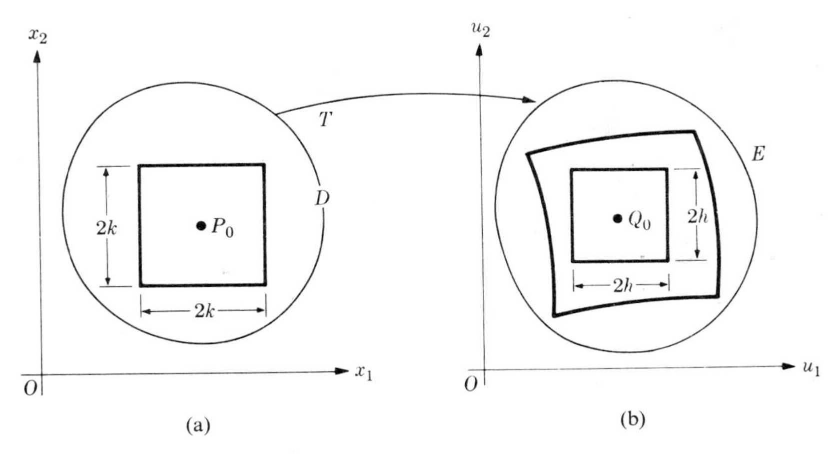

Fig. 7-14

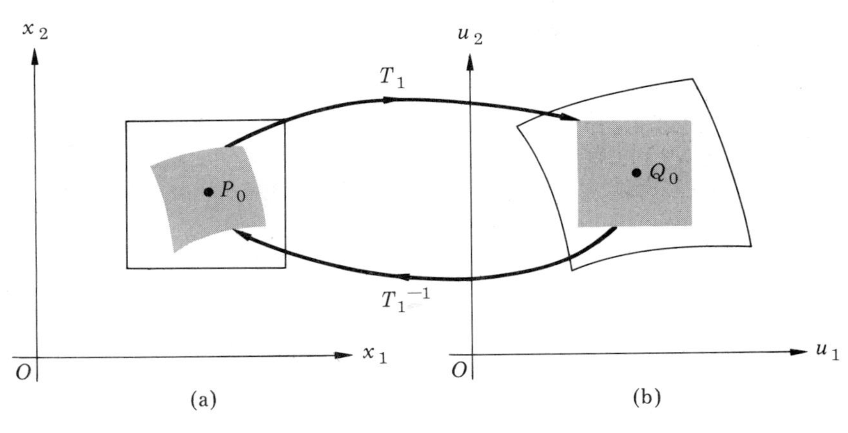

Fig. 7-15

locally) for each x_i and obtain

$$x_i = f_i(u_1, u_2, \ldots, u_n), \qquad i = 1, 2, \ldots, n,$$

with the f_i continuously differentiable.
iii) We define T_1 to be the transformation T *restricted* to those points P in the square $|x_i - x_i^0| < k$, $i = 1, 2$ which correspond to points Q lying in the image square defined by $|u_i - u_i^0| < h$, $i = 1, 2$. Then T_1 is one to one, and its inverse T_1^{-1} is just the transformation $x_i = f_i(u_1, u_2)$, $i = 1, 2$. (See Fig. 7-15.)
iv) The theorem says nothing about the sizes of h and k, and extreme caution must be used in the applications. It may happen that all other hypotheses of the theorem are satisfied, but if h and k are not taken sufficiently small the inversion from (1) to (3) may not be possible. [See part (d) of the examples below.]

v) The theorem fails, and we have no information when the Jacobian vanishes. Actually, if $J_0 = 0$ anything can happen, and we exhibit this fact with the following four illustrative examples.

a) The transformation

$$u = x^3, \qquad v = y$$

is one to one for all (x, y) in R^2, since the inversion formulas are functions given by

$$x = \sqrt[3]{u}, \qquad y = v.$$

However, the Jacobian

$$J\left(\frac{u, v}{x, y}\right) = 3x^2$$

vanishes along the entire y-axis.

b) The transformation $u = x^2$, $v = y$ is not one to one for all (x, y) in R^2, since for any $a \neq 0$ we have (a, b) and $(-a, b)$ transforming into (a^2, b). However, the Jacobian

$$J\left(\frac{u, v}{x, y}\right) = 2x$$

vanishes along the entire y-axis, exactly as in Example (a) above, which is one to one.

c) The transformation

$$u = x^2 - y^2, \qquad v = 2xy \tag{4}$$

has Jacobian

$$J\left(\frac{u, v}{x, y}\right) = 4(x^2 + y^2),$$

which vanishes only at the origin. Setting $w = u + iv$ and $z = x + iy$, we see that

$$u + iv = w = z^2 = (x + iy)^2 = x^2 - y^2 + 2ixy,$$

and so the transformation (4) is the same as the complex variable transformation

$$w = z^2.$$

Therefore, to each $w \neq 0$ there correspond two numbers z such that $w = z^2$ and the transformation is *not one to one* in the plane.

d) In Example (c) we select for domain D the annular ring between the circles $x^2 + y^2 = 1$ and $x^2 + y^2 = 9$. Then the Jacobian of (4) never vanishes in D. Yet the transformation is not one to one in D. [That is, $(2, 0)$ and $(-2, 0)$ both map into $(4, 0)$.] This example shows that the size of h and k must be suitably small before the theorem is valid. The theorem states that the transformation (4) is one to one in any sufficiently small

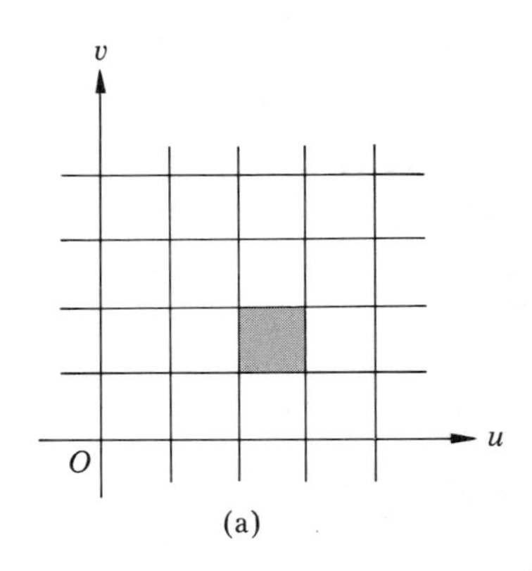

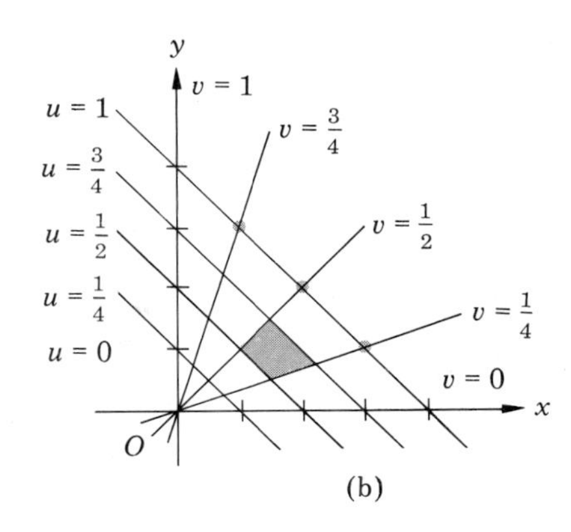

Fig. 7-16

rectangle in the annular ring. Actually, it is one to one within *any* rectangle whose interior is entirely within D.

In discussing transformations in R^2 it is desirable to find the images of the lines $x = \text{const}$ and $y = \text{const}$, or those of other convenient curves. We illustrate with examples.

EXAMPLE 1. Given the transformation

$$x = u - uv, \qquad y = uv.$$

Find

$$J\left(\frac{x, y}{u, v}\right)$$

and the inverse transformation. In the (x, y)-plane draw the images of the lines $u = 0, \frac{1}{4}, \frac{1}{2}, \frac{3}{4}, 1$ and $v = 0, \frac{1}{4}, \frac{1}{2}, \frac{3}{4}, 1$, and find the image of the square $S = \{(u, v) : \frac{1}{2} \leq u \leq \frac{3}{4}, \frac{1}{4} \leq v \leq \frac{1}{2}\}$ (Fig. 7-16).

SOLUTION. We have

$$J\left(\frac{x, y}{u, v}\right) = \begin{vmatrix} 1 - v & -u \\ v & u \end{vmatrix} = u - uv + uv = u.$$

Solving for u and v, we obtain

$$u = x + y, \qquad v = \frac{y}{x + y}.$$

The line $u = c$ corresponds to $x + y = c$ (if $c \neq 0$) and the line $v = d$ corresponds to $dx = (1 - d)y$. The various lines are drawn and the image of the rectangle is shaded in Fig. 7-16.

EXAMPLE 2. Discuss the transformation

$$u = e^x \cos y, \qquad v = e^x \sin y.$$

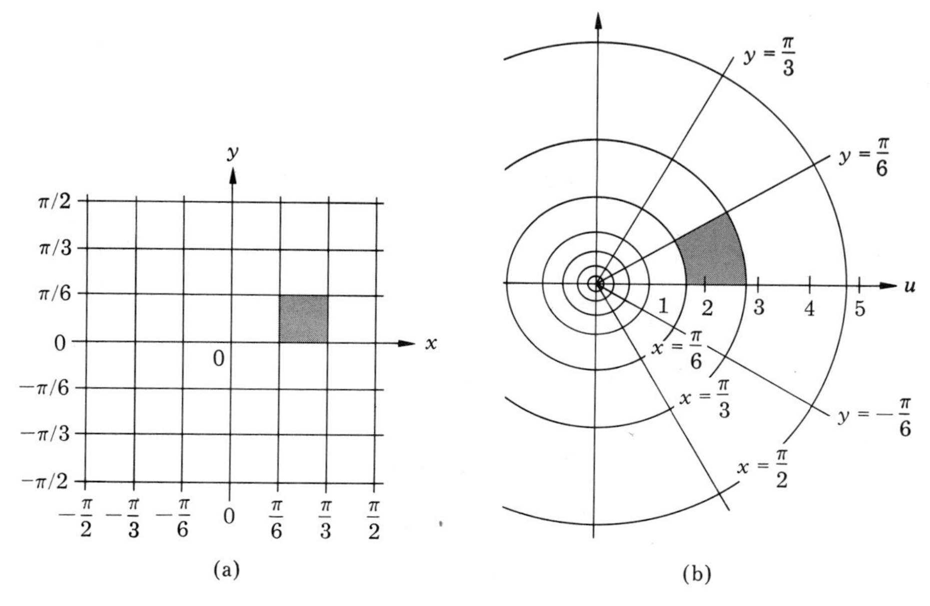

Fig. 7-17

Draw the lines $x = \pm k\pi/6$, $y = \pm k\pi/6$ for $k = 0, 1, 2, 3$ and draw their images in the (u, v)-plane. Find

$$J\left(\frac{u, v}{x, y}\right).$$

SOLUTION. If we set $w = u + iv$ and $z = x + iy$, the transformation becomes $w = e^z$. Introducing polar coordinates (ρ, ϕ) with $\rho > 0$ in the w-plane, we write

$$\rho = e^x, \qquad \phi = y.$$

Thus the transformation is periodic in y with period 2π. The lines $x = c$ correspond to circles $\rho = e^c$ and lines $y = d$ correspond to rays $\phi = d$. We obtain

$$J\left(\frac{u, v}{x, y}\right) = \begin{vmatrix} e^x \cos y & -e^x \sin y \\ e^x \sin y & e^x \cos y \end{vmatrix} = e^{2x}.$$

The lines and their images are drawn in Fig. 7-17.

PROBLEMS

In each of Problems 1 through 7, find the Jacobian

$$J\left(\frac{u, v}{x, y}\right)$$

and find the inverse transformation. In the uv-plane, draw the images of the lines $x = \frac{1}{4}, \frac{1}{2}, \frac{3}{4}, 1$ and $y = -\frac{1}{2}, -\frac{1}{4}, 0, \frac{1}{4}, \frac{1}{2}$.

1. $u = x, \quad v = y + x^2$

2. $u = x + xy, \quad v = y, \quad y > -1$

3. $u = 2x - 3y, \quad v = x + 2y$

4. $u = 2x + 3y, \quad v = x + 2y$

5. $u = x/(1 + x + y), \quad v = y/(1 + x + y), \quad x + y > -1$

6. $u = x^2 - y^2, \quad v = 2xy, \quad x > 0$

7. $u = x\cos(\pi y/2), \quad v = x\sin(\pi y/2), \quad x > 0, \quad -1 < y < 1$

8. Show that the transformation $u = x^2 - y^2$, $v = 2xy$ is one to one if the domain D is any rectangle situated in the half-plane $y > 0$.

9. Given the transformation

$$T: \quad u = \frac{x}{x^2 + y^2}, \qquad v = \frac{y}{x^2 + y^2},$$

show that boundary of any disk in the xy-plane with center at $(0, 0)$ is mapped into the boundary of a disk in the uv-plane. Compare the radii of the two disks.

10. Given the transformation

$$T: \quad u = -x + \sqrt{x^2 + y^2}, \qquad v = -x - \sqrt{x^2 + y^2}.$$

Show that T is not a one to one transformation. Decide whether or not T is one to one in the rectangle

$$R_1 = \{(x, y): 1 \le x \le 3, \quad -4 \le y \le 4\};$$

In the rectangle

$$R_2 = \{(x, y): -1 \le x \le 1, \quad 2 \le y \le 6\}.$$

11. Given the transformation

$$T: \quad u = f(x, y), \qquad v = g(x, y),$$

with Jacobian

$$J\left(\frac{u, v}{x, y}\right) \neq 0$$

at a point P_0. Show that if f and g are continuously differentiable near P_0, then

$$J\left(\frac{u, v}{x, y}\right) \cdot J\left(\frac{x, y}{u, v}\right) = 1.$$

12. Suppose that the transformation

$$T: \quad u = f(x, y), \qquad v = g(x, y)$$

is continuously differentiable and one to one. We have the relation

$$\frac{\partial u}{\partial x}\frac{\partial x}{\partial u} + \frac{\partial u}{\partial y}\frac{\partial y}{\partial u} = 1. \tag{*}$$

a) Denoting the inverse transformations $x = \phi(u, v)$, $y = \psi(u, v)$, differentiate the expression $v = g[\phi(u, v), \psi(u, v)]$ with respect to u to obtain

$$0 = \frac{\partial v}{\partial x}\frac{\partial x}{\partial u} + \frac{\partial v}{\partial y}\frac{\partial y}{\partial u}. \tag{**}$$

b) By differentiating (∗) and (∗∗) with respect to u (employing the chain rule) and then using Cramer's Rule, find expressions for $\partial^2 x/\partial u^2$ and $\partial^2 y/\partial u^2$.

13. a) Find the Jacobian of the transformation

$$u = \frac{x}{x^2 + y^2 + z^2}, \qquad v = \frac{y}{x^2 + y^2 + z^2}, \qquad w = \frac{z}{x^2 + y^2 + z^2}.$$

b) Show that the surface of a sphere in xyz-space with center at the origin transforms into a sphere in uvw-space.

14. a) Calculate the Jacobian of the transformation

$$u = \cos x \cosh y, \qquad v = \sin x \cosh y, \qquad w = \sinh z.$$

b) What is the image of the surface $\cosh^2 y - \sinh^2 z = 1$?

CHAPTER 8

Differentiation under the Integral Sign. Improper Integrals. The Gamma Function

1. Differentiation under the Integral Sign

We recall the elementary integration formula

$$\int_0^1 t^n\,dt = \left[\frac{1}{n+1}t^{n+1}\right]_0^1 = \frac{1}{n+1},$$

valid for any $n > -1$. Since n need not be an integer, we employ the variable x and write

$$\phi(x) = \int_0^1 t^x\,dt = \frac{1}{x+1}, \qquad x > -1. \tag{1}$$

Suppose we wish to compute the derivative $\phi'(x)$. We can proceed in two ways. Equating the first and last expressions in (1), we have

$$\phi(x) = \frac{1}{x+1}, \qquad \phi'(x) = -\frac{1}{(x+1)^2}.$$

On the other hand, we may try the following procedure:

$$\frac{d}{dx}\phi(x) = \frac{d}{dx}\int_0^1 t^x\,dt = \int_0^1 \frac{d}{dx}(t^x)\,dt = \int_0^1 t^x \log t\,dt. \tag{2}$$

Is it true that

$$\int_0^1 t^x \log t\,dt = -\frac{1}{(x+1)^2}, \tag{3}$$

at least for $x > -1$? In this section we shall determine conditions under which a process such as (2) is valid. To examine the validity of differentiation under the integral sign, as the process (2) is called, we first develop a property of continuous functions on R^2.

Let S be a region in R^2 and $f: S \to R^1$ a continuous function. We recall that f is continuous at a point $(x_0, y_0) \in S$ if for every $\varepsilon > 0$ there is a $\delta > 0$ such that

$$|f(x, y) - f(x_0, y_0)| < \varepsilon$$

whenever

$$|x - x_0| + |y - y_0| < \delta.$$

It is important to note that the size of δ depends not only on the size of ε but also on the particular point (x_0, y_0) at which continuity is defined. If the size of δ depends only on ε and not on the point (x_0, y_0), then f is said to be *uniformly continuous on S*. That is, f is *uniformly continuous* if for every $\varepsilon > 0$, there is a $\delta > 0$ such that

$$|f(x', y') - f(x'', y'')| < \varepsilon$$

for **all** (x', y'), (x'', y'') in S which satisfy the inequality

$$|x' - x''| + |y' - y''| < \delta.$$

In other words, the size of δ depends only on ε.

We denote the boundary of a region S in R^2 by ∂S. A region in R^2 is said to be **bounded** if it is contained in a sufficiently large disk. A region S is **closed** if it contains its boundary, ∂S. The basic theorem concerning uniformly continuous function states that *a function f which is continuous on a closed bounded region is uniformly continuous.* The same result holds in any number of dimensions. We omit the proof.

Suppose a function ϕ is given by the formula

$$\phi(x) = \int_c^d f(x, t)\,dt, \qquad a \le x \le b,$$

where c and d are constants. If the integration can be performed explicitly, then $\phi'(x)$ can be found by a computation. However, even when the evaluation of the integral is impossible, it sometimes happens that $\phi'(x)$ can be found. The basic formula is given in the next theorem, known as **Leibniz' Rule**.

Theorem 1. *Suppose that ϕ is defined by*

$$\phi(x) = \int_c^d f(x, t)\,dt, \qquad a \le x \le b, \tag{4}$$

where c and d are constants. If f and f_x are continuous in the rectangle

$$R = \{(x, t) : a \le x \le b, \quad c \le t \le d\},$$

then

$$\phi'(x) = \int_c^d f_x(x, t)\,dt, \qquad a < x < b. \tag{5}$$

That is, the derivative may be found by differentiating under the integral sign.

PROOF. We prove the theorem by showing that the difference quotient

$$[\phi(x+k) - \phi(x)]/k$$

tends to the right side of (5) as k tends to zero. If x is in (a, b) then, from (4), we have

$$\frac{\phi(x+k)-\phi(x)}{k} = \frac{1}{k}\int_c^d f(x+k, t)\,dt - \frac{1}{k}\int_c^d f(x,t)\,dt$$
$$= \frac{1}{k}\int_c^d [f(x+k,t) - f(x,t)]\,dt.$$

Since differentiation and integration are inverse processes, we can write

$$f(x+k,t) - f(x,t) = \int_x^{x+k} f_\xi(\xi, t)\,d\xi,$$

and so

$$\frac{\phi(x+k)-\phi(x)}{k} = \frac{1}{k}\int_c^d \int_x^{x+k} f_\xi(\xi,t)\,d\xi\,dt.$$

We note that f_x is uniformly continuous on R, since a function which is continuous on a bounded, closed set is uniformly continuous there. Therefore, using the comma notation for the derivative with respect to the first variable, if $\varepsilon > 0$ is given, there is a $\delta > 0$ such that

$$|f_{,1}(\xi,t) - f_{,1}(x,t)| < \frac{\varepsilon}{d-c}$$

for all t in $[c, d]$ and all ξ with $|\xi - x| < \delta$. We now wish to show that

$$\frac{\phi(x+k)-\phi(x)}{k} - \int_c^d f_{,1}(x,t)\,dt \to 0 \qquad \text{as } k \to 0.$$

We write

$$\int_c^d f_{,1}(x,t)\,dt = \frac{1}{k}\int_c^d \int_x^{x+k} f_{,1}(x,t)\,d\xi\,dt,$$

which is true because the integrand on the right does not contain ξ. Substituting this last expression in the one above, we find, for $0 < |k| < \delta$,

$$\left|\frac{\phi(x+k)-\phi(x)}{k} - \int_c^d f_{,1}(x,t)\,dt\right|$$
$$= \left|\int_c^d \left\{\frac{1}{k}\int_x^{x+k} [f_{,1}(\xi,t) - f_{,1}(x,t)]\,d\xi\right\} dt\right|$$
$$\le \int_c^d \left|\frac{1}{k}\int_x^{x+k} \frac{\varepsilon}{d-c}\,d\xi\right| dt = \frac{\varepsilon}{(d-c)}\cdot(d-c) = \varepsilon.$$

Since ε is arbitrary, the theorem follows.

Theorem 1 shows that the formula (3) is justified for $x > 0$, since the integrand $f(x, t)$ is then continuous in an appropriate rectangle. Later we shall examine more closely the validity of (3) when $-1 < x \leq 0$, in which case the integral is improper.

EXAMPLE 1. Find the value of $\phi'(x)$ if

$$\phi(x) = \int_0^{\pi/2} f(x, t)\, dt; \qquad f(x, t) = \begin{cases} \dfrac{\sin xt}{t} & \text{if } t \neq 0, \\ x & \text{if } t = 0. \end{cases}$$

SOLUTION. Since

$$\lim_{t \to 0} \frac{\sin xt}{t} = x \lim_{t \to 0} \frac{\sin xt}{xt} = x,$$

the integrand is continuous for $0 \leq t \leq \pi/2$ and for all x. Also, we have

$$f_x(x, t) = \begin{cases} \cos xt & \text{if } t \neq 0, \\ 1 = \cos xt & \text{if } t = 0, \end{cases}$$

so $f_x(x, t)$ is continuous everywhere. Therefore

$$\phi'(x) = \int_0^{\pi/2} \cos xt\, dt = -\left[\frac{1}{x} \sin xt\right]_0^{\pi/2} = -\frac{\sin(\pi/2)x}{x}, \qquad x \neq 0.$$

It is a fact that the integral expression for ϕ cannot be evaluated explicitly.

EXAMPLE 2. Evaluate

$$\int_0^1 \frac{du}{(u^2 + 1)^2}$$

by letting

$$\phi(x) = \int_0^1 \frac{du}{u^2 + x} = \frac{1}{\sqrt{x}} \arctan(1/\sqrt{x})$$

and computing $-\phi'(1)$.

SOLUTION.

$$\phi'(x) = -\int_0^1 \frac{du}{(u^2 + x)^2} = \frac{1}{\sqrt{x}} \frac{-\frac{1}{2}x^{-3/2}}{1 + (1/x)} - \frac{1}{2x\sqrt{x}} \arctan \frac{1}{\sqrt{x}}$$

and

$$-\phi'(1) = \int_0^1 \frac{du}{(u^2 + 1)^2} = \frac{1}{2}\left(\frac{1}{2} + \arctan 1\right) = \frac{1}{2}\left(\frac{1}{2} + \frac{\pi}{4}\right).$$

Leibniz' Rule may be extended to the case where the limits of integration

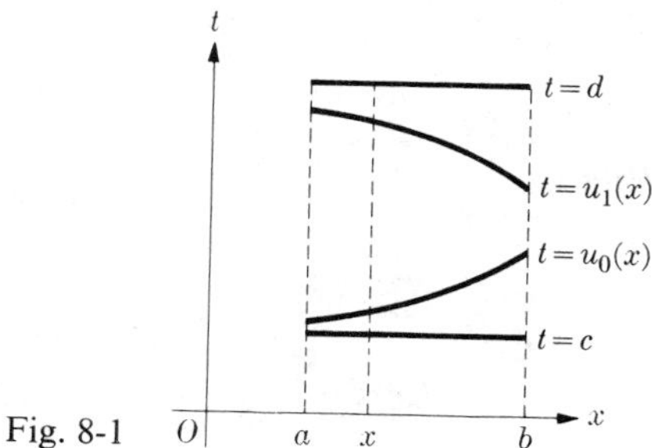

Fig. 8-1

also depend on x. We consider a function defined by

$$\phi(x) = \int_{u_0(x)}^{u_1(x)} f(x, t)\, dt, \tag{6}$$

where $u_0(x)$ and $u_1(x)$ are continuously differentiable functions for $a \leq x \leq b$. Furthermore, the ranges of u_0 and u_1 are assumed to lie between c and d (Fig. 8-1).

To obtain a formula for the derivative $\phi'(x)$, where ϕ is given by an integral such as (6), it is simpler to consider a new integral which is more general than (6). We define

$$F(x, y, z) = \int_y^z f(x, t)\, dt \tag{7}$$

and obtain the following corollary of Leibniz' Rule.

Theorem 2. *Suppose that f satisfies the conditions of Theorem* 1 *and that F is defined by* (7) *with $c < y, z < d$. Then*

$$\frac{\partial F}{\partial x} = \int_y^z f_{,1}(x, t)\, dt, \tag{8a}$$

$$\frac{\partial F}{\partial y} = -f(x, y), \tag{8b}$$

$$\frac{\partial F}{\partial z} = f(x, z). \tag{8c}$$

Proof. Formula (8a) is Theorem 1. Formulas (8b) and (8c) are precisely the Fundamental Theorem of Calculus, since taking the partial derivative of F with respect to one variable, say y, implies that x and z are kept *fixed.*

Theorem 3 (General Rule for Differentiation under the Integral Sign). *Suppose that f and $\partial f/\partial x$ are continuous in the rectangle*

$$R = \{(x, t) : a \leq x \leq b, \quad c \leq t \leq d\},$$

and suppose that $u_0(x)$, $u_1(x)$ are continuously differentiable for $a \leq x \leq b$

with the range of u_0 and u_1 in (c, d). If ϕ is given by

$$\phi(x) = \int_{u_0(x)}^{u_1(x)} f(x, t)\, dt,$$

then

$$\phi'(x) = f[x, u_1(x)]u_1'(x) - f[x, u_0(x)] \cdot u_0'(x) + \int_{u_0(x)}^{u_1(x)} f_x(x, t)\, dt. \tag{9}$$

PROOF. We observe that

$$F(x, u_0(x), u_1(x)) = \phi(x)$$

in Theorem 2. Applying the Chain Rule, we get

$$\phi'(x) = F_x + F_y u_0'(x) + F_z u_1'(x).$$

Inserting the values of F_x, F_y, and F_z from (8), we obtain the desired result (9).

EXAMPLE 3. Find $\phi'(x)$, given that

$$\phi(x) = \int_0^{x^2} \arctan \frac{t}{x^2}\, dt.$$

SOLUTION. We have

$$\frac{\partial}{\partial x}\left(\arctan \frac{t}{x^2}\right) = \frac{-2t/x^3}{1 + (t^2/x^4)} = -\frac{2tx}{t^2 + x^4}.$$

We use formula (9) and find

$$\phi'(x) = (\arctan 1) \cdot (2x) - \int_0^{x^2} \frac{2tx\, dt}{t^2 + x^4}.$$

Setting $t = x^2 u$ in the integral on the right, we obtain

$$\phi'(x) = \frac{\pi x}{2} - \int_0^1 \frac{2x^3 u \cdot x^2\, du}{x^4 u^2 + x^4} = \frac{\pi x}{2} - x \int_0^1 \frac{2u\, du}{u^2 + 1} = x\left(\frac{\pi}{2} - \log 2\right).$$

PROBLEMS

In each of Problems 1 through 5, express $\phi'(x)$ as a definite integral, using Leibniz' Rule.

1. $\phi(x) = \int_0^1 \frac{\sin xt\, dt}{1 + t}$
2. $\phi(x) = \int_0^2 \frac{e^{-xt}\, dt}{1 + t^2}$
3. $\phi(x) = \int_1^2 \frac{e^{-t}\, dt}{1 + xt}$
4. $\phi(x) = \int_0^1 \frac{t^2\, dt}{(1 + xt)^2}$

5. $\phi(x) = \int_0^1 \frac{t^x - 1}{\log t} dt, \qquad (f(x, 0) = 0, f(x, 1) = x, x > 0)$

In each of Problems 6 through 14, obtain expressions of the form (9) for $\phi'(x)$.

6. $\phi(x) = \int_1^x t^3 \, dt$

7. $\phi(x) = \int_1^{x^2} \cos(t^2) \, dt$

8. $\phi(x) = \int_x^1 e^{t^2} \, dt$

9. $\phi(x) = \int_{x^2}^x \sin(xt) \, dt$

10. $\phi(x) = \int_x^{\tan x} \frac{dt}{1 + xt}$

11. $\phi(x) = \int_{x^2}^{e^x} \tan(xt) \, dt$

12. $\phi(x) = \int_{\sin x}^x \log(1 + xt) \, dt, \quad 0 < x \le \pi$

13. $\phi(x) = \int_{\cos x}^{1+x^2} \frac{e^{-t} \, dt}{1 + xt}, \quad x > 0$

14. $\phi(x) = \int_{x^2}^{\sin x} e^{xt} \, dt$

15. Given that $\phi(x) = \int_0^{\pi/2} \cos xt \, dt$, obtain $\phi'(x)$ in two ways: (1) by integrating and then differentiating, and (2) by using Leibniz' Rule and then performing the integration.

16. Evaluate

$$\int_0^1 \frac{du}{(u^2 + 1)^3}$$

by using the methods and results of Example 2.

In each of Problems 17 through 25, find $\phi'(x)$ by first applying Theorem 1 or 3 and then integrating.

17. $\phi(x) = \int_{\pi/2}^{\pi} \frac{\cos xt}{t} dt$

18. $\phi(x) = \int_1^{x^2} \frac{e^{xt}}{t} dt$

19. $\phi(x) = \int_{x^2}^{x} \frac{\sin xt}{t} dt, \quad x > 0$

20. $\phi(x) = \int_x^{x^2} \frac{e^{-xt}}{t} dt, \quad x > 0$

21. $\phi(x) = \int_{x^m}^{x^n} \frac{dt}{x + t}, \quad x > 0$

22. $\phi(x) = \int_0^{\pi} \log(1 + x \cos t) \, dt, \quad |x| < 1$

23. $\phi(x) = \int_0^1 \frac{x \, dt}{\sqrt{1 - x^2 t^2}}, \quad |x| < 1$

24. $\phi(x) = \int_0^{x^2} f(x, t) \, dt, \quad f(x, t) = \begin{cases} t^{-1} \sin^2 xt & \text{if } t \neq 0 \\ 0 & \text{if } t = 0 \end{cases}$

25. $\phi(x) = \int_0^{\pi} \log(1 - 2x \cos t + x^2) \, dt, \quad |x| < 1$

26. Show that if m and n are positive integers, then

$$\int_0^1 t^n(\log t)^m\,dt = (-1)^m \frac{m!}{(n+1)^{m+1}}.$$

[*Hint*. Differentiate $\int_0^1 x^n\,dx = 1/(n+1)$ and use induction on m. Here we understand that $t^n(\log t)^m$ is defined to be 0 for $t = 0$, or else that the integral is improper.]

27. Suppose that $\phi(x, y, z) = \int_a^b f(z + x\cos t + y\sin t)\,dt$. Show that $\phi_{zz} = \phi_{xx} + \phi_{yy}$.

28. Show that $f(x) = 1/x$ is uniformly continuous on the interval $I = \{x : c \le x \le 1\}$ for any number $c > 0$, but is not uniformly continuous on $J = \{x : 0 < x \le 1\}$.

29. Let $f(x) = \sin 1/x$ on $I = \{x : 0 < x \le 1\}$. Is f uniformly continuous on I?

2. Tests for Convergence of Improper Integrals. The Gamma Function

Suppose that a function f is continuous in the half-open interval $a \le x < b$, and suppose that f tends to infinity as x tends to b. A typical example of such a function is shown in Fig. 8-2. Since f is continuous on the interval $a \le x \le c$ for every value of c between a and b, we define

$$\int_a^b f(x)\,dx = \lim_{\varepsilon\to 0} \int_a^{b-\varepsilon} f(x)\,dx$$

whenever the limit exists as ε tends to zero through positive values. If the limit does exist, we say the integral **converges**; otherwise it **diverges**. If the integrand becomes infinite at the left endpoint of an interval, convergence and divergence are defined similarly. If $f(x)$ is continuous for $a \le x < \infty$, we define

$$\int_a^\infty f(x)\,dx = \lim_{X\to\infty} \int_a^X f(x)\,dx$$

whenever the limit exists. The terms "convergence" and "divergence" are used to express the existence and nonexistence of the limit.

If f becomes infinite at several points in the interval of integration, we

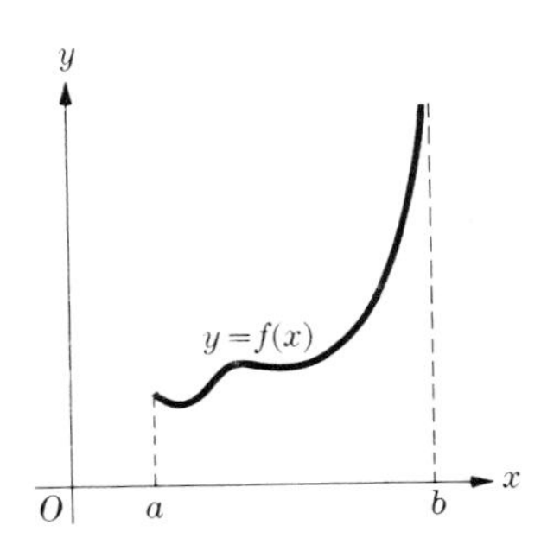

Fig. 8-2

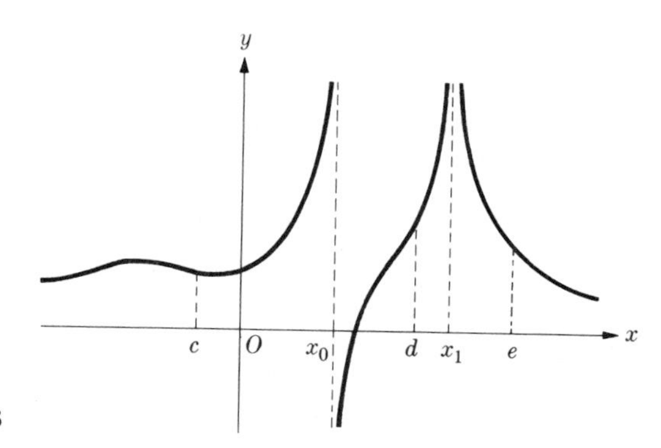

Fig. 8-3

decompose the interval and calculate each limit separately. Also, we compute the integral of a function f over the infinite interval $-\infty < x < \infty$ by selecting a convenient value C and calculating

$$\lim_{X \to -\infty} \int_X^C f(x)\,dx \qquad \text{and} \qquad \lim_{Y \to \infty} \int_C^Y f(x)\,dx.$$

The integral $\int_{-\infty}^{\infty} f(x)\,dx$ is said to converge if *both* limits above exist. For example, suppose that we wish to calculate $\int_{-\infty}^{\infty} f(x)\,dx$ for a function f which becomes infinite at the points x_0 and x_1, as shown in Fig. 8-3. We select the convenient values c, d, and e (Fig. 8-3) and evaluate

$$\lim_{X \to -\infty} \int_X^c f(x)\,dx, \qquad \lim_{\varepsilon_1 \to 0} \int_c^{x_0 - \varepsilon_1} f(x)\,dx, \qquad \lim_{\varepsilon_2 \to 0} \int_{x_0 + \varepsilon_2}^{d} f(x)\,dx,$$

$$\lim_{\varepsilon_3 \to 0} \int_d^{x_1 - \varepsilon_3} f(x)\,dx, \qquad \lim_{\varepsilon_4 \to 0} \int_{x_1 + \varepsilon_4}^{e} f(x)\,dx, \qquad \lim_{Y \to +\infty} \int_e^Y f(x)\,dx.$$

If all of these limits exist, their sum* yields the value of

$$\int_{-\infty}^{\infty} f(x)\,dx.$$

In the elementary study of convergence and divergence of integrals, we are usually able to obtain an indefinite integral and then evaluate the limit, either directly or by l'Hôpital's Rule. We now wish to determine methods for testing convergence which are useful even when the integrands are so complicated that we cannot perform the integrations. The following theorem, helpful in establishing convergence tests, is almost a corollary of the Axiom of Continuity.

Theorem 4. (a) *Suppose that F is nondecreasing in the half-open interval*

* Their sum can be shown to be independent of the particular values of c, d, and e as long as $c \in (-\infty, x_0)$, $d \in (x_0, x_1)$, and $e \in (x_1, \infty)$, and f is continuous except at x_0, x_1.

$a \le x < b$ *and that* $F(x) \le M$ *there. Then* (i) $F(x)$ *tends to a limit* L *as* x *tends to* b^-; (ii) $F(x) \le L$ *in* $[a, b)$; (iii) $L \le M$.
b) *Similarly, if* F *is nonincreasing in* $[a, b)$ *and if* $F(x) \ge m$ *for* $a \le x < b$, *then* (i) $F(x) \to l$ *as* $x \to b^-$; (ii) $F(x) \ge l$; (iii) $l \ge m$.

PROOF. We prove (a); the proof of (b) is the same. For each positive integer n, we define $x_n = b - (1/n)(b - a)$. Then $a \le x_n < b$ and $x_n < x_{n+1}$; also, $x_n \to b$ as $n \to \infty$. The numbers $F(x_n)$ form a nondecreasing bounded sequence and, by the Axiom of Continuity, tend to a limit L which is less than or equal to M; and $F(x_n) \le L$ for all n. Let $\varepsilon > 0$ be given. Then there is an N such that $F(x_n) > L - \varepsilon$ for all $n > N$. If x is in the interval (x_{N+1}, b), then $x_{N+1} < x < x_n$ for some n, so that

$$L - \varepsilon < F(x_{N+1}) \le F(x) \le F(x_n) \le L,$$

and the theorem is established.

The next theorem, known as the *comparison test*, is one of the basic tools used in establishing convergence. Note the analogy with the comparison test for series (Chapter 3, Section 3).

Theorem 5 (Comparison Test). *Suppose that* f *and* g *are continuous in the half-open interval* $[a, b)$, *that* $0 \le |f(x)| \le g(x)$, *and that* $\int_a^b g(x)\,dx$ *converges. Then* $\int_a^b f(x)\,dx$ *converges, and*

$$\left| \int_a^b f(x)\,dx \right| \le \int_a^b g(x)\,dx.$$

The same result holds if b *is replaced by* $+\infty$, *or if* $[a, b)$ *is replaced by* $(a, b]$, *or if the interval considered is* $(-\infty, b]$.

PROOF. We establish the result for $[a, b)$; the other cases are proved similarly. We first assume that $f(x) \ge 0$ and define

$$F(X) = \int_a^X f(x)\,dx, \qquad G(X) = \int_a^X g(x)\,dx.$$

Then F and G are nondecreasing on $[a, b)$ and, by hypothesis, $G(X)$ tends to a limit M as $X \to b$. Since, by hypothesis, $F(X) \le G(X) \le M$ on $[a, b)$, we find from Theorem 4 that $F(X) \to L \le M$.

If $f(x)$ is not always nonnegative, we define

$$f_1(x) = \frac{|f(x)| + f(x)}{2},$$

$$f_2(x) = \frac{|f(x)| - f(x)}{2}.$$

Then f_1 and f_2 are continuous on $[a, b)$ and nonnegative there. Furthermore,

$$f_1(x) + f_2(x) = |f(x)| \leq g(x), \qquad f_1(x) - f_2(x) = f(x).$$

Therefore the improper integrals of f_1, f_2, and $|f|$ all exist. From the elementary theorems on limits we conclude that

$$\left|\int_a^b f(x)\,dx\right| = \left|\int_a^b f_1(x)\,dx - \int_a^b f_2(x)\,dx\right|$$
$$\leq \int_a^b [f_1(x) + f_2(x)]\,dx \leq \int_a^b g(x)\,dx.$$

We have the following comparison test for divergence.

Theorem 6. *Suppose that f and g are continuous on $[a, b)$ where*

$$0 \leq g(x) \leq f(x),$$

and suppose that $\int_a^b g(x)\,dx$ diverges. Then $\int_a^b f(x)\,dx$ diverges. The same result holds if $[a, b)$ is replaced by $[a, +\infty)$, $(a, b]$, or $(-\infty, b]$.

PROOF. If $\int_a^b f(x)\,dx$ were convergent then, according to Theorem 5,

$$\int_a^b g(x)\,dx$$

would be also.

The comparison tests are useful if we have available a class or several classes of integrals which we *know* converge or diverge. Then these integrals may be used for comparison purposes. Two such classes are given in the next theorem. The proof is obtained by straightforward integration and evaluation of the resulting limit.

Theorem 7. (a) *The integrals*

$$\int_a^b (b - x)^{-p}\,dx, \qquad \int_a^b (x - a)^{-p}\,dx$$

converge if $p < 1$ and diverge if $p \geq 1$.
b) *The integrals*

$$\int_a^\infty x^{-p}\,dx \qquad \textit{and} \qquad \int_{-\infty}^{-b} |x|^{-p}\,dx$$

with $a > 0$, $b > 0$ converge if $p > 1$ and diverge if $p \leq 1$.

EXAMPLE 1. Test for convergence or divergence

$$\int_0^1 \frac{x^\alpha\,dx}{\sqrt{1 - x^3}},$$

where α is a positive constant.

SOLUTION. Taking

$$f(x) = \frac{x^\alpha}{\sqrt{1-x^3}}, \qquad g(x) = \frac{1}{\sqrt{1-x}},$$

we show that $|f(x)| \le g(x)$ for $0 \le x \le 1$. To see this, we observe that

$$f(x) = \frac{x^\alpha}{\sqrt{1-x^3}} = \frac{x^\alpha}{\sqrt{(1-x)(1+x+x^2)}} = \frac{x^\alpha}{\sqrt{1+x+x^2}} g(x).$$

Since $x^\alpha \le 1 \le \sqrt{1+x+x^2}$ for $0 \le x \le 1$ so long as $\alpha \ge 0$, we conclude that

$$|f(x)| \le g(x).$$

However,

$$\int_0^1 g(x)\,dx$$

converges by Theorem 7(a) with $p = \frac{1}{2}$. The Comparison Test shows that the given integral converges for all $\alpha \ge 0$.

EXAMPLE 2. Test for convergence or divergence:

$$\int_1^\infty \frac{\sqrt{x}}{1+x^{3/2}}\,dx.$$

SOLUTION. For $x \ge 1$, we see that

$$f(x) \equiv \frac{\sqrt{x}}{1+x^{3/2}} \ge \frac{\sqrt{x}}{x^{3/2}+x^{3/2}} = \frac{1}{2}\cdot\frac{1}{x} \equiv g(x).$$

However, according to Theorem 7(b), $\int_1^\infty g(x)\,dx$ diverges. Therefore the integral diverges.

EXAMPLE 3. Test for convergence or divergence:

$$\int_1^\infty \frac{\sin x}{x^{3/2}}\,dx.$$

SOLUTION. Since

$$\left|\frac{\sin x}{x^{3/2}}\right| \le \frac{1}{x^{3/2}},$$

we employ Theorem 7(b), with $p = \frac{3}{2}$, and the Comparison Test to conclude that the integral converges.

EXAMPLE 4. Test for convergence or divergence:

$$\int_0^\infty \frac{e^{-x}}{\sqrt{x}}\,dx.$$

Solution. Because the integrand is infinite at $x = 0$, we select a convenient value, say $x = 1$, and break the integral into two parts:

$$\int_0^\infty \frac{e^{-x}}{\sqrt{x}}\,dx = \int_0^1 \frac{e^{-x}}{\sqrt{x}}\,dx + \int_1^\infty \frac{e^{-x}}{\sqrt{x}}\,dx. \tag{1}$$

In the first integral on the right, we have

$$\frac{e^{-x}}{\sqrt{x}} \le \frac{1}{\sqrt{x}},$$

and $\int_0^1 (1/\sqrt{x})\,dx$ converges by Theorem 7(a) with $p = \frac{1}{2}$. In the second integral on the right, we have

$$\frac{e^{-x}}{\sqrt{x}} \le e^{-x}.$$

The integral

$$\int_1^X e^{-x}\,dx = [-e^{-x}]_1^X = e^{-1} - e^{-X} \to \frac{1}{e} \quad \text{as } X \to +\infty,$$

so that the second integral converges. Since both integrals on the right in (1) converge, the original integral is convergent.

We shall show that the integral

$$\int_0^\infty t^{x-1}e^{-t}\,dt \tag{2}$$

is convergent for $x > 0$. To do so, we notice that for $0 < x < 1$ the integrand becomes infinite at $t = 0$, and so we treat the integral as in Example 4. We write

$$\int_0^\infty t^{x-1}e^{-t}\,dt = \int_0^1 t^{x-1}e^{-t}\,dt + \int_1^\infty t^{x-1}e^{-t}\,dt. \tag{3}$$

In the first integral on the right, we obtain the inequality

$$t^{x-1}e^{-t} \le t^{x-1}.$$

The integral $\int_0^1 t^{x-1}\,dx$ converges for $x > 0$ (Theorem 7(a), $p < 1$). As for the second integral on the right in (3), we first note that

$$t^{x-1}e^{-t} = t^{-2}\cdot f(t)$$

where $f(t) = t^{x+1}e^{-t}$. We obtain an inequality by computing the maximum value of the function $f(t) = t^{x+1}e^{-t}$. The derivative $f'(t)$ is zero when $t = x + 1$; the maximum value of f occurs at $x + 1$ and is

$$f(x+1) = (x+1)^{x+1}e^{-(x+1)}.$$

Therefore

$$\int_1^\infty t^{x-1}e^{-t}\,dt = \int_1^\infty (t^{x+1}e^{-t})\frac{1}{t^2}\,dt \le (x+1)^{x+1}e^{-(x+1)}\int_1^\infty \frac{1}{t^2}\,dt.$$

The last integral on the right converges and, consequently, so does (2). This last device can be used to show the convergence of the integral (2) when $x \ge 1$.

Definition. The **Gamma function**, denoted by Γ, is defined by

$$\Gamma(x) = \int_0^\infty t^{x-1}e^{-t}\,dt, \qquad x > 0.$$

This function, which has many important applications in both mathematics and physics, has a number of interesting properties. The observation that

$$\Gamma(x+1) = \int_0^\infty t^x e^{-t}\,dt$$

and an integration by parts yield one of the most important properties. We have

$$\int_0^T t^x e^{-t}\,dt = [t^x(-e^{-t})]_0^T + x\int_0^T t^{x-1}e^{-t}\,dt, \qquad x \ge 1. \tag{4}$$

Letting $T \to +\infty$ and using l'Hôpital's Rule in the first term on the right, we obtain

$$\Gamma(x+1) = x\Gamma(x). \tag{5}$$

Since

$$\Gamma(1) = \int_0^\infty e^{-t}\,dt = \lim_{X\to\infty}[e^{-t}]_0^X = 1,$$

we find, by successive applications of (5), that

$$\Gamma(2) = 1\cdot\Gamma(1) = 1, \qquad \Gamma(3) = 2\cdot\Gamma(2) = 2,$$
$$\Gamma(4) = 3\cdot\Gamma(3) = 1\cdot 2\cdot 3,$$

and, by induction, that

$$\Gamma(n+1) = n!$$

In other words, the Gamma function, which is defined for all real numbers $x > 0$, is a generalization of the "factorial function" defined for positive integers.

The relation (5), which is a **recursion formula**, shows that if the Gamma function is known for any particular value of x, it may be found for all numbers of the form $x + n$, where n is a positive integer. For example, it can be shown that $\Gamma(\frac{3}{2}) = \frac{1}{2}\sqrt{\pi}$. From this fact it is easy to deduce that

$$\Gamma\left(\frac{2k+1}{2}\right) = \frac{1 \cdot 3 \cdot 5 \cdots (2k-1)}{2^k}\sqrt{\pi}$$

for any positive integer k.

Problems

In each of Problems 1 through 22, test the integral for convergence or divergence.

1. $\displaystyle\int_1^\infty \frac{dx}{(x+2)\sqrt{x}}$

2. $\displaystyle\int_2^\infty \frac{dx}{(x-1)\sqrt{x}}$

3. $\displaystyle\int_0^1 \frac{dx}{\sqrt{1-x^4}}$

4. $\displaystyle\int_0^1 \frac{dx}{\sqrt{(1-x)^4}}$

5. $\displaystyle\int_0^\infty \frac{dx}{\sqrt{x^3+1}}$

6. $\displaystyle\int_2^\infty \frac{dx}{\sqrt{x^3-1}}$

7. $\displaystyle\int_0^\infty \frac{x\,dx}{\sqrt{x^4+1}}$

8. $\displaystyle\int_1^2 \frac{dx}{x\sqrt{x^2-1}}$

9. $\displaystyle\int_3^4 \frac{\sqrt{16-x^2}}{x^2-x-6}\,dx$

10. $\displaystyle\int_{-1}^1 \frac{dx}{\sqrt{1-x^2}}$

11. $\displaystyle\int_{-1}^1 \frac{dx}{x^2}$

12. $\displaystyle\int_0^{\pi/2} \frac{\sqrt{x}\,dx}{\sin x}$

13. $\displaystyle\int_1^3 \frac{\sqrt{x}}{\log x}\,dx$

14. $\displaystyle\int_0^\infty \frac{(\arctan x)^2\,dx}{x^2+1}$

15. $\displaystyle\int_0^\infty e^{-x^2}\,dx$

16. $\displaystyle\int_0^\infty x^2 e^{-x^2}\,dx$

17. $\displaystyle\int_0^\infty \frac{dx}{(x+1)\sqrt{x}}$

18. $\displaystyle\int_0^1 \frac{dx}{\sqrt{x-x^2}}$

19. $\displaystyle\int_0^\infty \frac{\cos x\,dx}{1+x^2}$

20. $\displaystyle\int_0^\infty e^{-x}\sin x\,dx$

21. $\displaystyle\int_0^\pi \frac{\cos x\,dx}{\sqrt{x}}$

22. $\displaystyle\int_0^\pi \frac{\sin x}{x\sqrt{x}}\,dx$

23. Show that $\int_2^\infty x^{-1}(\log x)^{-p}\,dx$ converges if $p > 1$ and diverges if $p \leq 1$.

24. Show that $\int_0^1 |\log x|^p\,dx$ converges to $\Gamma(p+1)$ for each $p > 0$. [*Hint*. Consider $\int_\varepsilon^1 (\log(1/x))^p\,dx$, set $x = e^{-t}$, and let $\varepsilon \to 0$.]

25. Show that $\int_1^\infty e^{-x^2}\,dx = \frac{1}{2}\Gamma(\frac{1}{2}) = \Gamma(\frac{3}{2})$.

26. Show that

$$\int_0^\infty x^p e^{-x^2}\,dx = \tfrac{1}{2}\Gamma\left(\frac{p+1}{2}\right), \qquad p > -1.$$

27. Show how to extend the procedure in Eq. (4) to establish (5) for $x > 0$.

3. Improper Multiple Integrals

For functions of one variable we have considered two types of improper integrals: (i) those in which the integrand becomes infinite at some point in the interval of integration, and (ii) those in which the interval of integration becomes infinite.

Double and triple integrals have been defined only for bounded functions and for bounded regions of integration. We now take up the problem of defining a double integral when the integrand f becomes infinite at a single point in a bounded region of integration. For example, suppose we wish to integrate the function

$$f:(x,y) \to \frac{1}{[(x-1)^2 + y^2]^{1/2}}$$

over the rectangular region

$$R = \{(x,y) : |x| \le 3, |y| \le 2\}.$$

The function f becomes infinite at the point $P(1,0)$, which is in the region of integration. We construct a small region S containing the point P, and we observe that the function f is continuous in the region $R - S$ (Fig. 8-4). Therefore the integral

$$\iint\limits_{R-S} f(x,y)\,dA_{xy} \tag{1}$$

may be defined in the usual way if S is a small disk, square, triangle, or even a region of rather irregular shape, so long as P is interior to S (that is,

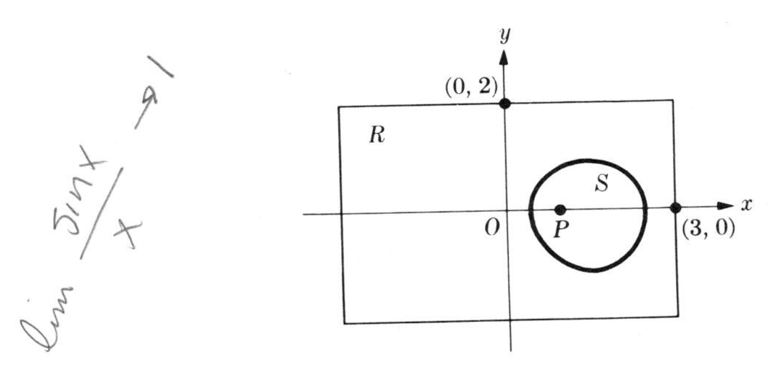

Fig. 8-4

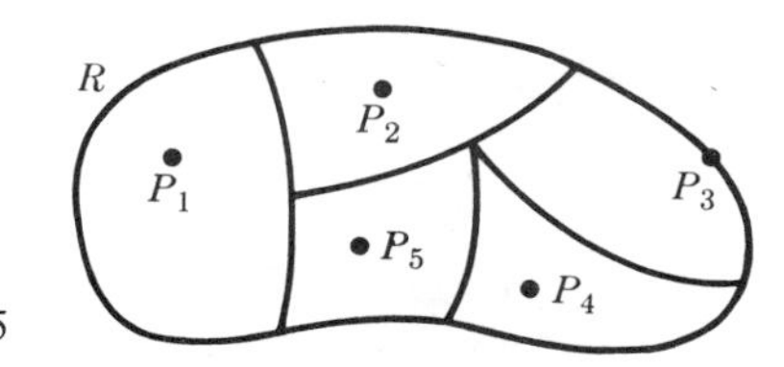

Fig. 8-5

so long as S contains a disk with center at P). For a set S we define the **diameter of** S as follows: Let $d(P, Q)$ be the distance between the points P and Q of S; then the diameter of S, denoted $d(S)$, is the least upper bound of $d(P, Q)$ as P and Q take on all possible values in S.

Let S_n be a sequence of closed regions, each containing P in its interior, and such that

$$d(S_n) \to 0 \qquad \text{as } n \to \infty.$$

Then, intuitively, we define the integral of f over R as the limit L of the integrals taken over $R - S_n$ whenever this limit exists. There are many different ways in which such a sequence S_n may be chosen, and of course the value L must be the same for all possible choices of the $\{S_n\}$. More precisely, we say that if for every $\varepsilon > 0$ there is a $\delta > 0$ such that

$$\left| \iint_{R-S} f(x, y)\, dA_{xy} - L \right| < \varepsilon$$

whenever S is a closed region containing P in its interior with diameter less than δ, then the **improper integral**

$$\iint_{R} f(x, y)\, dA_{xy}$$

exists and has the value L.

If a function becomes infinite at a finite number of points $P_1, P_2, \ldots, P_k$ within (or on the boundary of) the region of integration but is otherwise continuous, we subdivide R into a number of nonoverlapping pieces so that each portion contains one of the points P_i. (Figure 8-5 shows a typical situation.) Then, if the improper integral over each portion exists, we add the resulting values to obtain the improper integral over R. It can be shown that the value of the integral does not depend on how the region R is subdivided.

In any particular case, it is usually extremely difficult to verify the existence of an improper integral by use of the definition alone. We now establish theorems which are helpful not only in verifying the existence of improper integrals but also in their actual evaluation. Let R be a closed region and P a point in R. We define an **increasing sequence of regions closing down on** P as a sequence of closed regions $H_1, H_2, \ldots, H_n, \ldots$ with these properties: (i) every H_n is in $R - P$; (ii) for each n, $H_n \subset H_{n+1}$; and (iii) if R' is any closed set contained in $R - P$, then there is an integer n such that $R' \subset H_n$.

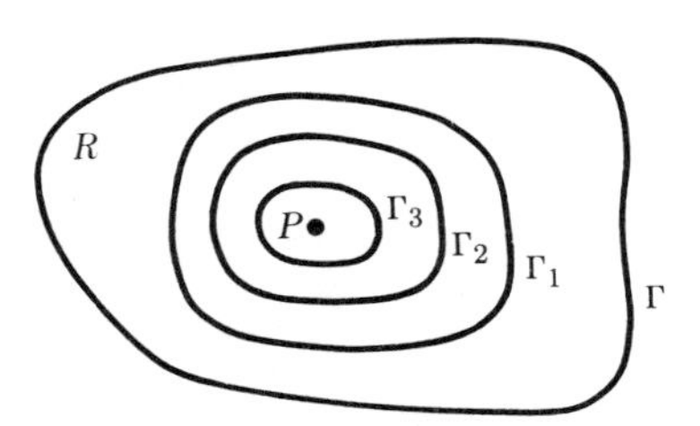

Fig. 8-6

Since H_n is closed and does not contain P, it also does not contain a small circle (of radius r_n, $r_n \to 0$) about P. In Fig. 8-6, for example, we could select H_1 as the portion of R between Γ and Γ_1, H_2 as the portion of R between Γ and Γ_2, H_3 as the portion of R between Γ and Γ_3, and so forth.

Theorem 8. *Suppose that $f(x, y) \geq 0$ in a closed, bounded region R, and suppose that f is continuous on $R - P$, where P is a point of R. If there is a number M such that*

$$\iint_{R'} f(x, y)\, dA_{xy} \leq M$$

for every closed region $R' \subset R - P$, then the improper integral of f over R exists. Moreover, suppose that $\{H_n\}$ is an increasing sequence of regions closing down on P, as defined above. Then we have

$$\lim_{n \to \infty} \iint_{H_n} f(x, y)\, dA_{xy} = \iint_R f(x, y)\, dA_{xy}.$$

PROOF. Let $\{H_n\}$ be any sequence of regions of the type specified. Since $f(x, y) \geq 0$, we see that for each n

$$\iint_{H_{n+1}} f(x, y)\, dA = \iint_{H_n} f(x, y)\, dA + \iint_{H_{n+1} - H_n} f(x, y)\, dA \geq \iint_{H_n} f(x, y)\, dA.$$

Therefore $\iint_{H_n} f(x, y)\, dA_{xy}$ is a nondecreasing sequence bounded by M. From the Axiom of Continuity, it follows that there is a number L such that

$$\lim_{n \to \infty} \iint_{H_n} f(x, y)\, dA_{xy} = L. \tag{2}$$

Now, let $\varepsilon > 0$ be given. There is an integer N such that

$$L - \varepsilon < \iint_{H_{N+1}} f(x, y)\, dA_{xy} \leq \iint_{H_n} f(x, y)\, dA_{xy} \leq L \qquad \text{for } n \geq N + 1.$$

Let R' be any closed region such that

$$H_{N+1} \subset R' \subset R - P.$$

Since R' does not contain P, there is an H_n such that $R' \subset H_n$ for some n (and so for all larger n). Consequently,

$$L - \varepsilon < \iint_{H_{N+1}} f(x, y)\, dA \le \iint_{R'} f(x, y)\, dA \le \iint_{H_n} f(x, y)\, dA \le L.$$

Since ε is arbitrary, it follows that

$$L = \iint_R f(x, y)\, dA.$$

Corollary. *If f, R, P are as in Theorem 8, if $\{H_n\}$ is an increasing sequence closing down on P, and if*

$$\iint_{H_n} f(x, y)\, dA_{xy} \le M$$

for all n, then the improper integral of f over R exists and (2) *holds.*

If an improper integral exists, we say that the integral is **convergent**; if it does not exist, we say that the integral is **divergent**.

EXAMPLE 1. Discuss the convergence or divergence of

$$\iint_R \frac{1}{\sqrt{(x^2 + y^2)^p}}\, dA_{xy}, \qquad p > 0,$$

where R is the disk $R = \{(x, y) : x^2 + y^2 \le 1\}$.

SOLUTION. The integrand becomes infinite at the origin. We select for H_n the ring $H_n = \{(x, y) : (1/n) \le \sqrt{x^2 + y^2} \le 1\}$. Then, changing to polar coordinates by setting $x = r\cos\theta$, $y = r\sin\theta$, we have

$$\iint_{H_n} \frac{1}{r^p}\, dA = \int_0^{2\pi} \int_{1/n}^1 r^{-p} r\, dr\, d\theta$$

$$= \begin{cases} \dfrac{2\pi}{2 - p}\left(1 - \dfrac{1}{n^{2-p}}\right) & \text{if } p < 2, \\ 2\pi \log n & \text{if } p = 2, \\ \dfrac{2\pi}{p - 2}(n^{p-2} - 1) & \text{if } p > 2. \end{cases}$$

By the Corollary to Theorem 8, the integral converges if $p < 2$. Since $r^{-p} > 0$, it follows that the definition of the existence of improper integral is not satisfied if $p \ge 2$.

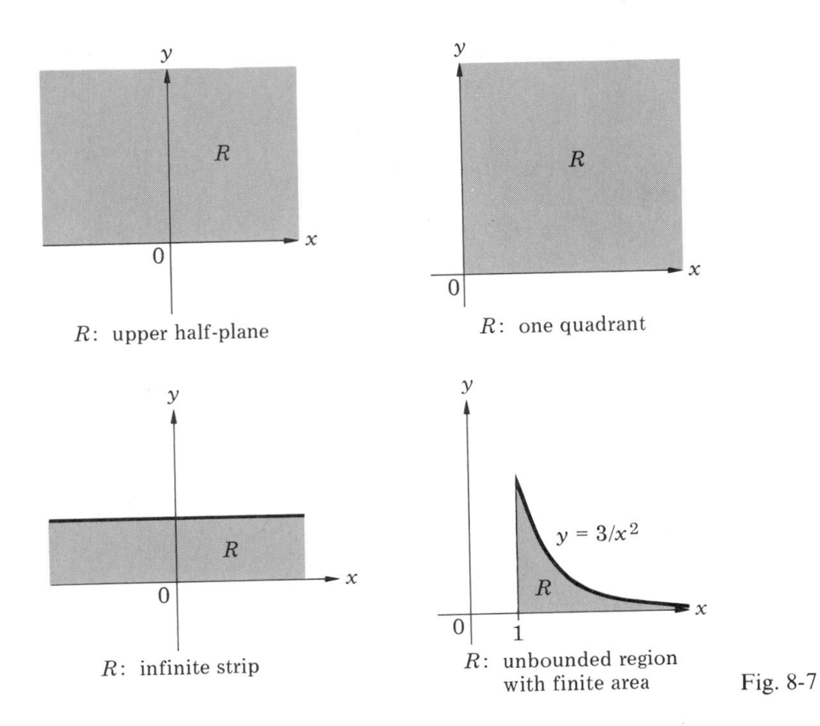

Fig. 8-7

So far we have considered integrals in which the integrand becomes infinite at one or several points. However, for functions of two or more variables the integrand may misbehave in many ways. For example, the function

$$f: (x, y) \to \frac{1}{(x^2 - y)^{1/3}}$$

becomes infinite all along the parabola $C = \{(x, y) : y = x^2\}$. It is possible to define an improper integral for functions which are infinite along an arc C. It is necessary to define an increasing sequence of regions closing down on C and to make an appropriate generalization of Theorem 8. This topic and various extensions to triple integrals will not be discussed in detail.

Suppose that R is an unbounded region in the plane and that f is a function on R. In defining the improper integral of f over R we must be aware that the situation is more complicated than it is for functions of one variable. For integrals along the x axis, the interval of integration could extend from some point a to $-\infty$, from a point a to $+\infty$, or from $-\infty$ to $+\infty$. As exhibited in Fig. 8-7, unbounded regions in the plane may be of many types. The reader can easily think of many other kinds of unbounded regions.

Definition. Suppose that R is an unbounded region and that f is continuous on R. We say that the **improper integral of f over R** exists if and only if there is a number K such that: for every $\varepsilon > 0$ there is a closed bounded region

$R_1 \subset R$ with the property that

$$\left| \iint_{R'} f(x, y)\, dA_{xy} - K \right| < \varepsilon$$

for every closed bounded region R' with $R_1 \subset R' \subset R$. The number K is the value of the improper integral, and we write

$$\iint_{R} f(x, y)\, dA_{xy} = K.$$

For unbounded regions R, we define an **increasing sequence of regions** $\{H_n\}$ **filling** R as a sequence of closed bounded regions in R with the properties: (i) for each n, $H_n \subset H_{n+1}$, and (ii) if R' is any bounded region in R, then there is an integer n such that $R' \subset H_n$. With this definition of the sequence $\{H_n\}$, Theorem 8 and its Corollary have obvious analogs for an improper integral taken over an unbounded region R.

Example 2. Evaluate $\int_0^\infty e^{-x^2}\, dx$.

Solution. We employ a device which has become classical in the study of improper multiple integrals but which is seldom useful otherwise. First we observe that

$$\int_0^\infty e^{-x^2}\, dx = \lim_{n \to \infty} \int_0^n e^{-x^2}\, dx$$

is a convergent integral (by comparison with $\int_0^\infty e^{-x}\, dx$, for example). Next we note that

$$\left(\int_0^n e^{-x^2}\, dx \right)^2 = \int_0^n \int_0^n e^{-(x^2+y^2)}\, dx\, dy = \iint_{R_n} e^{-(x^2+y^2)}\, dA,$$

where R_n is the square

$$R_n = \{(x, y) : 0 \le x \le n, \quad 0 \le y \le n\}.$$

Let R be the entire first quadrant and let R' be a closed bounded region in R. For n sufficiently large, $R' \subset R_n$ and, denoting

$$M = \left(\int_0^\infty e^{-x^2}\, dx \right)^2,$$

we see that

$$\iint_{R'} e^{-(x^2+y^2)}\, dA \le M.$$

The quantity M is also equal to

$$\lim_{n\to\infty} \iint_{G_n} e^{-(x^2+y^2)}\, dA,$$

where G_n is the quarter circle (polar coordinates)

$$G_n = \{(r, \theta) : 0 \leq r \leq n, \quad 0 \leq \theta \leq \pi/2\}.$$

But

$$\iint_{G_n} e^{-(x^2+y^2)}\, dA = \int_0^{\pi/2} \int_0^n e^{-r^2} r\, dr\, d\theta = \frac{\pi}{4}[-e^{-r^2}]_0^n \to \frac{\pi}{4} \qquad \text{as } n \to \infty.$$

Hence $M = \pi/4$ and, taking the square root, we conclude that

$$\int_0^\infty e^{-x^2}\, dx = \frac{\sqrt{\pi}}{2}.$$

Comparison theorems for determining the convergence and divergence of multiple integrals follow the same pattern as do those for integrals of functions of one variable. (See Section 2.)

Theorem 9. (a) *Suppose that f, R, P are as in Theorem* 8 *and that* $|f(x, y)| \leq g(x, y)$ *on* $R - P$. *If g is continuous on $R - P$ and the integral of g over R exists, then the integral of f over R does also, and*

$$\left| \iint_R f(x, y)\, dA \right| \leq \iint_R g(x, y)\, dA. \tag{3}$$

b) *Suppose that f and g are continuous on an unbounded region R. If*

$$|f(x, y)| \leq g(x, y)$$

on R and if the integral of g over R exists, then the integral of f does also, and the inequality (3) *holds.*

The proof of Theorem 9 is similar to that of Theorem 5 of the preceding section and is left to the reader.

Theorem 10. (a) *Suppose that f, g, R, P are as in Theorem* 9 *and that*

$$0 \leq g(x, y) \leq f(x, y).$$

If $\iint_R g(x, y)\, dA$ diverges, then $\iint_R f(x, y)\, dA$ does also.
b) *If R is unbounded and $\iint_R g(x, y)\, dA$ diverges, then $\iint_R f(x, y)\, dA$ diverges also.*

Theorem 10 follows directly from Theorem 9, since the supposition that $\iint_R f(x, y)\, dA$ converges implies that $\iint_R g\, dA$ does.

EXAMPLE 3. Test for convergence or divergence:

$$\iint_R \frac{\sin xy}{x^2(1+y^2)}\,dA,$$

where R is the half-infinite strip $R = \{(x, y): 1 \le x < \infty,\quad 0 \le y \le 1\}$.

SOLUTION. We define the domain $R_n = \{(x, y): 1 \le x \le n,\quad 0 \le y \le 1\}$. Setting

$$f(x, y) = \frac{\sin xy}{x^2(1+y^2)}, \qquad g(x, y) = \frac{1}{x^2(1+y^2)},$$

we have

$$\iint_{R_n} |f(x, y)|\,dA \le \iint_{R_n} g(x, y)\,dA = \int_1^n \int_0^1 \frac{1}{x^2} \cdot \frac{1}{1+y^2}\,dy\,dx.$$

However,

$$\iint_{R_n} g(x, y)\,dA = [\arctan y]_0^1 \int_1^n \frac{1}{x^2}\,dx = \frac{\pi}{4}\left[-\frac{1}{x}\right]_1^n \to \frac{\pi}{4} \qquad \text{as } n \to \infty.$$

Hence $\iint_R g\,dA$ is convergent, and so $\iint_R f\,dA$ is also.

The entire discussion of this section has an appropriate generalization to functions of three variable and to triple integrals. Improper integrals may be defined for functions $f(x, y, z)$ which become infinite at points, on curves, or on surfaces. Also, integrals over unbounded domains may be defined. We work an example.

EXAMPLE 4. Test for convergence or divergence:

$$\iiint_R \frac{1}{\sqrt{(x^2+y^2+z^2)^p}}\,dV, \qquad p > 0,$$

where R is the unit ball $R = \{(x, y, z): 0 \le x^2 + y^2 + z^2 \le 1\}$.

SOLUTION. We define the regions bounded by concentric spheres

$$R_n = \left\{(x, y, z): \frac{1}{n} \le \sqrt{x^2+y^2+z^2} \le 1\right\}.$$

Then, introducing spherical coordinates

$$x = \rho\cos\theta\sin\phi, \qquad y = \rho\sin\theta\sin\phi, \qquad z = \rho\cos\phi,$$

and using the change-of-variables formula for integration in spherical coordinates (see page 351), we find

$$\iiint_R \frac{1}{\rho^p} dV = \int_0^{2\pi} \int_0^{\pi} \int_{1/n}^{1} \frac{1}{\rho^p} \rho^2 \sin\phi \, d\rho \, d\phi \, d\theta$$

$$= \begin{cases} \dfrac{4\pi}{3-p}\left(1 - \dfrac{1}{n^{3-p}}\right), & 0 \le p < 3, \\ 4\pi \log n, & p = 3, \\ \dfrac{4\pi}{p-3}(n^{p-3} - 1), & p > 3. \end{cases}$$

We conclude that the integral is convergent for $p < 3$ and divergent for $p \ge 3$.

Problems

In each of Problems 1 through 10, the region R is the unit disk

$$R = \{(x, y) : 0 \le x^2 + y^2 \le 1\}.$$

Test the given integral for convergence or divergence. Specify the choice of the sequence $\{H_n\}$ and describe the set S where the integrand is singular.

1. $\displaystyle\iint_R \frac{x^2 \, dA}{(x^2 + y^2)^{3/2}}$

2. $\displaystyle\iint_R \frac{xy \, dA}{(x^2 + y^2)^{3/2}}$

3. $\displaystyle\iint_R \log\frac{1}{r} \, dA$

4. $\displaystyle\iint_R r^{-2}\left(\log\frac{2}{r}\right)^{-2} dA$

5. $\displaystyle\iint_R \left(\log\frac{1}{r}\right)^{-1} dA$

6. $\displaystyle\iint_R \frac{x^2 y^2}{(x^2 + y^2)^3} \, dA$

7. $\displaystyle\iint_R \frac{dA}{\sqrt{1 + x}}$

8. $\displaystyle\iint_R \frac{x \, dA}{(x^2 + y^2)\sqrt{1 - x}}$

9. $\displaystyle\iint_R \frac{dA}{1 - x}$

10. $\displaystyle\iint_R \frac{dA}{\sqrt[3]{(1 - 2x)(1 - 3x)(1 - 4x)}}$

In each of Problems 11 through 16, test for convergence or divergence. Specify your choice of $\{H_n\}$.

11. $\displaystyle\iint_R (x^2 + y^2)^{-p/2} \, dA, \quad R = \{(x, y) : x^2 + y^2 \ge 1\}; \quad p > 0$

12. $\displaystyle\iint_R \frac{y \, dA}{(x^2 + y^2)(1 - x)^{2/3}}, \quad R = \{(x, y) : 0 \le x \le 1, 0 \le y \le 1\}$

13. $\displaystyle\iint_R \frac{dA}{(x^2 + y^2)\sqrt{(x - 1)(y - 1)}}, \quad R = \{(x, y) : x \ge 1, y \ge 1\}$

14. $\iiint_R \rho^{-p}\,dV, \quad \rho = (x^2 + y^2 + z^2)^{1/2},$
$R = \{(x, y, z) : \rho \geq 1\}; \quad p > 0$

15. $\iiint_R \frac{x\,dV}{\rho^3}, \quad R = \{(x, y, z) : 0 \leq \rho \leq 1\}$

16. $\iiint_R \frac{dV}{\rho^2(1 - x)^{3/2}}, \quad R = \{(x, y, z) : 0 \leq \rho \leq 1\}$

17. Use the result of Example 2 to find $\Gamma(\frac{1}{2})$.

*18. (a) Define an improper double integral for a function f which becomes infinite along an arc C contained in a bounded region R. (b) State and prove the appropriate analog of Theorem 8 for improper integrals as defined in (a).

*19. (a) Define an improper triple integral for a function f which is continuous in a bounded region R except at one point where it becomes infinite. (b) State and prove the appropriate analog of Theorem 8 for improper integrals as defined in (a).

20. Define an improper triple integral for a function continuous in an unbounded domain R.

21. Write out a proof of Theorem 9.

22. State and prove the analog of Theorem 9 for triple integrals.

23. State the appropriate generalization of Theorem 9(a) for functions of two variables which become infinite along an arc.

4. Functions Defined by Improper Integrals

We recall that the Gamma function is defined by the formula

$$\Gamma(x) = \int_0^\infty t^{x-1}e^{-t}\,dt, \qquad x > 0.$$

Is it possible to differentiate under the integral sign? That is, is the formula (Leibniz' Rule)

$$\Gamma'(x) = \int_0^\infty \frac{\partial}{\partial x}(t^{x-1}e^{-t})\,dt$$
$$= \int_0^\infty t^{x-1}(\log t)e^{-t}\,dt$$

valid? Leibniz' Rule, given in Section 1, page 422, was established for proper integrals. Not only does the integral for the Gamma function have an infinite interval of integration, but also its integrand has a singularity at $t = 0$ if $0 < x < 1$. In order to extend Leibniz' Rule to functions given by improper

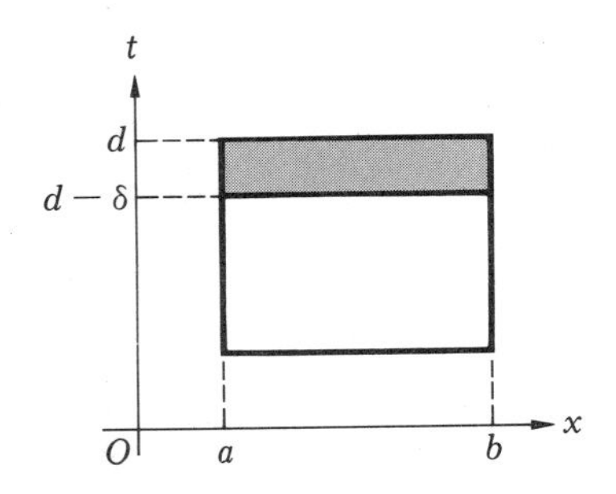

Fig. 8-8

integrals, we must first extend the notion of uniform convergence introduced in Chapter 3, page 154.

Suppose that $F(x, t)$ is continuous for $a \le x \le b$ and for $c \le t < d$. We wish to consider the limit of $F(x, t)$ as t tends to d (from values less than d). The limiting value will depend on x, and we write

$$\lim_{t \to d} F(x, t) = f(x).$$

Definition. We say that $F(x, t) \to f(x)$ **uniformly on** $[a, b]$ **as** $t \to d^-$ if and only if for each $\varepsilon > 0$ there is a $\delta > 0$ such that

$$|F(x, t) - f(x)| < \varepsilon \tag{1}$$

for all t in the interval $d - \delta < t < d$ and for all x on $[a, b]$.

Of course the size of δ will depend on ε. However, the crucial distinction between uniform limits (called **uniform convergence**) and ordinary limits lies in the fact that for uniform convergence, δ **is independent of** x. In other words, in uniform convergence, inequality (1) holds for all values of x and t in the shaded strip shown in Fig. 8-8.

Uniform convergence for a "continuous variable" may also be defined if the interval $c \le t < d$ is replaced by $c \le t < \infty$. We say that $F(x, t) \to f(x)$ **uniformly for** x **on** $[a, b]$ as $t \to +\infty$ if and only if for each $\varepsilon > 0$ there is a value T such that

$$|F(x, t) - f(x)| < \varepsilon$$

for all $t > T$. Once again, the value of T will depend on ε, but the uniformity condition requires that T **not** depend on x.

The next theorems, which are analogs of those for series given in Chapter 3, Section 11, are useful in establishing Leibniz' Rule for improper integrals.

Theorem 11. *Suppose that $F(x, t)$ is continuous in x on $[a, b]$ for each t,*

$$c \le t < d,$$

and suppose that $F(x, t) \to \phi(x)$ uniformly on $[a, b]$ as $t \to d^-$. Then $\phi(x)$ is continuous on $[a, b]$. The same result holds if d is replaced by $+\infty$.

PROOF. The proof follows the pattern of that given for Theorem 30 (Chapter 3, page 159). Let x_1, x_2 be two values of x in $[a, b]$. We write $\phi(x_1) - \phi(x_2)$ in the complicated form

$$\phi(x_1) - \phi(x_2) = \phi(x_1) - F(x_1, t) + F(x_1, t) - F(x_2, t) + F(x_2, t) - \phi(x_2).$$

Therefore, by the triangle inequality,

$$|\phi(x_1) - \phi(x_2)| \le |\phi(x_1) - F(x_1, t)| + |F(x_1, t) - F(x_2, t)| \\ + |F(x_2, t) - \phi(x_2)|.$$

Let $\varepsilon > 0$ be given. Then, from the uniform convergence of F, the first and third terms on the right may be made $< \varepsilon/3$ for all $t \in (d - \delta, d)$. The middle term on the right may be made $< \varepsilon/3$ if x_1 and x_2 are close enough, since F is continuous in x. Hence $|\phi(x_1) - \phi(x_2)| < \varepsilon$ if $|x_1 - x_2|$ is sufficiently small—that is, ϕ is continuous.

Theorem 12. *Suppose that the hypotheses of Theorem* 11 *hold and also, that* $F_x(x, t)$ *is continuous on* $[a, b]$ *for each* t. *If* $F_x(x, t) \to \psi(x)$ *uniformly for* x *on* $[a, b]$ *as* $t \to d^-$ *or as* $t \to +\infty$, *then* $\psi(x) = \phi'(x)$ *on* (a, b).

The proof follows the pattern of that given for Theorem 32, Chapter 3, page 161, and is left to the reader.

We now see that Leibniz' Rule for improper integrals, given below, is an immediate consequence of the two theorems above.

Theorem 13. *Suppose that* $f(x, \tau)$ *is continuous for* $a \le x \le b$ *and* $c \le \tau < d$; *we define*

$$F(x, t) = \int_c^t f(x, \tau)\, d\tau.$$

If the improper integral

$$\phi(x) = \int_c^d f(x, \tau)\, d\tau$$

exists for $a \le x \le b$, *and if* $F(x, t) \to \phi(x)$ *uniformly for* x *on* $[a, b]$ *as* $t \to d^-$, *then* ϕ *is continuous on* $[a, b]$. *The same result holds if* d *is replaced by* $+\infty$. (This result is a corollary of Theorem 11.)

Theorem 14 (Leibniz' Rule for improper integrals). *Suppose that the hypotheses of Theorem* 13 *hold, and suppose also that* f_x *is continuous. If* $F_x(x, t)$ *converges uniformly to* $\psi(x)$ *on* $[a, b]$ *as* $t \to d^-$ *(or* $+\infty$*), then*

$$\psi(x) = \phi'(x) = \int_c^d f_x(x, \tau)\, d\tau, \qquad a < x < b,$$

or, if d *is replaced by* $+\infty$,

$$\psi(x) = \phi'(x) = \int_c^{+\infty} f_x(x, \tau)\, d\tau, \qquad a < x < b.$$

PROOF. According to Leibniz' Rule, for each $t < d$ (or $< +\infty$), we have

$$F_x(x, t) = \int_c^t f_x(x, \tau)\, d\tau.$$

Hence the result follows from Theorem 12.

EXAMPLE 1. Show that if $\phi(x) = \int_0^\infty e^{-xt}\, dt$, then ϕ and ϕ' are continuous for $x > 0$ and

$$\phi'(x) = \int_0^\infty -te^{-xt}\, dt. \tag{2}$$

SOLUTION. We define

$$F(x, t) = \int_0^t e^{-x\tau}\, d\tau = \frac{1 - e^{-xt}}{x}.$$

Also,

$$F_x(x, t) = \frac{-1 + e^{-xt} + xte^{-xt}}{x^2}.$$

As $t \to \infty$, we see that $F(x, t) \to 1/x$ and $F_x(x, t) \to -1/x^2$. To show that the convergence is uniform, we note that for $h > 0$

$$\left| F(x, t) - \frac{1}{x} \right| = \frac{e^{-xt}}{x} < \frac{e^{-ht}}{h} \qquad \text{for all } x \geq h,$$

$$\left| F_x(x, t) - \left(-\frac{1}{x^2} \right) \right| = \frac{e^{-xt}(1 + xt)}{x^2} \leq \frac{e^{-ht}(1 + ht)}{h^2} \qquad \text{for } x \geq h.$$

Thus the convergence is uniform on any interval $x \geq h$ for positive h. Applying Leibniz' Rule for improper integrals, we conclude that (2) is valid.

In Example 1 it is possible to integrate the expression for $F(x, t)$ and so verify the uniform convergence directly. Since in most instances this direct approach is not possible, it is important to have some indirect tests for uniform convergence. The next theorem, which is a comparison test, is useful in that the direct evaluation of $F(x, t)$ is not required. The test may also be applied to $F_x(x, t)$ and in this way the applicability of Leibniz' Rule for improper integrals may be verified.

Theorem 15 (Comparison Test). (a) *Suppose that $f(x, t)$ is continuous for $a \leq x \leq b$ and $c \leq t < d$, and suppose that*

$$|f(x, t)| \leq g(t) \qquad \text{for } a \leq x \leq b, \quad c \leq t < d.$$

If $\int_c^d g(t)\,dt$ *converges, then the improper integral*

$$\phi(x) = \int_c^d f(x,t)\,dt$$

is defined for each x *on* $[a,b]$ *and* ϕ *is continuous on* $[a,b]$.
b) **(Leibniz Rule)** *Suppose, also, that* f_x *is continuous for* $a \le x \le b, c \le t < d$ *and that*

$$|f_x(x,t)| \le g_1(t) \qquad \text{for } a \le x \le b, \quad c \le t < d.$$

If $\int_c^d g_1(t)\,dt$ *converges, then*

$$\phi'(x) = \int_c^d f_x(x,t)\,dt, \qquad a < x < b.$$

The same results hold if d *is replaced by* $+\infty$.

PROOF. (a) That $\phi(x)$ is defined for each x follows from Theorem 5. We define

$$F(x,t) = \int_c^t f(x,\tau)\,d\tau.$$

Then

$$|\phi(x) - F(x,t)| = \left|\int_t^d f(x,\tau)\,d\tau\right| \le \int_t^d g(\tau)\,d\tau.$$

But since $\int_c^d g(\tau)\,d\tau$ converges, we know that for each $\varepsilon > 0$ there is a $\delta > 0$ such that for $t \in (d-\delta, d)$ we have $\left|\int_t^d g(\tau)\,d\tau\right| < \varepsilon$. Thus

$$|\phi(x) - F(x,t)| \to 0 \qquad \text{as} \qquad t \to d^-,$$

uniformly for x on $[a,b]$.

To prove (b), we note that

$$\left|\int_t^d f_x(x,\tau)\,d\tau\right| \le \int_t^d g_1(\tau)\,d\tau \to 0 \qquad \text{as } t \to d^-,$$

as above. The convergence is uniform and the result established.

Definition. If

$$F(x,t) = \int_c^t f(x,\tau)\,d\tau$$

and

$$\phi(x) = \int_c^d f(x,\tau)\,d\tau \tag{3}$$

and if $F(x,t) \to \phi(x)$ uniformly for x on $[a,b]$, we say that the **improper integral** (3) **converges uniformly for** x **on** $[a,b]$.

EXAMPLE 2. Show that the improper integral

$$\phi(x) = \int_1^\infty \frac{\sin t}{x^2 + t^2}\,dt$$

converges uniformly for all x.

SOLUTION. We have

$$\left|\frac{\sin t}{x^2 + t^2}\right| \le \frac{1}{t^2}$$

for all x and all $t \ge 1$. Since $\int_1^\infty (1/t^2)\,dt$ converges, Theorem 15 applies.

EXAMPLE 3. Given the integral

$$\phi(x) = \int_0^\infty \frac{e^{-xt} - e^{-t}}{t}\,dt.$$

Show that the integral for $\phi(x)$ and the integral for $\phi'(x)$ (obtained by differentiation under the integral sign) both converge uniformly for x on $[a, b]$ if $a > 0$. Evaluate $\phi(x)$ by this means.

SOLUTION. Setting $f(x, t) = (e^{-xt} - e^{-t})/t$, we have

$$f_x(x, t) = -e^{-xt}, \qquad |f_x(x, t)| \le \begin{cases} e^{-at}, & a \le 1, \\ e^{-t}, & 1 < a. \end{cases}$$

The last inequality may be written more compactly as

$$|f_x(x, t)| \le e^{-ht} \qquad \text{where } h = \min(a, 1).$$

Now, using the Theorem of the Mean on f (as a function of x), we find that

$$\left|\frac{e^{-xt} - e^{-t}}{t}\right| = |e^{-\xi t}|\,|1 - x| \qquad \text{for } 0 \le t \le 1.$$

Also,

$$\left|\frac{e^{-xt} - e^{-t}}{t}\right| \le e^{-ht} \qquad \text{for } t \ge 1.$$

Thus the integrals for ϕ and ϕ' converge uniformly. We may apply Leibniz' Rule to get

$$\phi'(x) = \int_0^\infty -e^{-xt}\,dt = \lim_{t\to\infty}\left[\frac{e^{-xt}}{x}\right]_0^t = -\frac{1}{x}.$$

The equation $\phi'(x) = 1/x$ can now be integrated to give $\phi(x) = C - \log x$. Since

$$\phi(1) = \int_0^\infty \frac{e^{-t} - e^{-t}}{t}\,dt = 0,$$

we obtain

$$\phi(x) = -\log x = \int_0^\infty \frac{e^{-xt} - e^{-t}}{t}\,dt.$$

Problems

In each of Problems 1 through 8, show that the integrals for $\phi(x)$ and $\phi'(x)$ converge uniformly on the given intervals, and find $\phi'(x)$ by Leibniz' Rule.

1. $\phi(x) = \int_0^\infty \frac{e^{-xt}\,dt}{1+t}, \quad a \le x, \quad a > 0$
2. $\phi(x) = \int_0^1 \frac{e^{xt}}{\sqrt{t}}\,dt, \quad |x| \le A, \quad A > 0$
3. $\phi(x) = \int_0^\infty \frac{\cos xt}{1+t^3}\,dt, \quad |x| \le A, \quad A > 0$
4. $\phi(x) = \int_0^\infty \frac{e^{-t}\,dt}{1+xt}, \quad x \ge 0$
5. $\phi(x) = \int_0^1 \frac{\sin xt}{t}(\log t)\,dt, \quad |x| \le A, \quad A > 0$
6. $\phi(x) = \int_0^1 \frac{(\log t)^2}{1+xt}\,dt, \quad x \ge A, \quad A > -1$
7. $\phi(x) = \int_0^1 \frac{dt}{(1+xt)\sqrt{1-t}}, \quad x \ge A, \quad A > -1$
8. $\phi(x) = \int_0^\infty \frac{\sin xt}{t(1+t^2)}\,dt, \quad |x| \le A, \quad A > 0$
9. Using the complex exponential function (and Theorem 15), show that

$$\int_0^\infty e^{-ax}\cos bx\,dx = \frac{a}{a^2+b^2},$$

$$\int_0^\infty e^{-ax}\sin bx\,dx = \frac{b}{a^2+b^2}, \quad a > 0.$$

10. From the fact that

$$\int_0^1 t^x\,dt = \frac{1}{x+1}, \qquad x > -1,$$

deduce that

$$\int_0^1 t^x(-\log t)^m\,dt = \frac{m!}{(x+1)^{m+1}}, \quad x > -1.$$

11. From the fact that

$$\int_0^\infty \frac{t^{x-1}}{1+t}\,dt = \frac{\pi}{\sin \pi x}, \quad 0 < x < 1,$$

deduce that

$$\int_0^\infty \frac{t^{x-1}\log t}{1+t}\,dt = \frac{\pi^2\cos\pi x}{\cos^2\pi x - 1}.$$

12. Verify that

$$\Gamma'(x) = \int_0^\infty t^{x-1}(\log t)e^{-t}\,dt$$

and, more generally, that

$$\frac{d^n}{dx^n}(\Gamma(x)) = \int_0^\infty t^{x-1}(\log t)^n e^{-t}\,dt.$$

13. If

$$\phi(x) = \int_0^\infty \frac{e^{-xt}}{1+t}\,dt,$$

show that $\phi(x) - \phi'(x) = 1/x$. Justify your steps.

14. If

$$\phi(x) = \int_0^\infty \frac{e^{-xt}\,dt}{1+t^2},$$

show that $\phi(x) + \phi''(x) = 1/x$. Justify your steps.

15. Given that

$$\phi(x) = \int_0^\infty e^{-t}\left(\frac{1-\cos xt}{t}\right)dt.$$

Find $\phi'(x)$ by Leibniz' Rule. Evaluate the integral for $\phi'(x)$ and then find $\phi(x)$ by integration.

16. Given that

$$\phi(x) = \int_0^1 \frac{t^x - 1}{\log t}\,dt, \qquad x \geq a, \quad a > -1.$$

Find $\phi'(x)$ by Leibniz' Rule, evaluate, and find $\phi(x)$ by integration.

17. Given that

$$\phi(x) = \int_0^\infty \frac{e^{-t}(1-\cos xt)}{t^2}\,dt.$$

Find $\phi'(x)$ and $\phi''(x)$ by Leibniz' Rule. Find ϕ' and ϕ by integration.

18. Given that

$$\phi(x) = \int_0^\infty e^{-xt}\,dt = \frac{1}{x},$$

show that

$$\phi^{(n)}(x) = (-1)^n \int_0^\infty t^n e^{-xt}\,dt = (-1)^n \frac{n!}{x^{n+1}}, \quad x > 0.$$

19. Starting from the formula

$$\int_0^\infty \frac{dt}{t^2+x} = \frac{\pi}{2}x^{-1/2},$$

deduce the formula

$$\int_0^\infty \frac{dt}{(t^2+x)^{n+1}} = \frac{1\cdot 3\cdot 5\cdots(2n-1)}{2\cdot 4\cdot 6\cdots 2n}x^{-n-1/2} = \frac{(2n)!}{2^{2n}(n!)^2}x^{-n-1/2}.$$

20. Prove Theorem 12.

*21. Given that

$$P_n(x) = \frac{1}{n!\sqrt{x}}\int_{-\infty}^\infty e^{-(1-x^2)t^2}\left(-\frac{d}{dx}\right)^n (e^{-x^2t^2})\,dt,$$

show that

$$x\frac{d}{dx}(P_n(x)) - \frac{d}{dx}(P_{n-1}(x)) = nP_n(x).$$

[$P_n(x)$ is the Legendre polynomial.]

CHAPTER 9

Vector Field Theory

1. Vector Functions

A **vector function** $\mathbf{v}(P)$ assigns a specific vector to each element P in a given domain $\mathscr{D}$. The range of such a function is the collection of vectors which correspond to the points in the domain. In Chapter 2, Section 9, we discussed vector functions with domain a portion (or all) of R^1 and with range a collection of vectors in R^2 and R^3. For example, if a vector function is defined on the interval $a \leq t \leq b$, then we may represent such a function in the form

$$\mathbf{v}(t) = f(t)\mathbf{i} + g(t)\mathbf{j},$$

where $\mathbf{i}$ and $\mathbf{j}$ are the customary unit vectors. The real-valued functions f and g are defined on the interval $a \leq t \leq b$.

If the range of a vector function of one variable is a collection of vectors in three-space, we can write

$$\mathbf{v}(t) = f(t)\mathbf{i} + g(t)\mathbf{j} + h(t)\mathbf{k},$$

where $\mathbf{i}$, $\mathbf{j}$, and $\mathbf{k}$ are mutually perpendicular unit vectors in R^3. The functions f, g, and h are ordinary real-valued functions of the variable t. In Chapter 2, we discussed various properties of vector functions with range in R^3.

We now continue the study of vector functions by considering those with *domain* a portion of R^2 or R^3. A vector function with domain in R^2 and with range consisting of vectors in R^2 has the representation

$$\mathbf{v}(x, y) = f(x, y)\mathbf{i} + g(x, y)\mathbf{j}. \tag{1}$$

If the range consists of vectors in R^3, we write

$$\mathbf{v}(x, y) = f(x, y)\mathbf{i} + g(x, y)\mathbf{j} + h(x, y)\mathbf{k}. \tag{1'}$$

If the domain $\mathscr{D}$ is a region in three-space and the range is in R^2, so that we are dealing with functions of three variables, a vector function $\mathbf{u}(P)$ has the representation

$$\mathbf{u}(x, y, z) = f(x, y, z)\mathbf{i} + g(x, y, z)\mathbf{j}.$$

If the range of such a vector function is a collection of vectors in three-space, we write

$$\mathbf{u}(x, y, z) = f(x, y, z)\mathbf{i} + g(x, y, z)\mathbf{j} + (x, y, z)\mathbf{k}. \tag{2}$$

The vector functions described above are actually special cases of the general transformations described in Chapter 7, Section 3. In fact, we may easily define a vector function with domain $\mathscr{D}$ in some m-dimensional Euclidean space and with range in some n-dimensional vector space. The numbers m and n may be different. Although many of the results of this chapter are valid in quite general spaces, we shall state and prove theorems in two and three dimensions only. In this way we can take full advantage of our geometrical insight.

The usual properties of continuity, differentiation, and integration for vector functions of several variables are immediate generalizations of those for vector functions of one variable. A function $\mathbf{v}(x, y)$ as given by (1) is ***continuous*** if and only if the functions f and g are. A function $\mathbf{u}$ as in (2) has partial derivatives if and only if the functions f, g, and h do. For example, we have the formula

$$\frac{\partial \mathbf{u}}{\partial x} = \frac{\partial f(x, y, z)}{\partial x}\mathbf{i} + \frac{\partial g(x, y, z)}{\partial x}\mathbf{j} + \frac{\partial h(x, y, z)}{\partial x}\mathbf{k}.$$

We also use the comma notation

$$\mathbf{u}_{,1}(x, y, z) = f_{,1}(x, y, z)\mathbf{i} + g_{,1}(x, y, z)\mathbf{j} + h_{,1}(x, y, z)\mathbf{k}.$$

The formulas for partial derivatives with respect to y and z are obvious analogs.

EXAMPLE 1. Find $\partial \mathbf{u}/\partial x$ and $\partial^2 \mathbf{u}/\partial x \partial y$ if

$$\mathbf{u}(x, y) = (x^2 + 2xy)\mathbf{i} + (1 + x^2 - y^2)\mathbf{j} + (x^3 - xy^2)\mathbf{k}.$$

SOLUTION. We have

$$\frac{\partial \mathbf{u}}{\partial x} = 2(x + y)\mathbf{i} + 2x\mathbf{j} + (3x^2 - y^2)\mathbf{k},$$

$$\frac{\partial^2 \mathbf{u}}{\partial x \partial y} = 2\mathbf{i} + 0 \cdot \mathbf{j} - 2y\mathbf{k} = 2\mathbf{i} - 2y\mathbf{k}.$$

The formulas for the partial derivatives of the scalar and cross products of vector functions are similar to those for vector functions of variable. (See Chapter 2. Sections 6 and 7.) For example, if **u** and **v** are differentiable functions, then the reader may verify that

$$\frac{\partial}{\partial x}(\mathbf{u}\cdot\mathbf{v}) = \mathbf{u}\cdot\frac{\partial \mathbf{v}}{\partial x} + \frac{\partial \mathbf{u}}{\partial x}\cdot\mathbf{v},$$

$$\frac{\partial}{\partial x}(\mathbf{u}\times\mathbf{v}) = \mathbf{u}\times\frac{\partial \mathbf{v}}{\partial x} + \frac{\partial \mathbf{u}}{\partial x}\times\mathbf{v}.$$

Example 2. Given the vector functions

$$\mathbf{u} = (e^x \cos y)\mathbf{i} + (e^y \sin z)\mathbf{j} + (e^z \sin x)\mathbf{k}$$

and

$$\mathbf{v} = (\cos z)\mathbf{i} + (e^{2x} \sin y)\mathbf{j} + e^z\mathbf{k}.$$

Find $\partial(\mathbf{u}\cdot\mathbf{v})/\partial y$. Decide whether the vectors $\partial\mathbf{u}/\partial z$ and $\partial\mathbf{v}/\partial x$ are perpendicular at the point $P_0: x = 1,\ y = 0,\ z = \pi/4$.

Solution. We have

$$\frac{\partial \mathbf{u}}{\partial y} = (-e^x \sin y)\mathbf{i} + (e^y \sin z)\mathbf{j}, \qquad \frac{\partial \mathbf{v}}{\partial y} = e^{2x}\cos y\mathbf{j},$$

$$\begin{aligned}\frac{\partial}{\partial y}(\mathbf{u}\cdot\mathbf{v}) &= \mathbf{u}\cdot\frac{\partial \mathbf{v}}{\partial y} + \frac{\partial \mathbf{u}}{\partial y}\cdot\mathbf{v}\\ &= e^{2x+y}\cos y \sin z - e^x \sin y \cos z + e^{2x+y}\sin y \sin z.\end{aligned}$$

Computing the derivatives

$$\frac{\partial \mathbf{u}}{\partial z} = (e^y \cos z)\mathbf{j} + (e^z \sin x)\mathbf{k}, \qquad \frac{\partial \mathbf{v}}{\partial x} = (2e^{2x}\sin y)\mathbf{j},$$

we find

$$\frac{\partial \mathbf{u}}{\partial z}\cdot\frac{\partial \mathbf{v}}{\partial x} = 2e^{2x+y}\sin y \cos z.$$

At $P_0(1, 0, \pi/4)$, this scalar product is zero, and so the vectors are orthogonal there.

The integration of a vector function is given in terms of the integration of each of its components. If R is a region and **v** is a vector function as in (1′) with domain containing R, then the integral

$$\iint_R \mathbf{v}(x, y)\, dA$$

is defined by the formula

$$\iint_R \mathbf{v}(x, y)\, dA = \left(\iint_R f(x, y)\, dA\right)\mathbf{i} + \left(\iint_R g(x, y)\, dA\right)\mathbf{j} + \left(\iint_R h(x, y)\, dA\right)\mathbf{k}.$$

We see that the definite integral of a vector function is a vector. Similarly, triple integrals and line integrals of a vector function are defined in terms of the corresponding integrals of each of its components.

The parametric equations

$$x = f(t), \qquad y = g(t), \qquad a \le t \le b \tag{3}$$

represent a curve in the xy plane. Similarly, the equations

$$x = f(t), \qquad y = g(t), \qquad z = h(t), \qquad a \le t \le b$$

represent a curve in three-space. These curves may also be represented by vector functions. A vector is an equivalence class of directed line segments. If we consider the particular directed line segment which has its base at the origin, then the vector function

$$\mathbf{v}(t) = f(t)\mathbf{i} + g(t)\mathbf{j} \tag{4}$$

is a characterization of the curve given by (3). The heads of the directed line segments representing $\mathbf{v}(t)$ which have their base at the origin trace out the curve (3). This geometric interpretation was discussed in some detail in the study of vector functions of one variable.

The parametric equations

$$x = f(s, t), \qquad y = g(s, t), \qquad z = h(s, t) \tag{5}$$

represent a surface in three-space. The vector function

$$\mathbf{v}(s, t) = f(s, t)\mathbf{i} + g(s, t)\mathbf{j} + h(s, t)\mathbf{k}$$

characterizes the same surface when we consider the directed line segments with base at the origin which represent $\mathbf{v}$. The heads of these directed line segments form the surface (5).

EXAMPLE 3. Sketch the surface represented by the vector function

$$\mathbf{v}(s, t) = (s \cos t)\mathbf{i} + (s \sin t)\mathbf{j} + (\sqrt{3}s)\mathbf{k}, \qquad s \ge 0, \quad 0 \le t \le 2\pi.$$

SOLUTION. We write

$$x = s \cos t, \qquad y = s \sin t, \qquad z = \sqrt{3}s.$$

Squaring and adding to eliminate the parameters s and t, we find

$$3(x^2 + y^2) = z^2,$$

which we recognize as the equation of a cone. The domain of the vector function and the surface represented by it are shown in Fig. 9-1(a) and (b). Since $s \ge 0$, the given surface is the upper half of the cone only.

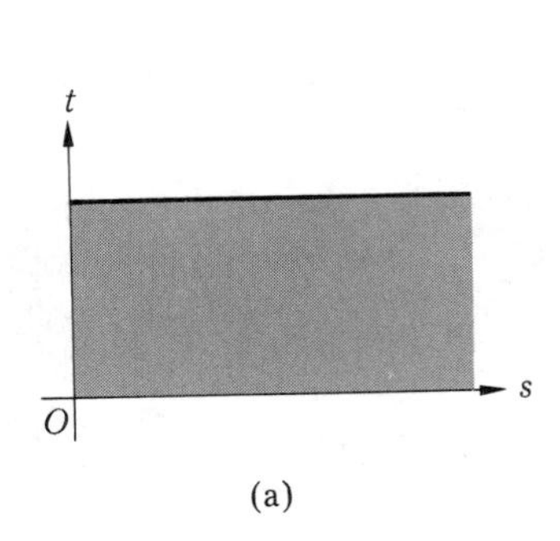

(a)

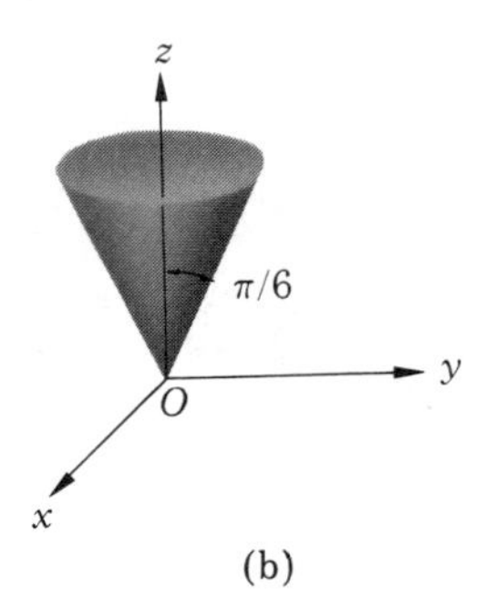

(b)

Fig. 9-1

PROBLEMS

In each of Problems 1 through 9, find the partial derivatives as indicated.

1. $\mathbf{u}(x, y) = (x^2 + 2xy)\mathbf{i} + (x^3 - y^3)\mathbf{j}$; $\quad \dfrac{\partial \mathbf{u}}{\partial x}, \dfrac{\partial \mathbf{u}}{\partial y}$

2. $\mathbf{u}(x, y) = (x \cos y)\mathbf{i} + (y \sin x)\mathbf{j}$; $\quad \dfrac{\partial \mathbf{u}}{\partial x}, \dfrac{\partial \mathbf{u}}{\partial y}$

3. $\mathbf{u}(x, y, z) = \dfrac{x}{x^2 + y^2}\mathbf{i} + \dfrac{y}{x^2 + z^2}\mathbf{j} + \dfrac{z}{x^2 + y^2}\mathbf{k}$; $\quad \dfrac{\partial \mathbf{u}}{\partial x}, \dfrac{\partial \mathbf{u}}{\partial z}$

4. $\mathbf{u}(x, y, z) = (e^{xy} \log z)\mathbf{i} + (e^{yz} \log x)\mathbf{j}$; $\quad \dfrac{\partial \mathbf{u}}{\partial y}, \dfrac{\partial \mathbf{u}}{\partial z}$

5. $\mathbf{v}(s, t) = (s \cos t)\mathbf{i} + (t \sin s)\mathbf{j} + (t^2 - s^2)\mathbf{k}$; $\quad \dfrac{\partial \mathbf{v}}{\partial s}, \dfrac{\partial \mathbf{v}}{\partial t}$

6. $\mathbf{v}(s, t) = (e^s \tan t)\mathbf{i} + (e^{-s} \cos t)\mathbf{j}$; $\quad \dfrac{\partial^2 \mathbf{v}}{\partial s^2}, \dfrac{\partial^2 \mathbf{v}}{\partial t^2}$

7. $\mathbf{v}(r, s, t) = (s^2 - t^2)\mathbf{i} + (t^2 - r^2)\mathbf{j} + (r^2 - s^2)\mathbf{k}$; $\quad \dfrac{\partial^2 \mathbf{v}}{\partial r\, \partial s}, \dfrac{\partial^2 \mathbf{v}}{\partial s\, \partial t}$

8. $\mathbf{u}(x, y) = (x + y)\mathbf{i} + x^2\mathbf{j} + (y - x)\mathbf{k}$; $\quad \dfrac{\partial}{\partial x}(|\mathbf{u}(x, y)|^2)$

9. $\mathbf{u}(x, y, z) = (\sin xy)\mathbf{i} + (\cos yz)\mathbf{j} + (\sin xz)\mathbf{k}$; $\quad \dfrac{\partial}{\partial y}(|\mathbf{u}(x, y, z)|^2)$

10. Find $\partial(\mathbf{u} \cdot \mathbf{v})/\partial x$ if

$$\mathbf{u} = (xy)\mathbf{i} + (x - y)\mathbf{j} + (x + y)\mathbf{k},$$
$$\mathbf{v} = (2x - y)\mathbf{i} - x\mathbf{j} + (x + 2y)\mathbf{k}.$$

11. Find $\partial(\mathbf{u} \cdot \mathbf{v})/\partial y$ if

$$\mathbf{u} = \log(x + y)\mathbf{i} + (x - y)\mathbf{j} + \log(x - y)\mathbf{k},$$
$$\mathbf{v} = e^{x-y}\mathbf{i} + (x + y)\mathbf{j} + \log(x + y)\mathbf{k}.$$

12. Find $\partial(\mathbf{u} \times \mathbf{v})/\partial x$ if

$$\mathbf{u} = (1 + x)\mathbf{i} + (y + 2)\mathbf{j} + (x + y)\mathbf{k},$$

$$\mathbf{v} = 2\mathbf{i} + (3 - x + y)\mathbf{j} + (x - 2y)\mathbf{k}.$$

13. Find $\partial(\mathbf{u} \times \mathbf{v})/\partial t$ if

$$\mathbf{u} = (e^t \cos s)\mathbf{i} + (e^{-t} \sin s)\mathbf{j} + e^t\mathbf{k},$$

$$\mathbf{v} = (e^{-t} \sin 2s)\mathbf{i} + (e^t \cos s)\mathbf{j} + 2\mathbf{k}.$$

14. Given the vector functions

$$\mathbf{u}(x, y) = (x + y)\mathbf{i} + (2x - y)\mathbf{j},$$

$$\mathbf{v}(x, y) = (x^2 - y)\mathbf{i} + (x + 2y^2)\mathbf{j}.$$

a) At what values of x and y are $\partial\mathbf{u}/\partial x$ and $\partial\mathbf{v}/\partial x$ orthogonal?
b) At what values of x and y are $\partial\mathbf{u}/\partial y$ and $\partial\mathbf{v}/\partial y$ orthogonal?
c) At what values of x and y do both (a) and (b) hold?

15. Given the vector functions

$$\mathbf{u}(x, y) = (x - 2y)\mathbf{i} + (x - y)\mathbf{j} + (x + 2y)\mathbf{k},$$

$$\mathbf{v}(x, y) = (2x - y)\mathbf{i} + (x + y)\mathbf{j} + (x + 3y)\mathbf{k}.$$

Find the values of x and y such that

$$\mathbf{u} \cdot (\mathbf{u}_{,1} \times \mathbf{v}_{,2}) = 0.$$

16. Derive a formula for

$$\frac{\partial}{\partial x}[\mathbf{u} \cdot (\mathbf{v} \times \mathbf{w})]$$

in terms of $\partial\mathbf{u}/\partial x$, $\partial\mathbf{v}/\partial x$, and $\partial\mathbf{w}/\partial x$.

17. Describe the surface represented by the vector function

$$\mathbf{v}(s, t) = (3 \cos s \cos t)\mathbf{i} + (3 \cos s \sin t)\mathbf{j} + (3 \sin s)\mathbf{k}$$

for $\{(s, t) : 0 \le s \le (\pi/2), 0 \le t \le 2\pi\}$.

18. Describe the surface represented by the vector function

$$\mathbf{v}(s, t) = (s + t)\mathbf{i} + (s + t)\mathbf{j} + (st)\mathbf{k}$$

for $\{(s, t) : -\infty < s < \infty, -\infty < t < \infty\}$.

19. Describe the surface represented by the vector function

$$\mathbf{v}(s, t) = (2s \cos t)\mathbf{i} + (3s \sin t)\mathbf{j} + s^2\mathbf{k}$$

for $\{(s, t) : 0 \le s < \infty, 0 \le t \le 2\pi\}$.

20. Show that if

$$\mathbf{v}(x, y, z) = \frac{\mathbf{w}(x, y, z)}{F(x, y, z)},$$

then

$$\mathbf{v}_{,1} = \frac{F\mathbf{w}_{,1} - \mathbf{w}F_{,1}}{F^2}.$$

2. Vector and Scalar Fields. Directional Derivative and Gradient

Vector functions occur frequently in applications. The vector velocity of the wind in the atmosphere is an example of a vector function. Other examples of vector functions are the vector velocity of the particles of fluid in a stream and the vector force of gravity exerted by the earth on an object in space.

We may represent a vector function graphically. At each point P of the domain D of a function $\mathbf{v}(P)$, we construct the directed line segment of $\mathbf{v}(P)$ having its base at P. If D is a plane region and if the range of $\mathbf{v}(P)$ is a collection of vectors in the plane, the graphical representation appears as in Fig. 9-2. The vectors in Fig. 9-2 form a field of vectors. More precisely, **vector field** is a synonym for vector function. In analogy, an ordinary function f which assigns a real number to each point of a region in the plane or in space is called a **scalar field** or a **scalar function**.

In Section 1 we defined vector fields in terms of the unit vectors **i**, **j**, and **k**. In other words, we introduced a rectangular coordinate system and used this system as a basis for various definitions. Many of the most important properties of vector fields are geometric in character and so are independent of any coordinate system. Therefore, whenever it is possible we shall define properties of vector functions without reference to a particular coordinate system. For example, in Section 1 we defined the continuity of a vector function $\mathbf{v}(x, y)$ given by

$$\mathbf{v}(x, y) = f(x, y)\mathbf{i} + g(x, y)\mathbf{j}$$

in terms of the continuity of f and g. It is also possible to proceed without such a reference to coordinate vectors **i** and **j**. Suppose that P_0 is a point in the domain D of a vector field $\mathbf{v}(P)$. Then we say that $\mathbf{v}(P)$ *is continuous at point* P_0 if and only if for every $\varepsilon > 0$ there is a $\delta > 0$ such that

$$|\mathbf{v}(P) - \mathbf{v}(P_0)| < \varepsilon$$

whenever $0 < |PP_0| < \delta$. Observe that this definition is valid if the domain of $\mathbf{v}(P)$ is a two- or three-dimensional region. Furthermore, since

$$|\mathbf{v}(P) - \mathbf{v}(P_0)|$$

is the length of the vector $\mathbf{v}(P) - \mathbf{v}(P_0)$, the definition is applicable for vector fields which have either a two- or a three-dimensional range.

It is possible to define the definite integral of a scalar or vector field without

Fig. 9-2

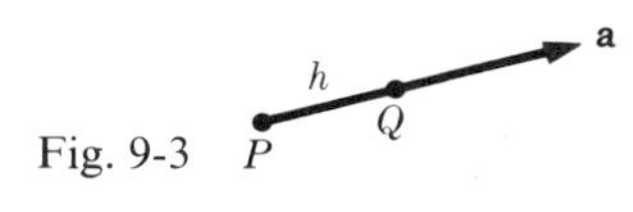

Fig. 9-3

reference to a coordinate system. If R is the region of integration, we subdivide R into a number of smaller regions $R_1, R_2, \ldots, R_n$. Let P_i be any point in R_i. Then we form the sums

$$\sum_{i=1}^{n} f(P_i)A(R_i) \qquad \text{and} \qquad \sum_{i=1}^{n} \mathbf{v}(P_i)A(R_i),$$

where $A(R_i)$ is the area (or volume if R is in three-space) of the region R_i. If the above expressions approach limits under the usual hypotheses imposed for the definition of integrals, then we say the integral exists. The first sum above will approach a numerical limit, while the limit of the second sum will be a vector. Note that no reference to a particular coordinate system is required in these definitions.

Suppose that f is a scalar field, that P is a point in the domain of f, and that $\mathbf{a}$ is a unit vector. We define the **directional derivative of f in the direction of a at the point P by**

$$\lim_{h \to 0} \frac{f(Q) - f(P)}{h},$$

where Q is a point at distance h from P in the direction of $\mathbf{a}$. The relation between P and Q is shown in Fig. 9-3. We denote this directional derivative by

$$D_{\mathbf{a}} f(P).$$

If we hold $\mathbf{a}$ and f fixed and consider P as variable, then $D_{\mathbf{a}} f(P)$ defines a scalar field which is analogous to a partial derivative. In fact, if we introduce a rectangular (x, y, z)-coordinate system and select $\mathbf{a} = \mathbf{i}$, the directional derivative is $\partial/\partial x$. Similarly, choosing $\mathbf{a} = \mathbf{j}$ and $\mathbf{a} = \mathbf{k}$ correspond to partial derivatives with respect to y and z, respectively.

Definition. A scalar field f is **continuously differentiable** on an open set $\mathscr{D}$ if and only if f and $D_{\mathbf{a}} f$ are continuous on $\mathscr{D}$ for each fixed $\mathbf{a}$.

The definition of directional derivative of a vector field is analogous to that of a scalar field.

Definition. The **directional derivative**, $D_{\mathbf{a}}\mathbf{w}(P)$, **of a vector field w in the direction of a at the point** P is given by

$$D_{\mathbf{a}}\mathbf{w}(P) = \lim_{h \to 0} \frac{\mathbf{w}(Q) - \mathbf{w}(P)}{h},$$

where the relation between Q and P is that shown in Fig. 9-3. We say that $\mathbf{w}$ is **continuously differentiable on** $\mathscr{D}$ if and only if $\mathbf{w}$ and $D_{\mathbf{a}}\mathbf{w}$ are continuous on $\mathscr{D}$ for every fixed $\mathbf{a}$.

REMARKS. For fixed $\mathbf{a}$ and $\mathbf{w}$, the directional derivative $D_{\mathbf{a}}\mathbf{w}$ defines a vector field on $\mathscr{D}$. If a rectangular (x, y, z)-coordinate system is introduced, then selecting $\mathbf{a}$ equal to $\mathbf{i}$, $\mathbf{j}$, and $\mathbf{k}$, in turn, gives the partial derivatives of $\mathbf{w}$ with respect to x, y, and z, respectively. The computation of these derivatives was discussed in Section 1.

While it is convenient to define properties of scalar and vector fields without reference to any coordinate axes, it is usually desirable to use some appropriate coordinate system when specific computations are to be made. Suppose that f is a scalar field in space and that a rectangular (x, y, z)-coordinate system τ is introduced. We use the symbol f to denote a scalar field without reference to any coordinate system and the symbol f_τ to denote the same scalar field in the τ-coordinate system. Then with f there is associated the unique function f_τ defined by the equations

$$f(P) = f[\tau(x_P, y_P, z_P)] \equiv f_\tau(x_P, y_P, z_P)$$

where (x_P, y_P, z_P) are the τ-coordinates of P in three-space. In general, two different coordinate systems τ_1 and τ_2 will associate two different functions f_{τ_1} and f_{τ_2} to the same scalar field f.

We shall now derive an expression for the directional derivative $D_{\mathbf{a}} f$ in terms of an (x, y, z)-coordinate system τ. Using the corresponding mutually orthogonal unit vectors $\mathbf{i}$, $\mathbf{j}$, $\mathbf{k}$, the unit vector $\mathbf{a}$ has the representation

$$\mathbf{a} = \lambda\mathbf{i} + \mu\mathbf{j} + \nu\mathbf{k}, \qquad \lambda^2 + \mu^2 + \nu^2 = 1.$$

If the τ-coordinates of P are (x_P, y_P, z_P) and those of Q are $(x_P + \lambda h, y_P + \mu h, z_P + \nu h)$, then the directional derivative is given by

$$D_{\mathbf{a}} f(P) = \lim_{h \to 0} \frac{f_\tau(x_P + \lambda h, y_P + \mu h, z_P + \nu h) - f_\tau(x_P, y_P, z_P)}{h}, \tag{1}$$

where f_τ is the coordinate function in R^3 corresponding to the scalar field f. Recalling the definition of directional derivative as defined on page 218 of Chapter 4, we find

$$D_{\mathbf{a}} f(P) = \lambda f_{\tau x}(x_P, y_P, z_P) + \mu f_{\tau y}(x_P, y_P, z_P) + \nu f_{\tau z}(x_P, y_P, z_P). \tag{2}$$

From this formula, we see that a function f is continuously differentiable if and only if its corresponding function with respect to every coordinate system is continuously differentiable.

A similar discussion holds for vector fields. If $\mathbf{u}$ is a vector field in space, then in each coordinate system there corresponds a triple of functions. More precisely, if (x, y, z) is a rectangular coordinate system and $\mathbf{i}$, $\mathbf{j}$, $\mathbf{k}$ are the customary unit vectors, then the components of $\mathbf{u}$ with respect to this system are defined by

$$\mathbf{u}(P) = f_\tau(x_P, y_P, z_P)\mathbf{i} + g_\tau(x_P, y_P, z_P)\mathbf{j} + h_\tau(x_P, y_P, z_P)\mathbf{k}.$$

The directional derivative $D_\mathbf{a}\mathbf{u}(P)$ in this coordinate system is

$$D_\mathbf{a}\mathbf{u}(P) = (D_\mathbf{a} f_\tau)\mathbf{i} + (D_\mathbf{a} g_\tau)\mathbf{j} + (D_\mathbf{a} h_\tau)\mathbf{k}.$$

The directional derivatives of the coordinate functions on the right may be calculated according to formula (2).

While it is important to distinguish between a scalar field f and its corresponding function f_τ given in a particular coordinate system, we shall usually, drop the τ notation. Since almost all computational work is done in some coordinate system, there is little danger of confusion.

EXAMPLE 1. Given the vector field

$$\mathbf{u}(P) = (x^2 - y + z)\mathbf{i} + (2y - 3z)\mathbf{j} + (x + z)\mathbf{k}$$

and the unit vector

$$\mathbf{a} = \frac{1}{\sqrt{6}}(2\mathbf{i} - \mathbf{j} + \mathbf{k}).$$

Find $D_\mathbf{a}\mathbf{u}(P)$.

SOLUTION. We have

$$D_\mathbf{a}(x^2 - y + z) = \frac{2}{\sqrt{6}}(2x) - \frac{1}{\sqrt{6}}(-1) + \frac{1}{\sqrt{6}}(1) = \frac{1}{\sqrt{6}}(4x + 2),$$

$$D_\mathbf{a}(2y - 3z) = \frac{2}{\sqrt{6}}(0) - \frac{1}{\sqrt{6}}(2) + \frac{1}{\sqrt{6}}(-3) = -\frac{5}{\sqrt{6}},$$

$$D_\mathbf{a}(x + z) = \frac{2}{\sqrt{6}}(1) - \frac{1}{\sqrt{6}}(0) + \frac{1}{\sqrt{6}}(1) = \frac{3}{\sqrt{6}}.$$

Therefore

$$D_\mathbf{a}\mathbf{u}(P) = \frac{1}{\sqrt{6}}(4x + 2)\mathbf{i} - \frac{5}{\sqrt{6}}\mathbf{j} + \frac{3}{\sqrt{6}}\mathbf{k}.$$

Let $f(P)$ be a scalar field, let $\tau : (x, y, z)$ be a rectangular coordinate system, and let $\mathbf{a}$ be a unit vector. We define the vector field

$$\mathbf{v}(P) = f_{\tau x}(x_P, y_P, z_P)\mathbf{i} + f_{\tau y}(x_P, y_P, z_P)\mathbf{j} + f_{\tau z}(x_P, y_P, z_P)\mathbf{k}, \tag{3}$$

where f_τ, $\mathbf{i}$, $\mathbf{j}$, and $\mathbf{k}$ have the significance described above. Then according to the definition of scalar product, formula (2) may be written

$$D_\mathbf{a} f(P) = \mathbf{v}(P) \cdot \mathbf{a}. \tag{4}$$

We now show that the vector $\mathbf{v}(P)$, as given in (4), is unique.

Theorem 1. *Suppose that*

$$\mathbf{v} \cdot \mathbf{a} = \mathbf{w} \cdot \mathbf{a}$$

for every unit vector **a** (*in the plane or in space*). *Then* $\mathbf{v} = \mathbf{w}$.

PROOF. The equation $\mathbf{v} \cdot \mathbf{a} = \mathbf{w} \cdot \mathbf{a}$ is equivalent to

$$(\mathbf{v} - \mathbf{w}) \cdot \mathbf{a} = 0. \tag{5}$$

If $\mathbf{v} - \mathbf{w} \neq \mathbf{0}$, we select **a** to be a unit vector in the direction of $\mathbf{v} - \mathbf{w}$. Then (5) would not hold.

Definition. The unique vector $\mathbf{v}(P)$ determined by Eq. (4) is called the **gradient of** f **at** P. We denote this vector by ∇f, which we read "del$f$." Then equation (4) becomes

$$D_{\mathbf{a}} f(P) = \nabla f \cdot \mathbf{a} \tag{6}$$

for any unit vector **a**. Since $D_{\mathbf{a}} f(P)$ and **a** are defined without reference to a coordinate system, the vector ∇f (which is uniquely determined) has a significance independent of coordinates.

To calculate ∇f when a rectangular (x, y, z)-system is introduced, we simply use the formula

$$\nabla f = f_{\tau x}\mathbf{i} + f_{\tau y}\mathbf{j} + f_{\tau z}\mathbf{k}.$$

The geometric significance of ∇f is easily seen with reference to the surface $f(x, y, z) = \text{const}$. If P is a point on the surface, the vector $\nabla f(P)$, whenever it is not zero, is perpendicular to the surface. To see this, observe that if **a** is any vector tangent to the surface f, then $D_{\mathbf{a}} f = 0$ since f is constant. Then (6) states that ∇f is perpendicular to **a**, i.e., perpendicular to the surface. To find a unit vector normal to a surface $f(P) = \text{const}$, we select

$$\mathbf{n} = \frac{\nabla f}{|\nabla f|}$$

evaluated at the desired point.

EXAMPLE 2. Given the function $f(x, y, z) = 3x^2 - y^2 + 2z^2$, find ∇f and a unit normal to the surface $f(x, y, z) = 17$ at the point $(1, -2, 3)$.

SOLUTION. We have

$$\nabla f = f_{\tau x}\mathbf{i} + f_{\tau y}\mathbf{j} + f_{\tau z}\mathbf{k} = 6x\mathbf{i} - 2y\mathbf{j} + 4z\mathbf{k},$$

$$\nabla f(1, -2, 3) = 6\mathbf{i} + 4\mathbf{j} + 12\mathbf{k} = 2(3\mathbf{i} + 2\mathbf{j} + 6\mathbf{k}),$$

$$\mathbf{n} = \pm\tfrac{1}{7}(3\mathbf{i} + 2\mathbf{j} + 6\mathbf{k}).$$

The proof of the next theorem is left to the reader.

Theorem 2. *Suppose that f, g, and u are continuously differentiable scalar fields. Then*

$$\left.\begin{aligned} \nabla(f+g) &= \nabla f + \nabla g, \qquad \nabla(fg) = f\nabla g + g\nabla f \\ \nabla\left(\frac{f}{g}\right) &= \frac{1}{g^2}(g\nabla f - f\nabla g) \qquad \textit{if} \quad g \neq 0 \end{aligned}\right\}, \tag{7}$$

$$\nabla f(u) = f'(u)\nabla u. \tag{8}$$

Problems

In each of Problems 1 through 6, find ∇f and $D_{\mathbf{a}} f$ at $P_0(x_0, y_0, z_0)$ as given.

1. $f(x, y, z) = x^2 - y^2 + 2yz + 2z^2$, $P_0(2, -1, 1)$, $\mathbf{a} = \dfrac{1}{\sqrt{14}}(2\mathbf{i} - \mathbf{j} + 3\mathbf{k})$
2. $f = e^x \cos y + xz^2$, $P_0(0, \pi/2, 1)$, $\mathbf{a} = \frac{1}{7}(3\mathbf{i} + 2\mathbf{j} - 6\mathbf{k})$
3. $f = \dfrac{x}{(x^2 + y^2 + z^2)^{3/2}}$, $P_0(2, 2, 1)$, $\mathbf{a} = \frac{1}{3}(\mathbf{i} - 2\mathbf{j} + 2\mathbf{k})$
4. $f = \log(x^2 + y^2)$, $P_0(3, 4, 2)$, $\mathbf{a} = \frac{1}{7}(6\mathbf{i} - 3\mathbf{j} - 2\mathbf{k})$
5. $f = \tan x + \log(y + z)$, $P_0(\pi/4, 1, 1)$, $\mathbf{a} = \frac{1}{3}(2\mathbf{i} + 2\mathbf{j} - \mathbf{k})$
6. $f = e^{xy}\cos z + e^{yz}\sin x$, $P_0(\pi/6, 0, \pi/3)$, $\mathbf{a} = \dfrac{1}{\sqrt{17}}(3\mathbf{i} + 2\mathbf{j} - 2\mathbf{k})$

In Problems 7 through 9, find $D_{\mathbf{a}}\mathbf{u}$ at $P_0(x_0, y_0, z_0)$ as given.

7. $\mathbf{u} = (x^2 + y^2)\mathbf{i} + 2xy\mathbf{j} + 3xz\mathbf{k}$, $P_0(1, 0, 2)$, $\mathbf{a} = \dfrac{1}{\sqrt{21}}(\mathbf{i} + 4\mathbf{j} + 2\mathbf{k})$
8. $\mathbf{u} = \dfrac{x}{x^2 + y^2}\mathbf{i} + \dfrac{y}{y^2 + z^2}\mathbf{j} + \dfrac{z}{x^2 + z^2}\mathbf{k}$, $P_0(1, 1, 2)$, $\mathbf{a} = \dfrac{1}{\sqrt{6}}(2\mathbf{i} + \mathbf{j} - \mathbf{k})$
9. $\mathbf{u} = \cos(xy)\mathbf{i} + \sin(z^2)\mathbf{j} + 2\mathbf{k}$, $P_0(0, \pi/4, 0)$, $\mathbf{a} = \dfrac{1}{\sqrt{29}}(2\mathbf{i} + 5\mathbf{k})$
10. Find the unit vector $\mathbf{a}$ such that $D_{\mathbf{a}} f(P_0)$ is a maximum, given that
$$f = 2x^2 - 3y^2 + z^2, \qquad P_0 = (2, 1, 3).$$
11. Find the unit vector $\mathbf{a}$ such that $D_{\mathbf{a}} f(P_0)$ is a minimum, given that
$$f = xyz, \qquad P_0 = (1, -3, -2).$$
12. Prove the laws stated in Eqs. (7). [*Hint.* Introduce a coordinate system.]
13. Prove the law (8).

14. Suppose that O is the origin of coordinates, P is a point, and $\mathbf{r}$ is the vector having the directed line segment $\overrightarrow{OP}$ as a representative. We write $\mathbf{r} = \mathbf{v}(\overrightarrow{OP})$. Show that

$$\nabla f(|\mathbf{r}|) = \frac{1}{|\mathbf{r}|} f'(|\mathbf{r}|)\mathbf{r}.$$

In each of Problems 15 through 20, find the unit normal to the surface at the given point P_0.

15. $x^2 - 2xy + 2y^2 - z^2 = 9, \quad P_0 = (2, -1, 1)$
16. $z = x^2 - y^2, \quad P_0 = (3, -2, 5)$
17. $z^2 = x^2 + y^2, \quad P_0 = (-3, 4, 5)$
18. $z = e^x \sin y, \quad P_0 = (1, \pi/2, e)$
19. $z = \arctan x + \log(1 + y), \quad P_0(1, 0, \pi/4)$
20. $x^3 + y^3 + z^3 - 3xyz = 14, \quad P_0(2, 1, -1)$

3. The Divergence of a Vector Field

Suppose τ is a rectangular coordinate system in space and $\mathbf{i}$, $\mathbf{j}$, and $\mathbf{k}$ are the corresponding orthogonal unit vectors. For any scalar function f, we may regard the gradient, ∇, as an *operator* (i.e., a transformation) which takes scalar fields into vector fields. In the coordinate system above, we write ∇ in the **symbolic form**

$$\nabla = \frac{\partial}{\partial x}\mathbf{i} + \frac{\partial}{\partial y}\mathbf{j} + \frac{\partial}{\partial z}\mathbf{k}. \tag{1}$$

If we make the convention that ∇f has the meaning

$$\nabla f = \frac{\partial f}{\partial x}\mathbf{i} + \frac{\partial f}{\partial y}\mathbf{j} + \frac{\partial f}{\partial z}\mathbf{k},$$

we see that ∇f is the gradient of f.

Suppose, now, that a vector field $\mathbf{w}$ in space has the representation

$$\mathbf{w}(P) = u(x_P, y_P, z_P)\mathbf{i} + v(x_P, y_P, z_P)\mathbf{j} + w(x_P, y_P, z_P)\mathbf{k}, \tag{2}$$

in the coordinate system above. We form the inner product of the symbolic vector ∇ and $\mathbf{w}$, denoted by $\nabla \cdot \mathbf{w}$, to get

$$\nabla \cdot \mathbf{w} = \frac{\partial u}{\partial x} + \frac{\partial v}{\partial y} + \frac{\partial w}{\partial z}. \tag{3}$$

The fact that the operator ∇ has a significance independent of the coordinates suggests that the expression on the right in (3) might have such a significance. We show this to be true in Theorem 3 below.

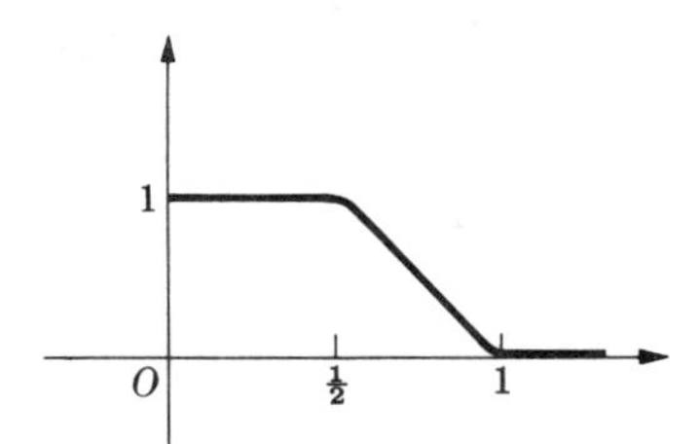

Fig. 9-4

Lemma. *Suppose that A is a continuous function defined in a region G in R^3, and that*

$$\iiint_G U(x,y,z)A(x,y,z)\,dA = 0$$

for all continuously differentiable functions U defined on G which vanish outside some ball contained in G. (The location and size of the ball may depend on U.) Then $A(x,y,z) \equiv 0$ in G.*

PROOF. Let $P_0(x_0, y_0, z_0)$ be a point of G and suppose $A(x_0, y_0, z_0) \neq 0$. We shall reach a contradiction. We may assume that $A(x_0, y_0, z_0) > 0$; otherwise we would replace A by $-A$. Since A is continuous in G, there is a closed ball with center at P_0 and radius $\delta > 0$ which is contained in G and such that $A(x,y,z) > 0$ in this ball. We denote this ball by $B(P_0, \delta)$. We now define the function ϕ by the formula

$$\phi(s) = \begin{cases} 1 & \text{for } 0 \le s \le \frac{1}{2}, \\ 4(4s^3 - 9s^2 + 6s - 1) & \text{for } \frac{1}{2} \le s \le 1, \\ 0 & \text{for } s \ge 1. \end{cases}$$

We observe that $\phi(\frac{1}{2}) = 1$, $\phi(1) = 0$, and $\phi'(\frac{1}{2}) = \phi'(1) = 0$, and so ϕ is continuously differentiable for all values of s. See Fig. 9-4. We choose $U(P)$ by setting $U(P) = \phi(|PP_0|/\delta)$. Then U is continuously differentiable in G and

$$UA > 0 \quad \text{in} \quad B(P_0, \delta).$$

Also, we have $UA \equiv 0$ outside $B(P_0, \delta)$. We conclude that

$$\iiint_G UA\,dV = \iiint_{B(P_0,\delta)} UA\,dV > 0,$$

which contradicts the hypotheses.

Theorem 3. *Suppose that* **w** *is a continuously differentiable vector field defined in a region G in three-space.*

* We recall that a function is **continuously differentiable** if all first partial derivatives are continuous.

a) *Then there exists a unique continuous scalar field A defined in G such that*

$$\iiint_G UA\,dV = -\iiint_G \nabla U\cdot\mathbf{w}\,dV \tag{4}$$

for every continuously differentiable scalar field U which vanishes outside some ball contained in G.

b) *Consider a rectangular coordinate system with* **i**, **j**, *and* **k** *corresponding orthogonal unit vectors. If the vector field* **w** *in part* (a) *is given by*

$$\mathbf{w}(x, y, z) = u(x, y, z)\mathbf{i} + v(x, y, z)\mathbf{j} + w(x, y, z)\mathbf{k}, \tag{5}$$

then the scalar field A in (4) *is given by*

$$A(x, y, z) = \frac{\partial u}{\partial x} + \frac{\partial v}{\partial y} + \frac{\partial w}{\partial z}. \tag{6}$$

PROOF. Suppose **w** is given by (5) in a rectangular coordinate system. We define A by the right side of (6). We must show that (4) holds and that A is unique.

Let $B(P_0, \delta)$ be a ball with center at P_0 and radius δ which together with its boundary is entirely in G. Suppose that U is any continuously differentiable scalar field on G which vanishes outside $B(P_0, \delta)$. Then (4) is equivalent to

$$\iiint_G (UA + \nabla U\cdot\mathbf{w})\,dV = 0 \Leftrightarrow \iiint_{B(P_0,\delta)} (UA + \nabla U\cdot\mathbf{w})\,dV = 0. \tag{7}$$

In terms of the rectangular coordinate system, we have

$$\begin{aligned} UA + \nabla U\cdot\mathbf{w} &= U\left(\frac{\partial u}{\partial x} + \frac{\partial v}{\partial y} + \frac{\partial w}{\partial z}\right) + \frac{\partial U}{\partial x}u + \frac{\partial U}{\partial y}v + \frac{\partial U}{\partial z}w \\ &= \frac{\partial}{\partial x}(Uu) + \frac{\partial}{\partial y}(Uv) + \frac{\partial}{\partial z}(Uw). \end{aligned} \tag{8}$$

We now integrate the first term on the right in (8). Denoting the coordinates of P_0 by (x_0, y_0, z_0), we find

$$\iiint_{B(P_0,\delta)} \frac{\partial}{\partial x}(Uu)\,dV_{xyz} = \iint_{C(y_0,z_0;\delta)} \left\{\int_{x_0-\sqrt{\delta^2-(y-y_0)^2-(z-z_0)^2}}^{x_0+\sqrt{\delta^2-(y-y_0)^2-(z-z_0)^2}} \frac{\partial}{\partial x}(Uu)\,dx\right\} dA_{yz} \tag{9}$$

where $C(y_0, z_0; \delta)$ is the circular disk in the yz-plane which has center at (y_0, z_0) and radius δ. See Fig. 9-5. Since $U \equiv 0$ on the boundary of $B(P_0, \delta)$, that is, when $x = x_0 \pm \sqrt{\delta^2 - (y - y_0)^2 - (z - z_0)^2}$, when we perform the x-integration in (9) and substitute the limits, the result is zero. In a similar way, when we integrate the second and third terms on the right in (8), we get zero. We conclude that (4) holds for every scalar field U with the required properties. To show that A is unique, we assume there is another continuous

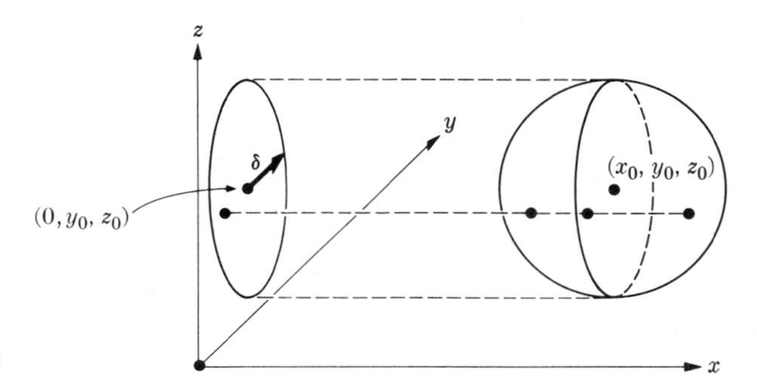

Fig. 9-5

scalar field A' which satisfies (4). Then we obtain

$$\iiint_G U(A - A')\,dV = 0$$

for every U which satisfies the conditions of the Lemma. Hence $A \equiv A'$ in G and the proof is complete.

Definition. If $\mathbf{w}$ is a continuously differentiable vector field in a region in three-space, we define the **divergence** of $\mathbf{w}$ as the scalar field A given in Theorem 3. We denote the divergence by

$$\operatorname{div} \mathbf{w} \qquad \text{or} \qquad \nabla \cdot \mathbf{w}.$$

It is clear from Theorem 3 that $\operatorname{div} \mathbf{w}$ is independent of the coordinate system. The formula for $\operatorname{div} \mathbf{w}$ in a rectangular coordinate system in which

$$\mathbf{w} = u(x, y, z)\mathbf{i} + v(x, y, z)\mathbf{j} + w(x, y, z)\mathbf{k}$$

is given by

$$\operatorname{div} \mathbf{w} = \frac{\partial u}{\partial x} + \frac{\partial v}{\partial y} + \frac{\partial w}{\partial z}.$$

The proof of the following simple properties of the divergence operator is left to the reader.

Theorem 4. *If* $\mathbf{u}$ *and* $\mathbf{v}$ *are vector fields and* f *is a scalar field, all continuously differentiable, then*

a) $\operatorname{div}(\mathbf{u} + \mathbf{v}) = \operatorname{div} \mathbf{u} + \operatorname{div} \mathbf{v}$,
b) $\operatorname{div}(f\mathbf{u}) = f \operatorname{div} \mathbf{u} + \nabla f \cdot \mathbf{u}$.

REMARKS. (i) The results of this section can be extended to vectors and scalars in n dimensional Euclidean space. (ii) The divergence operator, a differential operator which takes vector fields into scalar fields, has an important geometric interpretation. We exhibit this fact in the Divergence

Theorem which is discussed in Section 8 of Chapter 10. The divergence operator is also useful for many problems in mechanics, fluid flow, and electromagnetism. We give an application to mechanics in Example 2 below, and we develop the connection of the divergence operator with fluid flow problems in Section 8 of Chapter 10.

EXAMPLE 1. Given the vector

$$\mathbf{w} = \frac{x+z}{x^2+y^2+z^2}\mathbf{i} + \frac{y-x}{x^2+y^2+z^2}\mathbf{j} + \frac{z-y}{x^2+y^2+z^2}\mathbf{k},$$

find div $\mathbf{w}$. Evaluate div $\mathbf{w}$ at $P_0(1, 0, -2)$.

SOLUTION. We have

$$\frac{\partial}{\partial x}\left(\frac{x+z}{x^2+y^2+z^2}\right) = \frac{(x^2+y^2+z^2)-(x+z)(2x)}{(x^2+y^2+z^2)^2} = \frac{y^2+z^2-x^2-2xz}{(x^2+y^2+z^2)^2},$$

$$\frac{\partial}{\partial y}\left(\frac{y-x}{x^2+y^2+z^2}\right) = \frac{x^2+z^2-y^2+2xy}{(x^2+y^2+z^2)^2},$$

$$\frac{\partial}{\partial z}\left(\frac{z-y}{x^2+y^2+z^2}\right) = \frac{x^2+y^2-z^2+2yz}{(x^2+y^2+z^2)^2}.$$

Therefore

$$\operatorname{div}\mathbf{w} = \frac{x^2+y^2+z^2+2(xy-xz+yz)}{(x^2+y^2+z^2)^2}.$$

At P_0 we obtain div $\mathbf{w} = 2/5$.

EXAMPLE 2. Suppose R is the radius of the earth, O is its center, and g is the acceleration due to gravity at the surface of the earth. If P is a point in space near the surface, we denote by $\mathbf{r}$ the directed line segment $\overrightarrow{OP}$. The length $|\mathbf{r}|$ of the vector $\mathbf{r}$ we denote simply by r. From classical physics it is known that the vector field $\mathbf{v}(P)$ due to gravity (called the *gravitational field of the earth*) is given (approximately) by the equation

$$\mathbf{v}(P) = \frac{-gR^2}{r^3}\mathbf{r}.$$

Show that for $r > R$, we have div $\mathbf{v}(P) = 0$.

SOLUTION. A computation yields

$$\operatorname{div}\mathbf{v} = \operatorname{div}\left(\frac{-gR^2}{r^3}\mathbf{r}\right),$$

and using formula (b) of Theorem 4, we find

$$\begin{aligned}\operatorname{div} \mathbf{v} &= \frac{-gR^2}{r^3} \operatorname{div} \mathbf{r} + \nabla\left(\frac{-gR^2}{r^3}\right) \cdot \mathbf{r} \\ &= -gR^2\left[\frac{1}{r^3} \operatorname{div} \mathbf{r} + \frac{d}{dr}\left(\frac{1}{r^3}\right) \nabla r \cdot \mathbf{r}\right] \\ &= -gR^2(r^{-3} \operatorname{div} \mathbf{r} - 3r^{-4} \nabla r \cdot \mathbf{r}).\end{aligned}$$

Introducing a rectangular (x, y, z)-system with origin at O, we write

$$\mathbf{r} = x\mathbf{i} + y\mathbf{j} + z\mathbf{k},$$

from which we obtain $\operatorname{div} \mathbf{r} = 3$. Also, the length r of $\mathbf{r}$ is given by

$$r = \sqrt{x^2 + y^2 + z^2},$$

and so

$$\frac{\partial r}{\partial x} = \frac{x}{r}, \qquad \frac{\partial r}{\partial y} = \frac{y}{r}, \qquad \frac{\partial r}{\partial z} = \frac{z}{r}.$$

Hence

$$\nabla r = \frac{1}{r}(x\mathbf{i} + y\mathbf{j} + z\mathbf{k}) = \frac{\mathbf{r}}{r}$$

and

$$\nabla r \cdot \mathbf{r} = \frac{\mathbf{r} \cdot \mathbf{r}}{r} = r.$$

We conclude that

$$\operatorname{div} \mathbf{v} = -gR^2(3r^{-3} - 3r^{-4} \cdot r) = 0.$$

The force field in Example 2 is illustrative of a large class of vector fields which occur in mechanics, electrostatics, fluid dynamics, and so forth. A vector field $\mathbf{v}(P)$ is called **conservative** if there exists a scalar function $u(P)$ of which $\mathbf{v}$ is the gradient. That is, if there is a function u such that

$$\mathbf{v}(P) = \nabla u,$$

we say that u is the **potential function** which corresponds to the conservative force field $\mathbf{v}$. In fact, the potential function u which yields the force field in Example 2 above is

$$u(P) = \frac{gR^2}{r}.$$

A simple computation shows that

$$\nabla u = gR^2 \nabla\left(\frac{1}{r}\right) = -\frac{gR^2}{r^2} \nabla r = -\frac{gR^2}{r^3} \mathbf{r} = \mathbf{v}.$$

Since the result of Example 2 shows that $\nabla \cdot \mathbf{v} = 0$, we conclude that

$$\nabla \cdot \nabla u = \operatorname{div} \nabla u = 0. \tag{10}$$

Equation (10) is known as **Laplace's equation** and is often written

$$\nabla^2 u = 0 \qquad \text{or} \qquad \Delta u = 0.$$

The latter notation is used by mathematicians and the former by scientists in related fields. It is clear that if $\mathbf{v} = \nabla u$ with $\operatorname{div} \mathbf{v} = 0$, then u satisfies Laplace's equation, and conversely.

In any rectangular coordinate system we may write

$$\nabla u = \frac{\partial u}{\partial x}\mathbf{i} + \frac{\partial u}{\partial y}\mathbf{j} + \frac{\partial u}{\partial z}\mathbf{k},$$

and, therefore, the **Laplacian** has the form

$$\Delta u \equiv \nabla^2 u = \nabla \cdot \nabla u = \frac{\partial^2 u}{\partial x^2} + \frac{\partial^2 u}{\partial y^2} + \frac{\partial^2 u}{\partial z^2}. \tag{11}$$

The Laplacian is a differential operator from scalar fields to scalar fields. The next corollary is an immediate consequence of Theorem 3.

Corollary. *If u and ∇u are continuously differentiable and if (x, y, z) and (x', y', z') are two rectangular coordinate systems, then*

$$\frac{\partial^2 u}{\partial x^2} + \frac{\partial^2 u}{\partial y^2} + \frac{\partial^2 u}{\partial z^2} = \frac{\partial^2 u'}{\partial x'^2} + \frac{\partial^2 u'}{\partial y'^2} + \frac{\partial^2 u'}{\partial z'^2}.$$

That is, the form of the Laplace operator (11) *is the same in every rectangular coordinate system.*

Example 3. Let $\mathbf{r}(P)$ be the vector from the origin O to a point P in the xy plane. Define $r = |\mathbf{r}|$. Show that the plane scalar field

$$u(P) = \log r$$

satisfies the Laplace equation for $r > 0$.

Solution. We have

$$\nabla u = \nabla(\log r) = \frac{1}{r}\nabla(r) = \frac{1}{r^2}\mathbf{r}.$$

Therefore

$$\nabla \cdot \nabla u = \operatorname{div} \nabla u = \operatorname{div}\left(\frac{1}{r^2}\mathbf{r}\right) = \frac{1}{r^2}\operatorname{div}\mathbf{r} - 2r^{-3}\nabla r \cdot \mathbf{r}.$$

Since $\mathbf{r}$ is a vector in the plane, we have

$$\mathbf{r} = x\mathbf{i} + y\mathbf{j}, \qquad r = \sqrt{x^2 + y^2},$$

and so

$$\operatorname{div} \mathbf{r} = 2 \qquad \text{and} \qquad \nabla r = \frac{x}{r}\mathbf{i} + \frac{y}{r}\mathbf{j} = \frac{\mathbf{r}}{r}.$$

Therefore

$$\nabla \cdot \nabla u = \frac{1}{r^2}(2) - 2r^{-3}\frac{\mathbf{r}\cdot\mathbf{r}}{r} = \frac{2}{r^2} - 2r^{-3}(r) = 0.$$

PROBLEMS

In each of Problems 1 through 10, find div $\mathbf{v}$ in the given coordinate system.

1. $\mathbf{v} = (a_{11}x + a_{12}y + a_{13}z)\mathbf{i} + (a_{21}x + a_{22}y + a_{23}z)\mathbf{j} + (a_{31}x + a_{32}y + a_{33}z)\mathbf{k}$
2. $\mathbf{v} = (x^2 - y^2)\mathbf{i} + (x^2 - z^2)\mathbf{j} + (y^2 - z^2)\mathbf{k}$
3. $\mathbf{v} = (x^2 + 1)\mathbf{i} + (y^2 - 1)\mathbf{j} + z^2\mathbf{k}$
4. $\mathbf{v} = 4xz\mathbf{i} - 2yz\mathbf{j} + (2x^2 - y^2 - z^2)\mathbf{k}$
5. $\mathbf{v} = e^{xz}(\cos yz\,\mathbf{i} + \sin yz\,\mathbf{j} - \mathbf{k})$
6. $\mathbf{v} = y\log(1 + x)\mathbf{i} + z\log(1 + y)\mathbf{j} + x\log(1 + z)\mathbf{k}$
7. $\mathbf{v} = \nabla u, \quad u = x^3 - 3xy^2$
8. $\mathbf{v} = \nabla u, \quad u = a_{11}x^2 + a_{22}y^2 + a_{33}z^2 + 2a_{12}xy + 2a_{13}xz + 2a_{23}yz$
9. $\mathbf{v} = \nabla u, \quad u = e^x\cos y + e^y\cos z + e^z\cos x$
10. $\mathbf{v} = 2x\mathbf{i} + 3y\mathbf{j} + (\Delta u)\mathbf{k}, \quad \text{where } u = x^3 + y^3 + z^3$

In each of Problems 11 through 14, find div $\mathbf{v}$, assuming that $\mathbf{r} = x\mathbf{i} + y\mathbf{j} + z\mathbf{k}$ and that $r = |\mathbf{r}|$.

11. $\mathbf{v} = \mathbf{r}/r^n, \quad n > 0$
12. $\mathbf{v} = f(r)\mathbf{r}$
13. $\mathbf{v} = \nabla u, \quad u = \phi(r)$
14. $\mathbf{v} = \nabla u, \quad u = \phi(x + y + z)$
15. Prove Theorem 4.
16. Find $\operatorname{div}(f\nabla g - g\nabla f)$
17. Show that $\operatorname{div}(\mathbf{a} \times \mathbf{r}) = 0$ for any constant vector $\mathbf{a}$.
18. Show that there is a vector $\mathbf{w}$ such that

$$\operatorname{div}(\mathbf{v} \times \mathbf{a}) = \mathbf{w} \cdot \mathbf{a}$$

for any constant vector $\mathbf{a}$.

19. Suppose that a rigid body is rotating about the z axis with an angular velocity ω. Show that the velocity of a particle of the body which is at position $P(x, y, z)$ at

time t is given by

$$\mathbf{v} = \omega\mathbf{k} \times \mathbf{r}.$$

20. a) Compute the quantity Δu in cylindrical coordinates given by

$$x = r\cos\theta, \qquad y = r\sin\theta, \qquad z = z.$$

[*Hint*. Start with formula (11) and use the Chain Rule.]
b) Compute Δu in spherical coordinates:

$$x = \rho\cos\theta\sin\phi, \qquad y = \rho\sin\theta\sin\phi, \qquad z = \rho\cos\phi.$$

4. The Curl of a Vector Field

The cross or vector product of two vectors, as defined in Chapter 2, page 67, changes in sign when we shift from a right-handed system of coordinates to a left-handed one. Three-dimensional space is said to have an **orientation** when one of these two systems of coordinates is introduced. We shall use the term **positive orientation** for a right-handed coordinate system and **negative orientation** for a left-handed one.*

If $\mathbf{i}$, $\mathbf{j}$, $\mathbf{k}$ and $\mathbf{i}'$, $\mathbf{j}'$, $\mathbf{k}'$ are two sets of mutually perpendicular unit vectors, it can be shown that there is a relation between them of the form

$$\begin{aligned}\mathbf{i} &= c_{11}\mathbf{i}' + c_{12}\mathbf{j}' + c_{13}\mathbf{k}',\\ \mathbf{j} &= c_{21}\mathbf{i}' + c_{22}\mathbf{j}' + c_{23}\mathbf{k}',\\ \mathbf{k} &= c_{31}\mathbf{i}' + c_{32}\mathbf{j}' + c_{33}\mathbf{k}',\end{aligned}$$

where $C = (c_{ij})$ is a matrix with $\det C = \pm 1$. If both bases are positively oriented or if both are negatively oriented, then $\det C = +1$. If one system is positively oriented and the other negatively oriented, then $\det C = -1$. We shall always consider positively oriented systems in three-space and, when discussing transformations from one system to another, we shall assume that the new system has the same orientation as the old.

Suppose that $\mathbf{v}$ is a continuously differentiable vector field in space. In the last section we saw that the differential operator $\operatorname{div}\mathbf{v}$, which we also write $\nabla \cdot \mathbf{v}$, is independent of the axes chosen and so defines a scalar field. It might be expected—and indeed we shall show it to be the case—that the formal expression $\nabla \times \mathbf{v}$ defines a vector field.

* It can be shown that it is impossible for a right-handed triple of basis vectors $\mathbf{i}$, $\mathbf{j}$, $\mathbf{k}$ to be deformed continuously into a left-handed one without the vectors becoming linearly dependent at some time during the process. Furthermore, it can be shown that any triple of linearly independent vectors can be deformed continuously (always remaining linearly independent in the process) into either a right-handed set or a left-handed set and, in the light of the preceding sentence, not both.

Let **i**, **j**, **k** be an orthogonal set of unit vectors (positively oriented, of course), and let **u** and **v** be two vector fields. Then, in terms of the basis vectors, we may write

$$\mathbf{u} = u_1\mathbf{i} + u_2\mathbf{j} + u_3\mathbf{k},$$
$$\mathbf{v} = v_1\mathbf{i} + v_2\mathbf{j} + v_3\mathbf{k}.$$

We recall (Chapter 2, page 68) that the cross product $\mathbf{u} \times \mathbf{v}$ is given by

$$\mathbf{u} \times \mathbf{v} = (u_2v_3 - u_3v_2)\mathbf{i} + (u_3v_1 - u_1v_3)\mathbf{j} + (u_1v_2 - u_2v_1)\mathbf{k}. \tag{1}$$

This formula may be written more compactly in the symbolic form

$$\mathbf{u} \times \mathbf{v} = \begin{vmatrix} \mathbf{i} & \mathbf{j} & \mathbf{k} \\ u_1 & u_2 & u_3 \\ v_1 & v_2 & v_3 \end{vmatrix}, \tag{2}$$

where it is understood that the determinant is expanded in minors according to the first row. Note that (1) and (2) are identical. With the vector operator ∇ given by

$$\nabla = \frac{\partial}{\partial x}\mathbf{i} + \frac{\partial}{\partial y}\mathbf{j} + \frac{\partial}{\partial z}\mathbf{k},$$

we **define** the vector $\nabla \times \mathbf{v}$ by the **symbolic** determinant

$$\begin{aligned} \nabla \times \mathbf{v} &= \begin{vmatrix} \mathbf{i} & \mathbf{j} & \mathbf{k} \\ \dfrac{\partial}{\partial x} & \dfrac{\partial}{\partial y} & \dfrac{\partial}{\partial z} \\ v_1 & v_2 & v_3 \end{vmatrix} \\ &= \left(\frac{\partial v_3}{\partial y} - \frac{\partial v_2}{\partial z}\right)\mathbf{i} + \left(\frac{\partial v_1}{\partial z} - \frac{\partial v_3}{\partial x}\right)\mathbf{j} + \left(\frac{\partial v_2}{\partial x} - \frac{\partial v_1}{\partial y}\right)\mathbf{k}. \end{aligned} \tag{3}$$

Since $\nabla \times \mathbf{v}$ is defined in terms of a particular coordinate system, it is essential to show that $\nabla \times \mathbf{v}$ is a true vector field (if orientation is preserved).

For any three vectors **b**, **c**, **d**, it is not hard to establish the identity

$$(\mathbf{b} \times \mathbf{c}) \cdot \mathbf{d} = \mathbf{b} \cdot (\mathbf{c} \times \mathbf{d}). \tag{4}$$

Using the usual definitions, $\mathbf{b} = b_1\mathbf{i} + b_2\mathbf{j} + b_3\mathbf{k}$, etc., the reader may verify that

$$(\mathbf{b} \times \mathbf{c}) \cdot \mathbf{d} = \begin{vmatrix} b_1 & b_2 & b_3 \\ c_1 & c_2 & c_3 \\ d_1 & d_2 & d_3 \end{vmatrix}.$$

Identity (4) is now an immediate consequence of the properties of determinants.

The next theorem establishes the invariance of (3) under a change of basis.

Theorem 5. *Suppose that* $\mathbf{v}$ *is a continuously differentiable vector field. Then there exists a unique continuous vector field* $\mathbf{w}$ *such that*

$$\nabla \cdot (\mathbf{v} \times \mathbf{a}) = \mathbf{w} \cdot \mathbf{a}$$

for every constant vector $\mathbf{a}$. *In fact, if* $\mathbf{i}, \mathbf{j}, \mathbf{k}$ *is an orthonormal basis, then*

$$\mathbf{w} = \nabla \times \mathbf{v}$$

where the vector $\nabla \times \mathbf{v}$ *is given by* (3).

PROOF. Let $\mathbf{i}, \mathbf{j}, \mathbf{k}$ be an orthonormal basis. We define $\mathbf{a} = a_1\mathbf{i} + a_2\mathbf{j} + a_3\mathbf{k}$. Then

$$\mathbf{v} \times \mathbf{a} = (v_2a_3 - v_3a_2)\mathbf{i} + (v_3a_1 - v_1a_3)\mathbf{j} + (v_1a_2 - v_2a_1)\mathbf{k}.$$

The formula for the divergence yields

$$\operatorname{div}(\mathbf{v} \times \mathbf{a}) = a_3\frac{\partial v_2}{\partial x} - a_2\frac{\partial v_3}{\partial x} + a_1\frac{\partial v_3}{\partial y} - a_3\frac{\partial v_1}{\partial y} + a_2\frac{\partial v_1}{\partial z} - a_1\frac{\partial v_2}{\partial z}$$

or

$$\nabla \cdot (\mathbf{v} \times \mathbf{a}) = \left(\frac{\partial v_3}{\partial y} - \frac{\partial v_2}{\partial z}\right)a_1 + \left(\frac{\partial v_1}{\partial z} - \frac{\partial v_3}{\partial x}\right)a_2 + \left(\frac{\partial v_2}{\partial x} - \frac{\partial v_1}{\partial y}\right)a_3.$$

Taking into account the identity (3) and the definition of $\nabla \times \mathbf{v}$, we conclude that the last equation states the result of the theorem. The uniqueness of $\mathbf{w}$ follows from Theorem 1.

Definition. For any vector field $\mathbf{v}$ we define **curl** $\mathbf{v}$ by the formula

$$\operatorname{curl} \mathbf{v} = \nabla \times \mathbf{v},$$

where $\nabla \times \mathbf{v}$ is given by (3).

EXAMPLE 1. Given the vector field

$$\mathbf{v} = (x^2 - y^2 + 2xz)\mathbf{i} + (xz - xy + yz)\mathbf{j} + (z^2 + x^2)\mathbf{k},$$

find $\operatorname{curl} \mathbf{v}$. Show that the vectors given by $\operatorname{curl} \mathbf{v}$ evaluated at $P_0(1, 2, -3)$ and $P_1(2, 3, 12)$ are orthogonal.

SOLUTION. We have

$$\operatorname{curl} \mathbf{v} = \begin{vmatrix} \mathbf{i} & \mathbf{j} & \mathbf{k} \\ \dfrac{\partial}{\partial x} & \dfrac{\partial}{\partial y} & \dfrac{\partial}{\partial z} \\ x^2 - y^2 + 2xz & xz - xy + yz & z^2 + x^2 \end{vmatrix}$$

$$= -(x + y)\mathbf{i} + (z + y)\mathbf{k}.$$

At $P_0(1, 2, -3)$, we find $\operatorname{curl} \mathbf{v} = -3\mathbf{i} - \mathbf{k} \equiv \mathbf{v}_0$. At $P_1(2, 3, 12)$ we find $\operatorname{curl} \mathbf{v} = -5\mathbf{i} + 15\mathbf{k} \equiv \mathbf{v}_1$. Since $\mathbf{v}_0 \cdot \mathbf{v}_1 = 0$, the vectors are orthogonal.

The proofs of the following identities are left to the reader.

Theorem 6. *Suppose that* **u**, **v**, *and* f *are continuously differentiable fields. Then*

a) $\operatorname{curl}(\mathbf{u} + \mathbf{v}) = \operatorname{curl}\mathbf{u} + \operatorname{curl}\mathbf{v}$,
b) $\operatorname{curl}(f\mathbf{u}) = f\operatorname{curl}\mathbf{u} + \nabla f \times \mathbf{u}$,
c) $\operatorname{div}(\mathbf{u} \times \mathbf{v}) = \mathbf{v}\cdot\operatorname{curl}\mathbf{u} - \mathbf{u}\cdot\operatorname{curl}\mathbf{v}$.
d) *If* ∇f *is a continuously differentiable vector field, then*

$$\operatorname{curl}\nabla f = \mathbf{0}.$$

e) *If* **v** *is a twice continuousely differentiable vector field, then*

$$\operatorname{div}\operatorname{curl}\mathbf{v} = 0.$$

EXAMPLE 2. Given the scalar field $f = x^2 + y^2 + z^2$ and the vector field

$$\mathbf{u} = (x^2 + y^2)\mathbf{i} + (y^2 + z^2)\mathbf{j} + (z^2 + x^2)\mathbf{k},$$

compute $\operatorname{curl}(f\mathbf{u})$.

SOLUTION. We use formula (b) of Theorem 6 and the definition of curl as given in (3). Then

$$\operatorname{curl}\mathbf{u} = -2(z\mathbf{i} + x\mathbf{j} + y\mathbf{k}),$$

$$\nabla f = 2(x\mathbf{i} + y\mathbf{j} + z\mathbf{k}).$$

We have

$$\operatorname{curl}(f\mathbf{u}) = f\operatorname{curl}\mathbf{u} + \nabla f \times \mathbf{u}.$$

A computation yields

$$\begin{aligned}\operatorname{curl}(f\mathbf{u}) & \\ &= -2(x^2 + y^2 + z^2)(z\mathbf{i} + x\mathbf{j} + y\mathbf{k}) + \begin{vmatrix} \mathbf{i} & \mathbf{j} & \mathbf{k} \\ 2x & 2y & 2z \\ x^2 + y^2 & y^2 + z^2 & z^2 + x^2 \end{vmatrix} \\ &= -2[z(x^2 + 2y^2 + 2z^2) - y(x^2 + z^2)]\mathbf{i} \\ &\quad - 2[x(2x^2 + y^2 + 2z^2) - z(x^2 + y^2)]\mathbf{j} \\ &\quad - 2[y(2x^2 + 2y^2 + z^2) - x(y^2 + z^2)]\mathbf{k}.\end{aligned}$$

The curl of a vector is intimately connected with the notion of exact differential. We repeat the definition of exact differential as given in Chapter 4, page 275. Suppose that $P(x, y, z)$, $Q(x, y, z)$ and $R(x, y, z)$ are continuously differentiable functions on some region $\mathscr{D}$. We say that the expression

$$P\,dx + Q\,dy + R\,dz \tag{5}$$

is an **exact differential** if and only if there is a continuously differentiable function $f(x, y, z)$ such that

$$df = P\,dx + Q\,dy + R\,dz. \tag{6}$$

Whenever such a function f exists, the definition of the total differential of a function yields the relations

$$f_x = P, \qquad f_y = Q, \qquad f_z = R. \tag{7}$$

If the region $\mathscr{D}$ is a box in R^3 then a necessary and sufficient condition for the expression (5) to be an exact differential is that the three equations

$$\frac{\partial Q}{\partial x} = \frac{\partial P}{\partial y}, \qquad \frac{\partial P}{\partial z} = \frac{\partial R}{\partial x}, \qquad \frac{\partial R}{\partial y} = \frac{\partial Q}{\partial z} \tag{8}$$

hold. (See Chapter 4, page 279.)

We may formulate the notion of exact differential in vector terms. We introduce the usual rectangular coordinate system and define the vectors

$$\mathbf{v} = P\mathbf{i} + Q\mathbf{j} + R\mathbf{k}, \qquad \mathbf{r} = x\mathbf{i} + y\mathbf{j} + z\mathbf{k}, \qquad d\mathbf{r} = (dx)\mathbf{i} + (dy)\mathbf{j} + (dz)\mathbf{k}.$$

Suppose that f and ∇f are continuously differentiable on some domain $\mathscr{D}$ in three-space. Then Eqs. (6) and (7) are equivalent to

$$df = \mathbf{v} \cdot d\mathbf{r} \qquad \text{and} \qquad \nabla f = \mathbf{v}.$$

The necessary and sufficient conditions (8) become the simple vector condition

$$\operatorname{curl} \mathbf{v} = \mathbf{0}. \tag{9}$$

Whenever a vector $\mathbf{v}$ has the form ∇f, we see that (9) is a consequence of Theorem 6(d). Conversely, if (9) holds, then the results on exact differentials in Chapter 4, Section 13, show that there is a function f such that $\nabla f = \mathbf{v}$. We state this conclusion in the following theorem.

Theorem 7. *Suppose that* $\mathbf{v}$ *is a continuously differentiable vector field with* $\operatorname{curl} \mathbf{v} = \mathbf{0}$ *in some rectangular parallelepiped* $\mathscr{D}$ *in space. Then there exists a continuously differentiable scalar field* f *in* $\mathscr{D}$ *such that* $\nabla f = \mathbf{v}$. *Any two such fields differ by a constant.*

In Section 6 we shall state the above theorem for more general domains $\mathscr{D}$ (Theorem 13).

If we are given a specific vector field $\mathbf{v}$ with $\operatorname{curl} \mathbf{v} = \mathbf{0}$, then the method of determining f such that $\nabla f = \mathbf{v}$ is precisely the one described in Chapter 4, Section 13. We review the process by working an example.

EXAMPLE 3. Given the vector field

$$\mathbf{v} = 2xyz\mathbf{i} + (x^2z + y)\mathbf{j} + (x^2y + 3z^2)\mathbf{k},$$

verify that $\operatorname{curl} \mathbf{v} = \mathbf{0}$ and find the function f such that $\nabla f = \mathbf{v}$.

SOLUTION. We have

$$\operatorname{curl}\mathbf{v} = \begin{vmatrix} \mathbf{i} & \mathbf{j} & \mathbf{k} \\ \dfrac{\partial}{\partial x} & \dfrac{\partial}{\partial y} & \dfrac{\partial}{\partial z} \\ 2xyz & (x^2z + y) & (x^2y + 3z^2) \end{vmatrix}$$

$$= (x^2 - x^2)\mathbf{i} + (2xy - 2xy)\mathbf{j} + (2xz - 2xz)\mathbf{k} = \mathbf{0}.$$

We wish to find f such that

$$f_x = 2xyz, \qquad f_y = x^2z + y, \qquad f_z = x^2y + 3z^2.$$

Integrating the first equation, we get

$$f = x^2yz + C(y, z);$$

differentiating f with respect to y and z, we obtain

$$f_y = x^2z + C_y(y, z) = x^2z + y, \qquad f_z = x^2y + C_z(y, z) = x^2y + 3z^2.$$

The preceding equations yield

$$C_y = y, \qquad C_z = 3z^2.$$

Hence $C = \frac{1}{2}y^2 + K(z)$. Differentiation of C with respect to z shows that

$$C_z = K'(z) = 3z^2,$$

so that $K(z) = z^3 + C_1$, where C_1 is a constant. We conclude that

$$C = \tfrac{1}{2}y^2 + z^3 + C_1$$

and, finally, that

$$f = x^2yz + C = x^2yz + \tfrac{1}{2}y^2 + z^3 + C_1.$$

PROBLEMS

In each of Problems 1 through 8, find curl $\mathbf{v}$; in case curl $\mathbf{v} = \mathbf{0}$, find the function f such that $\nabla f = \mathbf{v}$.

1. $\mathbf{v} = (2x - y + 3z)\mathbf{i} + (-x + 3y + 2z)\mathbf{j} + (2x + 3y - z)\mathbf{k}$
2. $\mathbf{v} = (2xy + z^2)\mathbf{i} + (2yz + x^2)\mathbf{j} + (2xz + y^2)\mathbf{k}$
3. $\mathbf{v} = e^x(\sin y \cos z\,\mathbf{i} + \cos y \cos z\,\mathbf{j} - \sin y \sin z\,\mathbf{k})$
4. $\mathbf{v} = (x + 2y - z)\mathbf{i} + (x - y + z)\mathbf{j} + (-x + y + 2z)\mathbf{k}$
5. $\mathbf{v} = (x^2 + y^2)^{-1/2}[-xz\mathbf{i} - yz\mathbf{j} + (x^2 + y^2)\mathbf{k}]$
6. $\mathbf{v} = (x^2 + y^2 + z^2)^{-1}(x\mathbf{i} + y\mathbf{j} + z\mathbf{k}), \quad (x, y, z) \neq (0, 0, 0)$
7. $\mathbf{v} = x^2z\mathbf{i} + 0 \cdot \mathbf{j} + xz^2\mathbf{k}$
8. $\mathbf{v} = \dfrac{2x + y + z}{(x + y)(x + z)}\mathbf{i} + \dfrac{x + 2y + z}{(x + y)(y + z)}\mathbf{j} + \dfrac{x + y + 2z}{(y + z)(x + z)}\mathbf{k}$

9. Suppose that $G_1'(x) = g_1(x)$, $G_2'(y) = g_2(y)$, $G_3'(z) = g_3(z)$, and that
$$\mathbf{v} = [x^2(y^3 + z^3) + g_1(x)]\mathbf{i} + [y^2(x^3 + z^3) + g_2(y)]\mathbf{j} + [z^2(x^3 + y^3) + g_3(z)]\mathbf{k}.$$
Show that curl $\mathbf{v} = \mathbf{0}$ and find f such that $\nabla f = \mathbf{v}$.

10. Verify the identity $(\mathbf{b} \times \mathbf{c}) \cdot \mathbf{d} = \mathbf{b} \cdot (\mathbf{c} \times \mathbf{d})$ by introducing coordinate vectors and calculating each side separately.

11. Prove Theorem 6(a).

12. Prove Theorem 6(b).

13. Prove Theorem 6(c).

14. Prove Theorem 6(d).

15. Prove Theorem 6(e).

In Problems 16 through 18, suppose that $\mathbf{r} = x\mathbf{i} + y\mathbf{j} + z\mathbf{k}$, and let $r = |\mathbf{r}|$.

16. Show that
$$\operatorname{curl}\left(\frac{\mathbf{r}}{r}\right) = \mathbf{0}.$$

17. Find curl $[\phi(r)\mathbf{r}]$, where ϕ is a differentiable function.

18. Assuming p and $\mathbf{a}$ constant, find curl $(r^p\mathbf{a} \times \mathbf{r})$.

19. Verify Theorem 6(b) for $f = (x^2 + y^2 + z^2)^p$, $\mathbf{u} = z\mathbf{i} + x\mathbf{j} + y\mathbf{k}$.

20. Verify Theorem 6(c) for $\mathbf{u} = y\mathbf{i} + z\mathbf{j} + x\mathbf{k}$, $\mathbf{v} = z\mathbf{i} + x\mathbf{j} + y\mathbf{k}$.

21. Verify Theorem 6(d) with $f = (x^2 + y^2 + z^2)^{-1/2}x$.

22. Verify Theorem 6(e) with $\mathbf{v} = (y^2 - z^2)\mathbf{i} + (z^2 - x^2)\mathbf{j} + (x^2 - y^2)\mathbf{k}$.

23. If $\mathbf{v}$ is any vector of the form
$$\mathbf{v} = v_1\mathbf{i} + v_2\mathbf{j} + v_3\mathbf{k},$$
we define the Laplacian of $\mathbf{v}$, denoted $\Delta\mathbf{v}$ or $\nabla^2\mathbf{v}$, by the formula
$$\Delta\mathbf{v} = (\Delta v_1)\mathbf{i} + (\Delta v_2)\mathbf{j} + (\Delta v_3)\mathbf{k}.$$
For any vector field u (sufficiently differentiable), establish the identity
$$\operatorname{curl}\operatorname{curl}\mathbf{u} = \operatorname{grad}\operatorname{div}\mathbf{u} - \Delta\mathbf{u}. \tag{10}$$

24. Verify (10) with $\mathbf{u} = (y^2 + zx)\mathbf{i} + (z^2 + xy)\mathbf{j} + (x^2 + yz)\mathbf{k}$.

5. Line Integrals; Vector Formulation

An arc in three-space is the graph of the equations
$$x = f(t), \qquad y = g(t), \qquad z = h(t), \qquad a \le t \le b,$$
provided that f, g, and h are continuous and that no point on the graph

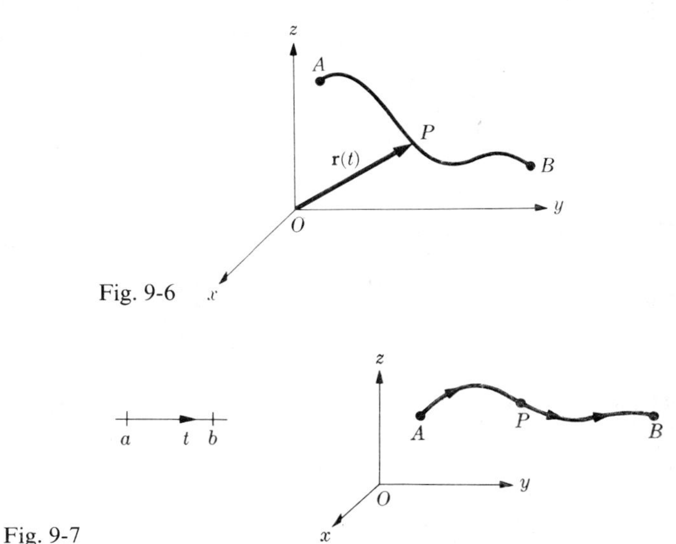

Fig. 9-6

Fig. 9-7

corresponds to two different values of t. In terms of vectors and transformations, we define an **arc** as the range of a one-to-one continuous transformation of the form (Fig. 9-6)

$$\mathbf{v}(\overrightarrow{OP}) = \mathbf{r}(t), \qquad a \le t \le b,$$

with the auxiliary condition that

$$\mathbf{r}(t') \ne \mathbf{r}(t'') \qquad \text{if } t' \ne t''.$$

The definition of an arc in a space of any number of dimensions is analogous to its definition in three-space.

Suppose that in a rectangular coordinate system an arc C is given by the transformation

$$C: \quad \mathbf{r} = \mathbf{r}(t) = f(t)\mathbf{i} + g(t)\mathbf{j} + h(t)\mathbf{k}, \qquad a \le t \le b.$$

As t increases from a to b, the point P moves along the arc from the point A, corresponding to $t = a$, to the point B, corresponding to $t = b$. Since the transformation is one to one, the point P moves along "without doubling back." We say that "P describes the arc in a certain sense." (See Fig. 9-7.)

Any arc C will have many different parametric representations. Suppose that two representations of C are

$$C: \quad x = f(t), \qquad y = g(t), \qquad z = h(t), \qquad a \le t \le b;$$

$$C: \quad x = F(\tau), \qquad y = G(\tau), \qquad z = H(\tau), \qquad c \le \tau \le d.$$

In vector notation, we write

$$C: \quad \mathbf{r} = \mathbf{r}(t), \quad a \le t \le b; \qquad \mathbf{r}(t') \ne \mathbf{r}(t'') \quad \text{if } t' \ne t''; \tag{1}$$

$$C: \quad \mathbf{r} = \mathbf{R}(\tau), \quad c \le \tau \le d; \qquad \mathbf{R}(\tau') \ne \mathbf{R}(\tau'') \quad \text{if } \tau' \ne \tau''. \tag{2}$$

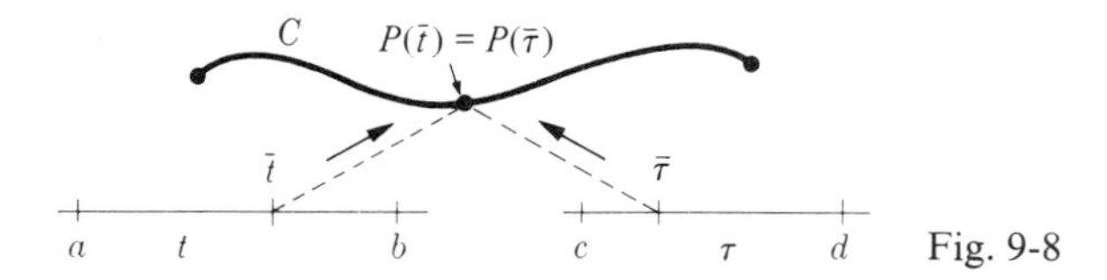

Fig. 9-8

Intuitively we see that as t increases from a to b, the arc C is described in a particular sense. As τ increases from c to d, the arc is described either in the same sense or in the opposite sense. (See Fig. 9-8.) There are no other possibilities. We make this conclusion precise in the next theorem, which is stated without proof.

Theorem 8. *Suppose that the arc C is the range of the continuous transformations* (1) *and* (2). *Then there are continuous functions $S_1(t)$, $S_2(\tau)$ defined on $[a, b]$ and $[c, d]$, respectively, and with ranges $[c, d]$ and $[a, b]$, respectively, such that*

$$\mathbf{R}[S_1(t)] = \mathbf{r}(t) \text{ for } t \text{ on } [a, b] \text{ and } \mathbf{r}[S_2(\tau)] = \mathbf{R}(\tau) \text{ for } \tau \text{ on } [c, d].$$

Either S_1 and S_2 are both increasing or they are both decreasing; moreover, each is the inverse of the other.

As Fig. 9-8 shows, the function $S_1(t)$ is the mapping obtained by going from a point $\bar{t}$ to the point $P(\bar{t})$ on C and then finding the unique point $\bar{\tau}$ in $[c, d]$ which corresponds to the same point P. Thus we get $\bar{\tau} = S_1(\bar{t})$. The function S_2 is obtained by reversing the process.

Theorem 8 implies that all parametric representations of an arc C fall into two classes: one class, in which S_1 and S_2 are both increasing, and the other, in which S_1 and S_2 are both decreasing. Two representations in the same class define the same ordering of the points of C in the sense that a point P' on C precedes P'' if and only if $t' < t''$.

Definitions. A **directed arc** $\vec{C}$ is an arc C together with one of the two orderings described above. If $\vec{C}$ is a directed arc, we denote the corresponding undirected arc by C or, if we wish to emphasize the undirected property, by $|\vec{C}|$. A transformation (1) is said to be a **parametric representation** of $\vec{C}$ if and only if it is a representation of $|\vec{C}|$ and establishes the given order on $\vec{C}$. The arc **oppositely directed** to $\vec{C}$ is denoted by $-\vec{C}$.

The definitions and basic properties of line integrals in the plane and in space were taken up in Chapter 4, page 282. We shall now see how vector notation may be used to simplify the statements and proofs of some of the elementary theorems on line integrals in space.

Let $\vec{C}$ be a directed arc from a point A to a point B. We make a **subdivision of** $\vec{C}$ by arranging the points $A = P_0, P_1, P_2, \ldots, P_{n-1}, P_n = B$ in order along the directed arc. As usual, we define the norm $\|\Delta\|$ of the subdivision

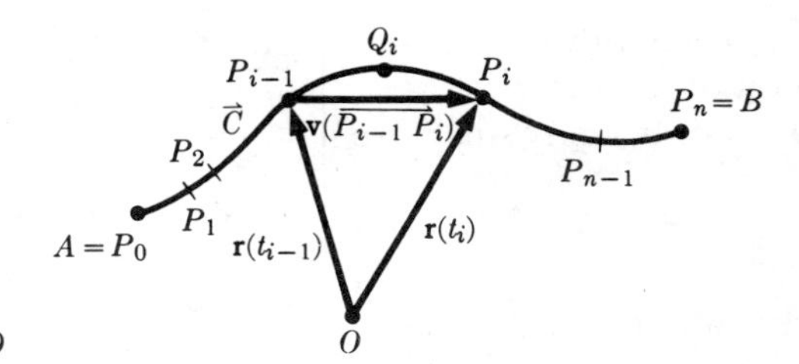

Fig. 9-9

as the length of the longest line segment connecting two successive points P_{i-1}, P_i in the subdivision (Fig. 9-9). Suppose we are given a vector field $\mathbf{w}$ defined on $|\vec{C}|$. For each i, we select a point Q_i on the subarc from P_{i-1} to P_i (Fig. 9-9) and form the sum of the scalar products

$$\sum_{i=1}^{n} \mathbf{w}(Q_i) \cdot \mathbf{v}(\overrightarrow{P_{i-1}P_i}) = \sum_{i=1}^{n} \mathbf{w}(Q_i) \cdot [\mathbf{r}(t_i) - \mathbf{r}(t_{i-1})]. \tag{3}$$

We may abbreviate this expression by denoting $\Delta_i \mathbf{r} = \mathbf{r}(t_i) - \mathbf{r}(t_{i-1})$ and then writing

$$\sum_{i=1}^{n} \mathbf{w}(Q_i) \cdot \Delta_i \mathbf{r}.$$

Definition. Suppose there is a number L with the property that for each $\varepsilon > 0$ there is a $\delta > 0$ such that

$$\left| \sum_{i=1}^{n} \mathbf{w}(Q_i) \cdot \Delta_i \mathbf{r} - L \right| < \varepsilon$$

for all subdivisions with norm less than δ and for all choices of the Q_i on the arc $\widehat{P_{i-1}P_i}$. Then we say that **the differential $\mathbf{w} \cdot d\mathbf{r}$ is integrable along $\vec{C}$.** We write

$$L = \int_{\vec{C}} \mathbf{w} \cdot d\mathbf{r}.$$

The next theorem establishes a few of the elementary properties of line integrals in three-space.

Theorem 9. (a) *There is at most one number L satisfying the conditions of the definition above.*

b) *When such a number L exists, it is independent of the choice for the origin O.*

c) *If $\mathbf{w} \cdot d\mathbf{r}$ is integrable along C, it is integrable along $-\vec{C}$, and*

$$\int_{-\vec{C}} \mathbf{w} \cdot d\mathbf{r} = -\int_{\vec{C}} \mathbf{w} \cdot d\mathbf{r}.$$

PROOF. (a) Suppose there are two numbers L_1 and L_2 satisfying the required conditions. We select $\varepsilon = (L_2 - L_1)/2$ (supposing that $L_2 > L_1$). Then, if $\|\Delta\|$ is sufficiently small, we have

$$\sum_{i=1}^{n} \mathbf{w}(Q_i) \cdot \Delta_i \mathbf{r} < L_1 + \varepsilon = L_1 + \frac{L_2 - L_1}{2} = \frac{L_1 + L_2}{2}. \tag{4}$$

On the other hand, since L_2 is also a limit, we see that

$$\frac{L_1 + L_2}{2} = L_2 - \varepsilon < \sum_{i=1}^{n} \mathbf{w}(Q_i) \cdot \Delta_i \mathbf{r}. \tag{5}$$

Since (4) and (5) are contradictory, $L_1 = L_2$.
b) This statement follows from the fact that each sum, as given on the left side of (3), is independent of O.
c) If $P_0, P_1, \ldots, P_n$ is a subdivision for $\vec{C}$, then $P_n, P_{n-1}, \ldots, P_0$ is a subdivision for $-\vec{C}$. Therefore a sum of the type (3) for $-\vec{C}$ is

$$\begin{aligned} &\mathbf{w}(Q_n) \cdot \mathbf{v}(\overrightarrow{P_n P_{n-1}}) + \mathbf{w}(Q_{n-1}) \cdot \mathbf{v}(\overrightarrow{P_{n-1} P_{n-2}}) + \cdots + \mathbf{w}(Q_1) \cdot \mathbf{v}(\overrightarrow{P_1 P_0}) \\ &\qquad = -\sum_{i=1}^{n} \mathbf{w}(Q_i) \cdot \Delta_i \mathbf{r}. \end{aligned} \tag{6}$$

The result of part (c) follows by letting $\|\Delta\| \to 0$ in (6).

Suppose that $\vec{C}$ is a directed arc and that $\mathbf{w}$ is a vector field defined on $|\vec{C}|$. We introduce a rectangular coordinate system (x, y, z) and basis vectors $\mathbf{i}, \mathbf{j}, \mathbf{k}$. We now write

$$C: \quad \mathbf{r}(t) = x(t)\mathbf{i} + y(t)\mathbf{j} + z(t)\mathbf{k}, \tag{7}$$

$$\mathbf{w} = P(x, y, z)\mathbf{i} + Q(x, y, z)\mathbf{j} + R(x, y, z)\mathbf{k}. \tag{8}$$

If $\mathbf{r}(t)$ is a continuously differentiable function and $\mathbf{w}$ is a continuous vector field, it can be shown that

$$\int_{\vec{C}} \mathbf{w} \cdot d\mathbf{r} = \int_{\vec{C}} P\,dx + Q\,dy + R\,dz,$$

where the line integral on the right is defined in terms of coordinates as in Chapter 4, Section 14. The evaluation of line integrals is reduced to that of ordinary integrals, as is shown in the next theorem.

Theorem 10. *Suppose that $\vec{C}$ has the parametric representation $\mathbf{r} = \mathbf{r}(t)$, with $\mathbf{r}$ continuously differentiable for $a \le t \le b$. Suppose that $\mathbf{w}$ is continuous on $|\vec{C}|$. Then $\mathbf{w} \cdot d\mathbf{r}$ is integrable along $\vec{C}$ and*

$$\int_{\vec{C}} \mathbf{w} \cdot d\mathbf{r} = \int_a^b \mathbf{w}(t) \cdot \mathbf{r}'(t)\,dt.$$

Furthermore, if $\mathbf{w}$ is given by (8), then

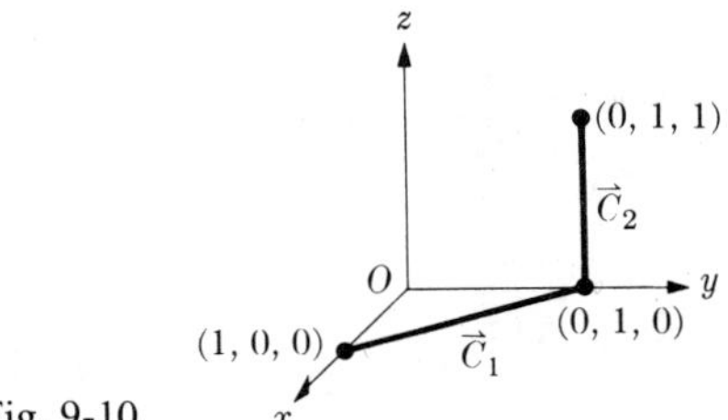

Fig. 9-10

$$\int_{\vec{C}} \mathbf{w} \cdot d\mathbf{r} = \int_a^b \{P[x(t), y(t), z(t)]x'(t) + Q[x(t), y(t), z(t)]y'(t) + R[x(t), y(t), z(t)]z'(t)\}\, dt.$$

REMARK. If $\mathbf{r}$ is continuous and only piecewise smooth, we may evaluate the line integral along each smooth subarc and add the results.

EXAMPLE. Compute $\int_{\vec{C}} \mathbf{w} \cdot d\mathbf{r}$, where

$$\mathbf{w} = xy\mathbf{i} + xz\mathbf{j} - y\mathbf{k},$$

$$\mathbf{r} = x\mathbf{i} + y\mathbf{j} + z\mathbf{k},$$

and $\vec{C}$ is the directed line segment $\vec{C}_1$ from $(1, 0, 0)$ to $(0, 1, 0)$, followed by $\vec{C}_2$, which is the segment from $(0, 1, 0)$ to $(0, 1, 1)$. (See Fig. 9-10.)

SOLUTION. Along $\vec{C}_1$ we have $z = 0$. Taking $t = y$, we may write the equation of the line segment

$$\mathbf{r}(y) = (1 - y)\mathbf{i} + y\mathbf{j}, \qquad 0 \le y \le 1.$$

Then

$$\mathbf{w} = (1 - y)y\mathbf{i} - y\mathbf{k}.$$

Therefore

$$\int_{\vec{C}_1} \mathbf{w} \cdot d\mathbf{r} = \int_0^1 \mathbf{w}(y) \cdot \mathbf{r}'(y)\, dy = \int_0^1 (-y + y^2)\, dy = -\tfrac{1}{6}.$$

Along $\vec{C}_2$ we have $x = 0$ and we take $t = z$. Then we find

$$\mathbf{r}(z) = \mathbf{j} + z\mathbf{k}, \qquad \mathbf{w}(z) = -\mathbf{k}, \qquad \mathbf{r}'(z) = \mathbf{k}.$$

Hence

$$\int_{\vec{C}_2} \mathbf{w} \cdot d\mathbf{r} = \int_0^1 -\mathbf{k} \cdot \mathbf{k}\, dz = -1.$$

The result is

$$\int_{\vec{C}_1} \mathbf{w} \cdot d\mathbf{r} + \int_{\vec{C}_2} \mathbf{w} \cdot d\mathbf{r} = -\tfrac{1}{6} - 1 = -\tfrac{7}{6}.$$

PROBLEMS

In each of Problems 1 through 10, evaluate $\int_{\vec{C}} \mathbf{w} \cdot d\mathbf{r}$. Sketch the arc $\vec{C}$ in each case.

1. $\mathbf{w} = xy\mathbf{i} - y\mathbf{j} + \mathbf{k}$; $\vec{C}$ is the segment going from $(0, 0, 0)$ to $(1, 1, 1)$.
2. $\mathbf{w} = xy\mathbf{i} - y\mathbf{j} + \mathbf{k}$; $\vec{C}$ is the arc given by $x = t$, $y = t^2$, $z = t^3$, $0 \le t \le 1$.
3. $\mathbf{w} = x\mathbf{i} - y\mathbf{j} + z\mathbf{k}$; $\vec{C}$ is the helical path $x = \cos\theta$, $y = \sin\theta$, $z = (1/\pi)\theta$, $0 \le \theta \le 2\pi$.
4. $\mathbf{w} = x\mathbf{i} - y\mathbf{j} + z\mathbf{k}$; $\vec{C}$ is the segment from $(1, 0, 0)$ to $(1, 0, 2)$.
5. $\mathbf{w} = 2x\mathbf{i} - 3y\mathbf{j} + z^2\mathbf{k}$; $\vec{C}$ is the path $x = \cos\theta$, $y = \sin\theta$, $z = \theta$, $0 \le \theta \le (\pi/2)$.
6. $\mathbf{w} = 2x\mathbf{i} - 3y\mathbf{j} + z^2\mathbf{k}$; $\vec{C}$ is the segment from $(1, 0, 0)$ to $(0, 1, \pi/2)$.
7. $\mathbf{w} = y^2\mathbf{i} + x^2\mathbf{j} + 0 \cdot \mathbf{k}$; $\vec{C}$ is the arc of the parabola $x = t$, $y = t^2$, $z = 0$, $1 \le t \le 2$.
8. $\mathbf{w} = z^2\mathbf{i} + 0 \cdot \mathbf{j} + x^2\mathbf{k}$; $\vec{C}$ is the segment $\vec{C}_1$ from $(1, 0, 1)$ to $(2, 0, 1)$ followed by the segment $\vec{C}_2$ from $(2, 0, 1)$ to $(2, 0, 4)$.
9. $\mathbf{w} = (x^2 - y^2)\mathbf{i} + 2xy\mathbf{j} + 0 \cdot \mathbf{k}$; $\vec{C}$ is the segment from $(2, 0, 0)$ to $(0, 2, 0)$.
10. $\mathbf{w} = 2yz\mathbf{j} + (z^2 - y^2)\mathbf{k}$; $\vec{C}$ is the circular arc given by: $x = 0$, $y^2 + z^2 = 4$ going from $(0, 2, 0)$ to $(0, 0, 2)$.
11. Show that

$$\int_{\vec{C}} \mathbf{w} \cdot d\mathbf{r} = \int_{\vec{C}} \sqrt{P^2 + Q^2 + R^2} \cos\theta \, ds,$$

where $\mathbf{w} = P\mathbf{i} + Q\mathbf{j} + R\mathbf{k}$, ds is the element of arc along $\vec{C}$, and θ is the angle made by the vector $d\mathbf{r}$ and the vector $\mathbf{w}$.

6. Path-Independent Line Integrals

A continuous transformation

$$\mathbf{v}(\overrightarrow{OP}) = \mathbf{r}(t), \qquad a \le t \le b$$

is called a **path** in space. Since a path is not necessarily one to one, we see that arcs are special cases of paths. As the examples in Fig. 9-11 show, a path may have loops and multiple intersections. Parametric representations are needed to distinguish the first two paths in Fig. 9-11, both of which start at A and end at B; they are identical in appearance. However, it is intuitively clear that if $\mathbf{r}(t)$ is a representation of a path and $t = S(\tau)$ is

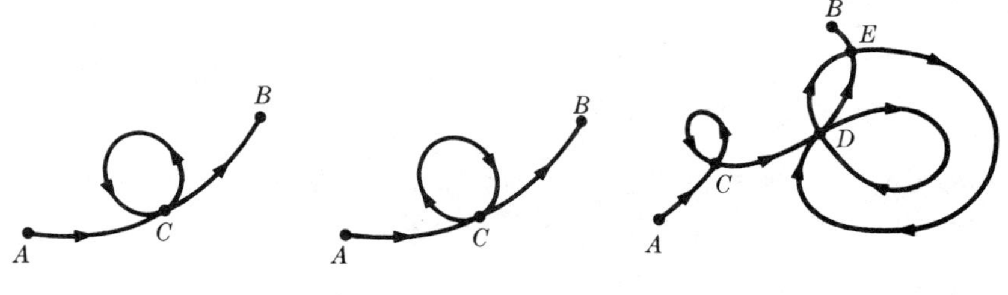

Fig. 9-11

an increasing function, then the representation $\mathbf{R}(\tau)$ determined by the relation

$$\mathbf{R}(\tau) = \mathbf{r}[S(\tau)]$$

describes the same path as $\mathbf{r}(t)$, and in the same order. Also, we denote **directed paths** by symbols such as $\vec{C}$, undirected paths by $|\vec{C}|$, and oppositely directed paths by $-\vec{C}$.

The theorems about line integrals which we established for piecewise smooth arcs are equally valid for piecewise smooth paths. As we saw in the preceding section, a line integral $\int_{\vec{C}} \mathbf{w} \cdot d\mathbf{r}$ depends not only on the vector field $\mathbf{w}$ and the endpoints A and B of the path but also on the path $\vec{C}$ itself. Whenever the value of a line integral depends only on the vector field $\mathbf{w}$ and on the endpoints A and B of the path but *not* on $\vec{C}$ itself, we say **the integral is independent of the path**. Such integrals were first discussed on page 291 of Chapter 4. The results given there are now stated in vector form so that the extension to higher-dimensional spaces becomes evident.

We say that D is a **connected region in space** if it has the property that any two points in D can be joined by a smooth arc which lies in D.

Theorem 11. *Suppose that u is a continuously differentiable scalar field on a connected region D and that A and B are in D. Then*

$$\int_{\vec{C}} \nabla u \cdot d\mathbf{r} = u(B) - u(A)$$

for any piecewise smooth path from A to B which is contained in D.

PROOF. Since each piecewise smooth path is the finite sum of smooth paths, it is sufficient to prove the theorem for a smooth path. Let (x, y, z) be a rectangular coordinate system with $\mathbf{i}$, $\mathbf{j}$, $\mathbf{k}$ the usual basis. The path $\vec{C}$ is given by

$$\mathbf{r} = \mathbf{r}(t) = x(t)\mathbf{i} + y(t)\mathbf{j} + z(t)\mathbf{k}, \qquad a \le t \le b,$$

and the vector field ∇u is

$$\nabla u = \frac{\partial u}{\partial x}\mathbf{i} + \frac{\partial u}{\partial y}\mathbf{j} + \frac{\partial u}{\partial z}\mathbf{k}.$$

Then

$$\int_{\vec{C}} \nabla u \cdot d\mathbf{r} = \int_{\vec{C}} \left\{ \frac{\partial u}{\partial x}\frac{dx}{dt} + \frac{\partial u}{\partial y}\frac{dy}{dt} + \frac{\partial u}{\partial z}\frac{dz}{dt} \right\} dt.$$

Along $\vec{C}$ we have $u = u[x(t), y(t), z(t)]$ which we denote by $g(t)$. Therefore (noting that $\mathbf{r}(a) = A$, $\mathbf{r}(b) = B$) the above integral is

$$\int_a^b \frac{d}{dt} u[x(t), y(t), z(t)]\, dt = \int_a^b g'(t)\, dt = g(b) - g(a) = u(B) - u(A),$$

which is the desired result.

Theorem 11 shows that, under appropriate hypotheses, every vector field which is the gradient of a scalar field has a path-independent line integral. The next theorem shows that, conversely, if a vector field leads to path-independent line integrals in a domain, it must be the gradient of some scalar field.

Theorem 12. *Suppose that* $\mathbf{v}$ *is a continuous vector field on a domain* D, *and suppose that for every ordered pair* (A, B) *of points in* D, *the integral*

$$\int_{\vec{C}} \mathbf{v} \cdot d\mathbf{r}$$

has the same value for every smooth path from A *to* B *with* $|\vec{C}|$ *in* D. *That is, suppose the integral is path-independent. Then there is a continuously differentiable scalar field* u *with domain* D *such that*

$$\mathbf{v} = \nabla u.$$

PROOF. Let A be a fixed point in D, and define

$$u(P) = \int_{\vec{C}} \mathbf{v} \cdot d\mathbf{r},$$

where $\vec{C}$ is any smooth path from A to a point P (which lies in D). We shall show that u is the desired scalar field. We introduce the customary rectangular coordinate system and let P_0 be any point in D with coordinates (x_0, y_0, z_0). We construct a ball with center at P_0 and radius ρ so small that the ball is entirely in D (Fig. 9-12). Let $\vec{C}_0$ be a directed path in D from A to P_0 which ends with a directed straight line segment $\overrightarrow{P_1 P_0}$, as shown in Fig. 9-12. The coordinates of P_1 are (x_1, y_1, z_1). We suppose for convenience that this segment is parallel to the x-axis. Let P be a point with coordinates $(x_0 + h, y_0, z_0)$, where $|h| < \rho$. Then we have

$$u(x_0 + h, y_0, z_0) = \int_{C_1} \mathbf{v} \cdot d\mathbf{r} + \int_{C_2} \mathbf{v} \cdot d\mathbf{r}, \tag{1}$$

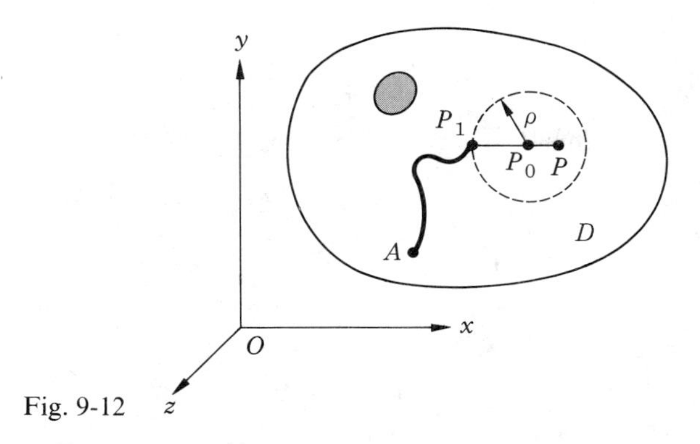

Fig. 9-12

where $\vec{C}_1$ is the directed path from A to P_1 and $\vec{C}_2$ is the directed segment $\overrightarrow{P_1P}$. Now we write

$$\mathbf{v}(x, y, z) = v_1(x, y, z)\mathbf{i} + v_2(x, y, z)\mathbf{j} + v_3(x, y, z)\mathbf{k},$$

$$\mathbf{r} = x\mathbf{i} + y\mathbf{j} + z\mathbf{k},$$

and we observe that, along $\overrightarrow{P_1P}$, we have $d\mathbf{r} = (dx)\mathbf{i}$. Then, defining

$$\phi(x) = \int_{x_1}^{x} v_1(\xi, y_0, z_0)\, d\xi,$$

we see from (1) that

$$u(x_0 + h, y_0, z_0) - u(x_0, y_0, z_0) = \phi(x_0 + h) - \phi(x_0).$$

We conclude from Leibniz' Rule that

$$u_x(x_0, y_0, z_0) = \lim_{h\to 0} \frac{\phi(x_0 + h) - \phi(x_0)}{h} = \phi'(x_0) = v_1(x_0, y_0, z_0).$$

Since the same arguments work in the y- and z-directions, we obtain

$$\nabla u(P_0) = \mathbf{v}(P_0).$$

But P_0 is an arbitrary point of D, and so the result is established.

The above theorem is intimately connected with Theorem 7 of Section 4, because any vector field $\mathbf{v}$ in a box in R^3 with the property that

$$\operatorname{curl} \mathbf{v} = \mathbf{0}$$

is the gradient of a scalar function. Thus $\int_{\vec{C}} \mathbf{v} \cdot d\mathbf{r}$ will be independent of the path in such a box. We would like to establish Theorem 7 for more general domains, but in doing so we must exercise extreme care, as the following illustration shows.

We consider the vector field

$$\mathbf{v} = -\frac{y}{x^2 + y^2}\mathbf{i} + \frac{x}{x^2 + y^2}\mathbf{j} + 0 \cdot \mathbf{k}.$$

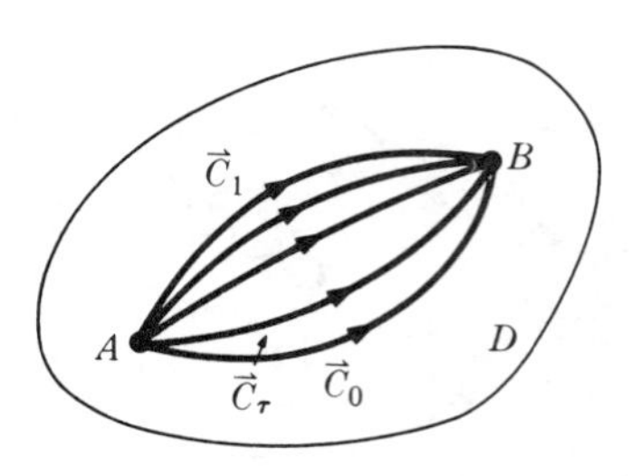

Fig. 9-13

The reader may easily verify that curl $\mathbf{v} = \mathbf{0}$ for all (x, y, z) so long as $(x, y) \neq (0, 0)$. However, when we select for $\vec{C}$ the circular path

$$\vec{C}: \quad x = \cos t, \qquad y = \sin t, \qquad z = 0, \qquad -\pi \leq t \leq \pi,$$

and for D any region containing $\vec{C}$ and excluding the z-axis, a calculation shows that

$$\int_{\vec{C}} \mathbf{v} \cdot d\mathbf{r} = \int_{\vec{C}} \frac{x\,dy - y\,dx}{x^2 + y^2} = \int_{-\pi}^{\pi} dt = 2\pi.$$

If $\mathbf{v} \cdot d\mathbf{r}$ were an exact differential in D, then $\mathbf{v} = \nabla f$ and the integral would be zero, according to Theorem 7. So we see that *some restriction* on the domain D is essential before Theorem 7 can be extended.

Suppose that A and B are points of a domain D which are connected by two paths $\vec{C}_0$ and $\vec{C}_1$, both lying in D. We shall define formally the concept which states: "$\vec{C}_0$ can be deformed smoothly into $\vec{C}_1$ without going outside D."

Definition. A domain D in the plane (or in space) is said to be **simply connected** if for each pair $\vec{C}_0$ and $\vec{C}_1$ of smooth directed paths in D joining the same points A and B, there exists a vector function $\mathbf{f}$ on R^2 (or R^3) with the following properties

i) $\mathbf{f}(t, \tau)$ is continuous for $a \leq t \leq b$, $0 \leq \tau \leq 1$.
ii) If $\mathbf{v}(\overrightarrow{OP})$ is the vector from the origin to a point P, then for $\tau = 0$, $\mathbf{f}(t, 0)$ describes $\vec{C}_0$. That is,

$$\vec{C}_0: \quad \mathbf{v}(\overrightarrow{OP}) = \mathbf{f}(t, 0), \qquad a \leq t \leq b.$$

iii) Similarly, for $\tau = 1$,

$$\vec{C}_1: \quad \mathbf{v}(\overrightarrow{OP}) = \mathbf{f}(t, 1), \qquad a \leq t \leq b.$$

iv) For each fixed τ on $[0, 1]$, $\mathbf{v}(\overrightarrow{OP}) = \mathbf{f}(t, \tau)$, $a \leq t \leq b$, is a directed path from A to B lying entirely in D, i.e., $\mathbf{f}(0, \tau) = A$, $\mathbf{f}(1, \tau) = B$ for $0 \leq \tau \leq 1$.

The definition of simple connectivity in the plane is illustrated in Fig. 9-13, where we denote by $\vec{C}_\tau$ the path $\mathbf{v}(\overrightarrow{OP}) = \mathbf{f}(t, \tau)$ for fixed τ going from A to B. We note that the paths vary smoothly from $\vec{C}_0$ to $\vec{C}_1$ as τ goes from 0 to 1.

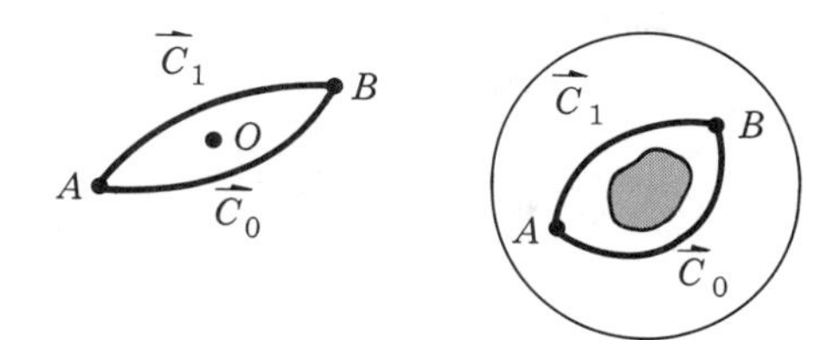

Fig. 9-14

It is intuitively clear that the domain D, consisting of the entire plane with a single point (denoted O) removed, is not simply connected. As Fig. 9-14 shows, it is impossible to deform $\vec{C}_0$ into $\vec{C}_1$ without some $\vec{C}_\tau$ passing through O. Similarly, the domain between two concentric circles (Fig. 9-14) is not simply connected.

In three-space the domain inside a sphere or ellipsoid is simply connected. Also, it can be shown (although we shall not do so) that, on one hand, the region between two concentric spheres *is* simply connected while, on the other, the region inside a torus (that is, the space inside a doughnut) is *not* simply connected. The region between two coaxial cylinders is another example of a region which is not simply connected.

The next theorem which is an extension to three dimensions of Theorem 7 of Section 4 uses a slightly stronger definition of simple connectivity. We say a domain is **strongly simply connected** if and only if it is simply connected and if the derivatives $\mathbf{f}_t$, $\mathbf{f}_\tau$, $\mathbf{f}_{t\tau}$, and $\mathbf{f}_{\tau t}$ of the function $\mathbf{f}$ given in the definition of simple connectivity are all continuous.

Theorem 13. *Suppose that* $\mathbf{v}$ *is continuously differentiable in a strongly simply connected region* D *in space and that curl* $\mathbf{v} = \mathbf{0}$ *in* D. *Then there is a continuously differentiable scalar field* u *on* D *such that* $\mathbf{v} = \nabla u$.

PROOF. In the light of Theorem 12, it is sufficient to prove that $\int_C \mathbf{v} \cdot d\mathbf{r}$ is independent of the path. It is convenient to use the summation notation, and so we denote the coordinates of a rectangular system by (x_1, x_2, x_3) instead of (x, y, z). We let $\vec{C}_0, \vec{C}_1, \vec{C}_\tau$ be paths from A to B as in the definition of simple connectivity, and we write

$$\mathbf{v}(x_1, x_2, x_3) = v_1(x_1, x_2, x_3)\mathbf{i} + v_2(x_1, x_2, x_3)\mathbf{j} + v_3(x_1, x_2, x_3)\mathbf{k},$$
$$\mathbf{f}(t, \tau) = f_1(t, \tau)\mathbf{i} + f_2(t, \tau)\mathbf{j} + f_3(t, \tau)\mathbf{k}.$$

We define

$$\phi(\tau) = \int_{\vec{C}_\tau} \mathbf{v} \cdot d\mathbf{r} = \int_a^b \sum_{i=1}^3 v_i[f_1(t, \tau), f_2(t, \tau), f_3(t, \tau)] \frac{\partial f_i}{\partial t} dt.$$

Then, using Leibniz' Rule (and the Chain Rule), we obtain

$$\phi'(\tau) = \int_a^b \sum_{i=1}^3 \left\{ v_i(f_1, f_2, f_3) \frac{\partial^2 f_i}{\partial t \partial \tau} + \sum_{j=1}^3 \frac{\partial v_i}{\partial x_j} \frac{\partial f_i}{\partial t} \frac{\partial f_j}{\partial \tau} \right\} dt.$$

Integrating by parts the terms in the first sum above, we eliminate the second derivatives of the f_i and find

$$\phi'(\tau) = \left[\sum_{i=1}^{3} v_i \frac{\partial f_i}{\partial \tau} \right]_{t=a}^{t=b} + \int_a^b \sum_{i,j=1}^{3} \frac{\partial v_i}{\partial x_j} \left(\frac{\partial f_i}{\partial t} \frac{\partial f_j}{\partial \tau} - \frac{\partial f_i}{\partial \tau} \frac{\partial f_j}{\partial t} \right) dt. \tag{2}$$

Denoting the coordinates of A and B by (x_1^0, x_2^0, x_3^0) and (x_1^1, x_2^1, x_3^1), respectively, we observe that for all τ in $[0, 1]$

$$f_i(a, \tau) = x_i^0 \qquad \text{and} \qquad f_i(b, \tau) = x_i^1, \qquad i = 1, 2, 3.$$

Thus the first term on the right in (2) vanishes. Moreover, if we interchange the indices i and j in the second sum in (2), we get

$$\phi'(\tau) = \int_a^b \sum_{i,j=1}^{3} \left(\frac{\partial v_i}{\partial x_j} - \frac{\partial v_j}{\partial x_i} \right) \frac{\partial f_i}{\partial t} \frac{\partial f_j}{\partial \tau} dt.$$

The condition that curl $\mathbf{v} = \mathbf{0}$ is equivalent to the condition

$$\frac{\partial v_i}{\partial x_j} - \frac{\partial v_j}{\partial x_i} = 0, \qquad i, j = 1, 2, 3$$

and so $\phi'(\tau) = 0$. Hence $\phi(\tau)$ is constant and therefore the integral is independent of the path.

REMARK. This proof generalizes to n dimensional space if we replace the condition curl $\mathbf{v} = \mathbf{0}$ by the condition

$$\frac{\partial v_i}{\partial x_j} - \frac{\partial v_j}{\partial x_i} = 0, \qquad i, j = 1, 2, \ldots, n \tag{3}$$

which can be shown to be independent of the coordinates. However, the equations (3) cannot be expressed in terms of vector operators for $n > 3$. More complicated objects called *alternating tensors* or *exterior differential forms* are employed. The specific details are usually discussed in courses in differential geometry.

A domain D in the plane or in three space is said to be **convex** if and only if the line segment $\overline{P_1 P_2}$ lies in D whenever P_1 and P_2 do. Figure 9-15 shows examples of convex and nonconvex domains. The next theorem establishes relationships between convex and simply connected domains.

Theorem 14. (a) *Any convex domain is simply connected.*
b) *Suppose D and D' are domains with D' simply connected. If there is a one-to-one twice continuously differentiable transformation with D as its domain and D' as its range, then D is simply connected.*

PROOF. (a) Suppose that $\vec{C}_0$ and $\vec{C}_1$ are paths in D. If $\vec{C}_0 : \mathbf{f}(t, 0)$, $\vec{C}_1 : \mathbf{f}(t, 1)$, we define

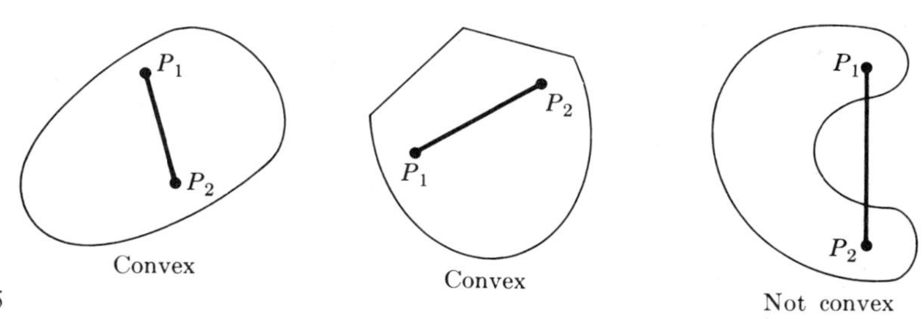

Fig. 9-15

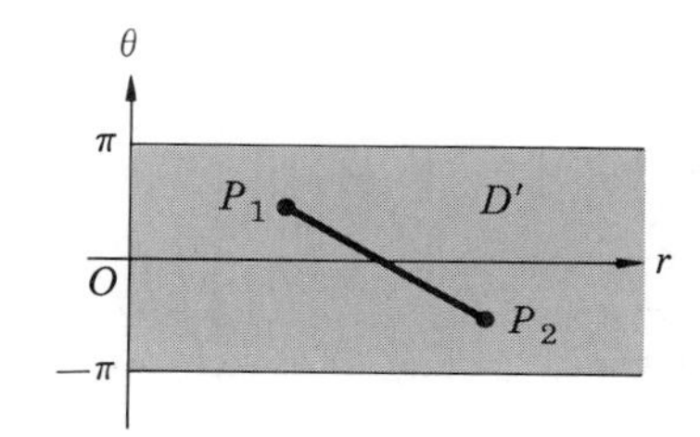

Fig. 9-16

$$\mathbf{f}(t, \tau) = (1 - \tau)\mathbf{f}(t, 0) + \tau\mathbf{f}(t, 1), \qquad 0 \le \tau \le 1.$$

As τ varies from 0 to 1, $\mathbf{f}(t, \tau)$ describes (for each t) the straight line segment joining the points $\mathbf{f}(t, 0)$ and $\mathbf{f}(t, 1)$. By convexity, this segment is in D.
b) If $\mathbf{f}(t, \tau)$ is the desired function in D' then, under the transformation, it would correspond to a function with similar properties in D.

Remark. The statement and proof of Theorem 14 are valid in a Euclidean space of any dimension.

Example. Let D be the set in the plane consisting of all points except the origin and the points on the negative x-axis. Show that D is simply connected.

Solution. The equations $x = r\cos\theta$, $y = r\sin\theta$ set up a one-to-one twice differentiable map of D onto the domain

$$D' = \{(r, \theta): r > 0, \quad -\pi < \theta < \pi\},$$

in the (r, θ)-plane. The inverse map is also twice differentiable. The domain D' is a half-infinite strip, and its convexity is easily verified. (See Fig. 9-16; see also Problem 14.) According to Theorem 14(a), D' is simply connected and then, by part (b), D is also.

Problems

In each of Problems 1 through 5, verify Theorem 11 by calculating the line integral $\int_{\vec{C}} \nabla u \cdot d\mathbf{r}$ for each of the given paths.

1. $u = x^2 - xy - y^2$;
   $\vec{C}_1 : x = \cos\theta, \quad y = \sin^2\theta, \quad z = 0, \quad 0 \le \theta \le \pi/2$;
   $\vec{C}_2$: straight segment from $(1, 0, 0)$ to $(0, 1, 0)$.

2. $u = x^2 - 2y^2 + xz + z^2$;
   $\vec{C}_1$: straight segment from $(1, 0, 1)$ to $(-1, 1, 2)$;
   $\vec{C}_2$: straight segment from $(1, 0, 1)$ to $(1, 0, 2)$ followed by straight segment from $(1, 0, 2)$ to $(-1, 1, 2)$.

3. $u = (x^2 + y^2 + z^2)^{-1/2}$;
   $\vec{C}_1 : x = \sin\theta\cos\theta, \quad y = \sin^2\theta, \quad z = \cos\theta, \quad 0 \le \theta \le \pi/2$;
   $\vec{C}_2$: straight segment from $(0, 0, 1)$ to $(0, 1, 1)$.

4. $u = z(x^2 + y^2)^{-1/2}$;
   $\vec{C}_1 : x = 3t, \quad y = -4t, \quad z = 5t^2, \quad 1 \le t \le 2$;
   $\vec{C}_2 : x = 3t, \quad y = -4t, \quad z = t^4 + 4, \quad 1 \le t \le 2$.

5. $u = \sin xy + \cos yz + \sin xz$;
   $\vec{C}_1$: straight segment from $(0, \pi/4, 1)$ to $(\pi/2, 1, 0)$;
   $\vec{C}_2$: straight segment from $(0, \pi/4, 1)$ to $(0, 1, \pi/4)$ followed by straight segment from $(0, 1, \pi/4)$ to $(\pi/2, 1, 0)$.

In each of Problems 6 through 10, decide whether Theorem 13 is valid in the given domain. If so, find the appropriate scalar field.

6. $\mathbf{v} = (2x + 8y - 2z)\mathbf{i} + (2y + 4z + 8x)\mathbf{j} + (2z - 2x + 4y)\mathbf{k}$; D is the interior of the ball $x^2 + y^2 + z^2 \le 1$.

7. $\mathbf{v} = 3(x^2 - yz)\mathbf{i} + 3(y^2 - xz)\mathbf{j} - 3xy\mathbf{k}$; D is the domain between the spheres $x^2 + y^2 + z^2 = 1, \quad x^2 + y^2 + z^2 = 9$.

8. $\mathbf{v} = (x^4 - 8y^2 + 2)\mathbf{i} + 2xyz\mathbf{j} + (y^2 - x^2)\mathbf{k}$; D is the interior of the ellipsoid $x^2 + 2y^2 + 3z^2 = 27$.

9. $$\mathbf{v} = \frac{x}{(x^2 + y^2 + z^2)^{2/3}}\mathbf{i} + \frac{y}{(x^2 + y^2 + z^2)^{2/3}}\mathbf{j} + \frac{z}{(x^2 + y^2 + z^2)^{2/3}}\mathbf{k};$$

   D is the domain between the cylinders

   $$x^2 + z^2 = 1 \quad \text{and} \quad x^2 + z^2 = 9.$$

10. $$\mathbf{v} = \frac{x}{x + y + z}\mathbf{i} + \frac{y}{x + y + z}\mathbf{j} + \frac{z}{x + y + z}\mathbf{k};$$

    D is the parallelepiped

    $$D = \{(x, y, z) : 1 \le x \le 2, \quad 1 \le y \le 3, \quad 2 \le z \le 4\}.$$

11. Prove that the intersection of any finite number of convex sets is a convex set.

12. Prove, using coordinates, that any half-plane (that is, the part of a plane on one side of a line) is convex.

13. Prove, using coordinates, that any half-space (that is, the portion of three-space on one side of a plane) is convex.

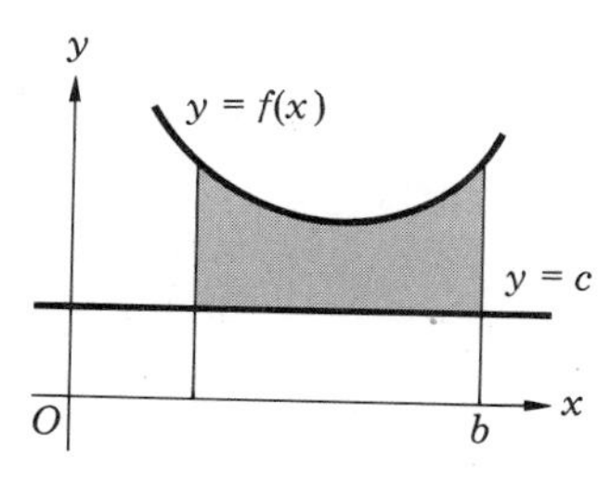

Fig. 9-17

Fig. 9-18

14. Using the results of Problems 11 and 12, show that the domain D' in Fig. 9-16 is convex.

15. Prove that any set of the form $a < x < b$, $c < y < f(x)$ in the plane, where f and f' are smooth in an interval containing $[a, b]$ in its interior, is simply connected. [*Hint*. Set up a transformation from this region onto a rectangle $a < \xi < b$, $0 < \eta < 1$; it is assumed that $f(x) > c$ for $a - h \leq x \leq b + h$ for some $h > 0$. (See Fig. 9-17.)]

16. Use Theorem 13 and the vector field

$$\mathbf{v} = (x^2 + y^2)^{-1}(-y\mathbf{i} + x\mathbf{j} + 0 \cdot \mathbf{k}),$$

$$\vec{C}: x = \cos\theta, \qquad y = \sin\theta, \qquad -\pi \leq \theta \leq \pi,$$

to show that the plane with the origin removed is not simply connected.

17. A torus is obtained by revolving a circle about an axis in its plane (provided the axis does not intersect the circle). (See Fig. 9-18.) The interior of the torus consists of all points whose cylindrical coordinates (r, θ, z) satisfy

$$(r - b)^2 + z^2 < a^2, \qquad 0 < a < b.$$

Using the example of Problem 16, show that the interior of a torus in three-space is not simply connected.

18. Show that the set of points (x, y) satisfying

$$\frac{x^2}{a^2} + \frac{y^2}{b^2} < 1$$

is convex. [*Hint*. Consider the function

$$\phi(t) = \frac{[x_1 + t(x_2 - x_1)]^2}{a^2} + \frac{[y_1 + t(y_2 - y_1)]^2}{b^2}, \qquad 0 \leq t \leq 1.$$

Draw a figure.]

19. Show that the totality of points in the plane which are not on the spiral $r = \theta$, $\theta \geq 0$, r and θ polar coordinates, is simply connected. Draw a figure.

20. Use the results of Problems 11 and 12 to show that the interior of every regular polygon in the plane is convex.

CHAPTER 10

Green's and Stokes' Theorems

1. Green's Theorem

The Fundamental Theorem of Calculus states that differentiation and integration are inverse processes. An appropriate extension of this theorem to double integrals of functions of two variables is known as Green's Theorem. Suppose that P and Q are smooth (i.e., continuously differentiable) functions defined in some region R of the plane. **A simple closed curve** is a curve that can be obtained as the union of two arcs which have only their endpoints in common. Thus a circle is the union of two half circles. Of course, any two points on a simple closed curve divide it into two arcs in this way. It is intuitively clear that a simple closed curve in the plane divides the plane into two regions, constituting the "interior" and the "exterior" of the curve. This fact which is surprisingly hard to prove, is not used in the proofs of any theorems. A **smooth simple closed curve** is one which has a parametric representation $x = x(t)$, $y = y(t)$, $a \leq t \leq b$, in which x, y, x', and y' are continuous and $[x'(t)]^2 + [y'(t)]^2 > 0$ and $x(b) = x(a)$, $x'(b) = x'(a)$, $y(b) = y(a)$, $y'(b) = y'(a)$. If Γ is a smooth simple closed curve which, together with its interior G, is in R, then the basic formula associated with Green's Theorem is

$$\iint_G \left(\frac{\partial Q}{\partial x} - \frac{\partial P}{\partial y}\right) dA = \oint_\Gamma (P\,dx + Q\,dy). \tag{1}$$

The symbol on the right represents the line integral taken in the counterclockwise sense, so that Γ is traversed with the interior of G always on the left.

We first establish Green's Theorem for regions which have a special

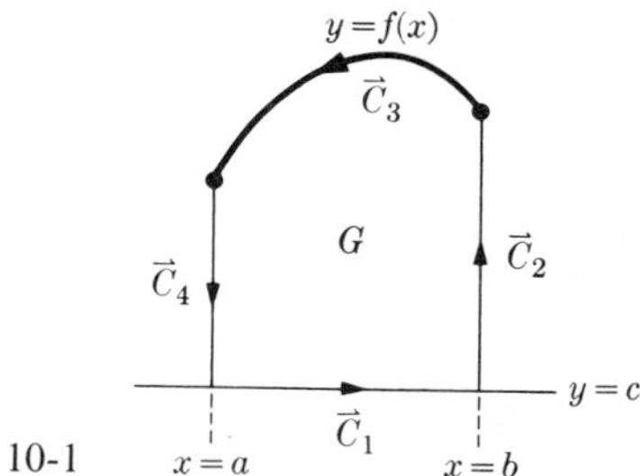

Fig. 10-1

shape. Then we show how the result for these special regions may be extended to yield the theorem for general domains.

Lemma 1. *Suppose that G is a region bounded by the straight lines $x = a$, $x = b$, $y = c$, and by an arc (situated above the line $y = c$) with equation*

$$y = f(x), \qquad a \le x \le b.$$

Assume that f is smooth (continuously differentiable). If $P(x, y)$ and $Q(x, y)$ are continuously differentiable in a region which contains G and its boundary then

$$\iint\limits_G \left(\frac{\partial Q}{\partial x} - \frac{\partial P}{\partial y}\right) dA = \oint\limits_{\partial G} (P\,dx + Q\,dy), \tag{2}$$

where the symbol ∂G denotes the boundary of G and the line integral is traversed in a counterclockwise sense.

PROOF. We establish the result by proving separately each of the formulas

$$-\iint\limits_G \frac{\partial P}{\partial y} dA = \oint\limits_{\partial G} P\,dx, \tag{2a}$$

$$\iint\limits_G \frac{\partial Q}{\partial x} dA = \oint\limits_{\partial G} Q\,dy. \tag{2b}$$

Figure 10-1 shows a typical region G with the arcs directed as shown by the arrows. To prove (2a), we change the double integral to an iterated integral and then employ the Fundamental Theorem of Calculus. We get

$$\begin{aligned} -\iint\limits_G \frac{\partial P}{\partial y} dA &= -\int_a^b \int_c^{f(x)} \frac{\partial P}{\partial y} dy\,dx \\ &= -\int_a^b \{P[x, f(x)] - P(x, c)\}\,dx. \end{aligned} \tag{3}$$

The right side of (3) may be written as the line integrals

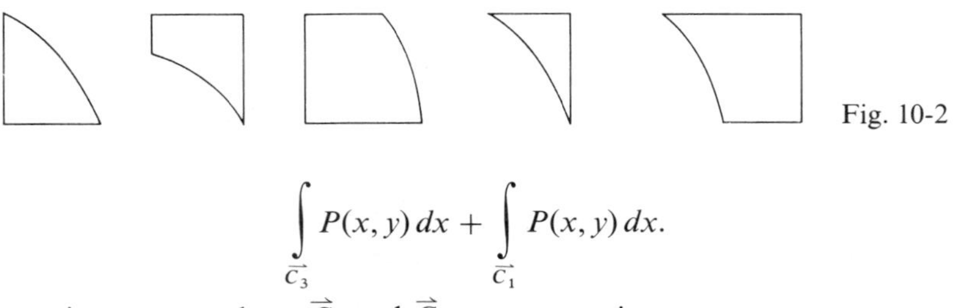

Fig. 10-2

$$\int_{\vec{C}_3} P(x, y)\, dx + \int_{\vec{C}_1} P(x, y)\, dx.$$

Since x is constant along $\vec{C}_2$ and $\vec{C}_4$, we may write

$$\int_{\vec{C}_2} P\, dx = \int_{\vec{C}_4} P\, dx = 0,$$

and so

$$-\iint_G \frac{\partial P}{\partial y}\, dA = \int_{\vec{C}_1+\vec{C}_2+\vec{C}_3+\vec{C}_4} P\, dx = \oint_{\partial G} P\, dx.$$

To prove (2b), we define

$$U(x, y) = \int_c^y Q(x, \eta)\, d\eta.$$

The formula for differentiating under the integral sign yields

$$U_x(x, y) = \int_c^y Q_x(x, \eta)\, d\eta, \qquad U_y(x, y) = Q(x, y), \qquad U_{yx} = Q_x = U_{xy}.$$

Therefore we may apply Theorem 11 of Chapter 9 to get

$$\int_{\partial G} (U_x\, dx + U_y\, dy) = 0 \Leftrightarrow \int_{\partial G} U_x\, dx = -\int_{\partial G} U_y\, dy = -\int_{\partial G} Q\, dy. \tag{4}$$

Now we use (2a) with $P = U_x$ to find

$$\iint_G Q_x\, dA = \iint_G U_{xy}\, dA = -\int_{\partial G} U_x\, dx.$$

Taking (4) into account, we conclude that

$$\iint_G Q_x\, dA = \int_{\partial G} Q\, dy.$$

By means of a simple change of coordinates or other minor adjustment, we see easily that the above lemma holds for all regions of the type shown in Fig. 10-2. The next important step is the observation that the result of Lemma 1 [i.e., Eq. (2)] may be established for any region which can be divided up into a finite number of regions, each of the type considered in the lemma. How this may be done is suggested in Fig. 10-3, which shows a region with a smooth simple closed curve as boundary divided by straight-

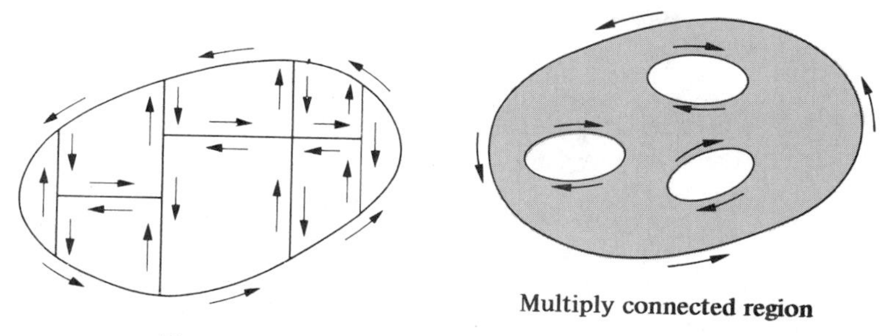

Multiply connected region

Fig. 10-3

Fig. 10-4

line segments into a number of regions of special type. It is clear that the double integral over the whole region is the sum of the double integral over the parts. In adding the line integrals we observe that the integrals over the interior segments must cancel, since each segment is directed in opposite senses when considered as part of the boundary of two adjacent special regions. Unfortunately, it is not true that every region for which formula (2) holds can be divided into special regions in the manner described above.

We shall not attempt to examine the most general type of domain for which formula (2) is valid, but shall note some examples of regions to which we may apply the lemma. The shaded domain shown in Fig. 10-4 is bounded by four smooth simple closed curves. It may be subdivided into regions to which the lemma applies. It is important to notice in such a case that the line integral as given in (2) must be traversed so that the region G always remains on the left—counterclockwise for the outer boundary curve and clockwise for the three inner boundary curves.

A simple closed curve is **piecewise smooth** if it is made up of a finite number of smooth arcs (i.e., having parametric representations as above) and if these arcs are joined at points called corners. A **corner** is the juncture of two smooth arcs which have limiting tangent lines *making a positive angle* (Fig. 10-5). For a piecewise smooth boundary with corners, it is possible to use Lemma 1, thus establishing formula (2) for any region which has a boundary consisting of a piecewise smooth simple closed curve. Since polygons have piecewise smooth boundaries, they are included in the collection of regions for which (2) is valid. We now state Green's Theorem in a form which is sufficiently general for most applications.

Theorem 1 (Green's Theorem in the Plane). *Suppose that G is a region with a boundary consisting of a finite number of piecewise smooth simple closed curves, no two which intersect. Suppose that P, Q are continuously differentiable functions defined in a region which contains G and ∂G. Then*

$$\iint_G \left(\frac{\partial Q}{\partial x} - \frac{\partial P}{\partial y}\right) dA = \oint_{\partial G} (P\,dx + Q\,dy), \tag{5}$$

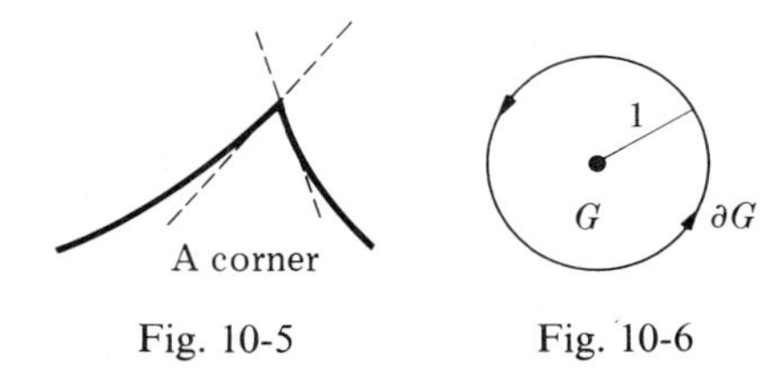

Fig. 10-5 Fig. 10-6

where the integral on the right is defined to be the sum of the integrals over the boundary curves, each of which is directed so that G is on the left.

The proof of Green's Theorem for all possible regions which may be subdivided into regions of special type has been described above in a discursive manner. In the next section we provide a proof which is sufficiently general to be used in most problems in advanced analysis.

Green's Theorem may be stated in vector form. We write

$$\mathbf{v} = P\mathbf{i} + Q\mathbf{j} + 0 \cdot \mathbf{k}$$

and denote by $\mathbf{r}(t)$ the vector from the origin O of a rectangular coordinate system in the plane to the boundary ∂G of G. We interpret

$$\operatorname{curl} \mathbf{v} = \left(\frac{\partial Q}{\partial x} - \frac{\partial P}{\partial y}\right)\mathbf{k}$$

as the scalar function $[(\partial Q/\partial x) - (\partial P/\partial y)]$ in the $\mathbf{i}$, $\mathbf{j}$-plane. We call this expression the **scalar curl of v**, although we use the same symbol. Then, under the hypotheses of Theorem 1, we have the formula

$$\iint_G \operatorname{curl} \mathbf{v}\, dA = \oint_{\partial G} \mathbf{v} \cdot d\mathbf{r}.$$

It is a simple matter to verify that this formula is identical with (5). However, the vector formulation has the advantage of exhibiting the invariance of the result under a change of coordinates. Furthermore, the nature of the extension to three-space is apparent from the vector formulation.

We now illustrate Green's Theorem in the plane with several examples.

EXAMPLE 1. Verify Green's Theorem when $P(x, y) = 2y$, $Q(x, y) = 3x$ and G is the unit disk,

$$G = \{(x, y) : x^2 + y^2 \leq 1\}$$

(see Fig. 10-6).

SOLUTION. We have

$$\frac{\partial Q}{\partial x} - \frac{\partial P}{\partial y} = 3 - 2 = 1.$$

Therefore

$$\iint_G \left(\frac{\partial Q}{\partial x} - \frac{\partial P}{\partial y}\right) dA = \iint_G dA = \pi \cdot 1^2 = \pi.$$

The boundary ∂G is given by $x = \cos\theta$, $y = \sin\theta$, $-\pi \le \theta \le \pi$. Hence

$$\oint_{\partial G} (P\,dx + Q\,dy) = \int_{-\pi}^{\pi} \{2(\sin\theta)\,d(\cos\theta) + 3(\cos\theta)\,d(\sin\theta)\}$$
$$= \int_{-\pi}^{\pi} (-2\sin^2\theta + 3\cos^2\theta)\,d\theta = -2\pi + 3\pi = \pi.$$

EXAMPLE 2. If G is the unit disk as in Example 1, use Green's Theorem to evaluate

$$\int_{\partial G} [(x^2 - y^3)\,dx + (y^2 + x^3)\,dy].$$

SOLUTION. Here, $P = x^2 - y^3$, $Q = y^2 + x^3$. Therefore

$$\int_{\partial G} [(x^2 - y^3)\,dx + (y^2 + x^3)\,dy] = \iint_G 3(x^2 + y^2)\,dA$$
$$= 3\int_0^{2\pi}\int_0^1 r^2 \cdot r\,dr\,d\theta = \frac{3\pi}{2}.$$

EXAMPLE 3. Let G be the region outside the unit circle which is bounded on the left by the parabola $y^2 = 2(x + 2)$ and on the right by the line $x = 2$. (See Fig. 10-7.) Use Green's Theorem to evaluate

$$\int_{\vec{C}_1} \left(\frac{-y}{x^2 + y^2}\,dx + \frac{x}{x^2 + y^2}\,dy\right),$$

where $\vec{C}_1$ is the oriented outer boundary of G as shown in Fig. 10-7.

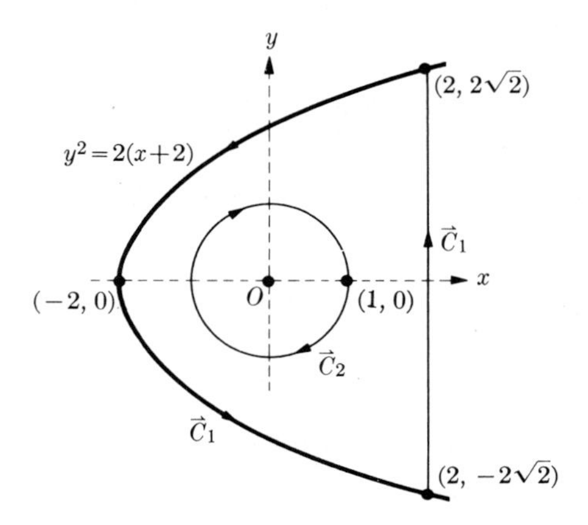

Fig. 10-7

SOLUTION. We write $P = -y/(x^2 + y^2)$, $Q = x/(x^2 + y^2)$ and observe that P and Q have singularities at the origin. Denoting the boundary of the unit disk oriented *clockwise* by $\overrightarrow{C_2}$ and noting that $(\partial Q/\partial x) - (\partial P/\partial y) = 0$, we use Green's Theorem to write

$$\int_{\overrightarrow{C_1}+\overrightarrow{C_2}} \left(-\frac{y}{x^2+y^2}dx + \frac{x}{x^2+y^2}dy\right) = 0.$$

Therefore

$$\int_{\overrightarrow{C_1}} (P\,dx + Q\,dy) = -\int_{\overrightarrow{C_2}} (P\,dx + Q\,dy) = \int_{-\overrightarrow{C_2}} (P\,dx + Q\,dy),$$

where $-\overrightarrow{C_2}$ is the unit circle oriented *counterclockwise*. Since $x = \cos\theta$, $y = \sin\theta$, $-\pi \le \theta \le \pi$ on the unit circle $-\overrightarrow{C_2}$, we obtain

$$\int_{\overrightarrow{C_1}} (P\,dx + Q\,dy) = \int_{-\pi}^{\pi} (\sin^2\theta\cos^2\theta)\,d\theta = 2\pi.$$

EXAMPLE 4. Let $\mathbf{v} = -\frac{1}{2}y\mathbf{i} + \frac{1}{2}x\mathbf{j}$ be defined in a region G with area A. Show that

$$A = \int_{\partial G} \mathbf{v}\cdot d\mathbf{r}.$$

SOLUTION. We apply Green's Theorem and obtain

$$\int_{\partial G} \mathbf{v}\cdot d\mathbf{r} = \iint_G \operatorname{curl}\mathbf{v}\,dA = \iint_G (\tfrac{1}{2} + \tfrac{1}{2})\,dA = A,$$

in which we have used the scalar interpretation of curl $\mathbf{v}$.

PROBLEMS

In each of Problems 1 through 8, verify Green's Theorem.

1. $P(x, y) = -y$, $Q(x, y) = x$; $G: 0 \le x \le 1, 0 \le y \le 1$
2. $P = 0$, $Q = x$; G is the region outside the unit circle, bounded below by the parabola $y = x^2 - 2$ and bounded above by the line $y = 2$.
3. $P = xy$, $Q = -2xy$; $G = \{(x, y): 1 \le x \le 2, 0 \le y \le 3\}$
4. $P = e^x \sin y$, $Q = e^x \cos y$; $G = \{(x, y): 0 \le x \le 1, 0 \le y \le \pi/2\}$
5. $P = \frac{2}{3}xy^3 - x^2y$, $Q = x^2y^2$; G is the triangle with vertices at $(0, 0)$, $(1, 0)$, and $(1, 1)$.
6. $P = 0$, $Q = x$; G is the region inside the circle $x^2 + y^2 = 4$ and outside the circles $(x - 1)^2 + y^2 = \frac{1}{4}$, $(x + 1)^2 + y^2 = \frac{1}{4}$.
7. $\mathbf{v} = (x^2 + y^2)^{-1}(-y\mathbf{i} + x\mathbf{j})$; G is the region between the circles $x^2 + y^2 = 1$ and $x^2 + y^2 = 4$.

8. $P = 4x - 2y$, $Q = 2x + 6y$; G is the interior of the ellipse $x = 2\cos\theta$, $y = \sin\theta$, $-\pi \le \theta \le \pi$.

In each of Problems 9 through 14, compute the area $A(G)$ of G by using the formula

$$A(G) = \int_{\partial G} x\,dy,$$

which is valid because of Green's Theorem.

9. G is the triangle with vertices at (1, 1), (4, 1), and (4, 9).
10. G is the triangle with vertices (2, 1), (3, 4), and (1, 5).
11. G is the region given by $G = \{(x, y) : 0 \le x \le y^2, 1 \le y \le 3\}$.
12. G is the region bounded by the line $y = x + 2$ and the parabola $y = x^2$.
13. G is the region in the first quadrant bounded by the lines $4y = x$ and $y = 4x$ and the hyperbola $xy = 4$.
14. G is the region interior to the ellipse

$$\frac{x^2}{16} + \frac{y^2}{9} = 1.$$

In each of Problems 15 through 21, compute $\int_{\partial G} \mathbf{v} \cdot d\mathbf{r}$, using Green's Theorem.

15. $\mathbf{v} = (\frac{4}{5}xy^5 + 2y - e^x)\mathbf{i} + (2xy^4 - 4\sin y)\mathbf{j}$; G: $1 \le x \le 2$, $1 \le y \le 3$
16. $\mathbf{v} = (2xe^y - x^2y - \frac{1}{3}y^3)\mathbf{i} + (x^2e^y + \sin y)\mathbf{j}$; G: $x^2 + y^2 \le 1$
17. $\mathbf{v} = 2xy^2\mathbf{i} + 3x^2y\mathbf{j}$; G is the interior of the ellipse

$$\frac{x^2}{a^2} + \frac{y^2}{b^2} = 1.$$

18. $\mathbf{v} = -y\mathbf{i} + x\mathbf{j}$; G is the interior of the circle $(x - 1)^2 + y^2 = 1$.
19. $\mathbf{v} = (\cosh x - 2)\sin y\mathbf{i} + \sinh x \cos y\mathbf{j}$;
    $G = \{(x, y) : 0 \le x \le 1, 0 \le y \le (\pi/2)\}$
20. $\mathbf{v} = 2\operatorname{Arctan}(y/x)\mathbf{i} + \log(x^2 + y^2)\mathbf{j}$;
    $G = \{(x, y) : 1 \le x \le 2, -1 \le y \le 1\}$
21. $\mathbf{v} = -3x^2y\mathbf{i} + 3xy^2\mathbf{j}$; $G = \{(x, y) : -a \le x \le a, \quad 0 \le y \le \sqrt{a^2 - x^2}\}$
22. Evaluate

$$\int_{\vec{C}} \frac{x\,dy - y\,dx}{x^2 + y^2},$$

where $\vec{C}$ consists of the arc of the parabola $y = x^2 - 1$, $-1 \le x \le 2$, followed by the straight segment from (2, 3) to $(-1, 0)$. Do this by applying Green's Theorem to the region G interior to $|\vec{C}|$ and exterior to a small circle of radius ρ centered at the origin.

23. Evaluate

$$\int_{\vec{C}} \frac{x\,dx + y\,dy}{x^2 + y^2}$$

where $\vec{C}$ is the path described in Problem 22.

24. Suppose that $f(x, y)$ satisfies the Laplace equation ($f_{xx} + f_{yy} = 0$) in a region G. Show that

$$\int_{\partial G^*} (f_y\,dx - f_x\,dy) = 0,$$

where G^* is any region interior to G.

25. If (x^*, y^*) is the location of the center of gravity of a plane region G of uniform density, show that

$$x^* = \frac{\int_{\partial G} x^2\,dy}{2\int_{\partial G} x\,dy}, \qquad y^* = \frac{\int_{\partial G} y^2\,dx}{2\int_{\partial G} y\,dx}.$$

26. If $f_{xx} + f_{yy} = 0$ in a region R and v is any smooth function, use the identity $(vf_x)_x = vf_{xx} + v_x f_x$ and a similar one for the derivative with respect to y, to prove that

$$-\int_{\partial G} v(f_y\,dx - f_x\,dy) = \iint_{\partial G} (v_x f_x + v_y f_y)\,dA,$$

where G is any region interior to R.

2. Proof of Green's Theorem

We define a function ϕ by the formula

$$\phi(s) = \begin{cases} 1, & 0 \le s \le 1, \\ (2s-1)(s-2)^2, & 1 \le s \le 2, \\ 0, & 2 \le s. \end{cases} \tag{1}$$

We extend the definition to negative values of s by making ϕ even: $\phi(-s) = \phi(s)$. It is a simple matter to verify that $\phi(1) = 1$, $\phi(2) = 0$ and, therefore,

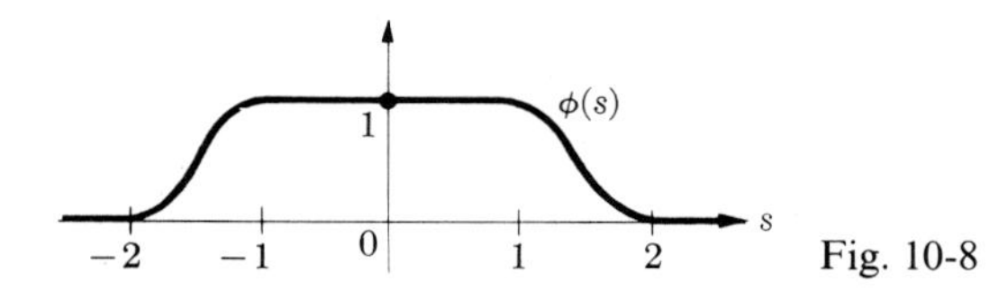

Fig. 10-8

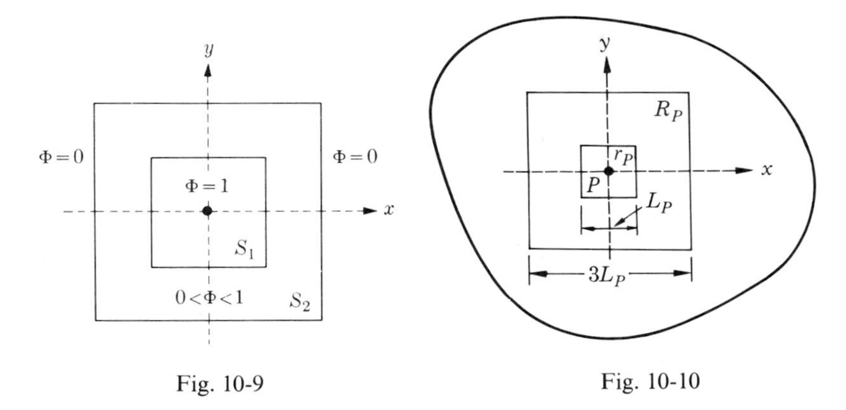

Fig. 10-9

Fig. 10-10

that ϕ is continuous everywhere. Furthermore, since $\phi'(1) = 0$, $\phi'(2) = 0$, we see that ϕ has a continuous first derivative everywhere. Its graph is shown in Fig. 10-8. The function $\phi(s/b)$ with $b > 0$ has the same general behavior as $\phi(s)$ except that the scale is changed. We note that $\phi(s/b)$ is 1 for $|s| \leq b$, is between 0 and 1 for $b \leq |s| \leq 2b$, and vanishes for $|s| \geq 2b$.

Functions of two variables with analogous properties are defined by taking products. The function

$$\Phi(x, y) = \phi(x)\phi(y)$$

is 1 in the square

$$S_1 = \{(x, y): -1 \leq x \leq 1, \quad -1 \leq y \leq 1\};$$

is 0 *outside* the square

$$S_2 = \{(x, y): -2 \leq x \leq 2, \quad -2 \leq y \leq 2\};$$

and is between 0 and 1 in the region between S_1 and S_2 (Fig. 10-9). The function Φ and its first partial derivatives are continuous everywhere. A change of scale shows that the function $\Phi(x/b, y/c)$ has the same properties as $\Phi(x, y)$ in a rectangle of width $2b$ and height $2c$.

The function Φ is the basic quantity in performing a decomposition of a region G in the plane. This decomposition will be used to prove Green's Theorem. Let a bounded region G have for its boundary a finite number of smooth arcs which may form corners at points where they meet. With each point P of G and its boundary ∂G, we associate both a rectangular coordinate system which has P as its origin and a function Φ of the type described above. We consider three cases:

1) If P is interior to G, we select any two perpendicular lines intersecting at P as coordinate axes (properly oriented, of course). We label these x and y. Choose any square with center at P, with sides parallel to the axes x and y, and situated entirely inside G. Denote this square by R_P and the length of one side by $3L_P$. The parallel square with side L_P and center at P

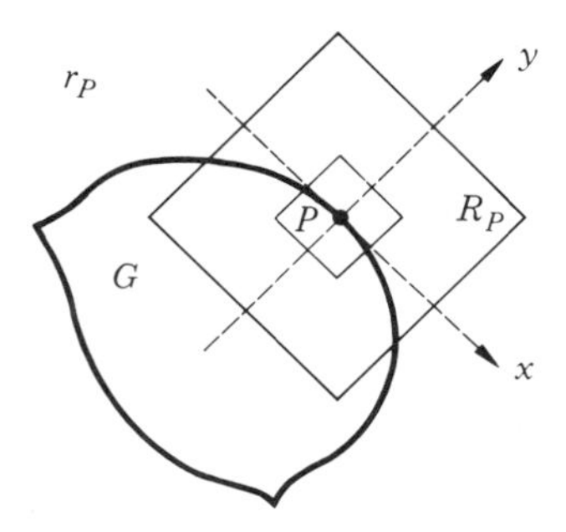

Fig. 10-11

we designate r_P (Fig. 10-10). The function

$$\Phi_P(x, y) = \Phi\left(\frac{x}{L_P}, \frac{y}{L_P}\right) = \phi\left(\frac{x}{L_P}\right) \cdot \phi\left(\frac{y}{L_P}\right)$$

has the property that it is smooth, is 1 in r_P, and is 0 outside of a square halfway between r_P and R_P. We may set up such a coordinate system and pairs of squares for each point P interior to G. Of course, if P is very near the boundary, the quantity $3L_P$ will be very small; nevertheless, the selections can be made and functions Φ_P formed.

2) If P is on the boundary of G but interior to a smooth arc, we choose the positive y axis in the direction of the *exterior* normal, as shown in Fig. 10-11. The x axis is then situated along the tangent, properly oriented. We choose a rectangle r_P of width L_P and height H_P parallel to the axes and centered at P. The rectangle of width $3L_P$ and height $3H_P$ is denoted R_P. The numbers L_P and H_P are taken so small that we may express the portion of the boundary in R_P by an equation

$$y = f(x),$$

where f is smooth. Furthermore, we make the rectangles so small, if necessary, that the conditions

$$|f(x)| < H_P \qquad \text{for } |x| < L_P,$$
$$|f(x)| < 3H_P \qquad \text{for } |x| < 3L_P,$$

are satisfied. In other words, the boundary arc must enter and leave the "sides" of the rectangles* r_P and R_P. With each such boundary point P, we associate the function

$$\Phi_P(x, y) = \phi\left(\frac{x_P}{L_P}\right)\phi\left(\frac{y_P}{H_P}\right),$$

* That such rectangles can always be found follows from the fact that a smooth arc in the plane has a representation $x = x(t)$, $y = y(t)$, $a \le t \le b$, in which x and y are smooth with $[x'(t)]^2 + [y'(t)]^2 > 0$. If $t = 0$ at P, we have $y'(0) = 0$, so $x'(0) \neq 0$ and the function x has a differentiable inverse T_P; thus we have $x(t) = x$ and $t = T_P(x)$. Hence $y = x[T_P(x)] = f(x)$ for $|x|$ small.

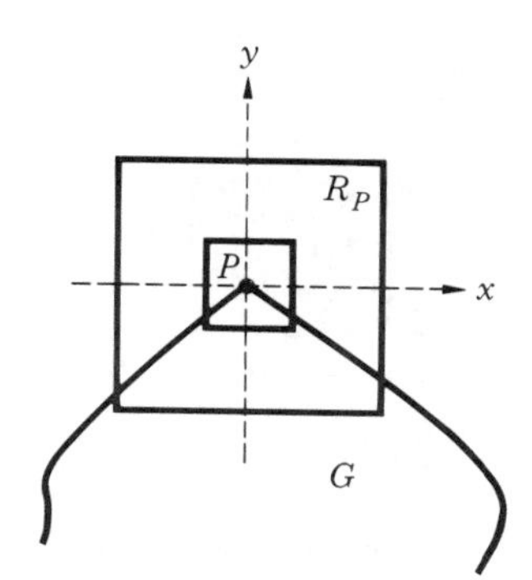

Fig. 10-12

which is 1 in the rectangle r_P and 0 outside a rectangle halfway between r_P and R_P.

3) If P is a corner point of the boundary, we choose the positive y axis along the bisector of the angle between the tangents at P and pointing outside of G; the x axis is selected accordingly (Fig. 10-12). The rectangles r_P and R_P are chosen as in case (2) above, and the function

$$\Phi_P(x, y) = \phi\left(\frac{x_P}{L_P}\right)\phi\left(\frac{y_P}{H_P}\right)$$

is defined as before.

The above description shows that with each point of the region G and its boundary ∂G we may associate a point P and a rectangle (or square) r_P. From this fact it is possible to conclude that a *finite number* of the interiors of the rectangles $\{r_P\}$ cover G and ∂G. We label these covering rectangles

$$r_1, r_2, \ldots, r_k$$

and we denote their centers $P_1, P_2, \ldots, P_k$. The associated functions are

$$\Phi_{P_1}, \Phi_{P_2}, \ldots, \Phi_{P_k}.$$

Let Q be any point in G. Then Q is in some r_i. According to the way we defined the functions Φ_P, we see that

$$\Phi_{P_i}(Q) = 1,$$

since Φ_{P_i} is identically equal to 1 in all of r_i. Thus, for every point Q of G, we have

$$\Phi_{P_1}(Q) + \Phi_{P_2}(Q) + \cdots + \Phi_{P_k}(Q) \geq 1.$$

We now define the function

$$\psi_i(Q) = \frac{\Phi_{P_i}(Q)}{\Phi_{P_1}(Q) + \Phi_{P_2}(Q) + \cdots + \Phi_{P_k}(Q)}.$$

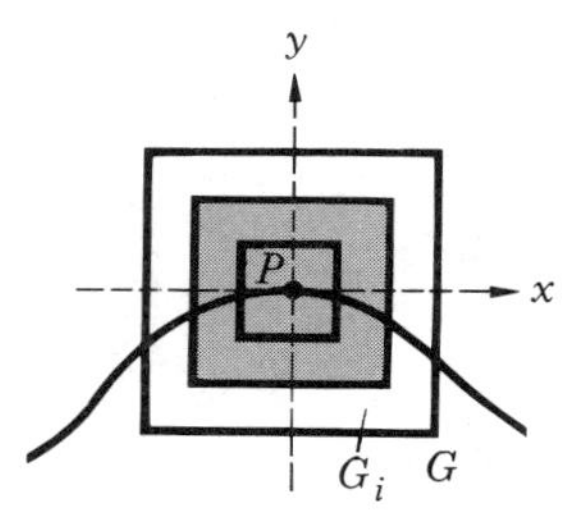

Fig. 10-13

Then each $\psi_i(Q)$ is smooth on all the r_i and, since $\Phi_{P_i}(Q)$ is zero outside a rectangle halfway between r_i and R_i, $\psi_i(Q)$ is also.

Definition. The sequence $\psi_1, \psi_2, \ldots, \psi_k$ is called a **partition of unity**. The term partition of unity comes from the formula

$$\sum_{i=1}^{k} \psi_i(Q) = 1,$$

valid for every point Q not only in G but also in any of the rectangles r_1, $r_2, \ldots, r_k$. We have defined a finite sequence of functions $\psi_1, \psi_2, \ldots, \psi_k$, which add up to 1 identically, and yet each member of the sequence vanishes, except for a small rectangle about a given point. This form of decomposition, is extremely useful not only in analysis but also in geometry and topology. It has the virtue of reducing certain types of global problems to local ones.

We now prove Green's Theorem i.e., Theorem 1 of Section 1. Let $\mathbf{v} = P(x, y)\mathbf{i} + Q(x, y)\mathbf{j}$ be a smooth vector field given in a region containing G. We define

$$\mathbf{v}_i = \psi_i \mathbf{v},$$

where $\psi_1, \psi_2, \ldots, \psi_k$ is a partition of unity, as described above. Then it is clear that for all points Q in G,

$$\mathbf{v}(Q) = \sum_{i=1}^{k} \mathbf{v}_i(Q).$$

Green's Theorem will then follow for $\mathbf{v}$ if we prove it for each vector field $\mathbf{v}_i$. Recalling that r_i and R_i are the rectangles associated with ψ_i, we define

$$G_i = G \cap R_i.$$

That is, G_i consists of those points of R_i contained in G. If P_i, the center of R_i, is interior to G, then so is R_i; in this case $G_i = R_i$. It is apparent that G_i is a region of the special type discussed in Lemma 1 on page 497. We have

$$\iint_{G_i} \operatorname{curl} \mathbf{v}_i \, dA = \int_{\partial G_i} \mathbf{v}_i \cdot d\mathbf{r} = 0,$$

because $\mathbf{v}_i = \psi_i \mathbf{v}$ is $\mathbf{0}$ near and on the boundary ∂G_i. Therefore (2) holds for all those R_i which have centers in G. Now suppose that P_i is on the boundary of G. We see that $G_i = G \cap R_i$ is again a region of special type, and so Lemma 1 of page 497 applies. Since $\mathbf{v}_i = \mathbf{0}$ outside G_i and on the part of ∂G_i (see Fig. 10-13) which is interior to G, we have

$$\iint_G \operatorname{curl} \mathbf{v}_i \, dA = \iint_{G_i} \operatorname{curl} \mathbf{v}_i \, dA = \int_{\partial G_i} \mathbf{v}_i \cdot d\mathbf{r} = \int_{\partial G} \mathbf{v}_i \cdot d\mathbf{r}.$$

The result holds in this case also, and so the theorem is established.

Problems

1. Given the function

$$\phi(s) = \begin{cases} 1, & 0 \le s \le 1, \\ (6s^2 - 9s + 4)(2 - s)^3, & 1 \le s \le 2, \\ 0, & 2 \le s < \infty, \end{cases}$$

and $\phi(-s) = \phi(s)$. Show that, for all s, ϕ is twice continuously differentiable and that $0 \le \phi \le 1$.

2. Using the function in Problem 1, construct a twice continuously differentiable function of two variables which is one in a given rectangle, zero outside a larger similarly placed rectangle, and between zero and one in the region between the rectangles.

3. Defining the function $F(x, y) = \phi(x^2 + y^2)$ where ϕ is the function given by (1), show that $F(x, y)$ is one in the unit disk, vanishes outside the circle $x^2 + y^2 = 2$, and is between zero and one otherwise.

4. Using the result of Problem 3, find a smooth function which is one in a disk of radius a, vanishes outside a concentric circle of radius $2a$, and is between zero and one otherwise.

5. Same as Problem 4, except that the function is to be twice continuously differentiable. (Use the function in Problem 1.)

6. Show how to construct a function $G(x, y, z)$ which is smooth, is one inside a rectangular box, is zero outside of a larger box, and is between zero and one in the region between boxes.

7. By considering an expression of the form

$$P(s)(2 - s)^4,$$

where P is a fourth-degree polynomial, show how to construct a function which is one for $|s| \le 1$, zero for $|s| \ge 2$, is between zero and one for $1 \le |s| \le 2$, and is three times continuously differentiable.

8. Show how to construct a smooth function which is one in the unit ball, vanishes outside the sphere $x^2 + y^2 + z^2 = 4$, and is between zero and one between the two spheres.

9. Show, by sketching the appropriate rectangles, how a partition of unity would be made for the triangle with vertices at $(0, 0)$, $(1, 0)$, $(0, 1)$. Do not construct the functions analytically.

10. Show, by sketching the appropriate rectangles, how a partition of unity would be made for the interior of the ellipse $4x^2 + 9y^2 = 36$.

11. Define the function

$$f(x) = \begin{cases} 0, & \text{if } 0 \le x \le 1, \\ e^{1/(x-2)} \cdot e^{1/(1-x)}, & \text{if } 1 < x < 2, \\ 0, & \text{if } 2 \le x. \end{cases}$$

Show that f has (continuous) derivatives of all orders.

12. Let f be the function of Problem 11. Define the function

$$F(x) = \frac{\displaystyle\int_x^2 f(t)\,dt}{\displaystyle\int_1^2 f(t)\,dt}.$$

Show that F is one for $0 \le x \le 1$, zero for $x \ge 2$, and between zero and one for $1 \le x \le 2$. Furthermore, show that F has derivatives of all orders.

13. Use the function in Problem 12 to find a function with derivatives of all orders which is one in a rectangle, vanishes outside a larger similarly placed rectangle, and is between zero and one in the region between the rectangles.

14. By considering $F(x^2 + y^2)$ where F is the function in Problem 12, show that F is one in a disk, zero outside a larger concentric disk, and between zero and one in the ring between the circles.

15. Describe a partition of unity with disks instead of rectangles.

16. Set $\mathbf{v} = x^2\vec{i} + (2y + x)j$. Set S be the ellipse $x^2 + 4y^2 = 16$. By making a partition of unity for the region interior to S, find the functions $\mathbf{v}_i$ as in the proof of Green's Theorem. Then verify the result of the Theorem for $\mathbf{v}$.

3. Change of Variables in a Multiple Integral

One of the principal techniques used in the evaluation of single integrals is the method of substitution. We do this by making a substitution or change of variable of the form $x = g(u)$ which enables us to transform an integral

$$\int f(x)\,dx \tag{1}$$

into

$$\int f[g(u)]g'(u)\,du. \tag{2}$$

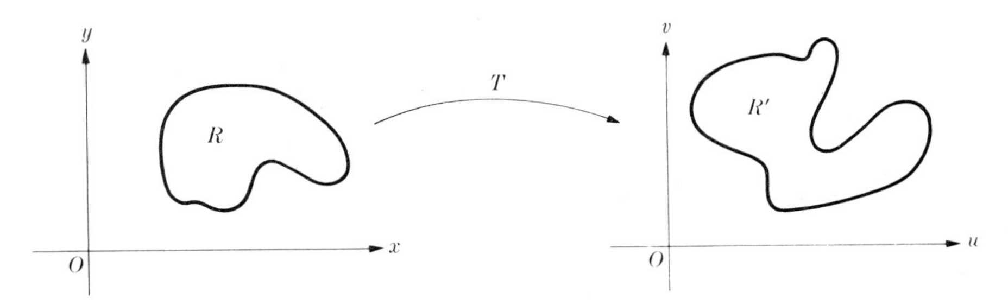

Fig. 10-14

It sometimes happens that (2) is simpler to evaluate than (1). If we start with a definite integral of the form

$$\int_a^b f(x)\,dx$$

then, after the change of variable $x = g(u)$, we obtain

$$\int_c^d f[g(u)]g'(u)\,du, \tag{3}$$

where $g(c) = a$ and $g(d) = b$. This method is always valid, provided that g and g' are continuous and f is defined and continuous for all the values of $g(u)$ for u on $[c, d]$.

In this section we apply Green's Theorem to obtain a general formula for a change of variables in a multiple integral. Let R be a bounded region in the xy plane and suppose that a transformation

$$T\colon\quad u = f(x, y), \qquad v = g(x, y) \tag{4}$$

takes R into a region R' in the uv plane. See Fig. 10-14. We assume that T is continuously differentiable throughout R and that it is one-to-one. Furthermore, we suppose that the Jacobian

$$J\left(\frac{u, v}{x, y}\right) = \begin{vmatrix} \dfrac{\partial u}{\partial x} & \dfrac{\partial u}{\partial y} \\[2mm] \dfrac{\partial v}{\partial x} & \dfrac{\partial v}{\partial y} \end{vmatrix}$$

does not vanish at any point of R. That is, J is always positive or always negative. We first establish a formula for the area of R' in terms of the area of R and the transformation T.

Theorem 2. *Suppose that T, given by* (4), *is a one-to-one, twice continuously differentiable transformation which takes a bounded region R with finite area $A(R)$ into a bounded region R' with area $A(R')$. Assume that the boundaries of R and R' consist of a finite number of piecewise smooth arcs, and that the*

Jacobian

$$J\left(\frac{u, v}{x, y}\right)$$

is never zero for $(x, y) \in R$. *Then*

$$A(R') = \iint_R \left| J\left(\frac{u, v}{x, y}\right) \right| dA_{xy}. \tag{5}$$

PROOF. Theorem 1 (Green's Theorem) applied to the region R' takes the form

$$\iint_{R'} \left(\frac{\partial Q}{\partial u} - \frac{\partial P}{\partial v}\right) dA_{uv} = \oint_{\partial R'} (P\,du + Q\,dv).$$

We choose $P \equiv 0$ and $Q = u$, and we find

$$A(R') = \iint_{R'} 1 \cdot dA_{uv} = \oint_{\partial R'} u\,dv.$$

Substituting from (4), we obtain

$$A(R') = \oint_{\partial R'} u\,dv = \oint_{\partial R} f(x, y)\left[\frac{\partial g}{\partial x} dx + \frac{\partial g}{\partial y} dy\right]. \tag{6}$$

The integral on the right is integrated in a counterclockwise direction if T preserves the orientation of the boundary of R; otherwise it is taken in a clockwise direction. Equivalently, the integration in (6) is counterclockwise or clockwise according as J is positive or negative in R. We now apply Green's Theorem to the integral on the right in (6) by setting

$$P = f\frac{\partial g}{\partial x} \qquad \text{and} \qquad Q = f\frac{\partial g}{\partial y}.$$

We find

$$A(R') = \oint_{\partial R} \left(f\frac{\partial g}{\partial x} dx + f\frac{\partial g}{\partial y} dy\right) = \pm \iint_R \left(\frac{\partial f}{\partial x}\frac{\partial g}{\partial y} - \frac{\partial g}{\partial x}\frac{\partial f}{\partial y}\right) dA_{xy}.$$

Since the area $A(R')$ is always positive, we obtain (5).

If T is the identity mapping, so that $u = x$ and $v = y$, then $J \equiv 1$ and $A(R') = A(R)$, as it should be. Hence the Jacobian is a measure of the distortion in the area which the transformation T introduces when a region R is mapped into a region R'.

In evaluating multiple integrals, we may use a transformation such as (4) for a change of variables of integration. The next theorem states the rule for the appropriate substitution in each case.

Theorem 3. *Suppose that the regions R and R' and the transformation T are as in Theorem 2. Let F be a continuous function defined on $R' \cup \partial R'$. We define $G(x, y) = F[f(x, y), g(x, y)]$. Then we have*

$$\iint_{R'} F(u, v)\, dA_{uv} = \iint_{R} G(x, y) \left| J\left(\frac{u, v}{x, y}\right) \right| dA_{xy}. \tag{7}$$

Sketch of Proof. We first divide the region R into subregions $R_1, R_2, \ldots, R_n$ as is customary in the definition of a double integral. The areas of these regions are denoted $A(R_1), \ldots, A(R_n)$. The transformation T maps the regions $R_1, R_2, \ldots, R_n$ into regions $R_1', R_2', \ldots, R_n'$ in the uv plane, and the latter regions have areas $A(R_1'), A(R_2'), \ldots, A(R_n')$. According to Theorem 2, we have

$$A(R_k') = \iint_{R_k} \left| J\left(\frac{u, v}{x, y}\right) \right| dA_{xy}, \qquad k = 1, 2, \ldots, n. \tag{8}$$

We apply the **Theorem of the Mean for double integrals** which states that if f is continuous over a region R in the plane with area $A(R)$ and if $m \le f(x, y) \le M$ for all (x, y) in R, then there is a point $(\bar{x}, \bar{y})$ in R such that

$$\iint_{R} f(x, y)\, dA_{xy} = f(\bar{x}, \bar{y}) A(R).$$

The proof is similar to that for single integrals. For the integral in (8), we find that there is a point $P_k(x_k, y_k)$ in R_k such that

$$A(R_k') = \left| J\left(\frac{u, v}{x, y}\right) \right|_{\substack{x = x_k \\ y = y_k}} A(R_k), \qquad k = 1, 2, \ldots, n.$$

The transformation T takes the point P_k into a point P_k' in R_k', and P_k' has coordinates which we denote by u_k, v_k; that is,

$$u_k = f(x_k, y_k), \qquad v_k = g(x_k, y_k).$$

From the definition of F and G, we have

$$F(u_k, v_k) = G(x_k, y_k),$$

and so we can form the sum

$$\sum_{k=1}^{n} F(u_k, v_k) A(R_k') = \sum_{k=1}^{n} G(x_k, y_k) \left| J\left(\frac{u, v}{x, y}\right) \right|_{\substack{x = x_k \\ y = y_k}} A(R_k). \tag{9}$$

We let $n \to \infty$ and the norm of the subdivision of R tend to zero, and employ the tools used in establishing the existence of double integrals for continuous functions. We recognize the left and right sides of (9) as sums which tend to the corresponding integrals in (7), a valid conclusion when the integrands are continuous.

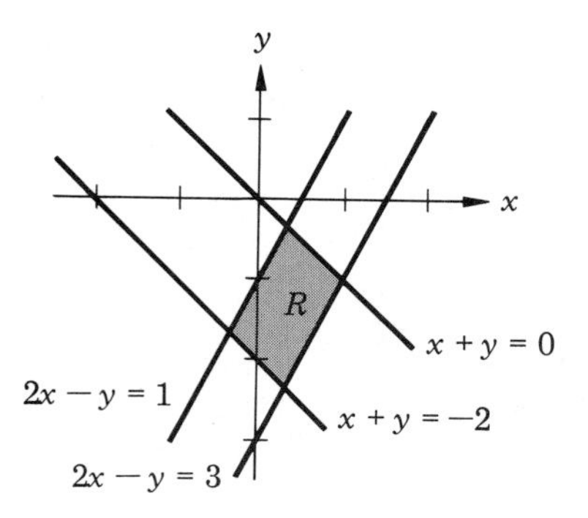

Fig. 10-15

REMARKS. (i) We note that formula (7) employs only the first derivatives of f and g, while in the proof we require that f and g have continuous second derivatives. By a more sophisticated argument it can be shown that the change-of-variables formula (7) is valid when f and g are once-continuously differentiable. (ii) We assumed in Theorems 2 and 3 that R', the image of R, is a region which has area. Actually, if T is continuously differentiable, then it can be proved that R' has area whenever R does.

EXAMPLE 1. Evaluate $\iint xy\,dA_{xy}$, where R is the parallelogram bounded by the lines $2x - y = 1$, $2x - y = 3$, $x + y = -2$, and $x + y = 0$. Draw a figure.

SOLUTION. See Fig. 10-15. Let $u = 2x - y$, $v = x + y$. Then R' is the region determined by the inequalities

$$1 \le u \le 3, \qquad -2 \le v \le 0,$$

and

$$x = \tfrac{1}{3}(u + v), \qquad y = \tfrac{1}{3}(-u + 2v).$$

We compute

$$f(x, y) = xy = \tfrac{1}{9}(-u^2 + uv + 2v^2),$$

$$J = \tfrac{1}{3}.$$

Therefore

$$\iint_R xy\,dA_{xy} = \tfrac{1}{27}\iint_{R'}(-u^2 + uv + 2v^2)\,dA_{uv}$$

$$= \tfrac{1}{27}\int_1^3\int_{-2}^0(-u^2 + uv + 2v^2)\,dv\,du = -\tfrac{44}{81}.$$

EXAMPLE 2. Evaluate $\iint_R x\,dA_{xy}$, where R is the region bounded by the curves

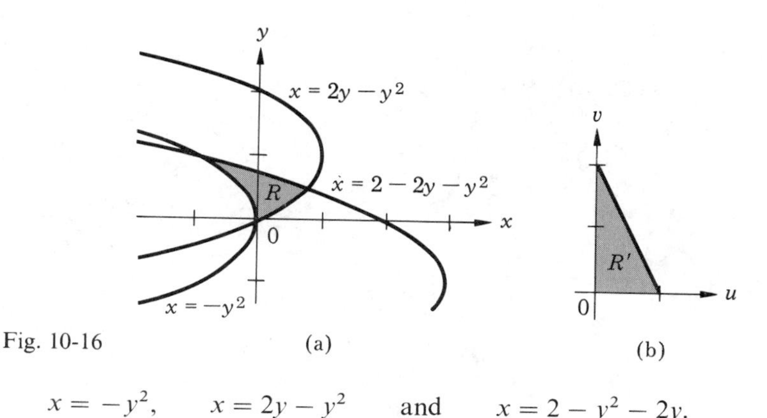

Fig. 10-16 (a) (b)

$$x = -y^2, \qquad x = 2y - y^2 \qquad \text{and} \qquad x = 2 - y^2 - 2y.$$

Perform the integration by introducing the new variables u, v:

$$x = u - \frac{(u+v)^2}{4}, \qquad y = \frac{u+v}{2}.$$

Draw R and the corresponding region R' in the uv-plane.

SOLUTION. The equations of the boundary curves for R' in the uv-plane are obtained by substitution. We find

$$x = -y^2 \to u = 0,$$
$$x = 2y - y^2 \to u = u + v \Leftrightarrow v = 0,$$
$$x + y^2 = 2 - 2y \to u = 2 - u - v \Leftrightarrow 2u + v = 2.$$

The regions R and R' are shown in Fig. 10-16. We have

$$J\left(\frac{x, y}{u, v}\right) = \begin{vmatrix} 1 - \dfrac{u+v}{2} & -\dfrac{u+v}{2} \\ \dfrac{1}{2} & \dfrac{1}{2} \end{vmatrix} = \frac{1}{2}.$$

Therefore

$$\iint_R x\,dA_{xy} = \iint_{R'} \left[u - \frac{(u+v)^2}{4}\right] \cdot \frac{1}{2} dA_{uv}$$
$$= \frac{1}{2}\int_0^1 \int_0^{2-2u} \left[u - \frac{(u+v)^2}{4}\right] dv\,du$$
$$= \frac{1}{2}\int_0^1 \left[uv - \frac{(u+v)^3}{12}\right]_0^{2-2u} du$$
$$= \frac{1}{2}\int_0^1 \left[2u - 2u^2 - \frac{(2-u)^3}{12} + \frac{u^3}{12}\right] du = \frac{1}{48}.$$

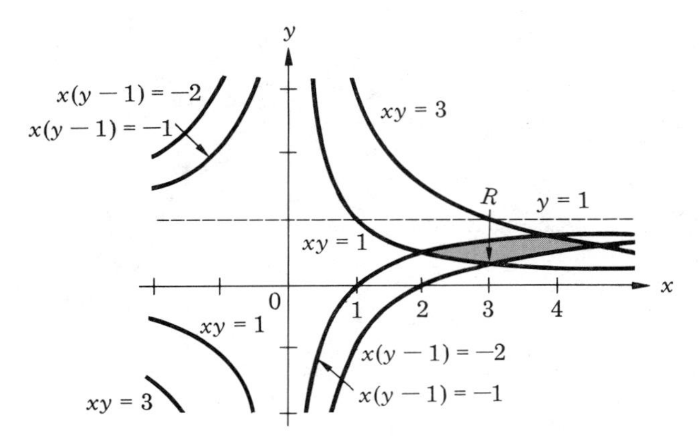

Fig. 10-17

EXAMPLE 3. Evaluate $\iint_R x\,dA_{xy}$, where R is the region bounded by the curves $x(1-y)=1$, $x(1-y)=2$, $xy=1$, and $xy=3$. The region R is shown in Fig. 10-17.

SOLUTION. Let $u = x(1-y)$, $v = xy$. Then

$$x = u + v, \qquad y = \frac{v}{u+v}, \qquad J\left(\frac{u,v}{x,y}\right) = x = u + v.$$

Since the Jacobian of the inverse of a one-to-one transformation is the reciprocal of the Jacobian of the transformation, we have

$$J\left(\frac{x,y}{u,v}\right) = \frac{1}{u+v}.$$

Therefore

$$\iint_R x\,dA_{xy} = \iint_{T(R)} (u+v)\left|J\left(\frac{x,y}{u,v}\right)\right| dA_{uv}$$

$$= \iint_{T(R)} dA_{uv} = \int_1^2 \int_1^3 dv\,du = 2.$$

The theorems of this section have appropriate generalizations to three or more dimensions. Furthermore, the specialization to one dimension shows that the ordinary method of substitution $x = g(u)$ is valid whenever g is continuously differentiable and $g' \neq 0$.

PROBLEMS

In each of Problems 1 through 6, evaluate $\iint_R x^2\,dA_{xy}$, where R is the parallelogram specified by the given inequalities. Draw a figure in each case. Use the method of Example 1.

1. $-1 \le x - y \le 1$
   $0 \le x + y \le 2$

2. $-1 \le 2x + y \le 2$
   $0 \le x + 2y \le 3$

3. $1 \le 2x - y \le 6$
   $-1 \le x + 2y \le 4$

4. $-1 \le 3x + 2y \le 3$
   $1 \le x + 2y \le 5$

5. $1 \le 3x + y \le 6$
   $2 \le x + 2y \le 7$

6. $2 \le 3x - y \le 9$
   $1 \le x + 2y \le 8$

7. Show that

$$J\left(\frac{x, y}{r, \theta}\right) = r \qquad \text{if } x = r\cos\theta,\ y = r\sin\theta.$$

8. Show that

$$J\left(\frac{x, y, z}{\rho, \theta, \phi}\right) = \rho^2 \sin\phi \text{ if } x = \rho\cos\theta\sin\phi,\ y = \rho\sin\theta\sin\phi,\ z = \rho\cos\phi.$$

9. Find

$$J\left(\frac{x, y, z}{u, v, w}\right) \qquad \text{given that } x = u(1 - v),\ y = uv,\ z = uvw.$$

10. Find

$$J\left(\frac{x_1, x_2, x_3, x_4}{u_1, u_2, u_3, u_4}\right)$$

if

$$\begin{aligned} x_1 &= u_1 \cos u_2, \\ x_2 &= u_1 \sin u_2 \cos u_3, \\ x_3 &= u_1 \sin u_2 \sin u_3 \cos u_4, \\ x_4 &= u_1 \sin u_2 \sin u_3 \sin u_4. \end{aligned}$$

In each of Problems 11 through 18, evaluate $\iint_R f(x, y)\, dA_{xy}$, where R is bounded by the curves whose equations are given. Perform the integration by introducing variables u and v as indicated. Draw a graph of R and the corresponding region R' in the uv-plane. Find the inverse of each transformation.

11. $f(x, y) = x^2$; R bounded by $y = 3x$, $x = 3y$ and $x + y = 4$; transformation: $x = 3u + v$, $y = u + 3v$.

12. $f(x, y) = x - y^2$; R bounded by $y = 2$, $x = y^2 - y$, $x = 2y + y^2$; transformation: $x = 2u - v + (u + v)^2$, $y = u + v$.

13. $f(x, y) = y$; R bounded by $x + y - y^2 = 0$, $2x + y - 2y^2 = 1$, $x - y^2 = 0$; transformation: $x = u - v + (u - 2v)^2$, $y = -u + 2v$.

14. $f(x, y) = x^2$; R bounded by $y = -x - x^2$, $y = 2x - x^2$, $y = \frac{1}{2}x - x^2 + 3$; transformation: $x = u - v$, $y = 2u + v - (u - v)^2$.

15. $f(x, y) = (x^2 + y^2)^{-3}$; R bounded by $x^2 + y^2 = 2x$, $x^2 + y^2 = 4x$, $x^2 + y^2 = 2y$, $x^2 + y^2 = 6y$; transformation: $x = u/(u^2 + v^2)$, $y = v/(u^2 + v^2)$.

16. $f(x, y) = 4xy$; R bounded by $y = x$, $y = -x$, $(x + y)^2 + x - y - 1 = 0$; transformation: $x = (u + v)/2$, $y = (-u + v)/2$ (assume that $x + y > 0$).

17. $f(x, y) = y$; R bounded by $x = e^{y/2} - \frac{1}{2}y$, $x = 5 + e^{y/2} - \frac{1}{2}y$, $x = 2y + e^{y/2}$, $3x = y + 3e^{y/2} - 5$; transformation: $u = x - e^{y/2} + (y/2)$, $v = y$.

18. $f(x, y) = x^2 + y^2$; R is the region in the first quadrant bounded by $x^2 - y^2 = 1$, $x^2 - y^2 = 2$, $2xy = 2$; $2xy = 4$; transformation: $u = x^2 - y^2$, $v = 2xy$.

19. Show that the integrals

$$\int_0^a \int_0^{a-y} f(x, y)\, dx\, dy \qquad \text{and} \qquad \int_0^1 \int_0^a f(u - uv, uv) u\, du\, dv$$

are equal if $u = x + y$, $v = y/(x + y)$.

20. Show that the integrals

$$\int_0^a \int_0^x f(x, y)\, dy\, dx \quad \text{and} \quad \int_0^1 \int_0^{a(1+u)} f\left[\frac{v}{1 + u}, \frac{uv}{1 + u}\right] \frac{v}{(1 + u)^2}\, dv\, du$$

are equal if $x = v/(1 + u)$, $y = uv/(1 + u)$.

21. Evaluate the integral

$$\iiint_R z\, dV_{xyz}$$

where R is the region $x^2 + y^2 \leq z^2$, $x^2 + y^2 + z^2 \leq 1$, $z \geq 0$, by changing to spherical coordinates.

22. By introducing polar coordinates, evaluate

$$\iint_R \frac{dA_{xy}}{(1 + x^2 + y^2)^2}$$

where R is the right-hand loop of the lemniscate:

$$(x^2 + y^2)^2 - (x^2 - y^2) = 0.$$

4. Surface Elements. Surfaces. Parametric Representation

So far we have interpreted the term "surface in three-space" as the graph in a given rectangular coordinate system of an equation of the form $z = f(x, y)$ of $F(x, y, z) = 0$. The definition and computation of areas of general surfaces were taken up in Section 7 of Chapter 5. Now we are interested in studying surfaces which have a structure much more complicated than those we have considered previously.

In order to simplify the study of surfaces, we decompose a given surface into a large number of small pieces and examine each piece separately. It

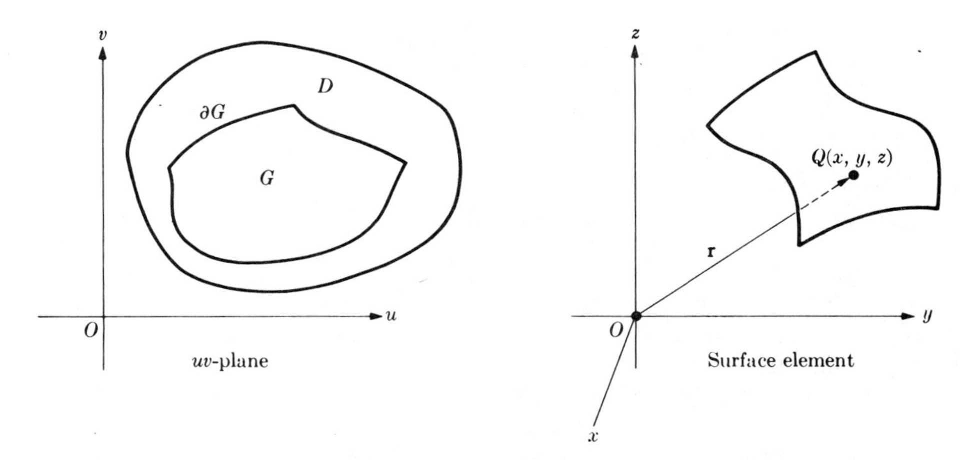

Fig. 10-18

may happen that a particular surface has an unusual or bizarre appearance and yet each small piece has a fairly simple structure. In fact, for most surfaces we shall investigate, the "local" behavior will be much like that of a piece of a sphere, a hyperboloid, a cylinder, or some other smooth surface.

Some surfaces have boundaries and others do not. For example, a hemisphere has a boundary consisting of its equatorial rim. An entire sphere, an ellipsoid, and the surface of a cube are examples of surfaces *without* boundary.

We now describe the "small pieces" into which we divide a surface. More precisely, a **smooth surface element** is the graph of a system of equations of the form

$$x = x(u, v), \qquad y = y(u, v), \qquad z = z(u, v), \qquad (u, v) \in G \cup \partial G, \tag{1}$$

in which x, y, z are continuously differentiable functions on a domain D containing G and its boundary ∂G. It is convenient to use vector notation, and we write the above system in the form

$$\mathbf{v}(\overrightarrow{OQ}) = \mathbf{r}(u, v), \qquad (u, v) \in G \cup \partial G, \qquad O \text{ is the origin of coordinates.} \tag{2}$$

We assume that G is a domain whose boundary consists of a finite number of piecewise smooth simple closed curves. See Fig. 10-18, in which ∂G consists of one piecewise smooth curve.

We shall suppose that

$$\mathbf{r}_u \times \mathbf{r}_v \neq \mathbf{0} \qquad \text{for } (u, v) \in D, \tag{3}$$

and that $\mathbf{r}(u_1, v_1) \neq \mathbf{r}(u_2, v_2)$ whenever $(u_1, v_1) \neq (u_2, v_2)$. In other words, the equations (1) (or (2)) define a one-to-one, continuously differentiable transformation from $G \cup \partial G$ onto the points of the surface element. We observe that since

$$\mathbf{r}(u, v) = x(u, v)\mathbf{i} + y(u, v)\mathbf{j} + z(u, v)\mathbf{k},$$

then

$$\mathbf{r}_u \times \mathbf{r}_v = (y_u z_v - y_v z_u)\mathbf{i} + (z_u x_v - z_v x_u)\mathbf{j} + (x_u y_v - x_v y_u)\mathbf{k},$$

which, in Jacobian notation, may be written

$$\mathbf{r}_u \times \mathbf{r}_v = J\left(\frac{y, z}{u, v}\right)\mathbf{i} + J\left(\frac{z, x}{u, v}\right)\mathbf{j} + J\left(\frac{x, y}{u, v}\right)\mathbf{k}. \tag{4}$$

We also say that the equations (1) or (2) form a **parametric representation** of the smooth surface element.

The next theorem establishes the relation between two parametric representations of the same smooth surface element.

Theorem 4. *Suppose that the transformation* (2) *satisfies the conditions above for* (u, v) *in a region* D *which contains* G *and* ∂G. *Let* S *and* S^* *denote the images under* (2) *of* $G \cup \partial G$ *and* D, *respectively. Let* (x, y, z) *be any rectangular coordinate system. Then*

a) *if* (u_0, v_0) *is any point of* $G \cup \partial G$, *there is a positive number* ρ *such that the part of* S^* *corresponding to the square*

$$|u - u_0| < \rho, \qquad |v - v_0| < \rho$$

is of one of the forms

$$z = f(x, y), \qquad x = g(y, z), \qquad \text{or} \quad y = h(x, z),$$

where f, g, *or* h *(as the case may be) is smooth near the point* (x_0, y_0, z_0) *corresponding to* (u_0, v_0).

b) *Suppose that another parametric representation of* S *and* S^* *is given by*

$$\mathbf{v}(\overrightarrow{OP}) = \mathbf{r}_1(s, t),$$

in which S *and* S^* *are the images of* $G_1 \cup \partial G_1$ *and* D_1, *respectively, in the* (s, t)*-plane. Suppose that* G_1, D_1, *and* $\mathbf{r}_1$ *have all the properties which* G, D, *and* $\mathbf{r}$ *have. Then there is a one-to-one continuously differentiable transformation*

$$T\colon \quad u = U(s, t), \qquad v = V(s, t), \qquad (s, t) \text{ on } D_1,$$

from D_1 *to* D *such that* $T(G_1) = G$, $T(\partial G_1) = \partial G$, *and*

$$\mathbf{r}[U(s, t), V(s, t)] = \mathbf{r}_1(s, t) \qquad \text{for } (s, t) \text{ on } D_1.$$

PROOF. (a) Since $\mathbf{r}_u \times \mathbf{r}_v \neq \mathbf{0}$ at a point (u_0, v_0), we conclude from the formula

$$\mathbf{r}_u \times \mathbf{r}_v = J\left(\frac{y, z}{u, v}\right)\mathbf{i} + J\left(\frac{z, x}{u, v}\right)\mathbf{j} + J\left(\frac{x, y}{u, v}\right)\mathbf{k} \tag{4}$$

that at least one of the three Jacobians is not zero at (u_0, v_0). Suppose,

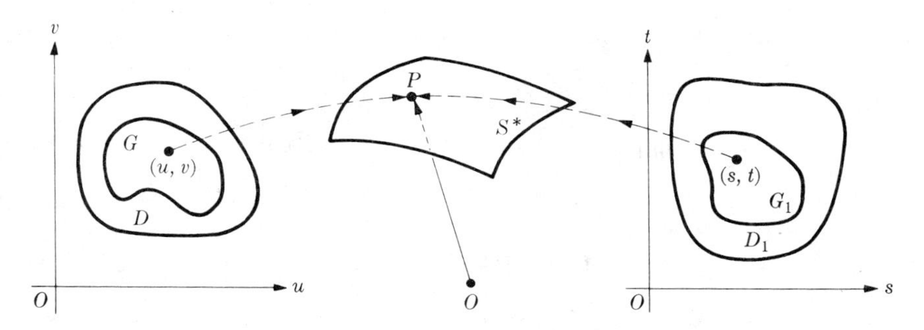

Fig. 10-19

for instance, that

$$J\left(\frac{x, y}{u, v}\right)$$

is not zero. We set

$$\mathbf{r}(u, v) = X(u, v)\mathbf{i} + Y(u, v)\mathbf{j} + Z(u, v)\mathbf{k},$$

and $x_0 = X(u_0, v_0)$, $y_0 = Y(u_0, v_0)$. Then, from the Implicit Function Theorem, it follows that there are positive numbers h and k such that all numbers x, y, u, v for which

$$|x - x_0| < h, \qquad |y - y_0| < h, \qquad |u - u_0| < k, \qquad |v - v_0| < k,$$

and $x = X(u, v)$, $y = Y(u, v)$ lie along the graph of $u = \phi(x, y)$, $v = \psi(x, y)$, where ϕ and ψ are smooth in the square $|x - x_0| < h$, $|y - y_0| < h$. In this case, the part of S^* near (x_0, y_0, z_0) is the graph of

$$z = Z[\phi(x, y), \psi(x, y)], \qquad |x - x_0| < h, \qquad |y - y_0| < h.$$

The conclusion (a) follows when we select $\rho > 0$ small enough so that the image of the square $|u - u_0| < \rho$, $|v - v_0| < \rho$ lies in the (x, y)-square above.
b) Since $\mathbf{r}(u, v)$ and $\mathbf{r}_1(s, t)$ are one-to-one, it follows that to each (s, t) in D_1 there corresponds a unique P on S^* which comes from a unique (u, v) in D (Fig. 10-19). If we define $U(s, t) = u$ and $V(s, t) = v$ by this correspondence, then T is one-to-one. To see that it is smooth, let (s_0, t_0) be any point in D_1 and let $(u_0, v_0) = T(s_0, t_0)$. Denote by P_0 the point of S corresponding to (s_0, t_0) and (u_0, v_0). At least one of the three Jacobians in (4) does not vanish. If, for example, the last one does not vanish, we can solve for u and v in terms of x and y as in part (a). We now set

$$\mathbf{r}_1(s, t) = X_1(s, t)\mathbf{i} + Y_1(s, t)\mathbf{j} + Z_1(s, t)\mathbf{k}.$$

Since there is a one-to-one correspondence between the points P near P_0 and the points (x, y) "below" them, we see that

$$U(s, t) = \phi[X_1(s, t),\ Y_1(s, t)], \qquad V(s, t) = \psi[X_1(s, t),\ Y_1(s, t)].$$

Hence U and V are smooth near (s_0, t_0), an arbitrary point of G_1.

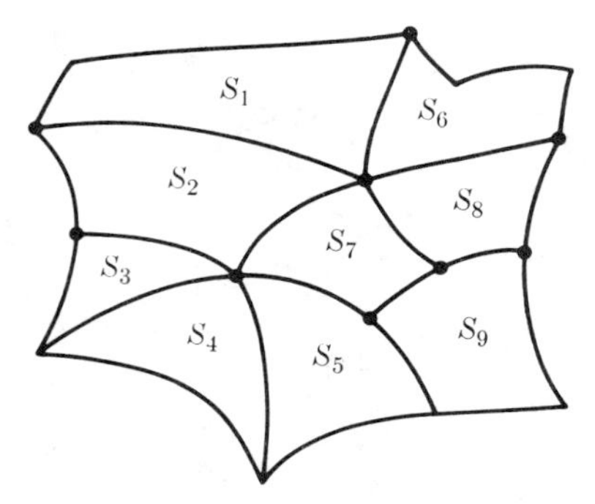

Fig. 10-20

REMARK. Of course, T^{-1}, the inverse transformation, has the same smoothness properties that T does.

Suppose that S is a smooth surface element given by

$$\mathbf{r}(u, v) = x(u, v)\mathbf{i} + y(u, v)\mathbf{j} + z(u, v)\mathbf{k}, \qquad (u, v) \in G \cup \partial G.$$

Then the **boundary** of S is the image of ∂G in the above parametric representation. From part (b) of Theorem 4, we see that the boundary of a smooth surface element is independent of the particular parametric representation. Also, it is not difficult to see that the boundary of a smooth surface element consists of a finite number of piecewise smooth, simple closed curves, no two of which intersect.

From Theorem 4, part (a), we observe that a neighborhood of any point of a smooth surface element is the graph of an equation of one of the forms

$$z = f(x, y), \qquad x = g(y, z), \qquad \text{or} \qquad y = h(x, z),$$

where f, g, or h is a smooth function on its domain. The set of points

$$\{(x, y, z) : z = f(x, y), \quad (x, y) \in G \cup \partial G\},$$

where f is smooth on a domain D containing $G \cup \partial G$, is a smooth surface element S. To see this, we observe that S is the graph of the equations

$$x = u, \qquad y = v, \qquad z = f(u, v), \qquad (u, v) \in G \cup \partial G.$$

Definition. A **piecewise smooth surface** S is the union $S_1 \cup S_2 \cup \cdots \cup S_n$ of a finite number of smooth surface elements $S_1, S_2, \ldots, S_n$ satisfying the following conditions:

i) no two of the S_i have common interior points;
ii) the intersection of the boundaries of two elements, $\partial S_i \cap \partial S_j$, is either empty, or a single point, or a piecewise smooth arc (see Fig. 10-20);
iii) the boundaries of any three elements have at most one point in common.
iv) any two points of S can be joined by an arc in S.

A piecewise smooth surface S may be decomposed into a finite union of smooth surface elements in many ways. A smooth arc which is part of the boundary of two of the S_i is called an **edge** of the decomposition. A corner of one of the boundaries of an S_i is called a **vertex**. The boundary ∂S of a piecewise smooth surface consists of all points P such that (i) P is on the boundary of exactly one S_i, or (ii) P is a limit point of such points. It can be shown that ∂S is either empty or the union of a finite number of piecewise smooth, simple closed curves no two of which intersect.

A piecewise smooth surface S is said to be a **smooth surface** if and only if for every point P not on ∂S a decomposition of S into piecewise smooth surface elements can be found such that P is an interior point of one of the surface elements. It can be shown that if F is a smooth scalar field on some domain and $F(P)$ and $\nabla F(P)$ are never simultaneously zero, then the graph of the equation $F(P) = 0$ is a smooth surface provided that the graph is a bounded, closed set. Thus, by choosing

$$F(P) = x^2 + y^2 + z^2 - a^2 \qquad \text{or} \qquad F(P) = \frac{x^2}{a^2} + \frac{y^2}{b^2} + \frac{z^2}{c^2} - 1,$$

we see that spheres and ellipsoids are smooth surfaces. Any polyhedron is a piecewise smooth surface.

EXAMPLE. Suppose that a smooth scalar field is defined in all of R^3 by the formula

$$F(P) = x^2 + 4y^2 + 9z^2 - 44.$$

Show that the set $S: F(P) = 0$ is a smooth surface.

SOLUTION. We compute the gradient:

$$\nabla F = 2x\mathbf{i} + 8y\mathbf{j} + 18z\mathbf{k}.$$

Then ∇F is zero only at $(0, 0, 0)$. Since S is not void [the point $(\sqrt{44}, 0, 0)$ is on it] and since $(0, 0, 0)$ is not a point of S, the surface is smooth.

5. Area of a Surface. Surface Integrals

If a surface S is given by

$$S = \{(x, y, z) : z = f(x, y), \quad (x, y) \in G \cup \partial G\},$$

we know that if f has continuous first derivatives, the area of S may be computed by the method described in Chapter 5, Section 7. We begin the study of area for more general surfaces by defining the area of a smooth surface element.

Let σ be a smooth surface element given by

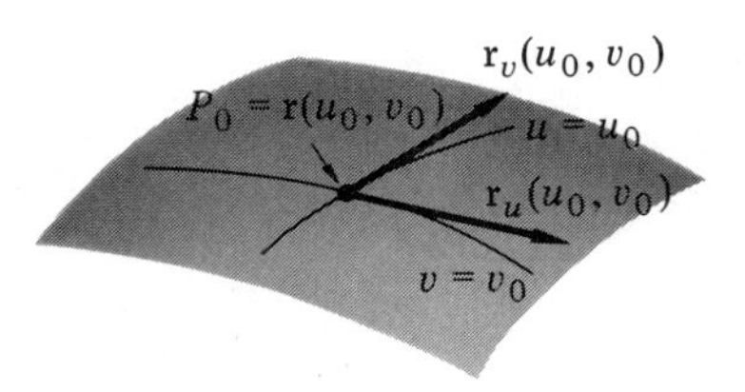

Fig. 10-21

$$\sigma: \quad \mathbf{v}(\overrightarrow{OQ}) = \mathbf{r}(u, v), \qquad \text{where } (u, v) \in G \cup \partial G, \tag{1}$$

with $\mathbf{r}$ and G satisfying the conditions stated in the definition of a surface element. If v is held constant, equal to v_0, say, then the graph of (1) is a smooth curve on the surface element σ. Therefore the vector $\mathbf{r}_u(u, v_0)$ is a vector tangent to the curve on the surface. Similarly, the vector $\mathbf{r}_v(u_0, v)$ is tangent to the smooth curve on the surface element σ obtained when u is set equal to the constant u_0 (Fig. 10-21). The vectors $\mathbf{r}_u(u_0, v_0)$ and $\mathbf{r}_v(u_0, v_0)$ lie in the plane tangent to the surface at the point P_0 corresponding to $\mathbf{r}(u_0, v_0)$. Since the cross product of two vectors is orthogonal to each of them, it follows that *the vector* $\mathbf{r}_u(u_0, v_0) \times \mathbf{r}_v(u_0, v_0)$ *is a vector normal to the surface of* P_0.

For convenience, we write $\mathbf{a} = \mathbf{r}_u(u_0, v_0)$, $\mathbf{b} = \mathbf{r}_v(u_0, v_0)$, and we consider the **tangent linear transformation** defined by

$$\mathbf{v}(\overrightarrow{OP}) = \mathbf{r}(u_0, v_0) + (u - u_0)\mathbf{a} + (v - v_0)\mathbf{b}. \tag{2}$$

We notice that the image of the rectangle

$$R = \{(u, v) : u_1 \le u \le u_2, \quad v_1 \le v \le v_2\}$$

under (2) is a parallelogram $ABCD$ in the plane tangent to the surface at the point $P_0 = \mathbf{r}(u_0, v_0)$ (Fig. 10-22). The points A, B, C, and D are determined by

$$\mathbf{v}(\overrightarrow{OA}) = \mathbf{r}(u_0, v_0) + (u_1 - u_0)\mathbf{a} + (v_1 - v_0)\mathbf{b},$$
$$\mathbf{v}(\overrightarrow{OB}) = \mathbf{r}(u_0, v_0) + (u_2 - u_0)\mathbf{a} + (v_1 - v_0)\mathbf{b},$$
$$\mathbf{v}(\overrightarrow{OC}) = \mathbf{r}(u_0, v_0) + (u_1 - u_0)\mathbf{a} + (v_2 - v_0)\mathbf{b},$$
$$\mathbf{v}(\overrightarrow{OD}) = \mathbf{r}(u_0, v_0) + (u_2 - u_0)\mathbf{a} + (v_2 - v_0)\mathbf{b}.$$

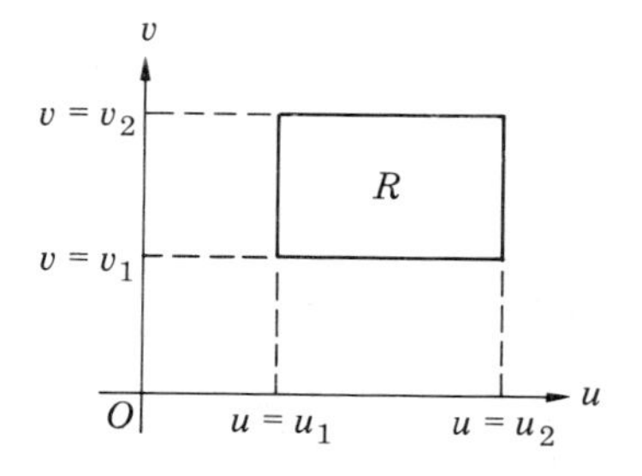

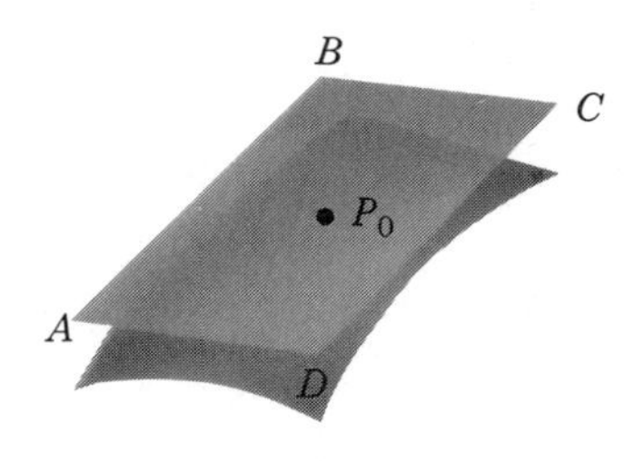

Fig. 10-22

Using vector subtraction, we find that

$$\mathbf{v}(\overrightarrow{AB}) = (u_2 - u_1)\mathbf{a}, \qquad \mathbf{v}(\overrightarrow{AD}) = (v_2 - v_1)\mathbf{b}.$$

Denoting the area of R by $A(R)$, we use the definition of cross product to obtain

$$\begin{aligned}\text{Area } ABCD &= |\mathbf{v}(\overrightarrow{AB}) \times \mathbf{v}(\overrightarrow{AD})| = |(u_2 - u_1)(v_2 - v_1)| \cdot |\mathbf{a} \times \mathbf{b}| \\ &= A(R)|\mathbf{r}_u(u_0, v_0) \times \mathbf{r}_v(u_0, v_0)|.\end{aligned}$$

In defining the area of a surface element, we would expect that for a very small piece near P_0, the area of $ABCD$ would be a good approximation to the (as yet undefined) area of the portion of the surface which is the image of R. We use this fact to define the area of a surface element σ. We subdivide the region G in the uv plane into a number of subregions $G_1, G_2, \ldots, G_k$ and define the norm $\|\Delta\|$ as the maximum diameter of any of the subregions. We define

$$\text{Area of } \sigma = \lim_{\|\Delta\| \to 0} \sum_{i=1}^{n} A(G_i)|\mathbf{r}_u(u_i, v_i) \times r_v(u_i, v_i)|,$$

where (u_i, v_i) is any point in G_i and where the limit has the usual interpretation as given in the definition of integral. Therefore the definition of area of σ is

$$A(\sigma) = \iint_G |\mathbf{r}_u(u, v) \times \mathbf{r}_v(u, v)|\, dA_{uv}. \tag{3}$$

We immediately raise the following question concerning the use of parameters. Suppose that the same smooth surface element σ has another parametric representation

$$\mathbf{r}' = \mathbf{r}'(s, t) \qquad \text{for } (s, t) \text{ in } G'.$$

Is it true that the formula

$$A(\sigma) = \iint_{G'} |\mathbf{r}'_s(s, t) \times \mathbf{r}'_t(s, t)|\, dA_{st} \tag{4}$$

gives the same value as formula (3)? Because of the rule for change of variable in a multiple integral (Theorem 3, page 513) and the rule for the product of Jacobians

$$J\left(\frac{x, y}{s, t}\right) = J\left(\frac{x, y}{u, v}\right) J\left(\frac{u, v}{s, t}\right),$$

it follows that (3) and (4) yield the same value. In fact, we have

$$|\mathbf{r}'_s(s, t) \times \mathbf{r}'_t(s, t)| = |\mathbf{r}_u(u, v) \times \mathbf{r}_v(u, v)| \cdot \left|J\left(\frac{u, v}{s, t}\right)\right|$$

If the representation is the simple one $z = f(x, y)$ discussed at the beginning of the section, we may set $x = u$, $y = v$, $z = f(u, v)$, and find

$$J\left(\frac{y, z}{u, v}\right) = -f_u, \qquad J\left(\frac{z, x}{u, v}\right) = -f_v, \qquad J\left(\frac{x, y}{u, v}\right) = 1.$$

Then (3) becomes

$$A(\sigma) = \iint_G \sqrt{1 + (f_u)^2 + (f_v)^2}\, dA_{uv}, \tag{5}$$

which is the formula we established in Section 7, Chapter 5.

Unfortunately, not every surface can be covered by a single smooth surface element, and so (5) cannot be used exclusively for the computation of surface area. In fact, it can be shown that even a simple surface such as a sphere cannot be part of a single smooth surface element. (See, however, Example 2 below.)

We define figures and their areas on more general piecewise continuous surfaces as follows: We say that a set F on a smooth surface element is a **figure** if and only if it is the image under (1) of a region E in $G \cup \partial G$ which has an area, and we define its area by the formula (3). If $r' = r'(s, t)$ is another representation, and E' is the corresponding set, it follows from Theorem 4(b) above and the rule for multiplying Jacobians, that E' is a figure and the area of F is given by (4) which is the same as (3). We say that a set F on a piecewise continuous surface is a **figure** $\Leftrightarrow F = F_1 \cup \cdots \cup F_k$ where each F_i is a figure contained in some smooth surface element σ_i and no two have common interior points; in this case we define

$$A(F) = A(F_1) + \cdots + A(F_k).$$

It is not difficult to see that two different decompositions of this sort yield the same result for $A(F)$.

If F is a closed figure (i.e., one which contains its boundary) on a piecewise smooth surface and f is a continuous scalar field on F, we define

$$\iint_F f\,dS = \sum_{i=1}^{k} \iint_{F_i} f\,dS, \tag{6}$$

where $F = \{F_1, \ldots, F_k\}$ is a subdivision of F as above; the terms in the right side of (6) are given by

$$\iint_{F_i} f\,dS = \iint_{E_i} f[\mathbf{r}_i(u, v)] \cdot |\mathbf{r}_{iu} \times \mathbf{r}_{iv}|\, du\, dv, \tag{7}$$

where F_i is the image of E_i under the transformation

$$\mathbf{v}(\overrightarrow{OQ}) = \mathbf{r}_i(u, v), \qquad (u, v) \in G_i \cup \partial G_i.$$

The result is independent of the subdivision and the parametric representations of the elements σ_i.

If a surface element σ has the representation $z = \phi(x, y)$ in a given co-

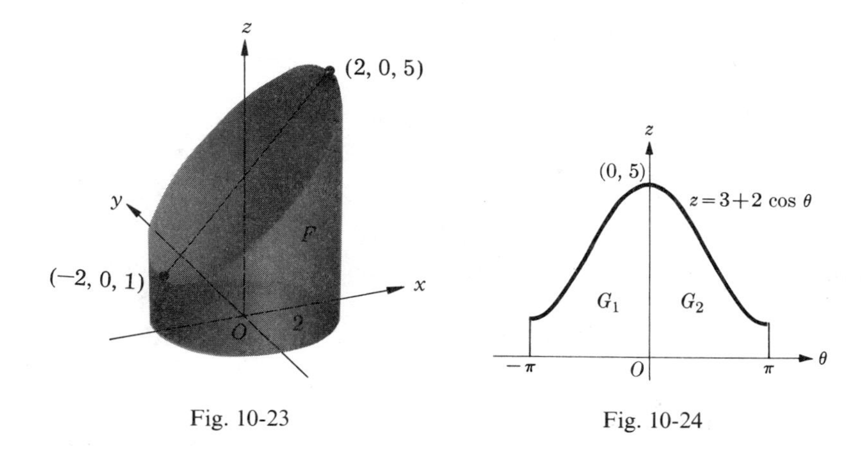

Fig. 10-23

Fig. 10-24

ordinate system, then (7) becomes

$$\iint_F f(Q)\,dS = \iint_G f[x, y, \phi(x, y)]\sqrt{1 + \phi_x^2 + \phi_y^2}\,dA_{xy}, \tag{8}$$

with corresponding formulas in the cases $x = \phi(y, z)$ or $y = \phi(x, z)$. The evaluation of the integral on the right in (8) follows the usual rules for evaluation of double integrals which we studied earlier. Some examples illustrate the technique.

Example 1. Evaluate $\iint_F z^2\,dS$, where F is the part of the lateral surface of the cylinder $x^2 + y^2 = 4$ between the planes $z = 0$ and $z = x + 3$.

Solution. (See Fig. 10-23.) If we transform to cylindrical coordinates r, θ, z then F lies on the surface $r = 2$. We may choose θ and z as parametric coordinates on F and write

$$F = \{(x, y, z): x = 2\cos\theta, \quad y = 2\sin\theta, \quad z = z, \quad (\theta, z) \text{ in } G\},$$

$$G = \{(\theta, z): -\pi \le \theta \le \pi, \quad 0 \le z \le 3 + 2\cos\theta\}.$$

See Fig. 10-24. The element of surface area dS is given by

$$dS = |\mathbf{r}_\theta \times \mathbf{r}_z|\,dA_{\theta z},$$

and since

$$\mathbf{r}_\theta \times \mathbf{r}_z = J\left(\frac{y, z}{\theta, z}\right)\mathbf{i} + J\left(\frac{z, x}{\theta, z}\right)\mathbf{j} + J\left(\frac{x, y}{\theta, z}\right)\mathbf{k}$$

$$= (2\cos\theta)\mathbf{i} + (2\sin\theta)\mathbf{j} + 0\cdot\mathbf{k},$$

we have

$$dS = 2\,dA_{\theta z}.$$

Then

$$\begin{aligned}\iint_F z^2\,dS &= 2\iint_G z^2\,dA_{\theta z} = 2\int_{-\pi}^{\pi}\int_0^{3+2\cos\theta} z^2\,dz\,d\theta \\ &= \tfrac{2}{3}\int_{-\pi}^{\pi}(3+2\cos\theta)^3\,d\theta \\ &= \tfrac{2}{3}\int_{-\pi}^{\pi}(27+54\cos\theta+36\cos^2\theta+8\cos^3\theta)\,d\theta \\ &= \tfrac{2}{3}(54\pi+36\pi) = 60\pi.\end{aligned} \tag{9}$$

REMARK. Strictly speaking, the whole of F is not a smooth surface element, since the transformation from G to F shows that the points $(-\pi, z)$ and (π, z) of G are carried into the same points on F. For a smooth surface element, the condition that the transformation be one to one is therefore violated. However, if we subdivide G into G_1 and G_2 as shown in Fig. 10-24, the images are smooth surface elements. The evaluation as given in (9) is unaffected.

Surface integrals can be used to make approximate computations of various physical quantities. The center of mass and the moment of inertia of a thin curvilinear plate are sometimes computable in the form of surface integrals. The potential due to a distribution of an electric charge over a surface may be expressed in the form of an integral. (See Problems 18 through 21 at the end of this section.) With each surface F we may associate a mass (assuming it to be made of a thin material), and this mass may be given by a density function δ. The density will be assumed continuous but not necessarily constant. The total mass $M(F)$ of a surface F is given by

$$M(F) = \iint_F \delta(P)\,dS.$$

The formula for the moment of inertia of F about the z axis becomes

$$I_z = \iint_F \delta(Q)(x_Q^2 + y_Q^2)\,dS,$$

with corresponding formulas for I_x and I_y. The formula for the center of mass is analogous to those which we studied earlier. (See Chapter 5, Section 6, and also Example 3 below.)

EXAMPLE 2. Find the moment of inertia about the x axis of the part of the surface of the unit sphere $x^2 + y^2 + z^2 = 1$ which is above the cone $z^2 = x^2 + y^2$. Assume $\delta = \text{const}$.

SOLUTION. (See Fig. 10-25.) In spherical coordinates (ρ, ϕ, θ), the portion F of the surface is given by $\rho = 1$. Then (ϕ, θ) are parametric coordinates and

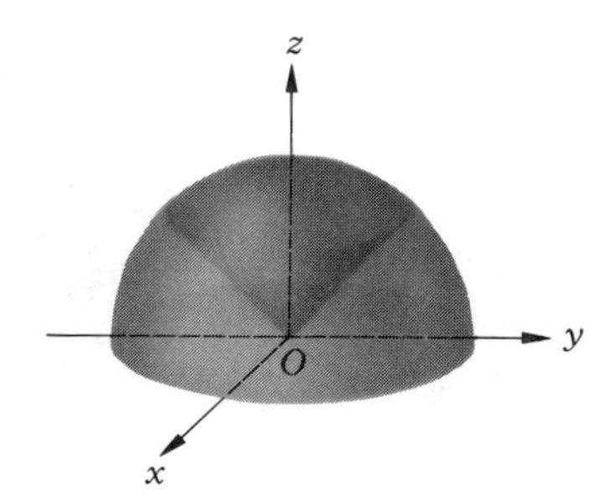

Fig. 10-25

$$F = \{(x, y, z) : x = \sin\phi\cos\theta, \quad y = \sin\phi\sin\theta, \quad z = \cos\phi, \quad (\phi, \theta) \in G\};$$

$$G = \{(\phi, \theta) : 0 \le \phi \le \pi/4, \quad 0 \le \theta \le 2\pi\}.$$

We compute $\mathbf{r}_\phi \times \mathbf{r}_\theta$, getting

$$\mathbf{r}_\phi \times \mathbf{r}_\theta = J\left(\frac{y, z}{\phi, \theta}\right)\mathbf{i} + J\left(\frac{z, x}{\phi, \theta}\right)\mathbf{j} + J\left(\frac{x, y}{\phi, \theta}\right)\mathbf{k}$$

$$= (\sin^2\phi\cos\theta)\mathbf{i} + (\sin^2\phi\sin\theta)\mathbf{j} + (\sin\phi\cos\phi)\mathbf{k}.$$

We obtain

$$dS = |\mathbf{r}_\phi \times \mathbf{r}_\theta|\, dA_{\phi\theta} = \sin\phi\, dA_{\phi\theta}.$$

Then

$$I_x = \iint_G (y^2 + z^2)\,\delta\sin\phi\, dA_{\phi\theta}$$

$$= \delta\int_0^{\pi/4}\int_0^{2\pi} (\sin^2\phi\sin^2\theta + \cos^2\phi)\sin\phi\, d\theta\, d\phi$$

$$= \pi\delta\int_0^{\pi/4} (\sin^2\phi + 2\cos^2\phi)\sin\phi\, d\phi$$

$$= \pi\delta\int_0^{\pi/4} (1 + \cos^2\phi)\sin\phi\, d\phi = \pi\delta\left[-\cos\phi - \tfrac{1}{3}\cos^3\phi\right]_0^{\pi/4}$$

$$= \frac{\pi\delta}{12}(16 - 7\sqrt{2}).$$

REMARK. The parametric representation of F by spherical coordinates does not fulfill the conditions required for such representations, as given in Theorem 4, since $|r_\phi \times \mathbf{r}_\theta| = \sin\phi$ vanishes for $\phi = 0$. However, the same surface, with a small hole cut out around the z axis, is of the required type. A limiting process in which the size of the hole tends to zero yields the above result for the moment of inertia I_x.

EXAMPLE 3. Let R be the region in three-space bounded by the cylinder

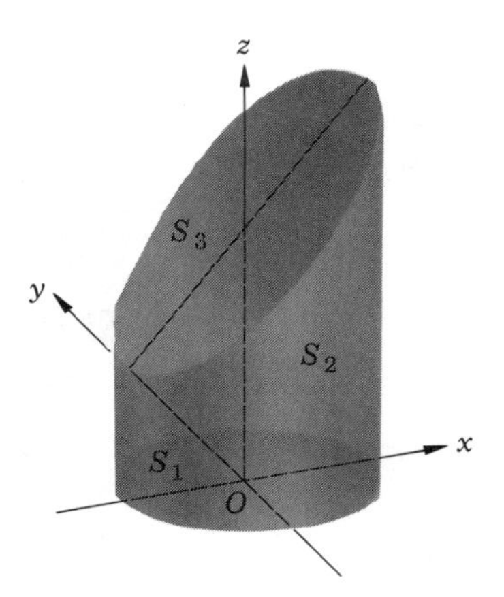

Fig. 10-26

$x^2 + y^2 = 1$ and the planes $z = 0$, $z = x + 2$. Evaluate $\iint_F x\,dS$, where F is the surface made up of the entire boundary of R (Fig. 10-26). If F is made of thin material of uniform density δ, find its mass. Also, find $\bar{x}$, the x-coordinate of the center of mass.

SOLUTION. As shown in Fig. 10-26, the surface is composed of three parts: F_1, the circular base in the xy plane; F_2, the lateral surface of the cylinder; and F_3, the part of the plane $z = x + 2$ inside the cylinder. We have

$$\iint_{F_1} x\,dS = \iint_{G_1} x\,dA_{xy} = 0,$$

where

$$G_1 = \{(x, y, z) : x^2 + y^2 \leq 1, \quad z = 0\}$$

is a disk. On F_3, we see that $z = f(x, y) = x + 2$, so that $dS = \sqrt{2}\,dA_{xy}$ and

$$\iint_{F_3} x\,dS = \sqrt{2} \iint_{G_1} x\,dA_{xy} = 0.$$

On F_2, we choose coordinates (θ, z) as in Example 1 and obtain $x = \cos\theta$, $y = \sin\theta$, $z = z$, $dS = dA_{\theta z}$. We write

$$\iint_{F_2} x\,dS = \iint_{G_2} \cos\theta\,dA_{\theta z} = \int_{-\pi}^{\pi} \int_0^{2+\cos\theta} \cos\theta\,dz\,d\theta,$$

where

$$G_2 = \{(\theta, z) : -\pi \leq \theta \leq \pi, \quad 0 \leq z \leq 2 + \cos\theta\}.$$

Therefore

$$\iint_{F_2} x\,dS = \int_{-\pi}^{\pi} (2\cos\theta + \cos^2\theta)\,d\theta = \pi.$$

We conclude that

$$\iint_F x\,dS = \iint_{F_1+F_2+F_3} x\,dS = \pi.$$

We observe that the surface F is a piecewise smooth surface without boundary. There are no vertices on F, but there are two smooth edges, namely the intersections of $x^2 + y^2 = 1$ with the planes $z = 0$ and $z = x + 2$. To compute the mass $M(F)$, we have

$$M(F) = \iint_F \delta\,dS = \delta A(F) = \delta[A(F_1) + A(F_2) + A(F_3)].$$

Clearly, $A(F_1) = \pi$, $A(F_3) = \pi\sqrt{2}$. To find $A(F_2)$, we write

$$A(F_2) = \iint_{G_2} dA_{\theta z} = \int_{-\pi}^{\pi}\int_0^{2+\cos\theta} dz\,d\theta = \int_{-\pi}^{\pi} (2 + \cos\theta)\,d\theta = 4\pi.$$

Therefore $M(F) = \delta[\pi + 4\pi + \pi\sqrt{2}] = \pi(5 + \sqrt{2})\,\delta$. We use the formula

$$\bar{x} = \frac{\iint_F \delta x\,dA}{M(F)} \quad \text{to get} \quad \bar{x} = \frac{\delta\pi}{\delta\pi(5 + \sqrt{2})} = \frac{1}{5 + \sqrt{2}}.$$

Problems

In each of Problems 1 through 9, evaluate

$$\iint_F f(x, y, z)\,dS.$$

1. $f(x, y, z) = x$, F is the part of the plane $x + y + z = 1$ in the first octant.
2. $f(x, y, z) = x^2$, F is the part of the plane $z = x$ inside the cylinder $x^2 + y^2 = 1$.
3. $f(x, y, z) = x^2$, F is the part of the cone $z^2 = x^2 + y^2$ between the planes $z = 1$ and $z = 2$.
4. $f(x, y, z) = x^2$, F is the part of the cylinder $z = x^2/2$ cut out by the planes $y = 0$, $x = 2$, and $y = x$.
5. $f(x, y, z) = xz$, F is the part of the cylinder $x^2 + y^2 = 1$ between the planes $z = 0$ and $z = x + 2$.
6. $f(x, y, z) = x$, F is the part of the cylinder $x^2 + y^2 = 2x$ between the lower and upper nappes of the cone $z^2 = x^2 + y^2$.
7. $f(x, y, z) = 1$, F is the part of the vertical cylinder erected on the spiral $r = \theta$, $0 \le \theta \le \pi/2$ (polar coordinates in the xy-plane), bounded below by the xy-plane and above by the cone $z^2 = x^2 + y^2$.
8. $f(x, y, z) = x^2 + y^2 - 2z^2$, F is the surface of the sphere

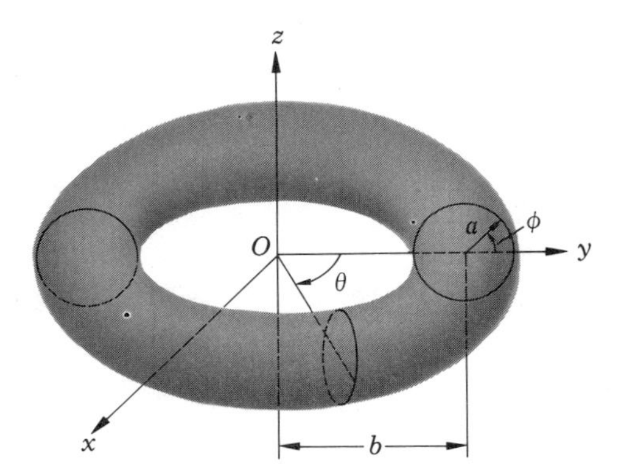

Fig. 10-27

$$x^2 + y^2 + z^2 = a^2.$$

9. $f(x, y, z) = x^2$, F is the total boundary of the region R in three-space bounded by the cone $z^2 = x^2 + y^2$ and the planes $z = 1$, $z = 2$. (See Problem 3.)

In each of Problems 10 through 14, find the moment of inertia of F about the indicated axis, assuming that $\delta = \text{const}$.

10. The surface F of Problem 3; x axis.

11. The surface F of Problem 6; x axis.

12. The surface F of Problem 7; z axis.

13. The total surface of the tetrahedron L bounded by the coordinate planes and the plane $x + y + z = 1$, having vertices $B(1, 0, 0)$, $E(0, 1, 0)$, $R(0, 0, 1)$, and $S(0, 0, 0)$; y axis.

14. The torus $(r - b)^2 + z^2 = a^2$, $0 < a < b$; z axis. Note that

$$r^2 = x^2 + y^2$$

and that we may introduce parameters

$$x = (b + a\cos\phi)\cos\theta,$$
$$y = (b + a\cos\phi)\sin\theta,$$
$$z = a\sin\phi,$$

with the torus swept out by

$$0 \le \phi \le 2\pi, \qquad 0 \le \theta \le 2\pi.$$

(See Fig. 10-27.)

In each of Problems 15 through 17, find the center of mass, assuming $\delta = \text{const}$.

15. F is the surface of Example 2.

16. F is the surface of Problem 13.

17. F is the part of the sphere $x^2 + y^2 + z^2 = 4a^2$ inside the cylinder $x^2 + y^2 = 2ax$.

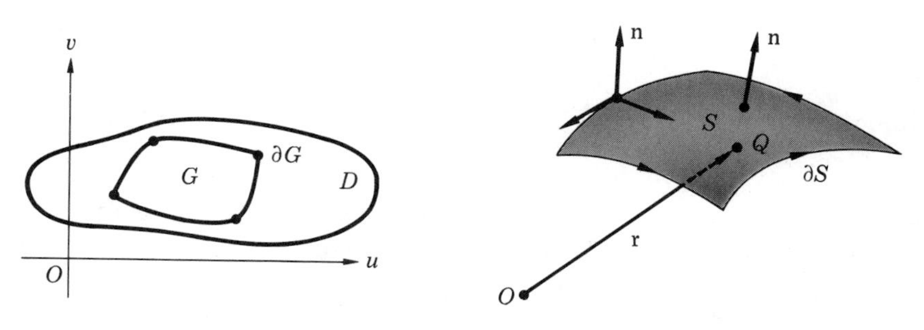

Fig. 10-28

The electrostatic potential $V(Q)$ at a point Q due to a distribution of charge (with charge density δ) on a surface F is given by

$$V(Q) = \iint_F \frac{\delta(P)\, dS}{|PQ|},$$

where $|PQ|$ is the distance from Q in space not on F to a point P on the surface F. In Problems 18 through 21, find $V(Q)$ at the point given, assuming δ constant.

18. $Q = (0, 0, 0)$, F is the part of the cylinder $x^2 + y^2 = 1$ between the planes $z = 0$ and $z = 1$.

19. $Q = (0, 0, c)$; F is the surface of the sphere $x^2 + y^2 + z^2 = a^2$. Do two cases: (i), $c > a > 0$ and (ii), $a > c > 0$.

20. $Q = (0, 0, c)$; F is the upper half of the sphere $x^2 + y^2 + z^2 = a^2$; $0 < c < a$.

21. $Q = (0, 0, 0)$; F is the surface of Problem 3.

6. Orientable Surfaces

Suppose S is a smooth surface element represented by the equation

$$\mathbf{v}(\overrightarrow{OQ}) = \mathbf{r}(u, v) \qquad \text{with } (u, v) \text{ in } G \cup \partial G \tag{1}$$

where G is a region in the (u, v)-plane with piecewise smooth boundary ∂G. According to the definition of smooth surface element, we know that $\mathbf{r}_u \times \mathbf{r}_v \neq \mathbf{0}$ for (u, v) in a domain D containing G and its boundary. Therefore we can define a **unit normal function to the surface** S by the formula (Fig. 10-28)

$$\mathbf{n} = \frac{\mathbf{r}_u \times \mathbf{r}_v}{|\mathbf{r}_u \times \mathbf{r}_v|}, \qquad (u, v) \text{ in } G. \tag{2}$$

Whenever S is a smooth surface element, the vector $\mathbf{n}$ is a continuous function of u and v. If we use Jacobian notation, we see that

$$\mathbf{n} = |\mathbf{r}_u \times \mathbf{r}_v|^{-1}\left[J\left(\frac{y,z}{u,v}\right)\mathbf{i} + J\left(\frac{z,x}{u,v}\right)\mathbf{j} + J\left(\frac{x,y}{u,v}\right)\mathbf{k}\right]. \tag{3}$$

Suppose now that

$$\mathbf{v}(\overrightarrow{OQ}) = \mathbf{r}_1(s,t), \qquad (s,t) \in G_1 \cup \partial G_1,$$

is another parametric representation of the same smooth surface element S. Then from part (b) of Theorem 4, it follows that there is a one-to-one continuously differentiable transformation

$$T\colon \quad u = U(s,t), \qquad v = V(s,t), \qquad (s,t) \in D_1,$$

from D_1 to D such that $T(G_1) = G$ and $T(\partial G_1) = \partial G$. Also, we have

$$\mathbf{r}[U(s,t),\ V(s,t)] = \mathbf{r}_1(s,t), \qquad (s,t) \in D_1.$$

Now we define

$$\mathbf{n}_1 = |\mathbf{r}_{1s} \times \mathbf{r}_{1t}|^{-1}\left[J\left(\frac{y,z}{s,t}\right)\mathbf{i} + J\left(\frac{z,x}{s,t}\right)\mathbf{j} + J\left(\frac{x,y}{s,t}\right)\mathbf{k}\right]. \tag{4}$$

Since Jacobians multiply according to the law

$$J\left(\frac{y,z}{s,t}\right) = J\left(\frac{y,z}{u,v}\right)J\left(\frac{u,v}{s,t}\right),$$

and similar laws when (y,z) is replaced by (z,x) or (x,y), we conclude from (3) and (4) that

$$\mathbf{n}_1(P) = +\mathbf{n}(P) \qquad \text{or} \qquad \mathbf{n}_1(P) = -\mathbf{n}(P)$$

for all P on S. The choice of sign depends upon whether $J\left(\frac{u,v}{s,t}\right)$ is positive or negative on $G_1 \cup \partial G_1$.

Definitions. A smooth surface S is **orientable** $\Leftrightarrow$ there exists a continuous unit normal function defined over the whole of S. Such a function is called an **orientation** of S.

Since any unit vector normal to S at a point P is either $\mathbf{n}(P)$ or $-\mathbf{n}(P)$, we see that *each smooth orientable surface possesses exactly two orientations each of which is the negative of the other*. We call a pair $(S, \mathbf{n})$, where $\mathbf{n}$ is an orientation of S, an **oriented surface** and denote it by $\vec{S}$. Suppose $\vec{S} = (S, \mathbf{n})$ is a smooth oriented surface and $\vec{\Sigma} = (\Sigma, \mathbf{n}')$, where Σ is a smooth surface element on S and $\mathbf{n}'$ is the restriction of $\mathbf{n}$ to Σ, and Σ is given by (1). Then the representation (1) of Σ is said to **agree with the orientation on S** $\Leftrightarrow$ $\mathbf{n}'$ is given by (2).

It is easy to see that smooth surfaces such as spheres, ellipsoids, tori (see Problem 14, Section 5), and so forth, are all orientable. However, there are smooth surfaces for which there is no way to choose a continuous normal function over the whole surface. One such surface is the **Möbius strip** drawn

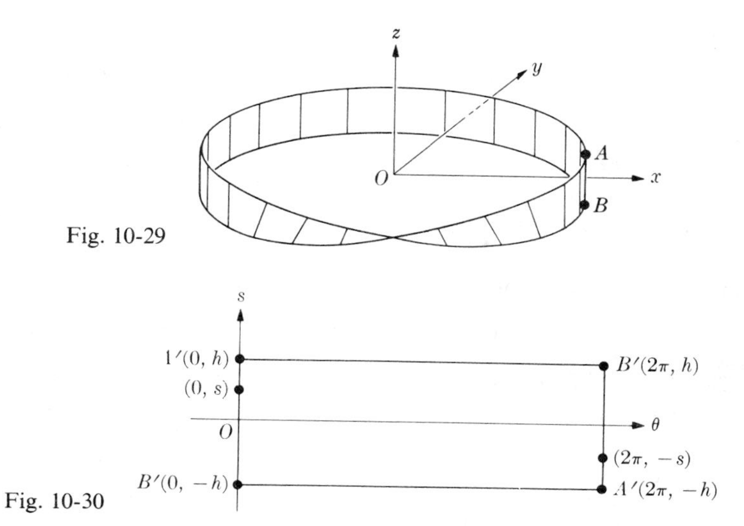

Fig. 10-29

Fig. 10-30

in Fig. 10-29. The reader can make a model of such a surface from a long, narrow, rectangular strip of paper by giving one end a half-twist and then gluing the ends together. A Möbius strip can be represented parametrically on a rectangle

$$R = \{(\theta, s) : 0 \le \theta \le 2\pi, \quad -h \le s \le h\}, \qquad (0 < h < a)$$

(see Fig. 10-30) by the equations

$$x = \left(a + s\sin\frac{\theta}{2}\right)\cos\theta, \qquad y = \left(a + s\sin\frac{\theta}{2}\right)\sin\theta, \qquad z = s\cos\frac{\theta}{2}. \tag{5}$$

To obtain the unit normal to the surface, we compute $\mathbf{r}_\theta \times \mathbf{r}_s$ from the formula

$$\mathbf{r}_\theta \times \mathbf{r}_s = J\left(\frac{y, z}{\theta, s}\right)\mathbf{i} + J\left(\frac{z, x}{\theta, s}\right)\mathbf{j} + J\left(\frac{x, y}{\theta, s}\right)\mathbf{k}.$$

We find

$$J\left(\frac{y, z}{\theta, s}\right) = \left(a + s\sin\frac{\theta}{2}\right)\cos\frac{\theta}{2}\cos\theta + \frac{s}{2}\sin\theta,$$

$$J\left(\frac{z, x}{\theta, s}\right) = \left(a + s\sin\frac{\theta}{2}\right)\cos\frac{\theta}{2}\sin\theta - \frac{s}{2}\cos\theta,$$

$$J\left(\frac{x, y}{\theta, s}\right) = -\left(a + s\sin\frac{\theta}{2}\right)\sin\frac{\theta}{2},$$

and

$$|\mathbf{r}_\theta \times \mathbf{r}_s|^2 = \left(a + s\sin\frac{\theta}{2}\right)^2 + \tfrac{1}{4}s^2.$$

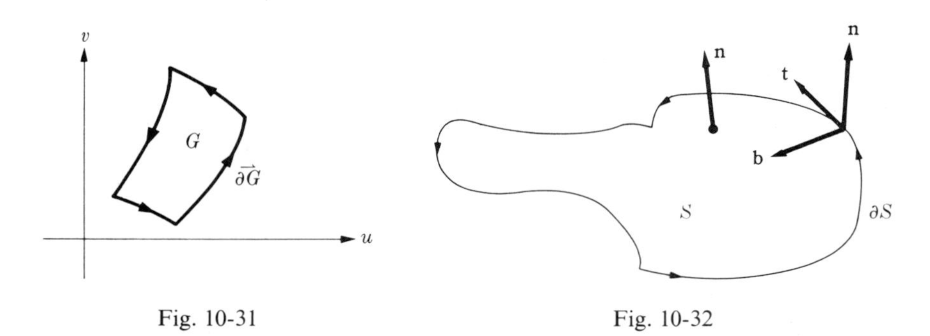

Fig. 10-31

Fig. 10-32

Therefore $\mathbf{r}_\theta \times \mathbf{r}_s$ is never $\mathbf{0}$, so that if we define $\mathbf{n}(\theta, s)$ by (2), we see that $\mathbf{n}(\theta, s)$ is continuous. However, $\mathbf{n}(0,0) = \mathbf{i}$, $\mathbf{n}(2\pi, 0) = -\mathbf{i}$, and the points $(0,0)$, $(2\pi, 0)$ in the parametric plane correspond to the same point on the surface. The transformation (5) is one to one, except that the points of R given by $(0, s)$ and $(2\pi, -s)$, $-h \le s \le h$, are always carried into the same point on the surface (Fig. 10-30). The Möbius strip, often called a "one-sided surface," is not a smooth surface element (defined in Section 4). If a pencil line is drawn down the center of the strip (corresponding to $s = 0$) then, after one complete trip around, the line is on the "opposite side." Two circuits are needed to "close up" the curve made by the pencil line. The Möbius strip is not an orientable surface.

A smooth surface element S which, according to (1) is the image of a plane region G, has a boundary (denoted ∂S) which is the image of the boundary ∂G of G. The closed curve ∂G, which is piecewise smooth, can be made into a directed curve by traversing it in a given direction. We say that ∂G is **positively directed** when we travel along it so that the interior of G is always on the left. We write $\partial\vec{G}$ for a positively directed closed curve. In Fig. 10-31 the curve ∂G is shown positively directed. The closed curve ∂S, the image of ∂G, becomes a directed curve in space, with the direction the one induced by $\partial\vec{G}$. We say the curve $\partial\vec{S}$ is **positively directed** when its direction corresponds to the positively directed curve $\partial\vec{G}$. Geometrically the curve $\partial\vec{S}$ is directed so that when one walks along $\partial\vec{S}$ in an upright position with his head in the direction of the positive normal $\mathbf{n}$ to the surface, then the surface is on his left. In terms of right- and left-handed coordinate systems, if $\mathbf{t}$ is tangent to $\partial\vec{S}$ pointing in the positive direction, if $\mathbf{n}$ is perpendicular to $\mathbf{t}$ and in the direction of the positive normal, and if $\mathbf{b}$ is perpendicular to the vectors $\mathbf{t}$ and $\mathbf{n}$ and pointing toward the surface, then the triple $\mathbf{t}$, $\mathbf{b}$, and $\mathbf{n}$ is a right-handed triple (Fig. 10-32). If ∂G and ∂S contain several closed curves, this argument holds for each curve.

We wish to extend the notion of orientable surface to *piecewise smooth surfaces*. Such surfaces have edges, and so a continuous unit normal vector field cannot be defined over the entirely of such a surface. However, a piecewise smooth surface F can be subdivided into a finite number of smooth surface elements $F_1, F_2, \ldots, F_k$. Each such surface element may be oriented

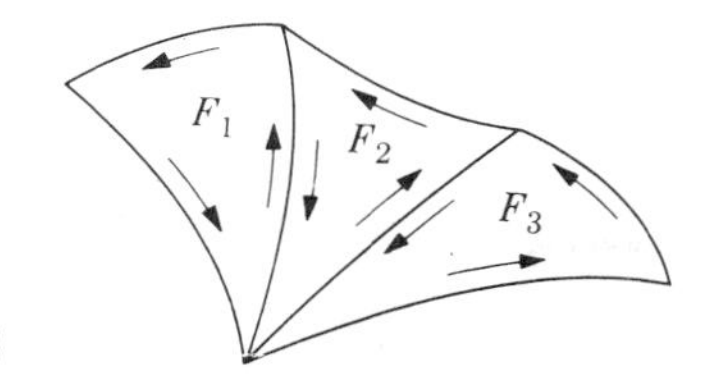

Fig. 10-33

and each boundary ∂F_i may be correspondingly positively directed. Suppose that γ_{ij} is a smooth arc which is the common boundary of two surface elements F_i and F_j. If the positive direction of γ_{ij} as part of $\partial \vec{F}_i$ is the opposite of the positive direction of γ_{ij} as part of $\partial \vec{F}_j$ for all arcs γ_{ij}, we say that the surface F is an **orientable, piecewise smooth surface**. In this case, the collection of orientations of the F_i form an orientation of F and the normal function defined on each $F_i^{(0)}$ (interior of F_i) is called the **positive normal**.

Figure 10-33 shows a piecewise smooth, orientable surface and its decomposition into smooth surface elements. The boundary $\partial \vec{F}$ is a positively directed, closed, piecewise smooth curve. Those boundary arcs of F_1, F_2, $\ldots$, F_k which are traversed only once comprise $\partial \vec{F}$. From the discussion of the last paragraph, it follows that if a piecewise smooth surface is orientable according to one decomposition into smooth surface elements, then it is orientable according to any other such decomposition. It can be shown that there is no way to subdivide a Möbius strip into smooth elements, with two adjacent elements always having oppositely directed common boundary arcs. In other words, a Möbius strip is not orientable even if it is treated as a piecewise smooth surface. On the other hand, the surface of a cube, which is piecewise smooth, is an orientable surface. We select the "outward" pointing normal on each face and traverse the boundary of any face in a counterclockwise direction as we view it from outside the cube. We easily verify that all edges are traversed twice, once in each direction.

7. Stokes' Theorem

Let $\vec{S}$ be a smooth oriented surface element and (x, y, z) a fixed rectangular coordinate system. We represent $\vec{S}$ parametrically:

$$S: \quad \mathbf{r}(u, v) = x(u, v)\mathbf{i} + y(u, v)\mathbf{j} + z(u, v)\mathbf{k}.$$

Then the positive unit normal $\mathbf{n}(u, v)$ is given by

$$\mathbf{n}(u, v) = \frac{1}{|\mathbf{r}_u \times \mathbf{r}_v|}\left[J\left(\frac{y, z}{u, v}\right)\mathbf{i} + J\left(\frac{z, x}{u, v}\right)\mathbf{j} + J\left(\frac{x, y}{u, v}\right)\mathbf{k}\right].$$

If $\mathbf{v}$ is a continuous vector field defined on S with coordinate functions

$$\mathbf{v}(x, y, z) = v_1(x, y, z)\mathbf{i} + v_2(x, y, z)\mathbf{j} + v_3(x, y, z)\mathbf{k},$$

we can compute the scalar product $\mathbf{v}\cdot\mathbf{n}$:

$$\mathbf{v}\cdot\mathbf{n} = \frac{1}{|\mathbf{r}_u \times \mathbf{r}_v|}\left[v_1 J\left(\frac{y,z}{u,v}\right) + v_2 J\left(\frac{z,x}{u,v}\right) + v_3 J\left(\frac{x,y}{u,v}\right)\right].$$

Since $\mathbf{v}\cdot\mathbf{n}$ is a continuous function on S, we may define the integral

$$\iint_S \mathbf{v}\cdot\mathbf{n}\,dS.$$

The surface element dS can be computed in terms of the parameters (u, v) according to the formula

$$dS = |\mathbf{r}_u \times \mathbf{r}_v|\,dA_{uv},$$

and so

$$\iint_S \mathbf{v}\cdot\mathbf{n}\,dS = \iint_G \left[v_1 J\left(\frac{y,z}{u,v}\right) + v_2 J\left(\frac{z,x}{u,v}\right) + v_3 J\left(\frac{x,y}{u,v}\right)\right] dA_{uv}, \tag{1}$$

where G is the domain in the (u, v)-plane which has $\vec{S}$ for its image.

If $\mathbf{r}_1(s, t)$ is another smooth representation of S, it follows from Theorem 4 that there is a one-to-one smooth transformation

$$u = U(s,t), \qquad v = V(s,t)$$

such that

$$\mathbf{r}[U(s,t),\ V(s,t)] = \mathbf{r}_1(s,t).$$

According to the rule for multiplying Jacobians,

$$J\left(\frac{y,z}{s,t}\right) = J\left(\frac{y,z}{u,v}\right) J\left(\frac{u,v}{s,t}\right)$$

[and the analogous equations for (z, x) and (x, y)], we see that the representation $\mathbf{r}_1(s, t)$ gives the same orientation as the representation $\mathbf{r}(u, v)$ if and only if

$$J\left(\frac{u,v}{s,t}\right) > 0.$$

In such a case, if we replace (u, v) by (s, t) in (1) and integrate over G_1 [the region in the (s, t)-plane whose image is S], we obtain the formula for

$$\iint_S \mathbf{v}\cdot\mathbf{n}\,dS$$

in terms of the parameters s and t.

When S is a piecewise smooth orientable surface, we can represent it as the union of a number of smooth surface elements $S_1, S_2, \ldots, S_k$. Then we define

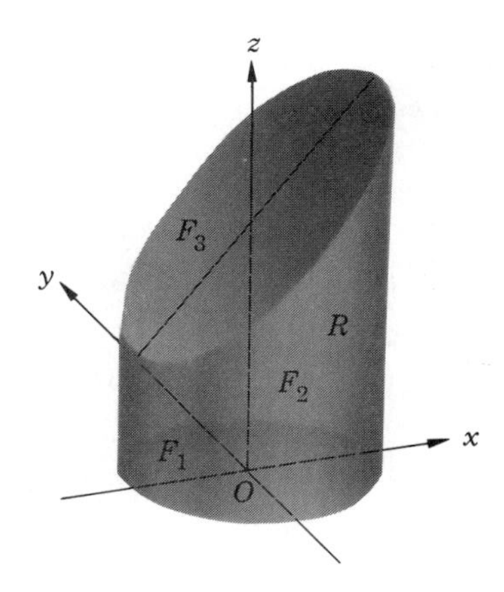

Fig. 10-34

$$\iint_S \mathbf{v}\cdot\mathbf{n}\,dS = \sum_{i=1}^{k} \iint_{S_i} \mathbf{v}\cdot\mathbf{n}_i\,dS_i.$$

Each integral on the right is evaluated as in (1).

EXAMPLE 1. Let R be the region bounded by the cylinder $x^2 + y^2 = 1$ and the planes $z = 0$ and $z = x + 2$ (Fig. 10-34). Let S be the entire boundary of R. Find the value of $\iint_S \mathbf{v}\cdot\mathbf{n}\,dS$ where $\mathbf{n}$ is the outward directed unit normal on S and

$$\mathbf{v} = 2x\mathbf{i} - 3y\mathbf{j} + z\mathbf{k}.$$

SOLUTION. The surface S is piecewise smooth and we label the smooth portions S_1, S_2, and S_3, as shown in Fig. 10-34. On S_1 we have

$$\mathbf{n} = -\mathbf{k}, \qquad \mathbf{v}\cdot\mathbf{n}\,dS = -z\,dA_{xy} = 0,$$

as S_1 is in the plane $z = 0$. Therefore we have $\iint_{S_1} \mathbf{v}\cdot\mathbf{n}\,dS = 0$. On S_3, we have $z = x + 2$ and

$$\mathbf{n} = \frac{1}{\sqrt{2}}(-\mathbf{i} + \mathbf{k}),$$

$$\mathbf{v}\cdot\mathbf{n} = \frac{1}{\sqrt{2}}(-2x + z), \qquad dS = \sqrt{2}\,dA_{xy},$$

$$\mathbf{v}\cdot\mathbf{n}\,dS = (-2x + z)\,dA_{xy}.$$

We obtain $\iint_{S_3} \mathbf{v}\cdot\mathbf{n}\,dS = \iint_{S_1} (-x + 2)\,dA_{xy}$. We see that $\iint_{S_1} x\,dA_{xy} = 0$, since the integrand is an odd function. Also, $2\iint_{S_1} dA_{xy} = 2\pi$.

On S_2, we select cylindrical coordinates

$$S_2\colon\ x = \cos\theta, \qquad y = \sin\theta, \qquad z = z, \qquad (\theta, z) \text{ in } F_2;$$

$$F_2 = \{(\theta, z) : -\pi \le \theta \le \pi,\ \ 0 \le z \le 2 + \cos\theta\}.$$

We find

$$\mathbf{n} = (\cos\theta)\mathbf{i} + (\sin\theta)\mathbf{j}, \qquad \mathbf{v}\cdot\mathbf{n} = 2\cos^2\theta - 3\sin^2\theta.$$

Therefore

$$\begin{aligned}\iint_{S_2} \mathbf{v}\cdot\mathbf{n}\,dS &= \iint_{F_2} (2\cos^2\theta - 3\sin^2\theta)\,dA_{\theta z}\\ &= \int_{-\pi}^{\pi}\int_0^{2+\cos\theta} (2 - 5\sin^2\theta)\,dz\,d\theta\\ &= \int_{-\pi}^{\pi} (4 - 10\sin^2\theta + 2\cos\theta - 5\sin^2\theta\cos\theta)\,d\theta\\ &= 8\pi - 10\pi + 0 + 0 = -2\pi.\end{aligned}$$

Hence

$$\iint_S \mathbf{v}\cdot\mathbf{n}\,dS = \iint_{S_1} \mathbf{v}\cdot\mathbf{n}\,dS + \iint_{S_2} \mathbf{v}\cdot\mathbf{n}\,dS + \iint_{S_3} \mathbf{v}\cdot\mathbf{n}\,dS = 0.$$

Suppose that $\vec{S}$ is an oriented, piecewise smooth surface with the boundary $\partial\vec{S}$ positively directed and made up of a finite number of smooth arcs $\vec{C}_1$ $\vec{C}_2, \ldots, \vec{C}_k$. If $\mathbf{v}$ is a continuous vector field defined on and near ∂S, we define

$$\int_{\partial\vec{S}} \mathbf{v}\cdot d\mathbf{r} = \sum_{i=1}^{k} \int_{\vec{C}_i} \mathbf{v}\cdot d\mathbf{r}.$$

With the aid of this definition we can state Stokes' Theorem.

Theorem 5 (Stokes' Theorem). *Suppose that $\vec{S}$ is a bounded, closed, orientable, piecewise smooth surface and that $\mathbf{v}$ is a smooth vector field defined in a region containing $\vec{S}$. Then*

$$\boxed{\iint_{\vec{S}} (\operatorname{curl}\mathbf{v})\cdot\mathbf{n}\,dS = \int_{\partial\vec{S}} \mathbf{v}\cdot d\mathbf{r}.} \tag{2}$$

Corollary. *Suppose that $\vec{S}$ is a bounded, closed, orientable, piecewise smooth surface without boundary and that $\mathbf{v}$ is a smooth vector field defined in a region containing S. Then*

$$\iint_{\vec{S}} (\operatorname{curl}\mathbf{v})\cdot\mathbf{n}\,dS = 0. \tag{3}$$

We require two lemmas to prove Stokes' Theorem.

Lemma 2. *Suppose that τ is a rectangular coordinate system in three-space with corresponding orthogonal unit vectors* **i**, **j**, **k**. *Suppose that $\vec{S}$ is a smooth oriented surface element with a parametric representation*

$$\mathbf{r}(u, v) = x(u, v)\mathbf{i} + y(u, v)\mathbf{j} + z(u, v)\mathbf{k}, \qquad (u, v) \in G \cup \partial G, \tag{4}$$

which agrees with the orientation of $\vec{S}$. Let **v** *be a continuous vector field defined in a region E containing S, and suppose* **v** *is given by*

$$\mathbf{v}(x, y, z) = v_1(x, y, z)\mathbf{i} + v_2(x, y, z)\mathbf{j} + v_3(x, y, z)\mathbf{k}, \qquad (x, y, z) \in E. \tag{5}$$

Then

$$\int_{\partial \vec{S}} \mathbf{v} \cdot d\mathbf{r} = \int_{\partial \vec{G}} [(v_1 x_u + v_2 y_u + v_3 z_u)\, du + (v_1 x_v + v_2 y_v + v_3 z_v)\, dv]. \tag{6}$$

Proof. We establish the result for the case that $\partial \vec{G}$ consists of a single piecewise smooth simple closed curve; the extension to several such curves is clear. Let

$$u = u(s), \qquad v = v(s), \qquad a \le s \le b,$$

be a parametric representation of the closed curve $\partial \vec{G}$. Then

$$x = x[u(s), v(s)], \qquad y = y[u(s), v(s)], \qquad z = z[u(s), v(s)], \qquad a \le s \le b$$

is a parametric representation of $\partial \vec{S}$. Using the Chain Rule, we find

$$\begin{aligned} d\mathbf{r} &= dx\,\mathbf{i} + dy\,\mathbf{j} + dz\,\mathbf{k} \\ &= (x_u u_s + x_v v_s)\, ds\,\mathbf{i} + (y_u u_s + y_v v_s)\, ds\,\mathbf{j} + (z_u u_s + z_v v_s)\, ds\,\mathbf{k}. \end{aligned}$$

Therefore

$$\int_{\partial \vec{S}} \mathbf{v} \cdot d\mathbf{r} = \int_a^b [(v_1 x_u + v_2 y_u + v_3 z_u) u_s + (v_1 x_v + v_2 y_v + v_3 z_v) v_s]\, ds. \tag{7}$$

Making the insertions

$$du = u_s\, ds, \qquad dv = v_s\, ds$$

in the right side of (6), we see that (6) and (7) are identical.

Lemma 3. *Suppose that $f(u, v)$ is smooth in a region D containing $G \cup \partial G$ where G has the usual piecewise smooth properties. Then there is a sequence $f_1, f_2, \ldots, f_n, \ldots$ such that each f_n, $\dfrac{\partial f_n}{\partial u}$, $\dfrac{\partial f_n}{\partial v}$ is smooth on G and*

$$f_n \to f, \qquad \frac{\partial f_n}{\partial u} \to \frac{\partial f}{\partial u}, \qquad \frac{\partial f_n}{\partial v} \to \frac{\partial f}{\partial v} \qquad \text{as } n \to \infty,$$

uniformly on G.

PROOF. Since G is bounded, there is no loss in generality in assuming that D is bounded. Since G is closed and contained in D, it follows that there is a $\rho > 0$ such that the disk $C(P, \rho)$ with center at P and radius ρ, is in D for P any point of G. We let G_0 be the set of all P in D such that $C(P, \rho/2)$ is in D; it can be shown that G_0 is closed. For (u_0, v_0) in G, it is then clear that the square

$$|u - u_0| \le h, \qquad |v - v_0| \le h$$

is in G_0 so long as $0 < h < \rho/2\sqrt{2}$. We now define the function

$$f_h(u, v) = \frac{1}{4h^2} \int_{u-h}^{u+h} \int_{v-h}^{v+h} f(s, t)\, dA_{st}$$

for all (u, v) in G with $0 < h < \rho/2\sqrt{2}$. Holding v fixed and applying Leibniz' Rule for differentiating under the integral sign, we obtain

$$\begin{aligned} \frac{\partial f_h(u, v)}{\partial u} &= \frac{1}{4h^2} \int_{v-h}^{v+h} [f(u + h, t) - f(u - h, t)]\, dt \\ &= \frac{1}{4h^2} \int_{v-h}^{v+h} \int_{u-h}^{u+h} \frac{\partial f(s, t)}{\partial s} dA_{st}. \end{aligned}$$

Applying Leibniz' Rule again, first with respect to u and then with respect to v, we get

$$\frac{\partial^2 f_h(u, v)}{\partial u^2} = \frac{1}{4h^2} \int_{v-h}^{v+h} \left[\frac{\partial f(u + h, t)}{\partial u} - \frac{\partial f(u - h, t)}{\partial u} \right] dt$$

$$\frac{\partial^2 f_h(u, v)}{\partial u\, \partial v} = \frac{1}{4h^2} \int_{u-h}^{u+h} \left[\frac{\partial f(s, v + h)}{\partial s} - \frac{\partial f(s, v - h)}{\partial s} \right] ds.$$

A similar argument shows that

$$\frac{\partial f_h(u, v)}{\partial v} = \frac{1}{4h^2} \int_{u-h}^{u+h} \int_{v-h}^{v+h} \frac{\partial f(s, t)}{\partial t} dA_{st}$$

and that $\dfrac{\partial f_h}{\partial v}$ is smooth for $0 < h < \rho/2\sqrt{2}$.

Since $\dfrac{\partial f}{\partial s}$ and $\dfrac{\partial f}{\partial t}$ are continuous on G_0, a closed bounded set, they are uniformly continuous there. Hence for any $\varepsilon > 0$, there is a δ with $0 < \delta < \rho/2\sqrt{2}$ such that

$$\left| \frac{\partial f(s', t')}{\partial s'} - \frac{\partial f(s'', t'')}{\partial s''} \right| < \varepsilon$$

if $|s' - s''| < \delta$ and $|t' - t''| < \delta$. A similar statement holds for $\partial f/\partial t$. Therefore if $0 < h < \delta$, then

$$\left| \frac{\partial f_h(u, v)}{\partial u} - \frac{\partial f(u, v)}{\partial u} \right| = \frac{1}{4h^2} \left| \int_{u-h}^{u+h} \int_{v-h}^{v+h} \left[\frac{\partial f(s, t)}{\partial s} - \frac{\partial f(u, v)}{\partial u} \right] dA_{st} \right| < \varepsilon,$$

and similarly for the derivative with respect to v. Now letting $\{h_n\}$ be any sequence tending to zero, we set $f_n = f_{h_n}$ and the result follows.

We now establish Stokes' Theorem (Theorem 5).

PROOF OF STOKES' THEOREM. We first prove this theorem for the case that $\vec{S}$ is a single smooth oriented surface element. Let τ be a rectangular coordinate system with corresponding vectors **i**, **j**, and **k**, and let **v** be given by (5) and let (4) be a parametric representation of $\vec{S}$ which agrees with the orientation on $\vec{S}$. Finally we assume that $x, y, z, x_u, y_u, z_u, x_v, y_v$, and z_v are all smooth on D. Then (6) holds on account of Lemma 2. Applying Green's Theorem to the integral on the right side of (6), we obtain

$$\int_{\partial\vec{S}} \mathbf{v}\cdot d\mathbf{r} = \iint_G \left[\frac{\partial}{\partial u}(v_1 x_v + v_2 y_v + v_3 z_v) - \frac{\partial}{\partial v}(v_1 x_u + v_2 y_u + v_3 z_u)\right] dA_{uv}$$

$$= \iint_G [x_v(v_{1x}x_u + v_{1y}y_u + v_{1z}z_u) + y_v(v_{2x}x_u + v_{2y}y_u + v_{2z}z_u) \quad (8)$$

$$+ z_v(v_{3x}x_u + v_{3y}y_u + v_{3z}z_u) - x_u(v_{1x}x_v + v_{1y}y_v + v_{1z}z_v)$$

$$- y_u(v_{2x}x_v + v_{2y}y_v + v_{2z}z_v) - z_u(v_{3x}x_v + v_{3y}y_v + v_{3z}z_v)]\, dA_{uv}.$$

It is easy to verify that all the terms in the right side of (8) involving the derivatives x_{uv}, y_{uv}, and z_{uv} cancel each other. Collecting terms in (8), we obtain

$$\int \mathbf{v}\cdot d\mathbf{r} = \iint_G \left[(v_{3y} - v_{2z})J\left(\frac{y,z}{u,v}\right) + (v_{1z} - v_{3x})J\left(\frac{z,x}{u,v}\right)\right.$$
$$\left. + (v_{2x} - v_{1y})J\left(\frac{x,y}{u,v}\right)\right] dA_{uv} \quad (9)$$

and the right side of (9) is the left side of (2), as is seen using (1).

Now, if we merely know that x, y, and z are smooth on D, it follows from Lemma 3 that there are sequences $\{x_n\}$, $\{y_n\}$, and $\{z_n\}$ such that $x_n, y_n, z_n, x_{nu}, y_{nu}, z_{nu}, x_{nv}, y_{nv}$, and z_{nv} are all smooth on an open set D' containing $G \cup \partial G$ such that all these quantities converge uniformly to x, y, z, and their first derivatives on $G \cup \partial G$. The formula (9) with x, y, z replaced by x_n, y_n, and z_n holds for each n. The uniform convergence implies that (9) holds in the limit.

Finally, if $\vec{S}$ is any piecewise smooth oriented surface, we may express $\vec{S} = \vec{S}_1 \cup \cdots \cup \vec{S}_k$, where each $\vec{S}_i$ is an oriented smooth surface element, oriented in such a way that any piecewise smooth arc on the boundary of $\vec{S}_i$ and $\vec{S}_j$ is directed oppositely as part of $\partial\vec{S}_i$ from the way it is directed as a part of $\partial\vec{S}_j$. Formula (2) follows for each i from the proof above. Adding these results, we obtain

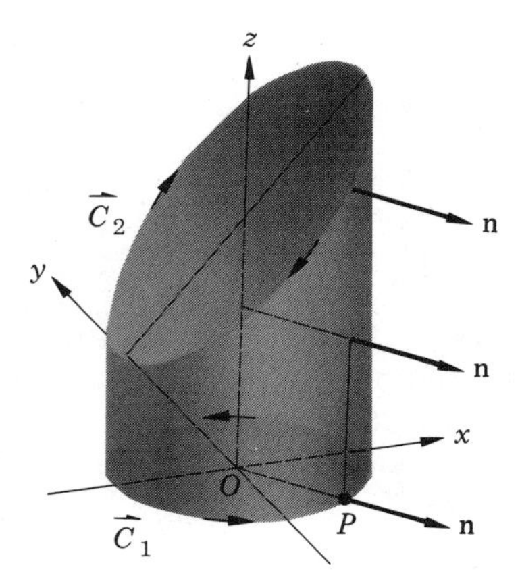

Fig. 10-35

$$\iint_{\vec{S}} (\operatorname{curl} \mathbf{v}) \cdot \mathbf{n}\, dS = \sum_{i=1}^{k} \int_{\partial \vec{S}_i} \mathbf{v} \cdot d\mathbf{r}. \tag{10}$$

In the sum on the right the integrals over arcs which are on the boundaries of two $\vec{S}_i$ cancel, leaving only the integrals over those arcs only on the boundary of one $\vec{S}_i$. These latter arcs comprise $\partial \vec{S}$.

EXAMPLE 2. Verify Stokes' Theorem, given that

$$\mathbf{v} = y\mathbf{i} + z\mathbf{j} + x\mathbf{k}$$

and that $\vec{S}$ is the part of the surface of the cylinder $x^2 + y^2 = 1$ between the planes $z = 0$ and $z = x + 2$, oriented with $\mathbf{n}$ pointing outward (Fig. 10-35).

SOLUTION. An examination of Fig. 10-35 shows that the curves $\vec{C}_1$ and $\vec{C}_2$ must be directed as exhibited. We choose cylindrical coordinates and write

$$\vec{S} = \{(x, y, z) : x = \cos\theta, \quad y = \sin\theta, \quad z = z, \quad (\theta, z) \text{ on } F\};$$

$$F = \{(\theta, z) : -\pi \le \theta \le \pi, \quad 0 \le z \le 2 + \cos\theta\}.$$

We think of S as made up of two smooth surface elements corresponding to F_1 and F_2, as shown in Fig. 10-36. We must make this subdivision because the representation of $\vec{S}$ by F is not one-to-one. The formulas

$$J\left(\frac{y, z}{\theta, z}\right) = \cos\theta,$$

$$J\left(\frac{z, x}{\theta, z}\right) = \sin\theta,$$

$$J\left(\frac{x, y}{\theta, z}\right) = 0,$$

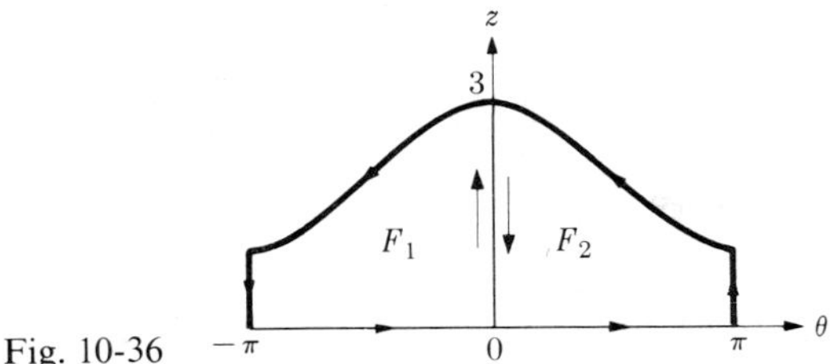

Fig. 10-36

$$\mathbf{n} = (\cos\theta)\mathbf{i} + (\sin\theta)\mathbf{j},$$

show that the representation agrees with the given orientation. Therefore the orientation of the boundary of $\vec{S}$ is determined by the orientations of the boundaries of F_1 and F_2. We note that when we form the boundary integral, those parts taken over vertical segments cancel. Now

$$\operatorname{curl}\mathbf{v} = -\mathbf{i} - \mathbf{j} - \mathbf{k}, \qquad \mathbf{n} = (\cos\theta)\mathbf{i} + (\sin\theta)\mathbf{j}, \qquad \text{and} \qquad dS = dA_{\theta z}.$$

We obtain

$$\begin{aligned}\iint_{\vec{S}} (\operatorname{curl}\mathbf{v}\cdot\mathbf{n})\,dS &= \int_{-\pi}^{\pi}\int_0^{2+\cos\theta} (-\cos\theta - \sin\theta)\,dz\,d\theta \\ &= -\int_{-\pi}^{\pi} [(2\cos\theta + \cos^2\theta) + (2+\cos\theta)\sin\theta)]\,d\theta \qquad (11) \\ &= -\pi.\end{aligned}$$

For the boundary integral, we have

$$\int_{\partial\vec{S}_i} \mathbf{v}\cdot d\mathbf{r} = \int_{\vec{C}_1} \mathbf{v}\cdot d\mathbf{r} - \int_{-\vec{C}_2} \mathbf{v}\cdot d\mathbf{r},$$

in which the integrals on the right are taken in a counterclockwise direction. We find:

$$\text{On } \vec{C}_1: \quad \mathbf{v} = (\sin\theta)\mathbf{i} + (\cos\theta)\mathbf{k}, \qquad d\mathbf{r} = (-\sin\theta\mathbf{i} + \cos\theta\mathbf{j})\,d\theta.$$

$$\text{On } -\vec{C}_2: \quad \mathbf{v} = (\sin\theta)\mathbf{i} + (2+\cos\theta)\mathbf{j} + (\cos\theta)\mathbf{k},$$

$$d\mathbf{r} = (-\sin\theta\mathbf{i} + \cos\theta\mathbf{j} - \sin\theta\mathbf{k})\,d\theta.$$

Taking the scalar products, we get

$$\int_{\vec{C}_1} \mathbf{v}\cdot d\mathbf{r} = \int_{-\pi}^{\pi} (-\sin^2\theta)\,d\theta = -\pi, \qquad (12)$$

$$\int_{\vec{C}_2} \mathbf{v}\cdot d\mathbf{r} = \int_{-\pi}^{\pi} (-\sin^2\theta + 2\cos\theta + \cos^2\theta - \sin\theta\cos\theta)\,d\theta = 0. \qquad (13)$$

A comparison of (11) with (12) and (13) verifies Stokes' Theorem.

Problems

In each of Problems 1 through 7, compute $\iint_{\vec{S}} \mathbf{v} \cdot \mathbf{n}\, dS$.

1. $\mathbf{v} = (x+1)\mathbf{i} - (2y+1)\mathbf{j} + z\mathbf{k}$; $\vec{S}$ is the triangle with vertices $(1, 0, 0)$, $(0, 1, 0)$, and $(0, 0, 1)$, with $\mathbf{n}$ pointing away from the origin.
2. $\mathbf{v} = x\mathbf{i} + y\mathbf{j} + z\mathbf{k}$; $\vec{S}$ is the part of the paraboloid $2z = x^2 + y^2$ inside the cylinder $x^2 + y^2 = 2x$ with $\mathbf{n} \cdot \mathbf{k} > 0$ ($\mathbf{n}$ pointing upward).
3. $\mathbf{v} = x^2\mathbf{i} + y^2\mathbf{j} + z^2\mathbf{k}$; $\vec{S}$ is the part of the cone
$$z^2 = x^2 + y^2 \qquad \text{for which} \qquad 1 \le z \le 2,$$
with $\mathbf{n} \cdot \mathbf{k} > 0$.
4. $\mathbf{v} = xy\mathbf{i} + xz\mathbf{j} + yz\mathbf{k}$; $\vec{S}$ is the part of the cylinder $y^2 = 2 - x$ cut out by the cylinders $y^2 = z$ and $y = z^3$.
5. $\mathbf{v} = y^2\mathbf{i} + z\mathbf{j} - x\mathbf{k}$; $\vec{S}$ is the part of the cylinder $y^2 = 1 - x$ between the planes $z = 0$ and $z = x$; $x \ge 0$, with $\mathbf{n} \cdot \mathbf{i} > 0$.
6. $\mathbf{v} = 2x\mathbf{i} - y\mathbf{j} + 3z\mathbf{k}$; $\vec{S}$ is the part of the cylinder $z^2 = x$ to the left of the cylinder $y^2 = 1 - x$ and $\mathbf{n} \cdot \mathbf{i} > 0$, $\mathbf{n}$ pointing to the right.

*7. $\mathbf{v} = x\mathbf{i} + y\mathbf{j} - 2z\mathbf{k}$; $\vec{S}$ is the part of the cylinder $x^2 + y^2 = 2x$ between the two nappes of the cone $z^2 = x^2 + y^2$, $\mathbf{n}$ pointing outward.

In each of Problems 8 through 13, verify Stokes' Theorem.

8. $\mathbf{v} = z\mathbf{i} + x\mathbf{j} + y\mathbf{k}$; $\vec{S}$ is the part of the paraboloid $z = 1 - x^2 - y^2$ for which $z \ge 0$ and $\mathbf{n} \cdot \mathbf{k} > 0$.
9. $\mathbf{v} = y^2\mathbf{i} + xy\mathbf{j} - 2xz\mathbf{k}$; $\vec{S}$ is the hemisphere $x^2 + y^2 + z^2 = a^2$, $z \ge 0$ with $\mathbf{n} \cdot \mathbf{k} > 0$.
10. $\mathbf{v} = -yz\mathbf{i}$; $\vec{S}$ is the part of the sphere $x^2 + y^2 + z^2 = 4$ outside the cylinder $x^2 + y^2 = 1$, $\mathbf{n}$ pointing outward.
11. $\mathbf{v} = -z\mathbf{j} + y\mathbf{k}$; $\vec{S}$ is the part of the vertical cylinder $r = \theta$ (cylindrical coordinates), $0 \le \theta \le \pi/2$, bounded below by the (x, y)-plane and above by the cone $z^2 = x^2 + y^2$, $\mathbf{n} \cdot \mathbf{i} > 0$ for $\theta > 0$.
12. $\mathbf{v} = y\mathbf{i} + z\mathbf{j} + x\mathbf{k}$; $\vec{S}$ is the part of the surface $z^2 = 4 - x$ to the right of the cylinder $y^2 = x$, $\mathbf{n} \cdot \mathbf{i} > 0$, $\mathbf{i}$ pointing to the right.
13. $\mathbf{v} = z\mathbf{i} - x\mathbf{k}$; $\vec{S}$ is the part of the cylinder $r = 2 + \cos\theta$ above the (x, y)-plane and below the cone $z^2 = x^2 + y^2$, $\mathbf{n}$ pointing outward.

In each of Problems 14 through 16, compute $\int_{\partial\vec{S}} \mathbf{v} \cdot d\mathbf{r}$, using Stokes' Theorem.

14. $\mathbf{v} = r^{-3}\mathbf{r}$, $\mathbf{r} = x\mathbf{i} + y\mathbf{j} + z\mathbf{k}$, $r = |\mathbf{r}|$; $\vec{S}$ is the surface of Example 2.
15. $\mathbf{v} = (e^x \sin y)\mathbf{i} + (e^x \cos y - z)\mathbf{j} + y\mathbf{k}$; $\vec{S}$ is the surface of Problem 3.
16. $\mathbf{v} = (x^2 + z)\mathbf{i} + (y^2 + x)\mathbf{j} + (z^2 + y)\mathbf{k}$; $\vec{S}$ is the part of the sphere $x^2 + y^2 + z^2 = 1$ above the cone $z^2 = x^2 + y^2$; $\mathbf{n} \cdot \mathbf{k} > 0$.
17. Show that if $\vec{S}$ is given by $z = f(x, y)$ for $x^2 + y^2 \le 1$, where f is smooth and if

$\mathbf{v} = (1 - x^2 - y^2)\mathbf{w}(x, y, z)$, where $\mathbf{w}$ is any smooth vector field defined on an open set containing S, then $\iint_{\vec{S}}(\text{curl } \mathbf{v} \cdot \mathbf{n})\, dS = 0$.

18. Suppose that $\mathbf{v} = r^{-3}(y\mathbf{i} + z\mathbf{j} + x\mathbf{k})$, where $r = |\mathbf{r}| = |x\mathbf{i} + y\mathbf{j} + z\mathbf{k}|$ and $\vec{S}$ is the unit sphere with $\mathbf{n}$ directed outward. Show by direct calculation that

$$\iint_{\vec{S}} (\text{curl } \mathbf{v}) \cdot \mathbf{n}\, dS = 0.$$

8. The Divergence Theorem

Stokes' Theorem, which relates an integral over a surface in space to a line integral over the boundary of the surface, is a generalization of Green's Theorem. Another type of generalization, known as the Divergence Theorem, establishes a connection between an integral over a three-dimensional domain and an integral over the surface which forms the boundary of the domain. It can be shown that the Divergence Theorem and the theorems of Green and Stokes are all special cases of a general formula which connects an integral over a set of points in some n-dimensional space with another integral over the boundary of that set of points.

Theorem 6 (The Divergence Theorem). *Suppose that a bounded domain G in three-space is bounded by one or more disjoint piecewise smooth, orientable surfaces without boundary and suppose that $\mathbf{v}$ is a smooth vector field defined on an open set containing G and ∂G. Then*

$$\iiint_G \operatorname{div} \mathbf{v}\, dV = \iint_{\partial G} \mathbf{v} \cdot \mathbf{n}\, dS, \tag{1}$$

where the boundary ∂G is oriented by taking $\mathbf{n}$ as the exterior normal.

We prove the Divergence Theorem (with an added condition that the boundary is not too irregular) later in this section. However, certain special cases which give the principal content of the result will be established first.

Lemma 4. *Suppose that $\mathbf{v}$ and G are such that there exists a rectangular coordinate system in which $\mathbf{v}(x, y, z) = R(x, y, z)\mathbf{k}$ and G is of the form*

$$G = \{(x, y, z) : (x, y) \in D, \quad c < z < f(x, y)\},$$

where f is piecewise smooth (as indicated in Fig. 10-37). If R and R_z are continuous on an open set containing G and ∂G, then (1) holds.*

* That is, the graph S of the equation $z = f(x, y)$, $(x, y) \in D \cup \partial D$ is a piecewise smooth surface and $z = f(x, y)$, $(x, y) \in D_i \cup \partial D_i$ is a parametric representation of each smooth part S_i.

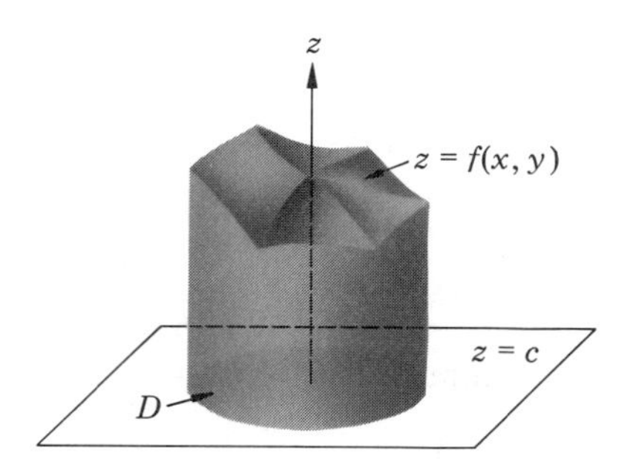

Fig. 10-37

PROOF. We note that $\operatorname{div} \mathbf{v} = R_z$ and, therefore, that

$$\iiint_G \operatorname{div} \mathbf{v}\, dV = \iiint_G R_z\, dV_{xyz} = \iint_D \int_c^{f(x,y)} R_z\, dz\, dA_{xy}.$$

Upon performing the integration with respect to z, we find

$$\iiint_G \operatorname{div} \mathbf{v}\, dV = \iint_D \{R[x, y, f(x, y)] - R(x, y, c)\}\, dA_{xy}. \tag{2}$$

Let $\partial G = S_1 \cup S_2 \cup S_3$, where S_1 is the domain D in the plane $z = c$, S_2 is the lateral surface of the cylinder, and S_3 is the top surface:

$$S_3 = \{(x, y, z) : (x, y) \in D, \quad z = f(x, y)\}.$$

We have

$$\text{On } S_2: \quad \mathbf{n} \cdot \mathbf{k} = 0, \qquad \iint_{S_2} \mathbf{v} \cdot \mathbf{n}\, dS = 0. \tag{3}$$

$$\text{On } S_1: \quad \mathbf{n} = -\mathbf{k}, \qquad \iint_{S_1} \mathbf{v} \cdot \mathbf{n}\, dS = -\iint_D R(x, y, c)\, dA_{xy}. \tag{4}$$

$$\begin{aligned} \text{On } S_3: \quad & \mathbf{n} = (1 + f_x^2 + f_y^2)^{-1/2}(-f_x\mathbf{i} - f_y\mathbf{j} + \mathbf{k}), \\ & dS = (1 + f_x^2 + f_y^2)^{1/2}\, dA_{xy}; \\ & \iint_{S_3} \mathbf{v} \cdot \mathbf{n}\, dS = \iint_D R[x, y, f(x, y)]\, dA_{xy}. \end{aligned} \tag{5}$$

A comparison of (2) with (3), (4), and (5) yields the result.

Lemma 5. *Suppose that the hypotheses of Lemma* 4 *hold, except that* $\mathbf{v}$ *has the form*

$$\mathbf{v} = P(x, y, z)\mathbf{i} + Q(x, y, z)\mathbf{j}$$

with P, Q smooth on a region containing G and ∂G. Then (1) *holds.*

PROOF. We define the functions U, V by the formulas

$$U(x, y, z) = \int_c^z Q(x, y, t)\,dt, \qquad V(x, y, z) = -\int_c^z P(x, y, t)\,dt.$$

Also, we set

$$\mathbf{w} = U\mathbf{i} + V\mathbf{j} \quad \text{and} \quad -R = V_x - U_y, \quad \mathbf{u} = R\mathbf{k}.$$

Then $\mathbf{w}$ is smooth and R, R_z are continuous, so that

$$\begin{aligned} \operatorname{curl}\mathbf{w} &= -V_z\mathbf{i} + U_z\mathbf{j} + (V_x - U_y)\mathbf{k} = \mathbf{v} - \mathbf{u}, \\ \operatorname{div}\mathbf{v} &= P_x + Q_y = R_z = \operatorname{div}\mathbf{u}. \end{aligned} \tag{6}$$

We apply the Corollary to Stokes' Theorem and obtain

$$\iint_{\partial G} (\operatorname{curl}\mathbf{w})\cdot\mathbf{n}\,dS = 0 = \iint_{\partial G} (\mathbf{v} - \mathbf{u})\cdot\mathbf{n}\,dS.$$

Therefore

$$\iint_{\partial G} \mathbf{v}\cdot\mathbf{n}\,dS = \iint_{\partial G} \mathbf{u}\cdot\mathbf{n}\,dS = \iiint_G \operatorname{div}\mathbf{u}\,dV; \tag{7}$$

the last equality holds because Lemma 4 may be applied to $\mathbf{u} = R\mathbf{k}$. Taking (6) and (7) into account, we get

$$\iint_{\partial G} \mathbf{v}\cdot\mathbf{n}\,dS = \iiint_G \operatorname{div}\mathbf{v}\,dV.$$

The Divergence Theorem holds also for cylindrical domains which are parallel to the x- or y-axes. By addition, the result is valid for smooth vector fields $\mathbf{v}$ defined over regions G which are the sum of cylindrical domains of the kind just described. These regions may be quite general. However, the proof we now present uses the partition of unity, and therefore avoids the difficulty of describing such regions in detail.

We suppose that G is a region in R^3, the boundary of which consists of a finite number of disjoint piecewise smooth surfaces. Each of these surfaces has the property: For any point P of such a surface S, we can associate a coordinate system (x, y, z) with P as the origin, and a function f which is piecewise smooth as in Lemma 4. That is, we have a domain

$$D = \{(x, y) : x^2 + y^2 < r^3\}$$

and a cylindrical region

$$\gamma = \{(x, y, z) : (x, y) \in D, \quad -K < z < K\}$$

such that the part of G in γ is

$$\{(x, y, z) : (x, y) \in D, \quad -K < z < f(x, y)\}.$$

We assume that $f(0, 0) = 0$. When each point P has the above property, we say that G has a **regular boundary**.

We now establish the Divergence Theorem with the additional condition that G has a regular boundary.

PROOF. Let P be a point of the boundary of G. We assume D, γ, and f are as described above. We also introduce the cylinder Γ which is similar to γ but has radius $3r$ instead of r. We take r so small that Γ is in G.

Let ϕ be the function defined in Section 2 (Equation 1) in the proof of Green's Theorem. We set

$$\alpha(x, y, z) = \phi\left(\frac{\sqrt{x^2 + y^2}}{r}\right)\phi\left(\frac{z}{K}\right). \tag{8}$$

Let $\gamma_1, \gamma_2, \ldots, \gamma_k$ be a set of cylinders of the above type which covers G, and $\alpha_1, \alpha_2, \ldots, \alpha_k$ the corresponding functions of type (8). We next define

$$\psi_i(Q) = \frac{\alpha_i(Q)}{\alpha_1(Q) + \alpha_2(Q) + \cdots + \alpha_k(Q)}$$

for each Q in R^3. We note that the ψ_i form a smooth partition of unity on a region containing G and its boundary. Setting

$$\mathbf{v}_i = \psi_i \mathbf{v}$$

we have

$$\mathbf{v} = \mathbf{v}_1 + \mathbf{v}_2 + \cdots + \mathbf{v}_k,$$

and it suffices to prove the Divergence Theorem for each $\mathbf{v}_i$. If for some i, the cylinder Γ_i is interior to G, then $\mathbf{v}_i = \mathbf{0}$ outside Γ_i as well as on and near its boundary. Applying Lemmas 4 and 5 to $\mathbf{v}_i$ in Γ_i we find that

$$\iiint_G \operatorname{div} \mathbf{v}_i \, dV \equiv \iiint_{\Gamma_i} \operatorname{div} \mathbf{v}_i \, dV = \iint_{\partial\Gamma_i} \mathbf{v}_i \cdot \mathbf{n}_i \, dS = 0 = \iint_{\partial G} \mathbf{v}_i \cdot \mathbf{n} \, dS.$$

When P_i is on the boundary of G, we still have $\mathbf{v}_i = \mathbf{0}$ in the exterior of Γ_i, so that Lemmas 4 and 5 yield

$$\iiint_G \operatorname{div} \mathbf{v}_i \, dV = \iiint_{\Gamma_i} \operatorname{div} \mathbf{v}_i \, dV = \iint_{\partial\Gamma_i} \mathbf{v}_i \cdot \mathbf{n} \, dS = \iint_{\partial G} \mathbf{v}_i \cdot \mathbf{n} \, dS.$$

The last inequality holds because $\mathbf{v}_i = \mathbf{0}$ on that portion of $\partial\Gamma_i$ which is not part of the boundary of G. The proof is complete.

In two dimensions the Divergence Theorem is a direct consequence of Green's Theorem. To see this we choose the usual coordinate system in the xy plane and suppose that $\mathbf{v} = P\mathbf{i} + Q\mathbf{j}$. We define $\mathbf{u} = Q\mathbf{i} - P\mathbf{j}$. As Fig.

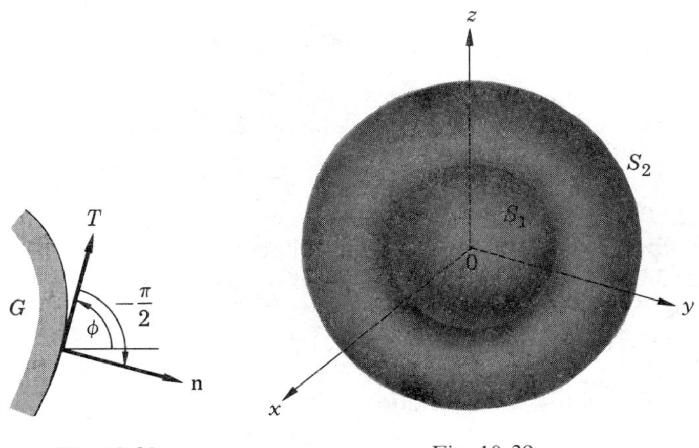

Fig. 10-38 Fig. 10-39

10-38 shows, the exterior normal **n** to a region G makes an angle of $-\pi/2$ with the tangent vector **T**, directed so that the interior of G is always on the left as we proceed around the boundary. If ϕ is the angle the vector **T** makes with the positive x-direction, we may write

$$\mathbf{T} = (\cos\phi)\mathbf{i} + (\sin\phi)\mathbf{j}, \qquad \mathbf{n} = (\sin\phi)\mathbf{i} - (\cos\phi)\mathbf{j},$$

$$\mathbf{v}\cdot\mathbf{T} = \mathbf{u}\cdot\mathbf{n}, \qquad \operatorname{div}\mathbf{u} = Q_x - P_y.$$

The Divergence Theorem (Equation (1)) applied to **u** is, when translated in terms of **v** (or P, Q), a restatement of Green's Theorem.

EXAMPLE 1. Verify the Divergence Theorem, given that G is the domain between the concentric spheres S_1 and S_2 of radius 1 and 2, respectively, and center at O; the vector **v** is $\mathbf{v} = \mathbf{r}/r^3$, with $\mathbf{r} = \mathbf{v}(\overrightarrow{OP})$, P a point in the domain, and $r = |\mathbf{r}|$.

SOLUTION. (See Fig. 10-39.) Let $\vec{S}_1$ and $\vec{S}_2$ be the spheres, both oriented with **n** pointing outward from O. Then

$$\iint_{\partial\vec{G}} \mathbf{v}\cdot\mathbf{n}\,dS = \iint_{\vec{S}_2} \mathbf{v}\cdot\mathbf{n}\,dS - \iint_{\vec{S}_1} \mathbf{v}\cdot\mathbf{n}\,dS,$$

since the normal on $\vec{S}_1$, as part of $\partial\vec{G}$, points toward O when the Divergence Theorem is used. According to Example 2 in Section 3 of Chapter 9, we easily find that

$$\operatorname{div}\mathbf{v} = 0$$

and hence

$$\iiint_G \operatorname{div}\mathbf{v}\,dV = 0.$$

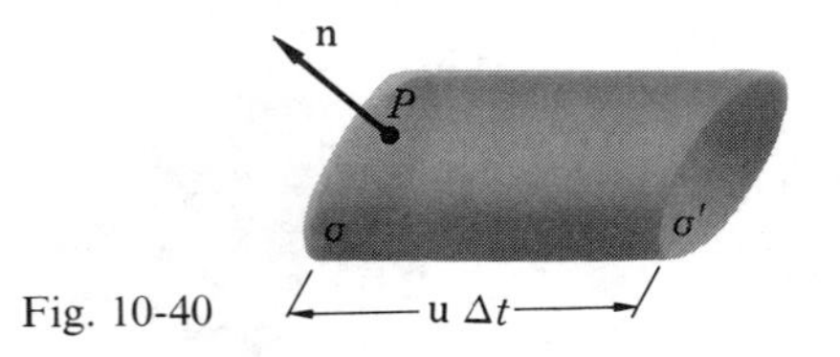

Fig. 10-40

Now, on both $\vec{S}_1$ and $\vec{S}_2$, we have $\mathbf{n} = r^{-1}\mathbf{r}$, and so

$$\iint_{\vec{S}_2} \mathbf{v}\cdot\mathbf{n}\,dS - \iint_{\vec{S}_1} \mathbf{v}\cdot\mathbf{n}\,dS = \tfrac{1}{4}A(S_2) - 1\cdot A(S_1) = 0.$$

EXAMPLE 2. Given that G is the domain inside the cylinder

$$x^2 + y^2 = 1$$

and between the planes $z = 0$ and $z = x + 2$, and given that

$$\mathbf{v} = (x^2 + ye^z)\mathbf{i} + (y^2 + ze^x)\mathbf{j} + (z^2 + xe^y)\mathbf{k},$$

use the Divergence Theorem to evaluate $\iint_{\partial G} \mathbf{v}\cdot\mathbf{n}\,dS$.

SOLUTION. We have $\operatorname{div}\mathbf{v} = 2x + 2y + 2z$. Therefore, letting F denote the unit disk, we obtain (Fig. 10-35)

$$\begin{aligned}\iint_{\partial G} \mathbf{v}\cdot\mathbf{n}\,dS &= \iiint_G 2(x+y+z)\,dV = 2\iint_F \left[(x+y)z + \tfrac{1}{2}z^2\right]_0^{x+2} dA_{xy}\\ &= \iint_F [2y(x+2) + 2x^2 + 4x + (x^2 + 4x + 4)]\,dA_{xy}\\ &= \iint_F (3x^2 + 4)\,dA_{xy} = \int_0^1\int_0^{2\pi} [(3r^3\cos^2\theta) + 4r]\,dr\,d\theta\\ &= \pi\int_0^1 (3r^3 + 8r)\,dr = \tfrac{19}{4}\pi.\end{aligned}$$

The Divergence Theorem has an important physical interpretation in connection with problems in fluid flow. We suppose that a fluid (liquid or gas) is flowing through a region is space. In general, the density ρ and the velocity vector $\mathbf{u}$ will depend not only on the point P in space but also on the time t. We select a point P in space, a value of t, and a small plane surface σ through P (Fig. 10-40). For a moment we suppose that ρ and $\mathbf{u}$ are constant. Then, after a short time Δt, all the particles on σ at time t would be in the shaded region σ' shown in Fig. 10-40. The particles sweep out a small cylindrical region as they travel from σ to σ'. The total mass of fluid flowing across σ in time Δt is just enough to fill up the oblique cylinder between σ

and σ'. We denote this cylindrical region by G, its boundary by ∂G, and the outward normal on the boundary by $\mathbf{n}$. If $\mathbf{u} \cdot \mathbf{n} > 0$ on σ, then the fluid flow is *out* of G across σ, while if $\mathbf{u} \cdot \mathbf{n} < 0$ on σ, the flow is *into* G across σ. The amount of fluid leaving (or entering, if the value is negative) G across σ is then given by

$$(\rho \mathbf{u})\,\Delta t \cdot \mathbf{n} A(\sigma).$$

The net *rate* of flow of mass out of G across σ (per unit time) is just

$$\rho \mathbf{u} \cdot \mathbf{n} A(\sigma).$$

If ρ and $\mathbf{u}$ are continuous on ∂G, a piecewise smooth boundary, and if we divide up ∂G into small, smooth surface elements such as σ and add the results, we obtain (in the limit)

$$\iint_{\partial G} (\rho \mathbf{u}) \cdot \mathbf{n}\, dS$$

for the rate of flow of the total amount of mass out of G. We denote by $M(G, t)$ the total mass in the region G at time t. Using the definition of density, we may write

$$M(G, t) = \iiint_G \rho(P, t)\, dV_P$$

and, using Leibniz' Rule (extended to triple integrals), we find

$$\frac{d}{dt} M(G, t) = \iiint_G \rho_t(P, t)\, dV_P.$$

Since $(d/dt)M$ is the amount of mass flowing into G, we obtain

$$\iiint_G \rho_t(P, t)\, dV_P = -\iint_G (\rho \mathbf{u}) \cdot \mathbf{n}\, dS.$$

If ρ and $\mathbf{u}$ are smooth, we may apply the Divergence Theorem to the boundary integral and get the equation

$$\iiint_G [\rho_t(P, t) + \operatorname{div}(\rho \mathbf{u})]\, dV = 0.$$

Since G is arbitrary, we can divide the above equation by $V(G)$, the volume of G, and let G shrink to a point P. The result is the **equation of continuity**

$$\rho_t + \operatorname{div}(\rho \mathbf{u}) = 0.$$

If the fluid is an incompressible liquid, so that $\rho = \text{const}$, the equation of continuity becomes

$$\operatorname{div} \mathbf{u} = 0.$$

Problems

In each of Problems 1 through 10, verify the Divergence Theorem by computing separately each side of Eq. (1).

1. $\mathbf{v} = xy\mathbf{i} + yz\mathbf{j} + zx\mathbf{k}$; G is bounded by the coordinate planes and the plane $x + y + z = 1$.
2. $\mathbf{v} = x^2\mathbf{i} - y^2\mathbf{j} + z^2\mathbf{k}$; G is bounded by: $x^2 + y^2 = 4$, $z = 0$, $z = 2$.
3. $\mathbf{v} = 2x\mathbf{i} + 3y\mathbf{j} - 4z\mathbf{k}$; G is the ball $x^2 + y^2 + z^2 \le 4$.
4. $\mathbf{v} = x^2\mathbf{i} + y^2\mathbf{j} + z^2\mathbf{k}$; G is bounded by: $y^2 = 2 - x$, $z = 0$, $z = x$.
5. $\mathbf{v} = x\mathbf{i} + y\mathbf{j} + z\mathbf{k}$; G is the domain outside $x^2 + y^2 = 1$ and inside
$$x^2 + y^2 + z^2 = 4.$$
6. $\mathbf{v} = x\mathbf{i} - 2y\mathbf{j} + 3z\mathbf{k}$; G is bounded by $y^2 = x$ and $z^2 = 4 - x$.
7. $\mathbf{v} = r^{-3}(z\mathbf{i} + x\mathbf{j} + y\mathbf{k})$; G is the domain outside $x^2 + y^2 + z^2 = 1$ and inside $x^2 + y^2 + z^2 = 4$.
8. $\mathbf{v} = 3x\mathbf{i} - 2y\mathbf{j} + z\mathbf{k}$; G is bounded by $x^2 + z^2 = 4$, $y = 0$,
$$x + y + z = 3.$$
9. $\mathbf{v} = 2x\mathbf{i} + y\mathbf{j} + z\mathbf{k}$; G is bounded by $z = x^2 + y^2$ and $z = 2x$.
10. $\mathbf{v} = x\mathbf{i} + y\mathbf{j} + z\mathbf{k}$; G is bounded by $x^2 + y^2 = 4$ and $x^2 + y^2 - z^2 = 1$.

In each of Problems 11 through 13, evaluate $\iint_{\partial G} \mathbf{v} \cdot \mathbf{n}\, dS$, using the Divergence Theorem.

11. $\mathbf{v} = ye^z\mathbf{i} + (y - ze^x)\mathbf{j} + (xe^y - z)\mathbf{k}$; G is the interior of the torus $(r - b)^2 + z^2 \le a^2$, $0 < a < b$; r, z cylindrical coordinates.
12. $\mathbf{v} = x^3\mathbf{i} + y^3\mathbf{j} + z^3\mathbf{k}$; G is the ball $x^2 + y^2 + z^2 \le 1$.
13. $\mathbf{v} = x^3\mathbf{i} + y^3\mathbf{j} + z\mathbf{k}$; G is bounded by: $x^2 + y^2 = 1$, $z = 0$, and
$$z = x + 2.$$
14. Suppose that G is a region in three-space with a boundary ∂G for which the Divergence Theorem is applicable. Prove the following formula for integration by parts if u and $\mathbf{v}$ are smooth on a region D containing G and ∂G:
$$\iiint_G u \operatorname{div} \mathbf{v}\, dV = \iint_{\partial G} u\mathbf{v} \cdot \mathbf{n}\, dS - \iiint_G \nabla u \cdot \mathbf{v}\, dV.$$
15. Suppose that D and G are as in Problem 14 and that u, ∇u, and v are smooth in D. Let $\partial/\partial n$ denote the directional derivative on ∂G in the direction of $\mathbf{n}$. Show that
$$\iiint_G v\nabla^2 u\, dV = \iint_{\partial G} v\frac{\partial u}{\partial n}\, dS - \iiint_G \nabla v \cdot \nabla u\, dV.$$
16. Suppose that u satisfies Laplace's equation in a region G of the type described in Problem 14. Show that

$$\iint_{\partial G} \frac{\partial u}{\partial n} \, dS = 0.$$

[*Hint*. Use the formula in Problem 15.]

17. Suppose that u satisfies Laplace's equation in a region G of the type described in Problem 14. Show that if $u = 0$ on ∂G, then $u \equiv 0$ in G. [*Hint*. Set $v = u$ in the formula of Problem 15.]

18. Evaluate $\iint_S \mathbf{v} \cdot \mathbf{n} \, dS$, where S is the torus

$$S = \{(r, \theta, z) : (r - 3)^2 + z^2 = 1\} \qquad ((r, \theta, z) \text{ cylindrical coordinates});$$

$\mathbf{n}$ is the outward normal on S; $\mathbf{R} = (x - 3)\mathbf{i} + y\mathbf{j} + z\mathbf{k}$; $R = |\mathbf{R}|$, $\mathbf{v} = R^{-3}\mathbf{R}$. [*Hint*. Use the Divergence Theorem, with G as the part of the interior of the torus which is outside a small sphere of radius ρ and center at $(3, 0, 0)$.]

APPENDIX 1

Matrices and Determinants

1. Matrices

A **matrix** is a rectangular array of numbers enclosed in parentheses. For example, the array

$$\begin{pmatrix} 3 & 2 & 0 & -1 \\ 1 & 4 & 6 & 2 \\ -3 & 1 & 3 & 0 \end{pmatrix}$$

is a matrix with three rows and four columns. A matrix with m rows and n columns is written

$$\begin{pmatrix} a_{11} & a_{12} & a_{13} & \cdots & a_{1n} \\ a_{21} & a_{22} & a_{23} & \cdots & a_{2n} \\ \vdots & & & & \vdots \\ a_{m1} & a_{m2} & a_{m3} & \cdots & a_{mn} \end{pmatrix}$$

The individual entries in the matrix are called its **elements**; the quantity a_{ij} in the above matrix is the element in the ith row and jth column. The subscripts used to indicate the elements will always denote the row first and the column second. If the number of rows of a matrix is the same as the number of columns, it is said to be **a square matrix**. We shall use capital letters such as A, B, C, ... to denote matrices. The corresponding lower-case letters with subscripts, such as a_{ij}, b_{ij}, c_{ij}, etc., will be used to denote the elements. We use the expression "m by n matrix" and write "$m \times n$ matrix" for a matrix with m rows and n columns.

A matrix with one row and n columns is called a **row vector**. For example, the matrix

$$(3, \quad -1, \quad 2, \quad 0, \quad 5, \quad 4)$$

is a row vector with 6 columns. A matrix with m rows and one column is called a **column vector**. The array

$$\begin{pmatrix} -3 \\ 0 \\ 2 \\ 4 \\ 5 \end{pmatrix}$$

is a column vector with 5 rows. We say that two matrices are of the **same size** if and only if they have the same number of rows and the same number of columns.

The most important properties of matrices are contained in the rules of operation which we now define.

Rule I (Multiplication by a Constant). *If A is a matrix and c is a number, then cA is the matrix obtained by multiplying each element of A by the number c. For example, if $c = 3$ and*

$$A = \begin{pmatrix} 3 & -2 & 0 \\ 1 & 4 & 2 \\ -1 & -3 & 1 \end{pmatrix}, \quad \text{then} \quad 3A = \begin{pmatrix} 9 & -6 & 0 \\ 3 & 12 & 6 \\ -3 & -9 & 3 \end{pmatrix}.$$

We use the symbol $-A$ for the matrix $(-1)A$.

Rule II (Addition of Matrices). *If A and B are of the same size and have elements a_{ij} and b_{ij}, respectively, we define their* **sum** *$A + B$ as the matrix C with elements c_{ij} such that*

$$c_{ij} = a_{ij} + b_{ij}$$

for each i and j.

For example, if

$$A = \begin{pmatrix} 3 & 2 & -1 \\ 4 & 6 & 0 \end{pmatrix}, \qquad B = \begin{pmatrix} -2 & 1 & 5 \\ 0 & 4 & -2 \end{pmatrix},$$

then

$$C = A + B = \begin{pmatrix} 1 & 3 & 4 \\ 4 & 10 & -2 \end{pmatrix}.$$

It is important to remember that addition of matrices can be defined only for matrices of the *same size*. Subtraction of two matrices is defined in terms of addition. We have $A - B = A + (-1)B$.

Rule III (Equality of Matrices). Two matrices are equal *if and only if all their corresponding elements are equal. That is, in order that $A = B$ we must have A and B the same size and $a_{ij} = b_{ij}$ for each i and j; in other words A and B denote the same matrix.*

EXAMPLE 1. Solve the following matrix equation for A:

$$2A - 3\begin{pmatrix} 2 & -1 & 3 \\ -3 & 2 & 1 \end{pmatrix} = \begin{pmatrix} -2 & 3 & -3 \\ 1 & 4 & 3 \end{pmatrix}.$$

SOLUTION. In order to make sense, A must be a 2×3 matrix. We may use the rules for adding matrices to write

$$\begin{aligned} 2A &= \begin{pmatrix} -2 & 3 & -3 \\ 1 & 4 & 3 \end{pmatrix} + 3\begin{pmatrix} 2 & -1 & 3 \\ -3 & 2 & 1 \end{pmatrix} \\ &= \begin{pmatrix} -2 & 3 & -3 \\ 1 & 4 & 3 \end{pmatrix} + \begin{pmatrix} 6 & -3 & 9 \\ -9 & 6 & 3 \end{pmatrix} = \begin{pmatrix} 4 & 0 & 6 \\ -8 & 10 & 6 \end{pmatrix}. \end{aligned}$$

Therefore

$$A = \begin{pmatrix} 2 & 0 & 3 \\ -4 & 5 & 3 \end{pmatrix}.$$

We define a **zero matrix** as any matrix with all elements zeros.

Rule IV (Multiplication of Matrices). *Let A be an $m \times n$ matrix and B an $n \times p$ matrix.* **The product** *AB is that $m \times p$ matrix C with elements c_{ij} given by*

$$c_{ij} = \sum_{k=1}^{n} a_{ik} b_{kj}, \qquad i = 1, 2, \ldots, m; \qquad j = 1, 2, \ldots, p.$$

It is extremely important to note that the product of two matrices is defined *only* when the number of columns in the first matrix is equal to the number of rows in the second matrix.

EXAMPLE 2. Compute AB, given that

$$A = \begin{pmatrix} 2 & -1 & 3 \\ 1 & -2 & -1 \end{pmatrix}, \qquad B = \begin{pmatrix} 3 & -1 \\ 1 & 2 \\ -1 & 1 \end{pmatrix}.$$

SOLUTION.

$$\begin{aligned} AB &= \begin{pmatrix} 2\cdot 3 + (-1)\cdot 1 + 3\cdot(-1) & 2\cdot(-1) + (-1)\cdot 2 + 3\cdot 1 \\ 1\cdot 3 + (-2)\cdot 1 + (-1)(-1) & 1\cdot(-1) + (-2)\cdot 2 + (-1)\cdot 1 \end{pmatrix} \\ &= \begin{pmatrix} 2 & -1 \\ 2 & -6 \end{pmatrix}. \end{aligned}$$

REMARKS. Note that A is a 2×3 matrix and B is a 3×2 matrix. Therefore the product can be formed, and the result is a 2×2 matrix. The product BA can also be formed, the result being a 3×3 matrix. It is clear that $AB \neq BA$, since AB is a 2×2 matrix and BA is a 3×3 matrix. In general, **multiplication of matrices is not commutative**. This is true even for square matrices. We observe that if

$$A = \begin{pmatrix} 1 & 2 \\ -1 & 1 \end{pmatrix} \quad \text{and} \quad B = \begin{pmatrix} 2 & 0 \\ 1 & -1 \end{pmatrix},$$

then

$$AB = \begin{pmatrix} 4 & -2 \\ -1 & -1 \end{pmatrix}, \quad BA = \begin{pmatrix} 4 & 2 \\ 2 & 1 \end{pmatrix}.$$

The reader can easily establish the following result, which is stated in the form of a theorem.

Theorem 1. (a) *If A is an $m \times n$ matrix and B and C are $n \times p$ matrices, then*

$$A(B + C) = AB + AC.$$

b) *If A and B are $m \times n$ matrices and C is an $n \times p$ matrix, then*

$$(A + B)C = AC + BC.$$

c) *If A is an $m \times n$ matrix, B is an $n \times p$ matrix, and c is a number, then*

$$(cA)B = A(cB) = c(AB).$$

If the matrix C is the product of the matrices A and B, the elements of C may be expressed in terms of the inner or scalar product of two vectors. The element in the ith row and jth column of C is the inner product of the ith row vector of A with the jth column vector of B. The formula

$$c_{ij} = a_{i1}b_{1j} + a_{i2}b_{2j} + \cdots + a_{in}b_{nj}$$

verifies this fact.

A system of linear equations is easily written in matrix form. For example, the system

$$\begin{aligned} a_{11}x_1 + a_{12}x_2 + \cdots + a_{1n}x_n &= b_1 \\ a_{21}x_1 + a_{22}x_2 + \cdots + a_{2n}x_n &= b_2 \\ \vdots \qquad & \quad \vdots \\ a_{m1}x_1 + a_{m2}x_2 + \cdots + a_{mn}x_n &= b_m \end{aligned}$$

has m equations and the n unknowns $x_1, x_2, \ldots, x_n$; it may be written in the form

$$AX = B,$$

where A is the $m \times n$ matrix with elements a_{ij} and where X and B are the column vectors with n rows and m rows, respectively:

$$X = \begin{pmatrix} x_1 \\ x_2 \\ \vdots \\ x_n \end{pmatrix}, \qquad B = \begin{pmatrix} b_1 \\ b_2 \\ \vdots \\ b_m \end{pmatrix}.$$

A precise definition of an $m \times n$ matrix may be given in terms of functions. The domain of a matrix function consists of all ordered pairs (i, j) of integers with $1 \le i \le m$ and $1 \le j \le n$. The range of the function is the real number system. The particular nature of the function is determined by the rules of operation which we described in terms of rectangular arrays.

Problems

In Problems 1 through 7, solve for A.

1. $A = 2\begin{pmatrix} 1 & -2 \\ 2 & 3 \end{pmatrix} - 3\begin{pmatrix} 2 & 1 \\ -1 & 4 \end{pmatrix}$

2. $A = 3\begin{pmatrix} 2 & -3 & 1 \\ 0 & 1 & -1 \end{pmatrix} - 2\begin{pmatrix} 1 & -1 & 2 \\ 2 & 3 & -1 \end{pmatrix}$

3. $2A - \begin{pmatrix} 1 & 2 & 3 \\ 2 & -1 & 2 \end{pmatrix} = 3\begin{pmatrix} 3 & 0 & 5 \\ 2 & 1 & 4 \end{pmatrix}$

4. $3A + \begin{pmatrix} 1 & -1 \\ 2 & 3 \\ -1 & 2 \end{pmatrix} = 2\begin{pmatrix} 2 & 1 \\ 1 & -3 \\ 1 & 4 \end{pmatrix}$

5. $\begin{pmatrix} 2 & -1 & 3 \\ -1 & 3 & 2 \\ 1 & 2 & -1 \end{pmatrix} - 2A = \begin{pmatrix} 2 & 3 & 1 \\ 3 & -1 & 2 \\ 1 & -2 & 1 \end{pmatrix}$

6. $\begin{pmatrix} 2 & 1 \\ 0 & 3 \end{pmatrix} A = \begin{pmatrix} 3 & 0 \\ -3 & -6 \end{pmatrix}$

7. $A\begin{pmatrix} 3 & 1 \\ -2 & 2 \end{pmatrix} = \begin{pmatrix} 5 & 7 \\ -5 & 9 \end{pmatrix}$

8. If A is a matrix and c and d are numbers, show that $c(dA) = (cd)A$.

9. If A and B are matrices of the same size and c and d are numbers, show that $(c + d)A = cA + dA$ and $c(A + B) = cA + cB$.

In Problems 10 through 13, solve simultaneously for A and B.

10. $A - 2B = \begin{pmatrix} 1 & 2 \\ -1 & 1 \end{pmatrix}, \quad A - B = \begin{pmatrix} 2 & 1 \\ 1 & -1 \end{pmatrix}$

11. $A + 2B = \begin{pmatrix} 2 & 1 & 0 \\ 1 & -1 & 2 \end{pmatrix}, \quad 2A + 3B = \begin{pmatrix} 1 & 2 & -1 \\ 2 & 0 & 1 \end{pmatrix}$

12. $A - B = \begin{pmatrix} 1 & -2 \\ -1 & 3 \end{pmatrix}, \qquad A + B = \begin{pmatrix} 3 & 0 \\ 3 & 1 \end{pmatrix}$

13. $2A - B = \begin{pmatrix} 3 & -3 & 0 \\ 3 & 3 & 2 \end{pmatrix}, \qquad -A + 2B = \begin{pmatrix} 0 & 3 & 3 \\ -3 & 0 & -4 \end{pmatrix}$

In Problems 14 through 21, compute AB when possible and compute BA when possible.

14. $A = \begin{pmatrix} 2 & -1 \\ 1 & 2 \end{pmatrix}, \qquad B = \begin{pmatrix} 1 & 0 \\ 2 & 1 \end{pmatrix}$

15. $A = \begin{pmatrix} 1 & -1 \\ 2 & 3 \end{pmatrix}, \qquad B = \begin{pmatrix} 2 & -1 \\ 1 & 3 \end{pmatrix}$

16. $A = \begin{pmatrix} 1 & 0 \\ 0 & 0 \end{pmatrix}, \qquad B = \begin{pmatrix} 0 & 0 \\ 1 & 1 \end{pmatrix}$

17. $A = \begin{pmatrix} 1 & 0 \\ 0 & 1 \end{pmatrix}, \qquad B = \begin{pmatrix} 0 & 1 \\ 0 & 1 \end{pmatrix}$

18. $A = \begin{pmatrix} 1 & 0 & 2 \\ -2 & 1 & 3 \end{pmatrix}, \qquad B = \begin{pmatrix} 2 & 1 \\ -1 & 1 \\ 1 & 2 \end{pmatrix}$

19. $A = \begin{pmatrix} 1 & 2 & -1 \\ -2 & 1 & 3 \end{pmatrix}, \qquad B = \begin{pmatrix} 2 & -1 & 1 \\ 1 & 3 & -2 \\ -1 & 2 & 1 \end{pmatrix}$

20. $A = \begin{pmatrix} 2 & 1 \\ -1 & 2 \\ -2 & 3 \end{pmatrix}, \qquad B = \begin{pmatrix} 1 & 2 & -1 & -2 \\ -1 & 3 & 1 & -1 \end{pmatrix}$

21. $A = \begin{pmatrix} 1 & 1 & 0 \\ 0 & 0 & 0 \\ 0 & 1 & 0 \end{pmatrix}, \qquad B = \begin{pmatrix} 0 & 0 & 0 \\ 0 & 0 & 0 \\ 1 & 0 & 0 \end{pmatrix}$

22. Prove Theorem 1.

23. Let

$$A = \begin{pmatrix} 0 & 0 \\ 0 & 0 \end{pmatrix} \quad \text{and} \quad B = \begin{pmatrix} 1 & 0 \\ 0 & 1 \end{pmatrix}$$

Show that A and B commute with all 2×2 matrices. Are there any other matrices which commute with all 2×2 matrices?

24. Find two 3×3 matrices A and B with the property that $AB = 0$, and $BA = 0$ and $A \neq 0$, $B \neq 0$.

25. If

$$A = \begin{pmatrix} 1 & 0 \\ 3 & 2 \end{pmatrix} \quad \text{and} \quad B = \begin{pmatrix} -1 & 1 \\ 2 & 3 \end{pmatrix},$$

solve for X: $$AX = B.$$

Also, solve for Y: $$YA = B.$$

26. If

$$A = \begin{pmatrix} 2 & 1 & 0 \\ 1 & 3 & 1 \\ -2 & 1 & 1 \end{pmatrix} \quad \text{and} \quad B = \begin{pmatrix} 2 & 1 & 1 \\ 1 & 0 & 2 \\ -1 & 1 & 3 \end{pmatrix},$$

solve for X:

$$AX = B.$$

27. If A, B, and C are 3×3 matrices, prove that

$$(AB)C = A(BC).$$

2. Matrices, Continued. Double Sums and Double Sequences

If A is an $m \times n$ matrix, the **transpose of** A is that $n \times m$ matrix obtained from A by interchanging its rows and columns. We use the symbol A^t for the transpose of A. The element in the ith row and jth column of A^t, denoted a^t_{ij}, is given by

$$a^t_{ij} = a_{ji}, \qquad i = 1, 2, \ldots, n; \qquad j = 1, 2, \ldots, m.$$

For example, if

$$A = \begin{pmatrix} 2 & -1 & 3 \\ 1 & -2 & -1 \end{pmatrix}, \qquad \text{then} \qquad A^t = \begin{pmatrix} 2 & 1 \\ -1 & -2 \\ 3 & -1 \end{pmatrix}.$$

The following theorem follows from the definition of transpose.

Theorem 2. (a) *If A and B are $m \times n$ matrices and c is a constant, then*

$$(A + B)^t = A^t + B^t, \qquad (cA)^t = cA^t.$$

b) *If A is an $m \times n$ matrix and B is an $n \times p$ matrix, then*

$$(AB)^t = B^t A^t.$$

PROOF. Part (a) can be performed by the reader. To prove (b), we let $C = AB$. Then C is an $m \times p$ matrix, and C^t is a $p \times m$ matrix. Using a_{ij}, b_{ij}, and c_{ij} for elements of A, B, and C, respectively, we have

$$c_{ij} = \sum_{k=1}^{n} a_{ik} b_{kj}, \qquad i = 1, 2, \ldots, m; \qquad j = 1, 2, \ldots, p.$$

Also

$$c_{ij}^t = c_{ji} = \sum_{k=1}^{n} a_{jk} b_{ki}, \qquad i = 1, 2, \ldots, p; \qquad j = 1, 2, \ldots, m.$$

We must show that the elements of $B^t A^t$ are precisely c_{ij}^t. We let $D = B^t A^t$. Since B^t is a $p \times n$ matrix and since A^t is an $n \times m$ matrix, the matrix D is $p \times m$, and we have

$$b_{ij}^t = b_{ji}, \qquad i = 1, 2, \ldots, p, \qquad j = 1, 2, \ldots, n,$$

$$a_{ij}^t = a_{ji}, \qquad i = 1, 2, \ldots, n, \qquad j = 1, 2, \ldots, m,$$

$$d_{ij} = \sum_{k=1}^{n} b_{ik}^t a_{kj}^t = \sum_{k=1}^{n} b_{ki} a_{jk} = \sum_{k=1}^{n} a_{jk} b_{ki} = c_{ji} = c_{ij}^t.$$

Definitions. If A is a square matrix, we call those elements of the form a_{ii} **diagonal elements**, and we call the totality of diagonal elements the **diagonal of the matrix**. A **diagonal matrix** is one in which all elements are zero, except possibly those on the diagonal. That is,

$$a_{ij} = 0 \qquad \text{if } i \neq j.$$

If, in a diagonal matrix A, we have $a_{ii} = 1$ for all i, the matrix is called the **identity matrix** and is denoted by I. A square matrix is said to be **triangular** if and only if all the elements on one side of the diagonal are zero, i.e.,

$$a_{ij} = 0 \quad \text{for} \quad i > j \qquad \text{or} \qquad a_{ij} = 0 \quad \text{for} \quad i < j.$$

Examples of 4×4 diagonal, identity, and triangular matrices are shown below.

$$D = \begin{pmatrix} 3 & 0 & 0 & 0 \\ 0 & 2 & 0 & 0 \\ 0 & 0 & 1 & 0 \\ 0 & 0 & 0 & 4 \end{pmatrix} \qquad I = \begin{pmatrix} 1 & 0 & 0 & 0 \\ 0 & 1 & 0 & 0 \\ 0 & 0 & 1 & 0 \\ 0 & 0 & 0 & 1 \end{pmatrix}$$

Diagonal Identity

$$T_1 = \begin{pmatrix} 3 & -1 & 0 & 4 \\ 0 & 2 & 1 & 5 \\ 0 & 0 & 3 & 4 \\ 0 & 0 & 0 & 5 \end{pmatrix} \qquad T_2 = \begin{pmatrix} 2 & 0 & 0 & 0 \\ 1 & 3 & 0 & 0 \\ 2 & 5 & 1 & 0 \\ 0 & 4 & 6 & 2 \end{pmatrix}$$

$$a_{ij} = 0 \quad \text{if} \quad i > j \qquad\qquad a_{ij} = 0 \quad \text{if} \quad i < j$$

Triangular Triangular

It is useful to introduce the **Kronecker delta** symbol δ_{ij}, which is defined by the relation

$$\delta_{ij} = \begin{cases} 1 & \text{if} \quad i = j, \\ 0 & \text{if} \quad i \neq j. \end{cases}$$

If B is a square matrix with elements b_{ij}, then the relation $b_{ij} = \delta_{ij}$ implies that the matrix B is the identity matrix I. We also note that the Kronecker delta allows simplifications of the type

$$\sum_{j=1}^{n} \delta_{ij} x_j = x_i, \qquad \sum_{i=1}^{n} \delta_{ij} y_i = y_j.$$

These formulas are used frequently.

Theorem 3. (a) *If A and B are each $n \times n$ diagonal matrices, then $AB = BA$ and this is a diagonal matrix.* (b) *If A is any $n \times n$ matrix and I is the $n \times n$ identity matrix, then $IA = AI = A$.*

The proofs are left to the reader.

We recall that a sequence is a succession of numbers such as

$$a_1, \quad a_2, \quad a_3, \quad \ldots, \quad a_{14}.$$

If there is a first and last element, the sequence is called **finite**; otherwise it is **infinite**. We may consider a sequence as a function with domain consisting of a portion of the integers (i.e., the subscripts) and with the range (the numbers a_i) in the real number system.

A **double sequence** is a function the domain of which is some set S of ordered pairs (i, j) of integers and with range consisting of a portion of the real number system. A matrix is a special type of double sequence in which the domain S consists of pairs (i, j) in which $1 \leq i \leq m$ and $1 \leq j \leq n$. In general, the domain S of a double sequence is not restricted in this way; in fact, S may be infinite, in which case we say that the double sequence is **infinite**.

A finite **double sum** is an expression of the form

$$\sum_{(i,j) \in S} a_{ij},$$

in which the a_{ij} form a finite double sequence with domain S and with the sum extending over all elements of S. The symbol $(i, j) \in S$, which we read "(i, j) belongs to S," indicates this fact. The general commutative property for addition of ordinary numbers shows that the order in which we add the terms of a finite double sum is irrelevant.

Suppose the domain of a double sequence consists of all pairs (i, j) for which $1 \leq i \leq m$ and $1 \leq j \leq n$. Then we can easily conclude that

$$\sum_{(i,j) \in S} a_{ij} = \sum_{i=1}^{m} \left[\sum_{j=1}^{n} a_{ij} \right] = \sum_{j=1}^{n} \left[\sum_{i=1}^{m} a_{ij} \right]. \tag{1}$$

The latter two sums are called **iterated sums**, since we sum first with respect to one index and then with respect to the other. The fact that double sums and iterated sums are identical leads to many conveniences, and we shall use this fact on a number of occasions. For example, it follows that

$$\left(\sum_{i=1}^{m} a_i\right)\left(\sum_{j=1}^{n} b_j\right) = \sum_{(i,j)\in S} a_i b_j, \qquad S = \{(i,j) : 1 \le i \le m, \quad 1 \le j \le n\}.$$

There is a convenient notation which consists of the symbol

$$1 \le i < j \le n.$$

This is used to represent the set S, which consists of all (i,j) for which $1 \le i < n$, $1 < j \le n$ and $i < j$. Note that i cannot have the value n, since i must be strictly less than j and j can be at most n. Similarly, j cannot have the value 1. We write

$$\sum_{1 \le i < j \le n} a_{ij} \qquad \text{to mean} \qquad \sum_{(i,j)\in S} a_{ij},$$

in which S is the set described above. We may also conclude on the basis of the equivalence of double sums and iterated sums that

$$\sum_{1 \le i < j \le n} a_{ij} = \sum_{i=1}^{n-1} \sum_{j=i+1}^{n} a_{ij}. \tag{2}$$

It can also be shown that

$$\sum_{1 \le i < j \le n} a_{ij} = \sum_{j=2}^{n} \sum_{i=1}^{j-1} a_{ij}. \tag{3}$$

The set S for $n = 6$ is shown schematically below.

j						
6	*	*	*	*	*	
5	*	*	*	*		
4	*	*	*			
3	*	*				
2	*					
1			$1 \le i < j \le 6$			
	1	2	3	4	5	6 $\quad i$

EXAMPLE 1. Write out the iterated sum

$$\sum_{i=1}^{4} \sum_{j=1}^{i} a_{ij}.$$

SOLUTION. We have

$$\sum_{i=1}^{4}\sum_{j=1}^{i} a_{ij} = \left[\sum_{j=1}^{1} a_{1j}\right] + \left[\sum_{j=1}^{2} a_{2j}\right] + \left[\sum_{j=1}^{3} a_{3j}\right] + \left[\sum_{j=1}^{4} a_{4j}\right]$$
$$= a_{11} + (a_{21} + a_{22}) + (a_{31} + a_{32} + a_{33})$$
$$+ (a_{41} + a_{42} + a_{43} + a_{44}).$$

EXAMPLE 2. Verify formulas (2) and (3) for $n = 3$.

SOLUTION. The symbol

$$\sum_{1 \le i < j \le 3} a_{ij} \qquad \text{means} \qquad \sum_{(i,j) \in S} a_{ij},$$

where S consists of all (i, j) for which $1 \le i < 3$, $1 < j \le 3$ and $i < j$. Therefore

$$\sum_{1 \le i < j \le 3} a_{ij} = a_{12} + a_{13} + a_{23}.$$

From formulas (2) and (3) we have

$$\sum_{i=1}^{2}\sum_{j=i+1}^{3} a_{ij} = \sum_{j=2}^{3} a_{1j} + \sum_{j=3}^{3} a_{2j} = (a_{12} + a_{13}) + a_{23},$$
$$\sum_{j=2}^{3}\sum_{i=1}^{j-1} a_{ij} = \sum_{i=1}^{1} a_{i2} + \sum_{i=1}^{2} a_{i3} = a_{12} + (a_{13} + a_{23}).$$

Theorem 4 (Associative Law for Multiplication of Matrices). *Suppose that A, B, and C are $m \times n$, $n \times p$, and $p \times q$ matrices, respectively. Then*

$$(AB)C = A(BC).$$

PROOF. We define $D = AB$ and $F = DC = (AB)C$. Also, we define $E = BC$ and $G = AE = A(BC)$. We must show that $F = G$. Using lower-case letters for the elements of the matrix with the corresponding capital letters, we have

$$d_{ij} = \sum_{k=1}^{n} a_{ik}b_{kj}, \qquad f_{ij} = \sum_{l=1}^{p} d_{il}c_{lj}.$$

We substitute the expression for d_{ij} on the left into that for f_{ij} (changing subscripts in the process), and we find

$$f_{ij} = \sum_{l=1}^{p}\sum_{k=1}^{n} a_{ik}b_{kl}c_{lj}.$$

Similarly, we may write

$$e_{ij} = \sum_{l=1}^{p} b_{il}c_{lj}, \qquad g_{ij} = \sum_{k=1}^{n} a_{ik}e_{kj}.$$

A straight substitution shows that the expressions for f_{ij} and g_{ij} are identical.

Problems

In each of Problems 1 through 6, compute the quantities $(AB)^t$ and B^tA^t for the given matrices A and B. Verify that the expressions are equal.

1. $A = \begin{pmatrix} 2 & 3 & 1 \\ -1 & 0 & 2 \end{pmatrix}, \quad B = \begin{pmatrix} 1 & 1 \\ 0 & 1 \\ 1 & 1 \end{pmatrix}$

2. $A = \begin{pmatrix} 3 & 1 \\ 2 & -1 \end{pmatrix}, \quad B = \begin{pmatrix} 3 & 2 \\ -1 & -2 \end{pmatrix}$

3. $A = \begin{pmatrix} 1 & 2 & -1 \\ 3 & -2 & 1 \end{pmatrix}, \quad B = \begin{pmatrix} 2 & 1 & 0 \\ -1 & 1 & 2 \\ 0 & 1 & 0 \end{pmatrix}$

4. $A = \begin{pmatrix} 4 \\ 1 \\ 2 \\ 3 \end{pmatrix}, \quad B = (2 \quad 5 \quad -1 \quad 4)$

5. $A = \begin{pmatrix} 3 & -1 & 2 & 5 \\ 4 & 1 & 2 & 4 \\ -1 & 0 & 2 & 0 \\ 6 & 1 & 5 & 4 \end{pmatrix}, \quad B = \begin{pmatrix} 3 & -2 & 1 & 4 \\ -2 & 0 & 5 & -2 \\ 1 & 5 & 1 & 3 \\ 4 & -2 & 3 & 2 \end{pmatrix}$

6. $A = \begin{pmatrix} 3 & 2 & -1 & 4 \\ 1 & 6 & 8 & 5 \end{pmatrix}, \quad B = \begin{pmatrix} 4 & 2 \\ 1 & 6 \\ 8 & 5 \\ -1 & 0 \end{pmatrix}$

7. Show that $IA = A$ if I is the $m \times m$ identity and A' is any $m \times n$ matrix.

8. Prove part (a) of Theorem 2.

9. Prove part (a) of Theorem 3.

10. Prove part (b) of Theorem 3.

11. Verify formula (1) of this section for $m = 2$ and $n = 3$.

12. Verify formula (1) of this section for $m = 4$ and $n = 3$.

13. Verify the formula

$$\left(\sum_{i=1}^{m} a_i\right)\left(\sum_{j=1}^{n} b_j\right) = \sum_{(i,j)\in S} a_i b_j, \; S = \{(i,j) : 1 \le i \le m; \quad 1 \le j \le n\},$$

for $m = 3$ and $n = 4$.

14. Verify the formula

$$\sum_{1 \le i < j \le n} a_{ij} = \sum_{i=1}^{n-1} \sum_{j=i+1}^{n} a_{ij} \text{ for } n = 4.$$

15. Verify the formula

$$\sum_{1 \le i < j \le n} a_{ij} = \sum_{j=2}^{n} \sum_{i=1}^{j-1} a_{ij}$$

for $n = 5$.

In each of Problems 16 through 18, write out and evaluate the following double sums.

16. $\sum_{(i,j) \in S} (i + j), \qquad S = \{(i,j) : 1 \le i \le 3, \quad 1 \le j \le 4\}.$

17. $\sum_{(i,j) \in S} ij, \qquad S = \{(i,j) : 1 \le i \le 4, \quad 1 \le j \le 4, \quad j \le i\}.$

18. $\sum_{(i,j) \in S} (3i + 2j), \qquad S = \{(i,j) : 0 \le i \le 4, \quad 0 \le j \le 4, \quad 0 \le i + j \le 4\}.$

In each of Problems 19 through 23, write out and evaluate the given iterated sums.

19. $\sum_{i=1}^{4} \sum_{j=1}^{3} (i - j)$

20. $\sum_{i=1}^{4} \sum_{j=1}^{5-i} (i^2 j)$

21. $\sum_{i=0}^{3} \sum_{j=i+1}^{i+3} (i + j + 1)$

22. $\sum_{i=-1}^{4} \sum_{j=0}^{5} \frac{i}{j + 2}$

23. $\sum_{i=0}^{3} \sum_{j=0}^{i+1} ij$

24. Write the iterated sum in Example 1 as an iterated sum in the other order (i.e., with j in the "outside" summation).

25. Write the double sum in Problem 18 as an iterated sum with j in the outside summation.

26. In the sum in Problem 18, let $p = i + j$ be the index in the outer sum and verify that

$$\sum_{(i,j) \in S} (3i + 2j) = \sum_{p=0}^{4} \sum_{j=0}^{p} [3(p - j) + 2j] = \sum_{p=0}^{4} \sum_{j=0}^{p} (3p - j).$$

27. Given

$$A = \begin{pmatrix} 2 & x \\ -1 & y \end{pmatrix} \quad \text{and} \quad B = \begin{pmatrix} 5 & 0 \\ 0 & 5 \end{pmatrix}.$$

Find x and y so that $AA^t = B$.

28. A square matrix $A = (a_{ij})$ is called **symmetric** if for all i and j we have $a_{ij} = a_{ji}$. Show that for any square matrix B, it is always true that BB^t is symmetric.

29. If A, B, C, and D are matrices, prove that

$$(AB)(CD) = A(BC)D,$$

assuming that all multiplications are appropriate.

30. Name necessary and sufficient conditions in order that the formula

$$(A + B)(A - B) = A^2 - B^2$$

should be true for two matrices A and B.

3. Determinants

With each $n \times n$ square matrix we associate a real number called its **determinant**. If the matrix is A, we denote its determinant by **det** A. If the matrix A is written out as a square array, its determinant is denoted by the same array between vertical bars. For example, if

$$A = \begin{pmatrix} 2 & -1 & 3 \\ 3 & 1 & 2 \\ -1 & 2 & -3 \end{pmatrix}, \qquad \text{then} \qquad \det A = \begin{vmatrix} 2 & -1 & 3 \\ 3 & 1 & 2 \\ -1 & 2 & -3 \end{vmatrix}.$$

Definitions. The **order** of an $n \times n$ square matrix is the integer n. A **submatrix** of a given matrix is any matrix obtained by deleting certain rows and columns from the original matrix and consolidating the remaining elements.

For example, the matrix

$$\begin{pmatrix} 2 & 0 & 1 & 4 \\ -3 & \frac{1}{2} & 2 & 5 \\ \sqrt{6} & 7 & 8 & 1 \\ 0 & 5 & 4 & -6 \end{pmatrix}$$

has the submatrix

$$\begin{pmatrix} 2 & 0 & 4 \\ \sqrt{6} & 7 & 1 \\ 0 & 5 & -6 \end{pmatrix},$$

obtained by deleting the second row and third column. Another submatrix, obtained by striking out the third and fourth rows and the fourth column, is

$$\begin{pmatrix} 2 & 0 & 1 \\ -3 & \frac{1}{2} & 2 \end{pmatrix}.$$

Definitions. A **determinant** is a function the domain of which is the collection of square matrices; its range is the real number system. To define the determinant of a matrix, we proceed inductively with respect to its order n. For $n = 1$, we define

$$\det (a_{11}) = a_{11}.$$

Assuming, for $n > 1$, that we have defined determinants of order $\leq (n-1)$, we define those of order n by the formula

$$\det A \equiv \begin{vmatrix} a_{11} & a_{12} & \cdots & a_{1n} \\ a_{21} & a_{22} & \cdots & a_{2n} \\ \vdots & & & \vdots \\ a_{n1} & a_{n2} & \cdots & a_{nn} \end{vmatrix} = \sum_{i=1}^{n} (-1)^{i+n} a_{in} M_{in}. \tag{1}$$

In this formula M_{in} is the determinant of the $(n-1) \times (n-1)$ submatrix of A, obtained by deleting its ith row and nth column. The determinant M_{in} is called the **minor** of the element a_{in}.

To illustrate formula (1), we obtain for $n = 2$

$$\begin{vmatrix} a_{11} & a_{12} \\ a_{21} & a_{22} \end{vmatrix} = (-1)^{1+2}a_{12}a_{21} + (-1)^{2+2}a_{22}a_{11} = a_{11}a_{22} - a_{12}a_{21};$$

for $n = 3$,

$$\begin{vmatrix} a_{11} & a_{12} & a_{13} \\ a_{21} & a_{22} & a_{23} \\ a_{31} & a_{32} & a_{33} \end{vmatrix} = (-1)^{1+3}a_{13}\begin{vmatrix} a_{21} & a_{22} \\ a_{31} & a_{32} \end{vmatrix} + (-1)^{2+3}a_{23}\begin{vmatrix} a_{11} & a_{12} \\ a_{31} & a_{32} \end{vmatrix} + (-1)^{3+3}a_{33}\begin{vmatrix} a_{11} & a_{12} \\ a_{21} & a_{22} \end{vmatrix}$$

$$= a_{13}(a_{21}a_{32} - a_{31}a_{22}) - a_{23}(a_{11}a_{32} - a_{31}a_{12}) + a_{33}(a_{11}a_{22} - a_{21}a_{12}).$$

In formula (1) it is convenient to consolidate the quantity $(-1)^{i+n}$ and the minor M_{in}. We define the **cofactor** A_{ij} of the element a_{ij} in det A of (1) by the formula

$$A_{ij} = (-1)^{i+j}M_{ij}. \tag{2}$$

In terms of cofactors we obtain the basic expansion theorem for determinants:

Theorem 5. *If A is an $n \times n$ matrix with $n \geq 2$, then*

a) $\det A = \sum_{i=1}^{n} a_{ij}A_{ij}$ *for each fixed j,* $1 \leq j \leq n$;

b) $\det A = \sum_{j=1}^{n} a_{ij}A_{ij}$ *for each fixed i,* $1 \leq i \leq n$.

We postpone the proof until the next section.

The formula in (a) is called the **expansion of det A according to its ith column**; that in (b) is called the **expansion of det A according to its ith row**. *From* (2) *it follows that the signs preceding M_{ij} alternate as one proceeds along a row or column, so that* (2) *is needed only to get the first sign correct.*

EXAMPLE 1. Evaluate the determinant of A by expanding it according to its second column and then evaluating the 2×2 determinants, given that

$$A = \begin{pmatrix} 1 & -2 & 3 \\ 2 & 1 & -1 \\ -2 & -1 & 2 \end{pmatrix}.$$

SOLUTION. Expanding according to the second column, we have

$$\det A = -(-2)\begin{vmatrix} 2 & -1 \\ -2 & 2 \end{vmatrix} + 1\begin{vmatrix} 1 & 3 \\ -2 & 2 \end{vmatrix} - (-1)\begin{vmatrix} 1 & 3 \\ 2 & -1 \end{vmatrix}$$

$$= 2(4-2) + 1(2+6) + 1(-1-6) = 5.$$

EXAMPLE 2. Write out the expansion of the determinant

$$\det A = \begin{vmatrix} -1 & 2 & 3 & -4 \\ 4 & 2 & 0 & 1 \\ -1 & 1 & 2 & 3 \\ -5 & 1 & 6 & 2 \end{vmatrix}$$

according to the fourth column. Do not evaluate.

SOLUTION.

$$\det A = -(-4)\begin{vmatrix} 4 & 2 & 0 \\ -1 & 1 & 2 \\ -5 & 1 & 6 \end{vmatrix} + 1\begin{vmatrix} -1 & 2 & 3 \\ -1 & 1 & 2 \\ -5 & 1 & 6 \end{vmatrix} - 3\begin{vmatrix} -1 & 2 & 3 \\ 4 & 2 & 0 \\ -5 & 1 & 6 \end{vmatrix}$$

$$+ 2\begin{vmatrix} -1 & 2 & 3 \\ 4 & 2 & 0 \\ -1 & 1 & 2 \end{vmatrix}.$$

PROBLEMS

In each of Problems 1 through 4, evaluate the given determinant by expanding it according to (a) the second row, and (b) the third column.

1. $\begin{vmatrix} 3 & 2 & -1 \\ -1 & 0 & 1 \\ 2 & 1 & -2 \end{vmatrix}$

2. $\begin{vmatrix} 1 & 0 & 3 \\ 2 & -1 & -2 \\ 1 & 3 & 2 \end{vmatrix}$

3. $\begin{vmatrix} 2 & 3 & 1 \\ 1 & 2 & -2 \\ -2 & 1 & 3 \end{vmatrix}$

4. $\begin{vmatrix} 1 & \frac{1}{2} & 5 \\ 2 & 1 & 0 \\ 3 & -6 & \frac{1}{3} \end{vmatrix}$

In Problems 5 through 7, expand in each case according to the second row. Do not evaluate.

5. $\begin{vmatrix} 2 & 0 & 2 & 3 \\ -1 & 3 & 6 & -1 \\ 1 & -1 & -2 & 4 \\ 0 & 4 & 8 & 2 \end{vmatrix}$

6. $\begin{vmatrix} 0 & 1 & -1 & 2 \\ 4 & 4 & -4 & 2 \\ 3 & -1 & 2 & 3 \\ 1 & -2 & 3 & 4 \end{vmatrix}$

7. $\begin{vmatrix} 8 & 0 & 2 & 5 \\ 6 & -1 & -1 & 4 \\ 0 & 2 & 5 & 1 \\ 4 & 4 & 0 & 0 \end{vmatrix}$

In each of Problems 8 through 10, expand according to the fourth column. Do not evaluate.

8. $\begin{vmatrix} 6 & 1 & 2 & 0 \\ 0 & 1 & -1 & -1 \\ -1 & -1 & 3 & 2 \\ 0 & 2 & 5 & 1 \end{vmatrix}$

9. $\begin{vmatrix} -1 & 1 & 6 & -3 & -5 \\ -2 & 2 & 0 & -1 & 6 \\ 1 & 0 & 6 & 2 & 1 \\ -3 & 3 & 2 & 4 & 1 \\ 3 & 1 & 4 & 1 & 2 \end{vmatrix}$

10. $\begin{vmatrix} 3 & -1 & 2 & 4 & 6 \\ 7 & 8 & 2 & 0 & 5 \\ 4 & -1 & -2 & 0 & \frac{1}{2} \\ 6 & 2 & 8 & 5 & 1 \\ 7 & 4 & 3 & 0 & 5 \end{vmatrix}$

11. Prove the identity

$$\begin{vmatrix} 1 & a & a^2 \\ 1 & b & b^2 \\ 1 & c & c^2 \end{vmatrix} = (b - a)(c - a)(c - b).$$

12. Show that in the plane the equation of a line through the points (x_0, y_0) and (x_1, y_1) is given by

$$\begin{vmatrix} x & y & 1 \\ x_0 & y_0 & 1 \\ x_1 & y_1 & 1 \end{vmatrix} = 0.$$

13. Expand a general 4×4 determinant (a) according to the third column, and (b) according to the second row. Do not evaluate.

14. Show that if $\mathbf{u} = a_1\mathbf{i} + a_2\mathbf{j} + a_3\mathbf{k}$ and $\mathbf{v} = b_1\mathbf{i} + b_2\mathbf{j} + b_3\mathbf{k}$ then (formally)

$$\mathbf{u} \times \mathbf{v} = \begin{vmatrix} \mathbf{i} & \mathbf{j} & \mathbf{k} \\ a_1 & a_2 & a_3 \\ b_1 & b_2 & b_3 \end{vmatrix}.$$

15. Show that the determinant of any triangular matrix is the product of the diagonal elements.

16. Show that if

$$A = (a_{ij}) \qquad \text{and} \qquad B = (b_{ij})$$

are both triangular matrices with $a_{ij} = 0$ for $i < j$, $b_{ij} = 0$ for $i < j$, then

$$\det(AB) = \sum_{i=1}^{n} a_{ii}b_{ii}.$$

17. A square matrix $A = (a_{ij})$ is called **skew** if $a_{ij} = -a_{ji}$ for all i and j. If A is a skew matrix, show that A^2 is symmetric.

4. Properties of Determinants

The evaluation of determinants of high order by expansion in rows or columns is a tedious and lengthy process. We now establish a number of important theorems which not only are of theoretical interest but also lead to rapid methods of evaluating determinants.

Theorem 6. *If A is an $n \times n$ matrix, then*

$$\det A^t = \det A.$$

PROOF. We proceed by induction on the order n. For $n = 1$ the result is obvious. Now let $n > 1$ and denote $B = A^t$. Assume that the result holds for all matrices of order $\leq (n-1)$. We wish to show that it holds for a matrix of order n. Using lower-case letters in the usual way, we write

$$b_{ij} = a_{ij}^t = a_{ji}.$$

The cofactor of b_{ij} is denoted B_{ij} and the cofactor of a_{ji} is the determinant A_{ji}. The determinants A_{ji} and B_{ij} come from two $(n-1) \times (n-1)$ matrices, each of which is the transpose of the other. According to the induction hypothesis,

$$B_{ij} = A_{ji}.$$

Expanding $\det B$ according to its ith row, we obtain

$$\det B = \sum_{j=1}^{n} b_{ij} B_{ij} = \sum_{j=1}^{n} a_{ji} A_{ji} = \det A.$$

Theorem 7. (a) *If all the elements in the kth row or kth column* $(1 \leq k \leq n)$ *of a matrix A are zero, then* $\det A = 0$.
b) *If a matrix A' is obtained from A by multiplying the elements of the kth row or column by a constant c*, then

$$\det A' = c \det A.$$

c) *If each element a_{kj} of the kth row of a matrix A equals $a'_{kj} + a''_{kj}$, then*

$$\det A = \det A' + \det A'',$$

where A' and A'' are obtained from A by replacing a_{kj} by a'_{kj} and by a''_{kj}, respectively. The analogous result holds for the kth column.

PROOF. (a) Expanding according to the kth row (or kth column), we introduce the factor zero in each term.

b) Expanding according to the kth row (or kth column), we introduce the factor c in each term of the expansion.
c) Expanding according to the kth row and setting $a_{kj} = a'_{kj} + a''_{kj}$, we get expansions of A' and A'' in terms of the kth row.

As an example of part (c) of the theorem, we have

$$\begin{vmatrix} a_{11} & a_{12} & a_{13} \\ a_{21} & a_{22} & a_{23} \\ a'_{31} + a''_{31} & a'_{32} + a''_{32} & a'_{33} + a''_{33} \end{vmatrix} = \begin{vmatrix} a_{11} & a_{12} & a_{13} \\ a_{21} & a_{22} & a_{23} \\ a'_{31} & a'_{32} & a'_{33} \end{vmatrix} + \begin{vmatrix} a_{11} & a_{12} & a_{13} \\ a_{21} & a_{22} & a_{23} \\ a''_{31} & a''_{32} & a''_{33} \end{vmatrix}.$$

Theorem 8. (a) *If A' is obtained from A by interchanging two rows or two columns, then*

$$\det A' = -\det A.$$

b) *If two rows or two columns of A are proportional, then* $\det A = 0$.

PROOF. (a) We proceed by induction on n. If $n = 1$, there is nothing to prove. For $n = 2$, the result follows at once by inspection of the formula

$$\begin{vmatrix} a_{11} & a_{12} \\ a_{21} & a_{22} \end{vmatrix} = a_{11}a_{22} - a_{21}a_{12}.$$

We let $n > 2$, suppose that two rows are interchanged, and assume that the result holds for $n - 1$. Now we expand A according to the ith row, where the ith row is *not* one of those being interchanged. Then $a'_{ij} = a_{ij}$, and each cofactor A'_{ij} is obtained from A_{ij} by interchanging two rows. Invoking the induction hypothesis, we have

$$A'_{ij} = -A_{ij}.$$

Therefore

$$\det A' = \sum_{j=1}^{n} a'_{ij}A'_{ij} = -\sum_{j=1}^{n} a_{ij}A_{ij} = -\det A.$$

The proof is identical for the case when two columns are interchanged.
b) If two rows (or columns) are proportional, then either one row consists of zeros so that the determinant is zero, or else one row (or column) is a constant c times the other row (or column). Using part (b) of Theorem 7, we see that the determinant $D = cD'$ where D' has two identical rows (or columns). Interchanging the two identical rows and employing part (a) (which was just established), we find $D' = -D'$, so that D' is zero; therefore $D = 0$.

By combining part (c) of Theorem 7 and part (b) of Theorem 8, we obtain the next extremely useful result.

Theorem 9. *If A' is obtained from A by multiplying the kth row by the constant c and adding the result to the ith row where $i \neq k$, then*

$$\det A' = \det A.$$

The same result holds for two columns.

PROOF. The element a'_{ij} of A' is of the form $a_{ij} + ca_{kj}$, so that [by Theorem 7(c)]

$$\det A' = \det A + c \det A'',$$

where A'' is obtained from A by replacing the ith row by the kth row. But then the ith and kth rows of A'' are identical, so that $\det A'' = 0$. The same proof holds for columns.

Corollary. *If A' is obtained from A by multiplying the kth row by c_i and adding the result to the ith row for $i = 1, 2, \ldots, k-1, k+1, \ldots, n$ in turn, then $\det A' = \det A$. The same is true for columns.*

PROOF. Each step of the process is one for which Theorem 9 applies and which leaves $\det A$ unchanged. Therefore a succession of such steps will not alter the determinant.

The next example shows how to use the results of this section to simplify and evaluate a determinant.

EXAMPLE 1. Simplify by using the Corollary, and use the expansion theorem to evaluate the determinant

$$D = \begin{vmatrix} 2 & -1 & 1 & 0 \\ -3 & 0 & 1 & -2 \\ 1 & 1 & -1 & 1 \\ 2 & -1 & 5 & -1 \end{vmatrix}.$$

SOLUTION. By adding the third row to the first and fourth rows in turn, we find that

$$D = \begin{vmatrix} 3 & 0 & 0 & 1 \\ -3 & 0 & 1 & -2 \\ 1 & 1 & -1 & 1 \\ 3 & 0 & 4 & 0 \end{vmatrix}.$$

Expanding according to the second column, we obtain

$$D = (-1) \begin{vmatrix} 3 & 0 & 1 \\ -3 & 1 & -2 \\ 3 & 4 & 0 \end{vmatrix}.$$

Multiplying the first row by 2 and adding the result to the second row, we get

$$D = -\begin{vmatrix} 3 & 0 & 1 \\ 3 & 1 & 0 \\ 3 & 4 & 0 \end{vmatrix} = -\begin{vmatrix} 3 & 1 \\ 3 & 4 \end{vmatrix} = -9.$$

The 2×2 determinant was obtained by expansion according to the third column.

EXAMPLE 2. Show that

$$D = \begin{vmatrix} x & x^2 & x^3 \\ y & y^2 & y^3 \\ z & z^2 & z^3 \end{vmatrix} = xyz(y-x)(z-y)(z-x).$$

SOLUTION. We may factor out an x from the first row, a y from the second row, and a z from the third row, to obtain

$$D = xyz\begin{vmatrix} 1 & x & x^2 \\ 1 & y & y^2 \\ 1 & z & z^2 \end{vmatrix}.$$

Subtracting the third row from the first and second in turn, and then expanding in terms of the first column, we get

$$D = xyz\begin{vmatrix} 0 & x-z & x^2-z^2 \\ 0 & y-z & y^2-z^2 \\ 1 & z & z^2 \end{vmatrix} = xyz\begin{vmatrix} x-z & x^2-z^2 \\ y-z & y^2-z^2 \end{vmatrix}$$

$$= xyz(x-z)(y-z)\begin{vmatrix} 1 & x+z \\ 1 & y+z \end{vmatrix} = xyz(y-x)(z-x)(z-y).$$

The next theorem is employed in Section 5.

Theorem 10. *For any square matrix A, we have*

a) $\sum_{i=1}^{n} a_{ij}A_{ik} = \delta_{jk}(\det A)$, b) $\sum_{k=1}^{n} a_{ik}A_{jk} = \delta_{ij}(\det A)$.

PROOF. (a) If $j = k$, then $\delta_{jk} = 1$ and the formula (a) is the statement of the expansion theorem (Theorem 5) according to the kth column. If $j \neq k$, then $\delta_{jk} = 0$ and the right side of (a) is zero. As for the left side of (a), we introduce the matrix A' obtained from A by replacing the kth column with the jth column. Then the left side of (a) is the expansion of A' according to the jth column. But since A' has two columns alike (jth and kth), we have $\det A' = 0$, and so the left side of (a) vanishes. The proof of (b) is the same.

We conclude this section with a proof of Theorem 5, stated in Section 3.

PROOF OF THEOREM 5. Our first aim is to establish the formula

$$\det A = \sum_{i=1}^{n} a_{ij}A_{ij} \qquad \text{for each fixed } j, \qquad 1 \le j \le n. \tag{1}$$

If $j = n$, the above formula is just the definition of $\det A$. It must be shown that all other expansions, $1 \le j \le (n-1)$, yield the same result.

We proceed by induction on n. If $n = 1$, there is nothing to prove. If $n = 2$, the result is easily verified by inspection of the expansion of a 2×2 determinant. So we let $n > 2$ and assume the result is true for all determinants of order $\le (n-1)$.

Choose j, $1 \le j \le (n-1)$, and let $M_{ik,jn}$ denote the determinant formed from A by deleting its ith and kth rows and its jth and nth columns. $M_{ik,jn}$ is the determinant of an $(n-2) \times (n-2)$ matrix. Recalling that M_{in} is the minor of the element a_{in}, we obtain from the definition of determinant

$$D \equiv \det A = \sum_{i=1}^{n} (-1)^{i+n} a_{in} M_{in}. \tag{2}$$

Now, using our induction hypothesis, we may expand each M_{in} according to its jth column. The determinant of A is shown below with lines through the ith row and nth column so that the remaining terms form M_{in}.

$$\begin{vmatrix} a_{11} & a_{12} & \cdots & a_{1j} & \cdots & a_{1n} \\ \vdots & & & & & \\ a_{k1} & a_{k2} & \cdots & a_{kj} & \cdots & a_{kn} \\ \vdots & & & & & \\ a_{i1} & a_{i2} & & a_{ij} & & a_{in} \\ \vdots & & & & & \\ a_{n1} & a_{n2} & \cdots & a_{nj} & \cdots & a_{nn} \end{vmatrix}$$

If $k < i$ (as shown above), then a_{kp} is in the kth row of M_{in}, but if $k > i$, then each a_{kp} is in the $(k-1)$st row of M_{in}. Therefore, if $1 < i < n$, we have the expansion

$$M_{in} = \sum_{k=1}^{i-1} (-1)^{k+j} a_{kj} M_{ik,jn} + \sum_{k=i+1}^{n} (-1)^{k-1+j} a_{kj} M_{ik,jn}. \tag{3}$$

If $i = 1$, the first sum in (3) is missing, while if $i = n$, the second sum is absent. Substituting M_{in} from (3) into (2), we see that

$$D = \sum_{i=2}^{n} \sum_{k=1}^{i-1} (-1)^{i+k+j+n} a_{in} a_{kj} M_{ik,jn} - \sum_{i=1}^{n-1} \sum_{k=i+1}^{n} (-1)^{i+k+j+n} a_{in} a_{kj} M_{ik,jn}. \tag{4}$$

If we interchange the indices i and k in the first sum and make use of the double-sum notation of Section 3, we can combine the two iterated sums in (4) into the single double sum

$$D = \sum_{1 \le i < k \le n} (-1)^{i+k+j+n}(a_{kn}a_{ij} - a_{in}a_{kj})M_{ik,jn}. \tag{5}$$

The relation $M_{ik,jn} = M_{ki,jn}$, which follows directly from the definition of these four-subscript determinants, was used to obtain (5).

To establish (1), it is sufficient to show that the expansion

$$\sum_{k=1}^{n} (-1)^{k+j} a_{kj} M_{kj} \tag{6}$$

is equal to (5). To do so, we expand M_{kj} according to its $(n-1)$st column, noting that the $(n-1)$st column of M_{kj} is the nth column of D with a_{kn} removed. If $i < k$, then the elements a_{ip} are in the ith row of M_{kj} but, if $i > k$, then the a_{ip} are in the $(i-1)$st row of M_{kj}. If we then write out the expansion of M_{kj}, collect terms, and substitute in (6), we find an expression which is identical with (5).

The proof of the formula

$$\det A = \sum_{k=1}^{n} a_{ik} A_{ik} \qquad \text{for each fixed } i, \qquad 1 \le i \le n,$$

is obtained by first showing that an expansion according to the nth row is equal to the expansion according to the nth column. This proof is omitted. Then the proof that the expansion according to two different rows yields the same result is analogous to the proof for columns.

Problems

In each of Problems 1 through 12, simplify and evaluate the determinant.

1. $\begin{vmatrix} 2 & -1 & 3 \\ 2 & -1 & 1 \\ 1 & 3 & -2 \end{vmatrix}$

2. $\begin{vmatrix} 3 & 1 & -1 \\ 1 & 3 & -2 \\ -2 & 1 & 3 \end{vmatrix}$

3. $\begin{vmatrix} 1 & 2 & 3 \\ 1 & -1 & 1 \\ 2 & 4 & -1 \end{vmatrix}$

4. $\begin{vmatrix} 5 & 0 & -1 & 1 \\ 1 & -1 & 1 & -2 \\ 4 & -4 & 2 & 1 \\ -2 & 1 & 1 & 0 \end{vmatrix}$

5. $\begin{vmatrix} 2 & 3 & 6 & 3 \\ 0 & 1 & 3 & 1 \\ -1 & -2 & 0 & 4 \\ 1 & 2 & 4 & -1 \end{vmatrix}$

6. $\begin{vmatrix} 2 & -1 & 3 & 2 \\ 1 & -2 & -2 & 3 \\ -1 & 3 & 2 & 0 \\ 3 & 2 & -2 & 1 \end{vmatrix}$

7. $\begin{vmatrix} -2 & 3 & 2 & 4 \\ -3 & 1 & -2 & 3 \\ 2 & 2 & 3 & -2 \\ 4 & -3 & 2 & 1 \end{vmatrix}$

8. $\begin{vmatrix} 1 & 2 & -3 & 4 \\ 3 & -4 & 2 & -1 \\ 2 & -2 & 3 & 4 \\ 1 & 2 & -2 & 3 \end{vmatrix}$

9. $\begin{vmatrix} 3 & 2 & -1 & 4 \\ 2 & -3 & 4 & 1 \\ -4 & 2 & 0 & 3 \\ 2 & 4 & -1 & 2 \end{vmatrix}$

10. $\begin{vmatrix} 2 & -1 & 3 & 4 & 0 \\ -1 & 2 & 1 & 0 & -1 \\ -3 & 0 & 0 & 1 & 0 \\ 0 & 1 & 0 & -1 & 2 \\ 0 & -2 & 0 & 2 & -1 \end{vmatrix}$

11. $\begin{vmatrix} 1 & 2 & 0 & -1 & 2 \\ 2 & 3 & -1 & 0 & 1 \\ 0 & -1 & 2 & 4 & -2 \\ -1 & 0 & 4 & -1 & 0 \\ 1 & 2 & -1 & 0 & 1 \end{vmatrix}$

12. $\begin{vmatrix} 2 & -1 & 0 & 4 & 1 & -3 \\ 2 & 1 & 2 & 1 & 3 & 2 \\ 4 & -1 & 0 & 2 & -2 & 3 \\ 1 & 5 & 4 & 0 & 2 & 0 \\ 6 & 2 & 1 & 4 & -3 & 0 \\ -1 & 2 & 5 & -3 & 4 & 2 \end{vmatrix}$

13. Show that

$$\begin{vmatrix} 1 & x & x^2 & x^3 \\ 1 & y & y^2 & y^3 \\ 1 & z & z^2 & z^3 \\ 1 & w & w^2 & w^3 \end{vmatrix} = (x-y)(y-z)(z-w)(x-z)(x-w)(y-w).$$

[*Hint:* Use the relation $a^3 - b^3 = (a-b)(a^2+ab+b^2)$.]

14. Show that the equation of a plane through the three points

$$P_0(x_0, y_0, z_0), \qquad P_1(x_1, y_1, z_1), \qquad \text{and} \qquad P_2(x_2, y_2, z_2)$$

is given by

$$\begin{vmatrix} 1 & x & y & z \\ 1 & x_0 & y_0 & z_0 \\ 1 & x_1 & y_1 & z_1 \\ 1 & x_2 & y_2 & z_2 \end{vmatrix} = 0.$$

15. Given the matrix

$$A = \begin{pmatrix} 3 & -1 & 2 \\ 1 & 4 & -1 \\ 2 & 0 & 5 \end{pmatrix},$$

verify Theorem 10 by computing $a_{11}A_{13} + a_{21}A_{23} + a_{31}A_{33}$, and showing that the result vanishes.

16. Given the matrices

$$A = \begin{pmatrix} 1 & 2 & -1 \\ 3 & 1 & 1 \\ -2 & 0 & 5 \end{pmatrix}, \qquad B = \begin{pmatrix} 3 & -1 & 2 \\ 4 & 0 & 1 \\ -2 & 1 & 5 \end{pmatrix},$$

show that $\det(AB) = (\det A)(\det B)$.

17. Let A be a square $n \times n$ matrix which is **skew** symmetric; that is,

$$a_{ij} = -a_{ji} \qquad \text{for all} \quad i \text{ and } j$$

Prove that if n is an odd integer, then $\det A = 0$.

18. Suppose A is a 4×4 matrix such that every 2×2 submatrix has determinant zero. Prove that $\det A = 0$.

19. Show that when

$$A = \begin{pmatrix} 1 & x_1 & x_1^2 & \dots & x_1^{n-1} \\ 1 & x_2 & x_2^2 & \dots & x_2^{n-1} \\ \vdots & & & & \\ 1 & x_n & x_n^2 & \dots & x_n^{n-1} \end{pmatrix},$$

then

$$\begin{aligned} \det A = {} & [(x_2 - x_1)] \cdot [(x_3 - x_2)(x_3 - x_1)] \\ & \times [(x_4 - x_3)(x_4 - x_2)(x_4 - x_1)] \cdot \cdots \\ & \times [(x_n - x_{n-1})(x_n - x_{n-2}) \cdot \cdots \cdot (x_n - x_1)]. \end{aligned}$$

5. Cramer's Rule

With the aid of determinants, we are able to give a formula for the solution of n simultaneous linear equations in n unknowns. The resulting theorem is known as **Cramer's Rule**.

Theorem 11. *If* $\det A \neq 0$, *the system of equations*

$$\begin{aligned} a_{11}x_1 + a_{12}x_2 + \cdots + a_{1n}x_n &= b_1 \\ a_{21}x_1 + a_{22}x_2 + \cdots + a_{2n}x_n &= b_2 \\ \vdots \qquad\qquad\qquad & \quad \vdots \\ a_{n1}x_1 + a_{n2}x_2 + \cdots + a_{nn}x_n &= b_n \end{aligned} \tag{1}$$

has a unique solution given by

$$x_k = \frac{1}{D}\begin{vmatrix} a_{11} & \dots & a_{1,k-1} & b_1 & a_{1,k+1} & \dots & a_{1n} \\ \vdots & & & & & & \vdots \\ a_{n1} & \dots & a_{n,k-1} & b_n & a_{n,k+1} & \dots & a_{nn} \end{vmatrix}, \quad \begin{array}{l} D = \det A, \\ k = 1, 2, \dots, n, \end{array} \tag{2}$$

where, if $k = 1$ *or* n, *the column of* b'*s is in the first or nth column, respectively.*

Proof. We must show that (1) implies (2) and that (2) implies (1). Suppose first that $x_1, x_2, \dots, x_n$ are numbers which satisfy (1). Multiply the ith equation of (1) by A_{ik}, the cofactor of a_{ik}, where k is a fixed number between 1 and n. We get

$$\sum_{j=1}^{n} a_{ij}A_{ik}x_j = b_iA_{ik}.$$

Now we add all such equations; that is, we sum on the index i. After interchanging the order of the iterated sum, we obtain

$$\sum_{j=1}^{n}\left[\sum_{i=1}^{n} a_{ij}A_{ik}\right]x_j = \sum_{i=1}^{n} b_i A_{ik}. \tag{3}$$

The sum on the right in (3) is the expansion according to the kth column of the determinant displayed in (2). As for the left side of (3), the quantity in brackets is precisely the expression which appears in Theorem 10 of the last section and is equal to δ_{jk} (det A). Therefore, using the property of the Kronecker δ (defined on page APP-8), we find that

$$\sum_{j=1}^{n} \delta_{jk}(\det A)x_j = Dx_k = \sum_{i=1}^{n} b_i A_{ik},$$

and the formula (2) follows.

If we assume formula (2) holds, we see that

$$Dx_k = \sum_{i=1}^{n} b_i A_{ik}.$$

We multiply this expression by a_{jk} and sum with respect to k to get (after interchanging the order of summation)

$$D\sum_{k=1}^{n} a_{jk}x_k = \sum_{i=1}^{n}\left[\sum_{k=1}^{n} a_{jk}A_{ik}\right]b_i.$$

Using Theorem 10 of Section 4 again, we obtain

$$D\sum_{k=1}^{n} a_{jk}x_k = \sum_{i=1}^{n} \delta_{ij}(\det A)b_i = Db_j,$$

from which (1) follows.

Example. Solve, using Cramer's Rule,

$$3x_1 - 2x_2 + 4x_3 = 5, \qquad x_1 + x_2 + 3x_3 = 2, \qquad -x_1 + 2x_2 - x_3 = 1.$$

Solution. We have

$$D = \begin{vmatrix} 3 & -2 & 4 \\ 1 & 1 & 3 \\ -1 & 2 & -1 \end{vmatrix} = \begin{vmatrix} 5 & -2 & 10 \\ 0 & 1 & 0 \\ -3 & 2 & -7 \end{vmatrix} = \begin{vmatrix} 5 & 10 \\ -3 & -7 \end{vmatrix} = -5,$$

$$x_1 = -\tfrac{1}{5}\begin{vmatrix} 5 & -2 & 4 \\ 2 & 1 & 3 \\ 1 & 2 & -1 \end{vmatrix} = -\tfrac{1}{5}\begin{vmatrix} 9 & 0 & 10 \\ 2 & 1 & 3 \\ -3 & 0 & -7 \end{vmatrix} = -\tfrac{1}{5}\begin{vmatrix} 9 & 10 \\ -3 & -7 \end{vmatrix} = \tfrac{33}{5}.$$

In a similar way, we find that

$$x_2 = -\tfrac{1}{5}\begin{vmatrix} 3 & 5 & 4 \\ 1 & 2 & 3 \\ -1 & 1 & -1 \end{vmatrix} = \tfrac{13}{5}, \qquad x_3 = -\tfrac{1}{5}\begin{vmatrix} 3 & -2 & 5 \\ 1 & 1 & 2 \\ -1 & 2 & 1 \end{vmatrix} = -\tfrac{12}{5}.$$

Cramer's Rule is useful in many theoretical investigations, since it gives an explicit formula for the solution. However, because of the enormous work of evaluating $n + 1$ determinants when n is large, Cramer's Rule is rather poor for numerical computations. Furthermore, there are many chances for error in the large number of multiplications and additions which must be performed. Techniques for solving linear systems which proceed by iteration and elimination are computationally superior not only because there are fewer arithmetical operations, but also because numerical errors are frequently self-correcting.

Problems

Solve the systems in Problems 1 through 10, using Cramer's Rule.

1. $\begin{aligned} 2x_1 - x_2 + 3x_3 &= 1 \\ 3x_1 + x_2 - x_3 &= 2 \\ x_1 + 2x_2 + 3x_3 &= -6 \end{aligned}$

2. $\begin{aligned} 2x_1 - x_2 + x_3 &= -3 \\ x_1 + 3x_2 - 2x_3 &= 0 \\ x_1 - x_2 + x_3 &= -2 \end{aligned}$

3. $\begin{aligned} x_1 + 3x_2 - 2x_3 &= 4 \\ -2x_1 + x_2 + 3x_3 &= 2 \\ 2x_1 + 4x_2 - x_3 &= -1 \end{aligned}$

4. $\begin{aligned} 2x_1 + x_2 - x_3 &= 1 \\ x_1 - 2x_2 + 3x_3 &= 0 \\ 2x_1 - 3x_2 + 4x_3 &= 0 \end{aligned}$

5. $\begin{aligned} 2x_1 - x_2 + 2x_3 &= 11 \\ x_1 + 2x_2 - x_3 &= -3 \\ 3x_1 - 2x_2 - 3x_3 &= -1 \end{aligned}$

6. $\begin{aligned} 2x_1 - x_2 - 2x_3 &= 0 \\ -x_1 + 2x_2 - 3x_3 &= 11 \\ 3x_1 - 2x_2 + 4x_3 &= -15 \end{aligned}$

7. $\begin{aligned} x_1 - 2x_2 + 2x_3 &= -1 \\ 2x_1 - 3x_2 - 3x_3 &= 1 \\ 3x_1 + x_2 + 2x_3 &= 3 \end{aligned}$

8. $\begin{aligned} x_1 \qquad + x_3 - 2x_4 &= 3 \\ x_2 + 2x_3 - x_4 &= 2 \\ 2x_1 + 3x_2 - 2x_3 \qquad &= -1 \\ x_1 - x_2 \qquad - 4x_4 &= 0 \end{aligned}$

9. $\begin{aligned} 3x_1 + 2x_2 \qquad - 4x_4 &= 0 \\ x_2 - 2x_3 + x_4 &= -1 \\ 2x_1 + 3x_2 \qquad &= 1 \\ x_1 \qquad + 4x_3 - 2x_4 &= 2 \end{aligned}$

10. $\begin{aligned} 2x_1 + x_2 - 2x_3 + 3x_4 - 4x_5 &= 0 \\ 4x_1 - x_2 + x_3 - 3x_4 + 2x_5 &= 1 \\ -2x_1 + x_2 + 2x_3 + 6x_4 - 2x_5 &= -2 \\ -4x_1 + 3x_2 - 5x_3 - 6x_4 + 4x_5 &= 13 \\ 6x_1 - 3x_2 + 4x_3 + 9x_4 - 6x_5 &= -13 \end{aligned}$

11. Given a system of two equations in three unknowns:

$$a_{11}x_1 + a_{12}x_2 + a_{13}x_3 = b_1,$$

$$a_{21}x_1 + a_{22}x_2 + a_{23}x_3 = b_2.$$

State conditions under which this system has (i) no solutions, (ii) one solution, (iii) infinitely many solutions.

12. Given a system of three equations in two unknowns:

$$a_{11}x_1 + a_{12}x_2 = b_1,$$
$$a_{21}x_1 + a_{22}x_2 = b_2,$$
$$a_{31}x_1 + a_{32}x_2 = b_3.$$

State conditions under which this system has (i) no solutions, (ii) one solution, (iii) infinitely many solutions.

6. The Rank of a Matrix. Elementary Transformations

Cramer's Rule applies when the number of equations is the same as the number of unknowns and when the determinant of the coefficients is not zero. In order to treat systems in which the number of equations is different from the number of unknowns, it is necessary to introduce a quantity called the *rank* of a matrix.

Definitions. A square matrix is said to be **nonsingular** if its determinant is not zero. The **rank** of an $m \times n$ matrix is the largest integer r for which a nonsingular $r \times r$ submatrix exists. The rank of any **matrix of zeros** is zero.

For example, the matrix

$$\begin{pmatrix} 4 & 0 & 0 & 0 \\ 1 & 0 & 3 & 0 \\ 2 & 0 & 0 & 0 \end{pmatrix}$$

is of rank 2, since the 2×2 submatrix

$$\begin{pmatrix} 4 & 0 \\ 1 & 3 \end{pmatrix},$$

obtained by deleting the third row and the second and fourth columns, is nonsingular, and since every 3×3 submatrix has zero determinant.

We note that the rank of an $m \times n$ matrix can never exceed the smaller of the numbers m and n.

Definition. An **elementary transformation** of a matrix is a process of obtaining a second matrix from the given matrix in one of the following ways:

a) interchanging two rows or two columns,
b) multiplying a row or column by a nonzero constant,

c) multiplying one row (or column) by a constant and adding it to another row (or column).

We observe that transformations of type (c) are just those which we used for simplifying and evaluating determinants.

Theorem 12. *If A' is obtained from A by an elementary transformation, the rank of A' equals the rank of A.*

PROOF. For transformations of type (a) and (b), the result is immediate from the definition. To prove the result for type (c), let r be the rank of A. If A is an $m \times n$ matrix and r is the smaller of m and n (i.e., the rank of A is as large as possible), then

$$\operatorname{rank} A' \leq \operatorname{rank} A. \tag{1}$$

We show first that (1) holds regardless of the rank of A. Suppose that r is smaller than m and n; let D' be a $k \times k$ submatrix of A' with $k > r$. To be specific, suppose that A' is obtained from A by multiplying the first row of A by c and adding the result to the second row. If D' contains both the first and second rows of A', then $\det D' = \det D$, where D is the corresponding matrix in A. But $\det D = 0$, since $k > r$ and since A is of rank r. The same result holds if D' contains neither the first nor the second row. If D' contains only the first row, we again have $\det D' = \det D$. The only remaining case occurs when D' contains the second row but not the first. But then $\det D'$ is a linear combination of two determinants of A of order k. Since all determinants of A of order k are zero, we conclude that $\det D' = 0$. Thus the determinant of every $k \times k$ submatrix of A' with $k > r$ is zero, and therefore (1) is established for every rank r. The argument just given works for any two matrices which are related to each other by a transformation of type (c). But since A can be obtained from A' in this manner, we conclude that

$$\operatorname{rank} A \leq \operatorname{rank} A'. \tag{2}$$

Combining (1) and (2), we get $\operatorname{rank} A = \operatorname{rank} A'$.

We say that two matrices are **equivalent** if and only if it is possible to pass from one to the other by applying a finite number of elementary transformations. We write $A \cong B$ when A and B are equivalent. From Theorem 12 we conclude that *equivalent matrices have the same rank*.

To compute the rank of an $m \times n$ matrix A directly from the definition, we must evaluate the determinant of every square submatrix of A. If m and n are large, this task is laborious, unless A has many zero entries. However, with the aid of elementary transformations and without an undue amount of computation, we can find a matrix equivalent to A whose rank can be determined by inspection. We first illustrate the technique with an example and then establish the appropriate theorem.

EXAMPLE 1. Determine the rank of the 4×5 matrix

$$A = \begin{pmatrix} -2 & -3 & -1 & 1 & 0 \\ 0 & 1 & 7 & 1 & -4 \\ 1 & 2 & 4 & 0 & -2 \\ -2 & -2 & 6 & 2 & -4 \end{pmatrix}.$$

SOLUTION. Interchanging the first and third rows, we find that

$$A \cong \begin{pmatrix} 1 & 2 & 4 & 0 & -2 \\ 0 & 1 & 7 & 1 & -4 \\ -2 & -3 & -1 & 1 & 0 \\ -2 & -2 & 6 & 2 & -4 \end{pmatrix}.$$

Multiplying the first row of this new matrix by 2 and adding the result to the third and fourth rows, we obtain

$$A \cong \begin{pmatrix} 1 & 2 & 4 & 0 & -2 \\ 0 & 1 & 7 & 1 & -4 \\ 0 & 1 & 7 & 1 & -4 \\ 0 & 2 & 14 & 2 & -8 \end{pmatrix}.$$

Multiplying the second row by 1 and 2 and subtracting the results from the third and fourth rows, respectively, we get

$$A \cong \begin{pmatrix} 1 & 2 & 4 & 0 & -2 \\ 0 & 1 & 7 & 1 & 4 \\ 0 & 0 & 0 & 0 & 0 \\ 0 & 0 & 0 & 0 & 0 \end{pmatrix} = A'.$$

By inspection, A' is seen to have rank 2. Therefore A has rank 2.

We now state a theorem which describes the process developed in the above example.

Theorem 13. *By a succession of elementary transformations of type* (a) *and* (c) *operating on rows only, any $m \times n$ matrix A can be reduced to an equivalent matrix A' in which*

$$a'_{ij} = 0 \qquad \text{for} \quad j < j_i, \quad i = 1, 2, \ldots, m,$$

where

$$\begin{aligned} 1 \le j_1 < j_2 < \cdots < j_r \le n, \\ j_i = n + 1 \qquad \text{for} \quad i = r + 1, r + 2, \ldots, m. \end{aligned} \tag{3}$$

The rank r is the integer equal to the number of rows which do not consist entirely of zeros. A corresponding result holds for columns.

Before proceeding with the proof, we give an example of the content of (3).

$$
\begin{matrix} j_1 = 1 \\ j_2 = 2 \\ j_3 = 4 \\ j_4 = 5 \\ \vdots \\ j_m = n+1 \end{matrix}
\quad
\begin{pmatrix}
3 & 2 & 3 & \ldots & & & & & \\
0 & 1 & 2 & 5 & \ldots & & & & \\
0 & 0 & 0 & 2 & -1 & 4 & \ldots & & \\
0 & 0 & 0 & 0 & 1 & 3 & \ldots & & \\
\vdots & \vdots & \vdots & & & & & & \\
0 & 0 & 0 & 0 & 0 & \ldots & & 0 & 0
\end{pmatrix}
$$

In other words, by means of elementary transformations we introduce as many zeros as possible in the bottom row, the second largest number of zeros in the second from bottom row, and so on until, in the first row, we introduce the fewest zeros (or perhaps none).

PROOF. If A is the zero matrix, then $j_1 = j_2 = \cdots j_m = n + 1$, and A is already in the desired form. The rank is zero. Otherwise we let the j_1st column be the first column which does not consist entirely of zeros. If a nonzero element occurs in the first column, then $j_1 = 1$. We interchange rows, if necessary, so that the element $a_{1,j_1} \neq 0$. Then, by elementary transformations of type (c), we make the remainder of that column all zeros; of course, all the numbers (if any) to the left of the j_1st column are still 0. If all the numbers below the first row are zero, then $j_2 = j_3 = \cdots = j_m = n + 1$, and the matrix is in the desired form; then $r = 1$. Otherwise, we let the j_2nd column (of course $j_2 > j_1$) be the first one which contains a nonzero element below the first row. By interchanging rows and performing elementary transformations of type (c), we arrange that $a_{2,j_2} \neq 0$ but that all the numbers $a_{i,j_2} = 0$ if $i > 2$; of course, all $a_{ij} = 0$ if $i \geq 2$ and $j < j_2$. If all the a_{ij} with $i > 2$ are zero, then $j_3 = \cdots j_m = n + 1$, and the matrix has been reduced to the desired form with $r = 2$. In any case, the process will stop after some step, say the rth, where $r \leq m$; if $r < m$, all the rows below the rth will consist of zeros. If, now, we select the submatrix of A' consisting of the first r rows and the $j_1, \ldots, j_r$ columns, this $r \times r$ submatrix has the form

$$
\begin{pmatrix}
a_{1j_1} & * & * & * & \ldots & * \\
0 & a_{2j_2} & * & * & \ldots & * \\
0 & 0 & \cdot & & & \cdot \\
0 & \cdot & & \cdot & & \cdot \\
\vdots & \cdot & & & \cdot & * \\
0 & 0 & \ldots & 0 & & a_{rj_r}
\end{pmatrix}.
$$

The diagonal elements are all nonzero and the asterisks stand for numbers which may or may not be zero. The matrix is triangular and nonsingular. The rank of A is r.

The proof for columns is the same.

Corollary 1. *If A is an $n \times n$ matrix, the process described in the proof of Theorem* 13 *transforms A to a triangular matrix with all zeros located below the diagonal.*

Corollary 2. *If A is a nonsingular $n \times n$ matrix, it can be reduced by means of elementary transformations, as in Theorem* 13, *to a matrix having nonzero elements along the diagonal and zeros elsewhere.*

PROOF. By Corollary 1, we make A into the triangular matrix A'. The value of the determinant is then the product of the diagonal elements; hence all diagonal elements are nonzero. Now, starting with the last column and the element in the lower right-hand corner, we use elementary transformations of type (c) to transform to zero all the elements above the bottom one in the last column. Proceeding to the second-from-last column, we transform to zero all the elements above the diagonal element. Continuing this process, working from right to left, we get a diagonal matrix which is equivalent to A.

Corollary 3. *The basic process as described in the proof of Theorem* 13 *leads to the simultaneous reduction to a similar form of each of the submatrices A_l, which are obtained by deleting all but the first l columns of A. The rank of A_l is the integer equal to the number of rows in A_l' which have nonzero elements.*

The proof of the theorem shows that, in principle, elementary transformations of type (b) are not needed to carry out the reduction from A to A'. However, if we have matrices with integers or if, in carrying out the reduction, we wish to avoid arithmetical difficulties, transformations of type (b) are helpful. The next example illustrates the use of type (b) transformations in performing the reduction.

EXAMPLE 2. Given the matrix

$$A = \begin{pmatrix} 3 & 2 & -2 & 3 \\ 2 & 3 & -3 & 4 \\ -2 & 4 & 2 & 3 \\ 5 & -2 & 4 & 2 \\ 3 & 4 & 2 & 3 \end{pmatrix},$$

determine the rank of A, as in Example 1. Use transformations of type (b) when convenient.

SOLUTION. We begin by interchanging the first two rows and then multiplying the second, fourth, and fifth rows by 2. In this way we avoid fractions. The reduction then proceeds as follows:

$$A \cong \begin{pmatrix} 2 & 3 & -3 & 4 \\ 6 & 4 & -4 & 6 \\ -2 & 4 & 2 & 3 \\ 10 & -4 & 8 & 4 \\ 6 & 8 & 4 & 6 \end{pmatrix} \cong \begin{pmatrix} 2 & 3 & -3 & 4 \\ 0 & -5 & 5 & -6 \\ 0 & 7 & -1 & 7 \\ 0 & -19 & 23 & -16 \\ 0 & -1 & 13 & -6 \end{pmatrix}$$

$$\cong \begin{pmatrix} 2 & 3 & -3 & 4 \\ 0 & -1 & 13 & -6 \\ 0 & -5 & 5 & -6 \\ 0 & 7 & -1 & 7 \\ 0 & -19 & 23 & -16 \end{pmatrix} \cong \begin{pmatrix} 2 & 3 & -3 & 4 \\ 0 & -1 & 13 & -6 \\ 0 & 0 & -60 & 24 \\ 0 & 0 & 90 & -35 \\ 0 & 0 & -224 & 98 \end{pmatrix}$$

$$\cong \begin{pmatrix} 2 & 3 & -3 & 4 \\ 0 & -1 & 13 & -6 \\ 0 & 0 & -5 & 2 \\ 0 & 0 & 18 & -7 \\ 0 & 0 & -16 & 7 \end{pmatrix}.$$

Multiplying the third row of the last matrix by 3 and combining with the fourth and fifth rows, we get

$$A \cong \begin{pmatrix} 2 & 3 & -3 & 4 \\ 0 & -1 & 13 & -6 \\ 0 & 0 & -5 & 2 \\ 0 & 0 & 3 & -1 \\ 0 & 0 & -1 & 1 \end{pmatrix} \cong \begin{pmatrix} 2 & 3 & -3 & 4 \\ 0 & -1 & 13 & -6 \\ 0 & 0 & -1 & 1 \\ 0 & 0 & 3 & -1 \\ 0 & 0 & -5 & 2 \end{pmatrix} \cong \begin{pmatrix} 2 & 3 & -3 & 4 \\ 0 & -1 & 13 & -6 \\ 0 & 0 & -1 & 1 \\ 0 & 0 & 0 & 2 \\ 0 & 0 & 0 & -3 \end{pmatrix}.$$

Finally, we obtain

$$A \cong \begin{pmatrix} 2 & 3 & -3 & 4 \\ 0 & -1 & 13 & -6 \\ 0 & 0 & -1 & 1 \\ 0 & 0 & 0 & 2 \\ 0 & 0 & 0 & 0 \end{pmatrix} = A',$$

and A is of rank 4.

The next example shows how the reduction proceeds when only elementary transformations involving columns are used.

EXAMPLE 3. Given the matrix

$$A = \begin{pmatrix} 2 & 1 & -1 & -2 \\ 4 & 2 & -2 & -4 \\ -5 & -2 & 3 & 2 \\ 1 & -1 & -2 & 8 \\ 8 & 3 & -2 & 1 \end{pmatrix},$$

reduce A to the form described in Theorem 13, using only elementary transformations involving columns. Find the rank of A.

SOLUTION. We begin by interchanging the first two columns; we then introduce zeros in the first row.

$$A \cong \begin{pmatrix} 1 & 2 & -1 & -2 \\ 2 & 4 & -2 & -4 \\ -2 & -5 & 3 & 2 \\ -1 & 1 & -2 & 8 \\ 3 & 8 & -2 & 1 \end{pmatrix} \cong \begin{pmatrix} 1 & 0 & 0 & 0 \\ 2 & 0 & 0 & 0 \\ -2 & -1 & 1 & -2 \\ -1 & 3 & -3 & 6 \\ 3 & 2 & 1 & 7 \end{pmatrix}$$

$$\cong \begin{pmatrix} 1 & 0 & 0 & 0 \\ 2 & 0 & 0 & 0 \\ -2 & -1 & 0 & 0 \\ -1 & 3 & 0 & 0 \\ 3 & 2 & 3 & 3 \end{pmatrix} \cong \begin{pmatrix} 1 & 0 & 0 & 0 \\ 2 & 0 & 0 & 0 \\ -2 & -1 & 0 & 0 \\ -1 & 3 & 0 & 0 \\ 3 & 2 & 3 & 0 \end{pmatrix}.$$

The rank is 3.

PROBLEMS

In each of Problems 1 through 9, reduce A to a matrix A', using elementary transformations involving rows, in accordance with the procedure of Theorem 13. Determine the rank of A.

1. $\begin{pmatrix} 1 & 4 & 6 & 1 \\ 1 & -1 & 2 & 3 \\ 1 & -11 & -6 & 7 \end{pmatrix}$

2. $\begin{pmatrix} 1 & 4 & -2 & 3 \\ 1 & 5 & 0 & 1 \\ -1 & -1 & 8 & -6 \\ 2 & 10 & 0 & 7 \end{pmatrix}$

3. $\begin{pmatrix} 1 & -1 & 0 & 3 \\ 2 & 1 & 0 & 1 \\ 2 & -2 & 0 & 6 \\ 1 & -2 & 1 & 2 \end{pmatrix}$

4. $\begin{pmatrix} 1 & 0 & -2 & -5 & -1 \\ 0 & -1 & 3 & 1 & 2 \\ 0 & 1 & 0 & 0 & 3 \\ 1 & 1 & -1 & 0 & 4 \end{pmatrix}$

5. $\begin{pmatrix} 1 & 1 & -2 & 0 \\ 2 & 1 & -3 & 0 \\ -4 & 2 & 2 & 0 \\ 6 & -1 & -5 & 0 \\ 7 & -3 & -4 & 1 \end{pmatrix}$

6. $\begin{pmatrix} 1 & 2 & -1 & 3 & 1 \\ -3 & -5 & -1 & 0 & -2 \\ 1 & 1 & 3 & -6 & 0 \\ 4 & 7 & 0 & 3 & 3 \\ 1 & 0 & 7 & -15 & -1 \end{pmatrix}$

7. $\begin{pmatrix} 3 & -3 & 2 & -2 \\ 2 & -1 & 3 & -2 \\ -2 & 3 & -2 & 1 \\ 4 & -2 & 1 & 3 \end{pmatrix}$

8. $\begin{pmatrix} 2 & 3 & -3 & -2 \\ -3 & 2 & 2 & 3 \\ 4 & -2 & 3 & 2 \\ 5 & 4 & -2 & 2 \end{pmatrix}$

9. $\begin{pmatrix} 3 & \frac{2}{3} & 1 & \frac{1}{2} & 6 \\ 2 & -\frac{1}{4} & 3 & -2 & 1 \\ 6 & -\frac{2}{5} & 3 & -\frac{1}{3} & 5 \end{pmatrix}$

In each of Problems 10 through 14, reduce A to a matrix A' as above, but use elementary transformations involving columns only. Determine the rank of A in each case.

10. A is the matrix in Problem 3.

11. A is the matrix in Problem 4.

12. A is the matrix in Problem 7.

13. A is the matrix in Problem 8.

14. A is the matrix in Problem 9.

In Problems 15 through 17, in each case reduce the given matrix to diagonal form as in Corollary 2.

15. $\begin{pmatrix} 1 & -2 & -4 & 2 \\ 1 & -3 & -2 & 1 \\ 2 & -2 & -11 & 4 \\ -1 & 3 & 3 & 0 \end{pmatrix}$

16. $\begin{pmatrix} 2 & -1 & -2 & 3 \\ 1 & -2 & -1 & 2 \\ 3 & -1 & 2 & 1 \\ -1 & 2 & -3 & 2 \end{pmatrix}$

17. $\begin{pmatrix} 1 & 3 & -1 & 2 & 1 \\ 4 & 1 & 5 & -1 & 2 \\ 3 & 2 & 1 & 6 & 5 \\ 2 & -1 & 4 & -2 & 1 \\ 3 & 1 & -5 & 6 & 3 \end{pmatrix}$

18. Show that every $n \times n$ matrix may be reduced by elementary row transformations to triangular form, with all elements above the diagonal consisting of zeros.

19. Let A be an $m \times n$ matrix with $m \le n$. If A is of rank r with $r < m$, describe a process using elementary row transformations which reduces A to a form having the first $(m - r)$ rows consisting entirely of zeros.

20. Show that every $n \times n$ nonsingular matrix A may be reduced by row transformations to a form in which the only nonzero elements are

$$a_{n1}, \quad a_{n-1,2}, \quad a_{n-2,3}, \quad \ldots, \quad a_{2,n-1}, \quad a_{1n}.$$

21. Let A and B be $n \times n$ matrices. Show that

$$\operatorname{rank}(A + B) \leq \operatorname{rank} A + \operatorname{rank} B.$$

22. Show that for any matrix A, the rank of A is the same as the rank of A^t.

7. General Linear Systems

We consider a system of m linear equations in n unknowns in which m may be different from n. In a system such as

$$\left\{\begin{array}{l} a_{11}x_1 + a_{12}x_2 + \cdots + a_{1n}x_n = b_1 \\ a_{21}x_1 + a_{22}x_2 + \cdots + a_{2n}x_n = b_2 \\ \vdots \qquad\qquad\qquad\qquad \vdots \quad \vdots \\ a_{m1}x_1 + a_{m2}x_2 + \cdots + a_{mn}x_n = b_m \end{array}\right\}, \tag{1}$$

we call the $m \times n$ matrix

$$A = \begin{pmatrix} a_{11} & a_{12} & \cdots & a_{1n} \\ a_{21} & a_{22} & \cdots & a_{2n} \\ \vdots & \vdots & & \vdots \\ a_{m1} & a_{m2} & \cdots & a_{mn} \end{pmatrix}$$

the **coefficient matrix** of the system (1). The $m \times (n + 1)$ matrix

$$B = \begin{pmatrix} a_{11} & a_{12} & \cdots & a_{1n} & b_1 \\ a_{21} & a_{22} & \cdots & a_{2n} & b_2 \\ \vdots & \vdots & & \vdots & \vdots \\ a_{m1} & a_{m2} & \cdots & a_{mn} & b_m \end{pmatrix}$$

is called the **augmented matrix** of (1).

As an illustration, suppose we wish to solve the system of equations

$$\begin{aligned} 3x + 2y - 2z &= 3, \\ 2x + 3y - 3z &= 4, \\ -2x + 4y + 2z &= 3, \\ 5x - 2y + 4z &= 2, \\ 3x + 4y + 2z &= 3. \end{aligned} \tag{2}$$

The reader, noticing that there are more equations than there are unknowns, would not expect a solution. However, sometimes such systems do have solutions. Geometrically, the above system represents five planes in three-dimensional space. If it should accidentally happen that the five planes have a point in common, then (2) has a solution. We shall show that the existence of a solution of (2) depends on the behavior of both the coefficient matrix and the augmented matrix.

Suppose now that we proceed to reduce an augmented matrix of a given system such as (1), using the methods of elementary transformations as described in the preceding section. First we note that the coefficient matrix is reduced simultaneously. Second, the interchange of two rows of a matrix corresponds to an interchange of two rows of the system of equations and does not affect the solution. Multiplication of a row by a constant is the same as multiplication of an equation by a constant and also has no effect on the solution. A transformation of type (c) is the same as multiplication of one equation by a constant and addition of the resulting equation to another equation. Again, the solution is unaffected. Thus to each elementary transformation of the augmented matrix involving only rows, there corresponds an operation which we call the **corresponding elementary transformation of the given system of equations**. By now, the following important theorem is evident.

Theorem 14. *Suppose B is the augmented matrix of a certain system of equations and B' is obtained from B by applying a finite number of elementary transformations involving only rows. Then B' is the augmented matrix of that system of equations which is obtained from the original system by performing the corresponding elementary transformations. The coefficient matrix of the transformed system is obtained from B' by omitting the last column. The systems corresponding to B and B' have the same solutions. We say* **the systems are equivalent**.

The augmented matrix of (2) is

$$\begin{pmatrix} 3 & 2 & -2 & 3 \\ 2 & 3 & -3 & 4 \\ -2 & 4 & 2 & 3 \\ 5 & -2 & 4 & 2 \\ 3 & 4 & 2 & 3 \end{pmatrix}.$$

We recognize this matrix as the one in Example 2 of the preceding section. The result of reduction shows that

$$\begin{pmatrix} 3 & 2 & -2 & 3 \\ 2 & 3 & -3 & 4 \\ -2 & 4 & 2 & 3 \\ 5 & -2 & 4 & 2 \\ 3 & 4 & 2 & 3 \end{pmatrix} \cong \begin{pmatrix} 2 & 3 & -3 & 4 \\ 0 & -1 & 13 & -6 \\ 0 & 0 & -1 & 1 \\ 0 & 0 & 0 & 2 \\ 0 & 0 & 0 & 0 \end{pmatrix}.$$

We now use Theorem 14 to conclude that the system (2) is equivalent to the system

$$\begin{aligned} 2x + 3y - 3z &= 4 \\ - y + 13z &= -6 \\ z &= 1 \\ 0 &= 2 \\ 0 &= 0, \end{aligned}$$

which has no solution. However, if the column of numbers on the right in (2) is replaced by the column

$$\begin{array}{r} -3 \\ -7 \\ -2 \\ 15 \\ 3, \end{array}$$

so that the system becomes

$$\begin{aligned} 3x + 2y - 2z &= -3 \\ 2x + 3y - 3z &= -7 \\ -2x + 4y + 2z &= -2 \\ 5x - 2y + 4z &= 15 \\ 3x + 4y + 2z &= 3, \end{aligned} \tag{3}$$

we see that exactly the same elementary transformations as before bring about an equivalent system. We obtain

$$\begin{aligned} 2x + 3y - 3z &= -7 \\ -y + 13z &= 27 \\ -z &= -2 \\ 0 &= 0 \\ 0 &= 0, \end{aligned}$$

which has the solution $x = 1$, $y = -1$, $z = 2$. The reader can easily verify that this set of numbers satisfies all equations in (3).

EXAMPLE. Solve, if possible, the following system of equations by using elementary row transformations on the augmented matrix to reduce the system to a simpler equivalent form:

$$
\begin{aligned}
x_1 + 2x_2 - 2x_3 + 3x_4 - 4x_5 &= -3,\\
2x_1 + 4x_2 - 5x_3 + 6x_4 - 5x_5 &= -1,\\
-x_1 - 2x_2 - 3x_4 + 11x_5 &= 15.
\end{aligned}
$$

SOLUTION. The augmented matrix is

$$
B = \begin{pmatrix} 1 & 2 & -2 & 3 & -4 & \vdots & -3 \\ 2 & 4 & -5 & 6 & -5 & \vdots & -1 \\ -1 & -2 & 0 & -3 & 11 & \vdots & 15 \end{pmatrix},
$$

and the part of B to the left of the dotted vertical line is the coefficient matrix. We multiply the first row by 2 and subtract from the second row. Then, adding the first row to the third row, we obtain

$$
B \cong \begin{pmatrix} 1 & 2 & -2 & 3 & -4 & -3 \\ 0 & 0 & -1 & 0 & 3 & 5 \\ 0 & 0 & -2 & 0 & 7 & 12 \end{pmatrix} \cong \begin{pmatrix} 1 & 2 & -2 & 3 & -4 & -3 \\ 0 & 0 & -1 & 0 & 3 & 5 \\ 0 & 0 & 0 & 0 & 1 & 2 \end{pmatrix}.
$$

Therefore, the given system is equivalent to the system

$$
\begin{aligned}
x_1 + 2x_2 - 2x_3 + 3x_4 - 4x_5 &= -3,\\
- x_3 + 3x_5 &= 5,\\
x_5 &= 2.
\end{aligned}
$$

From the last two equations we get at once $x_5 = 2$ and $x_3 = 1$. Solving the first equation for x_1, we find that

$$x_1 = 7 - 2x_2 - 3x_4.$$

In other words, regardless of the values we assign to x_2 and x_4, the remaining values of x_1, x_3, and x_5 will satisfy the given system. There is an infinite number of solutions.

We now give a criterion (in terms of the ranks of both the coefficient matrix and the augmented matrix) which determines when a general linear system has a solution. However, the actual solution of a problem is most easily obtained as in the example above.

Theorem 15 (Cramer's General Rule). (a) *A system of m linear equations in n unknowns has a solution if and only if the rank r of the augmented matrix equals that of the coefficient matrix.*
b) *If the two matrices have the same rank r and $r = n$, the solution is unique.*
c) *If the two matrices have the same rank r and $r < n$, then at least one set of r of the unknowns can be solved in terms of the remaining $(n - r)$ unknowns, and there are infinitely many solutions.*

PROOF. Let A be the $m \times n$ coefficient matrix and B be the $m \times (n + 1)$ augmented matrix of the given system. Suppose that B is reduced to the matrix B' as in Theorem 13 (using rows only). The coefficient matrix of the equivalent system corresponding to A is the submatrix A' of B' obtained by deleting the last column. If A and B are not of the same rank, neither are A' and B'. Then the last column of B' has at least one nonzero element b' in which all remaining elements of that row are zeros. The corresponding equation of the equivalent system is

$$0 = b',$$

which is clearly impossible.

If A' and B' are of the same rank r and $r = n$, then there are as many equations as unknowns and Cramer's Rule applies. There is a unique solution.

If $r < n$, we select an $r \times r$ submatrix of A' which is nonsingular, and we apply Cramer's Rule to solve for r of the unknowns in terms of the remaining $(n - r)$ unknowns. Thus part (c) of the theorem is established.

A system of linear equations is said to be homogeneous if and only if the numbers on the right [i.e., $b_1, b_2, \ldots, b_m$ in (1)] are all zero. In this case, it is clear that the rank of the augmented matrix equals that of the coefficient matrix. Every homogeneous system has the solution

$$x_1 = x_2 = \cdots = x_n = 0.$$

Moreover, if $x_1, x_2, \ldots, x_n$ and $y_1, y_2, \ldots, y_n$ are solutions of a homogeneous system and c is a constant, then

$$x_1 + y_1, \quad x_2 + y_2, \quad \ldots, \quad x_n + y_n \qquad \text{and} \qquad cx_1, \quad cx_2, \quad \ldots, \quad cx_n$$

are solutions. If the rank of a homogeneous system is n, then the solution is unique and $x_1 = x_2 = \cdots = 0$ is the only solution. We have just derived the following corollary.

Corollary. *A homogeneous system of m equations in n unknowns has a solution $x_1, x_2, \ldots, x_n$ in which not all the x_j are zero if and only if the rank r of the coefficient matrix is less than n. When $r < n$, some group of r of the x_j can be expressed in terms of the remaining x_j.*

PROBLEMS

In Problems 1 through 16, find in each case all the solutions, if any, of the system of equations. Begin by reducing the augmented matrix (and thus the coefficient matrix) to the simplified form, using only row transformations. Find the rank r of the coefficient matrix and the rank r^* of the augmented matrix.

1. $\begin{aligned} x_1 - x_2 + 2x_3 &= -2 \\ 3x_1 - 2x_2 + 4x_3 &= -5 \\ 2x_2 - 3x_3 &= 2 \end{aligned}$

2. $\begin{aligned} x_1 + x_2 - 5x_3 &= 26 \\ x_1 + 2x_2 + x_3 &= -4 \\ x_1 + 3x_2 + 7x_3 &= -34 \end{aligned}$

3. $$\begin{aligned} 2x_1 + 3x_2 - x_3 &= -15 \\ 3x_1 + 5x_2 + 2x_3 &= 0 \\ x_1 + 3x_2 + 3x_3 &= 11 \\ 7x_1 + 11x_2 &= -30 \end{aligned}$$

4. $$\begin{aligned} x_1 - x_3 + x_4 &= -2 \\ -x_2 + 2x_3 + x_4 &= 5 \\ x_1 - x_3 + 2x_4 &= 3 \\ 2x_1 + x_2 - x_3 &= -6 \end{aligned}$$

5. $$\begin{aligned} 3x_1 - x_2 + 2x_3 &= 3 \\ 2x_1 + 2x_2 + x_3 &= 2 \\ x_1 - 3x_2 + x_3 &= 4 \end{aligned}$$

6. $$\begin{aligned} 4x_1 - 6x_2 + 7x_3 &= 8 \\ x_1 - 2x_2 + 6x_3 &= 4 \\ 8x_1 - 10x_2 - 3x_3 &= 8 \end{aligned}$$

7. $$\begin{aligned} 2x_1 - x_2 + x_4 &= 2 \\ -3x_1 + x_3 - 2x_4 &= -4 \\ x_1 + x_2 - x_3 + x_4 &= 2 \\ 2x_1 - x_2 + 5x_3 &= 6 \end{aligned}$$

8. $$\begin{aligned} x_1 + x_2 + 2x_3 - x_4 &= 3 \\ 2x_1 - x_2 + x_3 + x_4 &= 1 \\ x_1 - 5x_2 - 4x_3 + 5x_4 &= -7 \\ 4x_1 - 5x_2 - x_3 + 5x_4 &= -3 \end{aligned}$$

9. $$\begin{aligned} 2x_1 - 7x_2 - 6x_3 &= 0 \\ 3x_1 + 5x_2 - 2x_3 &= 0 \\ 4x_1 - 2x_2 - 7x_3 &= 0 \end{aligned}$$

10. $$\begin{aligned} x_1 - x_2 - 5x_3 &= 0 \\ 2x_1 + 3x_2 &= 0 \\ 4x_1 - 5x_2 - 22x_3 &= 0 \end{aligned}$$

11. $$\begin{aligned} 2x_1 + 3x_2 - x_3 - x_4 &= 0 \\ x_1 - x_2 - 2x_3 - 4x_4 &= 0 \\ 3x_1 + x_2 + 3x_3 - 2x_4 &= 0 \\ 6x_1 + 3x_2 - 7x_4 &= 0 \end{aligned}$$

12. $$\begin{aligned} x_1 - 2x_2 + 10x_3 - 4x_4 &= 0 \\ 3x_1 - x_2 &= 0 \\ -2x_1 + x_2 + 5x_3 - 2x_4 &= 0 \\ 2x_1 - 3x_2 - 5x_3 + 2x_4 &= 0 \end{aligned}$$

13. $$\begin{aligned} x_1 + x_2 - 3x_3 + x_4 &= 1 \\ 2x_1 - 4x_2 + 2x_4 &= 2 \\ 3x_1 - 4x_2 - 2x_3 &= 0 \\ x_1 - 2x_3 + 3x_4 &= 3 \end{aligned}$$

14. $$\begin{aligned} x_1 - 2x_2 + 2x_3 + 3x_4 &= 1 \\ -x_1 + 4x_2 - x_3 - 5x_4 &= 2 \\ 2x_1 - 2x_2 + 5x_3 + 4x_4 &= 5 \\ -x_1 + 6x_2 - x_4 &= 12 \end{aligned}$$

15. $$\begin{aligned} x_1 + 3x_2 - 2x_3 - 3x_4 + 2x_5 &= 4 \\ -x_1 - 3x_2 + 4x_3 + 4x_4 + 4x_5 &= -1 \\ -x_1 - 3x_2 + 4x_3 + 4x_4 - x_5 &= -2 \\ -2x_1 - 6x_2 + 10x_3 + 9x_4 - 4x_5 &= 1 \end{aligned}$$

16. $$\begin{aligned} 2x_1 + x_2 - 3x_3 + x_4 + x_5 &= 0 \\ x_1 + 2x_2 - x_3 + 4x_4 + 2x_5 &= 1 \\ 2x_1 - 3x_2 + 2x_3 - x_4 + 3x_5 &= -6 \\ -x_1 + 2x_3 + 3x_4 - x_5 &= 8 \\ 2x_2 + x_3 + 2x_4 + 3x_5 &= -7 \\ 3x_1 - 4x_2 + 5x_3 - 2x_4 + x_5 &= 0 \end{aligned}$$

17. Let A be an $m \times n$ matrix. Let B be the matrix formed by adjoining p new columns, thus forming an $m \times (n + p)$ matrix. Show that rank $A \leq$ rank B.

18. Consider a system of m equations in n unknowns. Show that if this system has at least two solutions, then it has infinitely many.

19. The **trace** of a square matrix is the sum of the diagonal elements. Show that if A is an $m \times n$ matrix, then A is the zero matrix if and only if the trace of $AA^t = 0$.

APPENDIX 2

Proofs of Theorems 6, 10, 16 and 17 of Chapter 2

In this appendix we give statements and proofs of several of the more difficult theorems on vectors in three dimensions. The proofs make liberal use of the material on determinants given in Appendix 1.

Theorem 6. *Suppose that A, B, C, and D are points in space and that a Cartesian coordinate system is introduced in space. Denote the coordinates of A, B, C, and D by (x_A, y_A, z_A), (x_B, y_B, z_B), and so forth.* (i) *If the coordinates satisfy the equations*

$$x_B - x_A = x_D - x_C, \qquad y_B - y_A = y_D - y_C, \qquad z_B - z_A = z_D - z_C, \quad (1)$$

then $\overrightarrow{AB} \approx \overrightarrow{CD}$. (ii) *Conversely, if $\overrightarrow{AB} \approx \overrightarrow{CD}$, the coordinates satisfy the equations in* (1).

PROOF. (i) We assume that the equations in (1) hold. Then also,

$$x_C - x_A = x_D - x_B, \qquad y_C - y_A = y_D - y_B, \qquad z_C - z_A = z_D - z_B. \quad (2)$$

From (1) and Corollary 2 on p. 10, it follows that either $AB \| CD$ or A, B, C, and D are on a line. From Eqs. (2), we conclude that either $AC \| BD$ or A, B, C, and D are on a line. Thus, either $ACDB$ is a parallelogram or A, B, C, and D are on a line $\vec{L}$, which we may assume is directed.

If $ACDB$ is a parallelogram then $\overrightarrow{AB} \approx \overrightarrow{CD}$ by definition. If A, B, C, and D are on a line L, let $\vec{L}$ have the parametric equations

$$x = x_0 + t\cos\alpha, \qquad y = y_0 + t\cos\beta, \qquad z = z_0 + t\cos\gamma \quad (3)$$

and let A, B, C, and D have t coordinates t_A, t_B, t_C, and t_D, respectively. Thus $x_A = x_0 + t_A\cos\alpha$, $x_B = x_0 + t_B\cos\alpha$, etc. Subtracting, we get

$$\left.\begin{aligned} x_B - x_A &= (t_B - t_A)\cos\alpha, & x_D - x_C &= (t_D - t_C)\cos\alpha,\\ y_B - y_A &= (t_B - t_A)\cos\beta, & y_D - y_C &= (t_D - t_C)\cos\beta,\\ z_B - z_A &= (t_B - t_A)\cos\gamma, & z_D - z_C &= (t_D - t_C)\cos\gamma. \end{aligned}\right\} \tag{4}$$

From the equations in (1) and (4), we conclude that

$$\left.\begin{aligned} (t_B - t_A)\cos\alpha &= (t_D - t_C)\cos\alpha,\\ (t_B - t_A)\cos\beta &= (t_D - t_C)\cos\beta,\\ (t_B - t_A)\cos\gamma &= (t_D - t_C)\cos\gamma. \end{aligned}\right\} \tag{5}$$

Since $\cos\alpha$, $\cos\beta$, and $\cos\gamma$ are never simultaneously zero (as $\cos^2\alpha + \cos^2\beta + \cos^2\gamma = 1$), it follows from (5) that $t_B - t_A = t_D - t_C$, so that $\overline{AB} = \overline{CD}$ and hence $\overrightarrow{AB} \approx \overrightarrow{CD}$ in this case also.

ii) To prove the converse, we assume that $\overrightarrow{AB} \approx \overrightarrow{CD}$. Then either $ACDB$ is a parallelogram or A, B, C, and D are on a directed line $\vec{L}$. Let us first assume the former; we wish to show that the equations in (1) hold. Suppose they do not. It is clear that there are unique numbers x_E, y_E, z_E, coordinates of a point $E \neq D$ such that

$$x_E - x_C = x_B - x_A, \qquad y_E - y_C = y_B - y_A,$$

and

$$z_E - z_C = z_B - z_A.$$

Then, by part (i), we know that $ACEB$ is a parallelogram (since C is not on line AB because $ACDB$ is a parallelogram). But then D and E must coincide, thus contradicting the fact above that $D \neq E$. Accordingly, the equations in (1) must hold.

To consider the other case, let $\vec{L}$ have the parametric equations (3). If we use our previous notation, we conclude that the equations in (4) hold. But, since $\overrightarrow{AB} \approx \overrightarrow{CD}$, we know by definition that $\overline{AB} = \overline{CD}$, i.e., that $t_B - t_A = t_D - t_C$. But then the equations in (1) follow from those in (4), and the proof is complete.

We know that the vectors **i**, **j**, and **k** corresponding to any given coordinate system in space are linearly independent. Consequently Theorem 10 is a special case of the following Theorem 10′:

Theorem 10′. *Suppose that* $\mathbf{u}_1$, $\mathbf{v}_1$, *and* $\mathbf{w}_1$ *are linearly independent and suppose that*

$$\begin{aligned} \mathbf{u}_2 &= a_{11}\mathbf{u}_1 + a_{12}\mathbf{v}_1 + a_{13}\mathbf{w}_1,\\ \mathbf{v}_2 &= a_{21}\mathbf{u}_1 + a_{22}\mathbf{v}_1 + a_{23}\mathbf{w}_1,\\ \mathbf{w}_2 &= a_{31}\mathbf{u}_1 + a_{32}\mathbf{v}_1 + a_{33}\mathbf{w}_1, \end{aligned} \qquad D = \begin{vmatrix} a_{11} & a_{12} & a_{13}\\ a_{21} & a_{22} & a_{23}\\ a_{31} & a_{32} & a_{33} \end{vmatrix}. \tag{6}$$

Then the set $\{\mathbf{u}_2, \mathbf{v}_2, \mathbf{w}_2\}$ *is linearly dependent* $\Leftrightarrow$ $D = 0$.

PROOF. (a) Suppose the set is linearly dependent. Then there are constants c_1, c_2, and c_3, not all zero, such that

$$c_1\mathbf{u}_2 + c_2\mathbf{v}_2 + c_3\mathbf{w}_2 = \mathbf{0}. \tag{7}$$

If we substitute (6) into (7), we obtain

$$\begin{aligned}(c_1a_{11} + c_2a_{21} + c_3a_{31})\mathbf{u}_1 + (c_1a_{12} + c_2a_{22} + c_3a_{32})\mathbf{v}_1 \\ + (c_1a_{13} + c_2a_{23} + c_3a_{33})\mathbf{w}_1 = \mathbf{0}.\end{aligned} \tag{8}$$

Since $\mathbf{u}_1$, $\mathbf{v}_1$, and $\mathbf{w}_1$ are linearly independent, their coefficients must all vanish. That is, we must have

$$\begin{aligned} a_{11}c_1 + a_{21}c_2 + a_{31}c_3 &= 0, \\ a_{12}c_1 + a_{22}c_2 + a_{32}c_3 &= 0, \\ a_{13}c_1 + a_{23}c_2 + a_{33}c_3 &= 0. \end{aligned} \tag{9}$$

But if (9) holds with c_1, c_2, and c_3 not all zero, the determinant D' of the coefficients must vanish according to the Corollary to Cramer's Rule (Theorem 11, Appendix 1). But D' is obtained from D by interchanging rows and columns. Accordingly, $\det D = \det D' = 0$.
b) Now suppose $D = 0$. If all the cofactors A_{ij} are zero, any two rows of D and hence any two of the vectors $\mathbf{u}_2$, $\mathbf{v}_2$, and $\mathbf{w}_2$ are proportional and the set is linearly dependent. Otherwise, some $A_{pq} \neq 0$. By interchanging the order of the vectors, if necessary, we may assume that $p = 3$. From the expansion theorem (Theorem 5, Appendix 1), we conclude that

$$\begin{aligned} a_{11}A_{1q} + a_{21}A_{2q} + a_{31}A_{3q} &= 0, \\ a_{12}A_{1q} + a_{22}A_{2q} + a_{32}A_{3q} &= 0, \\ a_{13}A_{1q} + a_{23}A_{2q} + a_{33}A_{3q} &= 0. \end{aligned} \tag{10}$$

Since $A_{3q} \neq 0$, we can solve equations (10) for the a_{3j}, obtaining

$$\left.\begin{aligned} a_{31} = ka_{11} + la_{21}, \qquad a_{32} = ka_{12} + la_{22}, \qquad a_{33} = ka_{13} + la_{23}, \\ k = -A_{1q}/A_{3q}, \qquad l = -A_{2q}/A_{3q}. \end{aligned}\right\} \tag{11}$$

In this case it follows from (11) and (6) that $\mathbf{w}_2 = k\mathbf{u}_2 + l\mathbf{v}_2$, and the vectors $\mathbf{u}_2$, $\mathbf{v}_2$, and $\mathbf{w}_2$ are linearly dependent.

Theorem 16. *Suppose that* $\mathbf{u}$ *and* $\mathbf{v}$ *are any vectors, that* $\{\mathbf{i}, \mathbf{j}, \mathbf{k}\}$ *is a right-handed coordinate triple, and that* t *is any number. Then*

i) $\mathbf{v} \times \mathbf{u} = -\mathbf{u} \times \mathbf{v}$,
ii) $(t\mathbf{u}) \times \mathbf{v} = t(\mathbf{u} \times \mathbf{v}) = \mathbf{u} \times t(\mathbf{v})$,
iii) $\mathbf{i} \times \mathbf{j} = -\mathbf{j} \times \mathbf{i} = \mathbf{k}$,
$\mathbf{j} \times \mathbf{k} = -\mathbf{k} \times \mathbf{j} = \mathbf{i}$,
$\mathbf{k} \times \mathbf{i} = -\mathbf{i} \times \mathbf{k} = \mathbf{j}$,
iv) $\mathbf{i} \times \mathbf{i} = \mathbf{j} \times \mathbf{j} = \mathbf{k} \times \mathbf{k} = \mathbf{0}$.

PROOFS. (i) By definition $|\mathbf{v} \times \mathbf{u}| = |\mathbf{u} \times \mathbf{v}|$ and $\mathbf{v} \times \mathbf{u}$ and $\mathbf{u} \times \mathbf{v}$ are both orthogonal to both $\mathbf{u}$ and $\mathbf{v}$ (or are both zero if $\mathbf{u}$ and $\mathbf{v}$ are proportional). Thus $\mathbf{v} \times \mathbf{u} = \pm\mathbf{u} \times \mathbf{v}$. If we let $\mathbf{w} = \mathbf{u} \times \mathbf{v}$, then $\{\mathbf{u}, \mathbf{v}, \mathbf{w}\}$ and $\{\mathbf{v}, \mathbf{u}, -\mathbf{w}\}$ are right-handed (see Theorem 15, Chapter 2), so $\mathbf{v} \times \mathbf{u}$ must equal $-\mathbf{w}$.
ii) If $t = 0$ or $\mathbf{u}$ and $\mathbf{v}$ are proportional, (ii) certainly holds. Otherwise, let us set $\mathbf{w} = \mathbf{u} \times \mathbf{v}$. Then (ii) follows since all the terms in (ii) have the same magnitude, all are orthogonal to both $\mathbf{u}$ and $\mathbf{v}$, and $\{t\mathbf{u}, \mathbf{v}, t\mathbf{w}\}$ and $\{\mathbf{u}, t\mathbf{v}, t\mathbf{w}\}$ are right-handed by Theorem 15, Chapter 2.
iv) This follows, since we must have $\mathbf{i} \times \mathbf{i} = -\mathbf{i} \times \mathbf{i} = \mathbf{0}$, etc.
iii) To prove (iii), we note that, since $\{\mathbf{i}, \mathbf{j}, \mathbf{k}\}$ is right-handed, $\theta = \pi/2$, $|\mathbf{i}| = |\mathbf{j}| = |\mathbf{k}| = 1$, and $\mathbf{k}$ is orthogonal to both $\mathbf{i}$ and $\mathbf{j}$, it follows that $\mathbf{i} \times \mathbf{j} = \mathbf{k}$. That $\mathbf{j} \times \mathbf{i} = -\mathbf{k}$ follows from this and from (i). Since $\{\mathbf{i}, \mathbf{j}, \mathbf{k}\}$ is a coordinate triple, it follows as above that $\mathbf{j} \times \mathbf{k} = \pm\mathbf{i}$. Setting

$$\mathbf{u}_1 = \mathbf{i}, \quad \mathbf{v}_1 = \mathbf{j}, \quad \mathbf{w}_1 = \mathbf{k}, \quad \mathbf{u}_2 = \mathbf{j}, \quad \mathbf{v}_2 = \mathbf{k}, \quad \mathbf{w}_2 = \mathbf{i},$$

we see that

$$\begin{aligned} \mathbf{u}_2 &= 0 \cdot \mathbf{u}_1 + 1 \cdot \mathbf{v}_1 + 0 \cdot \mathbf{w}_1, \\ \mathbf{v}_2 &= 0 \cdot \mathbf{u}_1 + 0 \cdot \mathbf{v}_1 + 1 \cdot \mathbf{w}_1, \\ \mathbf{w}_2 &= 1 \cdot \mathbf{u}_1 + 0 \cdot \mathbf{v}_1 + 0 \cdot \mathbf{w}_1. \end{aligned}$$

Since $\{\mathbf{i}, \mathbf{j}, \mathbf{k}\}$ was given as right-handed, it follows from the discussion in Chapter 2, Section 7, that $\{\mathbf{u}_2, \mathbf{v}_2, \mathbf{w}_2\}$ is right-handed since

$$D = \begin{vmatrix} 0 & 1 & 0 \\ 0 & 0 & 1 \\ 1 & 0 & 0 \end{vmatrix} = +1.$$

The proof that $\mathbf{k} \times \mathbf{i} = \mathbf{j}$ is similar.

Theorem 17 (Distributive law). *If* $\mathbf{u}$, $\mathbf{v}$, *and* $\mathbf{w}$ *are any vectors,*

i) $\mathbf{u} \times (\mathbf{v} + \mathbf{w}) = (\mathbf{u} \times \mathbf{v}) + (\mathbf{u} \times \mathbf{w})$ *and*
ii) $(\mathbf{v} + \mathbf{w}) \times \mathbf{u} = (\mathbf{v} \times \mathbf{u}) + (\mathbf{w} \times \mathbf{u})$.

PROOF. Part (ii) follows from part (i) and part (i) of Theorem 11, for

$$\begin{aligned} (\mathbf{v} + \mathbf{w}) \times \mathbf{u} &= -[\mathbf{u} \times (\mathbf{v} + \mathbf{w})] = -[(\mathbf{u} \times \mathbf{v}) + (\mathbf{u} \times \mathbf{w})] \\ &= [-(\mathbf{u} \times \mathbf{v})] + [-(\mathbf{u} \times \mathbf{w})] = (\mathbf{v} \times \mathbf{u}) + (\mathbf{w} \times \mathbf{u}). \end{aligned}$$

It is clear that (i) holds if $\mathbf{u} = \mathbf{0}$. Otherwise, let $\{\mathbf{i}', \mathbf{j}', \mathbf{k}'\}$ be a right-handed coordinate triple such that $\mathbf{u} = |\mathbf{u}|\mathbf{i}'$ (i.e., $\mathbf{i}'$ is the unit vector in the direction of $\mathbf{u}$). Suppose that

$$\mathbf{v} = a_1\mathbf{i}' + b_1\mathbf{j}' + c_1\mathbf{k}', \qquad \mathbf{u} \times \mathbf{v} = \mathbf{V} = A_1\mathbf{i}' + B_1\mathbf{j}' + C_1\mathbf{k}'.$$

We first find A_1, B_1, C_1 in terms of a_1, b_1, and c_1.

Since $\mathbf{V}$ is orthogonal to both $\mathbf{u}$ and $\mathbf{v}$, we must have

$$\mathbf{V}\cdot\mathbf{u} = A_1|\mathbf{u}| = 0, \qquad \mathbf{V}\cdot\mathbf{v} = A_1 a_1 + B_1 b_1 + C_1 c_1 = 0.$$

Thus

$$A_1 = 0 \quad \text{and} \quad b_1 B_1 + c_1 C_1 = 0 \qquad \text{so that} \qquad B_1 = -kc_1, \quad C_1 = kb_1$$

for some k. Moreover,

$$|\mathbf{V}| = |\mathbf{u}|\cdot\sqrt{a_1^2 + b_1^2 + c_1^2}\sin\theta$$

and

$$\mathbf{u}\cdot\mathbf{v} = |\mathbf{u}|\cdot\sqrt{a_1^2 + b_1^2 + c_1^2}\cos\theta = |\mathbf{u}|a_1.$$

Since $0 \le \theta \le \pi$ and $\cos\theta = a_1/|\mathbf{v}|$, it follows that

$$\sin\theta = \sqrt{b_1^2 + c_1^2}/|\mathbf{v}|.$$

Thus

$$|\mathbf{V}| = |k|\cdot\sqrt{b_1^2 + c_1^2} = |\mathbf{u}|\cdot\sqrt{b_1^2 + c_1^2} \qquad \text{so} \quad k = \pm|\mathbf{u}|.$$

Finally $\{\mathbf{u}, \mathbf{v}, \mathbf{V}\}$ must be right-handed, so that

$$\begin{vmatrix} |\mathbf{u}| & 0 & 0 \\ a_1 & b_1 & c_1 \\ 0 & -kc_1 & kb_1 \end{vmatrix} = k|\mathbf{u}|\cdot(b_1^2 + c_1^2) > 0$$

(unless $\mathbf{v} = \mathbf{0}$ or $\mathbf{v}$ is proportional to $\mathbf{u}$). Hence $k = +1$ and

$$\mathbf{V} = |\mathbf{u}|\cdot(-c_1\mathbf{j}' + b_1\mathbf{k}'). \tag{12}$$

The result in (12) evidently holds also if $\mathbf{v} = \mathbf{0}$ or is proportional to $\mathbf{u}$.

In like manner, if we let

$$\mathbf{w} = a_2\mathbf{i}' + b_2\mathbf{j}' + c_2\mathbf{k}', \qquad \mathbf{W} = \mathbf{u}\times\mathbf{w}, \qquad \mathbf{X} = \mathbf{u}\times(\mathbf{v}+\mathbf{w}),$$

we see that

$$\mathbf{W} = |\mathbf{u}|\cdot(-c_2\mathbf{j}' + b_2\mathbf{k}'), \qquad \mathbf{X} = |\mathbf{u}|\cdot[-(c_1 + c_2)\mathbf{j}' + (b_1 + b_2)\mathbf{k}']$$

from which (i) follows.

APPENDIX 3

Introduction to the Use of a Table of Integrals

We recall some of the methods of integration with which the reader should be familiar. These devices, together with the integrals listed below, enable the reader to perform expeditiously any integration that is required in order to work the problems in this text.

1. Substitution in a Table of Integrals

EXAMPLE. Letting $u = \tan x$, $du = \sec^2 x\,dx$, we find that $\int e^{\tan x} \sec^2 x\,dx = \int e^u\,du = e^u + C = e^{\tan x} + C$.

2. Certain Trigonometric and Hyperbolic Integrals

We illustrate with trigonometric integrals; the corresponding hyperbolic forms are treated similarly.

a) $\int \sin^m u \cos^n u\,du$.

i) n **an odd positive integer**, m **arbitrary**. Factor out $\cos u\,du$ and express the remaining cosines in terms of sines.

EXAMPLE

$$\int \sin^4 2x \cos^3 2x\,dx = \tfrac{1}{2}\int \sin^4 2x(1 - \sin^2 2x)\cdot(2\cos 2x\,dx)$$
$$= \tfrac{1}{2}\int (\sin^4 2x - \sin^6 2x)d(\sin 2x)$$
$$= \tfrac{1}{10}\sin^5 2x - \tfrac{1}{14}\sin^7 2x + C.$$

ii) m **an odd positive integer,** n **arbitrary.** Factor out $\sin u\,du$ and express the remaining sines in terms of cosines.

iii) m **and** n **both even integers** ≥ 0**.** Reduce the degree of the expression by the substitutions

$$\sin^2 u = \frac{1 - \cos 2u}{2}, \qquad \cos^2 u = \frac{1 + \cos 2u}{2}.$$

EXAMPLE

$$\begin{aligned}\int \sin^4 u\,du &= \tfrac{1}{4}\int (1 - \cos 2u)^2\,du \\ &= \tfrac{1}{4}\int (1 - 2\cos 2u)\,du + \tfrac{1}{8}\int (1 + \cos 4u)\,du \\ &= \frac{3u}{8} - \frac{\sin 2u}{4} + \frac{\sin 4u}{32} + C.\end{aligned}$$

b) $\int \tan^m u \sec^n u\,du$.

i) n **an even positive integer,** m **arbitrary.** Factor out $\sec^2 u\,du$ and express the remaining secants in terms of $\tan u$.

EXAMPLE

$$\begin{aligned}\int \frac{\sec^4 u\,du}{\sqrt{\tan u}} &= \int (\tan u)^{-1/2}(1 + \tan^2 u)\cdot(\sec^2 u\,du) \\ &= \int [(\tan u)^{-1/2} + (\tan u)^{3/2}]\,d(\tan u) \\ &= 2(\tan u)^{1/2} + \tfrac{2}{5}(\tan u)^{5/2} + C.\end{aligned}$$

ii) m **and odd positive integer**, n **arbitrary**. Factor out $\sec u \tan u\,du$ and express the remaining tangents in terms of the secants.

EXAMPLE

$$\begin{aligned}\int \frac{\tan^3 u\,du}{\sqrt[3]{\sec u}} &= \int (\sec u)^{-4/3}\tan^2 u\cdot(\sec u \tan u\,du) \\ &= \int [(\sec u)^{2/3} - (\sec u)^{-4/3}]\,d(\sec u) \\ &= \tfrac{3}{5}(\sec u)^{5/3} + 3(\sec u)^{-1/3} + C.\end{aligned}$$

c) $\int \cot^m u \csc^n u\,du$. These are treated like those in (b).

3. Trigonometric and Hyperbolic Substitutions

a) If $\sqrt{a^2 - u^2}$ occurs (or $a^2 - u^2$ occurs in the denominator), set $u = a \sin\theta$, $\theta = \arcsin(u/a)$, $du = a\cos\theta\, d\theta$, $\sqrt{a^2 - u^2} = a\cos\theta$.

EXAMPLE

$$\begin{aligned}\int \sqrt{a^2 - u^2}\, du &= a^2 \int \cos^2\theta\, d\theta = \frac{a^2}{2}\int (1 + \cos 2\theta)\, d\theta \\ &= \frac{a^2}{2}(\theta + \sin\theta\cos\theta) + C \\ &= \frac{a^2}{2}\arcsin\frac{u}{a} + u\sqrt{a^2 - u^2} + C.\end{aligned}$$

b) If $\sqrt{a^2 + u^2}$ occurs, set $u = a\tan\theta$, etc.

EXAMPLE

$$\int u^3\sqrt{a^2 + u^2}\, du = a^5 \int \tan^3\theta \sec^3\theta\, d\theta.$$

The last integral is of type (2)(b)(ii) above.

c) If $\sqrt{u^2 - a^2}$ occurs, set $u = a\sec\theta$, etc.

EXAMPLE

$$\begin{aligned}\int \frac{\sqrt{u^2 - a^2}}{u}\, du &= \int \frac{a\tan\theta}{a\sec\theta}\cdot a\sec\theta\tan\theta\, d\theta \\ &= \int a\tan^2\theta\, d\theta = a\int (\sec^2\theta - 1)\, d\theta \\ &= a(\tan\theta - \theta) + C \\ &= \sqrt{u^2 - a^2} - a\,\mathrm{arcsec}\,(u/a) + C.\end{aligned}$$

A hyperbolic substitution is sometimes more effective:

EXAMPLE. If we let $u = a\sinh v$, then

$$\begin{aligned}\int \sqrt{a^2 + u^2}\, du &= \int a^2\cosh^2 v\, dv = \frac{a^2}{2}\int (1 + \cosh 2v)\, dv \\ &= \frac{a^2}{2}(v + \sinh v\cosh v) + C \\ &= \frac{a^2}{2}\,\mathrm{argsinh}\left(\frac{u}{a}\right) + u\sqrt{a^2 + u^2} + C.\end{aligned}$$

4. Integrals Involving Quadratic Functions

Complete the square in the quadratic function and introduce a simple change of variable to reduce the quadratic to one of the forms $a^2 - u^2$, $a^2 + u^2$, or $u^2 - a^2$.

EXAMPLE

$$\int \frac{(2x-3)\,dx}{x^2+2x+2} = \int \frac{(2x-3)\,dx}{(x+1)^2+1}.$$

Let $u = x + 1$. Then $x = u - 1$, $dx = du$, and

$$\int \frac{(2x-3)\,dx}{(x+1)^2+1} = \int \frac{(2u-5)\,du}{u^2+1} = \log(u^2+1) - 5\arctan u + C$$
$$= \log(x^2+2x+2) - 5\arctan(x+1) + C.$$

5. Integration by Parts: $\int u\,dv = uv - \int v\,du$.

EXAMPLE. If we let $u = x$ and $v = e^x$, then the formula for integration by parts gives

$$\int xe^x\,dx = \int u\,dv = uv - \int v\,du = xe^x - \int e^x\,dx = xe^x - e^x + C.$$

6. Integration of Rational Functions (Quotients of Polynomials)

If the degree of the numerator $\geq$ that of the denominator, divide out, thus expressing the given function as a polynomial plus a "proper fraction." Each proper fraction can be expressed as a sum of simpler "proper partial fractions":

$$\frac{P(x)}{Q(x)} = \frac{P_1(x)}{Q_1(x)} + \cdots + \frac{P_n(x)}{Q_n(x)},$$

in which no two Q_i have common factors and each Q_i is of the form $(x-a)^k$ or $(ax^2+bx+c)^k$; of course $Q = Q_1 \cdot Q_2 \cdots Q_n$. Each of these fractions can be expressed uniquely in terms of still simpler fractions as follows:

$$\frac{P_i(x)}{(x-a)^k} = \frac{A_1}{x-a} + \frac{A_2}{(x-a)^2} + \cdots + \frac{A_k}{(x-a)^k},$$

$$\frac{P_i(x)}{(ax^2+bx+c)^k} = \frac{A_1x+B_1}{ax^2+bx+c} + \frac{A_2x+B_2}{(ax^2+bx+c)^2} + \cdots$$
$$+ \frac{A_kx+B_k}{(ax^2+bx+c)^k}.$$

Each of these simplest fractions can be integrated by methods already described. The constants are obtained by multiplying up the denominators and either equating coefficients of like powers of x in the resulting polynomials or by substituting a sufficient number of values of x to determine the coefficients.

Example 1. Integrate

$$\int \frac{x^2+2x+3}{x(x-1)(x+1)}\,dx.$$

According to the results above, there exist constants A, B, and C such that

$$\frac{x^2+2x+3}{x(x-1)(x+1)} = \frac{A}{x} + \frac{B}{x-1} + \frac{C}{x+1}.$$

Multiplying up, we see that we must have

$$A(x-1)(x+1) + Bx(x+1) + Cx(x-1) \equiv x^2+2x+3.$$

The constants are most easily found by substituting $x = 0$, 1, and -1 in turn in this identity, yielding $A = -3$, $B = 3$, $C = 1$.

Example 2. Integrate

$$\int \frac{3x^2+x-2}{(x-1)(x^2+1)}.$$

There are constants A, B, and C such that

$$\frac{3x^2+x-2}{(x-1)(x^2+1)} = \frac{A}{x-1} + \frac{Bx+C}{x^2+1},$$

or

$$A(x^2+1) + Bx(x-1) + C(x-1) \equiv 3x^2+x-2.$$

Setting $x = 1$, 0, and -1 in turn, we find that $A = 1$, $C = 3$, $B = 2$.

Example 3. Show how to break up the fraction

$$\frac{2x^6-3x^5+x^4-4x^3+2x^2-x+1}{(x-2)^3(x^2+2x+2)^2} \equiv \frac{P(x)}{Q(x)}$$

into simplest partial fractions. Do not determine the constants.

SOLUTION.

$$\frac{P(x)}{Q(x)} = \frac{P_1(x)}{(x-2)^3} + \frac{P_2(x)}{(x^2+2x+2)^2} = \frac{A_1}{x-2} + \frac{A_2}{(x-2)^2}$$

$$+ \frac{A_3}{(x-2)^3} + \frac{A_4 x + A_5}{x^2+2x+2} + \frac{A_6 x + A_7}{(x^2+2x+2)^2}.$$

7. Three Rationalizing Substitutions

a) If the integrand contains a single irrational expression of the form $(ax+b)^{p/q}$, the substitution $z = (ax+b)^{1/q}$ will convert the integral into that of a rational function of z.

EXAMPLE. If we let $z = (x+1)^{1/3}$ so that $x = z^3 - 1$ and $dx = 3z^2\,dz$, then

$$\int \frac{\sqrt[3]{x+1}}{x}\,dx = \int \frac{z}{z^3-1}\cdot 3z^2\,dz = \int 3\,dz + \int \frac{3\,dz}{(z-1)(z^2+z+1)}.$$

b) If a single irrational expression of the form $\sqrt{a^2-x^2}$, $\sqrt{a^2+x^2}$, or $\sqrt{x^2-a^2}$ occurs with an odd power of x outside, the substitution $z = \sqrt{a^2-x^2}$ (or etc.) reduces the given integral to that of a rational function.

EXAMPLE. If we let $z = \sqrt{a^2-x^2}$, then

$$x^2 = a^2 - z^2, \qquad x\,dx = -z\,dz,$$

$$\int \frac{\sqrt{a^2-x^2}}{x^3}\,dx = -\int \frac{z^2\,dz}{(z^2-a^2)^2}.$$

c) In case a given integrand is a rational function of trigonometric functions, the substitution

$$t = \tan(\theta/2), \qquad \theta = 2\arctan t, \qquad d\theta = \frac{2\,dt}{1+t^2},$$

$$\cos\theta = \frac{1-t^2}{1+t^2}, \qquad \sin\theta = \frac{2t}{1+t^2},$$

reduces the integral to one of a rational function of t.

EXAMPLE. Making these substitutions, we obtain

$$\int \frac{d\theta}{5-4\cos\theta} = \int \frac{(2\,dt)/(1+t^2)}{5-4[(1-t^2)/(1+t^2)]} = \int \frac{2\,dt}{1+9t^2}.$$

A Short Table of Integrals. The constant of integration is omitted.

Elementary formulas

1. $\displaystyle\int u^n\,du = \frac{u^{n+1}}{n+1}, \quad n \neq -1$

2. $\displaystyle\int \frac{du}{u} = \log|u|$

3. $\displaystyle\int e^u\,du = e^u$

4. $\displaystyle\int a^u\,du = \frac{a^u}{\ln a}, \quad a > 0, \quad a \neq 1$

5. $\displaystyle\int \sin u\,du = -\cos u$

6. $\displaystyle\int \cos u\,du = \sin u$

7. $\displaystyle\int \sec^2 u\,du = \tan u$

8. $\displaystyle\int \csc^2 u\,du = -\cot u$

9. $\displaystyle\int \sec u \tan u\,du = \sec u$

10. $\displaystyle\int \csc u \cot u\,du = -\csc u$

11. $\displaystyle\int \sinh u\,du = \cosh u$

12. $\displaystyle\int \cosh u\,du = \sinh u$

13. $\displaystyle\int \operatorname{sech}^2 u\,du = \tanh u$

14. $\displaystyle\int \operatorname{csch}^2 u\,du = -\coth u$

15. $\displaystyle\int \operatorname{sech} u \tanh u\,du = -\operatorname{sech} u$

16. $\displaystyle\int \operatorname{csch} u \coth u\,du = -\operatorname{csch} u$

17. $\displaystyle\int \frac{du}{\sqrt{a^2 - u^2}} = \arcsin\frac{u}{a}, \quad a > |u|$

18. $\displaystyle\int \frac{du}{a^2 + u^2} = \frac{1}{a}\arctan\frac{u}{a}, \quad a \neq 0$

19. $\displaystyle\int \frac{du}{u\sqrt{u^2 - a^2}} = \frac{1}{a}\operatorname{arcsec}\frac{u}{a}, \quad |u| > a$

20. $$\int \frac{du}{\sqrt{a^2 + u^2}} = \begin{cases} \operatorname{argsinh}\dfrac{u}{a} \\ \log\left(u + \sqrt{a^2 + u^2}\right) \end{cases}$$

21. $$\int \frac{du}{a^2 - u^2} = \begin{cases} \dfrac{1}{a}\operatorname{argtanh}\dfrac{u}{a}, & |u| < a \\ \dfrac{1}{a}\operatorname{argcoth}\dfrac{u}{a}, & |u| > a \end{cases}$$

22. $\displaystyle\int \frac{du}{u\sqrt{a^2 - u^2}} = -\frac{1}{a}\operatorname{argsech}\frac{u}{a}, \quad 0 < u < a$

23. $\displaystyle\int \frac{du}{|u|\sqrt{u^2 + a^2}} = -\frac{1}{a}\operatorname{argcsch}\frac{u}{a}, \quad u \neq 0$

Elementary formulas (*cont.*)

24. $$\int \frac{du}{\sqrt{u^2 - a^2}} = \begin{cases} \operatorname{argcosh} \dfrac{u}{a} \\ \log |u + \sqrt{u^2 - a^2}| \end{cases} \quad |u| > a > 0$$

Algebraic forms

25. $$\int \frac{u\,du}{a + bu} = \frac{u}{b} - \frac{a}{b^2} \log (a + bu)$$

26. $$\int \frac{du}{u(a + bu)} = \frac{1}{a} \log \left| \frac{u}{a + bu} \right|$$

27. $$\int \frac{u\,du}{(a + bu)^2} = \frac{a}{b^2} \left(\frac{1}{a + bu} + \frac{1}{a} \log |a + bu| \right)$$

28. $$\int \frac{du}{u(a + bu)^2} = \frac{1}{a(a + bu)} + \frac{1}{a^2} \log \left| \frac{u}{a + bu} \right|$$

29. $$\int \frac{du}{u\sqrt{a + bu}} = \frac{1}{\sqrt{a}} \log \left| \frac{\sqrt{a + bu} - \sqrt{a}}{\sqrt{a + bu} + \sqrt{a}} \right|$$

30. $$\int u\sqrt{a + bu}\,du = \frac{2(3bu - 2a)\sqrt{(a + bu)^3}}{15b^2}$$

31. $$\int \frac{\sqrt{a + bu}}{u} du = 2\sqrt{a + bu} + a \int \frac{du}{u\sqrt{a + bu}}$$

32. $$\int \frac{u\,du}{\sqrt{a + bu}} = \frac{2(bu - 2a)}{3b^2} \sqrt{a + bu}$$

33. $$\int \sqrt{a^2 - u^2}\,du = \frac{u}{2}\sqrt{a^2 - u^2} + \frac{a^2}{2} \operatorname{Arcsin} \frac{u}{a}, \quad |u| < a$$

34. $$\int \sqrt{u^2 \pm a^2}\,du = \frac{u}{2}\sqrt{u^2 \pm a^2} \pm \frac{a^2}{2} \log |u + \sqrt{u^2 \pm a^2}|$$

35. $$\int \frac{\sqrt{a^2 \pm u^2}}{u} du = \sqrt{a^2 \pm u^2} - a \log \left| \frac{a + \sqrt{a^2 \pm u^2}}{u} \right|$$

36. $$\int \frac{\sqrt{u^2 - a^2}}{u} du = \sqrt{u^2 - a^2} - a \operatorname{Arccos} \frac{a}{u}, \quad 0 < a < |u|$$

Trigonometric forms

37. $$\int \tan u\,du = -\log |\cos u|$$

38. $$\int \sec u\,du = \log |\sec u + \tan u|$$

39. $$\int \sin^2 u\,du = \tfrac{1}{2}u - \tfrac{1}{4}\sin 2u$$

40. $$\int \sin^n u\,du = -\frac{\sin^{n-1} u \cos u}{n} + \frac{n-1}{n}\int \sin^{n-2} u\,du$$

41. $$\int \cos^n u\,du = \frac{\cos^{n-1} u \sin u}{n} + \frac{n-1}{n}\int \cos^{n-2} u\,du$$

42. $$\int \frac{du}{\sin^n u} = -\frac{\cos u}{(n-1)\sin^{n-1} u} + \frac{n-2}{n-1}\int \frac{du}{\sin^{n-2} u}, \quad n \neq 1$$

43. $$\int \sin mu \sin nu\,du = \frac{\sin (m-n)u}{2(m-n)} - \frac{\sin (m+n)u}{2(m+n)}, \quad m \neq \pm n$$

44. $$\int \cos mu \cos nu\,du = \frac{\sin (m-n)u}{2(m-n)} + \frac{\sin (m+n)u}{2(m+n)}, \quad m \neq \pm n$$

45. $$\int \sin mu \cos nu\,du = -\frac{\cos (m-n)u}{2(m-n)} - \frac{\cos (m+n)u}{2(m+n)}, \quad m \neq \pm n$$

46. $$\int u^n \sin u\,du = -u^n \cos u + n\int u^{n-1} \cos u\,du$$

47. $$\int u^n \cos u\,du = u^n \sin u - n\int u^{n-1} \sin u\,du$$

48. $$\int \operatorname{Arcsin} u\,du = u \operatorname{Arcsin} u + \sqrt{1-u^2}$$

49. $$\int \operatorname{Arccos} u\,du = u \operatorname{Arccos} u - \sqrt{1-u^2}$$

50. $$\int u \operatorname{Arcsin} u\,du = \tfrac{1}{4}[(2u^2-1)\operatorname{Arcsin} u + u\sqrt{1-u^2}]$$

Logarithmic and exponential forms

51. $$\int \log|u|\,du = u(\log|u| - 1)$$

52. $$\int (\log|u|)^2\,du = u(\log|u|)^2 - 2u\log|u| + 2u$$

53. $$\int u^n \log|u|\,du = \frac{u^{n+1}}{n+1}\log|u| - \frac{u^{n+1}}{(n+1)^2}, \quad n \neq -1$$

54. $$\int u^n e^u\,du = u^n e^u - n\int u^{n-1} e^u\,du$$

55. $$\int e^{au} \sin bu\,du = \frac{e^{au}(a \sin bu - b \cos bu)}{a^2 + b^2}$$

56. $$\int e^{au} \cos bu\,du = \frac{e^{au}(a \cos bu + b \sin bu)}{a^2 + b^2}$$

Miscellaneous forms

57. $\int (a + bu)^n \, du = \frac{(a + bu)^{n+1}}{b(n + 1)}, \quad n \neq -1$

58. $\int \frac{du}{u^2(a + bu)} = -(au)^{-1} + ba^{-2}[\log (a + bu) - \log u]$

59. $\int \frac{du}{(a + bu)^{1/2}(c + du)^{3/2}} = \frac{2}{bc - ad}\left(\frac{a + bu}{c + du}\right)^{1/2}, \quad bc - ad \neq 0$

60. $\int \frac{du}{u(a + bu^n)} = (an)^{-1}[\log (u^n) - \log (a + bu^n)], \quad n \neq 0$

61. $\int \frac{u \, du}{(a + bu)^{1/2}} = \frac{2(bu - 2a)}{3b^2}(a + bu)^{1/2}$

62. $\int \frac{du}{u^2(u^2 + a^2)^{1/2}} = -(a^2 u)^{-1}(u^2 + a^2)^{1/2}$

63. $\int \frac{(a^2 - u^2)^{1/2}}{u^2} du = -u^{-1}(a^2 - u^2)^{1/2} - \arcsin \frac{u}{a}$

64. $\int \sin^4 u \, du = \frac{3u}{8} - \frac{\sin 2u}{4} + \frac{\sin 4u}{32}$

65. $\int \cot^3 u \, du = -\frac{1}{2}\cot^2 u - \log \sin u$

Answers to Odd-Numbered Problems

Chapter 1

Section 1

1. $|AB| = \sqrt{30}$; $|BC| = \sqrt{19}$; $|AC| = \sqrt{21}$; scalene
3. $|AB| = \sqrt{17}$; $|AC| = \sqrt{18}$; $|BC| = \sqrt{35}$; right triangle
5. $|AB| = \sqrt{21}$; $|AC| = \sqrt{51}$; $|BC| = \sqrt{126}$
7. $(-4, \frac{3}{2}, \frac{9}{2})$
9. $(\frac{9}{4}, -\frac{3}{2}, \frac{7}{4})$, $(\frac{7}{2}, 3, \frac{7}{2})$, $(\frac{19}{4}, \frac{15}{2}, \frac{21}{4})$
11. $\frac{1}{2}\sqrt{61}$, $\frac{1}{2}\sqrt{61}$, $\frac{1}{2}\sqrt{10}$
13. $\frac{1}{2}\sqrt{74}$, $\frac{1}{2}\sqrt{74}$, $\frac{1}{2}\sqrt{26}$
15. $P_2(4, 5, -6)$, $Q(1, 3, 0)$
17. Not on a line
19. Not on a line
21. Block extending indefinitely in x and z directions, bounded by planes $y = -2$ and $y = 5$.
23. Interior of sphere of radius 1
27. $6x + 4y - 8z - 5 = 0$; plane

Chapter 1

Section 2

1. Dir. nos.: 1, 4, -6; dir. cosines: $\dfrac{1}{\sqrt{53}}, \dfrac{4}{\sqrt{53}}, \dfrac{-6}{\sqrt{53}}$
3. Dir. nos.: 2, -6, -3; dir. cosines: $\dfrac{2}{7}, \dfrac{-6}{7}, \dfrac{-3}{7}$
5. $(3, 5, 2)$
7. $(0, 4, 2)$
9. On line
11. Not on line
13. Parallel
15. Not parallel
17. Perpendicular
19. $\sqrt{2}/3$
21. $41/3\sqrt{190}$
23. $4\sqrt{10}$

Chapter 1

Section 3

1. $x = 1 + t,\ y = 3 - 4t,\ z = 2 + 2t$
3. $x = 4 - t,\ y = -2 + 4t,\ z = -t$
5. $\dfrac{x-1}{2} = \dfrac{y}{1} = \dfrac{z+1}{-3}$
7. $\dfrac{x-4}{2} = \dfrac{y}{-1} = \dfrac{z}{-3}$
9. $\dfrac{x-3}{2} = \dfrac{y+1}{0} = \dfrac{z+2}{0}$
11. Perpendicular
13. Not perpendicular
15. $\dfrac{x-4}{1} = \dfrac{y}{-2} = \dfrac{z-2}{0};\ \dfrac{x-3}{0} = \dfrac{y-1}{1} = \dfrac{z-4}{-2};\ \dfrac{x-2}{1} = \dfrac{y-5}{-3} = \dfrac{z}{2}$
17. $\dfrac{x-2}{a} = \dfrac{y+1}{b} = \dfrac{z-5}{-2a+3b}$
19. $\left(0, \dfrac{-5}{2}, \dfrac{-5}{2}\right), \left(\dfrac{-5}{3}, 0, \dfrac{10}{3}\right), \left(\dfrac{-5}{7}, \dfrac{-10}{7}, 0\right)$
21. $x = 3 - t,\ y = 1 + 3t,\ z = 5 + t$
23. $A'B' : \dfrac{x-2}{5} = \dfrac{y}{0} = \dfrac{z-6}{2};\ A'C' : \dfrac{x-2}{3} = \dfrac{y}{0} = \dfrac{z-6}{1};$

    $B'C' : \dfrac{x+3}{8} = \dfrac{y}{0} = \dfrac{z-4}{3}$

Chapter 1

Section 4

1. $3x + y - 4z + 1 = 0$
3. $3y - 2z - 4 = 0$
5. $2x - z = 1$
7. $9x + y - 5z = 16$
9. $2x + 3y - 4z + 11 = 0$
11. $3x - 2z = 10$
13. $\dfrac{x+2}{2} = \dfrac{y-3}{3} = \dfrac{z-1}{1}$
15. $\dfrac{x+1}{1} = \dfrac{y}{0} = \dfrac{z+2}{2}$
17. $3x + 2y - z = 0$
19. $x - 2y - 3z + 5 = 0$
21. $\dfrac{x-2}{3} = \dfrac{y+1}{-2} = \dfrac{z-3}{4}$
23. $\dfrac{x-1}{2} = \dfrac{y+2}{-1} = \dfrac{z}{4}$
25. $2x - 2y - z - 4 = 0$
27. $2x - 3y - 5z = 7$
29. $2x - 2y - z - 6 = 0$
33. $aA + bB + cC = 0$
35. $\begin{vmatrix} A_1 & B_1 & C_1 \\ A_2 & B_2 & C_2 \\ A_3 & B_3 & C_3 \end{vmatrix} = 0$

Chapter 1

Section 5

1. 8/21
3. $\sqrt{14/17}$

5. $x = 11 - 5t$, $y = -19 + 8t$, $z = t$
7. $x = -1 - 2t$, $y = t$, $z = 3$
9. $(3, 2, -1)$
11. $(-\frac{3}{5}, 0, -\frac{6}{5})$
13. 1
15. $8/\sqrt{29}$
17. $10x - 17y + z + 25 = 0$
19. $14x + 8y - 13z + 15 = 0$
21. $y + z = 1$
23. $(\frac{91}{57}, \frac{31}{57}, \frac{25}{57})$
25. No intersection
27. $x = 3 + 4t$, $y = -1 + 5t$, $z = 2 - t$
29. $x = 29t$, $y = 2 + t$, $z = 4 - 22t$
31. $5x + 2y + 3z - 3 = 0$
33. $\sqrt{14}$; $2\sqrt{14}/\sqrt{15}$

Chapter 1

Section 6

1. $x^2 + y^2 + z^2 - 2x - 8y + 4z + 12 = 0$
3. $x^2 + y^2 + z^2 - 2y - 8z - 19 = 0$
5. Sphere: $C(-1, 0, 2)$, $r = 2$
7. No graph
9. Sphere: $C(3, -2, -1)$, $r = 2$
11. $x^2 + y^2 + z^2 + 5x + 7y - 12z + 41 = 0$
13. Plane
15. Plane
17. Parabolic cylinder
19. Elliptic cylinder
21. Circular cylinder
23. Circular cylinder
25. Circle, $C(0, 0, 3)$, $r = 4$
27. No intersection

Chapter 1

Section 7

1. (a) $\left(3\sqrt{2}, \frac{\pi}{4}, 7\right)$ b) $(4\sqrt{5}, \arctan 2, 2)$ c) $\left(\sqrt{13}, \pi - \arctan\left(\frac{3}{2}\right), 1\right)$
3. (a) $\left(2\sqrt{3}, \frac{\pi}{4}, \arccos\frac{1}{\sqrt{3}}\right)$ b) $\left(2\sqrt{3}, \frac{-\pi}{4}, \arccos\frac{1}{\sqrt{3}}\right)$
   c) $\left(2\sqrt{2}, \frac{-\pi}{3}, \frac{\pi}{4}\right)$
5. (a) $\left(4, \frac{\pi}{3}, 0\right)$ b) $\left(1, \frac{2\pi}{3}, -\sqrt{3}\right)$ c) $\left(\frac{7}{2}, \frac{\pi}{2}, \frac{7}{2}\sqrt{3}\right)$
7. $r^2 + z^2 = 9$
9. $r^2 = 4z$
11. $r^2 = z^2$
13. $r^2 \cos 2\theta = 4$
15. $r = 4 \sin\theta$
17. $\rho = 4\cos\phi$
19. $\phi = \frac{\pi}{4}$
21. $\rho = \frac{\pm 2}{(1 \mp \cos\phi)}$

Chapter 2

Section 2

1. $\mathbf{v} = -2\mathbf{i} + 7\mathbf{j}$
3. $\mathbf{v} = -5\mathbf{i} + 4\mathbf{j}$
5. $\mathbf{v} = -12\mathbf{i} - 4\mathbf{j}$
7. $\mathbf{u} = -\frac{5}{13}\mathbf{i} - \frac{12}{13}\mathbf{j}$
9. $\mathbf{u} = \dfrac{-2}{\sqrt{29}}\mathbf{i} + \dfrac{5}{\sqrt{29}}\mathbf{j}$
11. $B(4, 6)$
13. $A(4, 7)$
15. $A(-3, \frac{1}{2})$, $B(-5, \frac{7}{2})$
17. $\pm 2\mathbf{i} + 2\mathbf{j}$
19. $-\mathbf{i} + \mathbf{j}$

Chapter 2

Section 3

1. $|\mathbf{v}| = 5$, $|\mathbf{w}| = 5$, $\cos\theta = -\frac{7}{25}$, proj. $\mathbf{v}$ on $\mathbf{w} = -\frac{7}{5}$
3. $|\mathbf{v}| = 5$, $|\mathbf{w}| = 13$, $\cos\theta = -\frac{33}{65}$, proj. $\mathbf{v}$ on $\mathbf{w} = -\frac{33}{13}$
5. $|\mathbf{v}| = \sqrt{13}$, $|\mathbf{w}| = \sqrt{13}$, $\cos\theta = 0$, proj. $\mathbf{v}$ on $\mathbf{w} = 0$
7. -1
9. $\dfrac{6}{\sqrt{17}}$
11. 0
13. $\cos\theta = \dfrac{+1}{\sqrt{5}}$, $\cos\alpha = \dfrac{1}{\sqrt{5}}$
15. $\cos\theta = \dfrac{2}{\sqrt{13}}$, $\cos\alpha = \dfrac{11}{\sqrt{130}}$
17. $\cos\theta = -1$, $\cos\alpha = 1$
19. $a = \frac{3}{2}$
21. Impossible
23. $a = \dfrac{(-240 + \sqrt{(240^2 + 69\cdot 407)})}{407}$
27. $|\overrightarrow{BC}| = \sqrt{93}$; proj. of $\overrightarrow{AB}$ on $\overrightarrow{BC} = -30/\sqrt{93}$; proj. of $\overrightarrow{AC}$ on $\overrightarrow{BC} = 63/\sqrt{93}$
29. Proj. of $\overrightarrow{AC}$ on $\overrightarrow{AB} = \dfrac{33}{5}$; proj. of $\overrightarrow{BC}$ on $\overrightarrow{AB} = \dfrac{-17}{5}$
31. $\mathbf{v}[\overrightarrow{DE}] = \frac{1}{2}\mathbf{v}[\overrightarrow{AC}] - \frac{2}{3}\mathbf{v}[\overrightarrow{AB}]$
33. $\mathbf{v}[\overrightarrow{EF}] = \frac{1}{3}\mathbf{v}[\overrightarrow{AC}] - \frac{1}{2}\mathbf{v}[\overrightarrow{AB}]$

Chapter 2

Section 4

1. $-\mathbf{i} + 4\mathbf{j} - 6\mathbf{k}$
3. $5\mathbf{i} - \mathbf{j} + 3\mathbf{k}$
5. $-2\mathbf{i} - 3\mathbf{k}$
7. $\dfrac{1}{\sqrt{29}}(3\mathbf{i} + 2\mathbf{j} - 4\mathbf{k})$
9. $\dfrac{1}{\sqrt{21}}(2\mathbf{i} - 4\mathbf{j} - \mathbf{k})$
11. $B: (3, 3, -4)$
13. $A: (-1, -2, 0)$
15. $A: (\frac{3}{2}, 0, 3)$, $B: (\frac{5}{2}, -2, 5)$
17. $A: (\frac{7}{4}, -\frac{3}{4}, \frac{7}{2})$, $B: (\frac{3}{4}, \frac{1}{4}, \frac{3}{2})$
19. $4\mathbf{i} - 5\mathbf{j} + \mathbf{k}$

Chapter 2

Section 5

1. Linearly independent
3. Linearly dependent

5. Linearly dependent
7. $13\mathbf{u} + 7\mathbf{v} - 2\mathbf{w}$
9. $\frac{1}{4}\mathbf{u} + \frac{3}{8}\mathbf{v} - \frac{9}{8}\mathbf{w}$
11. $-\frac{2}{5}\mathbf{u} + 3\mathbf{v} + \frac{8}{5}\mathbf{w}$
17. $\dfrac{x-1}{-1} = \dfrac{y+4}{4} = \dfrac{z}{7}$

Chapter 2

Section 6

1. $\dfrac{16}{7\sqrt{6}}$
3. $\dfrac{1}{\sqrt{3}}$
5. $\dfrac{-11}{29}$
7. $-\frac{2}{7}$
9. $-23/\sqrt{50}$
11. $\frac{1}{7}(2\mathbf{i} - 6\mathbf{j} + 3\mathbf{k})$
13. $\dfrac{1}{\sqrt{62}}(3\mathbf{i} - 2\mathbf{j} + 7\mathbf{k})$
15. $k = 2, h = \frac{2}{5}$
17. $k = 3, h = \frac{42}{145}$
21. $3g + 5h = 0$
23. $4g - 9h = 0$
25. $h = 1, g = -1$
27. Only if $\mathbf{u} = k\mathbf{v}$ for some $k > 0$
29. $\dfrac{2}{\sqrt{6}}$

Chapter 2

Section 7

1. $2\mathbf{i} - 3\mathbf{j} - 7\mathbf{k}$
3. $-4\mathbf{i} - 2\mathbf{j} - 5\mathbf{k}$
5. $20\mathbf{i} + 30\mathbf{j} - 16\mathbf{k}$
7. $\dfrac{7\sqrt{3}}{2}, x - y - z = 0$
9. $\dfrac{3\sqrt{35}}{2}, 11x + 5y + 13z - 30 = 0$
11. $\dfrac{\sqrt{421}}{2}, 9x + 12y + 14z = 32$
13. $\dfrac{107}{\sqrt{1038}}$
15. $\dfrac{x+1}{2} = \dfrac{y-3}{-11} = \dfrac{z-2}{-7}$
17. $\dfrac{x-1}{1} = \dfrac{y+2}{-1} = \dfrac{z-3}{1}$
19. $\dfrac{x-3}{-2} = \dfrac{y}{1} = \dfrac{z-1}{3}$
21. $\dfrac{x+2}{2} = \dfrac{y-1}{11} = \dfrac{z+1}{8}$
23. $8x + 14y + 13z + 37 = 0$
25. $2x - z + 1 = 0$
27. $5x - 3y - z = 6$
29. $4x - 3y - z + 9 = 0$
31. $x - y - z + 6 = 0$
33. $\dfrac{x-3}{2} = \dfrac{y+2}{1} = \dfrac{z}{2}$

Chapter 2

Section 8

1. $V = 20$
3. Plane: $x + 2y - 5 = 0$
7. $\mathbf{i} + 5\mathbf{j} - 2\mathbf{k}$
9. $8\mathbf{i} + 10\mathbf{j} - 14\mathbf{k}$
13. $\dfrac{x-2}{160} = \dfrac{y+1}{-45} = \dfrac{z-3}{37}$
15. $-16\mathbf{v} + 12\mathbf{w}$
17. $[(\mathbf{t} \times \mathbf{u}) \cdot \mathbf{w}]\mathbf{v} - [(\mathbf{t} \times \mathbf{u}) \cdot \mathbf{v}]\mathbf{w}$
19. $\mathbf{v} = |\mathbf{a}|^2 p\mathbf{a} - |\mathbf{a}|^2(\mathbf{a} \times \mathbf{b})$

Chapter 2

Section 9

1. $\mathbf{f}'(t) = 2t\mathbf{i} + (3t^2 - 3)\mathbf{j}$; $\mathbf{f}''(t) = 2\mathbf{i} + 6t\mathbf{j}$
3. $\mathbf{f}'(t) = 3\sec^2 3t\mathbf{i} - \pi \sin \pi t\mathbf{j}$; $\mathbf{f}''(t) = 18\sec^2 3t \tan 3t\mathbf{i} - \pi^2 \cos \pi t\mathbf{j}$
5. $\mathbf{f}'(t) = 2e^{2t}\mathbf{i} - 2e^{-2t}\mathbf{j}$; $\mathbf{f}''(t) = 4e^{2t}\mathbf{i} + 4e^{-2t}\mathbf{j}$
7. $\dfrac{2t(t^2+1)}{(t^2+2)^3} + 4t$
9. $\dfrac{2\sin 4t - 3\sin 6t}{2(\sin^2 2t + \cos^2 3t)^{1/2}}$
11. 0
13. $\dfrac{3t^2 + 40}{t^6}$
15. $\dfrac{d\theta}{dt} = -\dfrac{3}{13t^2 + 4t + 1}$ if $t > 0$;
    $\dfrac{d\theta}{dt} = \dfrac{3(2t+1)}{|2t+1|\cdot(13t^2+4t+1)}$ if $t < 0$, $t \neq -\frac{1}{2}$
19. Straight line, $x - y - 1 = 0$
21. $\mathbf{f}$ and $\mathbf{f}'$ are perpendicular (or $\mathbf{f} \perp \mathbf{f}'$), or the tangent is $\perp$ radius vector.

Chapter 2

Section 10

1. $\mathbf{v}(t) = 2t\mathbf{i} - 3\mathbf{j}$; $|\mathbf{v}(t)| = s'(t) = \sqrt{4t^2 + 9}$;
   $\mathbf{a}(t) = 2\mathbf{i}$; $|\mathbf{a}(t)| = 2$; $s''(t) = \dfrac{4t}{\sqrt{4t^2+9}}$
3. $\mathbf{v}(t) = 3\mathbf{i} + t^{-2}\mathbf{j}$; $|\mathbf{v}(t)| = s'(t) = \sqrt{9 + t^{-4}}$;
   $\mathbf{a}(t) = -2t^{-3}\mathbf{j}$; $|\mathbf{a}(t)| = |2t^{-3}|$; $s''(t) = \dfrac{-2}{t^3\sqrt{9t^4+1}}$
5. $\mathbf{v}(t) = \tan t\mathbf{i} + \mathbf{j}$; $|\mathbf{v}(t)| = s'(t) = |\sec t|$;
   $\mathbf{a}(t) = \sec^2 t\mathbf{i}$; $|\mathbf{a}(t)| = \sec^2 t$; $s''(t) = \sec t \tan t$
7. $\mathbf{v}(t) = 2e^t\mathbf{i} - 3e^{-t}\mathbf{j}$; $|\mathbf{v}(t)| = s'(t) = \sqrt{4e^{2t} + 9e^{-2t}}$;
   $\mathbf{a}(t) = 2e^t\mathbf{i} + 3e^{-t}\mathbf{j}$; $|\mathbf{a}(t)| = \sqrt{4e^{2t} + 9e^{-2t}}$; $s''(t) = \dfrac{4e^{2t} - 9e^{-2t}}{\sqrt{4e^{2t} + 9e^{-2t}}}$
9. $\mathbf{v}(1) = \mathbf{i}$; $\mathbf{a}(1) = 2\mathbf{i} + 2\mathbf{j}$; $s'(1) = 1$; $s''(1) = 2$
11. $\mathbf{v}\left(\dfrac{\pi}{6}\right) = 2\mathbf{i} + \frac{8}{3}\mathbf{j}$; $\mathbf{a} = \dfrac{10}{\sqrt{3}}\mathbf{i} + \dfrac{16}{3\sqrt{3}}\mathbf{j}$; $s' = \frac{10}{3}$, $s'' = \dfrac{154}{15\sqrt{3}}$
13. $\mathbf{v} = \frac{1}{3}\mathbf{i} - \frac{3}{4}\mathbf{j}$; $\mathbf{a} = -\frac{1}{9}\mathbf{i} + \frac{3}{4}\mathbf{j}$; $s' = \dfrac{\sqrt{97}}{12}$; $s'' = \dfrac{-259}{36\sqrt{97}}$
17. $\mathbf{T} = \dfrac{4e^{2t}\mathbf{i} - 3e^{-2t}\mathbf{j}}{(16e^{4t} + 9e^{-4t})^{1/2}}$

Chapter 2

Section 11

1. $\mathbf{f}' = 2t\mathbf{i} + 2t\mathbf{j} - 3\mathbf{k}$; $\mathbf{f}'' = 2\mathbf{i} + 2\mathbf{j}$
3. $\mathbf{f}' = -2(\sin 2t)\mathbf{i} + 2(\cos 2t)\mathbf{j} + 2\mathbf{k}$
   $\mathbf{f}'' = -4(\cos 2t)\mathbf{i} - 4(\sin 2t)\mathbf{j}$

5. $\mathbf{f}' = \dfrac{2t}{(t^2+1)^2}\mathbf{i} - \dfrac{2t}{(t^2+1)^2}\mathbf{j} + 2t\mathbf{k}$

$\mathbf{f}'' = \dfrac{-6t^2+2}{(t^2+1)^3}\mathbf{i} - \dfrac{-6t^2+2}{(t^2+1)^3}\mathbf{j} + 2\mathbf{k}$

7. $\mathbf{f}' = -(\sin t)\mathbf{i} + (\sec^2 t)\mathbf{j} + (\cos t)\mathbf{k}$

$\mathbf{f}'' = -(\cos t)\mathbf{i} + 2(\sec^2 t \tan t)\mathbf{j} - (\sin t)\mathbf{k}$

9. $f' = 9t^2 + 2t + 2$ 11. $f'(t) = 1$ 13. 14

15. $\dfrac{\pi}{8}\sqrt{\pi^2+8} + \log(\pi + \sqrt{\pi^2+8}) - \frac{3}{2}\log 2$

17. $\mathbf{v}(t) = (\sin t + t\cos t)\mathbf{i} + (\cos t - t\sin t)\mathbf{j} + \mathbf{k}$;
$s'(t) = \sqrt{t^2+2}$; $\mathbf{a}(t) = (2\cos t - t\sin t)\mathbf{i} - (2\sin t + t\cos t)\mathbf{j}$

Chapter 3

Section 1

1. $\frac{3}{4}$ 3. $\frac{3}{2}$ 5. $\frac{2}{3}$ 7. 0
9. 3 11. 4 13. $-\infty$ 15. $\frac{2}{3}$
17. ∞ 19. 0 21. -1 23. 0
25. 0 27. 2 29. 1 31. 0
33. $-\frac{2}{3}$ 35. 1 37. 1 39. e
41. 1 43. (c)0

Chapter 3

Section 2

1. $\dfrac{71}{99}$ 3. $\dfrac{13}{999}$ 5. $\dfrac{14}{99{,}000}$
7. $\dfrac{36140}{9999}$ 9. $\dfrac{9}{4}$ 11. 9
13. 12 meters 15. $\frac{1}{4} + \frac{4}{11} + \frac{3}{7} + \frac{8}{17} + \frac{1}{2} + \cdots$
17. $e - \dfrac{e^2}{8} + \dfrac{e^3}{27} - \dfrac{e^4}{64} + \dfrac{e^5}{125} - \cdots$

Chapter 3

Section 3

1. Convergent 3. Convergent 5. Divergent
7. Convergent 9. Divergent 11. Divergent
13. Convergent 15. Convergent 17. Divergent
19. Convergent 21. Convergent 23. Convergent
25. Convergent 27. Convergent 29. Convergent
31. Convergent if $p > 1$ 33. $q < 1$, and p arbitrary

Chapter 3

Section 4

1. Divergent
3. Divergent
5. Absolutely convergent
7. Conditionally convergent
9. Absolutely convergent
11. Absolutely convergent
13. Divergent
15. Absolutely convergent
17. Absolutely convergent
19. Conditionally convergent
21. Absolutely convergent
23. Divergent
25. Conditionally convergent
27. Conditionally convergent
29. p has no effect

Chapter 3

Section 5

1. Converges for $-1 < x < 1$
3. Converges for $-\frac{1}{2} < x < \frac{1}{2}$
5. Converges for $-1 < x < 1$
7. Converges for $-1 \leq x \leq 3$
9. Converges for $-3 \leq x < -1$
11. Converges for $-\infty < x < \infty$
13. Converges for $-\frac{2}{3} < x \leq \frac{2}{3}$
15. Converges for $0 < x < 2$
17. Converges for $-7 \leq x \leq -1$
19. Converges for $-6 < x < 10$
21. Converges for $-\frac{4}{3} < x < \frac{4}{3}$
23. Converges for $-1 \leq x < 1$
25. Converges for $-1 < x \leq 1$
27. Converges for $-\frac{3}{2} \leq x \leq \frac{3}{2}$
29. (a) Converges for $-1 < x < 1$
31. Converges for $-1 < x < 1$

Chapter 3

Section 6

1. $\displaystyle\sum_{n=0}^{\infty} \frac{x^n}{n!}$
3. $\displaystyle\sum_{n=1}^{\infty} \frac{(-1)^{n-1}x^n}{n}$
5. $\displaystyle\sum_{n=0}^{\infty} (n+1)x^n$
7. $\displaystyle 1 + \sum_{n=1}^{\infty} \frac{\left(\frac{1}{2}\right)\left(\frac{-1}{2}\right)\left(\frac{-3}{2}\right)\cdots\left(-n+\frac{1}{2}\right)x^n}{n!}$
9. $\displaystyle \log 3 + \sum_{n=1}^{\infty} \frac{(-1)^{n-1}(x-3)^n}{3^n \cdot n}$
11. $\displaystyle \frac{1}{2} - \frac{\sqrt{3}}{2}\left(x - \frac{\pi}{3}\right) - \frac{1}{2}\frac{(x-\pi/3)^2}{2!} + \frac{\sqrt{3}}{2}\frac{(x-\pi/3)^3}{3!} + \cdots$
13. $\displaystyle 2 + \tfrac{1}{4}(x-4) + 2\sum_{n=2}^{\infty} \frac{(-1)^{n-1}\cdot 1\cdot 3\cdots(2n-3)(x-4)^n}{2^n n!\, 4^n}$
15. $\displaystyle \cos\left(\tfrac{1}{2}\right) - x\sin\left(\tfrac{1}{2}\right) - \frac{x^2}{2!}\cos\left(\tfrac{1}{2}\right) + \frac{x^3}{3!}\sin\left(\tfrac{1}{2}\right) + \cdots$
17. $\displaystyle 1 - x^2 + \frac{x^4}{2!}$

19. $1 - x^2 + x^4$

21. $1 + x - \dfrac{x^3}{3} - \dfrac{x^4}{6}$

23. $x + \dfrac{x^3}{2\cdot 3} + \dfrac{3x^5}{2\cdot 4\cdot 5}$

25. $\dfrac{x^2}{2} + \dfrac{x^4}{12} + \dfrac{x^6}{45}$

27. $2 - 2\sqrt{3}\left(x - \dfrac{\pi}{6}\right) + 7(x - \pi/6)^2 - \dfrac{23\sqrt{3}(x - \pi/6)^3}{3} + \dfrac{305(x - \pi/6)^4}{12}$

29. $2 + 2\sqrt{3}\left(x - \dfrac{\pi}{3}\right) + 7\left(x - \dfrac{\pi}{3}\right)^2 + \dfrac{23\sqrt{3}(x - \pi/3)^3}{3} + \cdots$

31. $x + \dfrac{x^3}{3} + \dfrac{2}{15}x^5 + \dfrac{17}{315}x^7 + \cdots$

33. (c) Sum of Taylor's series $\equiv 0$

35. $(3 + x)^{-1/2} = \dfrac{1}{2} - \dfrac{1}{2\cdot 2^3}(x - 1) + \cdots$
$+ \dfrac{(-1)^k 1\cdot 3\cdots(2k-1)(x-1)^k}{2^k k!\, 2^{2k+1}} + \cdots.$
Interval of convergence: $-3 < x < 5$.

Chapter 3

Section 7

1. 0.81873
3. 1.22140
5. 0.87758
7. 0.1823
9. 0.36788
11. 1.01943
13. 0.96905
15. 1.97435
17. 0.95635
19. −0.22314
21. 0.017452
23. 0.99619
25. $|R_4| = |R_5| < 0.011$

Chapter 3

Section 8

1. $\displaystyle\sum_{n=0}^{\infty} \frac{(-1)^n x^{2n}}{(2n+1)!}$

3. $\displaystyle\sum_{n=0}^{\infty} \frac{x^n}{(n+1)!}$

5. $\displaystyle 2\sum_{n=0}^{\infty} \frac{x^{2n+1}}{2n+1}$

7. $\displaystyle \frac{1}{2}\sum_{n=0}^{\infty} (-1)^n(n+1)(n+2)x^n$

9. $\displaystyle\sum_{n=0}^{\infty} (-1)^n \frac{x^{2n+1}}{2n+1}$

11. $\displaystyle \frac{1}{2} - \frac{1}{2}\sum_{n=0}^{\infty} \frac{(-1)^n(2x)^{2n}}{(2n)!}$

13. $f(x) = (1 - x)^{-2}$

15. $\displaystyle x + \sum_{n=1}^{\infty} (-1)^n \frac{1\cdot 3\cdots(2n-1)}{2\cdot 4\cdots(2n)} \frac{x^{2n+1}}{2n+1}$

17. $x - \dfrac{x^3}{3} + \dfrac{x^5}{5\cdot 2!} + \dfrac{x^7}{3!\cdot 7} + \dfrac{x^9}{4!9}$; $F(1) = 0.747$, $0 > R > -\frac{1}{1320}$

19. $-x^{-2/3}\left[\tfrac{1}{3}\log(1 - x^{1/3}) - \tfrac{1}{6}\log(x^{2/3} + x^{1/3} + 1) + \dfrac{1}{\sqrt{3}}\arctan\left(\dfrac{2x^{1/3}+1}{\sqrt{3}}\right)\right.$
$\left. - \dfrac{1}{\sqrt{3}}\arctan\dfrac{1}{\sqrt{3}}\right]$

23. $\displaystyle \frac{1}{1+x^2} = \frac{1}{x^2(1 + (1/x^2))} = \sum_{n=0}^{\infty} (-1)^n \left(\frac{1}{x}\right)^{2n+2}$

Chapter 3

Section 9

1. $1 - \frac{3}{2}x + \dfrac{1\cdot 3\cdot 5}{2^2\cdot 2!}x^2 - \dfrac{1\cdot 3\cdot 5\cdot 7}{2^3\cdot 3!}x^3 + \dfrac{1\cdot 3\cdot 5\cdot 7\cdot 9}{2^4\cdot 4!}x^4 \cdots$
3. $1 - \frac{2}{3}x^2 + \dfrac{2\cdot 5}{3^2\cdot 2!}x^4 - \dfrac{2\cdot 5\cdot 8}{3^3\cdot 3!}x^6 + \dfrac{2\cdot 5\cdot 8\cdot 11}{3^4\cdot 4!}x^8 - \cdots$
5. $1 + 7x^3 + 21x^6 + 35x^9 + 35x^{12} + 21x^{15} + 7x^{18} + x^{21}$
7. $\dfrac{1}{3^3} - \dfrac{3}{3^4}x^{1/2} + \dfrac{3\cdot 4}{3^5\cdot 2!}x - \dfrac{3\cdot 4\cdot 5}{3^6\cdot 3!}x^{3/2} + \dfrac{3\cdot 4\cdot 5\cdot 6}{3^7\cdot 4!}x^2 - \cdots$
9. $0.9045243 + R$, where $0 < R < 1.4 \times 10^{-6}$
11. $0.4854018 + R$, where $0 > R > -9.5 \times 10^{-7}$
13. $1.3179019 + R$, where $0 < R < 3.4 \times 10^{-7}$
15. $0.3293897 + R$, where $0 < R < 10 \times 10^{-7}$
17. $0.5082641 + R$, where $0 < R < 4 \times 10^{-7}$
19. 0.69315

27. $\log 11 = \log 10 + 2\left[\dfrac{1}{21} + \dfrac{1}{3\cdot(21)^3}\right] + R,\ R < 4.9 \times 10^{-8}$

Chapter 3

Section 10

1. $x - x^2 + \frac{5}{6}x^3 - \frac{5}{6}x^4 + \frac{101}{120}x^5$
3. $1 - x + \dfrac{x^2}{2} - \dfrac{x^3}{2} + \dfrac{13x^4}{24} - \dfrac{13x^5}{24}$
5. $x - \dfrac{x^3}{24} + \dfrac{x^4}{24} - \dfrac{71x^5}{1920}$
7. $x - x^2 + \frac{23}{24}x^3 - \frac{11}{12}x^4 + \frac{563}{640}x^5$
9. $1 + x + x^2 + \frac{2}{3}x^3 + \dfrac{x^4}{2} + \dfrac{3x^5}{10}$
11. $1 + x - \dfrac{x^3}{3} + \dfrac{x^4}{6} + \dfrac{3}{10}x^5$
13. $x - x^2 + \frac{2}{3}x^3 - \frac{2}{3}x^4 + \frac{13}{15}x^5$
15. $x + \frac{2}{3}x^3 + \frac{11}{30}x^5$
17. $1 + \dfrac{x}{3} + \dfrac{11x^2}{9} + \dfrac{41x^3}{81} + \dfrac{8x^4}{243}$
19. $1 + \frac{3}{2}x^2 + \frac{1}{2}x^3 + \frac{3}{8}x^4$

Chapter 3

Section 11

1. Yes
3. Yes
5. Yes
7. Yes
9. Yes
11. Yes
13. No
17. f_n converges uniformly $\Leftrightarrow \beta > 2\alpha$
21. f_n' does not; F_n does
23. f_n' does not; F_n does not

Chapter 3

Section 12

1. $(0 <)h < 1$ 3. $0 < h < 1$ 5. $0 < h$
7. $0 < h < 2$ 9. $0 < h < 1$ 11. $0 < h \leq 1$
13. $h > 0$ 17. $0 < h < h_0, h_0 \log h_0 = 1$

Chapter 3

Section 13

1. $f(x) = \sum_{n=0}^{\infty} (n+1)x^{2n}$ 3. $f(x) = \sum_{n=0}^{\infty} \dfrac{ex^{2n}}{n!}$

5. $f(x) = \sum_{n=0}^{\infty} \dfrac{(-1)^n(n+1)(n+2)x^{3n}}{2! \cdot 3^{3n}}$

7. $f(x) = \sum_{n=0}^{\infty} \dfrac{(-1)^n(3x^2)^{2n+1}}{\pi^{2n+1} \cdot (2n+1)!}$ 9. $f(x) = \sum_{n=0}^{\infty} \dfrac{x^{2n}}{(n+1)!}$

Chapter 3

Section 14

3. Converges 7. Diverges
9. $1 - 2x + \frac{3}{2}x^2 - 3x^3 + \frac{85}{24}x^4 - \frac{49}{12}x^5 + \cdots$
11. $x + \frac{1}{2}x^2 + \frac{1}{3}x^3 + 0 \cdot x^4 + \frac{3}{40}x^5 + \cdots$
13. $1 + x + (\frac{1}{2}x^2 - \frac{1}{2}y^2) + (\frac{1}{6}x^3 - \frac{1}{2}xy^2) + \cdots$
15. $1 - x + (\frac{1}{2}x^2 + \frac{1}{2}y^2) - (\frac{1}{6}x^3 + \frac{1}{2}xy^2) + \cdots$
17. $1 + \cdots$

†19. Error $< (0.2)^5 \left(\dfrac{1}{116} + \dfrac{25}{24^2} + \dfrac{10}{57} + \dfrac{15}{28} + \dfrac{10}{9} + \dfrac{25}{16}\right) < 0.00111$

†21. Error $< (0.2)^5 \left(1 + 1 + \dfrac{1}{2} + \dfrac{1}{6} + \dfrac{1}{24} + \dfrac{1}{120}\right) < 0.00073$

Chapter 3

Section 15

13. (a) $e(\cos 1 + i \sin 1)$; (b) $-\frac{1}{2}(\sqrt{3} \cdot \cosh 1 + i \sinh 1)$;
  (c) $\frac{1}{2}(\cosh 2 + i\sqrt{3} \sinh 2)$

15. (a) $\log 4 + \pi i$; (b) $\frac{1}{2}\log 2 + \dfrac{\pi i}{4}$; (c) $-\dfrac{\pi i}{2}$

17. (a) $\frac{1}{2}(1 - \cos 2x \cosh 2y + i \sin 2x \sinh 2y)$
  (b) $\dfrac{\sin 2x + i \sinh 2y}{\cos 2x + \cosh 2y}$

19. (a) $\log 3 + \frac{1}{2}\log 2 + \dfrac{\pi i}{4}$
  (b) $e^{\sin(x^2 - y^2)\cosh(2xy)}(\cos\theta + i \sin\theta)$, $\theta = \cos(x^2 - y^2)\sinh(2xy)$

21. $1 + \frac{1}{2}z + \sum_{n=2}^{\infty} \frac{(-1)^{n-1} 1 \cdot 3 \cdots (2n-3)}{2 \cdot 4 \cdots (2n)} \qquad z^n, |z| < 1$

Chapter 4

Section 1

1. $f_x = 4x - 3y + 4; f_y = -3x$
3. $f_x = 3x^2 + 6xy - 3y^2; f_y = 3y^2 + 3x^2 - 6xy$
5. $f_x = x/\sqrt{x^2 + y^2}; f_y = y/\sqrt{x^2 + y^2}$
7. $f_x = 2x/(x^2 + y^2); f_y = 2y/(x^2 + y^2)$
9. $f_x = -y/(x^2 + y^2); f_y = x/(x^2 + y^2)$
11. $f_x = ye^{x^2+y^2}(2x^2 + 1); f_y = xe^{x^2+y^2}(2y^2 + 1)$
13. No solution
15. $f_x = \frac{1}{2}e^{\sqrt{2}/2}(4 + \sqrt{2}); f_y = (\pi/2)e^{\sqrt{2}/2}$
17. $f_x = 32 \log 2; f_y = 32(1 + \log 2)$
19. $f_x = 2xy - 4xz + 3yz + 2z^2;$
    $f_y = x^2 + 3xz - 2yz;$
    $f_z = -2x^2 + 3xy - y^2 + 4xz$
21. $f_x = f(x, y, z) \cdot (yz + y \cot xy - 2z \tan 2xz)$
    $f_y = (xz + x \cot xy) f(x, y, z)$
    $f_z = (xy - 2x \tan 2xz) f(x, y, z)$
23. $\frac{\partial w}{\partial x} = \frac{y^2 - x^2}{x(x^2 + y^2)}; \frac{\partial w}{\partial y} = \frac{x^2 - y^2}{y(x^2 + y^2)}$
25. $\frac{\partial w}{\partial x} = -\left(\frac{y}{x^2}\right) e^{\sin(y/x)} \cos\left(\frac{y}{x}\right); \frac{\partial w}{\partial y} = \frac{1}{x} \cos\left(\frac{y}{x}\right) e^{\sin(y/x)}$
35. f continuous at $(0, 0)$; g not continuous at $(0, 0)$

Chapter 4

Section 2

1. $\frac{\partial w}{\partial x} = \frac{1 - 6x}{12w}; \frac{\partial w}{\partial y} = -\frac{(1 + 4y)}{12w}$
3. $\frac{\partial w}{\partial x} = -\frac{x - y + w}{x + w}; \frac{\partial w}{\partial y} = \frac{x - 3y}{x + w}$
5. $\frac{\partial w}{\partial r} = \frac{2r \cos rw - w(r^2 + s^2) \sin rw}{1 + r(r^2 + s^2) \sin rw}$
   $\frac{\partial w}{\partial s} = \frac{2s \cos rw}{1 + r(r^2 + s^2) \sin rw}$
7. $\frac{\partial w}{\partial x} = \frac{y \cos 2xw(w \sin xy + \cos xy) - 2w \sin xy \sin 2xw}{\sin xy(2x \sin 2xw - xy \cos 2xw)}$
   $\frac{\partial w}{\partial y} = \frac{\cos 2xw(w \sin xy + \cos xy)}{\sin xy(2 \sin 2xw - y \cos 2xw)}$
9. $\frac{\partial w}{\partial x} = -\frac{z(y + 2x + w)}{xz - yz - 3w^2}; \frac{\partial w}{\partial y} = \frac{z(w - x - z)}{xz - yz - 3w^2}$

$$\frac{\partial w}{\partial z} = \frac{yw - xy - x^2 - xw - 2yz}{xz - yz - 3w^2}$$

11. $\dfrac{\partial w}{\partial x} = \dfrac{ye^{x(y-w)} + w^2}{y^2e^{w(y-x)} - xw - 1}$; $\dfrac{\partial w}{\partial y} = \dfrac{xe^{y(x-w)} - yw - 1}{y^2 - (1 + xw)e^{w(x-y)}}$

15. $\dfrac{\partial x}{\partial \rho} = \sin\phi\cos\theta$; $\dfrac{\partial x}{\partial \phi} = \rho\cos\phi\cos\theta$; $\dfrac{\partial x}{\partial \theta} = -\rho\sin\phi\sin\theta$

$\dfrac{\partial y}{\partial \rho} = \sin\phi\sin\theta$; $\dfrac{\partial y}{\partial \phi} = \rho\cos\phi\sin\theta$; $\dfrac{\partial y}{\partial \theta} = \rho\sin\phi\cos\theta$

$\dfrac{\partial z}{\partial \rho} = \cos\phi$; $\dfrac{\partial z}{\partial \phi} = -\rho\sin\phi$; $\dfrac{\partial z}{\partial \theta} = 0$

Chapter 4

Section 3

1. $\dfrac{\partial z}{\partial s} = 10s$; $\dfrac{\partial z}{\partial t} = 10t$

3. $\dfrac{\partial z}{\partial s} = 4s(s^2 + t^2)$; $\dfrac{\partial z}{\partial t} = 4t(t^2 + s^2)$

5. $\dfrac{\partial z}{\partial s} = \dfrac{y(2y - x)}{(x^2 + y^2)^{3/2}}$; $\dfrac{\partial z}{\partial t} = \dfrac{-y(2x + y)}{(x^2 + y^2)^{3/2}}$

7. $\dfrac{\partial w}{\partial s} = 10x + 13y - 4z$; $\dfrac{\partial w}{\partial t} = -5x + y + 2z$

9. $\dfrac{\partial w}{\partial r} = 4(2r^3 + 3r^2 - 6r + 1)$

11. $\dfrac{\partial w}{\partial r} = 2r(3u^2 + 4u - 3)$; $\dfrac{\partial w}{\partial s} = -2s(3u^2 + 4u - 3)$;

$\dfrac{\partial w}{\partial t} = 2t(3u^2 + 4u - 3)$

13. $\dfrac{\partial z}{\partial r} = 0$; $\dfrac{\partial z}{\partial \theta} = -4$

15. $\frac{1}{8}\sqrt{2}\pi(4 + \sqrt{2}) = \dfrac{dw}{dt}$

17. $\dfrac{\partial z}{\partial r} = 0$; $\dfrac{\partial z}{\partial \theta} = \dfrac{1}{2}$

19. $\dfrac{\partial w}{\partial s} = 0$; $\dfrac{\partial w}{\partial t} = -16$

Chapter 4

Section 4

1. $\dfrac{290\pi}{\sqrt{13}}$ cm^2/sec

3. $\dfrac{dT}{dt} = \dfrac{250{,}000}{R}$

5. $\dfrac{40\pi\sqrt{3}-63}{252}$

7. $-\frac{1}{15}(1+2\sqrt{10\pi})$

9. a) $\dfrac{\partial z}{\partial x}=f'(y/x)(-y/x^2)$; $\dfrac{\partial z}{\partial y}=f'(y/x)(1/x)$

11. a) $\dfrac{\partial z}{\partial r}=\dfrac{\partial z}{\partial x}\cos\theta+\dfrac{\partial z}{\partial y}\sin\theta$; $\dfrac{\partial z}{\partial \theta}=\dfrac{\partial z}{\partial x}(-r\sin\theta)+\dfrac{\partial z}{\partial y}(r\cos\theta)$

17. $1+2\sqrt{3}-\dfrac{6+3\sqrt{3}}{2\sqrt{12-3\sqrt{3}}}$

19. $1000e^8$

Chapter 4

Section 5

1. $d_\theta f=6\cos\theta+8\sin\theta$

3. $d_\theta f=\frac{1}{25}(-3\cos\theta+4\sin\theta)$

5. $d_\theta f=\frac{1}{2}(\cos\theta-\sqrt{3}\sin\theta)$

7. $d_\theta f=2\cos\theta+\sin\theta$; max if $\cos\theta=\dfrac{2}{\sqrt{5}}$, $\sin\theta=\dfrac{1}{\sqrt{5}}$

9. $d_\theta f=\frac{1}{2}\cos\theta+\dfrac{\sqrt{3}}{2}\sin\theta$; max if $\sin\theta=\dfrac{\sqrt{3}}{2}$, $\cos\theta=\frac{1}{2}$

11. $D_{\mathbf{a}}f=7\lambda+4\mu+2\nu$ where $\mathbf{a}=\lambda\mathbf{i}+\mu\mathbf{j}+\nu\mathbf{k}$

13. $D_{\mathbf{a}}f=-\lambda$ where $\mathbf{a}=\lambda\mathbf{i}+\mu\mathbf{j}+\nu\mathbf{k}$

15. $\frac{11}{3}$

17. $2/\sqrt{6}$

19. $d_\theta T(4,3)=400\cos\theta-300\sin\theta$; $\tan\theta=\frac{4}{3}$; slope of curve $=\frac{4}{3}$

21. $\nabla f=-4\mathbf{i}-\frac{4}{5}\mathbf{j}$

23. $\nabla f=(2e^2+2e^{-1}-e^{-2})\mathbf{i}+(e^{-2}+e^{-1}-3e^2)\mathbf{j}-(4e^2+e^{-2})\mathbf{k}$

25. $D_{\mathbf{a}}f=\dfrac{1}{\sqrt{14}}(-3e^2\cos 1-e^2\sin 1+2e)$

    $D_{\bar{\mathbf{a}}}f=e\sqrt{e^2+1}$ is max.

27. $D_{\mathbf{a}}f=\dfrac{1}{\sqrt{14}}(3-\cos 1-6\sin 1+2\cos 2)$

    $D_{\mathbf{a}}f=\sqrt{\cos^2 1+\cos^2 2+(1-2\sin 1)^2}$ is maximum

29. $\left(\dfrac{5}{2},\dfrac{-7}{2},\dfrac{1}{2}\right)$

Chapter 4

Section 6

1. $4x-4y-z-6=0$; $\dfrac{x-2}{4}=\dfrac{y+1}{-4}=\dfrac{z-6}{-1}$

3. $x-2y+z-2=0$; $\dfrac{x-2}{-1}=\dfrac{y+1}{2}=\dfrac{z+2}{-1}$

5. $ex-z=0$; $\dfrac{x-1}{e}=\dfrac{y-\pi/2}{0}=\dfrac{z-e}{-1}$

7. $3x-4y+25z=25(\log 5-1)$; $\dfrac{x+3}{3}=\dfrac{y-4}{-4}=\dfrac{z-\log 5}{25}$

9. $2x + 2y + 3z - 3 = 0; \dfrac{x-2}{2} = \dfrac{y-1}{2} = \dfrac{z+1}{3}$

11. $4x - 3z - 25 = 0; \dfrac{x-4}{4} = \dfrac{y+2}{0} = \dfrac{z+3}{-3}$

13. $3x + 6y + 2z - 36 = 0; \dfrac{x-4}{3} = \dfrac{y-1}{6} = \dfrac{z-9}{2}$

19. $\dfrac{x-4}{4} = \dfrac{y+2}{3} = \dfrac{z-20}{20}$

21. $\dfrac{x-4}{3} = \dfrac{y+3}{4} = \dfrac{z-16}{24}$

23. $(-1, \frac{3}{2}, -5)$

25. Value of constant is $a^{\alpha/(1-\alpha)}$

27. $(\sqrt{2}, 1 + \sqrt{2}, 1); (-\sqrt{2}, 1 - \sqrt{2}, 1)$

Chapter 4

Section 7

1. $df = -0.17; \Delta f = -0.1689$

3. $df = \dfrac{\pi}{2}(\pi - 5); \Delta f = -\sqrt{3} + \sin\dfrac{13\pi^2}{2}$

5. $df = -0.45; \Delta f = -0.456035$

7. $df = 0.05; \Delta f = 0.0499$

9. $df = -0.08; \Delta f = -0.080384$

11. $300x^{-1}h + 700y^{-1}k + 400z^{-1}l$

13. $V \approx 4309.92$

15. 1.4%

17. 3.30π

19. $\left(1 + \dfrac{5\pi\sqrt{3}}{18}\right)\%$

21. 0.111427

23. $[12t^5 + 15t^4 + 48t^3 - 15t^2 + 12t - 35]\,dt$

25. $(12r^3 - 4rs^2 + 24rs^{-2})h + (-4r^2s + 12s^3 - 24r^2s^{-3})k$

27. $G_1 = h, G_2 = 2k + 6h$

31. $T \approx \dfrac{6.28}{7} + .00185 = .89899$

Chapter 4

Section 8

1. $\dfrac{dy}{dx} = \dfrac{2x + 3y + 2}{-3x + 8y + 6}$

3. $\dfrac{dy}{dx} = -\dfrac{2x + (1 + x^2 + y^2)ye^{xy}}{2y + (1 + x^2 + y^2)xe^{xy}}$

5. $\dfrac{dy}{dx} = -\dfrac{y}{x}$

7. $\dfrac{dy}{dx} = \dfrac{y - 3x(x^2 + y^2)^{3/2}}{x + 3y(x^2 + y^2)^{3/2}}$

9. $\dfrac{\partial w}{\partial x} = \dfrac{3x^2 + 6xw + 2}{y^2 - 3x^2 - 4yw + 3}$

11. $\dfrac{\partial w}{\partial x} = -\dfrac{yw\cos(xyw) + 2x}{xy\cos(xyw) + 2w}$

13. $\dfrac{\partial w}{\partial y} = \dfrac{2y + 3x - 3}{2w - 4x + 3z}$

15. $\dfrac{dy}{dx} = -\dfrac{(2x + 2)}{y}; \dfrac{dz}{dx} = -(2x + 4)$

17. $\dfrac{dz}{dx} = -\dfrac{2xz + 3y}{4z^2 - 3y^2}; \dfrac{dy}{dx} = \dfrac{4z + 2xy}{4z^2 - 3y^2}$

19. $\dfrac{\partial u}{\partial x} = \dfrac{u}{2(u^2 + v^2)}; \dfrac{\partial u}{\partial y} = \dfrac{v}{2(u^2 + v^2)}; \dfrac{\partial v}{\partial x} = \dfrac{-v}{2(u^2 + v^2)}; \dfrac{\partial v}{\partial y} = \dfrac{u}{2(u^2 + v^2)}$

21. $\dfrac{\partial u}{\partial x} = \dfrac{2xv}{v + u}; \dfrac{\partial u}{\partial y} = \dfrac{1}{2(v + u)}; \dfrac{\partial v}{\partial x} = \dfrac{2xu}{(u + v)}; \dfrac{\partial v}{\partial y} = \dfrac{-1}{2(u + v)}$

23. $\dfrac{\partial u}{\partial x} = 0; \dfrac{\partial u}{\partial y} = \dfrac{y}{u}; \dfrac{\partial v}{\partial x} = -\dfrac{x}{v}; \dfrac{\partial v}{\partial y} = 0$

25. $\dfrac{\partial u}{\partial x} = \dfrac{g_v}{f_u g_v - f_v g_u}; \dfrac{\partial u}{\partial y} = \dfrac{-f_v}{f_u g_v - f_v g_u}; \dfrac{\partial v}{\partial x} = \dfrac{-g_u}{f_u g_v - f_v g_u}; \dfrac{\partial v}{\partial y} = \dfrac{f_u}{f_u g_v - f_v g_u}$

27. $\dfrac{\partial u_1}{\partial x_i} = \dfrac{F_{u_2} G_{x_i} - G_{u_2} F_{x_i}}{D}; \dfrac{\partial u_2}{\partial x_i} = \dfrac{G_{u_1} F_{x_i} - F_{u_1} G_{x_i}}{D}; D = F_{u_1} G_{u_2} - F_{u_2} G_{u_1}.$

Chapter 4

Section 9

1. $f_{xy} = 14x - 4y = f_{yx}$
3. $f_{xy} = -18x^2 y + 6xy^2 - 3y^2 = f_{yx}$
5. $e^{xy}(\sin x + xy \sin x + x \cos x + \cos y + xy \cos y - y \sin y)$
7. $\dfrac{2xy(x + y)}{[(x + y)^2 + x^2 y^2]^2}$
9. $\dfrac{-xy}{(x^2 + y^2 - z^2)^{3/2}}$
11. $u_{xy} = u_{yx} = -2xy/(x^2 + y^2 + z^2)^2; u_{xz} = u_{zx} = -2xz/(x^2 + y^2 + z^2)^2$
13. $u_{xy} = u_{yx} = -3z; u_{xz} = u_{zx} = -3y$
15. $u_{xy} = u_{yx} = e^{xy}[xy(x^2 + z^2) + z^2](x^2 + z^2)^{-3/2}$
    $u_{xz} = u_{zx} = ze^{xy}[3x - y(x^2 + z^2)](x^2 + z^2)^{-5/2}$
19. $\dfrac{\partial^2 z}{\partial r^2} = 2 \cos 2s$
21. $12r^2 - 24rs - 12s^2$
23. $\dfrac{\partial^2 u}{\partial y^2} = F_{yy} + 2F_{yz}\dfrac{\partial z}{\partial y} + F_{zz}\left(\dfrac{\partial z}{\partial y}\right)^2 + F_z \dfrac{\partial^2 z}{\partial y^2}$
27. $\dfrac{\partial^2 u}{\partial r\, \partial s} = F_x \dfrac{\partial^2 x}{\partial r\, \partial s} + F_y \dfrac{\partial^2 y}{\partial r\, \partial s} + F_{xx}\dfrac{\partial x}{\partial r}\dfrac{\partial x}{\partial s} + F_{xy}\left(\dfrac{\partial x}{\partial r}\dfrac{\partial y}{\partial s} + \dfrac{\partial x}{\partial s}\dfrac{\partial y}{\partial r}\right) + F_{yy}\dfrac{\partial y}{\partial r}\dfrac{\partial y}{\partial s}$
35. $\dfrac{\partial^2 u}{\partial s^2} = F_{xx}\left(\dfrac{\partial x}{\partial s}\right)^2 + 2F_{xy}\dfrac{\partial x}{\partial s}\dfrac{\partial y}{\partial s} + F_{yy}\left(\dfrac{\partial y}{\partial s}\right)^2 + F_x \dfrac{\partial^2 x}{\partial s^2} + F_y \dfrac{\partial^2 y}{\partial s^2}$

Chapter 4

Section 10

1. $10 + 13(x - 2) + 4(y - 1) + 6(x - 2)^2 + 2(x - 2)(y - 1)$
   $+ 2(y - 1)^2 + (x - 2)^3 + (x - 2)(y - 1)^2$
3. $x + y - \frac{1}{6}(x + y)^3$
5. $1 + (x + y) + \frac{1}{2}(x + y)^2 + \frac{1}{6}(x + y)^3 + \cdots$
7. $1 - \dfrac{x^2}{2!} - \dfrac{y^2}{2!} + \dfrac{x^2}{4!} + \dfrac{x^2 y^2}{2!2!} + \dfrac{y^4}{4!} +$

9. $3+4(x-1)+2(y-1)+z+(x-1)^2+z^2+2(x-1)(y-1)+(y-1)z$
11. $\theta''(0) = 2\lambda^2 + 8\lambda\mu + 2\mu^2$. No.
13. $\sum_{i=0}^{k}\sum_{j=0}^{i}\frac{i!k!}{i!j!(k-i)!(i-j)!}A^jB^{i-j}C^{k-i}$
17. $$f(x_1,x_2,\ldots,x_k)=f(a_1,a_2,\ldots,a_k)+\sum_{1\le r_1+\cdots+r_k\le p}\frac{\partial^{r_1+\cdots+r_k}f(a_1,\ldots,a_k)}{\partial x_1^{r_1}\partial x_2^{r_2}\cdots\partial x_k^{r_k}} \times\frac{(x_1-a_1)^{r_1}(x_2-a_2)^{r_2}\cdots(x_k-a_k)^{r_k}}{r_1!r_2!\cdots r_k!}+R_p$$
where $$R_p=\sum_{r_1+\cdots+r_k=p+1}\frac{\partial^{r_1+\cdots+r_k}f}{\partial x_1^{r_1}\partial x_2^{r_2}\cdots\partial x_k^{r_k}}\times\frac{(x_1-a_1)^{r_1}(x_2-a_2)^{r_2}\cdots(x_k-a_k)^{r_k}}{r_1!r_2!\cdots r_k!}$$

Chapter 4

Section 11

1. Rel. min. at $(2, -1)$
3. Rel. min. at $(1, -2)$
5. Rel. Min. $(\frac{27}{2}, 5)$; saddle point $(\frac{3}{2}, 1)$
7. No rel. max. or rel. min; $(1, \frac{2}{3})$, $(-1, -\frac{4}{3})$, saddle points
9. Rel. max. at $\left(\frac{\pi}{3}\pm 2n\pi, \frac{\pi}{3}\pm 2m\pi\right)$; rel. min. at $\left(-\frac{\pi}{3}\pm 2n\pi, -\frac{\pi}{3}\pm 2m\pi\right)$; test fails for $(\pi \pm 2n\pi, \pi \pm 2m\pi)$.
11. Test fails but $(0,0)$ is rel. max.
13. Test fails; critical points on lines $(x, n\pi)$
15. Critical point, $(-\frac{3}{2}, -\frac{1}{2}, -\frac{1}{4})$
17. $x = -\frac{9}{31}, y = \frac{55}{62}, z = \frac{18}{31}, t = -\frac{41}{31}$
19. $17/\sqrt{10}$
21. $\frac{1}{2}\sqrt{10}$
23. $\frac{6}{\sqrt{3}}, \frac{4}{\sqrt{3}}, \frac{8}{\sqrt{3}}$
25. $V = \frac{1}{4}\left(\sqrt{\frac{2D}{3}}\right)^3$
27. $x = \frac{P(2-\sqrt{3})}{2}, y = \frac{P(\sqrt{3}-1)}{2\sqrt{3}}$
29. (a) $f(x,y) = -(x-a)^4 - (y-b)^4$ (b) $f(x,y) = (x-a)^4 + (y-b)^4$ (c) $f(x,y) = (x-a)^3 + (y-b)^3$

Chapter 4

Section 12

1. $\frac{8}{7}$
3. $\frac{d^2}{a^2+b^2+c^2}$
5. Min. $\frac{134}{75}$ at $(\frac{16}{15}, \frac{1}{3}, -\frac{11}{15})$

7. $x = \pm\frac{1}{2}(1-\sqrt{5})\sqrt{50-10\sqrt{5}},\ y = \mp\sqrt{50-10\sqrt{5}}$
9. 2, 2, 1
11. $h = 2\sqrt{5};\ H = \dfrac{V}{25\pi} - \dfrac{4\sqrt{5}}{3}$
13. $\dfrac{17}{23}$ occurs when $x = \dfrac{10}{23},\ y = \dfrac{1}{23},\ z = \dfrac{-1}{23},\ t = \dfrac{17}{23}$
15. Closest: $\left(\pm\dfrac{1}{\sqrt{2}}, \pm\dfrac{1}{\sqrt{2}}\right)$; farthest: $(\pm\sqrt{2}, \mp\sqrt{2})$
17. $\dfrac{50}{3}, \dfrac{50}{3}, \dfrac{100}{3}$
19. If C is cost (in dollars) per square cm of material, then length = width = $\frac{1}{3}\sqrt{D/C}$; height $= \frac{1}{2}\sqrt{D/C}$.
21. $x = aA/(a+b+c);\ y = bA/(a+b+c);\ z = cA/(a+b+c)$
23. Max. value $= 1/k^k$
27. $x = \sqrt{750/29},\ y = 3x,\ z = 5x/3$

Chapter 4

Section 13

1. Exact: $(x^4/4) + x^3y + (y^4/4) + C$
3. Exact: $2xy - \log x + \log y + C$
5. Not exact
7. Exact: $e^{x^2}\sin y + C$
9. Exact: $\arctan(y/x) + C$
11. Not exact
13. Not exact
15. Not exact
17. Exact: $e^x \sin y \cos z + C$
19. (b) $f(x, y, z, t) = x^3 + z^3 - t^3 + y^2 + 2xz - yt + 3x - 2y + 4t + C$
23. $f(x, y) = xy + \frac{1}{2}x^2 + \log x + C$
25. $f(x, y) = (-2/xy) - \log x + \log y + C$
27. $I = (x^2+y^2)^{-2};\ f(x, y) = -\frac{1}{2}(x^2+y^2)^{-1} + \arctan(x/y) + C$

Chapter 4

Section 15

1. $\frac{7}{2}$
3. $\frac{231}{64}$
5. $\pi + 8$
7. $4 + 2\log(\frac{5}{4})$
9. $\arcsin\frac{4}{5}$
11. $\dfrac{508\sqrt{3}}{11}$
13. $\dfrac{64}{105}$
15. $$\frac{8}{9^4}\left\{(37)^{3/2}\left[\frac{(37)^3}{9} - \frac{3}{7}(37)^2 + \frac{3}{5}(37) - \frac{1}{3}\right] - (10)^{3/2}\left[\frac{(10)^3}{9} - \frac{3}{7}(10)^2 + \frac{3}{5}(10) - \frac{1}{3}\right]\right\}$$
17. $\frac{1}{2}(\sin 7 + \sin 1) + 2(\cos 3 - \cos 7)$
19. 24
21. $\frac{1}{2}\log 17$
23. 0

Chapter 4

Section 16

1. $\frac{101}{3}$
3. $-e + \cos 1$
5. $-\frac{7}{5}$
7. $\sin 3 - \cos 2 + e^{-6}\cos 6$
9. -9
11. $e^{-6}\cos 1 + \sin 2 - \cos 3$
13. $\frac{1}{2}\log\left(\frac{a^2 + b^2}{c^2 + d^2}\right)$

Chapter 5

Section 1

1. 0.9690
3. 3.5355
5. 3.8140
7. 3.2148
9. 0.182

Chapter 5

Section 2

1. Smallest $= 0$; largest $= 80$
3. Smallest $= 48$; largest $= 1680$
5. Smallest $= -27\pi\sqrt{2}$; largest $= 27\pi\sqrt{2}$
7. Smallest $= 9$; largest $= 9\sqrt{10}$

Chapter 5

Section 3

1. $\frac{423}{4}$
3. $-1498\frac{3}{14}$
5. $-\frac{18}{7}\sqrt{3} + \frac{16}{21}\sqrt{2} + \frac{559}{8}$
7. 0
9. $\frac{1}{4}(2\cos 4 - \cos 8 - 1)$
11. $40\frac{88}{105}$
13. $\frac{\pi}{4}(\sqrt{3} - 1)$
15. $(2\pi - 3\sqrt{3})/48$
17. $2e^{1/\sqrt{2}} - e\sqrt{2}$
19. 0
21. $(2\sqrt{2} - 1)/3$
23. $32\sqrt{2}/15$
25. $64/3$
27. $8/15$
29. 8π
31. $16/5$
37. $4\pi abc/3$

Chapter 5

Section 4

1. $\frac{5}{12}$
3. $\frac{2}{3} + 4\log\left(\frac{3}{2}\right)$
5. $\frac{9}{8}(3\pi + 2)$
7. $\frac{9}{20}$
9. $\frac{2}{5}$
11. $\frac{207}{10}$
13. $2a^3$

Chapter 5

Section 5

1. $4\pi/3$
3. 2π
5. $\dfrac{64\pi}{3} - \dfrac{256}{9}$
7. $\sqrt{2} - 1$
9. $\frac{64}{3}$
11. 4π
13. $32\pi(8 - 3\sqrt{3})/3$
15. $64/9$
17. $9\pi/2$
19. $4\pi\sqrt{3}$
21. $\pi/4$
23. $a^3(3\pi - 4)/9$
25. $a^3[\sqrt{2} + \log(1 + \sqrt{2})]/3$
27. $2\pi c^3/3$; $2c^3\,\phi/3$

Chapter 5

Section 6

1. $I = \rho a^4/3$;
3. $I = \frac{3}{35}\rho$;
5. $I = 4ka^5/15$;
7. $I = \rho(\pi^2 - 4)$;
9. $\frac{423}{28}\rho$
11. $3k/56$
13. $ka^5\pi/5$
15. $\dfrac{\rho\pi a^4}{16}$
17. $I = \rho\left(\dfrac{\pi}{8} - \dfrac{1}{5}\right)$
19. $\bar{x} = \frac{235}{112}$, $\bar{y} = \frac{25}{32}$
21. $\bar{x} = \frac{5}{9}$, $\bar{y} = \frac{4}{7}$
23. $\bar{x} = \bar{y} = 5a/8$
25. $\bar{x} = 5/3$, $\bar{y} = 0$
27. $\bar{x} = -2/(8\sqrt{3} - 1)$, $\bar{y} = 0$
33. $\pi\delta(r_2^4 - r_1^4)/2$; $R^4 = r_2^4 - r_1^4$

Chapter 5

Section 7

1. $\frac{2}{15}(1 + 9\sqrt{3} - 8\sqrt{2})$
3. $8a^2$
5. $2\pi\sqrt{2}$
7. $9\sqrt{2}$
9. $\pi\sqrt{2}/2$
11. $\frac{1}{9}(160 - 24\pi)$
13. $2a^2(\pi - 2)/3$
15. 16
17. $\dfrac{a^2}{3}(\pi + 6\sqrt{3} - 12)$
21. $\pi(17\sqrt{17} - 1)/6$
23. $2\sqrt{2}\pi$

Chapter 5

Section 8

1. $\frac{1}{8}$
3. $\frac{109}{1008}$
5. $(16 - 3\pi)/3$
7. $\frac{3}{2}$
9. $\frac{1}{3}$
11. $abc^2/24$
13. $a^4/840$
15. $243\pi/2$
17. $27a^5(2\pi + 3\sqrt{3})/2$
19. $\displaystyle\int_0^1 \int_{-\sqrt{x}}^{\sqrt{x}} \int_{-\sqrt{1-x}}^{\sqrt{1-x}} 2y^2\sqrt{x}\,dy\,dz\,dx$; $\displaystyle\int_0^1 \int_{-\sqrt{1-x}}^{\sqrt{1-x}} \int_{-\sqrt{x}}^{\sqrt{x}} 2y^2\sqrt{x}\,dz\,dy\,dx$

21. $\int_0^4 \int_{-\sqrt{4-x}}^{\sqrt{4-x}} \int_0^{y+2} (y^2+z^2)\,dz\,dy\,dx$; $\int_{-2}^{2} \int_0^{y+2} \int_0^{4-y^2} (y^2+z^2)\,dx\,dz\,dy$

23. $\int_{-\sqrt{12}}^{\sqrt{12}} \int_{-\sqrt{12-x^2}}^{\sqrt{12-x^2}} \int_2^{\sqrt{16-x^2-y^2}} f(x,y,z)\,dz\,dy\,dx$; $\int_{-\sqrt{12}}^{\sqrt{12}} \int_{-\sqrt{12-y^2}}^{\sqrt{12-y^2}} \int_2^{\sqrt{16-x^2-y^2}} f(x,y,z)\,dz\,dx\,dy$

$\int_2^4 \int_{-\sqrt{16-z^2}}^{\sqrt{16-z^2}} \int_{-\sqrt{16-y^2-z^2}}^{\sqrt{16-y^2-z^2}} f(x,y,z)\,dx\,dy\,dz$; $\int_{-\sqrt{12}}^{\sqrt{12}} \int_2^{\sqrt{16-y^2}} \int_{-\sqrt{16-y^2-z^2}}^{\sqrt{16-y^2-z^2}} f(x,y,z)\,dx\,dz\,dy$

$\int_2^4 \int_{-\sqrt{16-z^2}}^{\sqrt{16-z^2}} \int_{-\sqrt{16-x^2-z^2}}^{\sqrt{16-x^2-z^2}} f(x,y,z)\,dy\,dx\,dz$; $\int_{-\sqrt{12}}^{\sqrt{12}} \int_2^{\sqrt{16-x^2}} \int_{-\sqrt{16-x^2-z^2}}^{\sqrt{16-x^2-z^2}} f(x,y,z)\,dy\,dz\,dx$

25. $1/20$

Chapter 5

Section 9

1. $\pi\delta a^3(2-\sqrt{2})/3$
3. $\pi k(b^4-a^4)$
5. $12\pi k a^5\sqrt{3}/5$
7. $128ka^5(15\pi-26)/225$
9. $7\pi k a^4/6$
11. $7\pi\delta a^3/6$
13. $2\delta a^3(3\pi+20-16\sqrt{2})/9$
15. $\pi k a^4/4$

Chapter 5

Section 10

1. $2\delta a^5/3$
3. $ka^6/90$
5. $32\delta\left(\frac{4}{9}-\frac{1}{5}+\frac{1}{21}\right)=\frac{2944\delta}{315}$
7. $\frac{5\pi\delta}{16}$
9. $\frac{4\pi k}{9}(b^6-a^6)$
11. $\frac{4ka^6}{9}$
13. $4\pi\delta abc(a^2+b^2)/15$
15. $\frac{4\pi\delta}{15}[(c^2-a^2)^{3/2}(2c^2+3a^2)-(b^2-a^2)^{3/2}(2b^2+3a^2)]$
17. $\bar{x}=\bar{z}=2a/5;\ \bar{y}=a/5$
19. $\bar{y}=\bar{z}=0;\ \bar{x}=8/7$
21. $\bar{x}=\bar{y}=0;\ \bar{z}=1/3$
23. $\bar{y}=\bar{z}=0;\ \bar{x}=2a$
25. $\bar{x}=\bar{y}=0;\ \bar{z}=3a(2+\sqrt{2})/16$
27. $\bar{x}=\bar{y}=0;\ \bar{z}=9a/7$
29. $\bar{x}=\bar{y}=0;\ \bar{z}=\frac{4(1591-720\sqrt{3})}{7(391-192\sqrt{3})}$
31. $\bar{y}=\bar{z}=0;\ \bar{x}=a\sqrt{2}/2$

Chapter 6

Section 1

1. $f(x)=\sum_{k=1}^{\infty}\frac{\sin(2k-1)x}{2k-1}$
3. $f(x)=\frac{\pi^2}{3}-4\sum_{n=1}^{\infty}\frac{(-1)^{n-1}\cos nx}{n^2}$

5. $f(x) = \frac{2}{\pi}\sum_{k=1}^{\infty}\frac{\sin(2k-1)x}{2k-1} - \frac{2}{\pi}\sum_{k=1}^{\infty}\frac{\sin(4k-2)x}{2k-1}$

7. $f(x) = \frac{1}{2} - \frac{2}{\pi}\sum_{k=1}^{\infty}\frac{(-1)^{k-1}\cos(2k-1)x}{2k-1}$

9. $f(x) = \frac{2}{\pi} + \frac{4}{\pi}\sum_{k=1}^{\infty}\frac{(-1)^{k-1}\cos 2kx}{4k^2-1}$

11. $f(x) = \frac{2\sinh\pi}{\pi}\left[\frac{1}{2} + \sum_{n=1}^{\infty}\frac{(-1)^n(\cos nx - n\sin nx)}{n^2+1}\right]$

13. $f(x) = \frac{1}{2} - \frac{1}{2}\cos 2x$

15. $f(x) = -\frac{\pi}{4} + 3\sum_{k=1}^{\infty}\frac{\sin(2k-1)x}{2k-1} - \frac{2}{\pi}\sum_{k=1}^{\infty}\frac{\cos(2k-1)x}{(2k-1)^2} - \frac{1}{2}\sum_{k=1}^{\infty}\frac{\sin 2kx}{k}$

19. $f(x) = 2\sum_{n=1}^{\infty}\frac{(-1)^{n-1}\sin nx}{n} + \frac{\pi^2}{3} - 4\sum_{n=1}^{\infty}\frac{(-1)^{n-1}\cos nx}{n^2}$

Chapter 6

Section 2

1. $f(x) = \frac{1}{2} + \frac{2}{\pi}\left(\cos x - \frac{\cos 3x}{3} + \frac{\cos 5x}{5} - \cdots\right)$

3. $f(x) = \frac{2}{\pi} - \frac{4}{\pi}\sum_{k=1}^{\infty}\frac{\cos 2kx}{4k^2-1}$

5. $f(x) = \frac{\pi}{2} - \frac{4}{\pi}\sum_{k=1}^{\infty}\frac{\cos(2k-1)x}{(2k-1)^2}$

7. $f(x) = \frac{\pi^3}{4} + \sum_{k=1}^{\infty}\frac{3\pi}{2k^2}\cos 2kx + \frac{6}{\pi}\sum_{k=1}^{\infty}\left[\frac{4}{(2k-1)^4} - \frac{\pi^2}{(2k-1)^2}\right]\cos(2k-1)x$

9. $f(x) = \frac{2}{\pi}\sum_{k=1}^{\infty}\frac{\sin(2k-1)x}{2k-1} + \frac{2}{\pi}\sum_{k=1}^{\infty}\frac{\sin(4k-2)x}{2k-1}$

11. $f(x) = -\sum_{k=1}^{\infty}\frac{8k}{\pi(4k^2-1)}\sin 2kx$

13. $f(x) = 2\sum_{n=1}^{\infty}\frac{(-1)^{n-1}\sin nx}{n}$

15. $f(x) = 2\sum_{n=1}^{\infty}(-1)^{n-1}\left(\frac{\pi^2}{n} - \frac{6}{n^3}\right)\sin nx$

Chapter 6

Section 3

1. $f(x) = \frac{4}{\pi}\sum_{k=1}^{\infty}\frac{\sin(2k-1)x}{2k-1}$

3. $f(x) = \frac{2}{\pi} - \frac{4}{\pi}\sum_{n=1}^{\infty}\frac{\cos nx}{4n^2-1}$

5. $f(x) = \frac{1}{2} - \frac{2}{\pi}\sum_{n=1}^{\infty}\frac{(-1)^n}{n}\sin\frac{n\pi x}{2} - \frac{4}{\pi^2}\sum_{k=1}^{\infty}\frac{1}{(2k-1)^2}\cos\frac{(2k-1)\pi x}{2}$

7. $f(x) = \frac{1}{2} + \frac{4}{\pi^2}\sum_{k=1}^{\infty}\frac{\cos(2k-1)x}{(2k-1)^2}$

9. $f(x) = -\frac{4}{\pi}\sum_{k=1}^{\infty}\frac{\sin k\pi x}{k}$

$$- \frac{8}{\pi^3}\sum_{k=1}^{\infty}\left[-\frac{\pi^2}{2k-1} + \frac{4}{(2k-1)^3}\right]\sin\frac{(2k-1)\pi x}{2}$$

11. $f(x) = \frac{2}{\pi}\sum_{k=1}^{\infty}\frac{\sin 2k\pi x}{k}$

13. $f(x) = \frac{1}{4} + \frac{1}{\pi}\sum_{k=1}^{\infty}(-1)^{k-1}\frac{\cos(2k-1)x}{2k-1}$

$$+ \frac{1}{\pi}\sum_{k=1}^{\infty}\frac{5\sin(2k-1)x + \sin(4k-2)x}{2k-1}$$

Chapter 6

Section 5

1. $x^2 = \frac{\pi^2}{3} + 2\sum_{\substack{n=-\infty \\ n\neq 0}}^{\infty}\frac{(-1)^n}{n^2}e^{inx}$

3. $f(x) = \frac{2}{\pi i}\sum_{k=1}^{\infty}\frac{1}{2k-1}[e^{(2k-1)ix} - e^{-(2k-1)ix}]$

5. $e^{ax} = \sum_{n=-\infty}^{\infty}\frac{(-1)^n\sinh\pi a}{\pi(a-in)}e^{inx}$

7. $\sinh ax = \frac{\sinh\pi a}{\pi}\sum_{n=-\infty}^{\infty}\frac{(-1)^n in}{a^2+n^2}e^{inx}$

9. $f(x) = 1 - \sum_{\substack{n=-\infty \\ n\neq 0}}^{\infty}(-1)^n\frac{e^{inx}}{n}$

Chapter 7

Section 1

In Problems 1 through 10, $F(x_0, y_0) = 0$.

1. $F_y(x_0, y_0) = 1 \neq 0, f'(x_0) = -1$
3. $F_y(x_0, y_0) = 1, f'(x_0) = 0$
5. $F_y(x_0, y_0) = 2/3\sqrt{3}, f'(x_0) = -\sqrt{3}$
7. $F_y(x_0, y_0) = 2, f'(x_0) = -1$
9. $F_y(x_0, y_0) = 82, f'(x_0) = -\frac{39}{41}$

In Problems 11 through 16, $F(x_0, y_0, z_0) = 0$.

11. $F_z(x_0, y_0, z_0) = 9, f_x(x_0, y_0) = f_y(x_0, y_0) = \frac{1}{3}$

13. $F_z(x_0, y_0, z_0) = 1, f_x = 2, f_y = 0$
15. $F_z = 1, f_x = f_y = -1$
17. (a) No (b) No
19. $F_y = 0$ at $(2e^{-2}, 2)$, $F_x \neq 0$ there
21. $F_y = 0$ at $(0, 0)$, $F_x(0, 0) = 0$
23. $F_y = 0$ at $(0, 0)$ and $(\pm 2\sqrt{2}, 0)$
    $F_x(0, 0) = 0$, $F_x(\pm 2\sqrt{2}, 0) \neq 0$
25. Cannot solve for z in a full box about $(-1, 2)$. Can solve for y in a box about $x = -1$, $z = -4$.

Chapter 7

Section 2

In Problems 1 through 6, $F(x_0, y_0, z_0) = G(x_0, y_0, z_0) = 0$.

1. $F_y G_z - F_z G_y = 3 \neq 0$ at P_0

   Line: $\dfrac{x-1}{1} = \dfrac{y}{-1} = \dfrac{z}{1}$
3. $F_y G_z - F_z G_y = 8 \neq 0$ at P_0

   Tan line: $\dfrac{x-2}{2} = \dfrac{y-1}{1} = \dfrac{z+2}{-3}$
5. $F_y G_z - F_z G_y = 12 \neq 0$ at P_0

   Tan line: $\dfrac{x-2}{2} = \dfrac{y+1}{1} = \dfrac{z-1}{-3}$

In Problems 7 through 10, $F = G = 0$ at P_0.

7. $D_0 = F_u G_v - F_v G_u = 3 \neq 0$

   $f_x = -\frac{5}{3}$, $g_x = \frac{1}{3}$, $f_y = \frac{4}{3}$, $g_y = -\frac{5}{3}$
9. $D_0 = 6$ at P_0

   $f_x = \frac{2}{3}$, $f_y = \frac{4}{3}$, $g_x = -\frac{5}{3}$, $g_y = \frac{2}{3}$
11. $F_y G_z - F_z G_y \equiv D = -4y - 2z$

    $D = F = G = 0 \Leftrightarrow P = (2, -3, 6)$ or $(-10, -15, 30)$
13. $$\frac{\partial u_1}{\partial x_1} = \frac{1}{D}\begin{vmatrix} -3x_1^2 + 12x_2 - 3 & 2 - 12u_2 \\ -4x_1 - 3x_1^2 u_3 & 8u_1 - 4u_2 + 2u_3 \end{vmatrix}$$

    $$\frac{\partial u_2}{\partial x_1} = \frac{1}{D}\begin{vmatrix} 32u_1 & -3x_1^2 + 12x_2 - 3 \\ -12u_1 + 8u_2 + 8u_1u_3 & -4x_1 - 3x_1^2 u_3 \end{vmatrix}$$

    $$\frac{\partial u_3}{\partial x_1} = \frac{1}{6u_3}\left[8u_1\frac{\partial u_1}{\partial x_1} + 2\frac{\partial u_2}{\partial x_1} + 3x_1^2\right]$$

    $$D = \begin{vmatrix} 32u_1 & 2 - 12u_2 \\ -12u_1 + 8u_2 + 8u_1u_3 & 8u_1 - 4u_2 + 2u_3 \end{vmatrix}$$
17. The result will hold if, also,

    $$J\left(\frac{F, G}{x, y}\right) \neq 0,$$

    so that the equations define one-to-one transformations from (u, v) to (x, y).

Chapter 7

Section 3

1. $J\left(\frac{u, v}{x, y}\right) = 1;\ T^{-1}: \begin{array}{l} x = u \\ y = v - u^2 \end{array}$

3. $J\left(\frac{u, v}{x, y}\right) = 7;\ T^{-1}: \begin{array}{l} x = (2u + 3v)/7 \\ y = (-u + 2v)/7 \end{array}$

5. $J\left(\frac{u, v}{x, y}\right) = (1 + x + y)^{-3};\ T^{-1}: \begin{array}{l} x = u/(1 - u - v) \\ y = v/(1 - u - v) \end{array},\ u + v < 1$

7. $J\left(\frac{u, v}{x, y}\right) = \frac{\pi x}{2};\ T^{-1}:\ x = \sqrt{u^2 + v^2},\ y = \frac{2}{\pi}\arctan\frac{v}{u}$

9. $T^{-1}:\ x = \frac{u}{u^2 + v^2},\ y = \frac{v}{u^2 + v^2},\ x^2 + y^2 = r^2 \Leftrightarrow u^2 + v^2 = \frac{1}{r^2}$

13. (a) $J\left(\frac{u, v, w}{x, y, z}\right) = (x^2 + y^2 + z^2)^{-3}$

(b) $T^{-1}: x = \frac{u}{u^2 + v^2 + w^2},$

$y = \frac{v}{u^2 + v^2 + w^2},$

$z = \frac{w}{u^2 + v^2 + w^2}$

$x^2 + y^2 + z^2 = r^2 \Leftrightarrow u^2 + v^2 + w^2 = 1/r^2$

Chapter 8

Section 1

1. $\phi'(x) = \int_0^1 \frac{t \cos xt}{1 + t}\, dt$

3. $\phi'(x) = \int_1^2 -\frac{t}{(1 + xt)^2} e^{-t}\, dt$

5. $\phi'(x) = \int_0^1 t^x\, dt,\ x > 0$

7. $\phi'(x) = 2x \cos(x^4)$

9. $\phi'(x) = \sin(x^2) - 2x \sin(x^3) + \int_{x^2}^{x} t \cos(xt)\, dt$

11. $\phi'(x) = e^x \tan(xe^x) - 2x \tan(x^3) + \int_{x^2}^{e^x} t \sec^2(xt)\, dt$

13. $\phi'(x) = \frac{2x}{1 + x + x^3} e^{-(1+x^2)} + \frac{\sin x}{1 + x\cos x} e^{-\cos x} + \int_{\cos x}^{1+x^2} -\frac{t}{(1 + xt)^2} e^{-t}\, dt$

15. $\phi'(x) = \frac{\pi}{2x} \cos\left(\frac{\pi x}{2}\right) - \frac{1}{x^2} \sin\left(\frac{\pi x}{2}\right)$

17. $\phi'(x) = \frac{1}{x}\left[\cos(\pi x) - \cos\left(\frac{\pi x}{2}\right)\right]$

19. $\phi'(x) = \frac{1}{x}[2\sin(x^2) - 3\sin(x^3)]$

21. $\phi'(x) = \frac{x^{n-1} + nx^{n-2}}{1 + x^{n-1}} - \frac{x^{m-1} + mx^{m-2}}{1 + x^{m-1}}, x > 0$

23. $\phi'(x) = \frac{1}{\sqrt{1 - x^2}}$ 25. 0 29. No.

Chapter 8

Section 2

1. Converges 3. Converges 5. Converges 7. Diverges
9. Diverges 11. Diverges 13. Diverges 15. Converges
17. Converges 19. Converges 21. Converges

Chapter 8

Section 3

1. Converges 3. Converges 5. Diverges 7. Converges
9. Converges 11. Converges if $p > 2$, diverges if $p \leq 2$
13. Converges 15. Converges 17. $\Gamma(\frac{1}{2}) = \sqrt{\pi}$

Chapter 8

Section 4

1. $\phi'(x) = \int_0^\infty \frac{-te^{-xt}}{1 + t}\,dt$ 3. $\phi'(x) = -\int_0^\infty \frac{t\sin(xt)\,dt}{1 + t^3}$

5. $\phi'(x) = \int_0^1 (\log t)\cdot\cos(xt)\,dt$ 7. $\phi'(x) = -\int_0^1 \frac{t\,dt}{(1 + xt)^2\sqrt{1 - t}}$

15. $\phi(x) = \frac{1}{2}\log(1 + x^2)$

17. $\phi'(x) = \operatorname{Arctan} x$
$\phi(x) = x\operatorname{Arctan} x - \frac{1}{2}\log(1 + x^2)$

Chapter 9

Section 1

1. $\mathbf{u}_x = (2x + 2y)\mathbf{i} + 3x^2\mathbf{j}$
$\mathbf{u}_y = 2x\mathbf{i} - 3y^2\mathbf{j}$

3. $\mathbf{u}_x = \frac{-x^2 + y^2}{(x^2 + y^2)^2}\mathbf{i} - \frac{2xy}{(x^2 + z^2)^2}\mathbf{j} - \frac{2xz}{(x^2 + y^2)^2}\mathbf{k}$
$\mathbf{u}_z = -\frac{2yz}{(x^2 + z^2)^2}\mathbf{j} + \frac{1}{x^2 + y^2}\mathbf{k}$

5. $\mathbf{v}_s = (\cos t)\mathbf{i} + (t\cos s)\mathbf{j} - 2s\mathbf{k}$

$\mathbf{v}_t = (-s \sin t)\mathbf{i} + (\sin s)\mathbf{j} + 2t\mathbf{k}$

7. $\mathbf{v}_{rs} = \mathbf{0}$, $\mathbf{v}_{st} = \mathbf{0}$

9. $\dfrac{\partial}{\partial y}(|\mathbf{u}|^2) = x \sin(2xy) - z \sin(2yz)$

11. $\dfrac{\partial}{\partial y}(\mathbf{u} \cdot \mathbf{v}) = \dfrac{e^{x-y}}{x+y} - e^{x-y} \log(x+y) - 2y - \dfrac{\log(x+y)}{x-y} + \dfrac{\log(x-y)}{x+y}$

13. $\dfrac{\partial}{\partial t}(\mathbf{u} \times \mathbf{v}) = (-2e^{-t} \sin s - 2e^{2t} \cos s)\mathbf{i} + (-2e^t \cos s)\mathbf{j}$
$+ (2e^{2t} \cos^2 s + 2e^{-2t} \sin s \sin 2s)\mathbf{k}$

15. $\mathbf{u} \cdot (\mathbf{u}_{,1} \times \mathbf{v}_2) = 0 \Leftrightarrow y = 0$

17. S is the hemisphere $x^2 + y^2 + z^2 = 9$, $z \geq 0$

19. S is the paraboloid $z = \dfrac{x^2}{4} + \dfrac{y^2}{9}$

Chapter 9

Section 2

1. $\nabla f = 4\mathbf{i} + 4\mathbf{j} + 2\mathbf{k}$, $D_{\mathbf{a}} f = \dfrac{10}{\sqrt{14}}$
3. $\nabla f = \frac{1}{81}(-\mathbf{i} - 4\mathbf{j} - 2\mathbf{k})$, $D_{\mathbf{a}} f = \frac{1}{81}$
5. $\nabla f = 2\mathbf{i} + \frac{1}{2}\mathbf{j} + \frac{1}{2}\mathbf{k}$, $D_{\mathbf{a}} f = \frac{3}{2}$
7. $D_{\mathbf{a}} \mathbf{u} = \dfrac{2}{\sqrt{21}}(\mathbf{i} + 4\mathbf{j} + 6\mathbf{k})$
9. $D_{\mathbf{a}} \mathbf{u} = \mathbf{0}$
11. $\mathbf{a} = -\frac{1}{7}(6\mathbf{i} - 2\mathbf{j} - 3\mathbf{k})$
15. $\mathbf{n} = \pm \dfrac{1}{\sqrt{26}}(3\mathbf{i} - 4\mathbf{j} - \mathbf{k})$
17. $\mathbf{n} = \pm \dfrac{1}{5\sqrt{2}}(3\mathbf{i} - 4\mathbf{j} + 5\mathbf{k})$
19. $\mathbf{n} = \pm \frac{1}{3}(-\mathbf{i} - 2\mathbf{j} + 2\mathbf{k})$

Chapter 9

Section 3

1. $a_{11} + a_{22} + a_{33}$
3. $2x + 2y + 2z$
5. $e^{xz}(2z \cos yz - x)$
7. 0
9. 0
11. $r^{-n}(3 - n)$
13. $\phi''(r) + 2r^{-1}\phi'(r)$

Chapter 9

Section 4

1. $\operatorname{curl} \mathbf{v} = \mathbf{i} + \mathbf{j}$
3. $f(x, y, z) = e^x \sin y \cos z + C$

5. $\operatorname{curl} \mathbf{v} = \dfrac{2}{\sqrt{x^2+y^2}}(y\mathbf{i} - x\mathbf{j})$ 7. $\operatorname{curl} \mathbf{v} = (x^2 - z^2)\mathbf{j}$

9. $f = \frac{1}{3}(x^3y^3 + x^3z^3 + y^3z^3) + G_1(x) + G_2(y) + G_3(z) + C$

17. $\operatorname{curl}[\phi(r)\mathbf{r}] = \mathbf{0}$

Chapter 9

Section 5

1. $\frac{5}{6}$ 3. 2 5. $-\dfrac{5}{2} + \dfrac{\pi^3}{24}$ 7. $\frac{137}{10}$ 9. $\frac{8}{3}$

Chapter 9

Section 6

In Problems 1 through 5, we give the common answer.

1. -2 3. 0 5. $2 - \dfrac{1}{\sqrt{2}}$

7. Valid; $f = x^3 - 3xyz + y^3 + C$

9. Domain not simply connected but have $f = \frac{3}{2}(x^2 + y^2 + z^2)^{1/3} + C$ in D.

Chapter 10

Section 1

In Problems 1 through 8, we give the value of $\iint_G (Q_x - P_y)\, dA_{xy}$.

1. 2 3. $-\frac{27}{2}$ 5. $\frac{1}{4}$ 7. 0

9. 12 11. $\frac{26}{3}$ 13. $4 \log 4$ 15. $-\frac{988}{5}$

17. 0 19. 2 21. $\dfrac{3\pi a^4}{4}$ 23. 0

Chapter 10

Section 3

1. $\frac{5}{6}$ 3. $\frac{248}{15}$ 5. $\frac{10}{3}$

9. $J = u^2 v$ 11. $\frac{26}{3}$

13. 0; T^{-1}: $u = 2x + y - 2y^2$, $v = x + y - y^2$

15. $\frac{11}{5184}$; T^{-1}: $u = x/(x^2 + y^2)$, $v = y/(x^2 + y^2)$

17. $\frac{5}{6}$; T^{-1}: $x = u - v/2 + e^{v/2}$, $y = v$

21. $\pi/8$

Chapter 10

Section 5

1. $\dfrac{\sqrt{3}}{6}$ 3. $\dfrac{15\pi\sqrt{2}}{4}$ 5. 2π

7. $\frac{1}{3}[(1+\pi^2/4)^{3/2}-1]$ 9. $\dfrac{\pi}{4}(15\sqrt{2}+17)$

11. $\dfrac{1024\delta}{45}$ 13. $\dfrac{(2+\sqrt{3})\delta}{6}$

15. $\bar{z}=\dfrac{2+\sqrt{2}}{4}$, $\bar{x}=\bar{y}=0$ 17. $\bar{x}=\dfrac{4a}{3(\pi-2)}$, $\bar{y}=\bar{z}=0$

19. Case (i) $\dfrac{4\pi a^2\delta}{c}$, Case (ii) $4\pi\delta a$ 21. $2\pi\delta$

Chapter 10

Section 7

1. 0 3. $\dfrac{15\pi}{2}$ 5. $\frac{4}{15}$ 7. $\frac{64}{3}$

In Problems 8 through 13, we give the value of $\int_{\partial \vec{S}} \mathbf{v}\cdot d\mathbf{r}$.

9. 0 11. $2\left(\dfrac{\pi^2}{4}-1\right)$ 13. 0 15. 0

Chapter 10

Section 8

In Problems 1 through 10, we give the value of $\iiint_G \operatorname{div}\mathbf{v}\,dV$.

1. $\frac{1}{8}$ 3. $\dfrac{32\pi}{3}$ 5. $12\pi\sqrt{3}$ 7. 0 9. 2π 11. 0 13. 5π

Appendix 1

Section 1

1. $A=\begin{pmatrix}-4 & -7\\ 7 & -6\end{pmatrix}$ 3. $A=\begin{pmatrix}5 & 1 & 9\\ 4 & 1 & 7\end{pmatrix}$

5. $A=\begin{pmatrix}0 & -2 & 1\\ -2 & 2 & 0\\ 0 & 2 & -1\end{pmatrix}$ 7. $A=\begin{pmatrix}3 & 2\\ 1 & 4\end{pmatrix}$

11. $A=\begin{pmatrix}-4 & 1 & -2\\ 1 & 3 & -4\end{pmatrix}$, $B=\begin{pmatrix}3 & 0 & 1\\ 0 & -2 & 3\end{pmatrix}$

13. $A=\begin{pmatrix}2 & -1 & 1\\ 1 & 2 & 0\end{pmatrix}$, $B=\begin{pmatrix}1 & 1 & 2\\ -1 & 1 & -2\end{pmatrix}$

15. $AB = \begin{pmatrix} 1 & -4 \\ 7 & 7 \end{pmatrix}$, $BA = \begin{pmatrix} 0 & -5 \\ 7 & 8 \end{pmatrix}$

17. $AB = BA = \begin{pmatrix} 0 & 1 \\ 0 & 1 \end{pmatrix}$

19. $AB = \begin{pmatrix} 5 & 3 & -4 \\ -6 & 11 & -1 \end{pmatrix}$

21. $AB = \begin{pmatrix} 0 & 0 & 0 \\ 0 & 0 & 0 \\ 0 & 0 & 0 \end{pmatrix}$, $BA = \begin{pmatrix} 0 & 0 & 0 \\ 0 & 0 & 0 \\ 1 & 1 & 0 \end{pmatrix}$

25. $X = \begin{pmatrix} -1 & 1 \\ \frac{5}{2} & 0 \end{pmatrix}$, $Y = \begin{pmatrix} -\frac{5}{2} & \frac{1}{2} \\ -\frac{5}{2} & \frac{3}{2} \end{pmatrix}$

Appendix 1

Section 2

17. 65
19. 6
21. 72
23. 45
25. $\sum_{j=0}^{4} \sum_{i=0}^{4-j} (3i + 2j)$
27. $(x, y) = (1, 2)$, or $(-1, -2)$

Appendix 1

Section 3

1. -2
3. 24

Appendix 1

Section 4

1. 14
3. 21
5. -21
7. -69
9. 297
11. -22

Appendix 1

Section 5

1. $(1, -2, -1)$
3. $(-\frac{73}{79}, \frac{21}{19}, -\frac{43}{19})$
5. $(2, -1, 3)$
7. $(1, \frac{2}{3}, -\frac{1}{3})$
9. $(2, -1, \frac{1}{2}, 1)$

Appendix 1

Section 6

1. Rank = 2
3. Rank = 3
5. Rank = 3

7. Rank = 4
9. Rank = 3
11. Rank = 4
13. Rank = 4
15. Rank = 4

Appendix 1

Section 7

1. $(-1, 1, 0)$
3. $(2, -4, 7)$
5. $r = 2$, $r^* = 3$; inconsistent
7. $r = r^* = 3$; $(4 - 3x_3, 2 - x_3, x_3, 5x_3 - 4)$
9. $(0, 0, 0)$
11. $(\frac{5}{3}x_4, -x_4, -\frac{2}{3}x_4, x_4)$
13. $(0, 0, 0, 1)$
15. $r = 3$; $r^* = 4$; inconsistent

Index

Undergraduate Texts in Mathematics

continued from ii

Martin: The Foundations of Geometry and the Non-Euclidean Plane.

Martin: Transformation Geometry: An Introduction to Symmetry.

Millman/Parker: Geometry: A Metric Approach with Models.

Owen: A First Course in the Mathematical Foundations of Thermodynamics.

Prenowitz/Jantosciak: Join Geometrics.

Priestly: Calculus: An Historical Approach.

Protter/Morrey: A First Course in Real Analysis.

Protter/Morrey: Intermediate Calculus.

Ross: Elementary Analysis: The Theory of Calculus.

Scharlau/Opolka: From Fermat to Minkowski.

Sigler: Algebra.

Simmonds: A Brief on Tensor Analysis.

Singer/Thorpe: Lecture Notes on Elementary Topology and Geometry.

Smith: Linear Algebra. Second édition.

Smith: Primer of Modern Analysis.

Thorpe: Elementary Topics in Differential Geometry.

Troutman: Variational Calculus with Elementary Convexity.

Wilson: Much Ado About Calculus.

Table 3. Natural Logarithms of Numbers

n	$\log_e n$	n	$\log_e n$	n	$\log_e n$
0.0	*	4.5	1.5041	9.0	2.1972
0.1	7.6974	4.6	1.5261	9.1	2.2083
0.2	8.3906	4.7	1.5476	9.2	2.2192
0.3	8.7960	4.8	1.5686	9.3	2.2300
0.4	9.0837	4.9	1.5892	9.4	2.2407
0.5	9.3069	5.0	1.6094	9.5	2.2513
0.6	9.4892	5.1	1.6292	9.6	2.2618
0.7	9.6433	5.2	1.6487	9.7	2.2721
0.8	9.7769	5.3	1.6677	9.8	2.2824
0.9	9.8946	5.4	1.6864	9.9	2.2925
1.0	0.0000	5.5	1.7047	10	2.3026
1.1	0.0953	5.6	1.7228	11	2.3979
1.2	0.1823	5.7	1.7405	12	2.4849
1.3	0.2624	5.8	1.7579	13	2.5649
1.4	0.3365	5.9	1.7750	14	2.6391
1.5	0.4055	6.0	1.7918	15	2.7081
1.6	0.4700	6.1	1.8083	16	2.7726
1.7	0.5306	6.2	1.8245	17	2.8332
1.8	0.5878	6.3	1.8405	18	2.8904
1.9	0.6419	6.4	1.8563	19	2.9444
2.0	0.6931	6.5	1.8718	20	2.9957
2.1	0.7419	6.6	1.8871	25	3.2189
2.2	0.7885	6.7	1.9021	30	3.4012
2.3	0.8329	6.8	1.9169	35	3.5553
2.4	0.8755	6.9	1.9315	40	3.6889
2.5	0.9163	7.0	1.9459	45	3.8067
2.6	0.9555	7.1	1.9601	50	3.9120
2.7	0.9933	7.2	1.9741	55	4.0073
2.8	1.0296	7.3	1.9879	60	4.0943
2.9	1.0647	7.4	2.0015	65	4.1744
3.0	1.0986	7.5	2.0149	70	4.2485
3.1	1.1314	7.6	2.0281	75	4.3175
3.2	1.1632	7.7	2.0412	80	4.3820
3.3	1.1939	7.8	2.0541	85	4.4427
3.4	1.2238	7.9	2.0669	90	4.4998
3.5	1.2528	8.0	2.0794	95	4.5539
3.6	1.2809	8.1	2.0919	100	4.6052
3.7	1.3083	8.2	2.1041		
3.8	1.3350	8.3	2.1163		
3.9	1.3610	8.4	2.1282		
4.0	1.3863	8.5	2.1401		
4.1	1.4110	8.6	2.1518		
4.2	1.4351	8.7	2.1633		
4.3	1.4586	8.8	2.1748		
4.4	1.4816	8.9	2.1861		